Electrical Power Systems

(For B.E./B.Tech., AMIE, IETE, GATE, UPSC Engineering Services)
5th Revised Edition

Electrical Power Systems

(For B.E./B.Tech., AMIE, IETE, GATE, UPSC Engineering Services)

5th Revised Edition

Ashfaq Husain

Formerly, Reader in Electrical Engineering,
University Polytechnic, Aligarh Muslim University,
Aligarh (India)

CBS

CBS Publishers & Distributors Pvt. Ltd.

New Delhi • Bengaluru • Chennai • Kochi • Kolkata • Mumbai
Hyderabad • Uttarakhand • Nagpur • Patna • Pune • Jharkhand

Electrical Power Systems

ISBN: 978-81-239-1448-2

Copyright © Ashfaq Husain

Fifth Edition: 2007
Reprint: 2008, 2009, 2010, 2011, 2012, 2014, 2015, 2017, 2018, 2020. 2023
Third Edition: 1990
Reprint: 1992, 1993
Fourth Edition: 1994
Reprint: 1996, 1997, 1999, 2001, 2003, 2004, 2005, 2006

Published by Satish Kumar Jain and produced by Varun Jain for

CBS Publishers & Distributors Pvt Ltd
4819/XI Prahlad Street, 24 Ansari Road, Daryaganj, New Delhi 110 002, India
Ph: 011-23289259, 23266861   Website: www.cbspd.com
                                    e-mail: delhi@cbspd.com
Corporate Office: 204 FIE, Industrial Area, Patparganj, Delhi 110 092
Ph: 011-4934 4934   Fax: 011-4934 4935   e-mail: publishing@cbspd.com
                                                 publicity@cbspd.com

Branches

- **Bengaluru:** Seema House 2975, 17th Cross, KR Road, Banasankari 2nd Stage, Bengaluru 560 070, Karnataka, India
  Ph: +91-80-26771678/79   Fax: +91-80-26771680   e-mail: bangalore@cbspd.com
- **Chennai:** 7, Subbaraya Street, Shenoy Nagar, Chennai 600 030, Tamil Nadu, India
  Ph: +91-44-26680620, 26681266   Fax: +91-44-42032115   e-mail: chennai@cbspd.com
- **Kochi:** 42/1325, 1326, Power House Road, Opp KSEB, Power House, Ernakulam 682 018, India
  Ph: +91-484-4059061–65   Fax: +91-484-4059065   e-mail: kochi@cbspd.com
- **Kolkata:** 147, Hind Ceramics Compound, 1st Floor, Nilgunj Road, Belghoria, Kolkata 700 056, West Bengal, India
  Ph: +91-9096713055/56   e-mail: kolkata@cbspd.com
- **Lucknow:** Basement, Khushnuma Complex, 7-Meerabai Marg (behind Jawahar Bhawan), Lucknow 226 001, India
  Ph: +91-522-4000032   e-mail: tiwari.lucknow@cbspd.com
- **Mumbai:** PWD Shed. Gala no. 25/26, Ramchandra Bhatt Marg, Next to JJ Hospital Gate no. 2, Opp. Union Bank of India
  Noorbaug Mumbai 400 009, Maharashtra, India
  Ph: +91-22-66661880/89   e-mail: mumbai@cbspd.com
Representatives

• **Hyderabad**	0-9885175004	• **Jharkhand**	0-9811541605	• **Nagpur**	0-9421945513
• **Patna**	0-9334159340	• **Pune**	0-9923910676	• **Uttarakhand**	0-9716462459

Printed at Mudrak, Noida, UP, India

Preface to the Fifth Edition

This book is intended to serve as a textbook for the course 'Electrical Power Systems' for B.Tech./ B.E. degree students of Electrical Engineering. It will also serve as a text reference for students of Diploma in Electrical Engineering. The common topics included in the syllabi of almost all Engineering Institutions in India are covered in this book.

This text is also useful for candidates appearing in AMIE, GATE, UPSC Engineering services and other competitive examinations. The practising engineers would also find this book valuable as a concise reference for basic principles and applications in 'Electrical Power Systems'.

This book brings many widely varied topics together. It lays emphasis on the basic concepts and at the same time introduces modern methods of solution of power system problems.

The book is divided into twenty eight chapters. Each chapter is self-contained and is dealt with comprehensively. The subject matter in each chapter has been developed systematically from basic principles using the SI system of units. Generalized approach has been given in treatments.

The fifth edition has been thoroughly revised, further enlarged and updated.

In response to the suggestions from the teachers and students, the following new chapters are incorporated in this revised edition :

- Voltage stability
- Flexible AC Transmission System (FACTS)

Chapters on Line Insulators and Supports, Per Unit Representation, Long Transmission Lines, Economic Operation of Power Systems, Unsymmetrical Faults, Power System Stability, Overvoltage Protection, HVDC Transmission have been rewritten to include the recent trends in the subject.

The numerous illustrative examples enhance the utility of the text and make it easy for the beginner to thoroughly grasp the presentation of the theory. The worked-out examples are very carefully selected in the text to illustrate the practical applications. Most simplified methods of solving the problems are given. Step-by-step procedures of solving problems are provided. At the end of each chapter a large number of problems of wide variety have been included for practice.

With all these modifications the book has become well organised, up-to-date, lucid and very easy to follow for self study.

The author hopes that the new fifth edition will continue to help and will enhance the understanding of Electrical Power Systems in the same way that previous editions have done in the past.

It is hoped that the book will be useful to students and teachers alike.

The author would be grateful to readers for their suggestions for further improvement of the book.

Ashfaq Husain

Acknowledgements

I appreciate the patience, understanding and support of my wife Dr. Nigar Minhaj, Reader in Electronics, Women's Polytechnic, Aligarh Muslim University, Aligarh. Her valuable suggestions and comments have made it possible for me to complete this book.

I also appreciate the patience and concern of my sons Ahmad Ashfaq, M.Tech. (Environmental Engineering), Lecturer in Civil Engineering, University Polytechnic, Aligarh Muslim University, Aligarh; and Haroon Ashfaq, M.Tech. (Electrical Engineering), Lecturer in Electrical Engineering, Jamia Millia Islamia, New Delhi, who has helped me a lot in the revision of this book. Haroon Ashfaq has given valuable suggestions and made a lot of contribution to this book.

Ashfaq Husain

Preface to the First Edition

This book has been designed as a textbook for engineering degree, diploma, AMIE or equivalent examinations in Electrical Power Systems in India and abroad. It will also be useful to students preparing for various competitive examinations. It is equally helpful to practising engineers to understand the theoretical aspects of their profession.

The book includes the recent rapid developments that are taking place in the field of Electrical Power Systems which are not at present readily available in a single textbook because of the diversity of the topics. The present book brings many widely varied topics together to cater to the needs of revised syllabi of engineering and competitive examinations. Besides, this book bridges the gap between old and new concepts in Electrical Power Systems. The text lays emphasis on the basic concepts and at the same time introduces modern methods of solution of power system problems.

The book is divided into twenty-three chapters. Each chapter is self-contained and is dealt with comprehensively. The subject matter in each chapter has been developed systematically from basic principles using the S.I. system of units. Generalized approach has been given in treatments. Matrix analysis is used wherever necessary. The technical information available on the topics is up-to-date.

The chapters on Conductors, Insulators and Power Cables are written specially with a view to put before the reader recent developments in this important field which has not attracted much attention by previous authors. Clear diagrams and photographs are given for better understanding. The concepts of GMD and GMR which are necessary for handling calculations for lines having any conductor configuration are clearly discussed. Performance of short, medium and long lines is adequately explained. The chapter on General Network Constants gives the performance calculations of transmission lines by general methods. It emphasizes a unified approach to the problems instead of various partial solutions. The subject matter on Power Circle Diagrams covers analytical as well as graphical methods to demonstrate clearly the actual performance of lines.

In view of the importance of High Voltage Direct Current (HVDC) transmission in the present juncture of technological development in the subject, a chapter on Power Transmission by Direct Current is included in the text.

The mechanical performance of overhead lines under various operating conditions is fully discussed. The chapters on Load Flow Studies and Economic Operation of Power Systems are meant to introduce the basic aspects of the problems involved in this area.

Throughout the text, the importance of extra high voltage transmission of energy is discussed in detail. Recent developments regarding the choice of next higher voltages, standardization of transmission voltages are presented. One full chapter is devoted to the comprehensive treatment of Corona. Emphasis has been laid on generalized treatment of fault analysis instead of partial solution.

The chapters on Travelling Waves, and Power System Stability have been discussed exhaustively.

The numerous illustrative examples enhance the utility of the text and make it easy for the beginner to thoroughly grasp the presentation of the theory. The worked-out examples are very carefully selected in the text to illustrate the practical applications. Most simplified methods of solving the problems are given. In most cases, the worked-out examples illustrate the technique of solving problems. At the end of each chapter a large number of representative numerical examples of wide variety have been included for practice. Many of them have been selected from the past examination papers of standard institutions. The problems are of practical nature.

While utmost care has been exercised to eliminate misprints and errors, the author would feel obliged to have mistakes brought to his notice. He would welcome any comments and suggestions for the improvement of book.

Ashfaq Husain

Contents

Load Characteristics

1.1 INTRODUCTION

From the very outset energy has played a vital role in the development of civilization. There has been a universal basic drive towards better living through expanded utilization of energy. The history of civilization shows a close relationship between the utilization of energy and the progress of mankind. The degree of energy used is the symbol of the progress of a country. Energy consciousness in the people has created interests in them to tap new sources of energy from time to time. Of the various forms so far discovered the electrical energy has contributed a lot to the world's energy requirements.

1.2 ADVANTAGES OF ELECTRICAL ENERGY

Electrical energy is the most refined form of energy. The advantages derived from the electrical energy are many in number. Some of its important advantages over other forms of energy are :

1. It can be generated in large quantities at comparable cost with other types of energy.
2. It can be conveniently transmitted over long distances.
3. It can be utilized efficiently in a number of processes requiring energy.
4. It has got maximum flexibility and has most sensitive susceptibility control.

1.3 LOAD

A device which uses electrical energy is said to impose a load on the system. The term load is used in a number of ways.

- To indicate a device or collection of devices which consume electrical energy.
- To indicate power required from a given supply circuit.
- To indicate the current or power passing through a line or machine.

The load may be resistive, inductive, capacitive, or some combination of them.

Loads on power systems are divided into industrial, commercial, and residential. The industrial loads are composite loads, and induction motors form a high proportion of these loads. These composite loads are functions of voltage and frequency and form a major part of the system load. Commercial and residential loads consist largely of lighting, heating and cooling. These loads are independent of frequency and consume negligibly small reactive power.

1.4 CONNECTED LOAD

Connected load is the sum of continuous ratings of all loads connected to the system or any part thereof.

1.5 DEMAND

The demand of an installation of a system is the load that is drawn from the source of supply at the receiving terminals averaged over a suitable and specified interval of time.

The load may be given in kilowatts (kW), kilovars (kVAr), kilovoltamperes (kVA), or amperes (A).

1.6 DEMAND INTERVAL

Demand interval is the period over which the load is averaged. There are two demands :
   (a) Instantaneous demand
   (b) Sustained demand

The former is not very important because all the machines are designed for overloads. The sustained intervals are generally taken as 15 min, 30 min, or even longer. But 30 min is the basis time in India.

1.7 MAXIMUM DEMAND (MD) OR PEAK LOAD

The maximum demand of an installation or system is the greatest of all demands which have occurred during the specific period of time.

The maximum demand statement should also express the demand interval used to measure it. For example, the specified demand might be maximum of all demands such as daily, weekly, monthly, or annual.

Knowledge of maximum demand helps in determining the installed capacity of a generating station. The generating station must be capable of meeting the maximum demand. Hence the cost of plant and equipment increases with the increase in maximum demand.

1.8 DEMAND FACTOR DF

The demand factor is the ratio of the actual maximum demand of the system to the total connected load of the system. Therefore, the demand factor (DF) is

$$DF \triangleq \frac{\text{maximum demand}}{\text{total connected load}}$$

The demand factor can also be found for a part of the system. For example, an industrial or commercial consumer, instead of for the whole system.

In practice, consumers do not use all the devices at full load simultaneously. The maximum demand of each consumer is, therefore, less than his connected load. The demand factor depends upon the nature of load. Lighting loads have higher demand factors than power loads. The demand factor is usually less than 1.0.

1.9 AVERAGE LOAD OR AVERAGE DEMAND

It is the ratio of energy consumed in a given period of the time in hours.

$$\text{Average load} = \frac{\text{energy consumed in a given period}}{\text{hours in that time period}}$$

1.10 LOAD FACTOR

Load factor of a system is the ratio of the average load over a given period of time to the maximum demand (peak load) occurring in that period.

$$\text{Load factor} \triangleq \frac{\text{average load}}{\text{peak load}}$$

Multiplying the numerator and denominator by time T,

$$\text{Load factor} = \frac{\text{average} \times T}{\text{peak load} \times T} = \frac{\text{energy consumed during a time of } T \text{ hours}}{\text{peak load} \times T \text{ hours}}$$

This relation shows that the load factor can also be defined as the ratio of the actual energy consumed during a given period to the energy which would have been used if the maximum demand (peak load) had been continuously maintained throughout that period.

Depending upon the number of hours in days, weeks, months, or years we define different load factors. For daily load factor, the period of time T is taken as 24 hours and for annual load factor $T = 8760$ hours.

Mathematically,

$$\text{Daily load factor} = \frac{\text{total kWh during 24 h of the day}}{(\text{peak load in kW}) \times 24 \text{ h}}$$

$$\text{Monthly load factor} = \frac{\text{total kWh during the month}}{(\text{peak load in kW}) \times (\text{number of hours in the month})}$$

$$\text{Annual load factor} = \frac{\text{total kWh during the year}}{(\text{peak load in kW}) \times (8760 \text{ hours})}$$

Load factor plays an important role on the cost of generation per unit (kWh). The higher the load factor, the lesser will be the cost of generation per unit for the same maximum demand.

1.11 DIVERSITY FACTOR F_D

The maximum demands of the individual consumers of a group are not likely to occur simultaneously. Thus, there is a diversity in the occurrence of the loads. Due to this diverse nature of the load, power is never required to supply all connected loads to their full capacity at the same time.

Diversity factor is the ratio of the sum of the individual maximum demands of the various subdivisions of a system to the maximum demand of the whole system. Thus,

$$\text{Diversity factor} \triangleq \frac{\text{(sum of individual maximum demands)}}{\text{(coincident maximum demand of the whole system)}}$$

$$F_D \triangleq \frac{D_1 + D_2 + \ldots + D_n}{D_g}$$

or

$$F_D = \frac{\displaystyle\sum_{i=1}^{n} D_i}{D_g}$$

where D_i = maximum demand of the load i, irrespective of the time of occurrence.

$D_g = D_{(1+2+\ldots+n)}$ = coincident maximum demand of group of n loads

Diversity factor can be defined for loads, substations, feeders, and generating stations. Usually the maximum demand of various consumers do not occur at the same time and the simultaneous (coincident) maximum demand is less than their total maximum demand. The diversity factor can be equal to or greater than 1.0. The value of the diversity factor is generally greater than 1.0 with a high value representing a good diversity and 1.0 represents a poor diversity.

A large diversity factor has the effect of reducing the maximum demand. Consequently, lesser plant capacity is required. Thus, the capital investment on the plant is reduced and the cost of generation is also reduced.

A high diversity factor may be obtained by giving incentives to industries and farmers to use electrical energy at night or light-load periods.

1.12 LOAD DIVERSITY

It is the difference between the sum of the peaks of two or more individual loads and the peak of the combined load.

$$\text{Load diversity} \triangleq \left(\sum_{i=1}^{n} D_i \right) - D_g$$

1.13 UTILIZATION FACTOR F_u

It is the ratio of maximum demand of a system to the rated capacity of the system.

$$F_u \triangleq \frac{\text{maximum demand}}{\text{rated system capacity}}$$

The utilization factor can also be found for a part of the system.

1.14 PLANT FACTOR OR CAPACITY FACTOR

It is the ratio of the total actual energy produced or supplied over a specified period of time to the energy that would have been produced or supplied if the plant (or unit) had operated continuously at maximum rating. The maximum plant rating in the total installed plant capacity including the reserve capacity.

$$\text{Plant factor} \triangleq \frac{\text{(actual energy produced or supplied in time } T)}{\text{maximum plant rating} \times T}$$

Plant factor is mostly used in generation studies. For example,

$$\text{annual plant factor} = \frac{\text{actual annual energy generation}}{\text{maximum plant rating} \times 8760}$$

The capacity factor indicates the extent of the use of the generating station. If the plant is always run at its rated capacity, the capacity factor is 1.0 (100%). It is different from load factor because of the fact that the rated capacity of each plant is always greater than the maximum demand. The power plants have always some reserve capacity to take into account the future expansion, increase in load and maintenance.

It is to be noted that

$$\text{capacity factor} = \frac{\text{peak load}}{\text{plant capacity}} \times \text{load factor}$$

Thus, if the rate plant capacity equals peak load, the capacity factor and load factor become identical. That is, in absence of reverse capacity,

$$\text{capacity factor} = \text{load factor}$$

1.15 LOSS FACTOR F_{LS}

It is the ratio of the average power loss to the peak-load power loss during the specified period of time.

$$F_{LS} \overset{\Delta}{=} \frac{\text{average power loss}}{\text{power loss at peak load}}$$

This relationship is applicable for the copper losses of the system but not for the iron losses.

1.16 LOAD CURVE

Load curve (or chronological load curve) is a graphical representation between load in kW (or MW) in proper time sequence and time in hours. It shows the variation of load on the power station. When it is plotted for 24 hours a day, it is called *daily load curve*. If the time considered is one year (8760 hours) then it is called the *annual load curve*.

It is to be noted that the daily load curve of a system is not the same for all days. It differs from day-to-day and season-to-season. In practice, two types of curves are drawn — one for summer and the other for winter.

1.17 INFORMATIONS OBTAINED FROM LOAD CURVES

The following informations are obtained from load curves :

1. Load variation during different hours of the day.
2. The peak load indicated by the load curve gives the maximum demand on the power station.
3. The area under the load curve gives the total energy generated in the period under consideration.
4. The area under the load curve divided by the total number of hours gives the average load.

5. The ratio of the area under the load curve to the total area of the rectangle in which it is contained gives the load factor.

It would be ideal to have a flat load curve. But in practice, load curved are far from flat. For a flat load curve, the load factor will be higher. Higher load factor means more uniform load pattern with less variations in load. This is desirable from the point of view of maximum utilization of associated equipment which are selected on the basis of maximum demand.

1.18 UTILITY OF LOAD CURVES

On the basis of above informations load curves are useful as follows :
  (a) To decide the installed capacity of a power station.
  (b) To choose the most economical sizes of various generating units.
  (c) To estimate the generating cost.
  (d) To decide the operating schedule of the power station, that is, the sequence in which different generating units should run.

1.19 LOAD-DURATION CURVE

A load duration curve is also a graph between load and time in which the ordinates representing the load are plotted in the order of descending magnitude, that is, with the greatest load at the left, lesser loads towards the right and the lowest load at the time extreme right. The load duration curve is derived from the load curve and therefore, represents the same data as that of the load curve. The load duration curve is constructed by selecting the maximum peak points and connecting them by a curve.

1.20 PROCEDURE FOR PLOTTING THE LOAD-DURATION CURVE

  (a) From the data available from the load curve, determine the maximum load on the system and the duration for which it occurs.
  (b) Take the next lower load and determine the total time, during which this and the previous greater load occurs.
  (c) Plot the load against time during which it occurs.

The load-duration curve can be plotted for any duration of time, for example, a day or, a week, or a month, or a year. The abscissa of such a curve can also be chosen as per unit or percentage of time for which it occurs. The whole duration is taken as 1.0 pu or 100%.

The load duration curve plotted for 24 hours of day is called the *daily load duration curve*. Similarly, the load duration curve plotted for 8760 hours of a year is called the *annual load-duration curve*.

1.21 INFORMATIONS AVAILABLE FROM LOAD DURATION CURVE

  (a) It gives the minimum load present throughout the given period.
  (b) It enables the selection of base load and peak load power plants.
  (c) Any point on the load duration curve gives the total duration in hours for the corresponding load and all loads of greater value.

(d) The areas under load curve and corresponding load duration curve are equal. Both these areas represent the same associated energy during the period under consideration.

(e) The average demand during some specified time period such as a day, or month, or year can be obtained from the load duration curve as follows :

$$\text{Average demand} = \frac{\text{kWh (or MWh) consumed in a given time period}}{\text{hours in the time period}}$$

$$= \frac{\text{area under the load duration curve}}{\text{base of the load duration curve}}$$

Example 1.1 A consumer has the following connected load :

> 10 lamps each of 60 W
>
> 2 heaters each of 1000 W
>
> Maximum demand 1500 W

On the average he uses 8 lamps for 5 hours per day, each heater 3 hours per day. Find (a) average load, (b) monthly energy consumption, (c) load factor.

Solution

$$\text{Average load} = \frac{\text{actual energy consumed}}{\text{time duration}} = \frac{8 \times 60 \times 5 + 2 \times 1000 \times 3}{24} = 350 \text{ W}$$

Monthly energy consumption $= (8 \times 60 \times 5 + 2 \times 1000 \times 3) \times 30$ Wh $= 252$ kWh

$$\text{Load factor} = \frac{\text{average load}}{\text{maximum demand}} = \frac{350}{1500} = 0.2333$$

Example 1.2 There are four consumers of diversity having different load requirements at different timings.

Consumer 1

> Average load = 1 kW
>
> Maximum demand = 5 kW at 8 p.m.

Consumer 2

> Maximum demand = 2 kW at 9 p.m.
>
> Demand of 1.6 kW at 8 p.m.
>
> Daily load factor = 0.15

Consumer 3

> Maximum demand = 2 kW at 12 noon
>
> Load of 1 kW at 8 p.m.
>
> Average load of 500 W

Consumer 4

> Maximum demand = 10 kW at 5 p.m.
>
> Load of 5 kW at 8 p.m.
>
> Daily load factor = 0.25

The maximum demand of the system occurs at 8 p.m. Determine

(a) the diversity factor,

(b) average load and load factor of each consumer,

(c) average load and load factor of the combined load.

Solution

Consumer 1

Average load = 1 kW

$$\text{Load factor} = \frac{\text{average load}}{\text{maximum demand}} = \frac{1}{5} = 0.2$$

Consumer 2

Load factor = 0.15

Average load = load factor × maximum demand = $0.15 \times 2 = 0.3$ kW = 300 W

Consumer 3

Average load = 500 W

$$\text{Load factor} = \frac{500}{2000} = 0.25$$

Consumer 4

Load factor = 0.25

Average load = L.F. × M.D. = $0.25 \times 10 = 2.5$ kW

Combined average load = 1000 + 300 + 500 + 2500 = 4300 W = 4.3 kW

Combined maximum demand = 5 + 2 + 2 + 10 = 19 kW

At 8 p.m. the maximum demand = 5 + 1.6 + 1 + 5 = 12.6 kW

$$\text{Diversity factor} = \frac{\text{combined maximum demand}}{\text{actual maximum demand}} = \frac{19}{12.6} = 1.507$$

$$\text{Load factor} = \frac{\text{combined average load}}{\text{actual maximum demand}} = \frac{4.3}{12.6} = 0.341$$

Example 1.3 Determine the maximum value of a load which consumes 600 kWh per day at a load factor of 0.45. If the consumer increases the load factor to 0.65 without increasing the maximum demand, determine the consumption of energy in kWh.

Solution

$$\text{Load factor} = \frac{\text{energy consumed in 24 h}}{(\text{maximum demand in kW}) \times 24}$$

$$0.45 = \frac{600}{(\text{maximum demand}) \times 24}$$

$$\therefore \quad \text{maximum demand} = \frac{600}{24 \times 0.45} = 55.55 \text{ kW}$$

In the second case the load factor is 0.65.

Energy consumed in 24 h = (load factor) × (maximum demand in kW) × 24

$$= 0.65 \times 55.55 \times 24 = 866.6 \text{ kWh}$$

Example 1.4 Plot the load duration curve from the chronological load curve shown in Fig. 1.1 (a).

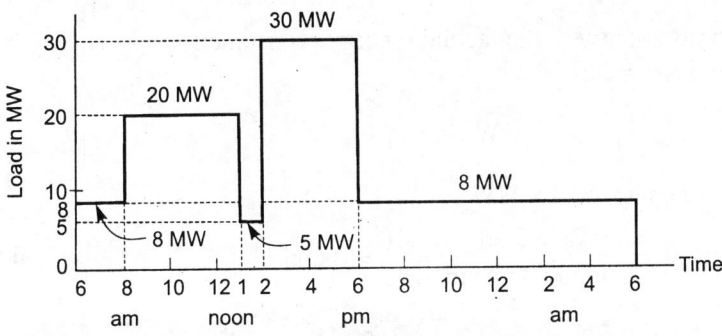

Fig. 1.1. (a) Chronological load curve.

Solution

The data available from the given chronological load curve are tabulated as follows. Here the total time is 24 hours or 100%.

Load in MW	Hours in a day	Time in percentage
30	4	$\dfrac{4}{24} \times 100 = 16.67\%$
20	4 + 5 = 9	$\dfrac{9}{24} \times 100 = 37.5\%$
8	2 + 4 + 5 + 12 = 23	$\dfrac{23}{24} \times 100 = 95.83\%$
5	4 + 5 + 2 + 12 + 1 = 24	$\dfrac{24}{24} \times 100 = 100\%$

The load duration curve is shown in Fig. 1.1 (b).

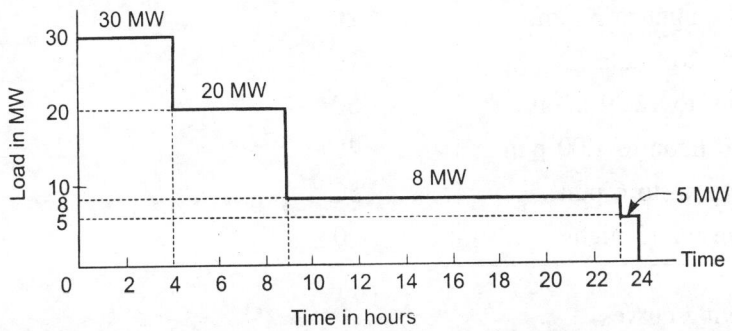

Fig. 1.1. (b) Load-duration curve.

Example 1.5 The load-duration curve for a system is shown in Fig. 1.2. Determine the load factor.

Solution

From the load-duration curve, the actual energy consumed
$= 15 \times 8 + 10 \times 8 + 5 \times 8 = 240$ MWh

$$\text{Average load} = \frac{240}{24} = 10 \text{ MW}$$

Maximum demand = 15 MW

$$\text{Load factor} = \frac{\text{average load}}{\text{maximum demand}} = \frac{10}{15} = 0.666$$

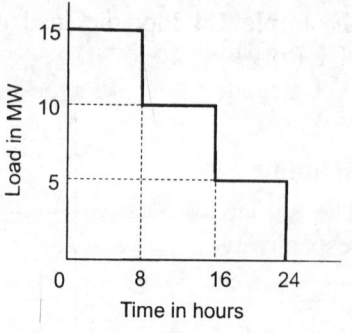

Fig. 1.2.

Example 1.6 The yearly load duration curve of a power plant is a straight line. The maximum load is 500 MW and the minimum load is 400 MW. The capacity of the plant is 750 MW. Find (a) plant capacity factor, (b) load factor, (c) utilization factor, (d) reserve capacity.

Solution

$$\text{Average annual load} = \frac{500 + 400}{2} = 450 \text{ MW}$$

$$\text{Capacity factor} = \frac{\text{average annual load}}{\text{capacity of the plant}} = \frac{450}{750} = 0.6$$

$$\text{Load factor} = \frac{\text{average load}}{\text{maximum demand}} = \frac{450}{500} = 0.9$$

$$\text{Utilization factor} = \frac{\text{maximum demand}}{\text{capacity of the plant}} = \frac{500}{750} = 0.667$$

Reserve capacity = plant capacity – maximum demand = 750 – 500 = 250 MW

Example 1.7 A power system had the daily load curve given by the following table :

Time	Load in MW
12.00 night to 2 a.m.	20
2 a.m. to 8 a.m.	10
8 a.m. to 12.30 noon	50
12.30 noon to 1.00 p.m.	40
1.00 p.m. to 6 p.m.	50
6 p.m. to 12 night	70

Plot the following curves :
(a) Chronological load curve
(b) Load-duration curve

(c) Load-energy curve

Calculate the load factor and the utilization factor of the plant if the installed capacity is 100 MW.

Solution

The chronological load curve and load-duration curve are plotted in Figs. 1.3 (a) and 1.3 (b) respectively.

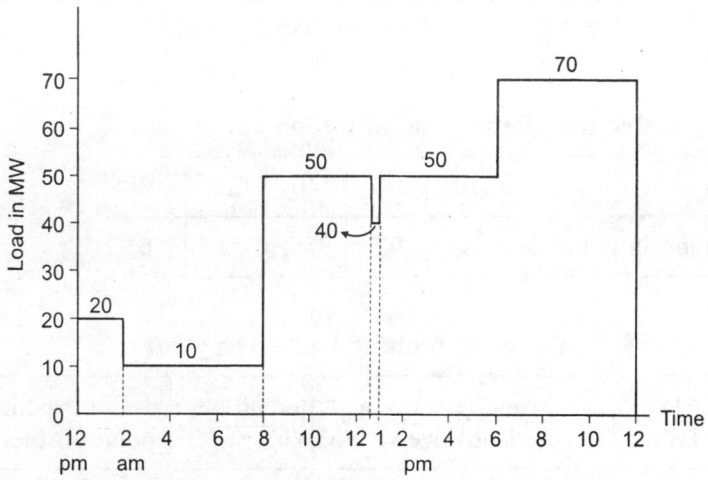

Fig. 1.3. (a) Chronological load curve.

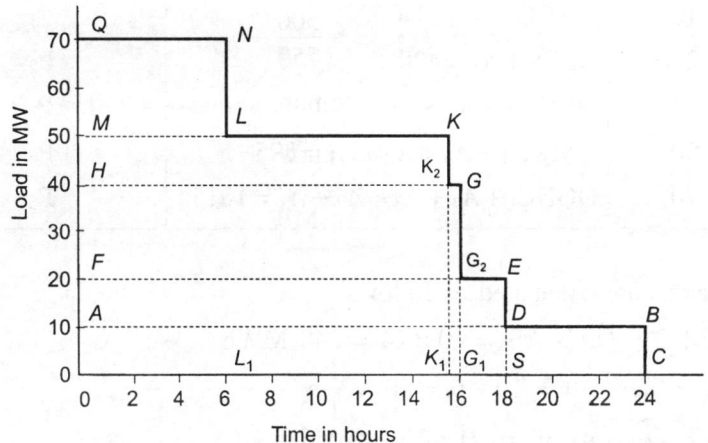

Fig. 1.3. (b) Load-duration curve.

Maximum demand of the system = 70 MW

Energy produced by the system in 24 hours = area under chronological load curve

$$= \text{area under load duration curve}$$
$$= 70 \times 6 + 50 \times 9.5 + 40 \times 0.5 + 200 \times 2 + 10 \times 6$$
$$= 1015 \text{ MWh}$$

$$\text{Daily load factor} = \frac{\text{total energy generated}}{\text{maximum demand} \times 24} = \frac{1015}{70 \times 24} = 0.604$$

$$\text{Utilization factor} = \frac{\text{maximum demand}}{\text{installed capacity}} = \frac{70}{100} = 0.7$$

Plotting of load energy curve

The load energy curve is the plot of the cumulative integration of the area under the load curve starting from zero load versus the particular load. The load energy curve is derived easily from the load-duration curve. The table is prepared by taking the area of the curve under successive loads as follows :

Table for plotting load-duration curve

Load	70	50	40	20	10
Duration in hours	6	9.5	0.5	2	6

Table for plotting load-energy curve

Load (MW)	Area (Energy at different load levels) (MWh)	Corresponding point on the graph
0	$0 = 0$	O
10	$OABC = A_1 = 240$	P_1
20	$OFES + A_1 = 420$	P_2
40	$OHGG_1 + A_2 + A_1 = 740$	P_3
50	$OMKK_1 + A_3 + A_2 + A_1 = 895$	P_4
70	$OQNL_1 + A_4 + A_3 + A_2 + A_1 = 1015$	P_5

The different areas are calculated as follows :

$$\text{Area OABC} = \text{OA} \times \text{AB} = 10 \times 24 = 240 \text{ MWh}$$

$$\text{Area } A_1 = \text{SD} \times \text{DB} = 10 \times 6 = 60 \text{ MWh}$$

$$\text{Area } A_2 = G_1G_2 \times G_2E = 20 \times 2 = 40 \text{ MWh}$$

$$\text{Area } A_3 = K_2K_1 \times K_2G = 40 \times 0.5 = 20 \text{ MWh}$$

$$\text{Area } A_4 = L_1L \times LK = 50 \times 9.5 = 475 \text{ MWh}$$

$$\text{Area } A_5 = \text{OQ} \times \text{QN} = 70 \times 6 = 420 \text{ MWh}$$

The load energy curve is plotted by taking load as abscissa and energy as ordinate. This curve is shown in Fig. 1.3 (c).

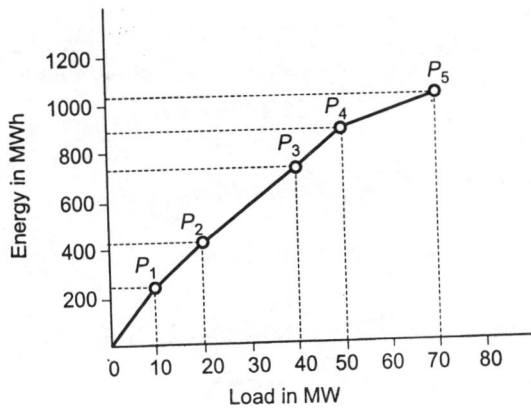

Fig. 1.3. (c) Load-energy curve.

Example 1.8 Find the diversity factor of a power station which supplies the following loads :

Load A : Motor load of 150 kW between 10 a.m. to 7 p.m.

Load B : Lighting load of 50 kW between 7 p.m. and 11 p.m.

Load C : Pumping load of 55 kW between 3 p.m. and 10 a.m.

Solution

Sum of individual maximum demands = 150 + 50 + 55 = 255 kW

The total load on the power station is as follows :

Time	Load
10 a.m. to 3 a.m.	150 kW
3 p.m. to 7 p.m.	150 + 55 = 205 kW
7 p.m. to 11 p.m.	50 + 55 = 105 kW
11 p.m. to 10 a.m.	55 kW

The above table shows that the coincident maximum demand of the whole system is 205 kW.

$$\therefore \quad \text{diversity factor} = \frac{\text{sum of individual maximum demands}}{\text{coincident maximum demand of the whole system}} = \frac{255}{205} = 1.2439$$

Example 1.9 A power station supplies the peak loads of 25 MW, 20 MW, and 30 MW to three localities. The annual load factor is 0.60 pu and the diversity of the load at the station is 1.65. Calculate (a) the maximum demand on the station, (b) the installed capacity, and (c) the energy supplied in a year.

Solution

(a) $\quad$ Diversity factor $= \dfrac{(\text{sum of individual maximum demands})}{(\text{maximum demand on the station})}$

$\therefore$ maximum demand on the station = $\dfrac{\text{sum of individual maximum demands}}{\text{diversity factor}}$

$$= \frac{(25 + 20 + 30)}{1.65} = 45.45 \text{ MW}$$

(b) Installed capacity = 25 + 20 + 30 = 75 MW

(c) Average load = load factor × maximum demand = $0.6 \times \dfrac{75}{1.65}$ MW

Energy supplied per year = $\dfrac{0.6 \times 75}{1.65} \times 8760 = 238909$ MWh

EXERCISES

1. Define the following terms :
   Connected load, maximum demand, demand factor, load factor.
   What is the effect of the load factor on the cost of generation?
2. Define the term diversity factor. Prove that the load factor of a power system is improved by an increase in diversity factor.
3. Define the terms plant capacity factor and plant use factor and explain their importance in an electrical power system.
4. The load curve of an electrical system is linear with the following values at different times of the day :

Time	12 midnight	4 a.m.	9 a.m.	12 noon	5 p.m.	8 p.m.	12 midnight
Load (MW)	40	40	100	100	120	150	40

Plot the following curves :
(a) Chronological load curve
(b) Load-duration curve
(c) Load-energy curve
Calculate the energy required by the system in one day and the system daily load factor.
5. The load on a power station on a typical day is as follows :

Time	Load (MW)
12 midnight to 6 a.m.	40
6 a.m. to 10 a.m.	60
10 a.m. to 6 p.m.	120
6 p.m. to 10 p.m.	180
10 p.m. to 12 midnight	40

Plot the chronological load curve and load duration curve. Determine the load factor of the power station and the energy supplied by the power station in 24 hours.

If the installed capacity of the plant is 200 MW, determine the capacity factor and the utilization factor.

6. The maximum demand of a power station is 100 kW. The capacity factor is 0.6 and the utilization factor is 0.8. Find (a) load factor, (b) plant capacity, (c) reserve capacity, (d) annual energy production.

7. The maximum demand on a power station is 200 MW. If the annual load factor is 0.55, calculate the total energy generated in a year.

8. A generating station has a connected had of 600 MW and the maximum demand is 450 MW. The energy generated per year is 2×10^9 kWh. Calculate the demand factor and the load factor.

9. The yearly load duration curve of a power station is a straight line from 50 MW to 10 MW. Three alternators each of 20 MW are installed to meet the demand. Determine (a) the installed capacity, (b) plant factor, (c) maximum demand, (d) load factor and (e) utilization factor.

ANSWERS

4. 1995 MWh; 0.554   **5.** 0.518, 2240 MWh, 0.4667, 0.9
6. (a) 0.75, (b) 125 MW, (c) 25 MW, (d) 657000 MWh
7. 963600 MWh   **8.** 0.75, 0.5074
9. (a) 60 MW, (b) 0.5, (c) 50 MW, (d) 0.6, (e) 0.833

2

Supply Systems

2.1 INTRODUCTION

Large amount of power is generated at the generating stations. The present trend is to install bigger size of alternators to generate large amount of power to cater the required increasing demand. The site of the power station depends upon the type of power station. The new thermal stations are being constructed at pit heads (near the coal mines) because of the higher cost of transportation of coal. Hydropower station sites are governed by the availability of water resources. The nuclear plants are also situated remote from the centres of consumption due to safety reasons. Thus, the difficulty of getting power station sites near the consuming centres make it inevitable to transfer bulk of electrical energy through longer distances. Long-distance, bulk power transfer is only possible by high voltage transmission systems. Extra high voltage (EHV) and ultra high voltage (UHV) transmission systems have been developed in most of the countries for transporting energy from remote power stations. *By EHV is meant the voltage above 220 kV. The voltages above 760 kV are called ultra high voltages (UHV).*

2.2 BASIC STRUCTURE OF AN AC POWER SYSTEM

Electrical energy generated at generating stations is transported to remote load centres. Between a generating station and a consumer we have transmission, sub transmission and distribution levels of voltage. Since the long distance transmission at high voltages is cheap and, low voltages are required for utility purposes, the voltage level goes on decreasing from the transmission system to the distribution system. An electrical power system may be divided into three main components, namely, the generating system, the transmission system, and the distribution system.

2.3 DISTRIBUTION VOLTAGE LEVEL

The component of an electrical power system connecting all the consumers in an area to the bulk power sources is called a *distribution system*. Bulk power stations are connected to generating stations by transmission lines. They feed a number of substations which are usually situated at convenient points near the load centres. A substation distributes the power to domestic, commercial and relatively small consumers. The consumers requiring larger blocks of power are usually supplied at subtransmission or even transmission levels. A bulk power station or a generating station usually supplies power to a subtransmission system.

2.4 SUBTRANSMISSION LEVEL

The operation of a subtransmission system is similar to that of a distribution system. It differs from a distribution system in the following manner :

1. A subtransmission system has a higher voltage level than a distribution system.
2. It supplies only bigger loads.
3. It supplies only few substations as compared to a distribution system which supplies a number of loads.

2.5 TRANSMISSION LEVEL

A transmission system is quite different from either a subtransmission or a distribution system. It has a higher voltage level than a subtransmission system. It supplies only large blocks of power to bulk power stations or very big consumers. The third and the main consideration in which a transmission system differs from either a subtransmission or a distribution system is that it interconnects the neighbouring generating stations into a *power pool*. Thus, a transmission line performs two functions : First, the transport of energy from generating stations to bulk receiving stations, and second, the interconnection of two or more generating stations. The interconnection of two neighbouring subtransmission systems is also done by transmission lines.

2.6 LAYOUT OF A POWER SUPPLY NETWORK

Fig. 2.1 shows the layout of a typical ac power supply network by a single-line diagram. Electrical energy is generated by three-phase synchronous generators (alternators). The generation voltages are usually 11 kV and 33 kV. This voltage is too low for transmission over long distances. It is therefore, stepped upto 132, 220, 400 kV, or more by means of step-up transformers. At that voltage, the electrical energy is transmitted to a bulk power substation (receiving station) where energy is supplied from several power stations. The voltage at these substations is stepped down to 66 kV in India and fed to the subtransmission system for onward transmission to distribution sub-stations. These substations are located in the vicinity of the load centres. The voltage is further stepped down to 33 kV and 11 kV. Large industrial consumers are supplied at the primary distribution level of 33 kV while smaller industrial consumers are supplied at 11 kV. The voltage is stepped down further by distribution transformers located in residential and commercial areas, where it is supplied to these consumers at the secondary distribution level of 400 V three phase and 230 V single phase.

It is to be noted that it is not necessary that every power system network should have all the stages shown in Fig. 2.1. For example, in a particular network there may be no secondary transmission. Similarly, for a smaller power network there is only distribution and no transmission.

2.7 SYSTEM INTERCONNECTION

Two or more generating stations are interconnected by tie lines. Interconnection of generating stations has the following important advantages :

1. It enables the mutual transfer of energy from surplus zone to deficit zone economically.

2. Lesser overall installed capacity to meet the peak demand.

3. Lesser standby reserve generating capacity.

4. The size of the biggest unit is not related to the peak load of an individual system but it relates to the peak load of the interconnected system. Thus, fewer but larger machines of greater efficiency are to be installed.

5. It permits the generation of energy at the most efficient and cheapest stations at every time.

6. It reduces the capital cost, operating cost and cost of energy generated.

7. If there is a major breakdown of a generating system unit in an interconnected system there is no interruption of power supply. Similarly, when a machine is taken out of service for its scheduled maintenance and inspection, the continuity of supply is maintained. In other, words the planning of plant outages for maintenance and repair work is facilitated.

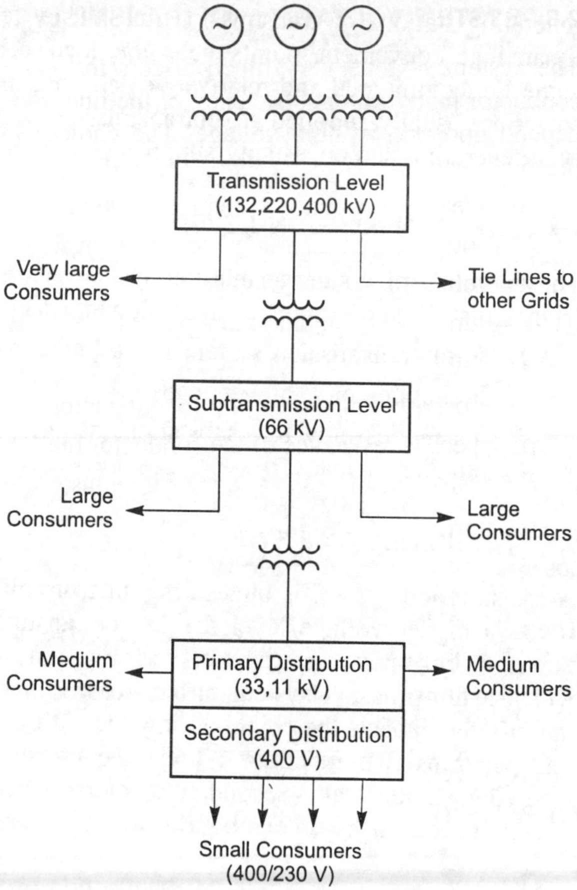

Fig. 2.1. A power system network showing various voltage levels.

Thus, interconnection provides the best use of power resources and ensures greater security of supply. It enables overall economic generation by optimum use of high capacity economical generating plants. Interconnection between networks is done either by HVAC links or HVDC links.

An interconnected power system covering a major portion of a country's territory (or state) is called a *grid*. The different grids may be interconnected through transmission lines (called *tie lines*) to form a *regional grid*. When the different regional grids are interconnected, they form a *national grid*.

Further details of interconnection are given in chapter 23.

2.8 SYSTEM VOLTAGE AND TRANSMISSION EFFICIENCY

The system voltage very much affects the capital cost of a transmission line. The weight of conductor material, the efficiency of the line, the voltage drop in the line and the system stability depend upon the system voltage. The choice of voltage, therefore, becomes a major factor in the line design.

Considering a n-phase system of transmission, let

P = power to be transmitted per phase in watts,

V = voltage to neutral in volts,

I = current in each phase in amperes,

l = length of the line in metres,

A = cross–sectional area of each conductor in m^2,

ρ = specific resistance of the conductor material in ohm–m,

R = resistance of each conductor in ohms,

α = current density in A/m^2, and

$\cos \varphi$ = power factor of the load.

Then, $P = VI \cos \varphi$; $I = \dfrac{P}{V \cos \varphi}$; $A = \dfrac{I}{\alpha} = \dfrac{P}{\alpha\, V \cos \varphi}$

The resistance of each conductor is given by

$$R = \rho \dfrac{l}{A} = \dfrac{\rho\, l\, \alpha\, V \cos \varphi}{P}$$

(a) Power Loss

The power loss in the line per phase, $p_L = I^2 R = \left(\dfrac{P}{V \cos \varphi}\right)^2 \cdot \dfrac{1}{P}\,(\rho\, l\, \alpha\, V \cos \varphi) = \dfrac{\alpha\, \rho\, l\, P}{V \cos \varphi}$ W

which shows that the power loss in the line is inversely proportional to both the system voltage and the power factor.

(b) Voltage Drop in Resistance

The voltage drop in resistance per phase $= IR = A\, \alpha\, \rho\, \dfrac{l}{A} = \alpha\, \rho\, l$

The percentage voltage drop in resistance in the line $= \dfrac{IR}{V} \times 100\% = \dfrac{\alpha\, \rho\, l}{V} \times 100\%$

The voltage drop in resistance per phase is, therefore, constant for a given value of current density and the percentage voltage drop in resistance decreases with the increase in the system voltage.

(c) Weight of Conductor Material

Let σ be the density of the material. The weight of conductor material required for each phase

$$= A\, l\, \sigma = \dfrac{P\, l\, \sigma}{\alpha\, V \cos \varphi}$$

This indicates that the weight of the conductor material required for the line will decrease with the increase in supply voltage and power factor.

(d) *Transmission Efficiency*

The efficiency of transmission, $\eta_T = \dfrac{\text{line output}}{\text{line output} + \text{line loss}} = \dfrac{P}{P + \dfrac{\alpha \rho l P}{V \cos \varphi}} = \dfrac{1}{1 + \dfrac{\alpha \rho l}{V \cos \varphi}}$

$$= \left(1 + \frac{\alpha \rho l}{V \cos \varphi}\right)^{-1} = 1 - \frac{\alpha \rho l}{V \cos \varphi} \quad \text{(approx.)}$$

The above relation shows that the efficiency of transmission also increases with the increase of supply voltage and power factor.

(e) *One more reason for using higher voltages is the enhancement of system stability.*

In the above discussion, for the effect of system voltages and power factor, P, l, ρ and α are assumed to be constants. The corona and leakage losses are neglected.

2.9 WORKING VOLTAGE

The above considerations indicate the desirability of using high voltages if the power is to be transmitted over a longer distance. It is also necessary that with the ac systems the load power factor should be as near to unity as possible for maximum economy. It is to be noted that heavy currents are more difficult to handle than high voltages. Considerable saving in the cost of the conductor material is possible when the voltage is high. Although much economy can be effected in the conductor material by employing extra high voltages, the cost of insulation of conductors, whether overhead or underground, increases appreciably. By adopting high voltages the electrical separation or clearances between the conductors is to be increased to avoid electrical discharge. The problems of mechanical supporting structures and right-of-way acquisition become more difficult and expensive.

The other problems encountered with high voltages are the insulation of the equipment, corona, radio and television interference. The insulation costs of transformers, switchgear and other terminal equipment increase tremendously. Corona, radio and television interference become very serious at extra high voltages. Thus, higher the voltage the more costly is the line. The voltage level of a system is, therefore, governed by the amount of power to be transmitted and the length of the line.

2.10 CHOICE OF THE NEXT HIGH VOLTAGE

The choice of the highest system voltage for a country is a matter of great significance. It is not merely the economic considerations to choose the next higher voltage but the site of power station, location and density of the load, and the technological developments of the times are also to be kept in view.

The next voltage level should be chosen on the basis of future requirements of load also. To determine the best system voltage the alternative plans should be compared by taking into account the power development over a period of about ten years. The long-term forecast should also include the development of power resources during that period.

The interval between the existing and the proposed level should not be too low as it will have a short life. At the same time too large an interval would lead to heavy expenses. It is, therefore, desirable to adopt the *skip-step basis* in which the next voltage chosen should be at least two steps higher than the existing one.

It is very necessary that a limited number of future EHV levels should be adopted and standardized so as to achieve the advantages of standardization more effectively without sacrificing the necessary flexibility of the existing systems and the proposed schemes.

2.11 STANDARDIZATION OF TRANSMISSION VOLTAGES

There is much variation in transmission voltages in different countries. A country adopts a voltage or a system of voltage levels to suit its requirements of load. Where long distance transmission is involved, the use of EHV becomes inevitable. Earlier, individual attempts were made to fix voltage levels for high power transmission but such an adoption of individual voltage levels resulted in waste of time. The designs due to their varied nature were costly. It was realized to standardize the transmission voltages for the following reasons :

(a) Standardization provides better facilities for research and development.

(b) The equipment can be manufactured economically with greater reliability.

(c) The maximum possible use of EHV for transmission and interconnection of EHV systems can be made.

The standardization would, therefore, avoid independent attacks to tackle EHV problems in different countries to find the optimum economic voltage and, thus, a lot of time is saved. By standardizing the equipment, the higher voltage can be adopted for a reasonable period of time before next change and, thus, the number of lines will also be lesser.

One of the difficulties encountered with in the standardization was to adopt the maximum continuous voltage or the nominal voltage to fix up a standard of system voltage. Again, there is a discrepancy in the matter of margin between these values in European and American practices. In the former case, the maximum value is ten per cent higher than the nominal value, while in the latter this difference is only five per cent. It was proposed to compensate the nominal values in such a way as to have the same maximum value in both the systems in order to avoid this margin. It was preferred to have the criterion of maximum value rather than the nominal value for the standardization purpose.

Earlier, standardization was very successful upto a voltage level of 230 kV. This voltage was the standard voltage used in many countries. Different opinions were put forward to standardize the system voltage above 230 kV and the international standardization was not so successful. The various voltages adopted by different countries above 230 kV are 275, 287, 345, 380, 400, 500, 735, 1100, 1200 kV, etc. The highest voltage used in India is 400 kV. The voltages above 765 kV are called ultra high voltages (UHV).

In India, a number of 400 kV lines are in operation since 1980. Even higher voltages of the order of 800 kV are needed in future. It is expected that by the year 2012, 5400 circuit kilometres of 800 kV and 48,000 circuit kilometres of 400 kV lines would be in operation.

2.12 CLASSIFICATION OF LINES

Lines may be classified either electrically or physically.

Electrical Classification

1. *Direct Current (DC) Systems*
   (a) DC 2-wire, 1-wire earthed.
   (b) DC 1-wire, earth return.
   (c) DC 2-wire, mid-point earthed.
   (d) DC 3-wire.
2. *Alternating Current (AC) Systems*
   (a) Single-phase, 2-wire, 1-wire earthed.
   (b) Single-phase, 1-wire, earth return.
   (c) Single-phase, 2-wire, mid-point earthed.
   (d) Single-phase, 3-wire.
   (e) Two-phase, 3-wire.
   (f) Two-phase, 4-wire.
   (g) Three-phase, 3-wire.
   (h) Three-phase, 4-wire.

All these systems are used in practice. Every system has got its own field of application. The line economics varies with the individual system. Direct current 2-wire, mid-point earthed system is used for transmitting large amounts of power over long distances.

Single-phase a.c. system is rarely used for transmission purposes. Three-phase 3-wire a.c. system finds its greatest use for transmission and primary distribution. For secondary distribution work three-phase, 4-wire system is common. The industrial and other big loads are supplied by three-phase, 4-wire system. The domestic and other small load consumers are supplied with single-phase power obtained from the individual phases of a three-phase supply.

Physical Classification

The other criterion of classification of lines is their physical configuration. The lines may be overhead, underground or underwater. Again, overhead lines may either have bare conductors or insulated conductors.

Most of the high voltage and extra high voltage lines have bare conductors since these are most economical. The conductors on overhead lines are sometimes insulated for safety or other considerations. The insulated aerial cables although costlier than the bare conductors are far cheaper than the underground cables. They are used in densely populated urban areas at relatively low voltages or for rural distribution work at medium voltages. Overhead lines and underground systems have been compared in Chapter 4.

For long river or sea crossing, it is not possible to erect overhead lines due to their longer spans. In such cases the need of underwater cables becomes inevitable. Such cables are very costly to manufacture and install.

2.13 COMPARISON OF CONDUCTOR COSTS IN VARIOUS SYSTEMS

The conductors in a line are insulated from each other and from the earth. The cost of insulation varies with the system. Since the conductor cost forms the bulk of expenditure of the line, it is necessary to compare it in various systems from that point of view. A fixed maximum voltage for the same insulation stress should always be specified to make such a comparison. With bare conductor overhead lines and single-core cables, the maximum voltage to earth is taken as the basis of comparing the conductor costs. With multicore belted type cables, it is the maximum voltage between the line conductors which forms the criterion of comparison.

Conductor efficacy is defined as the ratio of conductor material required for a given system to that required for the dc, 2-wire system with one wire earthed.

Let us now compare the volume of conductor material required for a given system to that required for the dc. 2-wire system with one wire earthed.

2.13.1 Criterion of Equal Maximum Voltage to Earth

This criterion is used for overhead lines and single-wire cables. The following assumptions are made for all the systems to be compared regarding the relative amount of conductor material required :

1. The same maximum voltage to earth, V.
2. The same power P to be transmitted.
3. The same route length l of the line.
4. The same efficiency of transmission, that is, the same I^2R losses for all the systems. Corona and leakage losses are neglected.
5. Balanced load so that there is no power loss in the neutral conductor.

General Procedure

1. Draw the circuit diagram of the given system. Mark the direction of currents and polarities of voltages. In ac systems the voltages and currents should be expressed in terms of their r.m.s. values.
2. Calculate the I^2R loss in *all* the conductors of the system and denote it by p_L.
3. Equate the I^2R loss of the system under consideration to the I^2R loss in the d.c. 2-wire system which is taken as the reference system.
4. Determine the volume of conductor material used in *all* the conductors of the system.
5. Determine the ratio of the volume of conductor found in step 4 to the volume of the conductor used in dc, 2-wire system with one-wire earthed.

Let us consider the following cases :

(a) *Direct current 2-wire system with one wire earthed*

The dc 2-wire system with one wire earthed is shown in Fig. 2.2.

Let I_1 = line current

R_1 = resistance of each conductor

A_1 = area of cross–section of each conductor

Power transmitted, $P = VI_1$

Line current $I_1 = \dfrac{P}{V}$

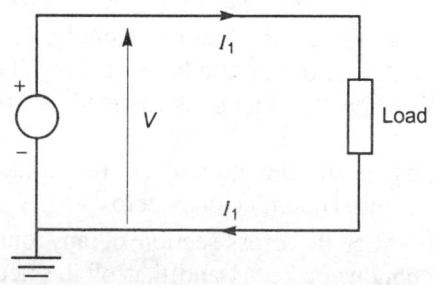

Fig. 2.2. Direct current 2-wire system with one wire earthed.

Total I^2R loss for both the conductors, $p_{L1} = 2I_1^2 R_1 = 2\left(\dfrac{P}{V}\right)^2 R_1$

Volume of conductor material used in both the conductors, $v_1 = 2lA_1$

(b) *Direct current 2-wire system with mid-point earthed*

The dc 2-wire system with mid-point earthed is called *dc 2-wire balanced system*. Such a system is shown in Fig. 2.3.

In this system,

line voltage $= 2V$

line current, $I_2 = \dfrac{P}{2V}$

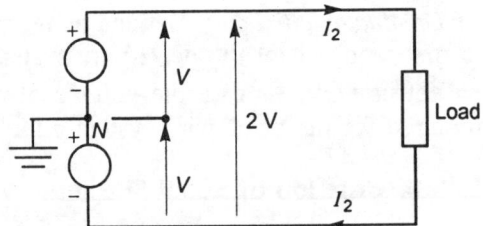

Fig. 2.3. Direct current 2-wire system with mid-point earthed.

If R_2 be the resistance per conductor, then the total I^2R loss for both conductors is

$$p_{L2} = 2I_2^2 R_2 = 2\left(\frac{P}{2V}\right)^2 R_2$$

Volume of conductor material used, $v_2 = 2lA_2$

Equating the I^2R loss for cases (a) and (b) for the same efficiency of transmission, we get

$$p_{L1} = p_{L2}$$

$$2\left(\frac{P}{V}\right)^2 R_1 = 2\left(\frac{P}{2V}\right)^2 R_2 \; ; \quad \frac{R_1}{R_2} = \frac{1}{4}$$

But $\quad R_1 = \rho\dfrac{l}{A_1}, \quad R_2 = \rho\dfrac{l}{A_2}$

$\therefore \quad \dfrac{R_1}{R_2} = \dfrac{A_2}{A_1} = \dfrac{1}{4} \; ; \quad \dfrac{v_2}{v_1} = \dfrac{2lA_2}{2lA_1} = \dfrac{A_2}{A_1} = \dfrac{1}{4} = 0.25$

Thus, the conductor material used with dc 2-wire mid point earthed system is 25 per cent of that required for dc 2-wire system with one wire earthed.

(c) *Direct current three-wire system*

Direct current three-wire system is shown in Fig. 2.4. In this system there are three conductors namely, two outers and one middle conductor. The middle conductor is earthed at the supply end. When the load is balanced, the current in the middle (neutral) conductor is zero.

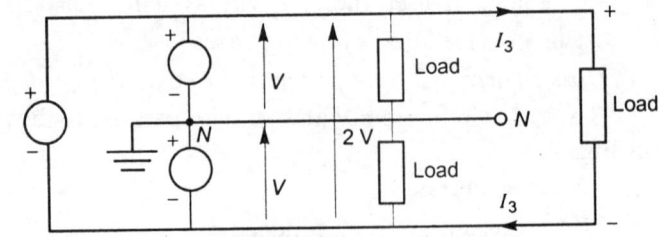

Fig. 2.4. Direct current 3-wire system.

Let the cross-section of any outer conductor be A_3 and that of the neutral wire be half that of any outer, that is $0.5A_3$.

Line voltage $= 2V$; Line current, $I_3 = \dfrac{P}{2V}$

I^2R loss in the system $= 2I_3^2 R_3 = 2\left(\dfrac{P}{2V}\right)^2 R_3$

Volume of conductor material required, $v_3 = 2lA_3 + 0.5lA_3 = 2.5lA_3$.

For the same efficiency of transmission in cases (a) and (c)

$p_{L1} = p_{L3}$

$2\left(\dfrac{P}{V}\right)^2 R_1 = 2\left(\dfrac{P}{2V}\right)^2 R_3$; $R_1 = \dfrac{1}{4} R_3$

$\rho \dfrac{l}{A_1} = \dfrac{1}{4} \rho \dfrac{l}{A_3}$; $\dfrac{A_3}{A_1} = \dfrac{1}{4}$

$\dfrac{v_3}{v_1} = \dfrac{2.5lA_3}{2lA_1} = \dfrac{2.5}{2 \times 4} = 0.3125$

Thus, the conductor material used with dc 3-wire system is 31.25 per cent of the material required for dc 2-wire system with one wire earthed.

(d) *Single-phase 2-wire system with one wire earthed*

This system is shown in Fig. 2.5.

Since V is the maximum voltage to earth, the rms voltage to earth is $V/\sqrt{2}$. If I_4 is the load current and $\cos\varphi$ the power factor of the load

$P = \dfrac{V}{\sqrt{2}} I_4 \cos\varphi; \quad I_4 = \dfrac{\sqrt{2}\,P}{V\cos\varphi}$

Let R_4 be the resistance of each conductor of cross section A_4.

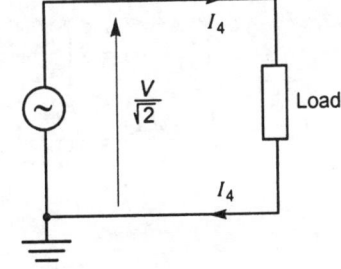

Fig. 2.5. Single-phase, 2-wire system with one wire earthed.

Line loss, $p_{L4} = 2I_4^2 R_4 = 2\left(\dfrac{\sqrt{2}\,P}{V\cos\varphi}\right)^2 R_4$

Volume of conductor material used, $v_4 = 2lA_4$

For the same efficiency of transmissions as in case (a)

$p_{L1} = p_{L4}$

$2\left(\dfrac{P}{V}\right)^2 R_1 = 2\left(\dfrac{\sqrt{2}\,P}{V\cos\varphi}\right)^2 R_4$; $\dfrac{R_1}{R_4} = \dfrac{2}{\cos^2\varphi}$

But $\dfrac{R_1}{R_4} = \dfrac{A_4}{A_1}$ $\left(R = \rho \dfrac{l}{A}\right)$

$\dfrac{v_4}{v_1} = \dfrac{2lA_4}{2lA_1} = \dfrac{2}{\cos^2\varphi}$

(e) *Single-phase, 2-wire mid-point earthed system*

A single-phase, 2-wire mid-point earthed system is shown in Fig. 2.6.

Maximum voltage between any conductor and earth = V

Maximum voltage between two line conductors = $2V$

Therefore, r.m.s. voltage between the line conductors $= \dfrac{2V}{\sqrt{2}} = \sqrt{2}\, V$

If I_5 be the line current, R_5 the resistance of each conductor of cross-section A_5, and $\cos \varphi$ the power factor of the load

$$P = (\sqrt{2}\, V)\, I_5 \cos \varphi \ ; \ I_5 = \frac{P}{\sqrt{2}\, V \cos \varphi}$$

Line loss, $p_{L5} = 2 I_5^2 R_5 = 2\left(\dfrac{P}{\sqrt{2}\cos \varphi}\right)^2 R_5$

Volume of conductor material used, $v_5 = 2 l A_5$

For the same efficiency transmission as in case (a)

$$p_{L1} = p_{L5}$$

$$2\left(\frac{P}{V}\right)^2 R_1 = 2\left(\frac{P}{\sqrt{2}\, V \cos \varphi}\right)^2 R_5 \ ; \quad \frac{R_1}{R_5} = \frac{0.5}{\cos^2 \varphi} = \frac{A_5}{A_1}$$

$$\frac{v_5}{v_1} = \frac{2 l A_5}{2 l A_1} = \frac{0.5}{\cos^2 \varphi}$$

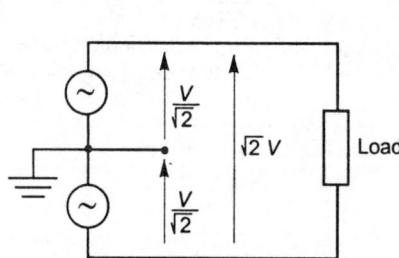

Fig. 2.6. Single-phase, 2-wire mid-point earthed system.

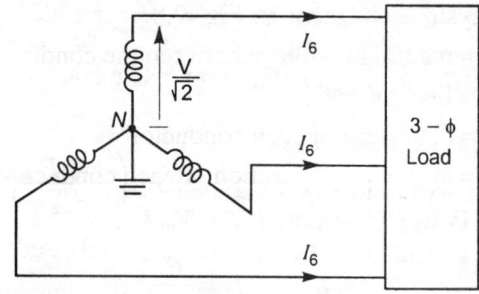

Fig. 2.7. Three-phase, 3-wire star connected system.

(f) *Three-phase, 3-wire star-connected system*

This system is shown in Fig. 2.7.

Maximum voltage to earth = V

Therefore, r.m.s. value of voltage per phase $= \dfrac{V}{\sqrt{2}}$

If I_6 is the rms phase current, R_6 the resistance of each conductor of cross section A_6, and $\cos \varphi$ the power factor of the load

$$P = 3V_{ph} I_{ph} \cos \varphi = 3 \times \frac{V}{\sqrt{2}} I_6 \cos \varphi \; ; \qquad I_6 = \frac{\sqrt{2}\, P}{3V \cos \varphi}$$

Line loss, $p_{L6} = 3I_6^2 R_6 = 3 \left(\dfrac{\sqrt{2}\, P}{3V \cos \varphi} \right)^2 R_6$

Volume of conductor material used, $v_6 = 3lA_6$

For the same efficiency of transmissions as in case (a)

$$p_{L1} = p_{L6}$$

$$2 \left(\frac{P}{V} \right)^2 R_1 = 3 \left(\frac{\sqrt{2}\, P}{3V \cos \varphi} \right)^2 R_6 \; ; \qquad \frac{R_1}{R_6} = \frac{1}{3 \cos^2 \varphi} = \frac{A_6}{A_1}$$

$$\frac{v_6}{v_1} = \frac{3lA_6}{2lA_1} = \frac{0.5}{\cos^2 \varphi}$$

2.13.2   Criterion of Equal Maximum Voltage between Line Conductors

We shall now compare the systems on the basis of the same maximum voltage between two line conductors. Such a criterion is important with multicore cables. Again, it shall be assumed that the the amount of power transmitted, route length and efficiency of transmission remain the same in all the systems. Corona and leakage losses are neglected and the system is assumed to be balanced so that there is no power loss in the neutral conductor. The ratio of conductor material required in any system to that required for the dc 2-wire system with one-wire earthed is found in the following cases. We shall use the same general procedure discussed in section 2.13.1.

(a) *DC 2-wire system*

    This system is shown in Fig. 2.8.

Let V_m = maximum voltage between the conductors

    I_1 = line current

    R_1 = resistance of each conductor

    A_1 = area of cross–section of each conductor

    Power transmitted, $P = V_m I_1$

$$I_1 = \frac{P}{V_m}$$

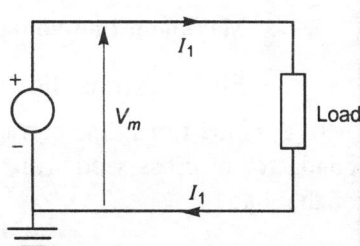

Fig. 2.8. Direct current, 2-wire system.

Line losses, $p_{L1} = 2I_1^2 R_1$

$$= 2 \left(\frac{P}{V_m} \right)^2 R_1$$

Volume of conductor material used, $v_1 = 2lA_1$

(b) *Direct-current, 2-wire mid-point earthed system*

    This system is shown in Fig. 2.9. Since this system is the same as (a), the conductor material used is also the same.

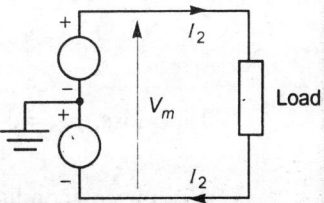

Fig. 2.9. Direct current, 2-wire mid-point earthed system.

(c) *Single-phase, 2-wire system*

This system is shown in Fig. 2.10.

Maximum voltage between line conductors = V_m

Effective (rms) line voltage = $V_m/\sqrt{2}$.

If I_3 is the r.m.s. line current, R_3 the resistance of each conductor of cross section A_3 and cos ϕ the power factor of the load

$$P = \frac{V_m}{\sqrt{2}} I_3 \cos \varphi, \quad I_3 = \frac{\sqrt{2}\, P}{V_m \cos \varphi}$$

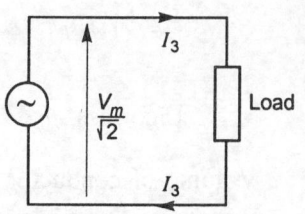

Fig. 2.10. Single-phase, 2-wire system.

Line loss, $p_{L3} = 2I_3^2 R_3 = 2\left(\dfrac{\sqrt{2}\, P}{V_m \cos \varphi}\right)^2 R_3$

Volume of conductor material used, $v_3 = 2lA_3$

For the same efficiency of transmission as in case (a),

$p_{L1} = p_{L3}$

$$2\left(\frac{P}{V_m}\right)^2 R_1 = 2\left(\frac{\sqrt{2}\, P}{V_m \cos \varphi}\right)^2 R_3$$

$$\frac{R_1}{R_3} = \frac{2}{\cos^2 \varphi} = \frac{A_3}{A_1}; \quad \frac{v_3}{v_1} = \frac{2lA_3}{2lA_1} = \frac{2}{\cos^2 \varphi}$$

(d) *Three-phase, 3-wire star-connected system*

Three-phase, 3-wire, star-connected system is shown in Fig. 2.11.

Maximum line voltage = V_m

Effective (rms) line voltage = $V_m/\sqrt{2}$

If I_4 is the r.m.s. line current, R_4 the resistance of each conductor of cross section A_4 and cos φ the power factor of the load

$$P = \sqrt{3}\, V_1 I_1 \cos \varphi = \sqrt{3} \times \frac{V_m}{\sqrt{2}} I_4 \cos \varphi$$

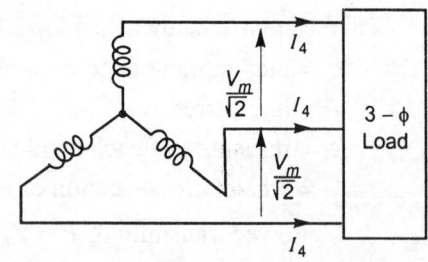

Fig. 2.11. Three-phase, 3-wire, star-connected system.

$$I_4 = \frac{\sqrt{2}\, P}{\sqrt{3}\, V_m \cos \varphi}$$

Line loss, $p_{L4} = 3I_4^2 R_4 = 3\left(\dfrac{\sqrt{2}\, P}{\sqrt{3}\, V_m \cos \varphi}\right)^2 R_4$

Volume of conductor material used, $v_4 = 3lA_4$

For the same efficiency of transmission as in case (a)

$p_{L1} = p_{L4}$

$$2\left(\frac{P}{V_m}\right)^2 R_1 = 3\left(\frac{\sqrt{2}\,P}{\sqrt{3}\,V_m\cos\varphi}\right)^2 R_4$$

$$\frac{R_1}{R_4} = \frac{1}{\cos^2\varphi} = \frac{A_4}{A_1}\;;\qquad \frac{v_4}{v_1} = \frac{3lA_4}{2lA_1} = \frac{1.5}{\cos^2\varphi}$$

(e) *Three-phase, 4-wire system*

This system is shown in Fig. 2.12. For a balanced load this system is the same as the three-phase, 3-wire system except that there is an additional neutral wire.

If the cross section of the neutral wire is half that of any line conductor, the ratio of the amount of conductor material required in this case and the d.c., 2-wire one wire earthed system is

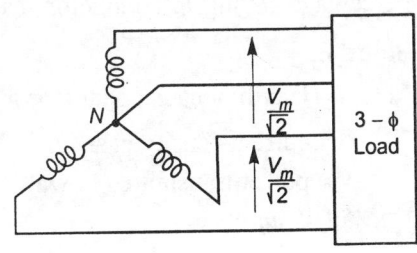

Fig. 2.12. Three-phase, 4-wire system.

$$\frac{3.5lA_4}{2lA_1} = \frac{1.75}{\cos^2\varphi}$$

In case the cross section of the neutral is taken to be equal to that of any line conductor, the ratio becomes

$$\frac{4lA_4}{2lA_1} = \frac{2}{\cos^2\varphi}$$

Table 2.1 gives the comparison of conductor material in various systems.

Table 2.1   Conductor Efficacies

System	Same maximum voltage to earth	Same maximum voltage between conductors
DC 2-wire, 1-wire earthed	1.0	1.0
DC 2-wire, mid-point earthed	0.25	1.0
DC 3-wire (neutral = ½ live)	0.3125	1.25
Single-phase, 2-wire	$2/\cos^2\varphi$	$2/\cos^2\varphi$
Single-phase, 2-wire, mid-point earthed	$0.5/\cos^2\varphi$	$2/\cos^2\varphi$
Single-phase, 2-wire (neutral = ½ live)	$0.625/\cos^2\varphi$	$2.5/\cos^2\varphi$
Three-phase, 3-wire	$0.5/\cos^2\varphi$	$1.5/\cos^2\varphi$
Three-phase, 4-wire (neutral = live)	$0.667/\cos^2\varphi$	$2/\cos^2\varphi$

From the Table 2.1, it is observed that on the basis of equal maximum voltage to earth the dc 2-wire mid-point earthed system is the cheapest on the basis of conductor cost alone. The conductor efficacies are the same in single-phase 2-wire and three-phase 3-wire systems. If equal maximum voltage between the conductors is made the basis of comparison, it is found that the dc balanced 2-wire system is the cheapest again and in ac three-phase, 3-wire star-connected system is more economical than single-phase system. It is also observed that the criterion of equal maximum voltage to earth is more economical and, therefore, the overhead lines will be cheaper.

Example 2.1 In a dc 2-wire system a feeder is working on 250 V supplying a constant load. If the supply voltage is increased to 400 V with the same power transmitted, calculate the percentage saving in conductor material.

Solution

Let l = length of each conductor in metres

α = current density in A/m^2

P = power transmitted in watts

(a) *250 V supply*

$$\text{Line current, } I_1 = \frac{P}{V_1} = \frac{P}{250}$$

$$\text{Cross-sectional area of conductor required, } A_1 = \frac{I_1}{\alpha} = \frac{P}{250\,\alpha} \text{ m}^2$$

$$\text{Volume of conductor material required, } v_1 = 2lA_1 = \frac{2lP}{250\,\alpha} = \frac{lP}{125\,\alpha} \text{ m}^3$$

(b) *400 V supply*

$$\text{Line current } I_2 = \frac{P}{V_2} = \frac{P}{400}$$

$$\text{Cross-sectional area of conductor required, } A_2 = \frac{I_2}{\alpha} = \frac{P}{400\,\alpha} \text{ m}^2$$

$$\text{Volume of conductor material required, } v_2 = 2lA_2 = \frac{2lP}{400\,\alpha} = \frac{lP}{200\,\alpha} \text{ m}^3$$

$$\frac{v_2}{v_1} = \frac{lP}{200\,\alpha} \div \frac{lP}{125\,\alpha} = \frac{125}{200} = 0.625$$

$$\text{Percentage saving in conductor material, } = \frac{v_1 - v_2}{v_1} \times 100 = \left(1 - \frac{v_2}{v_1}\right) \times 100$$

$$= (1 - 0.625) \times 100 = 37.5 \text{ per cent.}$$

Example 2.2 A three-phase 4-wire system is used for lighting. Compare the amount of conductor material required with that needed for a 2-wire dc system with same lamp voltage. Assume the same losses and balanced load. The neutral wire has half the cross-section of the outers.

Solution

(a) *DC 2-wire system*

Let V = voltage between conductors

P = power delivered

I_1 = line current

R_1 = resistance of each conductor

A_1 = conductor cross–section

$$\text{Line current } I_1 = \frac{P}{V} \; ; \qquad \text{Line loss } = 2I_1^2 R_1 = 2\frac{P^2}{V^2} R_1$$

Volume of conductor material required, $v_1 = 2lA_1$

(b) *AC three phase, 4-wire system*

Effective (rms) voltage between line and neutral = V.

If I_2 = line current

$\cos \varphi$ = power factor

R_2 = resistance of each conductor

A_2 = area of cross–section of each phase conductor

$P = 3VI_2 \cos \varphi$

For lighting load, $\cos \varphi = 1$

$\therefore \qquad I_2 = \dfrac{P}{3V}$

$$\text{Total line loss } = 3I_2^2 R_2 = 3\left(\frac{P}{3V}\right)^2 R_2 = \frac{P^2 R_2}{3V^2}$$

For equal line loss in both the systems

$$\frac{2P^2}{V^2} R_1 = \frac{P^2 R_2}{3V^2} \; ; \qquad \frac{R_1}{R_2} = \frac{1}{6}$$

But $\qquad \dfrac{R_1}{R_2} = \dfrac{A_2}{A_1} \; ; \quad \dfrac{A_2}{A_1} = \dfrac{1}{6}$

Volume of conductor material required, $v_2 = 3lA_2 + 0.5lA_2 = 3.5lA_2$

$$\frac{v_2}{v_1} = \frac{3.5lA_2}{2lA_1} = \frac{3.5}{2 \times 6} = 0.292$$

Example 2.3 A 50 km long transmission line supplies a load of 5 MVA at 0.8 power factor lagging at 33 kV. The efficiency of transmission is 90 per cent. Calculate the volume of conductor aluminium required for the line when

(a) single-phase, 2-wire system is used.

(b) three-phase, 3-wire system is used.

Take the resistivity of aluminium as 2.85×10^{-8} Ω-m.

Solution

Power transmitted = MVA $\cos \varphi = 5 \times 0.8$ MW $= 4 \times 10^6$ W

Efficiency of transmission = 90 per cent.

$\therefore \qquad$ line loss, p_L = 10 per cent of power transmitted $= \dfrac{10}{100} \times 4 \times 10^6 = 4 \times 10^5$ W

(a) *Single-phase system*

$$MVA = VI \times 10^{-6}$$

$$5 = 33 \times 10^3 \times I \times 10^{-6} ; \qquad I = 151.5 \text{ A}$$

Let A_1 be the area of cross-section of each conductor.

Line loss, $p_L = 2I^2 R = 2I^2 \rho \dfrac{l}{A_1}$

$$A_1 = \frac{2I^2 \rho l}{p_L} = \frac{2 \times (151.5)^2 \times 2.85 \times 10^{-8} \times 50 \times 10^3}{4 \times 10^5} = 1.635 \times 10^{-4} \text{ m}^2$$

Volume of conductor required, $v_1 = 2lA_1 = 2 \times 50 \times 10^3 \times 1.635 \times 10^{-4} = 16.35 \text{ m}^3$

(b) *Three-phase 3-wire ac system*

$$MVA = \sqrt{3} \ V_L I_L \times 10^{-6}$$

$$5 = \sqrt{3} \times 33 \times 10^3 \times I_L \times 10^{-6}$$

Line current, $I_L = \dfrac{5 \times 10^6}{\sqrt{3} \times 33 \times 10^3} = 87.5 \text{ A}.$

Let the area of each phase conductor be A_2.

Total line loss, $p_L = 3I_L^2 R_2 = 3I_L^2 \rho \dfrac{l}{A_2}$

$$A_2 = \frac{3I_L^2 \rho l}{p_L} = \frac{3 \times (87.5)^2 \times 2.85 \times 10^{-8} \times 50 \times 10^3}{4 \times 10^5} = 0.818 \times 10^{-4} \text{ m}^2$$

Volume of aluminium required, $v_2 = 3lA_2 = 3 \times 50 \times 10^3 \times 0.818 \times 10^{-4} = 12.27 \text{ m}^3$

Example 2.4 An existing single phase a.c. system comprising of two overhead conductors is to be converted into a three-phase, 3-wire system by providing an additional similar conductor, Calculate the percentage of additional load that can be transmitted by the three-phase system if the operating line voltage and percentage line losses remain the same in both the systems.

Solution

Let I_1 = single–phase current

       I_3 = three–phase current

       R = resistance of each conductor

$\cos \varphi$ = power factor in both the system

(a) *Single-phase system*

Power transmitted, $P_1 = VI_1 \cos \varphi$

Line loss $= 2I_1^2 R$

Percentage line loss $= \dfrac{2I_1^2 R}{VI_1 \cos \varphi} \times 100$

(b) *Three-phase system*

Power transmitted, $P_3 = \sqrt{3} \, VI_3 \cos \varphi$

Line loss $= 3I_3^2 R$

Percentage line loss $= \dfrac{3I_3^2 R}{\sqrt{3} \, VI_3 \cos \varphi} \times 100$

For the percentage line losses to be the same in both the systems, we have

$$\frac{2I_1^2 R}{VI_1 \cos \varphi} \times 100 = \frac{3I_3^2 R}{\sqrt{3} \, VI_3 \cos \varphi} \times 100$$

which gives

$$I_3 = \frac{2}{\sqrt{3}} I_1$$

$\therefore \qquad P_3 = \sqrt{3} \, V \left(\dfrac{2}{\sqrt{3}} I_1 \right) \cos \varphi = 2VI_1 \cos \varphi$

Additional load transmitted $= P_3 - P_1 = 2VI_1 \cos \varphi - VI_1 \cos \varphi = VI_1 \cos \varphi$

Percentage additional load $= \dfrac{P_3 - P_1}{P_1} \times 100 = \dfrac{VI_1 \cos \varphi}{VI_1 \cos \varphi} \times 100 = 100$

Thus, 100 per cent additional load can be transmitted by converting single-phase line to three-phase line.

Example 2.5 A given amount of power is to be transmitted by an overhead line. Compare the diameter and weight of aluminium conductor with those of copper for the same power loss in the line. The following data may be assumed :

Specific resistance of aluminium = 2.85 $\mu\Omega$-cm

Specific resistance of copper = 1.70 $\mu\Omega$-cm

Specific gravity of aluminium = 2.71

Specific gravity of copper = 8.89

Solution

For the same power loss in the line, the resistances of aluminium and copper will be the same. If the symbols have their usual meanings and the suffixes a and c be used for aluminium and copper

$$R_a = R_c \; ; \qquad \rho_a \frac{l}{A_c} = \rho_c \frac{l}{A_a} \; ; \qquad \frac{A_a}{A_c} = \frac{\rho_a}{\rho_c}$$

$$\frac{\dfrac{\pi}{4} d_a^2}{\dfrac{\pi}{4} d_c^2} = \frac{2.85 \times 10^{-6}}{1.70 \times 10^{-6}}$$

$$\frac{d_a}{d_c} = \sqrt{\frac{2.85}{1.70}} = 1.293$$

where d_a = diameter of aluminium conductor, and

d_c = diameter of copper conductor

If σ denotes the specific gravity,

$$\frac{\text{weight of aluminium}}{\text{weight of copper}} = \frac{\sigma_a A_c l}{\sigma_c A_a l} = \frac{2.71}{8.89} \times \frac{2.85}{1.70} = 0.51$$

This relation shows that the weight of an aluminium conductor is 51.1 per cent of the copper conductor of equal resistance.

EXERCISES

1. A 2-wire dc transmission line, 0.25 km long supplies a power of 125 kW to a distribution network. Calculate the volume of conductor copper required for the line when the potential difference (pd) maintained between the distributors is (a) 250 V and (b) 400 V, the current density in the copper in each case being 3A/mm². Calculate also in each case the power loss in the line, taking the resistivity of copper as 1.7×10^{-8} Ω-m.

2. Compare the relative weights of copper required for a distribution network on the dc 3-wire, and three-phase 4-wire systems. Assume in both cases the same voltage at consumers' terminals, the same copper losses, that the loads are balanced, and unity power factor in the three-phase case. Neglect the losses in the neutrals.

3. Calculate the ratio of the weights of copper used in two-phase 4-wire, and 2-wire dc systems. Assume the same power transmitted, the same maximum voltage between conductors and the same power loss. Take the power factor in ac system as $\cos \phi$.

4. Compare the weights of copper required in a cable transmission system utilizing (a) constant direct current, (b) single-phase (c) three-phase 3-wires. Assume the same transmitted power and maximum voltage between conductors, and the same percentage loss in each case. Assume also unity power factor and balanced load.

5. A 3-wire dc system is to be converted to a three-phase 4-wire ac system by addition of another conductor equal in section to one of the outers. If the percentage copper loss and the voltage at the consumer terminals are to be remain the same in the two cases, calculate the additional percentage load that can be supplied by the ac system. Assume the balanced load in the two cases and unity power factor in the case of ac system.

6. A single-phase load is transmitted by a pair of overhead conductors carried on similar insulators but with one line operated at earth potential. A third conductor of the same cross-section is added and a three-phase supply connected instead of the single-phase one. Calculate the percentage increase in power transmitted for the same loss. The voltage to earth for each conductor is to be the same as the voltage between lines in the single-phase case. Assume constant power factor.

7. Electric power of 50 MW at 0.85 power factor lagging is to be transmitted over a 220 kV, three-phase 3-wire 200 km transmission line. The efficiency of transmission is 0.9. Calculate the weight of conductor material required for the line in the following conductors :

(a) copper conductors, (b) aluminium conductors.

Resistivity of copper $= 1.70 \times 10^{-8}$ Ω-m

Resistivity of aluminium $= 2.85 \times 10^{-8}$ Ω-m

Specific gravity of copper $= 8.89$

Specific gravity of aluminium $= 2.71$.

ANSWERS

1. (a) 0.0833 m^3, 12.75 kW (b) 0.0521 m^3, 7.96 kW
2. 15 : 14
3. $2/\cos^2 \varphi$
4. a : b : c = 1 : 2 : 1.5
5. 50%
6. 144.9%
7. (a) 259339 kg (b) 132535 kg

3

Conductors

3.1 INTRODUCTION

The important factors taken into account in the selection of a conductor for a particular line are conductivity, tensile strength, fatigue strength, corona loss, local conditions and cost. The conductor materials mainly used are copper, aluminium and their alloys.

The conductors are usually stranded. Stranded conductors have not only greater flexibility but also greater mechanical strength than have single wires of the same cross-sectional area. Usually a central wire is surrounded by successive layers of wires containing 6, 12, 18, 24, ... wires. In practice the consecutive layers are spiralled in opposite directions to prevent unwinding and make the outer radius of one layer coincide with the inner radius of the next.

If there are n layers of strands of equal diameter in a circular strand formation with one central strand, the following general formulae are applicable :

Total number of conductors in a strand of n layers $= 1 + 3n(1 + n)$

Overall diameter of stranded conductor with n layers, $D = (1 + 2n)d$

where d is the diameter of each strand. Thus, a 7-strand conductor has a central strand with 6 outers; the 19-strand conductor has a central strand with 6 strands in the first layer and 12 strands in the next layer.

The conductor size is specified by its equivalent copper cross-sectional area and the number of strands with the diameter of each strand. For example, a conductor made up of 19 strands each of diameter 2.9 mm having an equivalent copper area of 130 mm^2 is designated as 130 mm^2, 19/2.9 mm conductor.

The overall diameter of this conductor can be calculated as follows :

Number of wires $= 19$

Therefore, $1 + 3n(1 + n) = 19$

$$n^2 + n - 6 = 0$$

$$(n + 3)\,(n - 2) = 0, \qquad n = 2$$

Overall diameter of the conductor, $D = (1 + 2n)\,d = (1 + 4) \times 29 = 14.5$ mm

The *equivalent cross-section* of a stranded conductor is the area of cross-section of a solid conductor of the same material and length as the stranded conductor and having the same resistance at the same temperature. For convenience, the conductors are identified by their code names assigned by the manufacturers. Usually names of some animals, birds or flowers are used.

3.2 TYPES OF CONDUCTORS

Hard-drawn copper, hard-drawn aluminium, and steel-cored aluminium conductors are most commonly used. In addition to these, various other materials are used for making conductors but their use is limited. Some of the important types of conductors are given in the discussion as follows.

3.2.1 Hard-Drawn Copper Conductors

Copper for overhead lines is hard-drawn to give a relatively high tensile strength. It has a high electrical conductivity, long life, and high scrap value. Other properties of hard-drawn copper are given in Table 3.1 along with the properties of hard-drawn aluminium. Copper conductor is most suitable for distribution work where spans are short and tappings are more.

Table 3.1. Electrical and Mechanical Characteristics of Hard-Drawn Aluminium and Copper Wires

	Hard-drawn aluminium	*Hard-drawn copper*
Conductivity at 20°C IACS*	61	97.4
Resistivity at 20°C (microhm-cm)	2.8264	1.774
Resistivity temperature coefficient (microhm-cm °C)	0.0115	0.00681
Constant mass temperature coefficient of resistance per °C at 20°C	0.00403	0.00381
Coefficient of linear expansion per °C	2.3×10^{-5}	1.7×10^{-5}
Density at 20°C (gm/cm^3)	2.703	8.89
Ultimate tensile strength (kgf/mm^2)	16.21	35-47
Final modulus of elasticity (kgf/mm^2)	7000 average	12700 average

* International Annealed Copper Standard

3.2.2 Cadmium Copper Conductor

The tensile strength of copper is increased by approximately 50 per cent by adding about 0.7 to 1.0 per cent cadmium to it. The conductivity is, however, reduced by about 15 to 17 per cent. The property of higher tensile strength enables it to be erected on longer spans with the higher tensile strength enables it to be erected on longer spans with the same sag. Like hard-drawn copper this alloy possesses the advantage of easy jointing, more resistance to atmospheric corrosion, better resistance to wear, easy machinability, etc. The temperature at which copper anneals and softens is also increased and temperature effects on stresses are less. The variations in sag due to changes in load and temperature are minimum. It has same Young's modulus of elasticity and coefficients of linear expansion as those of ordinary copper conductor. The supports required to carry these conductors are short as the conductors are subjected to low wind and ice loadings due to their smaller diameter. The smaller diameters render these conductors unsuitable to be used on high voltage lines where corona losses are serious. Usually, their use is confined to the lines posing no problems of corona.

3.2.3 Steel-Cored Copper Conductor (SCC)

One or more layers of copper strands surround a steel-core to make steel-cored copper conductors. The steel-core adds to the tensile strength of the conductors. The core is provided with bituminised cotton tape in order to protect the conductor from the galvanic action.

3.2.4 Copperweld Conductor

Copper is welded on to a steel wire by hot rolling and cold drawing a billet of steel coated with copper. It is ensured that the uniform thickness of copper is welded. The conductivity of copperweld conductor varies from 30 to 60 per cent of that a solid copper conductor with the same diameter. The conductivity of the standard grade is about 40 per cent. The modulus of elasticity is about 16800 kgf/mm^2, and coefficient of linear expansion is $1.296 \times 10^{-5}/^\circ$ C. Copperweld conductors may be used for longer spans such as river crossings.

3.2.5 Hard-Drawn Aluminium Conductor or All-Aluminium Conductor (AAC)

The increasing trend in the cost of copper has resulted in replacement of copper and adoption of aluminium for transmission work. At least 99.5 per cent electrolytically refined aluminium is rolled and hard drawn for conductor use. For a given resistance the cross-section of the aluminium conductor is 60 per cent greater than that of copper and its weight is only 48.3 per cent of that of copper conductor. Thus, handling, transportation and erection becomes economical. Corona effect is also reduced due to higher conductor diameter. Aluminium conductors are used for distribution lines in urban areas and short transmission lines with the lower voltages. Where high winds are frequent, aluminium conductors due to their lightness, large diameter and greater sag, are more likely to swing resulting in inter-phase faults. Also, line variations due to wind can cause failure of conductors through fatigue at the attachment points. Aluminium is somewhat inferior to copper in resisting fatigue.

3.2.6 Aluminium Conductor Steel Reinforced (ACSR)

Conductors made of all-aluminium are not sufficiently strong mechanically for construction of long span lines. This deficiencies in strength can be compensated by adding a steel core to the conductor. Such a conductor is called steel-cored aluminium (SCA) conductor or aluminium conductor steel reinforced ACSR. Fig. 3.1 shows the cross-section of a ACSR conductor. It has

7 steel strands forming a central core around which there are two layers of 30 aluminium strands. The conductor stranding is specified as 30 Al/7 St. For a given resistance, conductors of different strengths can be made by taking different proportions of steel and aluminium areas. Only the conductivity of aluminium conductor is taken for calculating current carrying capacity and resistance. The conductivity of steel-core is neglected for practical purposes.

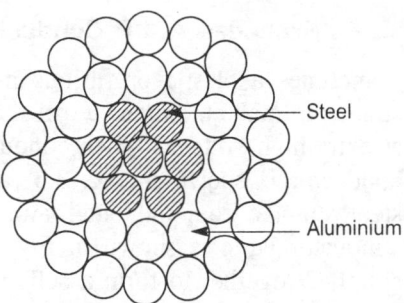

Fig. 3.1. ACSR conductor.

The important advantages of ACSR conductors are high tensile strength and light weight. The sag is, therefore, small and the line can be designed with shorter supports or longer spans for a given sag. The higher cross-section of the conductor enables to increase corona inception level. Heavier supports are required because of greater loading on them, but they are fewer in number. Insulators and other fittings needed are also less. As the supports are smaller in number the breakdown possibility is reduced.

The presence of steel in ACSR conductors creates a difficulty in making splices and dead ends. In urban construction this is particularly important. The other trouble is that of corrosion, which is due to an electrochemical action between aluminium and steel core. Corrosion rate depends on service conditions. In industrial and coastal areas this galvanic corrosion may be serious to affect the life of conductor. The application of grease on the steel-core helps in reducing the corrosion, but this practice is now avoided due to corona.

3.2.7 Smooth Body ACSR Conductor

A smooth body ACSR conductor is also called *Compacted* ACSR. Conventional ACSR conductor [Fig. 3.2 (a)] is pressed through dies to flatten the aluminium strands into segmental shape. The interstrand spaces are filled, and the diameter of the conductor is reduced without affecting its electrical or mechanical properties. The compacted ACSR conductor is shown in Fig. 3.2 (b). Thus for the same aluminium area the diameter of the steel core can be increased to increase the mechanical strength. Such a conductor can be made with different ratios of aluminium to steel. In Figs. 3.2 (a) and 3.2 (b) the ratio is 6 Al/1 St. It is possible to use these conductors on span lengths greater than those with conventional ACSR or given supports due to their reduced diameter and increased strength.

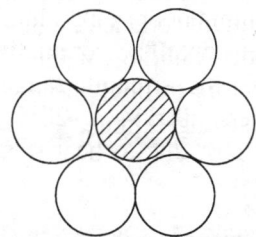

Fig. 3.2. (a) Conventional ACSR conductor.

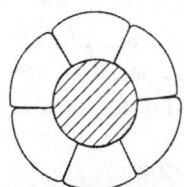

Fig. 3.2. (b) Compacted ACSR conductor.

3.2.8 Expanded ACSR Conductor

Sometimes a plastic or fibrous material is introduced between the steel core and aluminium strands to make the diameter of the conductor large to reduce corona loss and radio interference at extra high voltages. Such conductors are called *expanded conductors*. A typical form of such conductor is shown in Fig. 3.3 (a). It has a filler material such as paper separating the inner steel strands from the outer aluminium strands. Another design to increase the diameter of conductor large is shown in Fig. 3.3 (b). It consists of tongue and grooved copper segments spiralled together to form a self-supporting hollow tube.

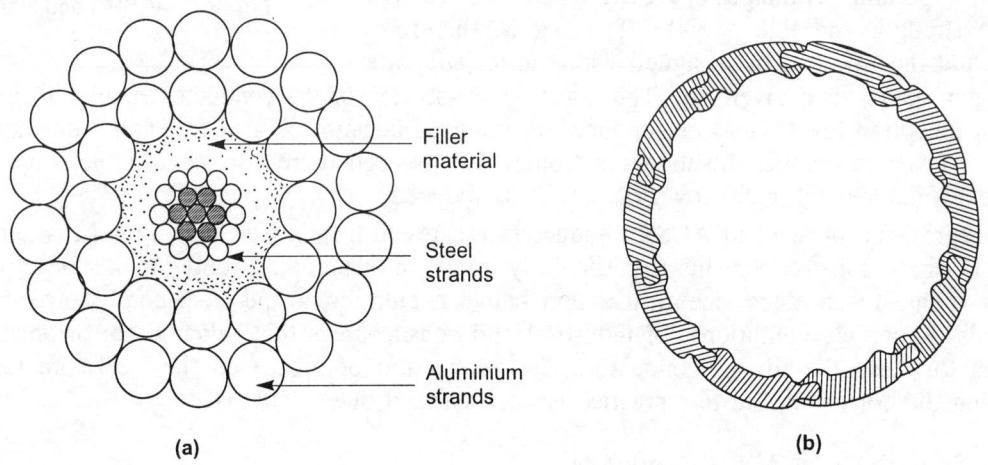

Filler material

Steel strands

Aluminium strands

(a)

(b)

Fig. 3.3. Expanded conductor.

3.2.9 All-Aluminium Alloy Conductor (AAAC)

Homogeneous aluminium alloy conductors find an application in urban areas. They provide a good combination of tensile strength and conductivity. These conductors are known by different trade names in different countries. One of these alloys is 'Silmalec', which contains 0.5 per cent silicon, 0.5 per cent magnesium and the remainder aluminium. The electrical conductivity is 53.5 per cent of the International Annealed Copper Standard, and ultimate tensile strength is 30 kgf/mm^2. These alloys are very costly as they are heat treated.

A standard commercial aluminium alloy popularly known as '5005' has been developed in America. It contains 0.8 per cent magnesium with the remainder being aluminium and impurities. The percentage of impurities does not exceed 0.55 per cent. The conductivity of '5005' is 53.5 per cent of International Annealed Copper Standard, the strength ranges between 21 to 25 kgf/mm^2 and the temperature coefficient of resistance 0.00354 at 20°C/1°C. '5005' alloy is cheaper than other aluminium alloys since additional cost due to heat treatment is eliminated.

3.2.10 ACAR Conductor

Aluminium conductor-alloy reinforced (ACAR) has a central core of alloy of aluminium surrounded by layers of conductor aluminium. It gives better conductance with a strength-weight ratio equal to ACSR construction of the same diameter. With the same electrical capacity ACAR conductor has a smaller size and lesser weight as compared to ACSR conductor.

3.2.11   Alumoweld Conductor

Aluminium powder is welded on to a high strength steel wire. About 75 per cent of area of conductor is covered by aluminium. Wires of different sizes are cold drawn from the steel rod on which aluminium is welded. It is more costly than SCA conductor. It has been used as earth wire and for making cores of SCA conductors.

3.2.12   Phosphor-Bronze Conductor

Phosphor-bronze is used as a conductor material for very long spans such as river crossings. It is more strong than copper conductor, but has got a low conductivity. The conductivity may be improved by using a cadmium-copper core. Phosphor-bronze is superior to aluminium bronze for atmospheres containing harmful gases such as ammonia.

3.2.13   Galvanized Steel Conductor

Galvanised steel conductors are used where high strength is desired. They are used on very long spans and particularly in rural areas where the load is small. In such cases the steel conductors may be replaced by SCA conductors to cope with the extra future load. The support and clearances need not be changed. The virtual modulus of elasticity for galvanised steel conductors is 18200 kgf/mm^2, and the coefficient of linear expansion is 1.08×10^{-5} per °C. Being a magnetic material, a galvanised steel conductor has large resistance, inductance and voltage drop. It has a comparatively short life.

3.3   RESISTANCE

The effective resistance of a conductor is the ratio of power loss in a conductor in watts to the square of the rms current in amperes. This is slightly greater than the dc resistance. The dc resistance is given by

$$R = \rho \frac{l}{A}$$

where ρ = resistivity of the conductor material

   l = length of the conductor

   A = area of cross–section of the conductor.

The resistivity of the conducting material is compared with the annealed copper having a conductivity of 100 per cent. The International Annealed Copper Standard (IACS) has a resistivity of 0.017241 ohm square millimeter per metre at 20°C is said to have a conductivity of 100 per cent. The conductivity of hard-drawn copper varies with the hardness. Hard-drawn copper used in practice has a conductivity of about 97 per cent at 20°C. The following formula may be used for hard-drawn copper :

$$P = \frac{T}{16}$$

where P is the percentage increase in the resistivity over annealed copper, and T is the tensile strength of hard-drawn copper in kgf/mm^2.

The formula gives sufficiently accurate results within range of 30-50 kgf/mm^2 tensile strength.

The average and maximum resistivity of hard-drawn aluminium may be taken as 2.845 $\mu\Omega$-cm and 2.873 $\mu\Omega$-cm respectively. The resistivity of conductors varies with the temperature.

It is given by the formula

$$\rho_t = \rho_{20} [1 + \alpha_{20} (t - 20)] \qquad \qquad ...(3.3.1)$$

where ρ_t = resistivity at $t°C$

ρ_{20} = resistivity at 20°C

α_{20} = temperature coefficient of resistance at 20°C

t = the operating temperature in °C

Similarly, $R_t = R_{20} [1 + \alpha_{20} (t - 20)]$ $\qquad \qquad ...(3.3.2)$

where R_t is resistance at $t°C$, and R_{20} is that at 20°C. Usually, the resistance of a conductor at 20°C is given in tables. The resistance at the operating temperature can be calculated from the above formula. α_{20} for hard-drawn copper of 97 per cent conductivity is 0.00381/°C. α_{20} for hard-drawn aluminium is 0.004/°C.

The resistance of stranded conductor is slightly greater than the resistance of the equivalent solid conductor because of increased length due to spiralling.

3.4 SKIN EFFECT

If d.c. is passed in a conductor, the current density is uniform over the cross-section of the conductor. In a conductor carrying a.c. there is a tendency of the current to crowd near the surface of the conductor. This phenomenon is called *skin effect*. A qualitative explanation of skin effect is given here.

Assume the conductor to be made up of a number of concentric cylinders. The magnetic flux linking a cylindrical element near the centre of the conductor is greater than that linking another element near the surface of conductor. This is due to the fact that the former element is surrounded by the internal as well as external flux, while the latter by the external flux only. The inner element will possess a greater self-inductance and, therefore, will offer a greater inductive reactance than the outer element. This difference in the inductive reactance gives a tendency to the current to crowd towards the surface or skin of the conductor. The current density is maximum at the surface and minimum at the centre of the conductor. The effect is virtually equivalent to a reduction of cross-sectional area of the conductor and, therefore, the effective resistance of the conductor is increased.

At low frequencies, such as 50 Hz, there is a small increase in the current density near the surface of conductor, but at high frequencies, such as with the radio, practically the whole of the current flows on the surface of the conductor.

Skin effect increases with the increase in frequency, conductor diameter and permeability. The concentration of current near the conductor surface has enabled the use of ACSR conductor. In the ACSR conductor the current flows mostly in the outer layer made of aluminium, while the steel near the centre carries practically no current and gives the high tensile strength to the conductor. The resistance of ACSR conductors is obtained from manufacturer's tables (see Appendix).

3.5 EQUIVALENT COPPER SECTION

The area of cross-section of a conductor made of a material other than copper is not specified by its actual value. Conductor sections are usually expressed and standardized in terms of equivalent copper sections. Consider the case of aluminium conductor. Let the suffixes a and c be used for aluminium and copper respectively. For the same length and resistance of aluminium and copper conductors :

$$R_a = R_c$$

$$\rho_a \frac{l}{A_a} = \rho_c \frac{l}{A_c} \; ; \quad A_c = \frac{\rho_c}{\rho_a} A_a \qquad\qquad\qquad ...(3.5.1)$$

The value of A_c is called the *equivalent copper section* for the corresponding cross-sectional area A_a of aluminium conductor.

Thus, equivalent copper section of aluminium conductor

$$= \frac{1}{(\rho_a/\rho_c)} \text{ (cross--sectional area of aluminium conductor)}$$

$$= \frac{1}{1.62} \times \text{cross--sectional area of aluminium conductor} \qquad\qquad \text{since } \frac{\rho_a}{\rho_c} = 1.62$$

3.6 KELVIN'S ECONOMY LAW

There are several factors considered in designing a line. Economy is also one of the considerations which is taken into account to select a conductor for the line. The cost of conductor material is a substantial part of the total line cost. It is, therefore, necessary to choose the most economic size of conductor. A design is considered to be most efficient if the total annual cost is a minimum. The total annual cost consists of two parts:

(a) The fixed standing charges, and

(b) the running charges.

The fixed charges consist of the interest on the capital cost of the conductor, the allowance for depreciation, and the maintenance cost. The running charges consist of cost of electrical energy wasted due to losses during operation. But the capital cost (and, therefore, the interest and depreciation on it) and cost of electrical energy wasted in the line are governed by the size of the conductor. A bigger size of conductor would be more costly, but due to its lesser resistance the cost of energy corresponding to I^2R loss will be smaller. On the other hand, if a smaller size of conductor is selected it will be cheaper, but its greater resistance will increase the cost of energy loss, and, therefore, the running charges. The cost of conductor and therefore, the standing charges, namely, the interest and depreciation on the initial investment will be directly proportional to the area of cross-section of the conductor. The cost of energy loss will be inversely proportional to the conductor section. Mathematically, they can be written as :

Annual interest and depreciation cost

$$C_1 \propto a \quad \text{or} \quad C_1 = k_1\, a$$

and annual cost of energy dissipated in the line

$$C_2 \propto 1/a \quad \text{or} \quad C_2 = \frac{k_2}{a}$$

where k_1 and k_2 are constants, and a represents the area of cross-section of conductor. The total annual cost may, therefore, be given by

$$C = C_1 + C_2 = k_1 a + \frac{k_2}{a}$$

For an economical design there will be one size of conductor at which the total cost is a minimum. For the most economical cross-section, the total annual cost is differentiated with respect to the cross-section and the result is equated to zero. That is,

$$\frac{dC}{da} = 0 \; ; \quad \frac{d}{da}\left(k_1 a + \frac{k_2}{a}\right) = 0$$

$$k_1 - \frac{k_2}{a^2} = 0 \; ; \quad k_1 a = \frac{k_2}{a}$$

i.e., $\qquad C_1 = C_2 \quad \text{and} \quad a = \sqrt{\frac{k_2}{k_1}}$ $\qquad\qquad$ ...(3.6.1)

Hence the most economical cross-sectional area of the conductor is that which makes the annual cost of energy loss equal to the annual interest and depreciation on the capital cost of the conductor material. This is known as *Kelvin's law*, after Lord Kelvin who first stated it in 1881. Kelvin's law is itself not sufficient to estimate the cross-section of the conductor. It gives the most economical current density. The most economical cross-section is given by Eq. (3.6.1).

3.7 MODIFIED KELVIN'S LAW

In our previous discussion we have not considered the cost of poles or towers, insulators, erection, etc. in an overhead line. Similarly, for an underground cable no account has been taken of the cost of the cable insulation and its laying. In practice, it is not true. As the size of conductor increases the mechanical stresses are increased, More strong towers and insulators are required. The cost of labour for erection also increases.

To achieve a close approximation to the true conditions G. Kapp assumed that the initial investment on the complete installation of the line may be divided into two parts, namely, (a) one part which is independent of the conductor size, and (b) the other part which is directly proportional to the conductor section.

The initial investment may, therefore, be written as

$$C'_1 = k'_0 + k'_1 a$$

where k'_0 is a constant and represents the part of the capital investment which is independent of the conductor size.

If x be the rate of annual interest and depreciation per unit, the annual cost of the line

$$C_1 = (k'_0 + k'_1 a) x$$

The annual cost of power loss

$$C_2 = \frac{k_2}{a}$$

The total annual cost

$$C = C_1 + C_2 = (k'_0 + k'_1\, a)\, x + \frac{k_2}{a}$$

For the cost to be a minimum

$$\frac{dC}{da} = 0 \; ; \quad k'_1\, x - \frac{k_2}{a^2} = 0$$

i.e., $\quad \dfrac{k_2}{a} = k'_1\, x\, a \quad$ and $\quad a = \sqrt{\dfrac{k_2}{x\, k'_1}}$

Hence the most economical cross-section is that for which the annual cost of energy loss is equal to the annual cost of interest and depreciation on that part of initial investment of the line which is proportional to the conductor area. This is the modification of Kelvin's law.

3.8 GRAPHICAL REPRESENTATION

Kelvin's law can be illustrated graphically as shown in Fig. 3.4. The annual cost C_1 of conductor is directly proportional to its cross-section. It can, therefore, be represented by a straight line passing through the origin. The variation of cost C_2 of energy wasted with the conductor size is represented by a rectangular hyperbola. The total cost C for any conductor section is the sum of the two component costs, C_1 and C_2, for that cross-section. The sum curve is represented by C in Fig. 3.4.

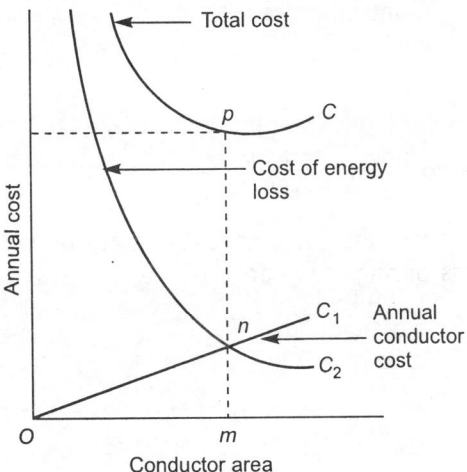

Fig. 3.4. Graphical representation of Kelvin's law.

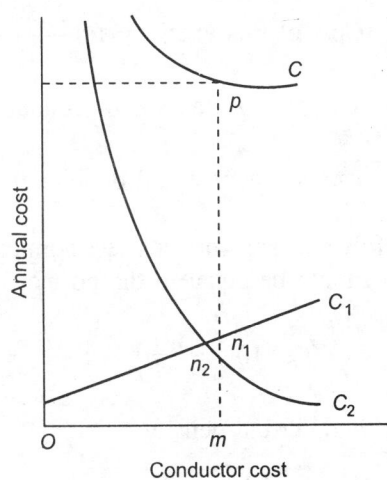

Fig. 3.5. Graphical representation of modified Kelvin's law.

The lowest point p on the total cost curve C gives the most economical area corresponding to the point of intersection of the two component cost curves C_1 and C_2. At the point of intersection the two component costs are equal. The most economical area is *om* and the minimum cost is *mp* in Fig. 3.4.

The modified Kelvin's law can also be shown graphically in Fig. 3.5. The cost C_1 is represented by a straight line. The intercept on the cost axis gives the annual cost of conductor which is not proportional to conductor area. The energy cost curve C_2 is represented by a hyperbola. The sum curve C gives the total cost. The lowest point on the total cost curve again gives the minimum cost and the most economical conductor section. In Fig. 3.5, *om* is the most economical section and *mp* the minimum cost. The length n_1n_2 is the cross-section of conductor cost which is not proportional to the area of cross section of conductor.

3.9 ECONOMIC CURRENT DENSITY

Let a = area of cross–section of the conductor in sq. mm.

R = resistance of the conductor of 1 sq. mm cross–section and 1 km length

I = rms value of current in conductor throughout the year.

W = weight of conductor of 1 sq. mm cross–section in kgf/km

t = number of working hours per year

x = cost of electrical energy wasted in rupees per kWh

y = cost of conductor per kgf in rupees

p = per cent annual interest and depreciation on capital cost.

Annual energy wasted in one kilometre of the conductor $= \dfrac{I^2 R t}{a} \times 10^{-3}$ kWh

Cost of this energy $= \text{Rs} \dfrac{I^2 R t x}{a} \times 10^{-3}$

The cost of one km of conductor $= \text{Rs } w\,a\,y$

Annual interest and depreciation on this cost $= \text{Rs } \dfrac{w\,a\,y\,p}{100}$

By Kelvin's law, for most economical cross-section, the fixed annual charges on conductor material should be equal to the cost of energy loss during the year.

$$\dfrac{I^2 R}{a} t x \cdot 10^{-3} = \dfrac{w\,a\,y\,p}{100}\ ; \quad \dfrac{I^2}{a^2} = \dfrac{10 w\,y\,p}{R t x}$$

Economic current density

$$\dfrac{I}{a} = \sqrt{\dfrac{10 w\,y\,p}{R t x}} \quad \text{A/mm}^2.$$

Example 3.1 A 500 V, 2-core feeder 0.8 km long is required to supply a constant load of 100 kW. The cost of the cable including installation charges is Rs $(6a + 1.3)$ per metre, where a is the cross-sectional area of each feeder in sq. cm. Interest and depreciation total 10 per cent. Determine the most economical size. Cost of energy is 12 paise per unit. Specific resistance of copper is 1.75×10^{-6} Ω per cm^2 cross sectional area and 1 cm long.

Solution

Full-load current, $I = \dfrac{100 \times 1000}{500} = 200$ A

Resistance of each core, $R = \rho \dfrac{l}{a} = \dfrac{1}{a} \times 1.75 \times 10^{-6} \times 0.8 \times 1000 \times 100 = \dfrac{0.14}{a}$ Ω

Power loss for the two cores of the cable, $= 2I^2 R = 2\,(200)^2 \times \dfrac{0.14}{a} = \dfrac{11200}{a}$ W

$$= \dfrac{11.2}{a} \text{ kW}$$

Annual energy loss $= \dfrac{11.2}{a} \times 365 \times 24$ kWh

Cost of annual energy loss $= $ Rs $\dfrac{11.2}{a} \times 365 \times 24 \times \dfrac{12}{100} = $ Rs $\dfrac{11773}{a}$

Total cost of the cable including installation cost $= $ Rs $(6a + 1.3) \times 800$

Annual interest and depreciation charges $= $ Rs $(6a + 1.3) \times 800 \times \dfrac{10}{100}$

$$= \text{Rs } (480a + 104)$$

By Kelvin's law the most economic cross-section is obtained when

$$480a = \dfrac{11773}{a}$$

$$a = \sqrt{\dfrac{11773}{480}} = 4.952 \text{ cm}^2 \qquad \dfrac{\pi}{4}d^2 = 4.952$$

$$d = \sqrt{\dfrac{4.952 \times 4}{\pi}} = 2.51 \text{ cm} = 25.1 \text{ mm}$$

Therefore, the most economical diameter of conductor is 25.1 mm.

Example 3.2 The daily load cycle of a three-phase, 33 kV, 10 km transmission line is as follows : 2500 kVA for 8 hours, 2000 kVA for 9 hours and 1500 kVA for 7 hours. Determine the most economical cross-section if the cost of line including erection is Rs $(7500 + 6000a)$ per km where a is the area of each conductor in sq. cm. The rate of interest and depreciation is 8 per cent and cost of energy is 15 paise per unit. The line is in use for 250 working days a year. The resistance per km and per sq. cm. is 0.173 ohm.

Solution

Cost of line $= $ Rs $(7500 + 6000a) \times 10$

Annual interest and depreciation on capital cost $= $ Rs $(7500 + 6000a) \times 10 \times \dfrac{8}{100}$

$$= \text{Rs } (6000 + 4800a)$$

Resistance of each conductor, $R = \dfrac{0.173 \times 10}{a} = \dfrac{1.73}{a}$

$\dfrac{\sqrt{3}\, V_L\, I_L}{1000} = \text{kVA}$

The load current at various loads are calculated by the above formula as follows :

At 2500 kVA, load current

$I_1 = \dfrac{2500 \times 1000}{\sqrt{3} \times 33 \times 1000} = 43.8 \text{ A}$

At 2000 kVA, load current

$I_2 = \dfrac{2000 \times 1000}{\sqrt{3} \times 33 \times 1000} = 35 \text{ A}$

At 1500 kVA, load current

$I_3 = \dfrac{1500 \times 1000}{\sqrt{3} \times 33 \times 1000} = 26.2 \text{ A}$

Daily energy loss $= 3I_1^2 R \times \dfrac{8}{1000} + 3I_2^2 R \times \dfrac{9}{1000} + 3I_3^2 R \times \dfrac{7}{1000} = \dfrac{3R}{1000}\,(8I_1^2 + 9I_2^2 + 7I_3^2)$

$= 3 \times \dfrac{1.73}{a} \times 10^{-3}\,[8\,(43.8)^2 + 9\,(35)^2 + 7\,(26.2)^2]$

$= \dfrac{79.8}{a} + \dfrac{57.2}{a} + \dfrac{24.9}{a} = \dfrac{161.9}{a} \text{ kWh}$

Annual energy loss $= \dfrac{161.9}{a} \times 250 \text{ kWh}$

Cost of energy loss per annum $= \text{Rs } \dfrac{161.9}{a} \times 250 \times \dfrac{15}{100} = \text{Rs } \dfrac{6071}{a}$

$\therefore$ by Kelvin's law, $\dfrac{6071}{a} = 4800a$ $\qquad a = \sqrt{\dfrac{6071}{4800}} = 1.124$

$\therefore$ the most economical cross-section $= 1.124 \text{ cm}^2 = 112.4 \text{ mm}^2$.

Example 3.3 Find the best current density for a three-phase overhead line which is in use for 2500 hours a year. Cost of copper per kgf = Rs 20, annual interest and depreciation = 12.5 per cent. Density of copper = 8.89 gf/cm^3. Resistance per conductor per km length and per sq. cm = 0.173 ohm. Cost of energy per unit = 16 paise.

Solution

Let $\quad l = $ length of each conductor in cm

$\quad a = $ area of cross–section in sq. cm

Volume of each conductor $= l\, a \text{ cm}^3$

Weight of each conductor $= l\, a \times 8.89 \text{ gf} = l\, a \times 8.89 \times 10^{-3} \text{ kgf}$

Capital cost of the conductor = Rs $20l\, a \times 8.89 \times 10^{-3}$

Interest and depreciation = Rs $20l\, a \times 8.89 \times 10^{-3} \times \dfrac{12.5}{100}$ = Rs $22.22 \times 10^{-3}\, l\, a$

Copper loss per conductor = $I^2 R \times 10^{-3}$ kW = $I^2 \left(1.73 \times 10^{-6} \times \dfrac{l}{a}\right) \times 10^{-3}$

$$= 1.73 \times 10^{-9}\, I^2 \dfrac{l}{a}\ \text{kW}$$

Cost of energy loss per year = Rs $1.3 \times 10^{-9} \times I^2 \dfrac{l}{a} \times 2500 \times \dfrac{16}{100}$

$$= \text{Rs } 6.92 \times 10^{-7}\, I^2 \dfrac{l}{a}$$

According to Kelvin's law, the two costs should be equal for the best current density.

$$22.22 \times 10^{-3}\, l\, a = 6.92 \times 10^{-7}\, I^2 \dfrac{l}{a}$$

$$\dfrac{I^2}{a^2} = \dfrac{22.22 \times 10^{-3}}{6.92 \times 10^{-7}} = 3.21 \times 10^4$$

The best current density $\dfrac{I}{a} = 179$ A/cm^2 = 1.79 A/mm^2

3.10 DETERMINATION OF LOSSES

It is difficult to estimate and evaluate the cost of $I^2 R$ loss. The load curves are drawn for different types of loads and the load factor is determined. The load factor will give an idea of average load current carried by the line during the whole year. The ohmic losses at any load are proportional to the square of the current. The annual losses are proportional to the mean value of the square of the current during the year. The square root of the mean value of the squares of the line currents throughout the year is called the *rms current*.

Load factor by itself is not sufficient to determine the rms current. To determine rms current flowing during the year average value of current obtained from load curve is multiplied by form factor.

If I_m = maximum full–load current of the line

K_l = annual load factor of the line = $\dfrac{\text{average load over a period}}{\text{maximum load over the period}}$

$$= \dfrac{\text{number of kWh generated per year}}{\text{maximum demand (kW)} \times 8760\ \text{(hrs)}} = \dfrac{I_{av}}{I_m} \qquad \qquad ...(3.10.1)$$

K_f = form factor of the load curve = $\dfrac{\text{root mean square current}}{\text{annual average current}} = \dfrac{I_{rms}}{I_{av}}$ $\qquad ...(3.10.2)$

$I_{rms} = K_f\, I_{av}$

$I_{rms} = K_l\, K_f\, I_m$

The rms current can be calculated to a fair degree of accuracy from the above expression. The approximate values of form factor for various load factors are given in Table 3.3.

Table 3.3 Approximate Values of Form Factor for Various Load Factors

Load factor $K_l\%$	Form factor $K_f\%$	Load factor $K_l\%$	Form factor $K_f\%$
10	2.25	50	1.20
15	1.88	60	1.13
20	1.68	70	1.08
25	1.55	80	1.04
30	1.45	90	1.02
40	1.30	100	1.00

The load factor of the losses is different from the load factor of the load current. The load factor of the losses is called the *loss factor* λ. It is equal to the ratio between the actual energy loss during a period and the energy loss assuming that maximum current is flowing during the whole period.

Annual loss may also be determined approximately with the help of loss factor. An empirical relation between the loss factor and the fractional load factor F is given by

$$\lambda = 0.2F + 0.8F^2 \qquad \qquad ...(3.10.4)$$

Two other similar expressions which have been used are

$$\lambda = 0.3F + 0.7F^2 \qquad \qquad ...(3.10.5)$$

and $\quad \lambda = 0.5F + 0.5F^2 \qquad \qquad ...(3.10.6)$

If loss factor is known the rms values of fluctuating current flowing during the year can be calculated.

Since $\quad$ loss factor, $\lambda = \dfrac{I^2_{rms}}{I^2_m}$

$\therefore \qquad I_{rms} = I_m \sqrt{\lambda} \qquad \qquad ...(3.10.7)$

Total annual loss per conductor

$$= I^2_{rms} R = \lambda \, I^2_m R \ \text{W} \qquad \qquad ...(3.10.8)$$

where R is the resistance of each conductor in ohms.

Example 3.4 Calculate the rms value of current for a three-phase, 33 kV line transmitting a maximum load of 7500 kVA for a load factor of 40 per cent.

Solution

$$\sqrt{3} \ VI_m = 7500$$

$$I_m = \frac{7500}{33 \sqrt{3}} = 131 \text{ A}$$

For a 40 per cent load factor

$K_f = 1.3$ (from Table 3.3)

$I_{rms} = K_f K_l I_m = 1.3 \times 0.4 \times 131 = 68.3 \text{ A}$

Example 3.5 Determine the most economical size of copper conductor for transmitting a maximum load of 7500 kVA over a three-phase, 33 kV overhead line for a load factor of 60 per cent. The cost of the line per km is Rs $20000a + 10000$, where a is the cross-sectional area of each conductor is sq. cm. The form factor K_f of the load curve for a load factor of 60 per cent is given by $K_f = I_{rms}/I_{av} = 1.13$. Cost of energy per kWh is 15 paise. Interest and depreciation is 10 per cent. Specific resistance of copper is 17.6 $\mu\Omega$-m^2/km.

Solution

$$\sqrt{3}\ VI_m = 7500$$

$$I_m = \frac{7500}{33 \sqrt{3}} = 131 \text{ A}$$

For a 60 per cent load factor,

$K_f = 1.13$

$I_{rms} = K_l K_f I_m = 0.6 \times 1.13 \times 131 = 88.9 \text{ A}$

$$\rho = \frac{17.6\ \mu\Omega\text{–m}^2}{\text{km}} = \frac{17.6\ \mu\Omega\text{–cm}^2 \times 10^4}{10^5\ \text{cm}} = 17.6\ \mu\Omega\text{–cm}$$

Resistance of one conductor per km, $R = \rho \dfrac{l}{a} = \dfrac{1.76 \times 10^{-6} \times 10^5}{a} = \dfrac{0.176}{a}\ \Omega$

Energy loss in the line per year = $3I_{rms}^2\ R\ t \times 10^{-3} = 3 \times (88.9)^2 \times \dfrac{0.176}{a} \times 8760 \times 10^{-3}$

$$= \frac{3.655}{a} \times 10^4 \text{ kWh}$$

Annual cost of energy wasted = Rs $\dfrac{3.655}{a} \times 10^4 \times 0.15 = $ Rs $\dfrac{5.48}{a} \times 10^3$

Annual cost due to interest and depreciation = Rs $0.1 \times 20000a = $ Rs $2000a$

By Kelvin's law for the most economical cross-section, the two annual charges should be equal.

$\therefore$ $2000a = \dfrac{5.48}{a} \times 10^3$

$a^2 = \dfrac{5.48}{2000} \times 1000 = 2.74$; $a = 1.65 \text{ cm}^2 = 165 \text{ mm}^2$

Example 3.6 The cost per km length of a three-phase overhead line is Rs $(25000a + 12000)$ where a is the area of cross-section of each conductor. The load factor for the load current is 50 per cent and that for the losses is given by $(0.2F + 0.8F^2)$, where F is the fractional load factor. If the combined rate of interest and depreciation is 9 per cent per annum, and the cost of energy wasted in the line is 12 paise per kWh, find the most economical current density of the line. Resistivity of conductors is 1.76 $\mu\Omega$-cm.

Solution

Consider 1 km length of the line.

Annual charges due to interest and depreciation $= 0.09 \times 25000a =$ Rs $2250a$

Load factor $= 0.2F + 0.8F^2 = 0.2 \times 0.5 + 0.8 \times (0.5)^2 = 0.10 + 0.20 = 0.30$

$$R = \rho \frac{l}{a} = \frac{1.76 \times 10^{-6} \times 10^5}{a} = \frac{0.176}{a} \ \Omega$$

Annual energy loss in the line $= 3I^2 R \, t\lambda \times 10^{-3} = 3I^2 \times \dfrac{0.176}{a} \times 8760 \times 0.3 \times 10^{-3}$

$$= 1.387 \frac{I^2}{a} \ \text{kWh}$$

Cost of energy loss $=$ Rs $\dfrac{1.387}{a} I^2 \times 0.12 =$ Rs $\dfrac{0.1665}{a} I^2$

By Kelvin's law two costs should be equal for the best current density, i.e.,

$$\frac{0.1665}{a} I^2 = 2250a$$

$$\frac{I^2}{a^2} = \frac{2250}{0.1665} = 1.35 \times 10^4$$

Best current density

$$\frac{I}{a} = 116 \text{ A/cm}^2 = 1.16 \text{ A/mm}^2.$$

3.11 LIMITATIONS OF KELVIN'S LAW

Kelvin's law is theoretically true. A number of difficulties are involved in its application. Of the numerous limitations some of the important ones are given below:

1. The amount of energy loss cannot be estimated accurately. This is due to the fact that the load factors of the losses and the load itself are different, and the future load conditions and load factors cannot be predicted accurately.

2. It is difficult to calculate the cost of energy loss. The cost of losses per unit is greater than the generating cost per unit. The prices of conductor material and the rates of interest are changing continuously.

3. The cost of energy and interest and depreciation are independent of resistance, voltage drop, temperature rise, etc. Two systems having identical demands but different energy costs and interest and depreciation will give two different sections.

4. The voltage drop may be beyond the permissible limits in some cases if economical conductor size is selected.

5. The most economical size of conductor may not be suitable to carry the required amount of current due to its thermal rating and temperature rise limits.

6. The economical section may not have the desired sufficient mechanical strength.

7. The problem of corona, leakage currents etc. prevent the use of economical section at extra high voltages.

Thus, it is found that Kelvin's law serves as a guide only to select the size of the conductor. More frequently the economic aspect is not considered due to its various limitations.

EXERCISES

1. A 2-conductor cable 1 km in length is required to supply a constant load of 250 A throughout the year. The cost of cable is Rs $(150a + 85)$ per metre, where a is the area of the conductor in sq. cm. Determine the most economical cross-section of the conductor if the cost of energy is 15 paise per kWh and interest and depreciation charges amount to 12 per cent. Specific resistivity of copper is 1.76×10^{-8} Ω-m.

2. The cost of a d.c. 2-wire overhead line is Rs 4000 per sq. cm. section per km length and 10 per cent of this is to be taken as annual cost. Cost of energy lost in the line is 12 paise per kWh. If full load is supplied for 40 per cent of the year (8760 hours) estimate the most economical current density. The resistance of the conductor material having a cross-sectional area of 1 mm^2 is 1/58 ohm for 1 m length.

3. The cost of a d.c. 2-wire overhead line is Rs $4000a + 5000$ per km where a is the area of conductor in sq. cm. The interest and depreciation per annum amounts to 12 per cent of capital cost. If the load is supplied for 60 per cent of the year estimate the most economical current density. Cost of energy is 15 paise per kWh. Take resistivity as 1.8 $\mu\Omega$-cm.

4. The cost of a three-phase overhead line is Rs $(150a + 10000)$ per km length where a is the cross-sectional area of the conductor in sq. mm. The rate of interest and depreciation per annum is 10 per cent. If the load is supplied 60 per cent of the year, estimate the most economical current density for the conductor. The cost of energy is 15 p. per kWh and resistivity 1.8 $\mu\Omega$-cm.

5. An 11 kV 3-core cable is to supply a factory which works 48 hours a week with a load of 500 kW at 0.9 power factor lagging. Capital cost of cable per core when laid is Rs $(40a + 250)$, per km where a is the cross-section of the conductor in sq. mm. The interest and depreciation charges are 14 per cent of the capital cost and the energy cost is 16 paise per kWh. Calculate the most economical area of cross-section of the conductor. Take the resistivity of copper as 1.724×10^{-8} Ω-m.

6. A factory takes through a 3-core underground cable a load of 500 kW at 11 kV and 0.8 power factor for 2500 hours per annum. Capital cost of cable per core when laid is

Rs $(35a + 400)$ per km, where a is cross-section of the conductor in sq. mm. The interest any depreciation charges are 12 per cent of the capital cost and energy cost of 20 paise per kWh. Calculate the most economical area of cross-section of the conductor. Take the resistivity of copper as 1.724×10^{-8} Ω-m.

7. The daily load cycle of the three-phase 110 kV transmission line is as follows : 15 MW at 0.8 power factor (p.f.) for 8 hours, 5 MW at unity p.f. for 10 hours, 16 MW at 0.85 p.f. for 6 hours. Determine the most economical cross-section if the cost of the line including erection is Rs $9000 + 6000a$ per km, where a is the cross-section of each conductor in sq. cm. The rate of interest and depreciation is 10 per cent and the cost of energy is 6 paise per kWh. The line is in use for whole of the year. The resistance per km of each conductor is $(0.176/a)$ ohm.

8. Find the best current density for a three-phase overhead line if it is used for 4000 hours per year. The conductor costs Rs 25 per kgf, has a resistivity of 1.8 $\mu\Omega$-cm and a density of 8.9 gf/cm^3. Energy costs 15 paise per kWh and annual interest and depreciation is 10 per cent of the capital cost of the conductor.

9. A 2-core cable 1 km long supplies a load of 125 kW for 12 hours, 40 kW for 8 hours and is on no load for the remainder of each day. The annual cost of the cable is Rs $0.6a + 30$ per metre, where a is the cross-sectional area in square millimetres. The supply voltage is 1 kV. Cost of energy per unit is 15 paise. Resistivity of the core material is 1.72×10^{-8} Ω-m. Determine the most economical core size and the annual cost of the cable.

10. A 2-core cable 1 km long supplies a load of 125 kW at a voltage of 2 kV and a power factor of 0.85. The load is switched on for 16 hours per day for 300 days of each year. The cost of the cable per metre to manufacture and lay is Rs $60a + 25$ per metre where a is cross-sectional area in sq. m. Interest and depreciation charges are 10 per cent and 5 per cent respectively per year. Cost of energy is 15 paise per kWh. Find

(a) the most economical size of the core,

(b) the maximum current density.

Resistivity of the aluminium core material is 2.8×10^{-8} Ω-m.

ANSWERS

1. 126.7 mm^2

2. 0.5253 A/mm^2

3. 0.4112 A/mm^2

4. 0.5934 A/mm^2

5. 56 mm^2

6. 47 mm^2

7. 1.66 cm^2

8. 143.5 A/cm^2

9. 27.46 mm^2, Rs 46476

10. (a) 0.4919 cm^2 (b) 149.4 A/cm^2

4

Power Cables

4.1 INTRODUCTION

Cables form the artery system for the transmission and distribution of electrical energy. The residential loads today have a trend towards their growing density. This requires rugged construction, greater service reliability, increased safety, and better appearance. The interference from external disturbances like storms, lightning, ice, trees, etc., should be reduced to a minimum. These difficulties are easily overcome by the use of underground cables and a trouble-free service is achieved under a variety of environmental conditions.

Earlier underground cables were mainly used in or near densely populated areas and were operated at low or medium voltages only, but the present day requirements seek to use them even at extra high voltages for longer distances. The underground system, although more costly as compared with an overhead system for the same voltage, is more acceptable to public from the point of view of its merits mentioned above. The additional advantage provided by the underground cables is lesser right-of-way requirements.

Underground cables are to be laid in areas where it is almost impracticable to use overhead lines, for example, in case of the transmission lines through sea, cross-over or terminal connections in substations or air-field crossings. The cross-channel link between France and England and the submarine link from Swedish mainland to the island of Gottland are the examples of high voltage power cables.

Increased working voltages of the overhead lines require, the cables to be insulated for such voltages in order to meet the requirements of the overhead lines. The design of power cables is, therefore, governed by the requirements of the overhead line.

The possibility of supply interruption due to lightning or other external influences is lesser with underground cables, but if a fault occurs due to any reason it is not easily located. For long-distance transmission, cables cannot be used due to their large charging currents.

4.2 CABLE CONSTRUCTION

A power cable consists of three main components, namely, conductor, dielectric, and sheath. The *conductor* provides the conducting path for the current. The *insulation* or *dielectric* withstands the service voltage, and isolates the conductor with other objects. The *sheath* does not allow the moisture to enter, and protects the cable from all external influences like chemical or electrochemical attack, fire, etc. It should be of a nonmagnetic material.

4.3 CONDUCTORS

Copper and aluminium are used as conductor materials in cables. As pointed out earlier, aluminium occupies a greater space than copper for a given conductance. Both copper and aluminium have been favoured for the conductor, materials because of their high electrical conductivity.

Solid or number of bare wires made of either copper or aluminium are used to make a power cable. For a conductor having more than three wires, the wires are arranged round a central wire such that there are six in the first layer, twelve in the second, eighteen in the third and so on. In this way the number of wires in conductors are 7, 19, 37, 61, 91, etc. The size of a conductor is represented by 7/*A*, 19/*B*, 37/*C*, etc., in which the first figures represent the number of strands and the second figures *A*, *B*, *C* etc., represent the diameters in cm or mm of the individual wires making the conductor.

Stranded conductors having more than one layer of wires are made such that the direction of lay of wires in adjacent layers are opposite to each other. The strands follow a helical path and, therefore, the individual strands are greater in length as compared to the length of the cable. Stranding thus increases the resistance of the cable.

The stranded wires are sometimes compacted by rollers to minimize the air spaces between the individual wires. Such a conductor is called a *segmental conductor* (Fig. 4.1). The conductor size is thus reduced for a given conductance. This consideration is kept in view in order to have control on the dimensions of the cable within reasonable limits. Another method adopted to reduce the diameter is to use *shaped conductors* (Fig. 4.2). The cables with sector-shaped core are economical.

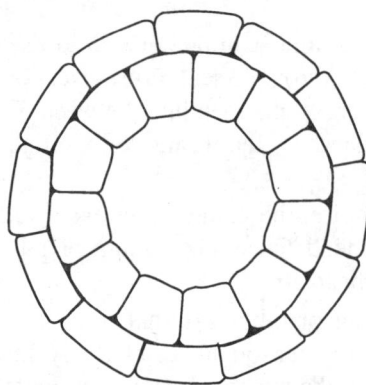

Fig. 4.1. Segmental conductor.

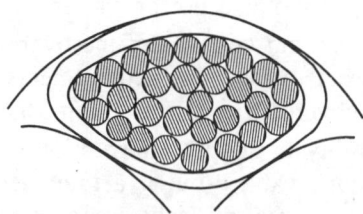

Fig. 4.2. Sector-shaped core conductor for 3-core cable.

4.4 INSULATION

The dielectric compounds as insulants for power cables should possess the following main properties:

1. High insulation resistance.
2. High dielectric strength.
3. Good mechanical properties.
4. Preferably non-hygroscopic, but if hygroscopic it should be provided with an economical water-tight covering or sheath.
5. Capable of being operated at high temperatures.
6. Low thermal resistance.
7. Low power factor.

The most commonly used dielectrics in power cables are impregnated paper, butyl rubber, polyvinyl chloride (PVC), polythene, cross-linked polyethylene (XLPE). Paper insulated cables, because of their relatively high current carrying capacity, long life and general reliability, are preferred. Other synthetic dielectrics are also being developed.

Polyvinyl chloride (PVC) cable is being used today for distribution purposes as an alternative to paper insulated cable. A polyvinyl chloride cable possesses several advantages over paper insulated cables. The important advantages of plastic cables are are :

1. Reduced cost.
2. Insulation is resistant to water.
3. No compound drainage trouble.
4. Simplified jointing.
5. Reduced weight.
6. Increased flexibility.
7. No plumbing required.

Polythene insulated power cables are found useful for special applications. Polythene, being non-hygroscopic, is used in cables for submarine use and damp soil. Polythene cables are comparatively lighter and have non-migratory dielectric. They are, therefore, suitable as aerial cables and for vertical installations. Additional advantages of polythene are low dielectric constant, low loss factor, and low thermal resistance. An 11 kV, 3-core, polythene insulated cable is shown in Fig. 4.3.

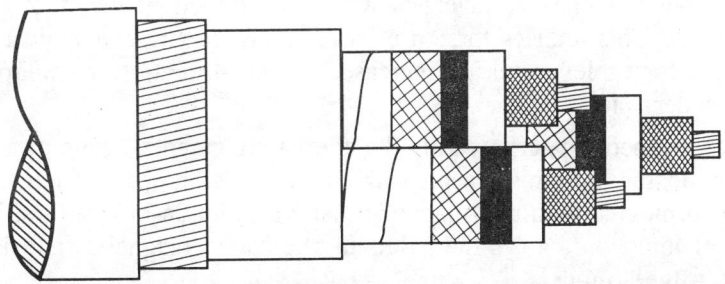

Fig. 4.3. 11 kV, 3-core, polythene cable.

4.5 SHEATH

Metal sheathing is required to provide an impervious layer to protect the cable from moisture, which would affect the insulation. Cable sheaths were made of lead in the beginning. Later lead alloys were used and they are also being used now. These new alloys can withstand the internal pressure of the pressurized cables.

Recently, aluminium is also being used a material for cable sheath. Aluminium sheath has a smaller weight and higher mechanical strength than the lead sheath. It has a greater conductivity. Aluminium sheath is cheaper than lead sheath. It is easy to manufacture and install. It has good screening properties for a.c. work. It eliminates the use of armour usually required in lead sheath cables. An aluminium sheath can withstand the required gas pressure without reinforcement. Owing to its greater conductivity the aluminium sheath of low voltage cables may be used as neutral conductor. Thus, there is no need of separate fourth neutral conductor.

Corrugated seamless aluminium (CSA) sheath is finding favour these days. It has better bending properties, **reduced** thickness and lesser weight. It is mainly used in high voltage oil-filled cables and telephone cables. The corrugated cable is very flexible and can be bent easily. Also, by repeated bending the sheath is not distorted unduly and, therefore, it is not damaged.

4.6 PROTECTIVE COVERING

Lead-sheathed cables are subject to mechanical damage, corrosion, and electrolytic action when laid direct in the ground. To protect them from these actions protective coverings are applied to the sheaths. For protecting against corrosion and electrolytic action bitumen and bituminised materials (paper, hessian, etc.) or polyvinyl chloride are used. Layers of fibrous materials (paper, hessian, etc.) or polyvinyl chloride are used. Layers of fibrous material permeated with waterproof compound applied to the exterior of the cable is called *serving*.

Armouring One or two layers of galvanized steel wires or two layers of metal tape *armouring* is applied over hessian or jute bedding to protect the sheath from mechanical damage. For longitudinal strength requirements, steel wire armour is preferred. The steel tapes are coated with preservative compound. The two tapes are wound helically in the same direction such that the outer layers covers the spaces between the turns of the inner layer. Single or double wire armour is used depending upon the degree of protection required. Single wire armour consists of wire applied over a compounded bedding. In case of double wire armour, a fabric tape is used as a separator between the two layers of wires. The directions of lay of the two layers are opposite to each other. This reduces torsion effect and gives extra mechanical strength. Double wire armour is used for cables requiring increased tensile strength, for example, along sloping routes, in mines, under water, etc.

The presence of magnetic material within the alternating magnetic field of a single-core cable produces excessive losses. For this reason single-core cables are either left unarmoured or, if necessary, they are armoured with non-magnetic materials like tin-bronze or silicon-bronze tapes or wires. In case of multicore cables the resultant alternating magnetic field is zero. There are no heating losses in the armour.

Aluminium has been used recently as an armour material due to its non-magnetic properties, high conductivity and mechanical strength. It is particularly useful for single-core cables working on ac. Aluminium or aluminium-alloy wires are used for armouring.

4.7 BELTED CABLE

Each of the three conductor cores is wrapped with oil impregnated paper. The cores are then assembled with a filler material. The assembly is enclosed by a paper insulating belt. A sheath is provided above the belt. The other things are done in the useful manner. A belted cable is shown in Fig. 4.4.

The electrostatic field of a 3-core belted type cable is not uniform along the radial direction but it is distorted. Due to this, a component of voltage acts in a direction tangential to the layers of the core insulation. The dielectric strength of impregnated paper along its laminations is about 6 per cent of the strength along the radial direction, and consequently a small tangential stress (Fig. 4.5) in the paper layers can cause breakdown of insulation.

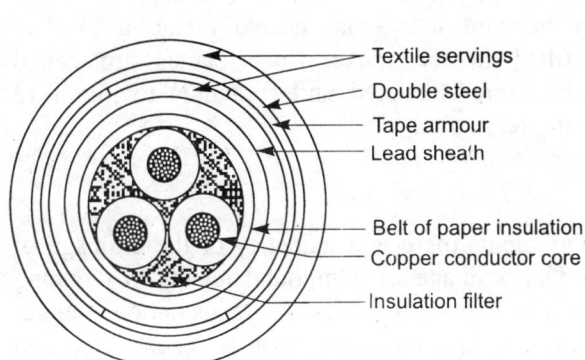

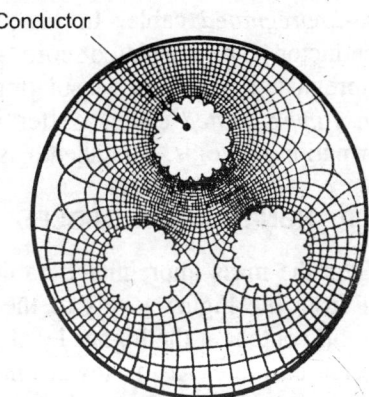

Conductor

Textile servings
Double steel
Tape armour
Lead sheath

Belt of paper insulation
Copper conductor core

Insulation filter

Fig. 4.4. Three-core belted cable.

Fig. 4.5. Tangential stress in a belted cable.

The original theory of tangential stress is now considered to be of lesser importance. The loose fillers between the cores situated in electric fields are electrically weak. With the variation of the load the cores expand and contract. This forms the vacuous spaces or *voids* in the fillers particularly in the space between the cores. The voids become the positions of discharges. If the discharges are sufficiently severe there may be local heating and charring of the paper with the result that there may be a breakdown of the cable insulation.

Belted cables are suitable up to and including 22 kV as they give satisfactory performance. Beyond this voltage a breakdown may occur due to the non-radial dielectric stresses associated with them and the formation of voids.

4.8 SCREENED CABLE

The defects of belted cable have been overcome in the screened, (shielded) cable (Fig. 4.6). Each core is insulated in the usual manner and is provided with a metallic screen. The screen consists of a metallised paper or a metal tape applied over each completed core insulation. The three screens are connected with the earthed metallic sheath. The screens are, therefore, at the earth potential. They eliminate the undesirable tangential stresses and make the stresses purely radial. The screened cable is also called H-type (after Hochstadter) cable. A construction in which individual insulated cores are metallised-paper screened and is separately lead sheathed is called HSL type cable.

Screened cables have greater breakdown strength, better heat dissipating properties and reduced risk of core-to-core faults. The current rating of screened cable is more than that of belted type. Such cables are generally used upto 66 kV.

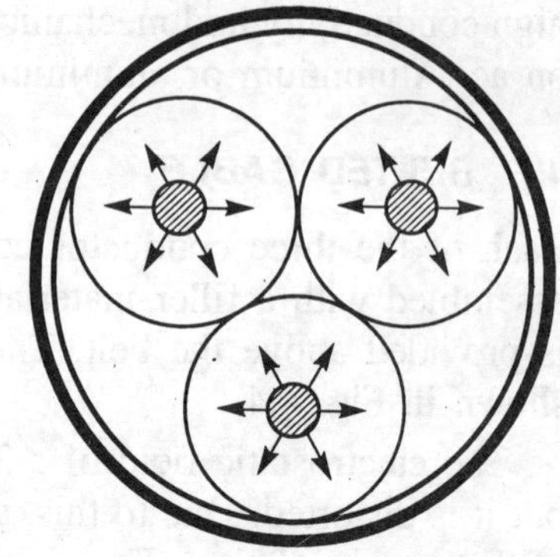

Fig. 4.6. Screened cable.

4.9 CABLE IMPREGNATION

Impregnation means application of an insulating oil compound on the paper used in cables. The cable in which the oil compound is applied to the dried paper before the paper is applied to the conductor is called a *pre-impregnated* cable. In the other type of construction the paper is first applied to the conductor. The insulated core is dried under vacuum in special heated tanks. It is then impregnated with the insulating compound filled in the tanks. Such cables are called *mass-impregnated* cables. After impregnation the cable is cooled under vacuum to minimize formation of voids. The sheath is applied after impregnation.

4.10 NON-DRAINED CABLES

When the mass-impregnated cables are laid on gradients there is a tendency of the drainage of oil from the higher levels to the lower levels. The drainage of compound results in reducing the oil content at the higher level positions. The voids may be formed and the insulation strength decreased there. At the lower end of the cable the impregnating compound is accumulated and builds up sufficient internal pressure which may burst the sheath.

To overcome this difficulty high viscosity insulating compound is used. This compound does not drain at normal operating temperature. Such cables are called *mass-impregnated non-draining* cables. They are also called *fully-impregnated non-draining* cables. Pre-impregnated cables are of non-draining type. They are shown in Fig. 4.7.

4.11 DIELECTRIC STRESS

If the electrostatic field be uniform, the stress on the dielectric may be found by dividing the applied voltage by the thickness of the dielectric. Since the field is not uniform the stress on the dielectric at any point in a single core cable (Fig. 4.8) is calculated as follows :

Let r = radius of conductor or inner radius of insulation

R = internal radius of sheath or outer radius of insulation

ε_0 = the permittivity of free space

ε_r = the relative permittivity of the dielectric

q = charge on the conductor per unit length in coulombs

V = operating phase to neutral voltage in volts

Flux passing through a cylinder of radius x and length one metre surrounding the core is q.

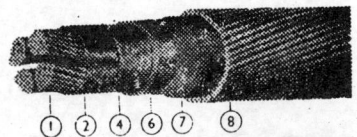

1.1. kV four-core lead-sheathed mass-impregnated non-draining cable, with single wire armour, left bare.

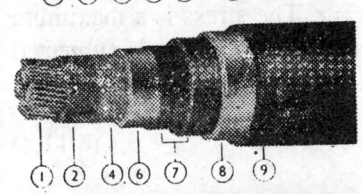

1.1 kV four-core lead-sheathed mass-impregnated non-draining cable, double steel tape armoured and served overall.

11 kV three-core screened lead-sheathed mass-impregnated non-draining cable, served overall.

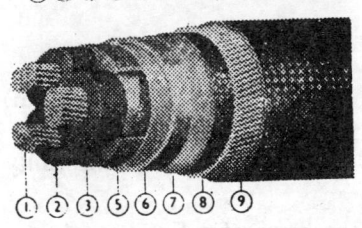

33 kV three-core screened lead-sheathed mass-impregnated non-draining cable, single wire armoured and served overall.

1. Stranded shaped copper conductor.
2. Paper insulation impregnated with special compound.
3. Core screening of metal tape (11 kV and higher voltages only).
4. Belt of impregnated paper insulation.

5. Copper woven fabric tape.
6. Sheath of lead or lead alloy.
7. Bedding for armour.
8. Armour, single layer of galvanized steel wires or double steel tape.
9. Overall serving.

Fig. 4.7. Mass-impregnated non-draining cables.

The electric flux density at a distance x from the centre is

$$D_x = \frac{q}{2\pi x} \ \text{C/m}^2$$

and the dielectric stress is given by

$$g_x = \frac{D_x}{\varepsilon_0\, \varepsilon_r} = \frac{q}{(2\pi\, \varepsilon_0\, \varepsilon_r)\, x} \ \text{V/m} \qquad \ldots(4.11.1)$$

The potential difference between the conductor and the sheath is equal to the work done to move a unit charge from the conductor to the sheath.

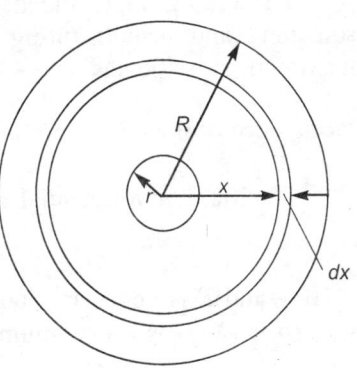

Fig. 4.8. Single-core cable.

Thus, $\quad V = \int_r^R g_x\, dx = \int_r^R \frac{q}{(2\pi\, \varepsilon_0\, \varepsilon_r)\, x}\, dx$

$$= \frac{q}{2\pi\, \varepsilon_0\, \varepsilon_r} \ln \frac{R}{r} \ \text{volts} \qquad \ldots(4.11.2)$$

Combination of Eqs. (4.11.1) and (4.11.2) gives

$$g_x = \frac{V}{x \ln (R/r)} \text{ V/m} \qquad \qquad ...(4.11.3)$$

The maximum stress will occur at the smallest radius, i.e., for $x = r$. The stress is a maximum at the surface of conductor, or in other words, the inner most layer of the dielectric is subjected to a maximum stress. The maximum stress is given by

$$g_{max} = \frac{V}{r \ln \dfrac{R}{r}} \text{ V/m} \qquad \qquad ...(4.11.4)$$

When $x = R$, the stress will be a minimum which indicates that stress has a minimum value at the sheath. The minimum value is given by

$$g_{min} = \frac{V}{R \ln \dfrac{R}{r}} \text{ V/m} \qquad \qquad ...(4.11.5)$$

Also, $\qquad \dfrac{g_{max}}{g_{min}} = \dfrac{R}{r} \qquad \qquad ...(4.11.6)$

The electric stress in a belted cable cannot be calculated accurately due to non-uniformity of the dielectric and the distortion in the electrostatic field.

The present day tendency is to design high voltage cables on the basis of a fixed maximum value of operating stress. The stress is usually expressed in kilovolts per millimetre. Eq. (4.11.4) is then utilized to determine the thickness of insulation necessary for a given diameter of conductor. It is clear from Eq. (4.11.4) that greater the value of permissible stress the lesser will be the insulation thickness. It is desirable to choose a higher value of operating stress in order to have a reduced thickness of insulation and, therefore, a reduced size of cable. Since a smaller cable size affords more economy, there is a tendency to increase the operating stresses to their highest values without the failure to cable either under actual operating conditions or during its approval specified tests.

Considerably improvements have been made in developing high strength paper, conductor screening and manufacturing techniques to achieve the objective of operating the cable at maximum stress levels.

Most Economical Size of Cable

$$\text{Maximum potential gradient, } g_{max} = \frac{V}{r \ln \dfrac{R}{r}}$$

If V and R are constant and r is made variable the expression for g_{max} has a minimum value when $r \ln (R/r)$ is a maximum. This occurs when

$$\frac{d}{dr} \left(r \ln \frac{R}{r} \right) = 0 \; ; \qquad \frac{d}{dr} (r \ln R - r \ln r) = 0$$

$$\ln R - r \cdot \frac{1}{r} - \ln r = 0 \; ; \qquad \ln \frac{R}{r} = 1 = \ln e \; ; \qquad \frac{R}{r} = e = 2.718$$

This is the condition that would make the voltage gradient at the conductor surface a minimum.

Example 4.1 A single-core cable, for working voltage of 6.5 kV (between core and sheath), has a conductor of 10 mm overall diameter, which is insulated with impregnated paper to a radial thickness of 7.5 mm and lead covered. Calculate the maximum electric stress on the insulation.

Solution

Diameter of conductor = 10 mm

Thickness of insulation, $t = 7.5$ mm

$r = 5$ mm, $R = r + t = 5 + 7.5 = 12.5$ mm

$$\ln \frac{R}{r} = \ln \frac{12.5}{5} = 0.916$$

By Eq. 4.11.4,

$$g_{max} = \frac{V}{r \ln \dfrac{R}{r}} = \frac{6.5}{5 \times 0.916} = 1.419 \text{ kV/mm (rms)}.$$

Example 4.2 Determine the overall diameter of a single-core cable and its most economical diameter when working on a three-phase, 275 kV system. The maximum permissible stress in the dielectric is not to exceed 15 kV/mm.

Solution

Effective (rms) value of phase voltage $= \dfrac{275}{\sqrt{3}}$ kV

Peak value of phase voltage $= \dfrac{275}{\sqrt{3}} \times \sqrt{2} = 224.5$ kV

For economical size of the cable, optimum ratio of sheath and core radii is given by

$$\frac{R}{r} = e$$

and $$g_{max} = \frac{V}{r \ln \dfrac{R}{r}} = \frac{V}{r}$$

Now, $g_{max} = 15$ kV/mm

∴ $$15 = \frac{224.5}{r} \; ; \qquad r = \frac{224.5}{15} = 14.96 \text{ mm}$$

Economical core diameter, $D = 2r = 2 \times 14.96 = 29.92$ mm

Also, for economical size of the cable

$R = e\, r = 2.718 \times 14.96 = 40.66$ mm

Inner diameter of the sheath $= 2R = 2 \times 40.66 = 81.32$ mm.

4.12 GRADING OF CABLES

We have seen that the stress is greatest at the surface of conductor or the inner-most part of the dielectric. It goes on decreasing as the outermost layer of dielectric is reached. It is minimum at the sheath. If the dielectric be chosen according to the maximum stress the thickness of the cable would increase considerably. Moreover, the benefit of the chosen dielectric would not be achieved fully at the outer layers. If all the dielectric were equally stressed the thickness of the insulation could be considerably reduced. This would reduce the cost of the cable. Moreover, such a construction would enable the heat generated in the conductor to be more easily conducted to the sheath, so increasing the current rating. The method of equalizing the stress in the dielectric of the cable is called *grading of cable*. There are two methods of grading the cables: (1) Capacitance grading, and (2) Intersheath grading.

Capacitance Grading or Dielectric Grading

The dielectric stress is given by Eq. (4.11.1) as

$$g_x = \frac{q}{2\pi\, \varepsilon_0\, \varepsilon_r\, x} = \frac{\text{constant}}{\varepsilon_r\, x} \qquad \qquad \ldots(4.12.1)$$

If it were possible to decrease ε_r as the radius x of the insulation increased in such a way that $\varepsilon_r\, x$ remained constant the stress would remain uniform and minimum insulation would be required. In *capacitance grading*, the homogeneous dielectric is replaced by layers of dielectrics having different values of relative permittivity such that $\varepsilon_r\, x$ remains constant. To get uniform stress infinite number of dielectrics will be required. In practice two or thee layers of insulation with suitable permittivities may be used for obtaining good results. Fig. 4.9 shows a cable having three dielectrics of relative permittivities ε_1, ε_2, and ε_3, such that $\varepsilon_3 < \varepsilon_2 < \varepsilon_1$. Let r_1, r_2, and R be the outer radii of dielectrics.

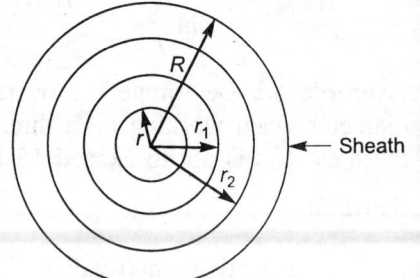

Fig. 4.9. Dielectric grading.

The potential difference across the inner layer

$$V_1 = \int_r^{r_1} g_x\, dx = \int_r^{r_1} \frac{q}{2\pi\, \varepsilon_0\, \varepsilon_1\, x}\, dx$$

$$= \frac{g}{2\pi\, \varepsilon_0\, \varepsilon_1} \ln \frac{r_1}{r} \qquad \qquad \ldots(4.12.2)$$

$$= \frac{q}{2\pi\, \varepsilon_0\, \varepsilon_1\, r} \cdot r \ln \frac{r_1}{r} = g_{\text{max }1}\, r \ln \frac{r_1}{r} \qquad \qquad \ldots(4.12.3)$$

Similarly, the potential difference between r_1 and r_2, i.e., across the middle layer

$$V_2 = g_{\text{max }2}\, r_1 \ln \frac{r_2}{r_1} \qquad \qquad \ldots(4.12.4)$$

and the potential difference between r_2 and R, i.e., across the the outer layer

$$V_3 = g_{max\ 3}\ r_2 \ln \frac{R}{r_2} \qquad \qquad \text{...(4.12.5)}$$

The total potential difference between core and earthed sheath

$$V = V_1 + V_2 + V_3$$

$$= g_{max\ 1}\ r \ln \frac{r_1}{r} + g_{max\ 2}\ r_1 \ln \frac{r_2}{r_1} + g_{max\ 3}\ r_2 \ln \frac{R}{r_2} \qquad \text{...(4.12.6)}$$

The capacitance of the cable

$$C = \frac{q}{V} = q \div \left[\frac{q}{2\pi\ \varepsilon_0} \left(\frac{1}{\varepsilon_1} \ln \frac{r_1}{r} + \frac{1}{\varepsilon_2} \ln \frac{r_2}{r_1} + \frac{1}{\varepsilon_3} \ln \frac{R}{r_2} \right) \right]$$

$$= \frac{2\pi\ \varepsilon_0}{\dfrac{1}{\varepsilon_1} \ln \dfrac{r_1}{r} + \dfrac{1}{\varepsilon_2} \ln \dfrac{r_2}{r_1} + \dfrac{1}{\varepsilon_3} \ln \dfrac{R}{r_2}} \qquad \text{...(4.12.7)}$$

The maximum stresses are given by

$$g_{max\ 1} = \frac{q}{2\pi\ \varepsilon_0\ \varepsilon_1\ r} \qquad \qquad \text{...(4.12.8)}$$

$$g_{max\ 2} = \frac{q}{2\pi\ \varepsilon_0\ \varepsilon_2\ r_1} \qquad \qquad \text{...(4.12.9)}$$

$$g_{max\ 3} = \frac{q}{2\pi\ \varepsilon_0\ \varepsilon_3\ R} \qquad \qquad \text{...(4.12.10)}$$

In order that the maximum stress is the same in each layer

$$g_{max\ 1} = g_{max\ 2} = g_{max\ 3} = g_{max} \ (\text{say}) \qquad \text{...(4.12.11)}$$

or, $$\varepsilon_1\ r = \varepsilon_2\ r_1 = \varepsilon_3\ R \qquad \qquad \text{...(4.12.12)}$$

The total applied voltage across the cable

$$V = g_{max} \left(r \ln \frac{r_1}{r} + r_1 \ln \frac{r_2}{r_1} + r_2 \ln \frac{R}{r_2} \right) \qquad \text{...(4.12.13)}$$

It is to be noted that g_{max} represents the peak value of electric stress and, therefore, all the voltages in the above relations are peak values and not rms values.

Intersheath Grading

In this method of grading, the same insulating material is used throughout the total thickness of the cable but it is divided into two or more layers by providing intersheaths. Intersheaths are *thin* metallic cylindrical sheaths concentric with the conductor and placed between the conductor and the outside sheath. These intersheaths are maintained at suitable potentials by connecting them to tappings from the supply transformer.

Consider a cable with one intersheath only as shown in Fig. 4.10.

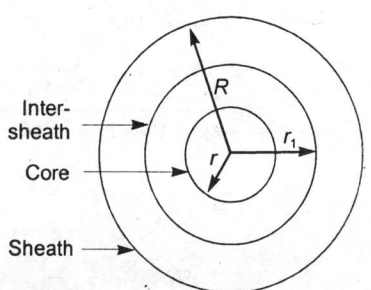

Fig. 4.10. Intersheath grading.

Let r = radius of the core

r_1 = radius of the intersheath

R = radius of the outer sheath

V_1 = voltage between the core and the intersheath

V_2 = voltage between the intersheath and outer sheath

V = applied voltage between the core and the sheath.

The maximum potential gradient in first layer

$$g_{max\ 1} = \frac{V_1}{r \ln \dfrac{r_1}{r}} \qquad \qquad \text{...(4.12.14)}$$

The maximum potential gradient in second layer

$$g_{max\ 2} = \frac{V_2}{r_1 \ln \dfrac{R}{r_1}} \qquad \qquad \text{...(4.12.15)}$$

If the two potential gradients are equal

$$\frac{V_1}{r \ln \dfrac{r_1}{r}} = \frac{V_2}{r_1 \ln \dfrac{R}{r_1}} = g_{max} \text{ (say)} \qquad \qquad \text{...(4.12.16)}$$

Also, $V = V_1 + V_2 = g_{max}\left(r \ln \dfrac{r_1}{r} + r_1 \ln \dfrac{R}{r_1} \right)$ \qquad \qquad \text{...(4.12.17)}

For economical size of the cable

$$\frac{r_1}{r} = e$$

and $g_{max} = \dfrac{V_1}{r}$ \qquad \qquad \text{...(4.12.18)}

$$r_1 = e\, r = \frac{e\, V_1}{g_{max}} \qquad \qquad \text{...(4.12.19)}$$

Also, $g_{max} = \dfrac{V_2}{r_1 \ln \dfrac{R}{r_1}} \; ; \qquad \dfrac{V_1}{r} = \dfrac{V_2}{r_1 \ln \dfrac{R}{r_1}}$

$$\ln \frac{R}{r_1} = \frac{V_2}{V_1} \cdot \frac{r}{r_1} = \frac{V_2}{e\, V_1} = \frac{V - V_1}{e\, V_1} = \frac{V}{e\, V_1} - \frac{1}{e} \; ; \qquad \frac{R}{r_1} = e^{\frac{V}{e\, V_1} - \frac{1}{e}}$$

$$R = r_1\, e^{\frac{V}{e\, V_1} - \frac{1}{e}} = \frac{e\, V_1}{g_{max}} e^{\frac{V}{e\, V_1} - \frac{1}{e}}$$

$$= \frac{V_1}{g_{max}} e^{1 - \frac{1}{e}} e^{\frac{V}{e\, V_1}} \qquad \qquad \text{...(4.12.20)}$$

$$= A\ V_1\ e^{\frac{V}{e\ V_1}}$$

where A is a constant equal to $\dfrac{1}{g_{max}}\ e^{1-\frac{1}{e}}$

For minimum value of R

$$\frac{dR}{dV_1} = 0\ ;\qquad A\ V_1 \left(\frac{-V}{e\ V_1^2}\right) e^{\frac{V}{e\ V_1}} + A\ e^{\frac{V}{e\ V_1}} = 0$$

$$A\ e^{\frac{V}{e\ V_1}} \left(1 - \frac{V}{e\ V_1}\right) = 0$$

$$V_1 = \frac{V}{e} \qquad\qquad\qquad\qquad\qquad\qquad\qquad\qquad\qquad ...(4.12.21)$$

From Eqs. (4.12.18) to (4.12.21)

$$r = \frac{V_1}{g_{max}} = \frac{V}{e\ g_{max}} \qquad\qquad\qquad\qquad\qquad ...(4.12.22)$$

$$r_1 = \frac{V_1\ e}{g_{max}} = \frac{V}{g_{max}} \qquad\qquad\qquad\qquad\qquad ...(4.12.23)$$

$$R = \frac{V}{g_{max}}\ e^{(1-1/e)} \qquad\qquad\qquad\qquad\qquad ...(4.12.24)$$

It should be remembered that $e = 2.718$

$$\therefore\qquad e^{(1-1/e)} = (2.718)^{(1-1/2.718)} = (2.718)^{1.718/2.718} = (2.718)^{0.632} = 1.881$$

$$R = 1.881\ \frac{V}{g_{max}} \qquad\qquad\qquad\qquad\qquad ...(4.12.25)$$

Limitations of Grading

Cable grading is of theoretical interest only. Practical difficulties are encountered in the application of grading methods. The main disadvantage of the capacitance grading is that the range of permittivity values of insulating materials available for cable insulation is limited. Moreover, the permittivities of the layers may not remain constant during the service period of the cable. Consequently, the stress distribution may change and may even cause insulation breakdown at the normal operating voltage.

The intersheaths, being very thin, are liable to be damaged during transportation or installation. High local stress may be developed at the points of damage leading to breakdown. Also, thin intersheaths are not able to carry the charging current of long cable lines and thus the current carrying capacity of the cable is reduced. Intersheath grading also presents difficulties in fixing the intersheath potentials. For these reasons the present trend is to avoid grading.

Example 4.3 A single-core, lead sheathed cable joint has a conductor of 10 mm diameter and two layers of different insulating materials, each 10 mm thick. The relative permittivities are 3 (inner) and 2.5 (outer). Calculate the potential gradient at the surface of the conductor when the potential difference between the conductor and the lead sheathing is 60 kV.

Solution

$$r = \frac{10}{2} = 5 \text{ mm}, \quad t = 10 \text{ mm}$$

$$r_1 = r + t = 5 + 10 = 15 \text{ mm}$$

$$R = r + 2t = 5 + 20 = 25 \text{ mm}$$

$$\varepsilon_1 = 3, \quad \varepsilon_2 = 2.5, \quad V = 60 \text{ kV}$$

$$g_{max\,1} = \frac{q}{2\pi \,\varepsilon_0 \,\varepsilon_1 \,r} \; ; \qquad g_{max\,2} = \frac{q}{2\pi \,\varepsilon_0 \,\varepsilon_2 \,r_1}$$

$$\therefore \qquad \frac{r_1 \, g_{max\,2}}{r \, g_{max\,1}} = \frac{\varepsilon_1}{\varepsilon_2}$$

$$V = V_1 + V_2 = g_{max\,1} \, r \ln \frac{r_1}{r} + g_{max\,2} \, r_1 \ln \frac{R}{r_1} = r \, g_{max\,1} \left(\ln \frac{r_1}{r} + \frac{g_{max\,2} \, r_1}{g_{max\,1} \, r} \ln \frac{R}{r_1} \right)$$

$$= r \, g_{max\,1} \left(\ln \frac{r_1}{r} + \frac{\varepsilon_1}{\varepsilon_2} \ln \frac{R}{r_1} \right)$$

$$60 = 5 g_{max\,1} \left(\ln \frac{15}{5} + \frac{3}{2.5} \ln \frac{25}{15} \right) = 5 g_{max\,1} \,(1.098 + 0.613) = 8.558 \, g_{max\,1}$$

$$g_{max\,1} = \frac{60}{8.558} = 7.01 \text{ kV/mm}$$

Example 4.4 A single-core, lead covered cable is to be designed for 66 kV to earth. Its conductor radius is 10 mm and its insulating materials A, B and C have relative permittivities of 5, 4 and 3 respectively and corresponding maximum permissible stresses of 3.8, 2.6 and 2.0 kV/mm (rms) respectively. Find the minimum diameter of the lead sheath.

Solution

$$g_{max\,A} = \frac{q}{2\pi \,\varepsilon_0 \,\varepsilon_A \,r} \; ; \qquad g_{max\,B} = \frac{q}{2\pi \,\varepsilon_0 \,\varepsilon_B \,r_1} \; ; \qquad g_{max\,C} = \frac{q}{2\pi \,\varepsilon_0 \,\varepsilon_C \,r_2}$$

$$\varepsilon_A \, r \, g_{max\,A} = \varepsilon_B \, r_1 \, g_{max\,B} = \varepsilon_C \, r_2 \, g_{max\,C}$$

$$5 \times 10 \times 3.8 = 4 r_1 \times 2.6 = 3 r_2 \times 2$$

$$r_1 = \frac{5 \times 10 \times 3.8}{4 \times 2.6} = 18.3 \text{ mm} \; ; \qquad r_2 = \frac{5 \times 10 \times 3.8}{3 \times 2} = 31.7 \text{ mm}$$

$$V_1 = g_{max\,A} \, r \ln \frac{r_1}{r} = 3.8 \times 10 \times \ln \frac{18.3}{10} = 38 \times 0.604 = 22.95 \text{ kV}$$

$$V_2 = g_{max\,B} \, r_1 \ln \frac{r_2}{r_1} = 2.6 \times 18.3 \ln \frac{31.7}{18.3} = 2.6 \times 18.3 \times 0.549 = 26.12 \text{ kV}$$

$$V_3 = g_{max\,C} \, r_2 \ln \frac{R}{r_2} = 2 \times 31.7 \ln \frac{R}{r_2} = 63.4 \ln \frac{R}{r_2}$$

Also, $\quad V = V_1 + V_2 + V_3$

$$66 = 22.95 + 26.12 + 63.4 \ln \frac{R}{r_2}$$

$$\ln \frac{R}{r_2} = \frac{66 - 22.95 - 26.12}{63.4} = \frac{16.93}{63.4} = 0.2670 \ ; \qquad \frac{R}{r_2} = 1.306$$

$R = 1.306 \ r_2 = 1.306 \times 31.7 = 41.4$ mm

Diameter of sheath $= 2R = 2 \times 41.4 = 82.8$ mm.

Example 4.5 A 60 kV (rms) single-core metal sheathed cable is to be graded by means of a metallic intersheath. The safe electric stress of the insulating material is 4 kV/mm (rms). (a) Calculate the diameter of the intersheath and the voltage at which it must be maintained in order to obtain minimum overall diameter. Calculate also the corresponding conductor diameter. (b) Compare the conductor diameter obtained in (a) with that of an ungraded cable working under the same conditions.

Solution

(a) $V = 60$ kV (rms), $g_{max} = 4$ kV/mm (rms)

$$V_1 = \frac{V}{e} = \frac{60}{2.718} = 22.1 \text{ kV}$$

$$r = \frac{V}{e \ g_{max}} = \frac{60}{2.718 \times 4} = 5.5 \text{ mm}$$

Diameter of core $= 2r = 2 \times 5.5 = 11$ mm

Radius of intersheath, $r_1 = \dfrac{V}{g_{max}} = \dfrac{60}{4} = 15$ mm

Diameter of intersheath, $d_1 = 2r_1 = 2 \times 15 = 30$ mm

$V_2 = V - V_1 = 60 - 22.1 = 37.9$ kV

$$R = 1.881 \frac{V}{g_{max}} = 1.881 \times \frac{60}{4} = 28.2 \text{ mm}$$

Minimum overall diameter of the cable, $D = 2R = 2 \times 28.2 = 56.4$ mm.

(b) *Cable without intersheath*

For economic cable size

$$\frac{R}{r} = e = 2.718 \ ; \qquad \ln \frac{R}{r} = 1$$

$$V = g_{max} \ r \ln \frac{R}{r} = g_{max} \ r$$

$$r = \frac{V}{g_{max}} = \frac{60}{4} = 15 \text{ mm}$$

Diameter of conductor $= 2r = 2 \times 15 = 30$ mm

$R = e \ r = 2.718 \times 15 = 40.77$ mm

$D = 2R = 2 \times 40.77 = 81.54$ mm.

4.13 CABLE CAPACITANCE

The capacitance of a cable transmission line is very much larger than that of an overhead line of the same length due to the following reasons :

1. The distance between the conductors is very small.
2. The distance between the cores and the earthed sheath is also small.
3. The permittivity of the cable insulation is usually 3 to 5 times greater than that of air insulation around the conductors of overhead line.

It is easier to calculate the capacitance of an overhead system accurately if its configuration be known, but for a cable system such a calculation is only approximate. Therefore, it is a practice to perform actual tests on a cable to find its capacitance.

The approximate method of cable capacitance is based on the assumption that the cable dielectric is perfectly homogeneous. The insulation of a cable is, however, far from being homogeneous or uniform. The capacitance of a single-core cable is found from the Eq. (4.11.2) as

$$C = \frac{q}{V} = \frac{2\pi\,\varepsilon_0\,\varepsilon_r}{\ln\dfrac{R}{r}} \quad \text{F/m} \qquad \qquad \text{...(4.13.1)}$$

$$= \frac{\varepsilon_r}{18 \times 10^9 \ln\dfrac{R}{r}} \quad \text{F/m} = \frac{\varepsilon_r}{18 \ln\dfrac{R}{r}} \quad \mu\text{F/km} \qquad \text{...(4.13.2)}$$

4.14 CHARGING CURRENT OR CAPACITIVE CURRENT

The capacitance of a cable determines the charging current, charging kVA, and the dielectric loss. The charging current restricts the use of cables on extra high voltage lines. For such lines additional reactive kVA is to be supplied to compensate this effect. The regulation and stability are affected by the charging current.

The current carrying capacity of an a.c. cable is also reduced by charging current. The maximum permissible load current goes on decreasing with greater lengths of an a.c. cable. With a d.c. cable there is negligible charging under normal operating conditions. The current carrying capacity of d.c. cable is, therefore, independent of its length.

4.15 CAPACITANCES IN A THREE-CORE BELTED CABLE

The conductors in a cable are separated from each other by the dielectric. Similarly, there is dielectric between the conductors and the sheath. When a potential difference is applied between the conductors the cable, in effect, is a combination of six capacitances. The capacitances between the conductors are represented by C_c, while those between conductors and sheath by C_s. Thus, a three-phase, belted-cable may be represented by a system of capacitances connected in star and delta as shown in Fig. 4.11.

The delta-connected capacitances C_c may be replaced by equivalent star-connected capacitances C_1 (Fig. 4.12). The capacitances between pairs of terminals will be the same in the two systems.

Capacitance between A and B in the delta system $= C_c + 0.5C_c = 1.5C_c$

and the capacitance between A and B in the star system $= 0.5C_1$.

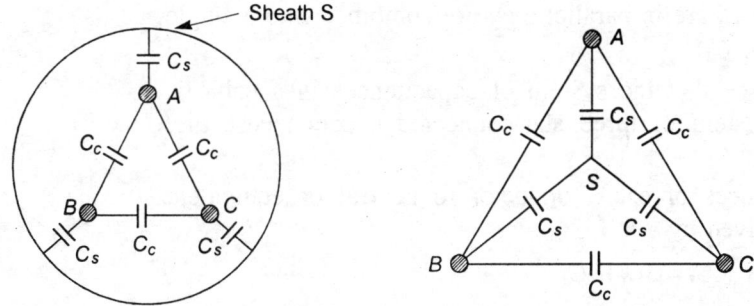

Fig. 4.11. Capacitances in three-core belted cable.

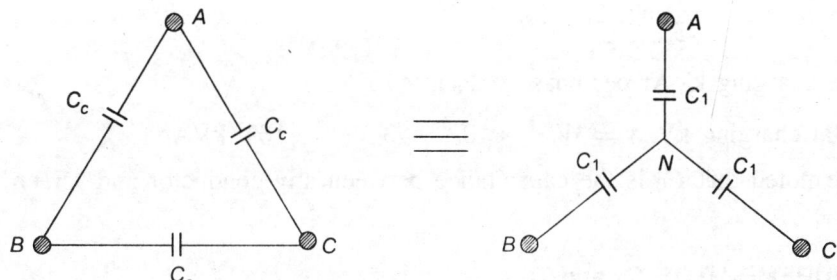

Fig. 4.12.

For the two systems to be equivalent

$$1.5C_c = 0.5C_1, \quad C_1 = 3C_c.$$

The cable may, therefore, be represented by Fig. 4.13. If the neutral point N of the system be earthed, and the sheath be also at zero potential, N and S will become equipotential, and Fig. 4.13 then becomes equivalent to that shown in Fig. 4.14.

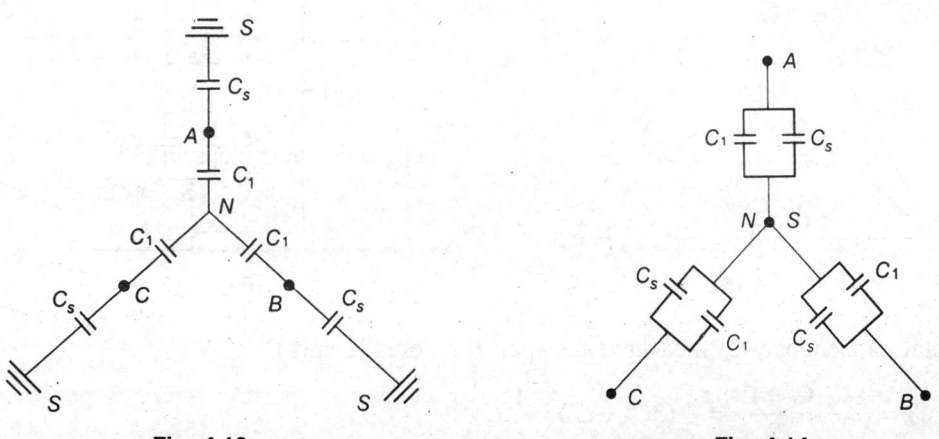

Fig. 4.13.   **Fig. 4.14.**

Since C_1 and C_s are in parallel they are combined into a single capacitance $(C_1 + C_s)$.

Finally, we see that the system of capacitances in a cable is reduced to a system of three star-connected capacitances (Fig. 4.15).

The capacitances of each conductor to neutral or equivalent capacitance is given by

$$C_0 = C_1 + C_s = 3C_c + C_s$$

If V_L = line voltage, V_p = phase voltage, the charging current per phase

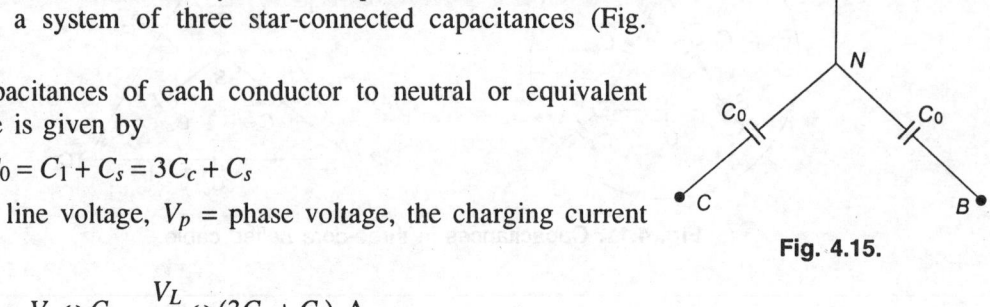

Fig. 4.15.

$$I_c = V_p \, \omega \, C_0 = \frac{V_L}{\sqrt{3}} \, \omega \, (3C_c + C_s) \text{ A}$$

The charging kVAr per phase $= V_p I_c \times 10^{-3}$

Total charging kVAr $= 3V_p I_c \times 10^{-3} = \sqrt{3} \, V_L I_c \times 10^{-3}$ kVAr

It is to be noted that C_0 is the capacitance between any conductor and screen for a 3-core screened cable.

4.16 MEASUREMENT OF C_c and C_s

Cable capacitance is determined from actual measurements instead of relying on the results obtained from the geometry of the cable. The non-uniformity of the insulation material, the variation in the shape of conductors and the use of fillers make it difficult to estimate the capacitance of a cable from its diameter. The following tests are generally performed :

(a) One conductor, say C, is connected to the sheath or insulated and the capacitance is measured between the remaining two conductors A and B. Fig. 4.16 then reduces to Fig. 4.17.

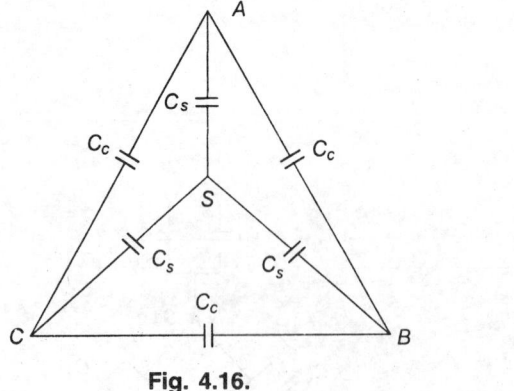

Fig. 4.16.

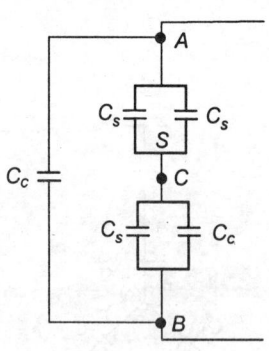

Fig. 4.17.

The total capacitance C_L measured between the cores A and B is

$$C_L = C_c + \frac{C_c + C_s}{2} = \frac{1}{2} (3C_c + C_s) = \frac{1}{2} C_0$$

The single measurement is sufficient for calculating the charging current per conductor.

(b) The three conductors are connected or bunched together (Fig. 4.18) and the capacitance is measured between this bunch and the sheath. Let it be denoted by C_b. Here C_c becomes zero and $C_b = C_s$.

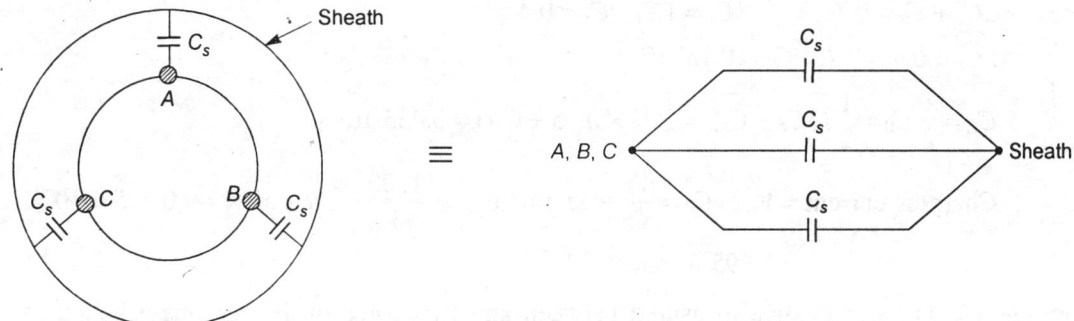

Fig. 4.18.

(c) Two conductors, say A and B, are joined together and the capacitance is measured between them and the remaining conductor. The arrangement then becomes as shown in Fig. 4.19.

The capacitance between B and $C = C_c + C_c + \dfrac{2}{3} C_s = \dfrac{2}{3} (3C_c + C_s) = \dfrac{2}{3} C_0$

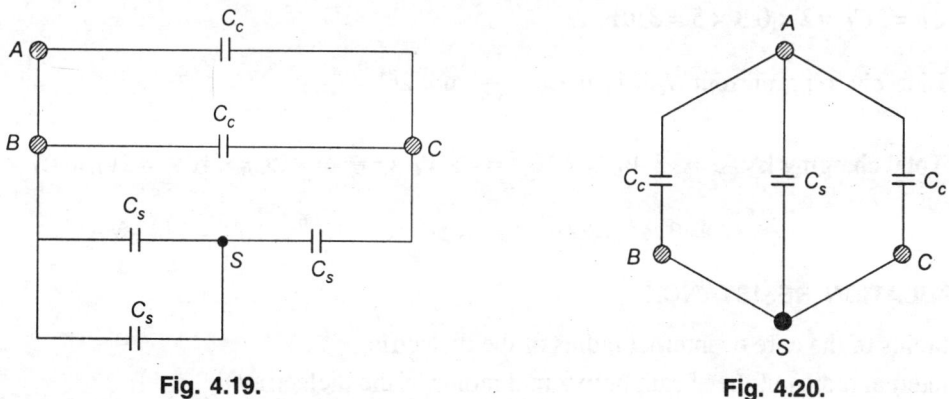

Fig. 4.19.   Fig. 4.20.

(d) Two conductors, say B and C, are connected to sheath and the capacitance is measured between these and the third conductor A. The capacitance arrangement of the system then reduces to Fig. 4.20.

The capacitance measured in this case

$$= C_s + C_c + C_c = 2C_c + C_s$$

From the above tests the value of C_c and C_s can also be determined separately.

Example 4.6 The test results for 1 km of a three-phase metal sheathed belted cable gave a measured capacitance of 0.7 μF between one conductor and the other two conductors bunched together with the earth sheath and 1.2 μF measured between the three bunched conductors and the sheath. Find (a) the capacitance between any pair of conductors, the sheath being isolated, and (b) the charging current when the cable is connected to 11 kV, 50 Hz supply.

Solution

Here $\qquad 2C_c + C_s = 0.7, \qquad 3C_s = 1.2, \quad C_s = 0.4$

$\qquad 2C_c + 0.4 = 0.7, \quad C_c = 0.15 \ \mu F$

$$C_L = \frac{1}{2} \ C_0 = \frac{1}{2} \ (3C_c + C_s) = \frac{1}{2} \ (3 \times 0.15 + 0.4) = 0.425 \ \mu F$$

$$\text{Charging current} = V_p \ \omega \ C_0 = \frac{V_L}{\sqrt{3}} \times 2\pi f \times 2C_L = \frac{11000}{\sqrt{3}} \times 2\pi \times 50 \times 2 \times 0.425 \times 10^{-6}$$

$$= 1.695 \ \text{A/phase}.$$

Example 4.7 The capacitance measured between any two cores of a three phase belted cable is 0.3 μF/km. Calculate the charging kVAr taken by a 5 km length of this cable when connected to an 11 kV, 50 Hz supply.

Solution

If C_L be the measured capacitance between any two cores

$$C_L = \frac{C_0}{2}$$

$$C_0 = 2C_L = 2 \times 0.3 \times 5 = 3 \ \mu F$$

Line charging current, $I_c = V_p \ \omega \ C_0 = \dfrac{V_L}{\sqrt{3}} \ \omega \times 2C_L$

$$\text{Total charging kVA,} = \sqrt{3} \ V_L \ I_c \times 10^{-3} = \sqrt{3} \ V_L \times \frac{V_L}{\sqrt{3}} \ \omega \times 2C_L \times 10^{-3} = 2V_L^2 \ \omega \ C_L \times 10^{-3}$$

$$= 3 \times (11000)^2 \times 2\pi \times 50 \times 3 \times 10^{-6} \times 10^{-3} = 114 \ \text{kVAr}.$$

4.17 INSULATION RESISTANCE

Let $\quad r =$ radius of the core or internal radius of the dielectric

$\qquad R =$ internal radius of the sheath or external radius of the dielectric

$\qquad l =$ cable length

$\qquad \rho =$ specific resistance of the dielectric.

The leakage current flows radially through the dielectric. Consider an element of the dielectric radius x and the thickness dx.

The length through which the leakage current flows $= dx$. The area of the cross-section through which it is flowing is $2\pi x l$. Therefore, the resistance of the element in the radial direction is

$$dR_i = \rho \ \frac{dx}{2\pi x l}$$

and the insulation resistance of the cable is

$$R_i = \int_r^R \frac{\rho \, dx}{2\pi x l} = \frac{\rho}{2\pi l} \, [\ln x]_r^R = \frac{\rho}{2\pi l} \ln \frac{R}{r} \qquad \qquad ...(4.17.1)$$

The above relation shows that the insulation resistance varies inversely as the length.

Example 4.8 A single-core cable 5 km long has an insulation resistance of 0.4 MΩ. The core diameter is 20 mm and the diameter of the cable over the insulation is 50 mm. Calculate the resistivity of the insulating material.

Solution

$$R_i = \frac{\rho}{2\pi l} \ln \frac{R}{r} \; ; \qquad \rho = \frac{2\pi l R_i}{\ln \frac{R}{r}}$$

Here $\quad R_i = 0.4 \times 10^6 \; \Omega, \quad l = 5 \text{ km} = 5 \times 10^3 \text{ m}$

$\qquad 2r = 20 \text{ mm}, \quad 2R = 50 \text{ mm}$

$\qquad \ln \dfrac{R}{r} = \ln \dfrac{50}{20} = 0.916$

$$\rho = \frac{2\pi \times 5 \times 10^3 \times 0.4 \times 10^6}{0.916} = 13.72 \times 10^9 \; \Omega\text{-m} = 13.72 \times 10^3 \text{ M}\Omega\text{-m}$$

4.18 DIELECTRIC LOSS

Power loss occurs in the dielectric of a cable due to three main causes. They are :

1. Conductivity of insulation.
2. Dielectric hysteresis or dielectric absorption.
3. Ionization or corona.

4.18.1 Conductivity of Insulation

The resistance of dielectric is very high but not infinite and therefore, a small current flows due to the conductivity of the dielectric. This current is called the *leakage current*. The leakage current through the insulation resistance gives rise to a power loss in the dielectric. This power loss is termed as *leakage loss*. The leakage loss occurs with ac and dc voltages.

4.18.2 Dielectric Hysteresis or Dielectric Absorption

If a constant d.c. voltage be suddenly applied across a cable the current in the beginning is very high. The current decreases gradually to a low steady value determined by the type of cable insulation. This steady-state value of current is called the *conduction current*. On removal of the applied voltage the current does not become zero instantaneously, but the cable is discharging by supplying a current of decreasing magnitude for practically the same length of time as it took previously to attain its steady-state value before the dielectric is fully discharged. However, the amount of charge released is somewhat lesser than that stored in dielectric during charge. A part of the absorbed charge recombines with the dielectric.

If an alternating voltage be applied to the cable, the dielectric is subjected to several cycles of charge and discharge per second. There is a molecular bipolar friction within the insulation when the stress is varied. The applied potential has to overcome this molecular friction and elasticity of dielectric. A power loss takes place in the material of the dielectric due to constant changing of absorbed charge. This loss due to absorption is much greater than that in the insulation resistance. The *dielectric absorption* is also called *dielectric hysteresis*. The phenomenon is similar to magnetic hysteresis. There is a time lag in storing and releasing charge in dielectric. The time lag is responsible for absorbing loss when a dielectric is subjected to several cycles of charge and discharge per second.

4.18.3 Corona or Gaseous Ionization in Cables

The spaces or voids in the cables are filled up with the vapours of the impregnating oil and air. The dielectric strength of the voids or gaseous spaces within the insulation is very small compared with that of the insulation. If the applied potential gradient in the insulation is large, there is a concentration of field intensity in the spaces resulting in the ionization or corona discharge. Thus, a breakdown of insulation may occur at relatively low stresses and temperatures. Another harmful effect of corona discharge within the cable insulation is that it produces ozone and oxides of nitrogen. These gases may deteriorate the insulation. The life of the cable is thus affected.

The electrical breakdown strength of impregnated insulation depends to a great extent upon the properties of its gas filled cavities and upon corona process occurring in them. The electrical breakdown stress is different for different insulations. One reason for this can be found in the difference of the properties of cavities occurring in the insulations. These properties are typical location, size, shape, gas pressure, kind of gas, and corona inception stress.

4.18.4 Dielectric Power Loss

In a perfect capacitor the applied voltage V lags the charging current by $90°$. A cable behaves as an imperfect capacitor. The equivalent circuit of a cable can be represented by a parallel combination of leakage resistance R_1 and a capacitance C_0 as shown in Fig. 4.21.

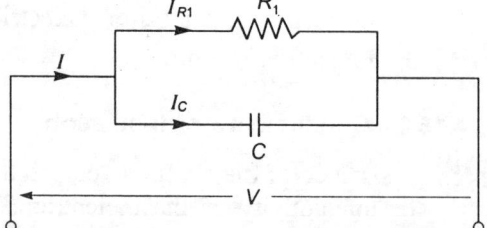

Fig. 4.21. Equivalent circuit of a cable.

Let V = line–to–neutral supply voltage

I_{R1} = current through R_1

I_c = current through the capacitor

= charging current

I = supply current

The phasor diagram is shown in Fig. 4.22. The current I leads the applied voltage V by an angle ϕ. The angle $(90 - \varphi) = \delta$ is called the dielectric loss angle.

For a single-phase line, the dielectric power loss

$$P_d = VI \cos \varphi = VI_{R1}$$

From the phasor diagram,

$$\frac{I_{R1}}{I_c} = \tan \delta$$

$$I_{R1} = I_c \tan \delta = (V \omega C_0) \tan \delta$$

∴ $$P_d = V I_{R1} = V^2 \omega C_0 \tan \delta$$

Since δ is usually very small,

$$\tan \delta = \delta$$

∴ $$P_d = V^2 \omega C_0 \delta \text{ watts} \qquad \qquad \dots(4.18.1)$$

where ω is the power supply frequency in rad/s and δ is in radians.

In a three-phase cable, if C_0 be the equivalent star capacitance or capacitance to neutral, the dielectric power loss is given by

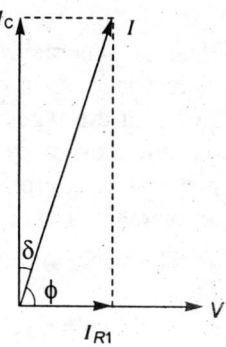

Fig. 4.22. Phasor diagram.

$$P_d = 3 V_p^2 \omega C_0 \delta \qquad \qquad \dots(4.18.2)$$

where V_p is the phase voltage.

Example 4.9 An 11 kV 50 Hz, single-phase cable has a diameter of 20 mm, and an internal sheath radius of 15 mm. If the dielectric has a relative permittivity of 2.4 and a loss angle of 0.031 radian, determine for 2.5 km length of the cable : (a) the capacitance, (b) the charging current, (c) the generated reactive voltamperes, (d) the dielectric loss, and (e) the equivalent insulation resistance.

Solution

(a) $$C = \frac{\varepsilon_r}{18 \ln \dfrac{R}{r}} \text{ } \mu\text{F/km} = \frac{2.4}{18 \ln \dfrac{15}{10}} = 0.329 \text{ } \mu\text{F/km}$$

Total cable capacitance $= 0.329 \times 2.5 = 0.8225 \text{ } \mu\text{F}$

(b)   Charging current, $I_c = \omega C V = 2\pi f C V = 2\pi \times 50 \times 0.8225 \times 10^{-6} \times 11 \times 10^3 = 2.84$ A

The no-load current may approximately be taken as charging current, i.e., $I_0 = I_c$.

(c)   Generated reactive voltamperes, $V I_c = 11 \times 10^3 \times 2.84$ VAr

$$= 11 \times 2.84 \text{ kVAr} = 31.24 \text{ kVAr}$$

(d)   Dielectric loss, $P_d = V I_0 \cos \varphi_0 = V I_c \cos \varphi_0 = V I_c \delta = 11 \times 10^3 \times 2.84 \times 0.031 = 968$ W

(e) If R_i is the equivalent insulation resistance

$$P_d = \frac{V^2}{R_i}$$

$$R_i = \frac{V^2}{P_d} = \frac{(11 \times 10^3)^2}{968} \text{ } \Omega = 0.125 \text{ M}\Omega$$

4.19 STRESS DISTRIBUTION IN A HVDC CABLE

Heat is generated in the dielectric medium of the cables both due to leakage currents and dielectric loss. In d.c. under steady-state operation, the loss is due to leakage currents only. In a.c. both the types of losses are there. In d.c. the current flows unidirectionally whereas in a.c. the direction of the current is changing every half cycle. Hence in a.c. the heat produced in the material is approximately constant and, therefore, the temperature gradient is zero. In d.c. the temperature is not so uniform and there is a greater temperature gradient.

The variation of insulation resistivity with temperature is governed by the relation.

$$\rho_\theta = \rho_0\, e^{-\alpha\theta} \qquad\qquad\qquad ...(4.19.1)$$

where ρ_θ = insulation resistivity at a temperature elevation of θ

ρ_0 = insulation resistivity at a standard reference temperature

α = a constant for the dielectric.

The above relation shows that the resistivity of the insulation decreases with the increase in temperature.

If it is assumed that the temperature is uniform in the whole dielectric (which is only possible at no load) the stress distribution is given by

$$g_x = \frac{V}{x \ln \dfrac{R}{r}}$$

The stress is maximum at the conductor surface and minimum at the sheath.

The stress distribution in a d.c. cable is different from that in an a.c. cable. In an a.c. cable permittivity decides the stress distribution. The permittivity is not affected by stress or temperature. The stress distribution in a cable subjected to d.c. is decided by the insulation resistivity. The insulation resistivity is a quantity depending upon a number of factors like temperature, humidity, non-homogeneity, and electrode configuration, etc.

Under normal operating conditions, the conductor is at a higher temperature than the sheath. When the cable is loaded the insulation near the conductor is heated up, but the sheath remains practically at a constant temperature governed by the ambient conditions. Thus, a temperature gradient is developed between the conductor surface and the sheath. For an unloaded cable the temperature gradient is zero. Since the resistivity of the insulation decreases with the increase in temperature, the stress decreases at the conductor surface and increases near the sheath when the cable is loaded. If the temperature gradient becomes large enough the stress near the sheath may become equal to or may exceed that near the conductor surface. It may exceed the design stress of the cable which is of course not desirable. Thus, the temperature gradient, and not the temperature, is responsible for stress inversion phenomenon. Therefore, a d.c. cable has two temperature limitations upon design, the first being the maximum permissible dielectric temperature and second is the maximum temperature gradient. The former limit is common to both a.c. and d.c. cables, but the latter is peculiar to a.c. only. It is very unlikely that both limits would be reached simultaneously.

4.20 SKIN EFFECT

The non-uniform distribution of current density in a conductor carrying a.c. due to greater inductance of the inner parts is called the *skin effect*. The apparent or d.c. ohmic resistance of the conductor is increased due to concentration of current on the surface or skin of the conductor.

4.21 PROXIMITY EFFECT

This is another electromagnetic effect which also results in the increment of the apparent resistance of the conductor due to the presence of other conductors carrying current in its vicinity. When two or more conductors are in proximity their electromagnetic fields interact with each other with the result that the current in each of them is redistributed such that the greater current density is concentrated in that part of the strand most remote from the interfering conductor. In each case, a reduced current rating results from the apparent increase of resistance.

The two effects assume importance only for conductor sizes greater than 125 mm^2. Correction factors are to be applied to take these facts into account.

If R_{dc} = uncorrected d.c. resistance of the core

$\quad y_s$ = skin effect factor, i.e., the fractional increment in resistance to allow for skin effect

$\quad y_p$ = proximity effect factor,

$\qquad$ i.e., the fractional increment in resistance to allow for proximity effect

$\quad R_e$ = effective or corrected ohmic resistance of the core,

and the allowance for skin and proximity effects is made, the a.c. resistance of the conductor becomes

$$R_e = R_{dc}\,(1 + y_s + y_p)$$

The resistance R_{dc} is known from standard tables.

4.22 CURRENT RATING OF CABLES

The current rating of power cables is very important. It decides the upper limit of power transfer by a cable. The cable rating is classified under three headings :

1. Normal maximum continuous current rating
2. Over-current rating
3. Short-circuit rating.

4.22.1 Normal or Safe Current Carrying Capacity

The normal or safe current carrying capacity depends upon a number of factors. Some of the important factors are: the minimum conductor operating temperature, heat dissipating properties of the cable and condition of installation. The cable may be installed in ground, air or ducts. There may be other cables also in the vicinity.

It is not proposed to discuss the problem of thermal rating of cables in detail. Only important aspects are given here.

Electrical analogy is helpful in solving heat flow problem. In such an analogy electrical resistance corresponds to thermal resistance, voltage difference represents the temperature difference and current flow represents heat flow.

In an electric circuit

$$I = \frac{\text{potential difference}}{\text{ohmic resistance}}$$

In a thermal circuit,

$$\text{heat dissipated} = \frac{\text{temperature difference}}{\text{total thermal resistance}}$$

Thermal resistance is expressed in thermal ohms. A *thermal ohm* is defined as the thermal resistance of a path through which a temperature difference of 1°C produces a heat flow of one watt.

Let W = heat dissipated in watts

θ = maximum permissible temperature drop (°C) between conductor and ambient

G_T = total thermal resistance to the passage of heat from conductor to ambient.

By application of Ohm's law to the thermal circuit

$$W = \frac{\theta}{G_T} .$$

Heat flows outward through the electric, sheath, serving, and then passes into the surrounding earth or air depending upon the method of installation.

The thermal resistivities of sheath and armour are neglected for calculating the current carrying capacity of the cable. Basic theory of heat transmission may be applied to calculate the current rating of the cables. In order to calculate the rating of a buried cable, it is assumed that heat generated per unit length of the cable is equal to heat transmitted by conduction from the outside surface of the cable to the ground surface.

To calculate the continuous current rating it is assumed and there is a stable equilibrium, i.e., the rates of generation and transmission of heat per unit length of cable are constant and equal. Further, the effects of prolonged heating and electrical stressing on the dielectric and the change in the soil thermal resistivity g due to the environment conditions are neglected.

If sheath losses are neglected for simplicity, the heat generated in the cable is due to losses in conductor and dielectric.

The heat generated in the cable due to various losses is conducted to the air or ground through various paths. These paths (Fig. 4.23) offer thermal resistance to the heat flow. In a three-phase cable all the three conductors are at the same temperature. The heat produced flows outwards through the dielectric in three parallel paths from the conductors to the sheath. The thermal resistances between the cores and the sheath may be assumed to be g_{c1}, g_{c2}, and g_{c3}. It then passes through the bedding of resistance g_b, metallic armouring, serving of resistance g'_s. Finally it passes into the surrounding air or ground depending upon the method of installation of cable. Let the thermal resistance of external heat flow path be g_e, i.e., g_e is the thermal resistance between the outer surface of the cable and ambient. The thermal resistance of metallic portion namely screens, sheath, and armouring are negligibly small. Heat is generated due to loss in the core.

$$\text{Total core loss} = n\, I^2\, R_i\, (1 + y_s + y_p)$$

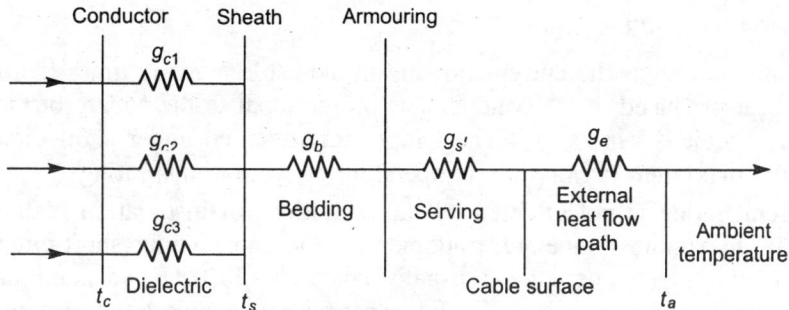

Fig. 4.23.

$$\text{Heat dissipated} = \frac{\text{temperature difference}}{\text{total thermal resistance}} = \frac{\theta}{G_T} \qquad \ldots(4.22.1)$$

where θ = temperature difference between maximum permissible temperature and ambient (°C),

and $G_T = g_d + (1 + \lambda)\,(g_b + g'_s + g_e)$ $\qquad \ldots(4.22.2)$

The maximum current rating is, therefore, given by

$$I = \left(\frac{\theta}{n\,R_\theta\,G_T}\right)^{1/2} \qquad \ldots(4.22.3)$$

where R_θ = a.c. resistance per unit length of conductor at the maximum operating temperature
including allowance for skin and proximity effects (Ω/cm)

n = number of loaded conductors in the cable

g_d = thermal resistance of the dielectric (°C cm/W)

g_b = thermal resistance of the bedding between sheath and armouring (°C cm/W)

g'_s = thermal resistance of serving (°C cm/W)

g_e = thermal resistance between the outer surface of the cable and ambient

λ = sheath loss factor, i.e., the fractional increment in a.c. resistance of the conductor to
allow for sheath loss

$$= \frac{\text{sheath loss}}{\text{core loss}} \qquad \ldots(4.22.4)$$

4.22.2   Over Current Rating

The rise in core temperature and the time of overcurrent should be known to calculate the overcurrent rating. The overcurrent rating depends upon the thermal conditions of the cable.

The values of maximum continuous current ratings of cable are supplied by the manufacturers. The ratings are valid for the specified conditions of installation (depth of laying, ground temperature, air temperature, etc.). Correction factors are to be applied for other conditions. The correction factors are also supplied by the manufacturers.

4.22.3 Short-Circuit Rating

Under short-circuit conditions the current flowing in the cable is many times the full load value of current. The heat produced in the conductor is proportional to the square of the current. The duration of short circuit is very small. The temperature attained under short circuit conditions is greater than the maximum permissible temperature for continuous rating.

The short-circuit rating is not affected by the conditions of installation of the cable, but it depends upon the maximum temperature attained by the cable under short-circuit conditions. The safe value of limiting temperature is usually taken as 120°C. For a maximum continuous operating conductor temperature of 80°C and a permissible temperature rise of 50°C during short circuit the approximate formula may be given as

$$I_{sc} = \frac{k A}{\sqrt{t}} \qquad\qquad ...(4.22.5)$$

where I_{sc} = short–circuit rms current

  A = cross–sectional area of conductor

  t = duration of short–circuit (sec)

  k = a constant.

The value of constant depends upon the rise of temperature.

4.23 THERMAL BREAKDOWN

If the amount of heat generated in a cable is equal to that dissipated, a thermal equilibrium is reached. For satisfactory operation of the cable the operating temperature should be lesser than the permissible maximum temperature. In case the amount of heat generated is more than the dissipated the cable temperature goes on rising till a thermal breakdown occurs when the insulation is damaged.

4.24 SOIL THERMAL RESISTIVITY

Heat is generated in the cable due to the load current in the conductor and the dielectric losses in the insulation. This heat should be dissipated under all conditions of load, season and soil. The thermal resistivity of the soil greatly affects the power transmission capability of the buried cables. *Soil thermal resistivity g* is defined as the temperature drop in degrees Celsius in one metre of thickness per watt heat flow per square metre of an area. It is expressed in the thermal ohm-metres, or deg C-m/W.

The value of g depends upon the type of soil, the density, the moisture content the depth of burial, and the temperature. Different values of g are used in different countries for the design of cable systems. For correct current rating of a cable it is necessary to measure the actual soil resistivity at suitable intervals along the route through which the cable is laid.

4.25 EXTRA HIGH VOLTAGE CABLES

A *solid cable* is an impregnated paper insulated cable in which no special provision is made to maintain the dielectric pressure. The main drawback with the solid cable is the formation of voids. They are suitable upto 22 kV. For 33 kV, metallised paper is applied over paper insulation to make the field the field radial. On loading the cable above 66 kV, the impregnating compound expands more than the paper. When the load is gradually removed the cable cools and the compound contracts. It will not return to its original position with the result that cavities or voids may be formed in the dielectric. Ionization of gas in these voids may deteriorate the insulation and ultimately cause a breakdown at relatively low values of stress and working temperature. Recent theories of stress have established that the ionization in void and, therefore, performance of a cable is decided by the size of void and not by the dielectric surrounding it.

The remedial measures to avoid gaseous ionization in the cable dielectric are either to prevent the formation of voids or to keep them under sufficient pressure to suppress ionization. The first method includes the use of low viscosity oil which fills the small spaces in oil impregnated paper under all conditions of varying load. Such cables are called *oil-filled cables*. The second method consists in introducing a gas at high pressure to suppress the ionization. Such cables are called *gas-pressure cables*.

4.26 OIL-FILLED CABLES

Low-viscosity oil is kept under pressure either within the cable sheath itself or within a containing pipe. The oil fills the voids in oil impregnated paper under all conditions of varying load. There are three main types of oil-filled (OF) cables.

1. Self-contained circular type cables.
2. Self-contained flat-type cable.
3. Pipe-line cables

1. Circular Oil-Filled Cables

These cables have oil ducts full of oil. The oil is kept under pressure. Figs. 4.24 (a) and 4.24 (b) show single-core cables with central oil ducts. Fig. 4.24 (c) shows a 66 kV, 3-core cable with oil ducts. A 33 kV 3-core, ductless cable is shown in Fig. 4.24 (d). In this type of cable, all the free space between the cores is available for oil flow.

2. Flat-Type Oil-Filled Cables

Flat type or Møllerhøj cable (Fig. 4.25) has three insulated cores laid up together horizontally. There is no filter material and the space is filled up by thin oil under pressure. The flat sides of lead sheath are reinforced wit h resilient metallic tapes or bands and binding wires. The resilient supporting bands are made corrugated to make the cable more flexible. On loading the cable the temperature rises and the oil expands. The flats sides of the sheath are deformed slightly. On decreasing the load oil contracts and deflections are reduced due to the presence of resilient bands. Thus the void formation in the dielectric during cooling is minimized.

Oil-storage tanks are placed along the cable route at suitable intervals to allow for thermal expansion and contraction. On loading the cable heat is produced and oil is driven from the cable in the oil-storage tanks and vice versa. Thus, the creation of voids is avoided.

3. Pipe-Type Oil-Filled Cables

A pipe type oil-filled cable consists of three separate paper insulated screened cores installed in a steel pipe. The pipe is filled with insulation oil kept at a pressure of 1.38×10^6 to 1.725×10^6 N/m^2. The high pressure oil prevents the formation of voids. It also removes heat from the cable. In this type of cable, the conductor oil duct is not required.

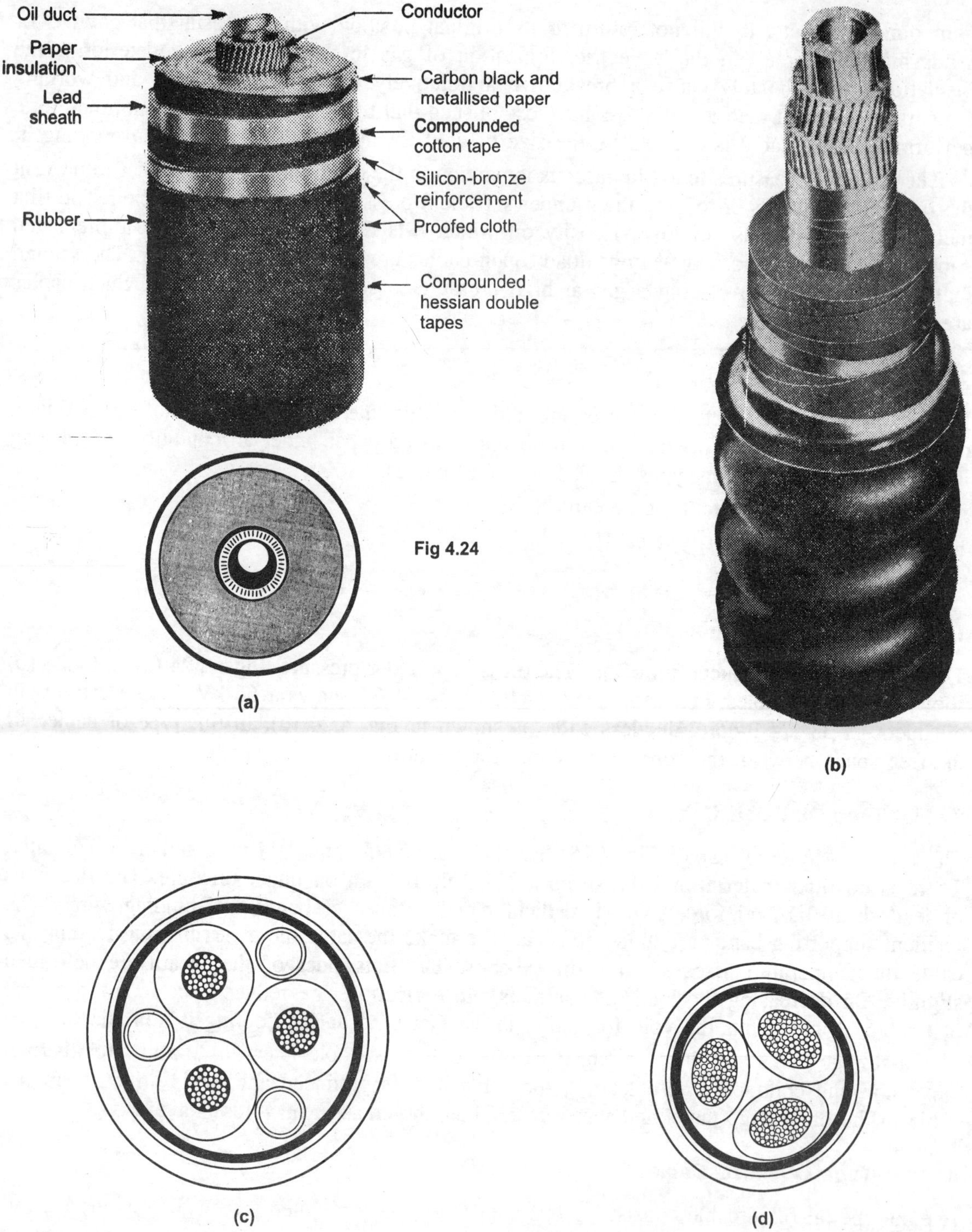

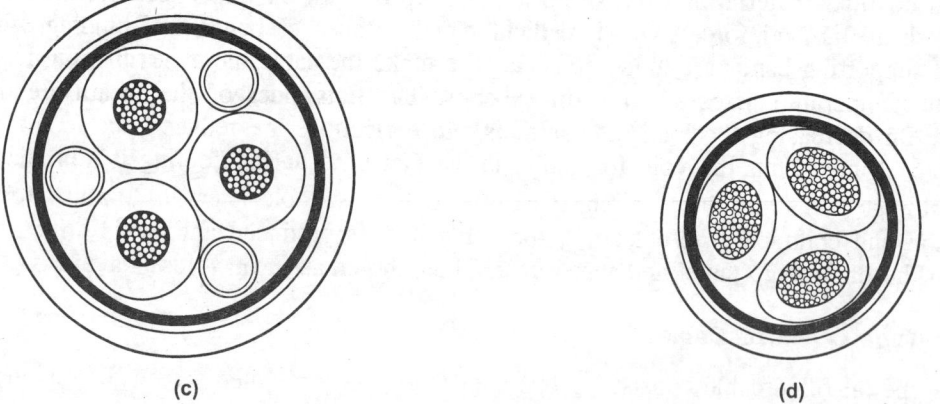

Oil duct — Conductor

Paper insulation

Lead sheath

— Carbon black and metallised paper

— Compounded cotton tape

— Silicon-bronze reinforcement

— Proofed cloth

Rubber —

— Compounded hessian double tapes

Fig 4.24

(a)

(b)

(c)

(d)

Fig. 4.24. Self-contained circular OF cables. (a) Single-core 300 kV; (b) Single-core, 132 kV; (c) Three-core, 66 kV cable with oil ducts; (d) Three-core, 33 kV ductless cable.

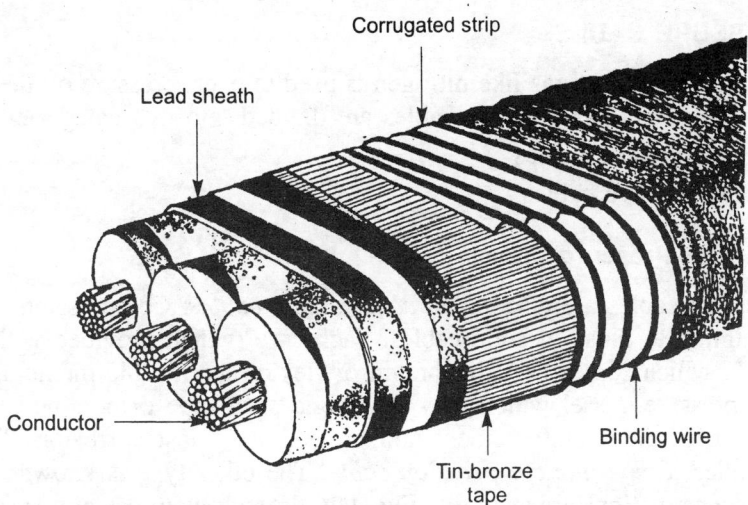

Fig. 4.25. Flat-type of cable (*Courtesy* : BICC, London).

4.26.1 Advantages of Self-Contained Oil-Filled Cables

The following are the advantages of self-contained type cables over pipe-type cables :

1. Smaller conductor size.
2. Cable installation is easier.
3. Cheaper.
4. No pumps are required, only oil tanks are to be used.
5. In case of damage, all the three phases are to be removed in case of pipe-type cables.

Self-contained oil-filled cables are very widely used for the merits mentioned above.

4.26.2 Advantages of Oil-Filled Cables

Oil-filed cables have the following advantages over solid cables:

1. Greater operating dielectric stress.
2. Greater working temperature and greater current carrying capacity.
3. Better impregnation.
4. Impregnation possible after sheathing.
5. No void formation.
6. Smaller size of cable due to reduced dielectric thickness.
7. A defect can easily be detected by oil leakage.

4.26.3 Oil For Cables

Different impregnation oils have been employed over the years. Most existing installations are using mineral oil. In 1960s dodecylbenzene (ddb) was very commonly used. More recently alkylates as, for example, linear decylbenzene (db) and branched nonylbenzene (nb) are becoming popular because of their lower viscosity and their ability to absorb water vapour liberated during ageing of cellulose.

4.27 GAS-PRESSURE CABLES

In gas pressure cables, an inert gas like nitrogen is used to exert pressure on the paper dielectric to prevent void formation. Gas pressure cables are divided into two categories.

1. External gas-pressure cables
2. Internal gas-pressure cables

4.27.1 External Gas-Pressure Cables

External gas pressure cables are also called *compression cables*. Compression cables have their insulated cores similar to those of solid cables. Each core is then provided with a polythene or a thin lead sheath, which serves as a diaphragm or flexible permeable membrane. The cable is then put in in a pressure vessel which may be a rigid steel pipe or common lead sheath over which is applied a metallic reinforcement and protection against corrosion. The first type of construction is called a *pipe-line compression cable*. The other type is known as *self-contained cable*. These two constructions are shown in Fig. 4.26. In both the types of construction, nitrogen gas is filled between the outer casing and the impermeable diaphragm. The gas is at a nominal pressure of 1.38×10^6 N/m^2 with a maximum pressure of 1.725×10^6 N/m^2. The indirect application of gas pressure through the flexible permeable membrane is done due to the fear of detrimental chemical reactions between gas and impregnated paper insulation.

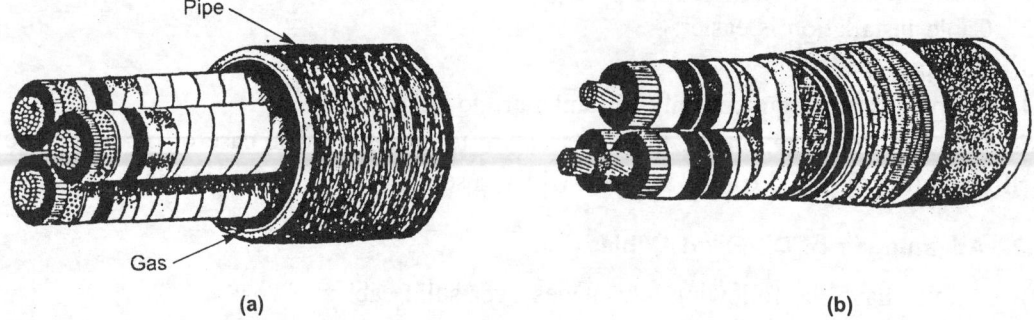

Pipe

Gas

(a)   (b)

Fig. 4.26. External gas-pressure cables. (a) Pipe-line compression cable; (b) Self-contained compression cable.

4.27.2 Internal Gas-Pressure Cables

In an internal gas-pressure cable, an inert gas like nitrogen is introduced within the lead sheath. The gas is in direct contact with the paper insulation thus suppressing ionization. The nominal gas pressure is 1.38×10^6 N/m^2 with a maximum pressure of 1.725×10^6 N/m^2.

For several reasons the use of gas-pressure assisted paper insulated cables appears to be coming to an end. Some of the reasons are as follows :

1. Lower dielectric operating stresses of gas pressure cables as compared to low-pressure oil-filled cables.
2. Costlier pressure-maintaining equipment
3. Steel enclosures can only be used with 3-core cables due to magnetic hysteresis losses.

4. Larger and stronger steel pipes are needed for bigger sizes of conductor. The pipes have to withstand gas pressures of the order of 1.725×10^6 N/m^2.

The pipe may as well be filled with oil, which although more costly than nitrogen gas, permits higher dielectric operating stresses than the cheaper gas filling.

4.28 COMPRESSED GAS INSULATED CABLES (GIC)

In a compressed gas insulted cable, high pressure sulphur hexafluoride (SF$_6$) gas fills the small spaces in oil-impregnated paper insulation and suppresses the ionization.

The conductors in gas insulated cables consist of hollow aluminium tubes rather than solid rods in order to have greater rigidity, lower electrical surface stress, and lower ac-dc resistance ratio. They are subject to severe mechanical, thermal and electrical stresses. It is, therefore, necessary to hold these conductors in position by spacers. Three types of spacers are used in rigid gas insulted cables. They are (a) disc type, (b) cone type, and (c) post type as shown in Fig. 4.27.

The spacers are made from epoxy resin insulating material. There are two possible configurations :

(a) Isolated phase GIC

(b) Single enclosure type three-phase GIC

In an *isolated phase GIC*, the three phases are enclosed in *separate enclosures*, filled with SF$_6$ gas at a pressure of 2 to 4 atmospheres (2×10^5 to 4×10^5 N/m^2) as shown in Fig. 4.28 (a). When all the three phases are put into the same enclosure (pipe or sheath), the configuration is called *single-enclosure three-phase GIC* [Fig. 4.28 (b)]. The enclosure is filled with SF$_6$ gas at a pressure of 2 to 4 atmospheres. The enclosure is made from an aluminium alloy for greater mechanical strength.

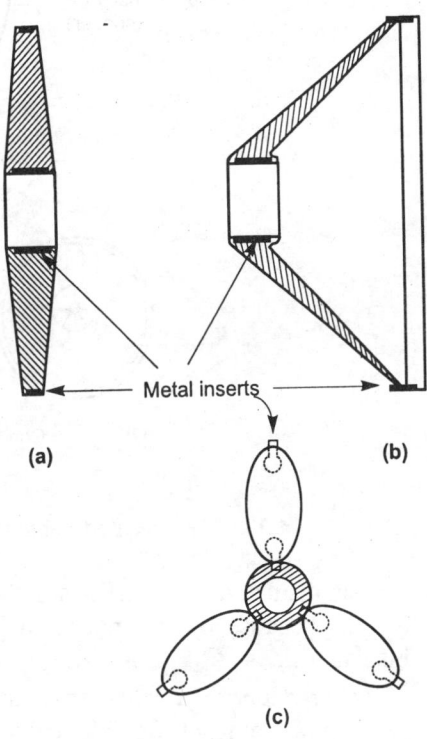

Fig. 4.27. Spacers used in rigid isolated phase gas insulated cables. (a) Disc, (b) Cone, (c) Post.

The choice between the GIC configurations is made on space and economic considerations. An isolated phase system requires (a) larger quantity of aluminium for enclosure pipes and (b) wider trenches. The rigid three-phase GIC system provide economies in metal consumption and trench width, but its current-carrying capacity is less than an isolated GIC system.

Most of the gas cables presently in use consist of three separate coaxial lines assembled from rigid pipes of aluminium alloy.

EHV/UHV lines insulated with sulphur hexafluoride (SF$_6$) gas are being used extensively for voltages above 132 kV upto 1200 kV.

GIC systems are very popular for short lengths, river crossings, and highway crossings, etc.

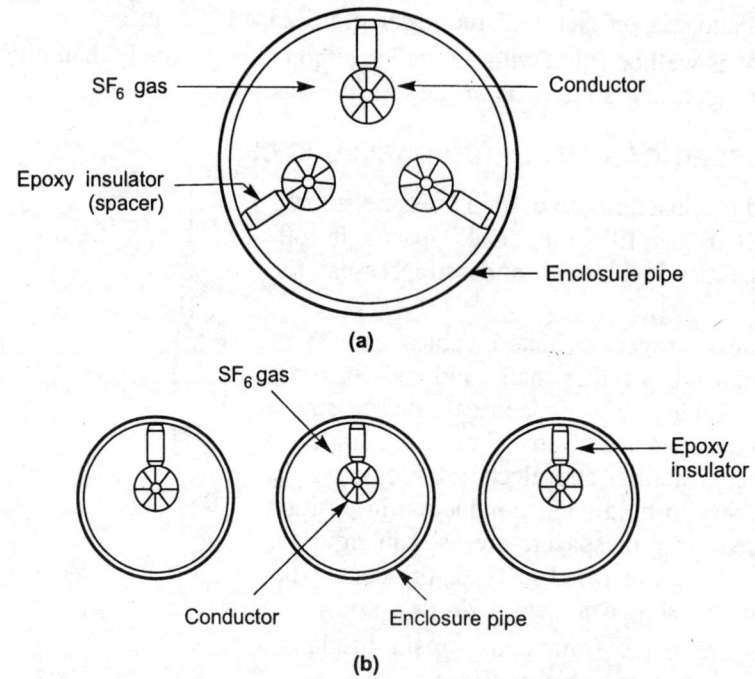

Fig. 4.28. SF$_6$ insulated cables. (a) Isolated phase GIC; (b) Single-enclosure type 3-ϕ GIC.

Gas insulated lines (GIL) are also known as *bus ducts*. Gas insulated lines are made in short lengths because of transportation limitations. Lengths upto 5 km are very common with gas insulated lines.

Isolated phase flexible gas insulated cables are also, becoming popular. The object is to develop a compact flexible cable of approximately the same diameter of a self-contained oil-paper insulated cable of similar transmission capability. This requires relatively high gas pressure of the order of 1.38×10^6 N/m^2. Rigid cables are operated at around 0.345×10^6 N/m^2.

4.29 ADVANTAGES OF GIC OVER OIL-FILLED CABLES

Gas insulated cables (GIC) have several advantages over oil-filled cables. Some of the important advantages are as follows :

1. The manufacturing process of wire drawing, stranding, paper taping, polythene extrusion, lead sheathing, etc. are not required for gas insulated cables.
2. The terminations of GIC cables are simpler and cheaper than those for oil-paper insulated cables.
3. Oil pressurization plant is not required.
4. More efficient heat transfer and therefore, GIC can carry more current.
5. Low capacitance and low dielectric losses.
6. Sulphur hexafluoride (SF$_6$) gas is nontoxic, chemically stable and nonflammable.

4.30 CROSS-LINKED POLYETHYLENE (XLPE) CABLES

Cross-linked polyethylene (XLPE) is a hydrocarbon thermoplastic insulating material. Its molecular chain is represented as

$$\left(\begin{array}{c} \text{C} \\ |\quad| \\ -\text{CH}_2-\text{C}-\text{CH}_2- \\ |\quad| \\ \text{C} \end{array} \right)_n$$

The number of cross-linked carbon bonds is usually between 7 and 10 per cent of the total bonds. Cross-linked polyethylene (XLPE) is being increasingly used for cable insulation for voltages of 3.3, 6.6, 11, 33, 66, 132, 220, 400 and 500 kV. Cross-linked polyethylene cables are being developed for voltages higher than 500 kV. XLPE cables are also used in HVDC systems. The main advantages of XLPE cables over other type of cables are :

1. Higher dielectric strength and higher impulse strength (400 kV/mm at 90°C).
2. Low power factor ($\tan \delta_i = 0.0005$) and low dielectric loss.
3. Good thermal properties.
4. Higher temperature withstand using normal loads and short duration overloads. Normal maximum operating temperature allowed is 90°C.
5. Resistant to deformation at higher temperatures.
6. Resists attack by chemicals.
7. Smaller size due to higher dielectric strength.
8. Moisture resistant.
9. Unaffected by partial discharges in the presence of moisture.

With the development of XLPE cables for higher voltages, the complex oil-filled and gas-pressure cables are becoming obsolete in the voltage range of 66 kV to 400 kV.

4.31 POWER CABLE INSTALLATION

The method of installation adopted for a power cable system is mainly governed by local conditions. The most important methods of installing power cables are :

1. Direct laying
2. Draw-in (duct) system
3. Cables installed above ground level in free air
4. Cables installed in tunnels

1. Direct Laying

This system, which is comparatively simple and cheaper in most cases, involves the digging of a trench in the ground. The depth of trench may vary between 0.5 m to 1.2 m. The bottom of the trench is levelled, freed from sharp edges, stones, rocks, etc. A layer of about 10 cm of sand is laid at the bottom of the trench. When more than one 3-core power cables are laid in the same trench, a horizontal interaxial spacing of 0.3 to 0.5 m is provided to reduce the effects of mutual heating and also to ensure that a fault occuring on one cable will not damage the adjacent cable. After laying the cable in the trench, it is covered with a layer of 10 cm of sand. Then a continuous layer of bricks or reinforced concrete slabs is provided for the protection of the laid cables against mechanical damage. This also provides a warning to future excavators of the presence of cables. To prevent damage, the dug out soil is filled in the trench as soon as possible after the cables have been laid.

2. Draw-in (duct) System

In congested areas where the excavation of a trench is both very expensive and inconvenient, the draw-in (duct) system is used. In this system, ducts or pipes are buried direct in the ground with manholes at positions from which the cables subsequently are drawn in. The duct system provides the following advantages :

(a) Repairs, alterations, or additions to the system can be made without reopening the ground.

(b) Cables can easily be withdrawn if any modifications to the route are required. The duct system suffers from the following drawbacks :

(i) Initial cost of installation is higher than burying the cables direct in the ground.

(ii) Lower current capacity of cables in ducts due to poor heat dissipation.

This method is also used for road crossings and under railway tracks.

3. Cables Installed in Free Air

Cables can be installed on cleats, or on racks or hangers. This is a very cheap method of power cable installation. This method is very commonly used in the installation of cables in power station, factories, large institutions, etc.

4. Cables Installed in Tunnels

Cable tunnels are commonly used in power stations and/or control rooms where a large number of cables is involved. The size of such channels should be large enough to accommodate present and future cables. The headroom should be at least 2 m with a walkway in the centre and racks or expanded metal trays on both sides. These tunnels, should be used only for cables and should be well ventilated for heat dissipation.

EXERCISES

1. A single-core cable has an inner diameter of 6 cm and a core diameter of 2 cm. Its paper dielectric has a working maximum dielectric stress of 60 kV/cm. Calculate the maximum permissible line voltage when such cables are used on a three-phase power system.

2. Find the most economical value of the diameter of single-core cable to be used on 66 kV three-phase system. Find also the overall diameter of the insulation if the maximum permissible stress is not to exceed 5 kV/mm.

3. A concentric cable has an inner conductor of radius 10 mm and an outer conductor of inside radius 30 mm. If the instantaneous electric stress in the dielectric is not to exceed 50 kV/cm, calculate the rms value of the allowable sinusoidal alternating potential difference which can be applied to the cable. If this potential difference has a frequency $f = 50$ Hz and the relative permittivity of the dielectric is 2.0, determine the charging current for a cable 1 km in length. (L.U.)

4. Show that, for a concentric cable of given dimensions and given maximum potential gradient in the dielectric the permissible voltage between the core and the sheath is independent of the permittivity of the insulating material. Find the maximum permissible rms voltage between the core and the sheath of a single-core cable having conductor radius 5 mm and a dielectric of radial thickness 4 mm if the maximum potential gradient is 2 kV/mm (rms). Find also the charging current per kilometre at this voltage given that the frequency is 50 Hz and the relative permittivity of the dielectric is 3.5. Derive the expression used to calculate the capacitance of the cable. (L.U.)

5. A single-phase concentric cable 2 km long has a capacitance of 0.2 μF per km, the relative permittivity of the dielectric being 3.5. The diameter of the inner conductor is 1.5 cm and the supply voltage is 33 kV at 50 Hz. Calculate the inner diameter of the outer conductor, the rms voltage gradient at the surface of the inner conductor and the rms value of the charging current. Derive any formula used for the capacitance. (L.U.)

6. The capacitance of a length of three-core lead sheathed belted cable are measured and found to be as follows :

   (a) Between one conductor and the other two connected together to the sheath, 4.8 μF.

   (b) Between three cores bunched together and the sheath, 7.2 μF.

   Calculate the capacitance to neutral and the total charging kVA when the cable is connected to an 11 kV, 50 Hz three-phase supply.

7. A three-core, three-phase metal sheathed cable 1.16 km in length gave the following results on test for capacitance : Capacitance between bunched conductors and sheath, 1.0 μF and capacitance between two conductors bunched with the sheath and the third conductor, 0.6 μF.

   With the sheath insulated, find the capacitance (a) between any two conductors, and (b) between any two bunched conductors and the third conductor. Calculate the charging current per conductor per km when connection is made to 11 kV, 50 Hz supply.

8. In a three-phase, three-core metal sheathed cable the measured capacitance between any two cores is 2 μF. Calculate the charging current and kVA taken by the cable when it is connected to 11 kV, 50 Hz supply.

9. In a coaxial cable the conductor diameter is 10 mm and the inner sheath diameter is 50 mm. There are two layers of insulation, the inner layer of dielectric constant 4 and a maximum working gradient of 6 kV/mm has a radial thickness of 4.6 mm; the outer layer has dielectric constant 2.5 and the maximum voltage gradient 5 kV/mm. Calculate the maximum working voltage for the cable.

10. A 85 kV, single-core, metal sheathed cable is to be graded by means of a metallic intersheath.

(a) Find the diameter Δ of the intersheath and the voltage at which it must be maintained in order to obtain the minimum overall cable diameter D. The insulating material can be worked at 60 kV/cm.

(b) Prove the formulae used, and compare the conductor and the outside diameter (d and D) with those of an ungraded cable of the same material under the same conditions.

11. A three-phase circuit, 5 km long, consists of 3 single-core cables. Each core has a diameter of 0.8 cm. The dielectric has a radial thickness of 0.8 cm and relative permittivity of 2.2. The loss angle of the dielectric is found to be 45 min of angle. Calculate the total dielectric loss in the circuit when connected to a 33 kV, 50 Hz supply.

12. A circuit, 10 km long, consisting of three single-core cables, is connected to a 33 kV, three-phase, 50 Hz supply. The core of each cable is 10 mm diameter and the dielectric of relative permittivity 2.25, has a radial thickness of 6 mm. If the total dielectric loss in the circuit is 10.5 kW and the capacitance to neutral of each cable is 2.55 µF, determine the loss angle of the dielectric.

13. A single core cable 1 km in length has a core diameter of 1.0 cm and a diameter under the sheath is 2.5 cm. The relative permittivity is 3.5. The power factor on open circuit is 0.03. Calculate (a) the capacitance of the cable, (b) its equivalent insulation resistance, (c) the charging current, and (d) the dielectric loss, when the cable is connected to 6600 V, 50 Hz busbars.

ANSWERS

1. 81 kV
2. 21.54 mm, 58.54 mm
3. 1.24 A
4. 1.95 A
5. 4 cm, 32 kV/cm, 4.2 A
6. µF, 228 kVAr
7. (a) 0.367 µF (b) 0.489 µF; 1.26 A
8. 8 A, 152.4 kVAr
9. 65.51 kV
10. (a) d = 1.04 cm, Δ = 2.84 cm, D = 5.34 cm, 53.8 kV to earth
    (b) d = 2.84 cm, D = 7.73 cm
11. 2489 W
12. 0.012 radian
13. (a) 0.212 µF (b) 0.5 MΩ (c) 0.44 A (d) 87.1 W

5

Line Insulators and Supports

5.1 INTRODUCTION

Overhead line insulators are used to separate line conductors from each other and from the supporting structures electrically. While designing an insulator the following considerations are made:

(a) The insulator should have high permittivity so that it can withstand high electrical stresses. That is, the dielectric strength of the insulating material should be high. The insulator should be able to withstand the overvoltages due to lightning, switching, or other causes under severe weather conditions in addition to the normal working voltage.

(b) It should possess high mechanical strength to bear the conductor load under worst loading conditions.

(c) It needs to have a high resistance to temperature changes to reduce damages from power flashover.

(d) The leakage of current to earth should be minimum to keep the corona loss and radio interference within reasonable limits.

(e) The insulator material should not be porous and should be impervious to gases in atmosphere and should be free from impurity and cracks which may lower the permittivity.

The electrical failure of insulators occurs either by *puncture* or *flashover*. In the case of a puncture the arc passes through the body of the insulator. Flashover is caused by an arc discharge between the conductor and earth through air surrounding the insulator. It is either due to line surges or due to the formation of wet conducting layer over the insulator surface. Normally, the insulator is not damaged by a flashover but it becomes useless after the puncture.

Sufficient thickness of material is provided in the insulator to prevent the puncture under surge conditions. Flashovers are reduced by increasing the resistance to leakage currents. The length of the leakage path is made large by constructing several layers called *petticoats* or *rainsheds.* They keep the inner surfaces relatively dry in wet weather and thus provide sufficient leakage resistance to prevent a flashover.

Accumulation of dirt, dust, salt, smoke, etc., on insulator surface in polluted area is kept minimum by providing semi-conducting glaze over the whole exposed surface of insulator. This reduces the surface deposition which may produce flashover at the operating voltage.

For satisfactory operation, the flashover should occur before puncture. The ratio of puncture voltage to flashover voltage, called the *factor of safety*, is kept as high as possible. The flashover voltage is reduced considerably by moisture and surface deposits.

For satisfactory operation, the rainsheds should have the shapes like those of equipotential surfaces and the insulator body should be constructed along the lines of electrostatic field around the pin. Also, the leakage resistance and capacitance of various rainsheds should be approximately equal. The flux distribution between pin and cap of a pin-type insulator is shown in Fig. 5.1 (a). Fig. 5.1 (b) shows the construction of a pin-type insulator based upon the principles given above.

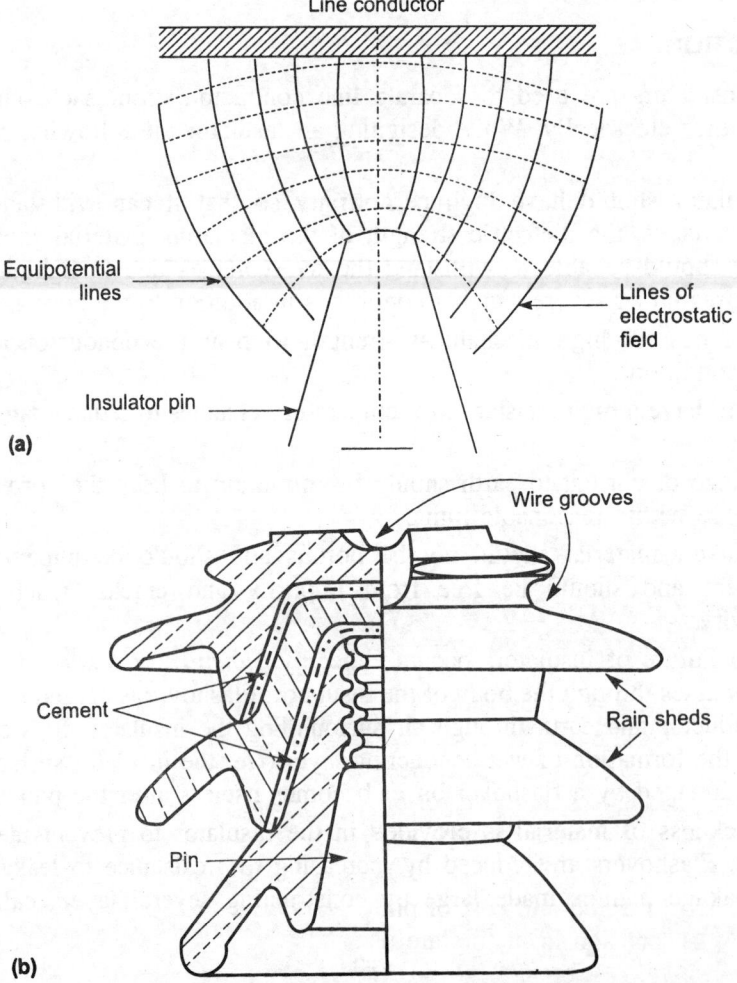

Fig. 5.1. (a) Flux distribution of a pin insulator; (b) Pin-type insulator.

5.2   TYPES OF INSULATOR

There are three main types of insulator used for overhead lines:
1. Pin type insulator
2. Suspension type insulator
3. Strain or tension type insulator.

5.2.1   Pin Type Insulator

The pin insulator is supported on a forged steel pin or bolt which is secured to the cross arm of the supporting structure. The conductor is tied to the insulator on the top groove on straight line positions and side groove in angle positions by annealed binding wire of the same material as conductor. A lead thimble is cemented into the insulator body to receive the pin. A pin insulator is shown in Fig. 5.1 (b).

Single piece type pin insulators are used for lower voltages, but for higher voltages two or more pieces are cemented together to provide sufficient thickness of porcelain and adequate leakage path or creepage path. Fig. 5.2 (a) shows a single piece insulator. Two-piece and three-piece insulators are shown in Figs. 5.2 (b) and 5.1 (b) respectively.

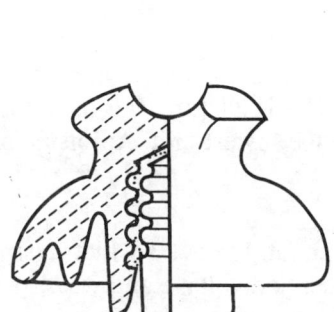

Fig. 5.2. (a) One piece pin insulator.

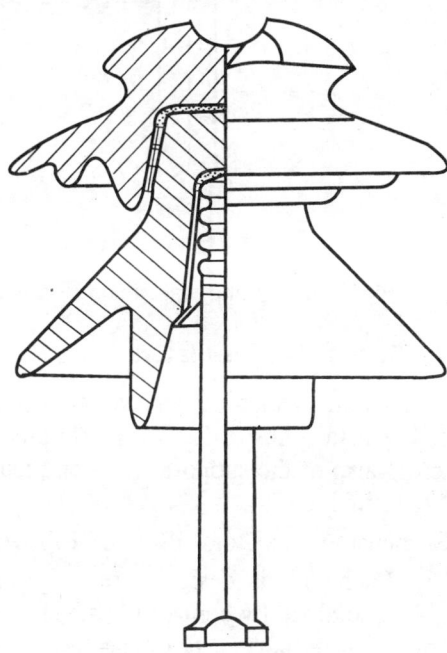

Fig. 5.2. (b) Two-piece pin insulator.

Fig. 5.3 shows dry and wet acting distances. As pointed out earlier, the flashover voltage for moist and dirty surfaces is less than that for dry and clean surfaces. The total dry arcing distance is the sum of all direct distances through air. In Fig. 5.3 it is shown by $(a + b + c)$. The total wet arcing distances is shown by $(A + B + C)$.

The increased size, weight, and cost of pin type insulator put a limit to its use above 66 kV and, therefore, the suspension insulators are used for high voltage work.

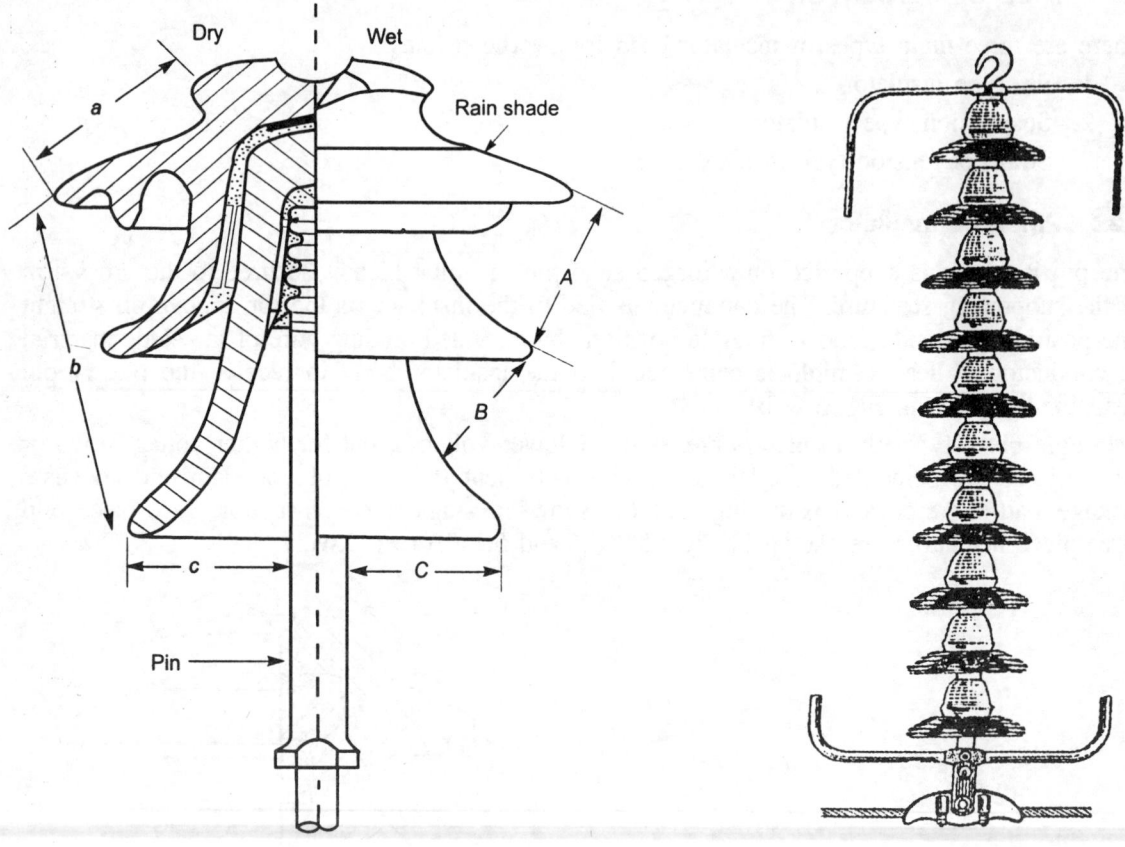

Fig. 5.3. Dry and wet arcing distances.   **Fig. 5.4.** Insulator string.

5.2.2 Suspension Insulators

A suspension insulator consists of a number of separate insulator units connected with each other by metal links to form a flexible chain or a string. The insulator string is suspended from the cross arm of the support. The conductor is attached to the lowest unit. An insulator assembly is shown in Fig. 5.4.

Suspension insulators offer the following advantages :

1. Each unit is designed for operating voltage of about 11 kV, so that a string can be assembled by connecting several units to suit the service voltage and weather conditions.

2. In case the line is to operate on a higher voltage in future to cope with the increasing load, additional units would be introduced to the same string. In case of damage to one of the units, only the damaged insulator, but not the whole string, is replaced by a new one.

3. The string is free to swing in any direction and, therefore, greater flexibility is provided. The tension in the successive spans are balanced. The lines can, therefore, be designed for longer spans and higher mechanical loading.

4. There is a decreased liability to lightning disturbances if the string is suspended from a metallic supporting structure, which works as a lightning shield for conductor.

Since the string is hung from the support, the tower height is to be increased. Greater spacing between the conductors is to be provided to allow for swinging.

The types of suspension insulators in use are :

1. Cap-and-pin type.
2. Hewlett or interlink type.

The first type is more common. A galvanised cast iron or forged-steel cap and a galvanised forged-steel pin are connected to porcelain in the cap-and-pin type construction. The units are joined together either by ball and socket or clevis-pin connections. The ball and socket type and pin and clevis type constructions are given in Fig. 5.5.

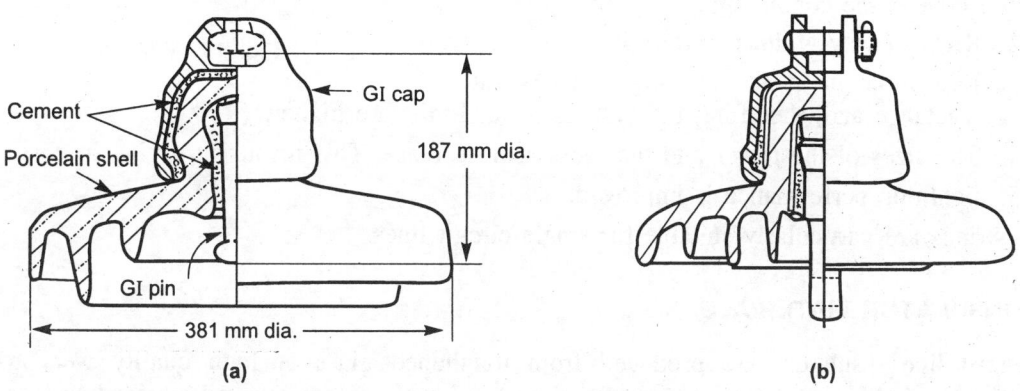

Fig. 5.5. Suspension insulators. (a) Pin and clevis type; (b) Ball and socket type.

The interlink type unit (Fig. 5.6) employs porcelain having two curved channels with planes at right angles to each other. U-shaped level covered steel links pass through these channels and serve to connect the units. Interlink type insulator is mechanically stronger than the cap-and-pin type unit. The metal links continue to support the line if the porcelain between the links breaks. Thus, the supply is not interrupted. The Hewlett type of insulator suffers from the disadvantage that the porcelain between the links is highly stressed electrically and, therefore, its puncture strength is lesser as compared to other types.

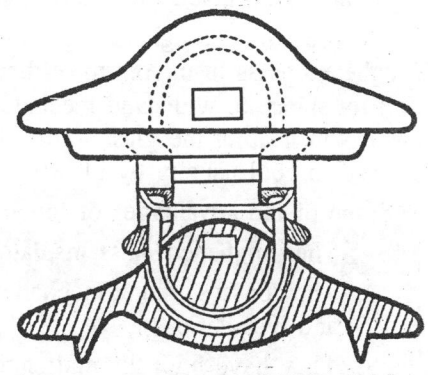

Fig. 5.6. Interlink type insulator.

5.2.3 Strain or Tension Insulators

Strain or tension insulators are designed for handling mechanical stresses at angle positions where there is a change in the direction of the line or at terminations of the line. Shackle and pin insulators serve the purpose for low voltage lines. For high voltage lines having longer spans and greater mechanical loading, suspension insulator strings are arranged in a horizontal position. In case a single string is not sufficient to take the load, two or more strings in parallel may be employed for higher conductor tensions.

5.3 V-STRINGS

A single string of insulator follows the conductor and sways like a pendulum in a strong side wind. V-strings are used to prevent conductor movement at towers. They find increased application in high voltage transmission systems. V-string construction (Fig. 5.7) offers the following advantages :

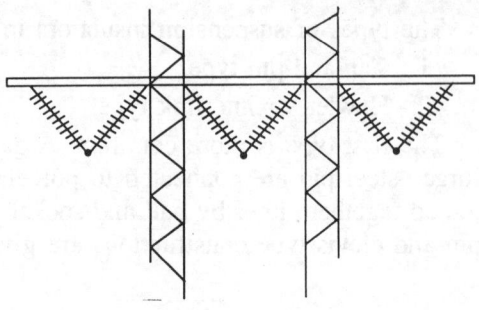

1. The insulator swing is reduced and, therefore, lesser spacing can be provided between the conductors.
2. Right-of-way width is reduced.
3. Reduction in phase spacing reduces line reactance and, therefore, the system power limits are higher.

Fig. 5.7. V-string.

4. The sizes of the tower and the cross arm decrease. This results in saving in cost.
5. lightning performance is improved.

V-strings are particularly suitable for single circuit lines.

5.4 INSULATOR MATERIALS

Overhead line insulators are produced from toughened glass or high quality wet process porcelain. Porcelain insulators are usually glazed in brown colour over all exposed surface, but sometimes creamglazed insulators are also used. Porcelain has been used from the very beginning as an insulator material and is still being used.

Prestressed or toughened glass has also been employed in constructing line insulators. Toughened glass insulators have their surface layers in state of high compression due to which their resistance to withstand mechanical and thermal stresses is greater. The toughening process consists in heating the glass uniformly to a temperature above its strain temperature. It is then allowed to cool usually by blowing air on its surface.

Some of the advantages of toughened glass insulators over porcelain insulators are :

1. The toughened glass insulators have greater puncture strength.
2. They posses greater mechanical strength and, therefore, there is less breakage in transport and installation.
3. They have high thermal shock resistance and, therefore, damage from power flashover is reduced.
4. If an insulator is damaged by electrical or mechanical cause, the outer shed breaks and falls on the ground. The cap and pin remain sufficiently strong to support the conductor in its position.
5. The life of a toughened glass insulator is long.

The glass insulator has the disadvantage that moisture readily condenses on its surface.

, Most of the lines use porcelain as an insulator material, but toughened glass insulators have also been used in large quantities in Great Britain at all voltages upto 275 kV and on the Swedish 380 kV system.

The performance of glass units is practically similar to that of porcelain units when tested for puncture strength in air with steep-fronted impulse waves.

Table 5.1 shows some characteristics of porcelain and glass for the purpose of comparison.

Table 5.1   Characteristics of Insulator Materials

Characteristic	Unit	Material	
		Glass	Porcelain
Tensile strength	kgf/mm^2	5.35-8.53	4.23-6.00
Crushing strength	kgf/mm^2	8.53-35.20	31.00-42.30
Modules of elasticity	kgf/mm^2	5000-8500	7000-10500
Coefficient of expansion	per °C	7.9×10^{-6} - 8.33×10^{-6}	3.3×10^{-6} - 6.6×10^{-6}
Density	gm/cc	2.49-3.46	2.21-2.35
Puncture strength	kV/mm	70-120	12.5-27.5
Dielectric constant		6.8	6.15

5.4.1 Polymer Insulator

A synthetic suspension insulator has been developed by General Electric under the trade name 'Gepol'. It is made up of combination of fibre glass and an epoxy polymer instead of porcelain. The polymer insulator is 70 per cent lighter than its equivalent porcelain unit. It is described as essentially puncture-proof with extremely high mechanical strength. It has high thermal shock resistance to reduce damage to flashover, excellent radio interference voltage performance, lesser hardware corrosion, and better performance in polluted atmosphere. The number of flexible parts to be handled in construction are lesser in number.

5.5 VOLTAGE DISTRIBUTION AND STRING EFFICIENCY

When a voltage is applied across a suspension insulator string, it is not equally distributed across the individual units. The unit (disc) near the line conductor is most highly stressed and takes the maximum percentage of voltage in absence of any stress control device. The determination of voltage distribution on an insulator string is helpful in determining the flashover voltage and the voltage at which localized corona and radio interference are produced.

The insulator material of each insulator, being between two metallic pins or knobs, forms a capacitor of capacitance C called the *self capacitance*. Also, the air between each pin and the tower, which is at earth potential, forms other set of capacitors. These capacitances are called *capacitances to earth*.

Let $k = \dfrac{\text{capacitance to earth}}{\text{self capacitance}}$

∴   capacitance to earth $= k \times$ self capacitance $= k\,C$ ...(5.5.1)

A string of insulators may, therefore, be treated as a combination of capacitances connected in series and parallel as shown in Fig. 5.8.

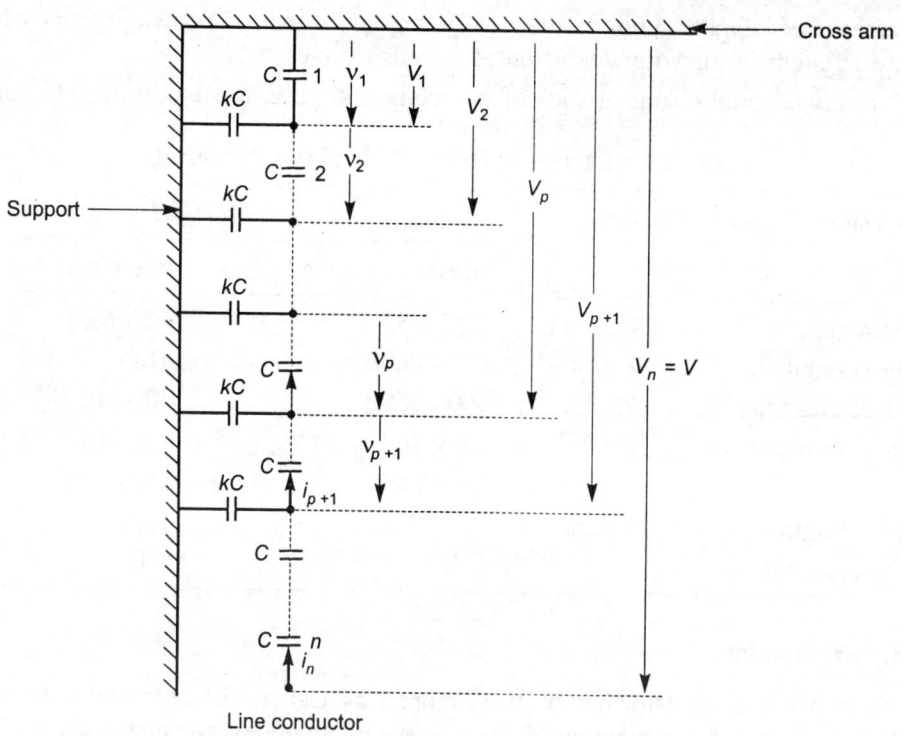

Fig. 5.8. Voltage distribution across the units of a string.

Let V be the total voltage across the string and v_1, v_2, ..., v_n the voltage drops across units, 1, 2, ..., n starting from the cross arm towards the line conductor. Suppose that V_n is the voltage across n units from the top.

By KCL at the node P,

$$i_{p+1} = i_p + I_p$$

$$v_{p+1} (j\,\omega\,C) = v_p (j\,\omega\,C) + V_p (j\,\omega\,k\,C)$$

$$v_{p+1} = v_p + k\,V_p \qquad\qquad ...(5.5.2)$$

Also, $v_1 + v_2 + ... + v_n = V$ 　　　　　　　　　...(5.5.3)

Equations (5.5.2) and (5.5.3) may be used to find the voltage across each unit (disc) of the insulator string. It is seen that the voltage across the insulator unit nearest to the line conductor is maximum and this unit is most highly stressed. The voltage goes on decreasing progressively all the way till the cross arm is reached.

It is to be noted that if the insulator unit nearest to the line conductor is electrically stressed to its safe operating value, then all the other units in the string would remain electrically understressed. Consequently, the insulator string as a whole is being inefficiently used. The string efficiently, η_s is defined as follows :

$$\text{String efficiency} \triangleq \frac{\text{voltage across the whole string}}{n \times (\text{voltage across the unit adjacent to line conductor})}$$

$$\eta_s = \frac{V}{n \, v_l} \qquad \qquad ...(5.5.4)$$

where V = voltage across the whole string

n = number of insulator discs in the string

v_l = voltage across the lowest unit connected to the line conductor

If the unit adjacent to the line conductor is about to flashover, then the whole string is about to flashover. Therefore, the string efficiency can also be expressed as follows :

$$\text{String efficiency} = \frac{\text{flashover voltage of string}}{n \times \text{flashover voltage of one unit}}$$

It is to be noted that the string efficiency decreases as the number of units increases.

It should be noted that the voltage across the string is equal to the phase voltage, that is, the voltage between the line and earth.

Therefore, line voltage = $\sqrt{3}$ × phase voltage = $\sqrt{3}$ × voltage across the string

In the above treatment we have made the following *assumptions.*

1. All the insulator discs in a string are identical.
2. The discs are at equal distances from the supporting structure which is at zero potential.
3. The dielectric losses in the discs are neglected and each disc is considered as a pure capacitance.
4. The capacitances between various connections and the line conductor are neglected.
5. The effect of corona and leakage across the insulator surface has also been neglected. Their effect is to equalize the voltage distribution.

5.5.1 Alternative Method

Based on hyperbolic function, the following method is used to calculate the voltage across any unit and the string efficiency. A string of insulators may be considered to be equivalent to a transmission line with receiving end short circuited and earthed. Suppose that there are n insulators in the string as shown in Fig. 5.9. Insulator 1 is the topmost unit connected to the cross arm (earth) and insulator n is the lowest unit attached to the line conductor.

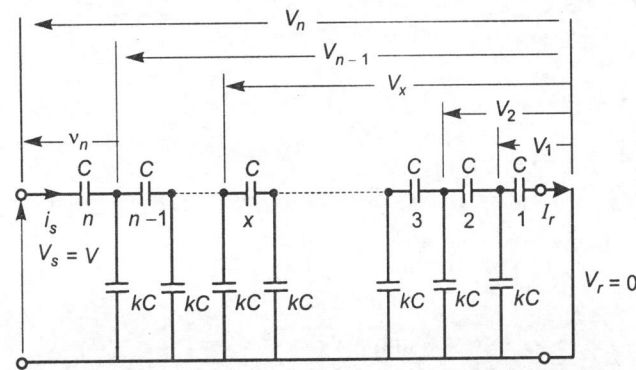

Fig. 5.9. Insulator string considered as a short-circuited line.

Let V_x be the voltage upto the xth unit from the top of the string. From Eq. (10.7.3)

$$V_x = V_r \cosh \gamma x + Z_0 I_r \sinh \gamma x \qquad \qquad ...(5.5.5)$$

where $\gamma = \sqrt{Z' Y'}$

Here, $Z' = \dfrac{1}{j \omega C}$, $\qquad Y' = j \omega (k C)$

$\therefore \qquad \gamma = \sqrt{Z' Y'} = \sqrt{\dfrac{1}{j \omega C} \times j \omega k C} = \sqrt{k}$

Also, for the line short circuited at the receiving end, $V_r = 0$. Therefore, Eq. (5.5.5) becomes

$$V_x = Z_0 I_r \sinh x \sqrt{k} \qquad \qquad ...(5.5.6)$$

At the sending end, $x = n$ and $V_x = V_n = V$. Therefore, Eq. (5.5.6) can be written as

$$V = Z_0 I_r \sinh n \sqrt{k} \qquad \qquad ...(5.5.7)$$

Dividing Eq. (5.5.7) by Eq. (5.5.6) we get

$$\frac{V_x}{V} = \frac{\sinh x \sqrt{k}}{\sinh n \sqrt{k}} \qquad \qquad ...(5.5.8)$$

For the mth disc, the total applied voltage

$$V_m = \frac{V \sinh m \sqrt{k}}{\sinh n \sqrt{k}}$$

For the $(m - 1)$th disc, the total applied voltage

$$V_{m-1} = V \frac{\sinh (m - 1) \sqrt{k}}{\sinh n \sqrt{k}}$$

Voltage drop across the mth disc

$$v_m = V_m - V_{m-1}$$

$$\text{String efficiency} = \frac{\text{voltage across the string}}{n \times \text{voltage across the lowest unit } n}$$

$$= \frac{V}{n (V_n - V_{n-1})} \qquad \qquad ...(5.5.9)$$

From Eq. (5.5.8)

$$V_n = V \frac{\sinh n \sqrt{k}}{\sinh n \sqrt{k}} = V \qquad \qquad ...(5.5.10)$$

$$V_{n-1} = V \frac{\sinh (n - 1) \sqrt{k}}{\sinh n \sqrt{k}} \qquad \qquad ...(5.5.11)$$

Therefore, string efficiency $= \dfrac{V}{n \left(V - \dfrac{V \sinh (n - 1) \sqrt{k}}{\sinh n \sqrt{k}} \right)}$

$$= \frac{\sinh n \sqrt{k}}{n [\sinh n \sqrt{k} - \sinh (n - 1) \sqrt{k}]} \qquad \qquad ...(5.5.12)$$

5.6 IMPROVING VOLTAGE DISTRIBUTION

For satisfactory performance it is essential that the voltage distribution across the units of the string should be uniform. Different methods have been attempted to get the uniform distribution of voltage along the insulators in order to exploit its insulation strength fully. The following methods may be utilized to achieve the uniformity :

1. Use of Long Cross Arm

It follows from the previous discussion that the non-uniformity in the potential distribution is due to stray capacitances from the line and ground to each unit of the string. The former, being very small, was neglected in calculating the potential distribution across various units.

It is, therefore, obvious that non-uniformity arises mainly due to stray capacitance of each unit to ground. The reduction is stray capacitance to ground would, therefore, equalize to some extent, the distribution. This can be achieved by using a longer cross arm. The conductor spacing and, therefore, the inductive reactance voltage drop increases. Moreover, economy does not permit the use of very long cross arms.

2. Capacitance Grading or Grading of Units

The second method consists in assembling a string with units of different capacitances for appropriate voltages. The line-end unit has got the greatest capacitance, while the top unit the smallest. By properly selecting the capacitances of the units the voltage distribution can be made uniform. Fig. 5.10 shows a string of suspension insulators and the various capacitances.

Let v = voltage across each unit of the string

C_x = self capacitance of x th unit from the top

C_{x+1} = self capacitance of $(x+1)$ th unit from the top.

Consider the pth link P from the top and apply KCL at node P

$$i_{x+1} = i_x + I_x$$

$$v \cdot j \omega \, C_{x+1} = v \cdot j \omega \, C_x + x v \cdot j \omega \, C_g$$

$$C_{x+1} = C_x + x \, C_g \qquad\qquad\qquad\qquad …(5.6.1)$$

Equation (5.6.1) may be used to find the capacitance of any unit if the capacitance of one unit is known.

Capacitance grading presents difficulties in selecting units for longer strings. Moreover, the stocks have to be kept for units for different sizes and ratings. This method is only used for very high voltage lines.

3. Use of Grading Rings or Static Shielding

The method of using grading or guard rings has been extensively used from the very beginning and has proved to be very effective by equalizing the voltage distribution. Fig. 5.11 shows one arrangement of a grading ring. The rings or shields are fitted to the bottom insulator hardware or to the clamp and connected to the line. They increase the stray capacitance to the line of the line-end units, and decrease the stray capacitances to earth, and thus help in making the voltage distribution uniform.

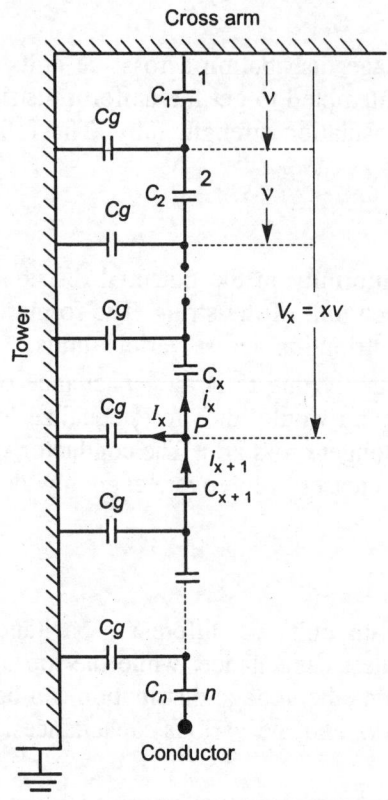

Fig. 5.10. Capacitance grading.

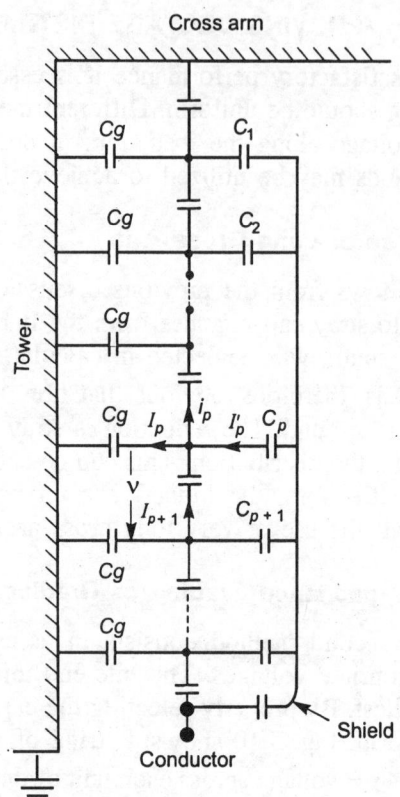

Fig. 5.11. Static shielding.

Consider pth link P from the top. Let C_p be the capacitance from the shield to the pth link. Applying KCL at P

$$I_p + i_p = I'_p + i_{p+1}$$

But since
$$i_p = i_{p+1} ; \qquad I_p = I'_p$$

$$pv \cdot j \omega C_g = (n-p) v \cdot j \omega C_p$$

$$C_p = \frac{p C_g}{n-p} \qquad \qquad ...(5.6.2)$$

Grading rings are generally circular or oval in shape and made of galvanized tube or pipe. Other shapes of ring have also been used.

A combination of an arcing horn at the top and grading ring at the bottom (Fig. 5.12) provides a very good protection for the string against power frequency and lightning flashovers. It diverts the power arc from the insulator surface to air, and thus reduces the risk of insulator damage by the heat of a flashover. These protective devices lower the flashover voltage value.

The voltage stress at the line-end units increases with the increase in transmission voltage. At voltages above 300 kV electrostatic overstress at the line-end units produces localised corona at the insulator hardware having sharp corners and projections. Localized corona produces radio

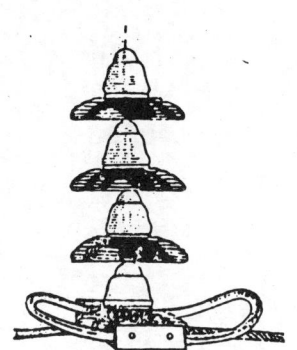

Fig. 5.12. Grading ring at the bottom of the string.

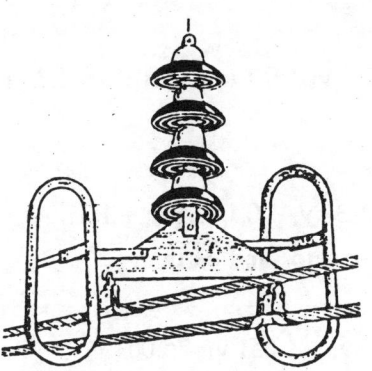

Fig. 5.13. Grading ring.

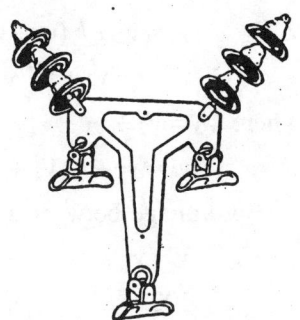

Fig. 5.14. Ringless hardware for triple conductor bundle for V-string suspension.

and television interference and corrosion damage to the insulator. It, therefore, becomes necessary to modify both insulators and suspension hardware for better performance, but the changes would present difficulties in installation and maintenance. The problem was, therefore, solved by using grading rings. The rings surrounding both line-end units and suspension fittings with an elementary Faraday cage are shown in Fig. 5.13. But the rings are costly. In the modern designs the rings have been eliminated. The suspension system has been modified to envelop the line-end units and supporting hardware in the electrical field of the bundled conductors. These insulator hardware systems reduce cost and give improved electrical performance of the line. The ringless hardware system for triple conductor bundle for V-string suspension is shown in Fig. 5.14.

Example 5.1 A string of suspension insulators consists of four units. The capacitance between each link pin and earth is one-tenth of the self capacitance of a unit. The voltage between the line conductor and earth is 100 kV. Find (a) the voltage distribution across each unit, and (b) the string efficiency.

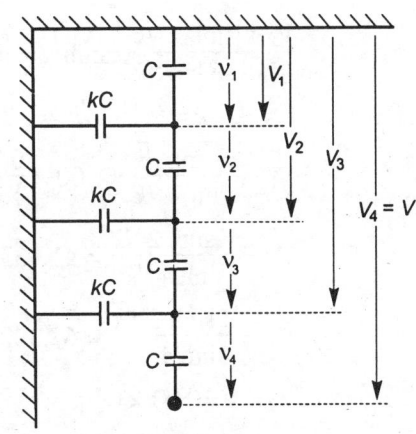

Fig. 5.15.

Solution

The arrangement of units is shown in Fig. 5.15. Here $k = 0.1$. By Eq. (5.5.2)

$$v_{p+1} = v_p + k\, V_p$$

This formula is applied repeatedly to calculate v_2, v_3, and v_4 in terms of v_1.

$$v_{1+1} = v_1 + k\, V_1$$

But $V_1 = v_1$

$\therefore$ $v_2 = v_1 + k\, v_1 = (1 + k)\, v_1 = (1 + 0.1)\, v_1 = 1.1\, v_1$

$$v_{2+1} = v_2 + k\, V_2$$

But $\quad V_2 = v_1 + v_2$

$\therefore \quad v_3 = v_2 + k(v_1 + v_2) = 1.1\, v_1 + 0.1\,(v_1 + 1.1\, v_1) = 1.31\, v_1$

$\quad\quad v_{3+1} = v_3 + k\, V_3$

where $\quad V_3 = V_2 + v_3$

$\therefore \quad v_4 = v_3 + k(V_2 + v_3) = 1.31\, v_1 + 0.1\,(2.1\, v_1 + 1.31\, v_1) = 1.651\, v_1$

But voltage between line conductor and earth

$\quad\quad V = v_1 + v_2 + v_3 + v_4$

$\quad\quad 100 = v_1 + 1.1\, v_1 + 1.31\, v_1 + 1.651\, v_1 = 5.061\, v_1$

$\quad\quad v_1 = \dfrac{100}{5.061} = 19.758 \text{ kV}$

$\quad\quad v_2 = 1.1\, v_1 = 1.1 \times 19.758 = 21.73 \text{ kV}$

$\quad\quad v_3 = 1.31\, v_1 = 1.31 \times 19.758 = 25.883 \text{ kV}$

$\quad\quad v_4 = 1.651\, v_1 = 1.651 \times 19.758 = 32.62 \text{ kV}$

Check

$\quad\quad v_1 + v_2 + v_3 + v_4 = 19.758 + 21.73 + 25.883 + 32.62 = 99.991 \simeq 100 \text{ kV}$

The maximum voltage is 32.62 kV. It is across the unit adjacent to line.

$$\text{String efficiency} = \frac{\text{voltage across the string}}{n \times \text{voltage across the line unit}} = \frac{V}{4v_4} = \frac{100}{4 \times 32.62} = 0.7664 \text{ or } 76.64\%$$

Alternative method

$$\frac{V_x}{V} = \frac{\sinh x \sqrt{k}}{\sinh n \sqrt{k}}, \quad \sinh x = \frac{e^x - e^{-x}}{2}$$

Here $\quad n = 4, \quad k = 0.1, \quad \sqrt{k} = 0.316227, \quad V = 100 \text{ kV}$

$$V_1 = \frac{V \sinh \sqrt{k}}{\sinh 4\sqrt{k}} = 100 \times \frac{0.3215246}{1.6302566} = 19.722 \text{ kV}$$

$$V_2 = \frac{V \sinh 2\sqrt{k}}{\sinh 4\sqrt{k}} = 41.43339 \text{ kV}$$

$$V_3 = \frac{V \sinh 3\sqrt{k}}{\sinh 4\sqrt{k}} = 67.322436 \text{ kV},$$

$\quad\quad V_4 = V = 100 \text{ kV}$

$\therefore \quad v_1 = V_1 = 19.722 \text{ kV}, \quad\quad v_2 = V_2 - V_1 = 21.711 \text{ kV}$

$\quad\quad v_3 = V_3 - V_2 = 25.8889, \quad\quad v_4 = V_4 - V_3 = 32.6778 \text{ kV}$

$$\text{String efficiency} = \frac{\sinh n \sqrt{k}}{n\,[\sinh n \sqrt{k} - \sinh (n-1)\sqrt{k}]}$$

$$= \frac{1.6302566}{4\,(1.6302566 - 1.0975284)} = 0.7650 \text{ per unit} = 76.5 \text{ per cent}$$

Example 5.2 In a 6-unit suspension insulator string, the capacitance between each link pin and earth is 0.1 of the self capacitance of each unit. Determine the voltage across the lowest unit as a percentage of the total voltage and also the string efficiency.

Solution

We have

$$\frac{V_x}{V} = \frac{\sinh x \sqrt{k}}{\sinh n \sqrt{k}}$$

$\therefore$ $\quad$ $\frac{V_n}{V} = \frac{\sinh n \sqrt{k}}{\sinh n \sqrt{k}}$ $\quad$ and $\quad$ $V_n = V$

Also, $\quad$ $\frac{V_{n-1}}{V} = \frac{\sinh (n-1) \sqrt{k}}{\sinh n \sqrt{k}}$

The voltage across the lowest unit is given by

$$v_n = V_n - V_{n-1} = V - \frac{\sinh (n-1) \sqrt{k}}{\sinh n \sqrt{k}}$$

For a six-unit string $n = 6$, therefore

$$v_n = V - \frac{V \sinh 5 \sqrt{k}}{\sinh 6 \sqrt{k}}$$

We have

$$\sinh x = \frac{1}{2} (e^x - e^{-x})$$

Here $\quad$ $k = 0.1, \ \sqrt{k} = 0.316227$

$$\sinh 5 \sqrt{k} = \frac{1}{2} (4.8604879 - 0.2057406) = 2.3273737$$

$$\sinh 6 \sqrt{k} = \frac{1}{2} (6.6683109) - 0.149963) = 3.259174$$

Now $\quad$ $\frac{v_n}{V} = 1 - \frac{\sinh 5 \sqrt{k}}{\sinh 6 \sqrt{k}} = 1 - \frac{2.3273737}{3.259174} = 0.2859$

Therefore, the voltage across the lowest unit as a percentage of the total voltage

$$= \frac{v_n}{V} \times 100 = 0.2859 \times 100 = 28.59\%$$

$$\text{String efficiency} = \frac{V}{n \, v_n} = \frac{1}{6 \times 0.2859} = 0.5829 \text{ pu or } 58.29\%$$

Example 5.3 Find the voltage distribution and string efficiency of a three unit suspension insulator string if the capacitances of the link pins to earth and to the line are respectively 20 per cent and 10 per cent of the self capacitance of each unit.

If a guard ring increases the capacitance to the line of lower link pin to 35 per cent of the self capacitance of each unit, find the redistribution of voltage and string efficiency.

Solution

Let C represent the self capacitance of each unit. The capacitances of the link pins to earth and to the line will, therefore, be $0.2C$ and $0.1C$ respectively. The capacitances, voltages and currents are shown in Fig. 5.16.

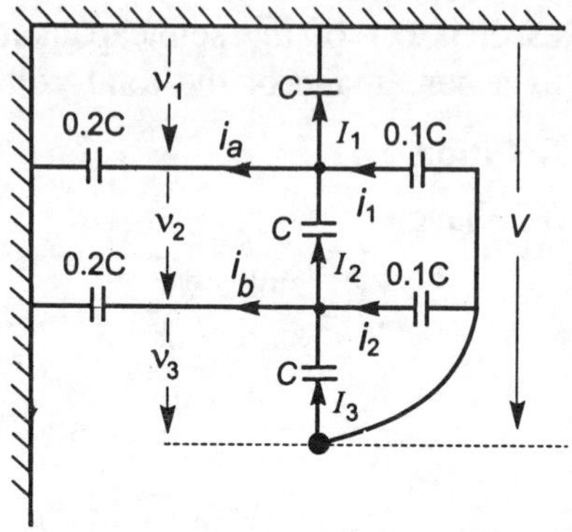

(a) We have

$$I_1 = v_1 (j \omega C), \quad I_a = v_1 (j \omega)(0.2C)$$

$$i_1 = (V - v_1) j \omega (0.1C)$$

$$I_2 = I_1 + I_a - i_1$$

$$= v_1 j \omega C + 0.2 v_1 j \omega C - 0.1 (V - v_1) j \omega C$$

$$= j \omega C (1.3 v_1 - 0.1V)$$

$$v_2 = \frac{I_2}{j \omega C} = 1.3 v_1 - 0.1V$$

$$I_b = (v_1 + v_2) \times 0.2 j \omega C = (v_1 + 1.3 v_1 - 0.1V) \times 0.2 j \omega C = j \omega C (0.46 v_1 - 0.02V)$$

$$i_2 = v_3 \times 0.1 j \omega C$$

$\therefore \quad I_3 = I_2 + I_b - i_2$

$$= j\omega C (1.3 v_1 - 0.1V) + j\omega C (0.46 v_1 - 0.02V) - 0.1 j\omega C v_3$$

$$= j\omega C (1.76 v_1 - 0.12V - 0.1 v_2)$$

$$v_3 = \frac{I_3}{j\omega C} = 1.76 v_1 - 0.12V - 0.1 v_3$$

$$v_3 = \frac{1.76 v_1}{1.1} - \frac{0.12V}{1.1} = 1.6 v_1 - 0.109V$$

Also, $\quad v_1 + v_2 + v_3 = V$

$$v_1 + 1.3 v_1 - 0.1V + 1.6 v_1 - 0.109V = V$$

$$3.9 v_1 = 1.209V, \quad v_1 = \frac{1.209}{3.9} V = 0.31V$$

$$v_2 = 1.3 v_1 - 0.1V = 1.3 \times 0.31V - 0.1V = 0.303V$$

$$v_3 = 1.6 v_1 - 0.109V = 1.6 \times 0.31V - 0.109V = 0.387V$$

$$\text{String efficiency} = \frac{V}{3 v_3} = \frac{V}{3 \times 0.387V} = 0.8613 \text{ pu}$$

(b) *With the grading ring*

The capacitance of the lower link pin now becomes $0.35C$ instead of $0.1C$. Therefore,

$$i_2 = v_3 \times 0.35 j\omega C$$

and $\quad I_3 = I_2 + I_b - i_2 = j\omega C (1.76 v_1 - 0.12V - 0.35 v_3)$

Fig. 5.16.

$$v_3 = \frac{I_3}{j\omega C} = 1.76v_1 - 0.12V - 0.35v_3$$

$$v_3 = \frac{1.76}{1.35}v_1 - \frac{0.12}{1.35}V = 1.3037v_1 - 0.0389V$$

Again, $v_1 + v_2 + v_3 = V$

$$v_1 + 1.3v_1 - 0.1V + 1.3037v_1 - 0.0889V = V$$

$$v_1(1 + 1.3 + 1.3037) = (1 + 0.1 + 0.0889)V$$

$$3.6037v_1 = 1.1889V$$

$$v_1 = \frac{1.1889}{3.6037} = 0.3299V$$

$$v_2 = 1.3v_1 - 0.1V = 0.32887V$$

$$v_3 = V - (v_1 + v_2) = V - (0.3299 + 0.32887)V = 0.34123V$$

It is found that v_1, v_2, and v_3 are practically equal to each other, that is, the voltage distribution becomes practically uniform.

$$\text{String efficiency} = \frac{V}{3v_3} \times 100 = 97.68\%$$

Example 5.4 A string of 6 suspension insulators is to be graded to obtain uniform distribution of voltage across the string. If the pin to earth capacitances are equal to C, and the self capacitance of the top insulator is $10C$, find the mutual capacitance of each unit in terms of C.

Solution

By Eq. (5.6.1)

$$C_{x+1} = C_x + x\, C_g$$

Here $C_g = C$

and $C_1 = 10C$

∴ $C_{1+1} = C_1 + 1 \times C$; $C_2 = 10C + C = 11C$

$C_{2+1} = C_2 + 2C$; $C_3 = 11C + 2C = 13C$

$C_{3+1} = C_3 + 3C$; $C_4 = 13C + 3C = 16C$

$C_{4+1} = C_4 + 4C$; $C_5 = 16C + 4C = 20C$

$C_{5+1} = C_5 + 5C$; $C_6 = 20C + 5C = 25C.$

Example 5.5 A string of 7 suspension insulators is to be fitted with a grading ring. If the pin to earth capacitances are equal to C, find the values of line to pin capacitances that would give a uniform voltage distribution over the string.

Solution

By Eq. (5.6.2), the capacitance of the pth link from the top is given by

$$C_p = \frac{p\, C_g}{n - p}$$

Here $n = 7$, $C_g = C$. Hence the line to pin capacitance from the top are given by

$$C_1 = \frac{1 \times C}{7-1} = \frac{1}{6} C ; \qquad C_2 = \frac{2C}{7-2} = \frac{2}{5} C ;$$

$$C_3 = \frac{3C}{7-3} = \frac{3}{4} C ; \qquad C_4 = \frac{4C}{7-4} = \frac{4}{3} C ;$$

$$C_5 = \frac{5C}{7-5} = \frac{5}{2} C ; \qquad C_6 = \frac{6C}{7-6} = 6C.$$

Example 5.6 Each conductor of a three-phase overhead line is suspended from a cross arm of a steel tower by a string of 4 suspension insulators. The voltage across the second unit is 14.2 kV and across the third 20 kV. Find the voltage between the conductors and the string efficiency.

Solution

By Eq. (5.5.2)

$$v_{p+1} = v_p + k V_p ; \qquad v_{1+1} = v_1 + k V_1$$

$$v_2 = v_1 + k v_1 = (1 + k) v_1 ; \qquad v_{2+1} = v_2 + k V_2$$

$$v_3 = v_2 + k (v_1 + v_2) = v_2 (1 + k) + k v_1$$

$$v_3 = (1 + k) (1 + k) v_1 + k v_1 = (1 + 3k + k^2) v_1$$

$$v_{3+1} = v_3 + k V_3$$

$$v_4 = (1 + 3k + k^2) v_1 + k (v_3 + v_2 + v_1) = (1 + 6k + 5k^2 + k^3) v_1$$

Here $v_2 = 14.2$ kV and $v_3 = 20$ kV

$$\frac{v_3}{v_2} = \frac{(1 + 3k + k^2) v_1}{(1 + k) v_1}, \qquad \frac{20}{14.2} = \frac{1 + 3k + k^2}{1 + k}$$

$$14.2 (1 + 3k + k^2) = 20 (1 + k), \qquad 14.2 k^2 + 22.6k - 5.8 = 0$$

$$k = \frac{1}{2 \times 14.2} \left[-22.6 \pm \sqrt{(22.6)^2 + 4 \times 5.8 \times 14.2} \right] = 0.22487.$$

We have neglected the negative value of k.

$$v_1 = \frac{v_2}{1 + k} = \frac{14.2}{1 + 0.22487} = 11.5931 \text{ kV}$$

$$v_4 = (1 + 6k + 5k^2 + k^3) v_1 = (1 + 1.34922 + 0.25283 + 0.011370) \times 11.5931 = 30.297 \text{ kV}$$

$$V = v_1 + v_2 + v_3 + v_4 = 11.5931 + 14.2 + 20 + 30.297 = 76.09 \text{ kV between conductor and earth.}$$

Line voltage, $V_l = \sqrt{3} \times 76.09 = 131.79$ kV

$$\text{String efficiency} = \frac{V}{4v_4} = \frac{76.09}{4 \times 30.297} = 0.6278 \text{ pu or } 62.78\%.$$

Example 5.7 In a string of three insulator units, the capacitance of each unit is C, from each conductor to ground is $C/3$, and from each connector to the line conductor is $C/5$. Calculate the voltage across each unit as a percentage of the total voltage.

To what value the capacitance between the connector of the bottom unit and the line has to be increased by a guard ring to make the voltage across it equal to that across the next higher unit?

Solution

By KCL at node A of Fig. 5.17,

$$I_1 + I_a = I_2 + I_x$$

$$v_1 (j\omega C) + v_1 \left(j\omega \frac{C}{3} \right) = v_2 (j\omega C) + (v_2 + v_3) \left(j\omega \frac{C}{5} \right)$$

$$\frac{4}{3} v_1 = \frac{6}{5} v_2 + \frac{1}{5} v_3$$

$$20v_1 = 18v_2 + 3v_3 \qquad \qquad \text{...(E5.7.1)}$$

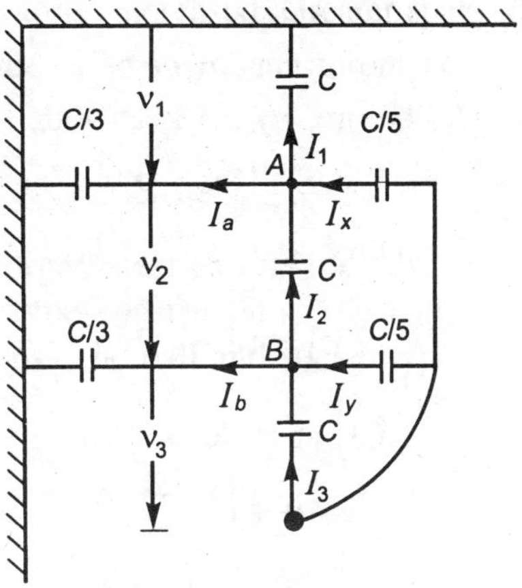

By KCL at node B,

$$I_2 + I_b = I_3 + I_y$$

$$v_2 (j\omega C) + (v_1 + v_2) \left(j\omega \frac{C}{3} \right) = v_3 (j\omega C) + v_3 \left(j\omega \frac{C}{5} \right)$$

$$\frac{4}{3} v_2 + \frac{1}{3} v_1 = \frac{6}{5} v_3$$

$$20v_2 + 5v_1 = 18v_3 \qquad \qquad \text{...(E5.7.2)}$$

Eliminating v_3 from Eqs. (E5.7.1) and (E5.7.2) by multiplying Eq. (E5.7.1) by 6 and subtracting Eq. (E5.7.2) from it,

$$120v_1 - 108v_2 - 20v_2 - 5v_1 = 0$$

$$115v_1 = 128v_2 ; \qquad v_2 = \frac{115}{128} v_1$$

Substituting this value of v_2 in Eq. (E5.7.2) we get

$$20 \times \frac{115}{128} v_1 + 5v_1 = 18v_3 ; \qquad v_3 = 1.276v_1$$

Total voltage across the string

$$v_1 + v_2 + v_3 = V ; \qquad v_1 + \frac{115}{128} v_1 + 1.276v_1 = V ; \qquad 3.1744v_1 = V$$

Voltage across first unit from top as a percentage of total voltage

$$= \frac{v_1}{V} \times 100 = \frac{100}{3.1744} = 31.5 \text{ per cent}$$

Voltage across second unit from top as a percentage of total voltage

$$= \frac{v_2}{V} \times 100 = \frac{115v_1 \times 100}{128V} = 28.30 \text{ per cent}$$

Voltage across bottom unit as a percentage of total voltage

$$= \frac{v_3}{V} \times 100 = \frac{1.276v_1}{3.1744v_1} \times 100 = 40.196 \text{ per cent}$$

Fig. 5.17.

With grading ring

Let the arrangement be as shown in Fig. 5.18.

As before, by KCL at node A

$$20v_1 = 18v_2 + 3v_3 \qquad \ldots(E5.7.1)$$

But since $v_2 = v_3$

$$20v_1 = 18v_2 + 3v_2, \quad v_2 = \frac{20}{21}v_1$$

By KCL at node B,

$$I_2 + I_b = I_3 + I_y$$

$$v_2 (j\omega C) + (v_1 + v_2)\left(j\omega \frac{C}{3}\right) = v_3 (j\omega C) + v_3 (j\omega C_y)$$

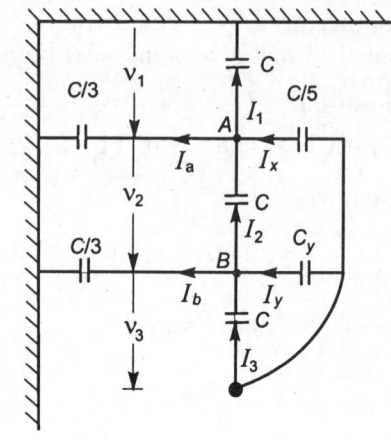

Fig. 5.18.

But $\quad v_2 = v_3 ; \qquad \dfrac{1}{3}(v_1 + v_2)\,C = v_2\,C_y$

$$C_y = \frac{1}{3}\left(\frac{v_1}{v_2} + 1\right)C = \frac{1}{3}\left(\frac{21}{20} + 1\right)C = 0.6833C$$

Thus, the capacitance between the connector of the bottom unit and the line has to be increased from $C/5$ to $0.6833C$ in order to make the voltage across the two bottom units equal.

5.7 SELECTION OF INSULATION

The insulation level affects the cost and reliability of high voltage system considerably. Normal operating and transient voltages are the primary considerations in selecting the insulation for a line. The transient voltages include switching and lightning surges. It is found that when sufficient insulation is provided for switching and lighting surges, power frequency voltage requirements are usually satisfied. The magnitude of switching surges is 2.5 to 3 times normal voltage. Switching surges assume greater importance above 230 kV insulation requirements. On lines up to 230 kV lightning is the major factor to decide the amount of insulation. The lightning insulation requirements depend on other factors also like tower footing resistance, number of ground wires, conductor configuration etc. in addition to the number of insulators. Final choice of the number of insulators for a line is also dictated by the atmospheric conditions, temperature, and altitude of the place.

Atmospheric conditions affect considerably the flashover voltages. Wet conditions cause flashovers at lower voltages. Dirty and wet surfaces of the insulators allow the leakage currents to flow over them resulting in flashovers. The flashover voltage is proportional to the length of leakage path. The leakage distances or creepage paths vary depending upon the quality and quantity of deposition on insulators. Usually about 25 to 28 mm of leakage distance per kV of line-to-ground voltage is provided. In heavily polluted areas either antifog type of units are used, or the number of units is increased in the string to increase the leakage path. In such cases the insulators should be washed periodically.

Temperature and Altitude

Correction factors are to be applied to take into account the effects of temperature and altitude on flashover values. At altitudes higher than 100 metres or 3300 feet the insulator strength is decreased due to lower relative air density and, therefore, the flashover value at normal voltage is multiplied by $\left[1 + \dfrac{0.03\,(h - 3300)}{1000}\right]$ for altitude h feet. The correction factor for places where temperature exceeds 40°C is $\left(\dfrac{273 + t}{313}\right)$ for service temperature $t°C$. More units are to be provided for higher insulation.

Flashover voltage is also affected by grading rings, bundle conductor field, and configuration of the string. Table 5.2 serves as a guide to show the present practice in selection of overhead line insulation under normal conditions.

5.8 LINE SUPPORTS

A line support has to perform the following functions :

1. To keep the proper spacing between the conductors.
2. To keep the conductors at prescribed distances from its grounded parts.
3. To maintain the specified ground clearance.

Table 5.2   Selection of Overhead Line Insulators

Line voltage	Suspension sets					Tension sets				
	Dia-meter	Centres	No. of units per string	String length	No. of strings in a set	Dia-meter	Centres	No. of units per string	String length	No. of strings in a set
kV	mm	mm	No.	mm	No.	mm	mm	No.	mm	No.
33	254	127	3	381	1	254	127	' 4	508	1
66	254	127	5	635	1	254	127	6	762	1
132	254	127	11	1397	1	254	127	11	1397	1
220	254	146	14	2044	1	254	146	16	2336	1
275	280	171	18	3078	1	280	178	16	2848	2
380	280	171	20	3342	1	280	171	25	4240	3
400	381	187	22	4114	2	330	190.5	22	4191	4

The air and ground clearances are decided by electrical and mechanical considerations. The main requirements of line supports are low initial cost, low maintenance expense, and long life. The final choice of the type of support is also governed by the availability of the material is addition to the above considerations. Overhead lines are carried on supports made of wood, concrete steel or aluminium. The supports may be in the form of poles or towers. Some of the line supports are described in the treatment to follow.

5.9 WOOD POLES

Of the various types of supports wood pole is the cheapest. When properly treated with a preservative such as creosote a very satisfactory service is obtained. These poles are suitable for lines where spans are short and tensions are low. A wood pole has the limitations of height and diameter.

Double pole structures of the 'A' or 'H' type are used where greater strength is required. The strength of these types of constructions varies from two to four times the strength of the single pole. In practice, it has been found that foundations fail before the pole is stressed to these values. H-type of construction is usually employed for terminal poles or those carrying switchgear and transformers.

Wood has got a natural insulating property and, therefore, lesser flashovers are likely to take place due to lightning. The drawbacks with wood poles are that their strength and durability cannot be predicted with certainty. Insects and birds are the problems for such poles. Typical wood pole constructions are shown in Fig. 5.19 (a) and (b).

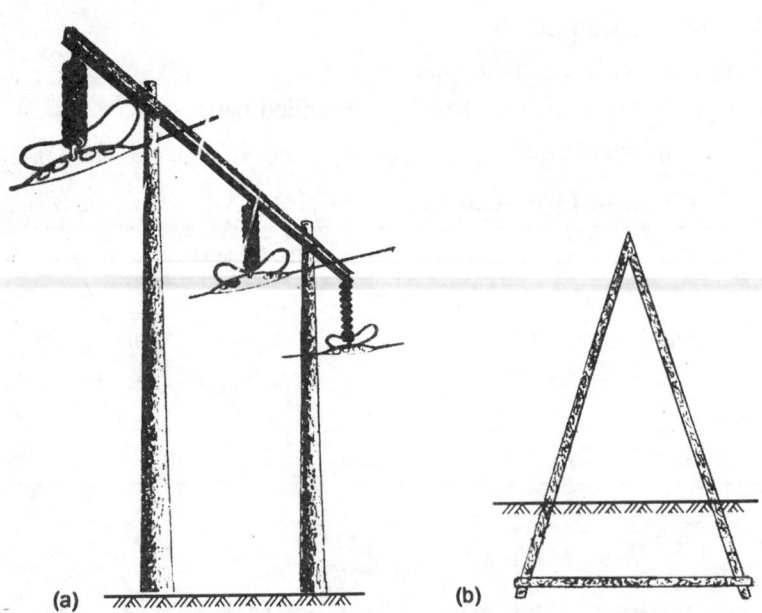

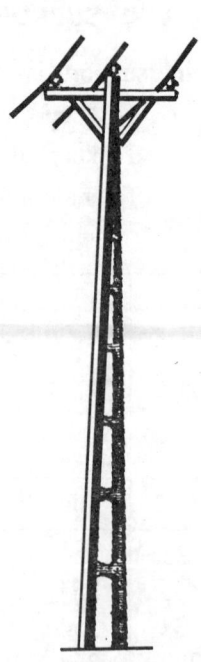

(a)   (b)

Fig. 5.19. (a) & (b) Typical wood pole constructions.

Fig. 5.20. Reinforced concrete pole.

5.10 CONCRETE POLES

Concrete pole (Fig. 5.20) is reinforced to give greater strength and is an alternative to a wood pole. It has a longer life than that of wood pole because of little deterioration. The maintenance cost is low. Reinforced concrete poles are very heavy and are liable to damage during loading, unloading, transportation and erection due to their brittle nature.

Handling and transportation difficulties are overcome by the use of pre-stressed concrete supports which can be manufactured in pieces and then assembled at the job sites. The weight of pre-stressed concrete pole is considerably less than that of reinforced concrete pole. The material used is less. Pre-stressed concrete poles are more durable than any other type of pole.

5.11 STEEL POLES

The use of tubular steel poles or girder steel masts is favoured for low and medium voltage distribution work. Longer spans are possible with steel poles. The poles need be galvanised or painted periodically to prevent them from corrosion. The maintenance expense is, therefore, more.

5.12 SUPPORTING TOWERS

High voltage and extra high voltage lines require large air and ground clearances. They have large mechanical loadings. Insulation costs are also considerable. Steel towers were developed for such lines where very long spans are essential. The long span construction cuts the insulation cost considerably as fewer supports are to be provided for. Moreover, the possibilities of breakdowns are reduced. The towers are either made of steel or aluminium. They are classified as follows:

1. Self-supporting towers.
2. Stayed or guyed towers.

5.12.1 Self-Supporting Towers

Self-supporting towers are divided into two categories, namely : wide-base and narrow-base towers. In a wide-base tower lattice type construction with bolted connections is adopted. Each leg has a separate foundation. The narrow-base design uses latticed construction of angle, channel or tubular steel section with bolted or welded connections.

Self-supporting towers are also classified as:

(a) tangent towers, and

(b) deviation towers.

Tangent towers are used for straight runs of the line. Suspension insulators are used with these towers.

Deviation towers are used at points where transmission line changes direction.

Strain insulators are used with these towers. They have broader base, stronger members, and are costlier as compared to tangent towers. Fig. 5.21 shows some typical self-supporting towers.

A narrow-base tower requires lesser steel or aluminium in comparison with a wide-base tower, but its cost of foundation is more. The selection between the two is to be made on the basis of costs of material, foundations, and right-of-way requirements.

5.12.2 Guyed or Stayed Towers

Guyed or stayed towers were adopted with a view to reduce weight and cost. Such towers have many identical members. This affords saving in fabrication and facilitates assembly and the erection as complete unit at the job site. Aluminium guyed towers due to their much lighter weight have been favoured for remote areas where transport is a problem. In some countries

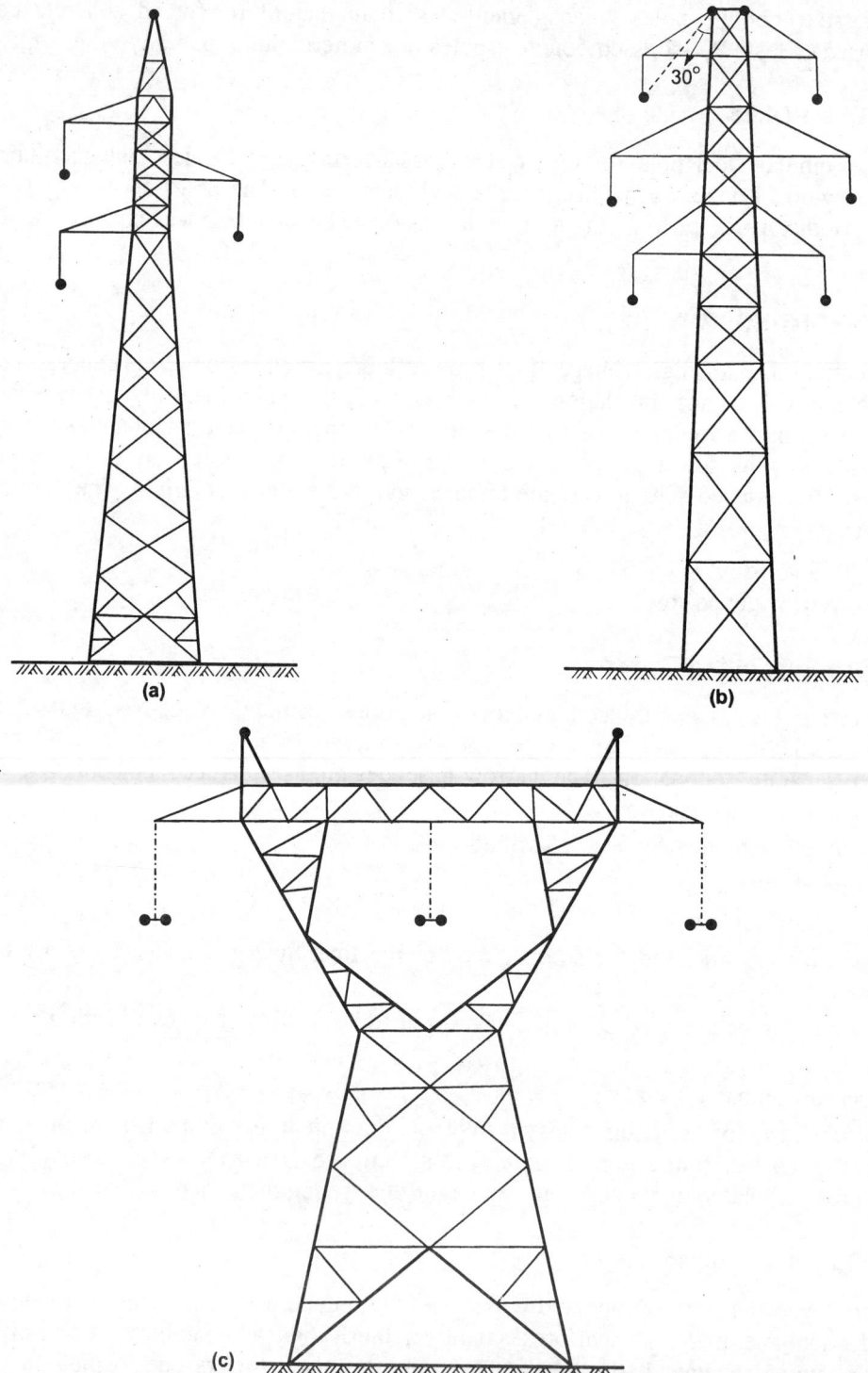

Fig. 5.21. Some typical self-supporting towers. (a) 132-kV single-circuit tower. (b) 132-kV double-circuit tower. (c) 400-kV single-circuit tangent tower.

like the USA and Canada, use is made of helicopter for transportation of such towers at sites which are not accessible by road or rail.

Guyed towers are either portal type or V-type. Both of them have got two masts connected at the top by a cross arm and provided with four guys. In case of portal type of structure each mast rests on its own foundation. The four guys are anchored to two double-acting guy anchors. A V-tower has two masts resting at an angle to one another on one thrust footing only which is of heavier type. Separate guy anchorages is to meet varying soil conditions and to take large uplift forces. The two types of guyed towers are shown in Fig. 5.22.

In some of the recent designs pre-stressed concrete has been utilized in construction of stayed towers.

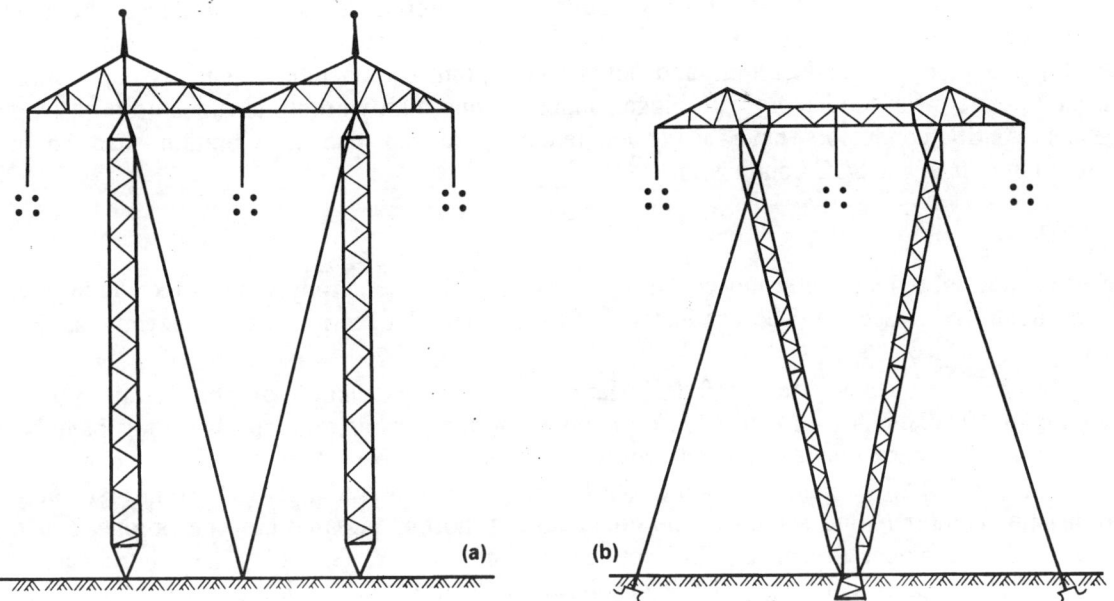

Fig. 5.22. Guyed towers. (a) Portal type, (b) V-type.

5.13   VIBRATION OF CONDUCTORS

The following types of mechanical vibration are common to line conductors :

1. Resonant vibration or aeolian vibration.
2. Galloping.
3. Dancing and sleet-jump.

5.13.1   Resonant Vibration

Resonant vibration of conductor arises from the eddies produced behind the conductor by the action of low-velocity winds. Due to the formation of eddies, the velocities of the wind at the sides towards and away from conductor become unequal. The unequal velocities result in unequal pressures, the pressure at the higher velocity side being lower. Air from the lower velocity side will, therefore, come to fill rarefied space behind the pressure of air. The conductor thus sets in motion in upward and downward direction depending on the existing conditions. During its motion the conductor leaves behind an empty space, which is then filled up by eddies from the higher velocity side and the direction of motion changes. Thus, oscillation of the conductor starts. When the frequency of eddies coincides with the natural frequencies of the conductor, resonant vibration of conductor results.

The change in velocity or direction of the wind damps the original vibration with the production of a new vibration. Such vibrations are in vertical plane, and have the characteristics of high frequency and low amplitudes with the formation of nodes and loops. The normal maximum amplitudes are of the order of 25 mm and the frequency ranges between 5-100 Hz for wind velocities of 3-25 km/h.

The high frequency oscillations may build up an amplitude, which produces alternating stresses large enough to cause fatigue failure. The failure occurs due to fatigue cracks caused by rapid bending of conductor up and down at the point of attachment. The risk of vibration trouble is more for conductors with bigger diameters and high working tensions used on longer spans. ACSR conductors have a greater tendency to fail due to vibration than copper, cadmium-copper, or SCC conductors.

5.13.2 Galloping

Self-excited vibrations are produced on line conductors by the aerodynamic forces acting upon non-circular cross-section. The conductor takes a non-circular section due to uneven coating of ice on its surface.

This low frequency ($\frac{1}{4}$ - $1\frac{1}{4}$ Hz), large amplitude oscillation of conductors is called *galloping*. Oscillations with amplitude of about 11 m and periods of about 8 seconds have been witnessed. Galloping conditions occur rarely. Galloping has been observed with wind velocities ranging between 15-75 km/h inclined to the line at an angle between 10-90°. Both torsional and translational motions are set up in the conductor. Abnormal stresses may be produced at the points of attachment to damage the conductors and fittings. Galloping also introduces the possibilities of electrical contact between phase conductors or from phase conductor to earth. The supply may thus be interrupted.

5.13.3 Dancing and Sleet-Jump

Ice falling from a conductor throws it into violent oscillations of large period and long amplitude. The oscillations die out quickly if the ends of the dancing span are dead ended, otherwise the oscillations are transmitted in the adjoining spans to considerable distance of line.

As such dancing is not harmful from mechanical damage point of view, but large amplitude vibrations may bring the conductors together resulting in short circuits and burning of conductors. Conductor clashing may be reduced by arranging the conductors in a horizontal configuration.

5.14 EFFECTS OF VIBRATION ON THE TRANSMISSION LINE

The following adverse effects are observed in the transmission line due to conductor vibrations :

1. Fatigue failure of strands or complete stranded conductor
2. Bird caging of stranded conductors

3. Breaking of insulator discs
4. Excessive sag
5. Collapse of supporting structures

5.15 PREVENTION OF VIBRATION

Of the various methods employed for minimising or eliminating vibration in a conductor, the two mentioned below have been found to be very effective.

1. Armour rods.
2. Stockbridge dampers.

5.15.1 Armour Rods

Armour rods consist of layer of wires of rods wrapped spirally around the conductor for a short distance on either side of the point of support. They provide a reinforcement of conductors at suspension points and reduce amplitude of vibration from 10 to 20 per cent. They relieve and distribute the stresses at the support point. They also serve as a protection against flashover burns. Fig. 5.23 shows a tapered armour rod fitted on a conductor.

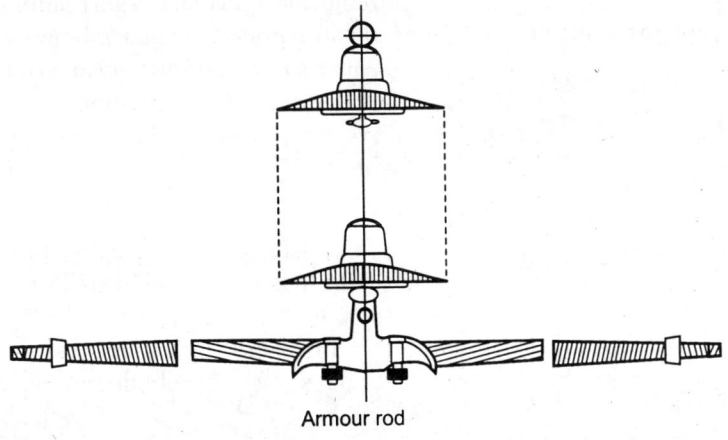

Armour rod

Fig. 5.23. Armour rod.

5.15.2 Stockbridge Damper

The stockbridge damper shown in Fig. 5.24 consists of two weights joined together by a flexible steel wire. It is provided with a clamp at its middle point to attach it to the conductor. Usually one damper is attached at each end of the span, for spans up to 300 m. The number may be increased for longer spans.

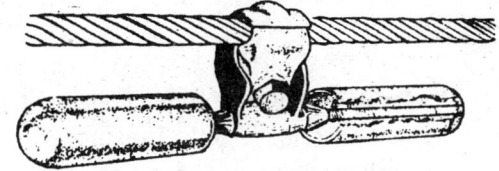

Fig. 5.24. Stockbridge damper.

V-strings have also been used to minimise vibrations of conductors. The adjustment of the system prevents the conductors from vibration

5.16 SPACING OF CONDUCTORS

Trouble free service requires proper spacing of conductors. The conductors should be so spaced that corona loss is minimum and that they do not clash during swinging or vibration. No exact rule can be derived to calculate the conductor spacing for a line. Various empirical formulae are in use in different countries giving widely different values of spacing for some lengths and conductor sizes. They have been deduced on the basis of experience in each country depending on the loading and local conditions of temperature, wind, ice, etc. Where the spacing is not governed by corona or vibration considerations, a simple rule of spacing is a function of voltage between the conductors, length of span and sag.

Some of the empirical formulae for conductor spacing S are given below :

$$S = 0.75 \sqrt{\delta} + \frac{V^2}{20000} \; ; \qquad S = 2\delta \sin \theta$$

$$S = 0.65 \sqrt{\delta} + 0.007 \; ; \qquad S = 0.75 \sqrt{\delta} + \frac{V}{150}$$

$$S = 0.8 \sqrt{\delta + l} + \frac{V}{150} \; ; \qquad S = \sqrt{\delta} + \frac{V}{150}$$

$$S = \sqrt{\delta} + 0.012V \; ; \qquad S = 0.25 + \frac{V}{100} + 0.7 \sqrt{\delta}$$

$$S = 0.75 \sqrt{(\delta + l) \sin \delta} + \frac{V}{125}$$

where δ = sag in metres

l = length of the insulator string in metres

θ = deflection of insulator string

V = voltage in kV.

These formulae may serve as a guide to calculate conductor spacing. It is obvious from the above formulae that it is not the voltage alone which decides the conductors spacing but other considerations such as sag, weight per unit length of conductor, span length, conductor diameter, length of insulator string, the angle of swing etc., are also taken into account to choose the conductor spacing. The local conditions of temperature, ice and wind also affect the final choice.

EXERCISES

1. Explain why the voltage across the insulators of a simple insulator strong are not equal and describe practical methods to improve this distribution.

2. If the voltage across the units in a 2-unit suspension insulator is 60 per cent and 40 per cent respectively of the line voltage, find the ratio of the capacitance of the insulator to that of its capacitance to earth. (C & G)

3. A string of suspension insulators consists of three similar units. The capacitance between the metal interlinks is 10 times the capacitance between each interlink and earth. The flashover voltage of one insulator is 100 kV. Calculate the voltage at which the string will flashover.   (C & G)

4. Find the voltage distribution and the string efficiency of a 3-unit suspension insulator if the capacitance of the link pins to earth and to the line are respectively 20 per cent and 10 per cent of the self-capacitance of each unit.   (L.U.)

5. A string of suspension insulators consists of 3 units. The capacitance between each link pin and earth is one-eighth of the self-capacitance of the unit. If the maximum peak voltage per unit is not to exceed 20 kV, find the greatest working voltage and the string efficiency.   (L.U.)

6. A string of 4 insulator units has self-capacitance equal to 5 times the pin to earth capacitances. Neglecting leakage, find (a) voltage distribution from top to the bottom insulator in percentages of the total voltage, (b) the string efficiency.

7. Find the potential difference across each unit of an overhead line suspension insulator consisting of four units all similar. The voltage between the line conductor and earth is 60 kV and the ratio of capacitance of each unit insulator to the capacitance to earth intermediate section of metal work is 5 : 1. It is assumed that no leakage takes place across the insulators.   (I.E.E.)

8. Determine the potential distribution and the voltage across a string of five suspension insulators when the potential across any unit cannot exceed 25 kV. The capacitance of each unit is eight times the capacitance between unit and ground. Also calculate the string efficiency.

9. A string of 5 suspension insulators is connected across a 100 kV line. Calculate the distribution of voltage on the insulator discs if the capacitance of each disc to earth is 0.1 of the capacitance of the insulator. Calculate also the string efficiency.

10. Explain why the voltage distribution along a string of suspension insulators is normally nonuniform.

    A voltage V is applied across a string of n cap and pin-type insulators suspended from an earthed cross-arm of a tower. The capacitance between the cap and pin of each insulator is C and the capacitance of each pin to earth is c. Show that the voltage across the insulator unit nearest the line terminal is

    $$V [\sinh n\gamma - \sinh (n-1)\,\gamma]/\sinh n\gamma$$
    where $\sinh \gamma/2 = \tfrac{1}{2}\sqrt{(c/C)}$

11. In a 3-unit suspension insulator string, the capacitance between each link pin and earth is 0.2 of the capacitance of the unit. Determine by how much the capacitance of the lowest unit should be increased to achieve a string efficiency of 90% the other two units being left unchanged.

    Describe (a) how such an increase in capacitance may be obtained in practice and (b) alternative methods of achieving a high string efficiency.

12. Define string efficiency. What is the necessity of having a high string efficiency? How can it be achieved?

A string of eight suspension insulators is to be fitted with a grading ring. If the pin-to-earth capacitances are all equal to C, find the values of line-to-pin capacitances that would give a uniform voltage distribution over the string.

13. The self capacitance of each unit in a string of three suspension units is C. The shunting capacitance of the connecting metal work of each insulator to earth is $0.25C$, while for the line it is $0.15C$. Find the percentage voltage distribution and string efficiency.

14. Write a brief note on vibration of conductors. How is the vibration minimized?

ANSWERS

2. $2 : 1$ **3.** 260 kV

4. From earth end 31%, 30.3%, 38.7%; efficiency 86%

5. 35.7 kV, 84.1%

6. (a) $v_1 = 16\%$, $v_2 = 19.2\%$, $v_3 = 26.3\%$, $v_4 = 38.5\%$ (b) 65%

7. $v_1 = 9.6$, $v_2 = 11.5$, $v_3 = 15.8$, $v_4 = 23.1$ kV

8. $v_1 = 10$, $v_2 = 11.25$, $v_3 = 13.906$, $v_4 = 18.301$, $v_5 = 25$ kV; 62.72%

9. $v_1 = 8$, $v_2 = 8.80$, $v_3 = 10.48$, $v_4 = 13.21$, $v_5 = 17.25$ kV; 67%

11. 26.5%   **12.** $\dfrac{C}{7}, \dfrac{C}{3}, \dfrac{3}{5}C, C, \dfrac{5}{3}C, 3C, 7C$

13. Top unit = 31.72%, second unit = 29.41%, bottom unit = 38.86%;
String efficiency = 85.77%

6

Sag and Tension

6.1 INTRODUCTION

The main mechanical considerations in the design and stringing of conductors on the supports are :

1. An adequate ground clearance of the conductors at maximum temperature and minimum loading conditions should be maintained.

2. The conductor should not break under most severe conditions of ice and wind loadings which are assumed to act on it at its place of erection. It should be strung with a predetermined tension so that under the conditions of maximum wind and, possibly, ice loading at minimum temperature, it is not stressed to the values greater than its ultimate tensile strength divided by the factor of safety.

To make sure that these conditions are satisfied, the lines are erected in accordance with the statutory regulations. These regulations are different in different countries because of variations in loading and other conditions.

6.2 SAG AND TENSION

Sag calculations in conductors made of homogeneous material, such as all copper, all aluminium etc., are easier. They become more involved for composite conductors such as steel cored aluminium conductors. With such conductors, the temperature coefficients and tensile strengths are different for the materials of which the conductors are made. Sag and tension charts supplied by the manufacturers should be used for composite conductors.

A line conductor of uniform cross-section and material, perfectly flexible but inelastic stretched between two supports, hanging freely under the influence of its own weight takes the form of curve called *catenary*. The horizontal distance between two adjacent supports is called the *span*. The vertical distance between the conductor at midpoint and the line joining the two adjacent level supports is known as *sag*. It should be measured along the resultant load on a conductor.

If the weight of the conductor is assumed to be distributed uniformly along a horizontal line, it is sufficiently accurate to assume that a freely suspended conductor hangs in the shape of a parabola. The error involved with the assumption of a parabolic shape increases with the increase in span. For level spans up to 300 metres and sags not exceeding 5 per cent of the span length, the error is negligible. For longer spans such as at river crossings and very steep slopes either correction factors should be applied or catenary method of solution should be adopted. The two methods of calculation of sag and tension are discussed separately.

6.3 PARABOLIC METHOD

As mentioned above, the parabolic method is applicable for spans up to 300 metres and there is not much difference in span and conductor lengths. Parabolic assumption gives accurate results except in special cases involving very long spans or very steep slopes.

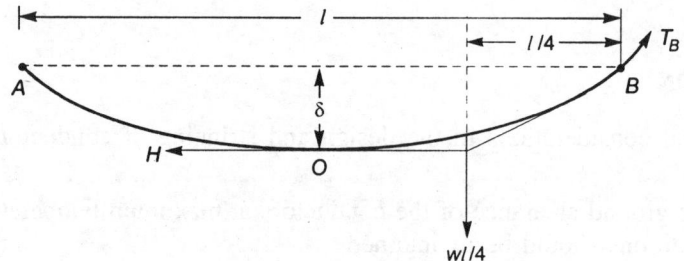

Fig. 6.1. Parabola.

Fig. 6.1 shows a conductor AOB suspended freely between supports A and B at the same level. Let its shape be parabolic. The lowest point is taken at O.

Let l = span length

   w = weight per unit length of the conductor

   δ = conductor sag

   H = tension in the conductor at the point of maximum deflection O

   T_B = tension in the conductor at the point of support B.

Unless otherwise indicated, the word tension shall denote the horizontal component of tension in a conductor.

Consider the equilibrium of the portion OB of the conductor. The forces acting on it are the horizontal tension H at O, the weight $(w \cdot OB)$ of the portion OB acting vertically downwards through the centre of gravity at a distance $l/4$ from B, and the tension T_B at the support B. Taking moments about B,

   $H \delta = (w \cdot OB) \, l/4$

Since OB is approximately equal to $l/2$,

$$H \cdot \delta = w \frac{l}{2} \cdot \frac{l}{4} \; ; \; \delta = \frac{w \, l^2}{8H}$$

$$...(6.3.1)$$

Equation (6.3.1) shows that the sag in a freely suspended conductor is directly proportional to the weight per unit length of the conductor, and to the square of the span length and inversely proportional to the horizontal tension H.

6.4 CATENARY METHOD

In this method it is assumed that the curve taken by the freely hanging conductor is a *catenary*. In Fig. 6.2, considering the equilibrium of the portion OP of the catenary, where P is a point on the curve such that the OP = s. The three forces acting on it are the horizontal tension H at the lowest point, the weight ws of OP acting through its centre of gravity, and the tension T at P along the tangent to the curve at P. For equilibrium these three force meet at a point M. The angle

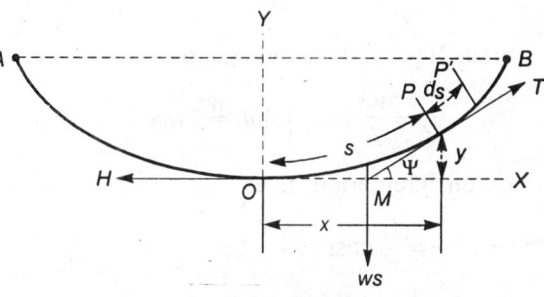

Fig. 6.2. Catenary.

which the tangent at P makes with the horizontal is assumed to be ψ. Resolving these forces horizontally and vertically.

$$T \cos \psi = H \qquad\qquad ...(6.4.1)$$

$$T \sin \psi = ws \qquad\qquad ...(6.4.2)$$

$$\tan \psi = ws/H \qquad\qquad ...(6.4.3)$$

If (x, y) be the coordinates of P, then for an element PP' of length ds of the conductor (Fig. 6.3)

$$(ds)^2 = (dx)^2 + (dy)^2$$

$$(ds/dx)^2 = 1 + (dy/dx)^2$$

$$dx = \frac{ds}{[1 + (dy/dx)^2]^{1/2}}$$

But $\qquad \dfrac{dy}{dx} = \tan \psi = ws/H \qquad\qquad ...(6.4.4)$

$$dx = \frac{ds}{[1 + (ws/H)^2]^{1/2}}$$

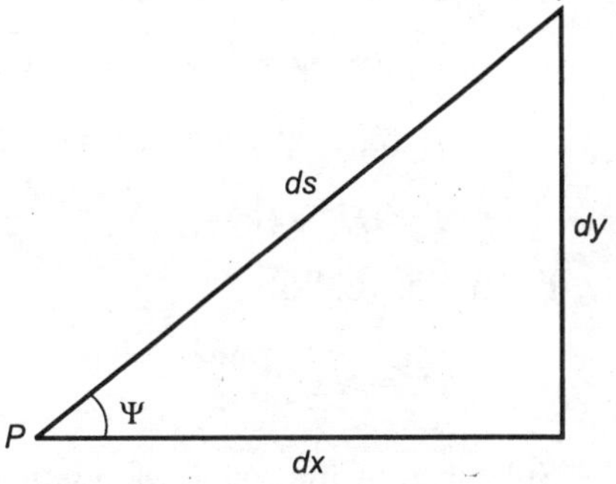

Integration on both sides gives

$$\int dx = \int \frac{ds}{[1 + (ws/H)^2]^{1/2}}$$

Fig. 6.3. Element of length ds shown enlarged.

or, $\qquad x = (H/w) \sinh^{-1}(ws/H) + C_1 \qquad\qquad ...(6.4.5)$

where C_1 is a constant of integration. This constant is evaluated by means of the conditions at the point O where $x = 0$ and $s = 0$.

Substituting these values in Eq. (6.4.5), we have

$$0 = \frac{H}{w} \sinh^{-1} (0) + C_1$$

which gives $C_1 = 0$

$$\therefore \qquad x = \frac{H}{w} \sinh^{-1} \frac{ws}{H} ; \qquad s = \frac{H}{w} \sinh \frac{wx}{H} \qquad \qquad ...(6.4.6)$$

From Eqs. (6.4.4) and (6.4.6)

$$dy = \frac{ws}{H} dx = \left(\sinh \frac{wx}{H} \right) dx$$

which on integration gives

$$y = \frac{H}{w} \cosh \frac{wx}{H} + C_2 \qquad \qquad ...(6.4.7)$$

where C_2 is another constant of integration.

At the point O, $x = 0$, $y = 0$

Substituting these values in Eq. (6.4.7),

$$0 = \frac{H}{w} \cosh 0 + C_2 ; \qquad 0 = \frac{H}{w} \cdot 1 + C_2 ; \qquad C_2 = -\frac{H}{w}$$

By substituting the value of C_2 in Eq. (6.4.7), it becomes

$$y = \frac{H}{w} \left(\cosh \frac{wx}{H} - 1 \right) \qquad \qquad ...(6.4.8)$$

This is the equation of a catenary.

Squaring and adding Eqs. (6.4.1) and (6.4.2)

$$T^2 \cos^2 \psi + T^2 \sin^2 \psi = H^2 + (ws)^2$$

$$T^2 (\cos^2 \psi + \sin^2 \psi) = H^2 + (ws)^2$$

$$T = [H^2 + (ws)^2]^{1/2} \qquad \qquad ...(6.4.9)$$

From Eq. (6.4.6)

$$ws = H \sinh \frac{wx}{H}$$

Substituting the value of ws in Eq. (6.4.9), the tension at any point is given by

$$T = \left(H^2 + H^2 \sinh^2 \frac{wx}{H} \right)^{1/2}$$

$$T = H \cosh \frac{wx}{H} \qquad \qquad ...(6.4.10)$$

The sag or dip δ, the tension T_B at the support, and the length of the conductor L between the supports are found by substituting the values of x and y at the supports in Eqs. (6.4.8), (6.4.10) and (6.4.6) respectively.

For level supports the maximum sag is the value of y at either support and the maximum tension occurs at the supports.

At the support B

$$x = \frac{l}{2}, \quad y = \delta, \quad s = \frac{L}{2}$$

$$\delta = \frac{H}{w}\left(\cosh\frac{wl}{2H} - 1\right) \qquad\qquad\qquad ...(6.4.11)$$

$$T_{max} = T_B = H\cosh\frac{wl}{2H} \qquad\qquad\qquad ...(6.4.12)$$

$$\frac{L}{2} = \frac{H}{w}\sinh\frac{wl}{2H}; \qquad L = \frac{2H}{w}\sinh\frac{wl}{2H} \qquad\qquad ...(6.4.13)$$

The Eqs. (6.4.11) to (6.4.13) give accurate results. The hyperbolic functions involved take much labour in calculation. For this reason approximations are made to calculate sag and tension, etc. For this purpose the hyperbolic functions are expanded and higher powers neglected. The results so obtained are fairly accurate if the sag does not exceed 5 per cent of the span length.

Approximate Results

Sag : Equation (6.4.8) on expansion gives

$$y = \frac{H}{w}\left[1 + \left(\frac{wx}{H}\right)^2\frac{1}{2!} + \left(\frac{wx}{H}\right)^4\frac{1}{4!} + ... - 1\right]$$

If the fourth- and higher-order terms are neglected,

$$y = \frac{H}{w}\cdot\frac{w^2x^2}{2H^2} \text{ (approx.)}; \; y = \frac{wx^2}{2H} \qquad\qquad ...(6.4.14)$$

This is an equation of a parabola.

At B, $x = \frac{l}{2}$; $\quad y = \delta$

$$\therefore \qquad \delta = \frac{w}{2H}\cdot\left(\frac{1}{2}\right)^2 = \frac{wl^2}{8H} \qquad\qquad\qquad ...(6.4.15)$$

This relation has been deduced earlier.

Length of conductor : By expanding Eq. (6.4.13), we get

$$L = \frac{2H}{w}\left[\frac{wl}{2H} + \left(\frac{wl}{2H}\right)^3\cdot\frac{1}{3!} + ...\right]$$

If fifth- and higher-order terms are neglected,

$$L = \frac{2H}{w}\left(\frac{wl}{2H} + \frac{w^3l^3}{8H^3}\cdot\frac{1}{6}\right) \text{ (approx.)}$$

$$L = l + \frac{w^2 l^3}{24H^2} \qquad \qquad ...(6.4.16)$$

But from Eq. (6.4.15)

$$H = \frac{wl^2}{8\delta} \; ; \; L = l + \frac{w^2 l^3}{24\left(\dfrac{wl^2}{8\delta}\right)^2}$$

$$L = l + \frac{8\delta^2}{3l} \qquad \qquad ...(6.4.17)$$

Equation (6.4.17) gives a relation between the conductor length, span and sag.

Maximum tension :   If Eq. (6.4.12) is expanded, we obtain

$$T_{max} = H\left[1 + \left(\frac{wl}{2H}\right)^2 \cdot \frac{1}{2!} + ...\right]$$

If third- and higher-order terms are neglected,

$$T_{max} = H + \frac{w^2 l^2}{8H} = H + w\delta \qquad \qquad ...(6.4.18)$$

Equation (6.4.18) gives a relation between the maximum tension T_{max} and minimum tension H. If sag is small, T_{max} is approximately equal to H, i.e., the tension remains constant throughout the length of the conductor if it is assumed to hang in the form of a parabola. The constant value of the tension is approximately equal to the tension H at the lowest point of the conductor. Unless otherwise indicated, the word tension denotes the horizontal component of tension in a conductor.

6.5 ACCURACY OF RESULTS

It should be noted that extreme accuracy for calculating sags may not be desirable if the following factors are taken into account :

  (a) arbitrary values of wind and ice loading selected in the design;
  (b) inaccuracies in the preparation and plotting of profile surveys;
  (c) intermediate effect of the longitudinal flexibility of supports on the sags in individual spans of a complete line;
  (d) variations in elastic properties of the conductors from those assumed in the calculations;
  (e) arbitrary values of erection temperatures;
  (f) inaccuracies in the adjustment of erection sags in the field.

6.6 LOADING ON CONDUCTORS

The following forces act on a conductor : (a) conductor weight, (b) ice loading, and (c) wind loading.

  Conductor weight   The weight of conductor acts vertically downwards and depends upon the type of conductor used. The weight of conductor per unit length is available from the table giving the mechanical characteristics of the conductors.

Ice loading   In areas having very cold winters and snowfall, ice deposits on conductors. The amount of deposition depends upon the severity of weather conditions. The weight per unit length of conductor increases by ice deposition. The wind blowing across the line exerts more pressure due to increased surface area. When the ice load is suddenly released there is a likelihood of flashover between the conductors due to jumping. The actual configuration of a conductor with ice coating is non-uniform. For the purpose of calculation, it is assumed that the coating of ice is of uniform thickness. Fig. 6.4 shows a conductor with ice coating.

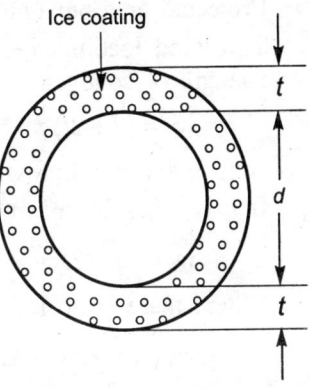

Fig. 6.4. Ice coated conductor.

Let d = diameter of conductor without ice coating in metres

$\quad\quad t$ = radial thickness of ice in metres.

Area of cross-section of the conductor = $\dfrac{\pi}{4} d^2$ m^2

Area of cross-section of the conductor when covered with ice = $\dfrac{\pi}{4} (d + 2t)^2$ m^2

Area of cross-section of the ice covering = $\dfrac{\pi}{4} (d + 2t)^2 - \dfrac{\pi}{4} d^2 = \pi t (d + t)$ m^2

Volume of ice per unit length of the conductor = $\pi t (d + t) \times 1 = \pi t (d + t)$ m^3

If ρ_i is the weight per unit volume or weight density of ice, the weight unit length of ice

$$w_i = \pi t (d + t) \rho_i \quad\quad\quad\quad\quad ...(6.6.1)$$

The weight per unit volume ρ_i is assumed to be 913.5 kgf per cubic metre or 8961 N/m^3 to calculate the ice loading.

$$w_i = \pi t (d + t) \times 913.5 \text{ kgf/m} \quad\quad\quad ...(6.6.2)$$

If d and t are given in mm

$$w_i = \pi t (d + t) \times 913.5 \times 10^{-6}$$

$$= 2.871 \, t (d + t) \times 10^{-3} \text{ kgf/m} \quad\quad\quad ...(6.6.3)$$

Wind loading   Wind exerts horizontal pressure on the exposed surface. The pressure of the wind is dependent upon its velocity. Wind velocities vary with the location and weather of a particular place. Empirical formulae connecting pressure and velocity of wind are used. Weather charts indicate ice and wind loading for various regions in any country. These loading conditions vary widely. In general, the loadings may be divided into three categories namely, *light loading*, *medium loading* and *heavy loading*. In India, there is no ice loading on the lines in most places. The classification is, therefore, based upon wind loading only.

Light loading = 75 kgf/m^2 = 736 N/m^2

Medium loading = 100 kgf/m^2 = 981 N/m^2

Heavy loading = 150 kgf/m^2 = 1472 N/m^2

Wind produces a transverse loading by exerting a pressure upon the projected area of conductor.

Projected area per unit length of ice covered conductor $= (d + 2t)$

The wind loading, i.e., the force exerted due to wind normal to the direction of span per unit length of conductor

$$w_h = (d + 2t) \, p \qquad \qquad \qquad \text{...(6.6.4)}$$

where p is the wind pressure per unit area acting in a direction normal to the direction of span.

If p is given in kgf/m^2 of projected area and d and t are in mm,

$$w_h = (d + 2t) \, p \times 10^{-3} \text{ kgf/m} \qquad \qquad \text{...(6.6.5)}$$

Where there is no ice

$$w_h = dp \times 10^{-3} \text{ kgf/m} \qquad \qquad \qquad \text{...(6.6.6)}$$

Usually, the wind pressure may be assumed to act on two-third of the projected area of the conductor. This is due to the fact that the effect of wind on flat surface of the same dimension will be more in comparison with the circular surface as in latter case the wind is normal to every point exposed to wind.

Effective loading The loads acting on the conductor are :

(i) Its own weight per unit length acting vertically downwards. It is represented by w_c.

(ii) The weight of ice per unit length acting vertically downwards. It is represented by w_i.

(iii) The wind loading per unit length acting horizontally, perpendicular to the line. It is represented by w_h.

These loads are represented as shown in Fig. 6.5.

The total vertical load $= (w_c + w_i)$

If w_r = effective or resultant loading per unit length of conductor,

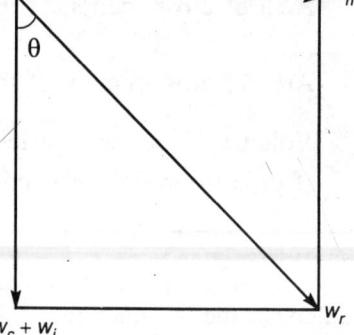

Fig. 6.5. Loads acting on the conductor.

then $\qquad w_r = [(w_c + w_i)^2 + w_h^2]^{1/2} \qquad \qquad \text{...(6.6.7)}$

The combined effect of ice and wind loading is to deflect the conductor at an angle θ to the vertical, but the shape of the conductor will remain the same. The sag so obtained will be the *deflected sag*.

The angle of deflection, θ, of the conductor from the vertical is given by

$$\tan \theta = \frac{w_h}{w_c + w_i} \qquad \qquad \qquad \text{...(6.6.8)}$$

The value of deflected sag is given by

$$\delta = \frac{w_r \, l^2}{8H} \qquad \qquad \qquad \text{...(6.6.9)}$$

The vertical component of the deflected sag δ, is due to vertical load only.

$$\delta_v = \delta \cos \theta = \frac{w_r \, l^2}{8H} \cdot \frac{(w_c + w_i)}{w_r}$$

or $\qquad \delta_v = \dfrac{(w_c + w_i) \, l^2}{8H}$ $\qquad\qquad\qquad\qquad\qquad\qquad$ ...(6.6.10)

Unless stated otherwise, the deflected sag should be calculated.

Factor of safety For working out the sag, the conductor is assumed to have a factor of safety (FOS). It is defined as the ratio of the ultimate tensile strength and working tension. The value of FOS is assumed on the basis of weather and loading conditions.

6.7 CONDUCTOR CLEARANCE FROM GROUND

An adequate clearance of a conductor from the ground under all loading conditions is to be maintained for safety reasons. The clearance distance depends upon the transmission voltage. The approximate value of conductor clearance from ground is given by

$\qquad$ GC = 6 m + 0.01 m per kV

The statutory regulations of a country govern the choice of minimum ground clearances depending upon the location and weather conditions. Indian Electricity (Supply) Act lays down a clearance of 17 feet (5.18 m) to be provided for 33 kV line and for every additional 33 kV or part thereof, additional 1 foot (0.3048 m) clearance is to be provided.

For example, for a 132 kV line

since $\qquad 132 = 33 + 3 \times 33$

$\therefore \qquad$ ground clearance $= 5.18 + 3 \times 0.3048 = 6.1$ m

For a 400 kV line, since $400 = 33 + 33 \times 11.12$

$\qquad$ ground clearance $= 5.18 + (0.3048) \times 11.12 = 8.57$ m

Example 6.1 Calculate the sag for a span of 200 m if the ultimate tensile strength of conductor is 5758 kgf. The weight of conductor is 604 kgf/km. Allow a factor of safety of 2.

Solution

$\qquad$ Maximum working tension, $H = \dfrac{\text{ultimate tensile strength}}{\text{factor of safety}} = \dfrac{5758}{2} = 2879$ kgf

Span $l = 200$ m

Weight of conductor, $w_c = 604$ kgf/km $= 0.604$ kgf/m

Sag is given by

$$\delta = \frac{w_c \, l^2}{8H} = \frac{0.604 \times (200)^2}{8 \times 2879} = 1.048 \text{ m.}$$

Example 6.2 A transmission line has a span of 275 metres between level supports. The conductor has a diameter of 19.53 mm, weighs 0.844 kgf/m and has an ultimate breaking strength of 7950 kgf. Each conductor has a radial covering of ice 9.53 mm thick and is subjected to a horizontal wind pressure of 40 kgf/m^2 of the ice covered projected area. If the factor of safety is 2, calculate the deflected sag and the vertical component of the sag. One cubic metre of ice weighs 913.5 kgf.

Solution

Weight of conductor, $w_c = 0.844$ kgf/m

Weight of ice per metre length, $w_i = \pi t (d + t) \times 913.5 \times 10^{-6}$

$$= \pi \times 9.53 (19.53 + 9.53) \times 913.5 \times 10^{-6} = 0.795 \text{ kgf}$$

Wind loading, $w_h = (d + 2t) p \times 10^{-3} = (19.53 + 19.06) \times 40 \times 10^{-3} = 1.544$ kgf/m

Resultant loading, $w_r = [(w_c + w_i)^2 + w_h^2]^{1/2} = [(0.844 + 0.795)^2 + (1.544)^2]^{1/2} = 2.252$ kgf/m

Maximum working tension, $H = \dfrac{\text{breaking load}}{\text{factor of safety}} = \dfrac{7950}{2} = 3975$ kgf

Length of span, $l = 275$ m

Deflected sag $= \dfrac{w_r l^2}{8H} = \dfrac{2.252 \times (275)^2}{8 \times 3975} = 5.36$ m

Vertical component of sag $= \dfrac{(w_c + w_i) l^2}{8H} = \dfrac{(0.844 + 0.795) \times (275)^2}{8 \times 3975} = 3.9$ m

6.8 ERECTION SAG AND TENSION

Sag and tension are dependent upon, and determine, each other. They are subject to variations due to change of temperature and loading. To ensure that the limits of maximum sag and tension are not exceeded it is necessary to calculate their values at stringing temperature also since the erection conditions are different from the most severe conditions of ice and wind. At the time of erection there are no ice or wind loads.

If suffix 1 is also used to refer to initial conditions of maximum loading and suffix 2 for conditions at erection, increase in length of the conductor for a rise of temperature from $t_1°$C to $t_2°$C

$$= \alpha (t_2 - t_1) L = \alpha tl \text{ (approx.)}$$

where $\alpha =$ coefficient of linear thermal expansion per degree C

$t =$ rise in temperature $= (t_2 - t_1)$

Since the load is decreased from w_1 to w_2, there will be a decrease in length due to reduced strain. By Hooke's law the decrease in length due to decrease in tension from T_1 to T_2

$$= \dfrac{(T_1 - T_2)}{EA} L = \dfrac{(T_1 - T_2)}{EA} l \text{ (approx.)}$$

where E is the modulus of elasticity, and A the area of cross-section of conductor.

The final length of conductor at erection temperature $t_2°$C = initial length of conductor at $t_1°$C + increase in length due to temperature rise of $t°$C – decrease in length due to decrease in tension

$$= L + \alpha tl - \dfrac{(T_1 - T_2)}{EA} l = \left(l + \dfrac{w_r^2 l^3}{24 T_1^2}\right) + \alpha tl - \dfrac{(T_1 - T_2)}{EA} l$$

The length of the conductor under new loading conditions at erection may be expressed as :

$$l + \frac{w_2^2 \, l^3}{24 T_2^2}$$

Equating these values of length of conductor

$$l + \frac{w_1^2 \, l^3}{24 T_1^2} + \alpha t l - \frac{T_1 - T_2}{EA} \, l = l + \frac{w_2^2 \, l^3}{24 T_2^2}$$

which, on simplification, gives

$$T_2^2 \left[T_2 - \left(T_1 - \frac{w_1^2 \, l^2 \, EA}{24 T_1^2} - \alpha \, tEA \right) \right] = \frac{w_2^2 \, l^2 \, EA}{24} \qquad \qquad ...(6.8.1)$$

The tension T_2 in still air and at temperature t_2 is calculated from the above cubic equation. To find the sag at erection temperature the value of T_2 is substituted in the relation

$$\delta = \frac{w_2 \, l^2}{8 T_2}$$

Equation (6.8.1) is of basic importance. It may be called the *behaviour equation* of the line.

6.8.1   Factors Affecting the Sag

The following factors affect the sag of a conductor :

1. *Conductor weight per unit length :*   The sag of a conductor is directly proportional to its weight per unit length. The weight of the conductor is increased due to ice loading.

2. *Span :*   The sag is directly proportional to the square of the span length. Longer span gives more sag.

3. *Tension :*   The sag is inversely proportional to the tension in the conductor. Higher tension reduces the sag but increases the stress in the insulators and supporting structures.

4. *Wind :*   Wind load increases sag in the inclined direction.

5. *Temperature :*   The sag is reduced at lower temperatures and is increased at higher temperatures.

Example 6.3 The following particulars refer to a 2-Camel ACSR conductor used on a 400 kV line :

Size of conductor 54/3.35 mm Al, 7/3.35 mm steel

Ultimate tensile strength = 14740 kgf

Area of cross section of conductor = 538.4 mm^2

Modulus of elasticity = 7000 kgf/mm^2

Weight of conductor = 1.805 kgf/m

Wind pressure = 100 kgf/m^2

Span = 335 m

Coefficient of linear expansion = 19.3×10^{-6}/°C.

The wind load on the conductor may be assumed to be acting on two-thirds of the exposed area. Ice loading is neglected. Determine the tension and sag

(a) at 0°C and full wind, factory safety = 2

(b) at 60°C with initial temperature of 0°C, no wind, factor of safety = 2

(c) at 60°C with initial temperature of 30°C, no wind, factor of safety = 4

Assume parabolic configuration.

Solution

Number of strands in the conductor = $54 + 7 = 61$

$\therefore \qquad 61 = 1 + 3n\,(1 + n)\,; \qquad n^2 + n - 20 = 0\,; \qquad (n + 5)\,(n - 4) = 0$

Thus $\qquad n = 4$

Overall diameter of the conductor, $d = (1 + 2n)\,d_s = (1 + 2 \times 4) \times 3.35 = 30.15$ mm

$$= 0.03015 \text{ m}$$

Wind load per metre, $w_h = \dfrac{2}{3} \times 1 \times$ wind load per unit area

$$= \frac{2}{3}\,dp = \frac{2}{3} \times 0.03015 \times 100 = 2.01 \text{ kgf}$$

(a) Sag at 0°C, full wind, and factor of safety equal to 2.

Effective weight of conductor, $w_r = [w_c^2 + w_h^2]^{1/2}$

$$= [(1.805)^2 + (2.01)^2]^{1/2} = \sqrt{7.31} = 2.70$$

Maximum working tension, $T = \dfrac{\text{breaking load}}{\text{factor of safety}} = \dfrac{14740}{2} = 7370$ kgf

Sag at 0°C and full wind, $\delta = \dfrac{w_r\,l^2}{8T} = \dfrac{2.7 \times (335)^2}{8 \times 7370} = 5.14$ m

(b) Sag and tension at 60°C with initial condition $t_0 = 0°C$, no wind (still wind), and factor of safety equal to 2.

Initial working tension, $T_1 = \dfrac{\text{breaking load}}{\text{factor of safety}} = \dfrac{14740}{2} = 7370$ kgf

If w_1 is the effective loading per metre length of conductor

$$w_1^2 = w_c^2 + w_h^2 = (1.805)^2 + (2.01)^2 = 3.26 + 4.04 = 7.30$$

Let $\quad T_2 =$ tension in conductor at 60°C

$w_2 =$ effective loading per metre length of conductor at 60°C and no wind

$$T_2^2\left[T_2 - \left(T_1 - \frac{w_1^2\,l^2\,EA}{24T_1^2}\right) + \alpha t EA\right] = \frac{w_2^2\,l^2\,EA}{24}$$

$$T_2^2\,(T_2 - K + \alpha t EA) = \frac{w_2^2\,l^2\,EA}{24}$$

where $\quad K = T_1 - \dfrac{w_1^2\,l^2\,EA}{24T_1^2}$

$$= 7370 - \frac{7.30 \times (335)^2 \times 7000 \times 538.4}{24 \times (7370)^2} = 7370 - 2368 = 5002$$

$$\alpha t EA = 19.3 \times 10^{-6} \times (60 - 0) \times 7000 \times 538.4 = 4364$$

When wind is still, $w_h = 0$; $w_2^2 = w_c^2 + w_h^2$; $w_2 = w_c = 1.805$

$$\frac{w_2^2 \, l^2 \, EA}{24} = \frac{1}{24} \times (1.805)^2 \times (335)^2 \times 7000 \times 538.4 = 57.4 \times 10^9$$

Again $T_2^2 (T_2 - K + \alpha t EA) = \dfrac{w_2^2 \, l^2 \, EA}{24}$

$$T_2^2 (T_2 - 5002 + 4364) = 57.4 \times 10^9$$

$$T_2^2 (T_2 - 638) = 57.4 \times 10^9$$

By trial $T_2 = 4083$ kgf

Sag at 60°

$$\delta_{60°} = \frac{w_c \, l^2}{8 T_2} = \frac{(1.805) \times (335)^2}{8 \times 4083} = 6.20 \text{ m}$$

(c) Sag and tension with initial temperature of 30°C, with no wind, final temperature of 60°C, no wind and factor of safety equal to 4

Here $w_h = 0$, $w_2 = w_c$

Initial working tension, $T_1 = \dfrac{14740}{4} = 3685$ kgf

$$K = T_1 - \frac{w_c^2 \, l^2 \, EA}{24 T_1^2} = 3685 - \frac{(1.805)^2 \times (335)^2 \times 7000 \times 538.4}{24 \times (3685)^2} = 3685 - 4228 = -543$$

At $t_1 = 60°C$ and $t_0 = 30°C$

$$t = 60 - 30 = 30°C$$

$$\alpha t EA = 19.3 \times 10^{-6} \times 30 \times 7000 \times 538.4 = 2182$$

$$\frac{w_2^2 \, l^2 \, EA}{24} = \frac{1}{24} \times (1.805)^2 \times (335)^2 \times 7000 \times 538.4 = 57.4 \times 10^9$$

$$T_2^2 (T_2 - K + \alpha t EA) = \frac{w_2^2 \, l^2 \, EA}{24}$$

$$T_2^2 (T_2 + 543 + 2182) = 57.4 \times 10^9$$

$$T_2^2 (T_2 + 2725) = 57.4 \times 10^9$$

By trial $T_2 = 3132$ kgf

Sag at 60°C and no wind

$$\delta_{60°} = \frac{w_c \, l^2}{8 T_2} = \frac{(1.805) \times (335)^2}{8 \times 3132} = 8.08 \text{ m}$$

Similar calculations can be made for other temperatures and loading conditions.

Example 6.4 An overhead line has a span of 160 m of stranded copper conductor between level supports. The sag is 3.96 m at –5.5°C with 9.53 mm thick ice coating and wind pressure of 40 kgf/m^2 of projected area. Calculate the temperature at which the sag will remain the same under conditions of no ice and no wind. The particulars of the conductor are as follows :

Size of conductor = 7/3.45 mm

Area of cross-section = 64.5 mm^2

Weight of conductor = 0.594 kgf/m

Modulus of elasticity = 12700 kgf/mm^2

Coefficient of linear expansion = 1.7×10^{-5}/°C.

Assume 1 m^3 of ice to weigh 913.5 kgf.

Solution

Diameter of conductor, $d = 3 \times 3.45 = 10.35$ mm

Weight of conductor, $w_c = 0.594$ kgf/m

Weight of ice, $w_i = \pi t (d + t) \times 913.5 \times 10^{-6} = \pi \times 9.53 (10.35 + 9.53) \times 913.5 \times 10^{-6}$

$$= 0.544 \text{ kgf/m}$$

Wind loading on ice covered conductor, $w_h = (d + 2t) p \times 10^{-3}$

$$= (10.35 + 2 \times 9.53) \times 40 \times 10^{-3} = 1.176 \text{ kgf/m}$$

Resultant loading at – 5.5°C, $w_r = [(w_c + w_i)^2 + w_h^2]^{1/2}$

$$= [(0.594 + 0.544)^2 (1.176)^2]^{1/2} = (1.295 + 1.383)^{1/2}$$

$$= (2.678)^{1/2} = 1.636 \text{ kgf/m}$$

If suffix 1 is used to refer to initial conditions and suffix 2 for conditions at erection

$$w_1 = w_r = 1.636 \text{ kgf/m}$$

Sag at initial temperature of – 5.5°C

$$\delta_1 = \frac{w_1 l^2}{8T_1} ; \qquad 3.96 = \frac{1.636 \times (160)^2}{8T_1}$$

$$T_1 = \frac{1.636 \times (160)^2}{8 \times 3.96} = 1322 \text{ kgf}$$

With no wind and no ice, effective loading at erection temperature t_2°C is

$$w_2 = w_c = 0.594 \text{ kgf/m}$$

Sag at erection temperature t_2°C

$$\delta_2 = \frac{w_2 l^2}{8T_2} = \frac{w_c l^2}{8T_2}$$

But $\delta_2 = \delta_1$; $\qquad \dfrac{w_1 l^2}{8T_1} = \dfrac{w_c l^2}{8T_2}$

$$T_2 = \frac{w_c}{w_1} T_1 = \frac{0.594}{1.636} \times 1322 = 480 \text{ kgf.}$$

Again $\dfrac{w_1}{T_1} = \dfrac{w_2}{T_2}$; $\dfrac{w_1^2}{T_1^2} = \dfrac{w_2^2}{T_2^2}$

$$\frac{w_1^2}{T_1^2} \cdot \frac{l^2 \, EA}{24} = \frac{w_2^2}{T_2^2} \cdot \frac{l^2 \, EA}{24}$$

The *behaviour equation* of the line may be written as :

$$T_2 - T_1 + \frac{w_1^2 \, l^2 \, EA}{24 T_1^2} + \alpha t EA = \frac{w_1^2 \, l^2 \, EA}{24 T_2^2}$$

$$\alpha t EA = T_1 - T_2 \; ; \qquad t = t_2 - t_1 = \frac{T_1 - T_2}{\alpha EA}$$

$$t_2 - (-5.5) = \frac{1322 - 480}{1.7 \times 10^{-5} \times 12700 \times 64.5}$$

$$t_2 = 60.46 - 5.5 = 54.96°\text{C}.$$

6.9 SPANS OF UNEQUAL LENGTH

The horizontal distance between two adjacent supports is called the *span*. The nature of the route decides the type of construction. In hilly areas, or sloping grounds, the supports will not be at the same level. Maximum benefit is taken of the elevated spots to erect the supports there. To achieve all this the spans between the tensioning points or dead ends are to be selected of unequal lengths and it becomes necessary to dead end some of the longer spans. Theoretically, the calculation of sags and tensions are to be made separately for individual spans, and during stringing of conductor, it is to be kept in mind that these values are not exceeded. Moreover, with the variation of temperature and mechanical loading the values of sags and tensions will alter. With suspension insulator strings this is not practicable since the insulator swinging allows the tension to be balanced in the adjacent spans. If the conductors are bound into pin-or post-type insulators the unbalanced longitudinal tension would be taken by the binders. This is not desirable.

For practical reasons, the sag and the tension-calculations are, therefore, made on the basis of a hypothetical span that will result in a constant tension throughout the whole section between tensioning points under any given conditions of temperature and loading. This hypothetical span is called the *equivalent span*, *basic span*, or *ruling span*. Sag and tension in individual spans of a dead ended section are then calculated.

Let, there be n spans of lengths l_1, l_2, l_3, ..., l_n in a section of line between dead ends or tensioning points and l_e the equivalent span of the section, between dead ends or tensioning points. The total length of the section, $(l_1 + l_2 + l_3 + ... + l_n)$ may be treated to be equal to the length of n spans each equal to l_e.

$$n \, l_e = l_1 + l_2 + l_3 + ... + l_n = \sum l$$

The total length of conductor under loading conditions of w and the tension T_0 is given by

$$\sum l + \frac{w^2 \sum l^3}{24 T_0^2}$$

Also, the length of conductor in n spans each of length l_e, under same condition of loading, is given by

$$n\left(l_e + \frac{w^2 l_e^3}{24 T_0^2}\right)$$

But these are the lengths of conductor in a section between two dead points. Therefore, they are equal, i.e.,

$$\sum l + \frac{w^2 \sum l^3}{24 T_0^2} = n\left(l_e + \frac{w^2 l_e^3}{24 T_0^2}\right); \qquad \sum l^3 = n l_e^3$$

$$l_e^2 = \frac{\sum l^3}{n l_e} = \frac{\sum l^3}{\sum l}$$

$$l_e = \left(\frac{\sum l^3}{\sum l}\right)^{1/2} = \left(\frac{l_1^3 + l_2^3 + l_3^3 + \ldots + l_n^3}{l_1 + l_2 + l_3 + \ldots l_n}\right)^{1/2} \qquad \ldots (6.9.1)$$

Thus, the basic or equivalent span is equal to the square root of the sum of the cubes of the individual span lengths divided by the length of the section considered.

The basic span is used to calculate the tension under any conditions of temperature and loading. These tensions remain constant throughout the section of the line irrespective of the length of any span. The uniform tension so obtained may be utilized to calculate the sag in any individual span by using the formula

$$\delta = \frac{w l^2}{8T} .$$

6.10 SAG AND TENSION CHARTS

Calculations can be made to determine the sags at different temperatures. The minimum and maximum temperatures of the proposed route are known from the records of the previous years. The sags are determined in this range of temperatures. Sag versus temperature and tension versus temperature curves are plotted. These curves are called *erection-sag curves* or *stringing charts* for the given conductor and loading conditions. Such a chart is shown in Fig. 6.6. These charts find their utility while stringing conductors so that they are adjusted to proper sag and tension and that the line will not break at any temperature to which it will be exposed in service.

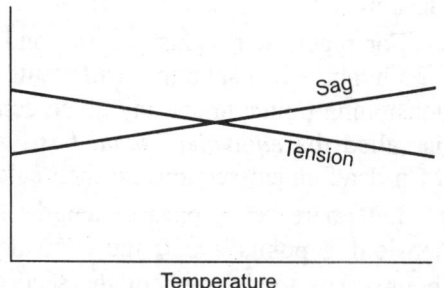

Fig. 6.6. Stringing chart.

Sag tension charts are usually supplied by the conductor manufacturers. These charts give the sags for ruling spans. To calculate the sag for any other span at any temperature the following formula is used :

$$\delta_2 = \delta_1 \left(\frac{l_2}{l_1} \right)^2$$

where δ_2 = sag for the required span and temperature

δ_1 = sag for the ruling span at the given temperature

l_2 = required span

l_1 = ruling span.

The above formula has been deduced on the basis of a constant tension at a given temperature throughout the section between dead ends. This uniform tension at a given temperature produces a definite sag on a definite basic span.

Table 6.1 gives the sags at 60°C for a 2-Camel ASCR conductor 538.4 mm^2 (54/3.35 Al + 7/3.35 steel) with a basic span of 335 m. The other data are as follows :

Constant erection tension = 3880 kgf

Sag at 60°C 335 m span = 6.20 m

Table 6.1

Span length in m	Sag in m
250	3.45
275	4.18
300	4.97
335	6.20
350	6.76
375	7.77
400	8.84
425	9.98
450	11.19

The sags in Table 6.1 have been calculated with the help of the formula

$$\delta_2 = \delta_1 \left(\frac{l_2}{l_1} \right)^2$$

For example, $\delta_{300} = \delta_{335} \left(\dfrac{300}{335} \right)^2 = 6.2 \left(\dfrac{300}{335} \right)^2 = 4.97$ m

6.11 SUPPORTS AT UNEQUAL LEVELS

In hilly areas or sloping grounds the supports are not usually at the same level.

Let h = difference in level between two supports A and B

l = horizontal span length between A and B

x = horizontal distance of A from the lowest point O

$l - x$ = horizontal distance of B from the lowest point O.

The conductor AOB is a portion of the catenary B'OB. The portions OA and OB may be treated as catenaries of half spans x and $(l-x)$ and sags δ_2 and δ_1 respectively. Then from Eq. (6.4.8)

$$\delta_1 = \frac{H}{w}\left[\cosh\frac{w(l-x)}{H} - 1\right] \qquad \ldots(6.11.1)$$

$$\delta_2 = \frac{H}{w}\left[\cosh\left(\frac{wx}{H}\right) - 1\right] \qquad \ldots(6.11.2)$$

The tension at A and B are given by Eq. (6.4.10)

$$T_A = H\cosh\frac{wx}{H} \qquad \ldots(6.11.3)$$

$$T_B = H\cosh\frac{w(l-x)}{H} \qquad \ldots(6.11.4)$$

The maximum tension would occur at B, since $(l-x)$ is greater than x as seen from the Fig. 6.7.

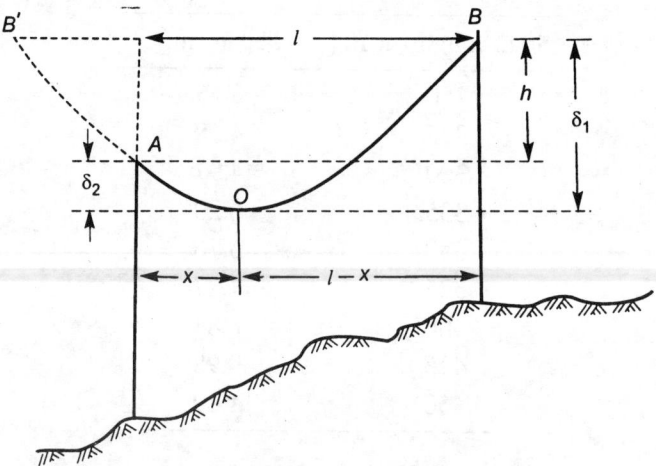

Fig. 6.7. Supports at unequal levels.

Vertical reaction at higher support $= w(l-x)$

Vertical reaction at lower support $= wx$

For approximation, AOB may be assumed parabolic. The calculations in this case become easier and are shown below :

$$\delta_1 = \frac{w[2(l-x)]^2}{8T} = \frac{w(l-x)^2}{2T} \qquad \ldots(6.11.5)$$

$$\delta_2 = \frac{w(2x)^2}{8T} = \frac{wx^2}{2T} \qquad \ldots(6.11.6)$$

But $\quad h = \delta_1 - \delta_2 = \dfrac{w(l-x)^2}{2T} - \dfrac{wx^2}{2T} = \dfrac{wl}{2T}(l-2x)$

$$\frac{2Th}{wl} = l - 2x$$

$$x = \frac{l}{2} - \frac{Th}{wl} \qquad \qquad ...(6.11.7)$$

and $\quad l - x = \frac{l}{2} + \frac{Th}{wl} \qquad \qquad ...(6.11.8)$

The value of x obtained above may be substituted in Eq. (6.11.5) to calculate δ_1. From Eq. (6.11.7), it is clear that if $\frac{l}{2} = \frac{Th}{wl}$, the lowest point O of the curve coincides with the lower support A [Fig. 6.8 (a)].

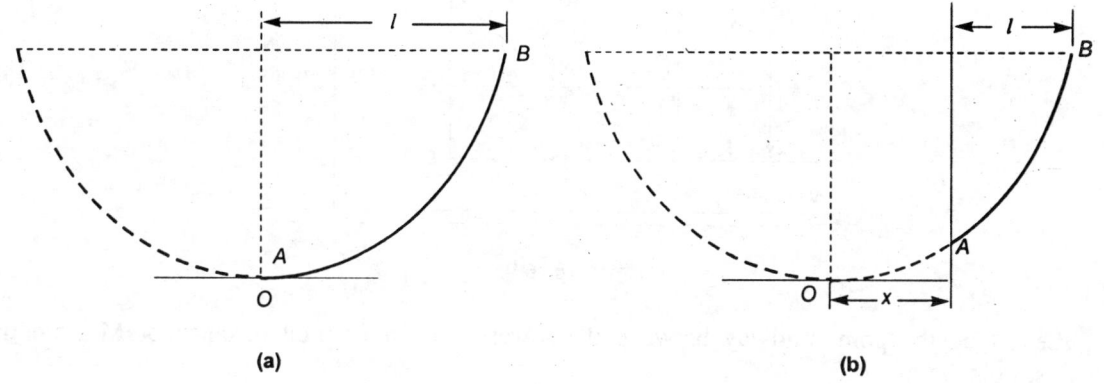

(a)   **(b)**

Fig. 6.8.

If $\frac{l}{2} < \frac{Th}{wl}$, x is negative, i.e., A is on right hand side of O. In this case, there will be a resultant upward pull on the lower support. The insulator at A will be subjected to an upward pull since the upward pull there is greater than the downward pull due to the weight of the wires in the next adjacent span. The amount of upward pull is equal to the weight of the conductor from A to 0.

It is obvious from Fig. 6.8 (b) that the lowest point lies outside the span AB. In this case the conductor in the span under consideration exerts an upward pull on the lower support. If the upward pull be greater than the downward load of the adjoining span uplift would be caused and the conductor tends to swing clear of the lower support.

Example 6.5 A transmission line conductor at a river crossing is supported from two towers at heights of 30 m and 90 m above water level. The horizontal distance between the towers is 270 m. If the tension in the conductor is 1800 kgf and the conductor weights 1.0 kgf per meter, find the clearance between the conductor and the water at a point midway between the towers. Assume parabolic configuration.

Solution

$l = 270$ m ; $\qquad T = 1800$ kgf ; $\qquad w = 1.0$ kgf ; $\qquad h = 90 - 30 = 60$ m

$$x = \frac{l}{2} - \frac{Th}{wl} = \frac{270}{2} - \frac{1800 \times 60}{1 \times 270} = -265 \text{ m}$$

The negative value of x shows that A and B (Fig. 6.9) are on the same side of the lowest point O of the conductor AB.

$$(l - x) = 270 - (-265) = 535 \text{ m}$$

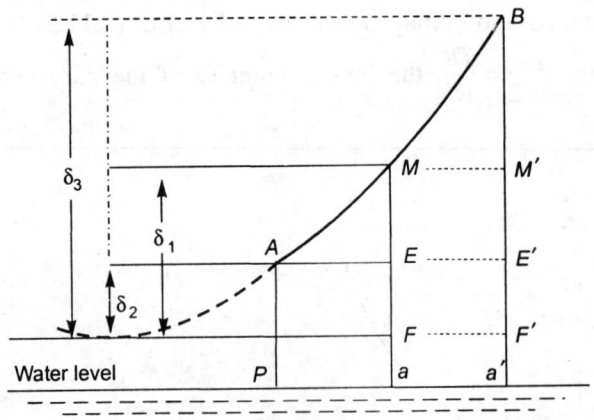

Fig. 6.9.

Let M be the point midway between the towers. The horizontal distance of M from the lowest point is given by

$$x_1 = 265 + \frac{270}{2} = 400 \text{ m}$$

Height of M above O, $\delta_1 = \dfrac{wx_1^2}{2T} = \dfrac{1.0 \times (400)^2}{2 \times 1800} = 44.44 \text{ m}$

Height of A above O, $\delta_2 = \dfrac{wx^2}{2T} = \dfrac{1 \times (265)^2}{2 \times 1800} = 19.05 \text{ m}$

Height of B above O, $= \dfrac{w(l-x)^2}{2T} = \dfrac{1 \times (535)^2}{2 \times 1800} = 79.50 \text{ m}$

Clearance between the conductor and water at point M

$$QM = QE + EM = QE + (FM - FE)$$
$$= AP + (\delta_1 - \delta_2) = 30 + (44.44 - 19.50) = 54.94 \text{ m}$$

Alternatively

$$QM = Q'M' = Q'B - M'B$$
$$= Q'B - (F'B - F'M') = Q'B - (\delta_3 - \delta_1) = 90 - (79.50 - 44.44) = 54.94 \text{ m}$$

Example 6.6 An overhead line over a hill side with a gradient of 1 in 20 is supported by two 30 m high towers. The horizontal distance between the towers is 300 m. The weight of each conductor is 1.492 kgf/m and the tension is 2200 kgf. The lowest conductor is fixed 6 m below the top of each tower. Find (a) the clearance of the lowest point of the conductor from the ground, and (b) the minimum ground clearance. Assume parabolic configuration.

Solution

The lowest conductor AOB and the profile of the ground CE are shown in Fig. 6.10. Take the origin at the lowest point O. Let OX and OY be the horizontal and vertical reference axes.

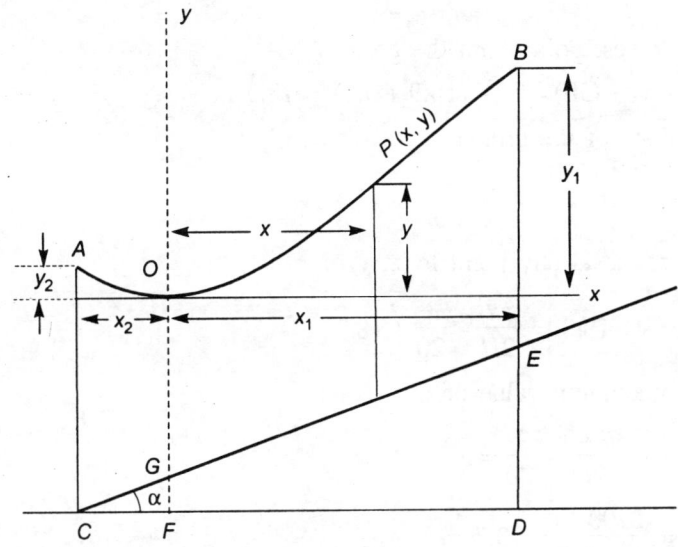

Fig. 6.10.

$$ED = CD \tan \alpha = 300 \times \frac{1}{20} = 15 \text{ m}$$

Height of the lowest conductor from the ground = AC = BE = 30 − 6 = 24 m

$$BD = BE + ED = 24 + 15 = 39 \text{ m}$$

The equation of the parabola AOB is given by

$$y = \frac{wx^2}{2H} \; ; \qquad y_1 = \frac{wx_1^2}{2H} \quad \text{and} \quad y_2 = \frac{wx_2^2}{2H}$$

Vertical distance between the supports

$$y_1 - y_2 = ED = 15 \text{ m}$$

Also, $$y_1 - y_2 = \frac{w}{2H} (x_1^2 - x_2^2) = \frac{w}{2H} (x_1 - x_2)(x_1 + x_2)$$

But $$w = 1.492 \text{ kgf/m}, \quad H = 2200 \text{ kgf}, \quad x_1 + x_2 = 300 \text{ m}$$

$$15 = \frac{1.492}{2 \times 2200} (x_1 - x_2) \times 300$$

$$x_1 - x_2 = \frac{15 \times 2 \times 2200}{1.492 \times 300} = 147.45$$

Again, $x_1 + x_2 = 300$

$$x_1 = \frac{1}{2}(147.45 + 300) = 223.7 \text{ m}$$

$$x_2 = 300 - 223.7 = 76.3 \text{ m} \qquad y_2 = 16.9 - 15 = 1.9 \text{ m}$$

$$GF = CF \tan \alpha = x_2 \tan \alpha = 76.3 \times \frac{1}{20} = 3.815 \text{ m}$$

Clearance of the lowest point from the ground

$$OG = BD - (y_1 + GF) = 39 - (16.9 + 3.8) = 18.3 \text{ m}$$

Equation of the slope of the ground

$$y = mx + C = x \tan \alpha - OG = \frac{1}{20}x - 18.3$$

Clearance from ground at any point P (x, y)

$$C_1 = y - \left(\frac{1}{20}x - 18.3\right) = \frac{wx^2}{2H} - \frac{x}{20} + 18.3$$

For minimum or maximum value of C_1,

$$\frac{dC_1}{dx} = 0 ; \qquad \frac{2wx}{2H} - \frac{1}{20} = 0$$

$$x = \frac{H}{20w} = \frac{2200}{20 \times 1.492} = 73.7 \text{ m}$$

Since $\dfrac{d^2 C_1}{dx^2} > 0$, $x = 73.7$ m is a condition for minimum.

Minimum clearance, $C_1 = \dfrac{wx^2}{2H} - \dfrac{x}{20} + 18.3$

$$= \frac{1.492 \times (73.7)^2}{2 \times 2200} - \frac{73.7}{20} + 18.3 = 1.84 - 3.68 + 18.3 = 16.46 \text{ m}$$

6.12 THE SAG TEMPLATE

The use of a sag template is essential to allocate the position and height of the supports correctly on the profile. It is usually made of transparent celluloid, perspex, or some times card board. The following curves are marked on it:

1. Hot template curve or hot curve
2. Ground clearance curve
3. Support foot or tower curve
4. Cold curve or uplift curve

Hot Curve

Hot curve is obtained by plotting the sags at maximum temperature against span lengths. It shows where the supports must be located in order to maintain the prescribed ground clearance.

Ground Clearance Curve

The clearance curve is below the hot curve. It is drawn parallel to the hot curve and at a vertical distance equal to the ground clearance as prescribed by the regulations for the given line.

Support Foot Curve

This curve is drawn for locating the position of the supports for tower lines. It shows the height from the base of the standard support to the point of attachment of the lowest conductor. For wood or concrete pole lines this curve is not required to be drawn since they can be put on any convenient position.

Cold Curve or Uplift Curve

Uplift curve is obtained by plotting the sags at minimum temperature without wind or ice against span lengths. Uplift curve is drawn to check whether uplift of conductors occurs on any support. The uplift conductors may occur at low temperatures when one support is much lower than either of the adjoining ones.

6.13 PREPARATION OF THE SAG TEMPLATE

The above mentioned curves are first drawn on a squared paper on the same scale as the line profile. Suitable scales are selected. With the help of sharp-pointed probe, the curves are then transferred to transparent celluloid or perspex. The celluloid or perspex is then cut along the line of maximum, sag, i.e., the hot curve.

Fig. 6.11 illustrates a typical sag template. New sag templates are to be prepared for lines with different ruling span, voltage and loading conditions.

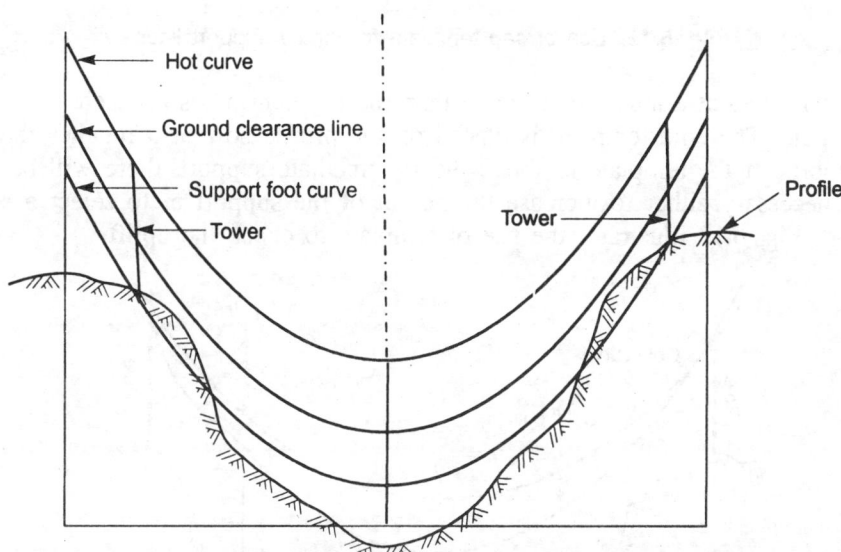

Fig. 6.11. Sag template.

6.14 METHOD OF USING THE TEMPLATE

The positions of the supports of normal height are marked on the profile plan. For single supports the template is then placed over the profile plane in such a way that the maximum sag line touches the tops of the single supports and the axis is kept vertical. The clearance line marked on the template should not cut any intermediate points between the supports on the profile in which case the necessary clearance is not maintained. In such a case the height of the support or supports should be increased. For tower line the support-foot curve gives the best position of a tower. The use of sag template for single support lines is shown in Fig. 6.12.

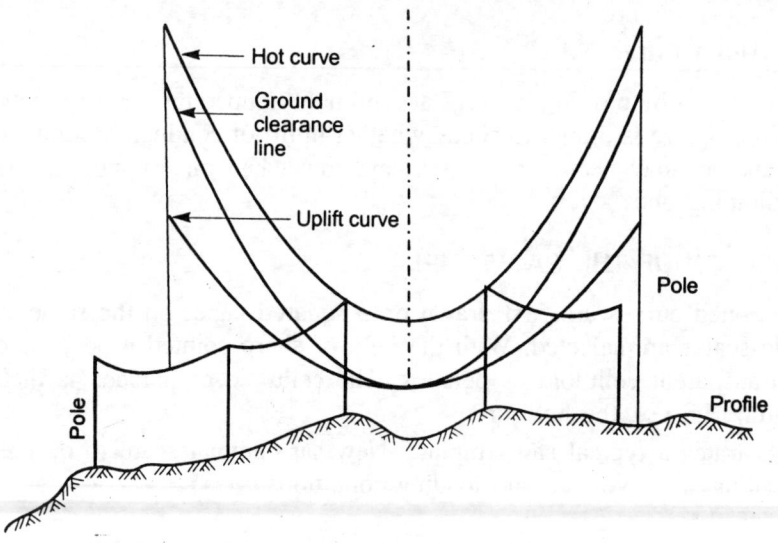

Fig. 6.12. Use of sag template for single support lines.

If a pole is located at a much lower level than the adjacent poles on each side there will be uplift on the pole. The cold template is placed on the profile so that it touches the tops of the alternate supports. If the template is above the intermediate support, there will be an uplift. It will then be necessary either to increase the height of the support or to select a new position of the support. Fig. 6.13 illustrates the use of template to check the uplift.

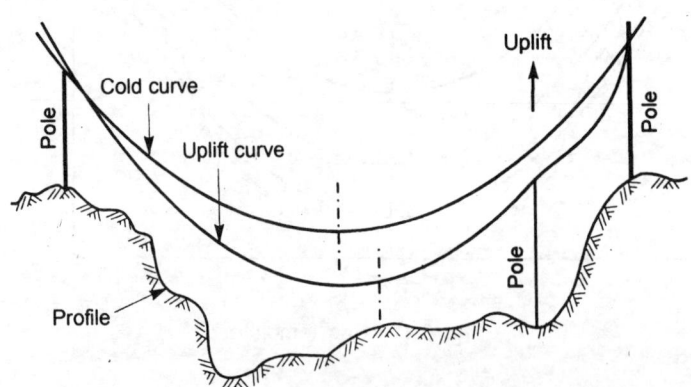

Fig. 6.13. Use of sag template to check the uplift.

6.15 ECONOMIC SPAN LENGTH

The following factors govern the choice of economic span length of transmission lines :

   (a) Size, material, and number of conductors;

   (b) Assumed wind and ice loading and temperature range;

   (c) Specified factors of safety;

   (d) Disposition of conductivity and earth wire and shielding angle;

   (e) Conductor clearance to ground;

   (f) Cost of erected hardware fittings and insulators;

   (g) Cost of erection of towers;

   (h) Cost of right of way requirements.

EXERCISES

1. An overhead transmission line has a span of 240 m between level supports. Calculate the maximum sag if the conductor weighs 727 kgf/km and has a breaking strength of 6880 kgf. Allow a factor of safety of 2. Neglect wind and ice loading.

2. A transmission line has a san of 180 m between level supports. The conductor has a cross-sectional area of 129 mm^2, weighs 1.17 kgf/m and has a breaking stress of 42 kgf/mm^2. Calculate the sag for a factor of safety of 5, allowing for a maximum wind pressure of 125 kgf/m^2 of projected surface.

3. Calculate the sag in an overhead conductor under the following conditions: length of span, 150 m, cross-sectional area of conductor 125 mm^2, breaking strength 42 kgf/mm^2, factor of safety 5, weight of conductor 0.859 kgf/m, maximum wind pressure, 150 kgf/m^2.

4. The line conductor of a transmission line has an overall diameter of 19.53 mm, weighs 0.844 kgf/m, and an ultimate breaking strength of 7950 kgf. If the factor of safety is to be 2 when the conductor has an ice load of 1 kgf/m and a horizontal wind pressure of 1.5 kgf/m, find approximately the vertical sag which corresponds to this loading for a 300 m span between level supports.

5. A transmission line with ACSR conductors has a span of 200 m between level supports. The conductor has a nominal copper area of 80 mm^2, weighs 0.604 kgf/m and has a breaking strength of 5758 kgf. Calculate the sag for a factor of safety of 4, allowing a maximum wind pressure of 125 kgf/m^2 of projected area. The overall diameter of conductor = 16.52 mm.

6. Calculate the vertical and deflected sag under the following conditions: length of span 458 m, cross-sectional area of conductor 538.4 mm^2, overall diameter of conductor 30.15 mm, weight of conductor per metre 1.805 kgf, ultimate tensile strength 14740 kgf, wind load 75 kgf/m^2 acting on two-thirds of exposed area. Factor of safety = 4.

7. The following particulars refer to an ACSR conductor: Nominal copper area 80 mm², overall diameter 15.84 mm, weight of conductor 493 kgf/km, ultimate breaking strength 4137 kgf. If the factor of safety is to be 2 when the conductor has an ice load of 0.792 kgf/m and a horizontal wind load of 0.917 kgf/m find the vertical sag which corresponds to this loading in a 200 m span between level supports.

8. An overhead line has a span of 160 m of copper conductor between level supports. The conductor diameter is 1.0 cm and has a breaking stress of 35 kgf/mm². Calculate (a) the deflected sag (b) the vertical sag. The line is subjected to a wind pressure of 40 kgf/m² of projected area and radial ice coating of 9.53 mm thickness. The weight of ice is 913.5 kgf/m³. Allow a factor of safety of 2 and take the density of copper as 8.9 g/cm³.

9. An overhead transmission line has a span of 180 m between level supports. Each conductor has a nominal copper area of 65 mm² and a diameter of 14.15 mm. The maximum tension in the conductor is not to exceed 2044 kgf/mm², allowing for ice coating of 9.53 mm radial thickness and wind pressure of 0.72 kgf/m. Calculate the sag to be allowed under these conditions. Weight of conductor = 394 kgf/km, weight of ice = 913 kgf/m³. The parabolic approximation may be used.

10. An overhead line has a span of 305 m between level supports. Each conductor has a diameter of 31.77 mm. Find the sag which must be allowed if the tension is not to exceed one-fourth of the ultimate strength of 16280 kgf in (a) still air, (b) with a wind pressure of 100 kgf/m² and an ice coating of 12.7 mm. In the latter case also find the vertical sag. One cubic metre of ice weighs 913.5 kgf. The weight of conductor = 2002 kgf/km.

11. A transmission lines has supports 275 m apart at the same level and each conductor 22.26 mm in diameter weighs 1097 kgf/m. Under severe weather conditions each conductor has a covering of ice 9.53 mm thick (radially) and is subjected to a horizontal wind pressure of 75 kgf/m² of the ice covered projected area. Calculate the deflected the vertical sags if the tension at the mid span is 5105 kgf. Assume 1 m³ of ice to weigh 913.5 kgf.

12. Calculate the vertical sag which must be allowed in a 245 m span of a conductor of a transmission line over a level ground. The overall conductor diameter is 25.79 mm. Maximum working tension in the conductor 6890 kgf. Allow for an ice coating of 9.53 radial thickness and a horizontal wind loading of 0.72 kgf/m. Weight of conductor 1.492 kgf/m, weight of ice 913 kgf/m³. The parabolic approximation may be used.

13. In a transmission line the line conductor has an effective diameter of 19.53 mm, weighs 844 kgf/km, and has an ultimate strength of 7950 kgf. Calculate the height above the ground at which a conductor with a span of 275 m must be supported in order that the total tension shall not exceed half the ultimate strength and with a 12.7 mm radial coating of ice and a horizontal with pressure of 380 N/mm² of projected area. The ground clearance required is 6.7 m. Weight of ice is 913.5 kgf/m³.

14. A transmission line has supports 150 m apart, at the same level and conductors 21.0 mm in diameter, weighing 0.976 kgf/m. Under severe weather conditions each conductor has a covering of ice 9.53 mm thick (radially) and is subjected to a horizontal wind pressure of 20 kgf/m² of the ice-covered projected area. Calculate the vertical component of the sag if the tension at the midspan is 4565 kgf. Assume 1 cubic metre of ice to weigh 913.5 kgf/m³.

15. A transmission line uses ACSR conductors. Each conductor has an effective diameter of 19.53 mm, weighs 844 kgf per km and has an ultimate breaking strength of 7950 kgf. Calculate the height above the ground at which a conductor with a span of 275 m must be supported in order that the total tension shall not exceed half the ultimate strength and with a 1.25 mm radial coating of ice and a horizontal wind pressure of 40 kgf/m^2 of projected area. The ground clearance is 6.7 m. Weight of ice 913.5 kgf/m^3.

16. An overhead line of ASCR conductor 110 mm^2 cross-section weighing 0.844 kgf/m has an overall diameter of 19.53 mm. The line is strung between two supports at equal heights and 275 m apart. The conductor was designed at a tension of 4000 kgf and at a temperature of 0°C, wind load of 1.75 kgf/m and ice load of 1.18 kgf/m. Calculate the tension and sag (a) at 0°C, full wind and ice loading; (b) at 50°C, no wind, and no ice. $[E = 7870$ kgf/mm^2; $\alpha = 17.73 \times 10^{-6}/°C]$

17. The following particulars refer to a 2-Moose ACSR conductor used on a 400 kV line; size of conductor 54/3.35 mm Al, 7/3.35 mm steel, ultimate tensile strength 16280 kgf, area of cross-section of conductor 597 mm^2, modulus of elasticity 7000 kgf/mm^2, weight of conductor 2.0015 kgf/m, wind pressure 100 kgf/m^2.

The wind load on the conductor may be assumed to be acting on two-thirds of the exposed area.

Coefficient of linear expansion = $19.3 \times 10^{-6}/°C$. Ice loading is neglected.

Determine the tensions and sags for spans of

(i) 458 m, (ii) 335 m, (iii) 244 m for the following conditions :

(a) 4.4°C, full wind, factor or safety 2.

(b) 60°C, with initial temperature of 32.2°C, no wind and factor of safety 4.

18. The following particulars refer to 3-Zebra ACSR conductor used on a 400 kV overhead transmission line: size of conductor 54/3.18 mm Al, 7/3.18 mm steel, ultimate tensile strength 13245 kgf, area of cross-section of conductor 482.9 mm^2, modulus of elasticity 7000 kgf/mm^2, weight of conductor 1.619 kgf/m, wind pressure 100 kgf/m^2.

The wind load on the conductor may be assumed to be acting on two-thirds of the exposed area. Coefficient of linear expansion = $19.3 \times 10^{-6}/°C$. Ice loading is neglected. Determine the tensions and sags for spans of (i) 458 m, (ii) 355 m, (iii) 244 m and the following conditions :

4.4°C, full wind and factor of safety 2.

19. An overhead line is supported between two towers having heights of 30 m and 70 m from the datum level. If the horizontal distance between them is 300 m, find the height of the conductor from the datum level between the supports. Assume maximum tension of 1720 kgf and weight per metre run is 0.727 kgf.

20. An overhead line at a river crossing is supported from two towers at heights of 50 m and 85 m above water level, the horizontal distance between the towers being 450 m. If the maximum tension is 3980 kgf and the conductor weighs 1.726 kgf/m, find (a) the minimum clearance, (b) the clearance between the conductor and the water level at a point midway between the towers, and (c) the clearance between the conductor and water at a point 200 m from the lower supporting tower.

21. An overhead transmission line conductor on a hillside is supported between two points separated by a horizontal distance of 400 m and at the heights of 1150 m and 900 m above sea level respectively. The weight of the conductor is 1.492 kgf/m and the tension is 3935 kgf. Determine the vertical clearance between the conductor and a point on the hillside at a height of 970 m and a horizontal distance of 175 m from the lower support. Assume parabolic configuration.

22. At a river crossing an overhead transmission line has a span of 560 m with the two supports of the lowest conductor at 15 m and 95 m above the water level. The weight of the conductor is 0.394 kgf/m. If the tension is adjusted to 1200 kgf determine clearance of the conductor above the adjusted to 1200 kgf determine clearance of the conductor above the water level at a point 215 m from the base of the higher tower.

ANSWERS

1. 1.52 m
2. 7.41 m
3. 5.56 m
4. 5.22 m
5. 7.47 m
6. 12.84 m, 16.73 m
7. 3.1 m
8. (a) 3.95 m (b) 2.87 m
9. 2.51 m
10. (a) 10.35 m (b) 19.34 m, 10.35 m
11. 6.8 m, 3.64 m
12. 2.68 m
13. 13.04 m
14. 1.11 m
15. 9.73 m
16. (a) 4000 kgf, 6.32 m (b) 2360.68 kgf, 3.38 m
17. (a) (i) 4973 kgf, 10.55 m (ii) 4700 kgf, 5.97 m (iii) 4438 kgf, 3.36 m
17. (b) (i) 3967 kgf, 14.2 m (ii) 3500 kgf, 8.02 m (iii) 3252 kgf, 4.58 m
18. (i) 3874 kgf, 10.92 m (ii) 3730 kgf, 6.09 m (iii) 3538 kgf, 3.41 m
19. 45.25 m
20. (a) 49.55 m (b) 56.53 m (c) 54.72 m
21. 32 m
22. 52.1 m

Line Parameters

7.1 INTRODUCTION

An electric transmission line has four parameters, namely *resistance, inductance, capacitance* and *shunt conductance*. The electrical design and performance of a line are dependent on these parameters. These four parameters are uniformly distributed along the whole line. Each line element has its own value, and it is not possible to concentrate or lump them at discrete points on the line. For this reason the line parameters are known as *distributed parameters*. Their values are given per unit length of line and they are denoted as R, L, C and G respectively. The line parameters are functions of the line-geometry, construction material and operational frequency. The line resistance and inductance form the *series impedance* of the line. The capacitance and conductance form the *shunt admittance* of the line.

7.2 LINE INDUCTANCE

When current flows in an electric circuit a magnetic flux is set up. With the variation of current in the circuit, the number of lines of flux also changes and an emf is induced in it. The magnitude of the self-induced emf is directly proportional to the rate of change of flux linkages, and its direction is such as to oppose the cause, i.e., the change of current which produces it.

Mathematically, the induced emf is given by

$$|e| = \frac{d}{dt}(\varphi N) = \frac{N d\varphi}{dt} \text{ V} \qquad \qquad ...(7.2.1)$$

where (φN) is the number of flux linkages of the circuit in weber-turns. *Flux linkage* means the sum of flux lines linking with each turn of the circuit, so that the number of flux linkage is equal to the product of the flux and the number of turns of the circuit linked.

The change in the circuit current causes a change in flux linkages proportionately provided that the permeability of the medium in which the magnetic field produced is assumed to be a constant. The self-induced emf will, therefore, be proportional to the rate of change of current, or

$$|e| = L \frac{di}{dt} \text{ volts} \qquad \qquad \qquad \qquad ...(7.2.2)$$

where L is a constant of proportionality and is known as the self-inductance of the circuit.

Equating the two values of induced emf from Eqs. (7.2.1) and (7.2.2),

$$N \frac{d\varphi}{dt} = L \frac{di}{dt} ; \qquad L = N \frac{d\varphi}{di} \text{ H} \qquad \qquad ...(7.2.3)$$

If the permeability of the magnetic circuit is assumed to be a constant,

$$\frac{d\varphi}{di} = \frac{\varphi}{i}$$

and

$$L = \frac{\varphi N}{i} \text{ H} \qquad \qquad \qquad \qquad ...(7.2.4)$$

which shows that the self-inductance of an electric circuit is numerically equal to the flux linkage of the circuit per unit of current.

$$L = \frac{\varphi}{i} \text{ H} \qquad \qquad \qquad \qquad ...(7.2.5)$$

because only one current path links the flux.

7.3 INDUCTANCE OF A CONDUCTOR

Consider a solid, round infinitely long conductor of radius r, situated in air and carrying a current of I amperes. The flux linking with the conductor consists of two parts, namely, the *internal flux* and the *external flux*. The internal flux is present inside the conductor due to the current in it. This flux does not link with the whole current, but only with a fraction of it. The external flux is produced around the conductor due to its own current and the currents of each of the other conductors in the vicinity. This flux is wholly outside the conductor and links with its entire cross-section. It is, therefore, necessary to determine the internal as well as the external flux linkages for calculations of inductance of a conductor. The inductance obtained by considering the total internal flux linkages due to all flux inside the conductor is called the *internal inductance*. It is denoted by L_{in}. The inductance due to all external flux linkages is known as the external inductance of the conductor. It is represented by L_{ex}. The total inductance L per metre length of the conductor is, therefore, given by

$$L_t = L_{in} + L_{ex} \text{ H/m} \qquad \qquad \qquad ...(7.3.1)$$

7.3.1 Internal Inductance

Let the return path for the current in the conductor be so far away that its magnetic field is not appreciably affected. It can, therefore, be assumed that the current distribution is uniform over the cross-section of the conductor. The cross-section of the conductor is shown in Fig. 7.1.

$$H_x = \frac{NI_x}{l_1} \qquad \qquad \qquad \qquad ...(7.3.2)$$

where H_x is the magnetic field intensity in ampere-turns/metre, I_x is the current enclosed by the flux around which H_x is measured, and l_1 is the length of the path of flux.

Here, $l_1 = 2\pi x$

Since only one turn or current path is linked, $N = 1$

$$H_x = \frac{I_x}{2\pi x} \qquad \qquad \dots(7.3.3)$$

If I is the total current in the conductor, the current density σ at any point on the cross-section of the conductor

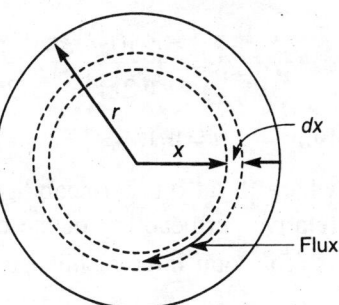

Fig. 7.1. Conductor cross-section.

$$\sigma = \frac{I}{\pi r^2} \ \text{A/m}^2 \qquad \qquad \dots(7.3.4)$$

This has a constant value at every point.

I_x = (current density) × (area enclosed)

$$= \frac{I}{\pi r^2} \cdot \pi x^2 = \frac{x^2}{r^2} I \ \ \text{A} \qquad \qquad \dots(7.3.5)$$

$$H_x = \frac{x^2 I}{r^2 \cdot 2\pi x} = \frac{xI}{2\pi r^2} \ \text{AT/m} \qquad \qquad \dots(7.3.6)$$

The flux density at any point inside the conductor distant x metres from its centre is B_x given by

$$B_x = \mu H_x = \frac{\mu x I}{2\pi r^2} \ \text{Wb/m}^2 \qquad \qquad \dots(7.3.7)$$

where μ is the permeability of the conductor.

The flux through a cylindrical shell distant x from the centre, with radial thickness dx and one metre in length is given by

$$d\varphi_x = (\text{the flux density at } x) \cdot (\text{elemental area normal to the flux path})$$

$$= B_x A_x = B_x \cdot (dx \cdot 1) = \frac{\mu I x}{2\pi r^2} \ dx \ \ \text{Wb} \qquad \qquad \dots(7.3.8)$$

This flux in the cylindrical shell links only the current I_x which is x^2/r^2 of the total current I, so that each weber of flux produces the fraction x^2/r^2 of the total flux linkages. Thus, the flux linkage $d\lambda_x$ due to flux $d\varphi_x$, is given by

$$d\lambda_x = \frac{x^2}{r^2} \ d\varphi_x = \frac{\mu I x^3}{2\pi r^4} \ dx \qquad \qquad \dots(7.3.9)$$

The total internal flux linkages λ_{in} due to all the flux inside the conductor may be obtained by integrating from the centre of the conductor to its surface.

At the centre $x = 0$, and at the surface $x = r$.

$$\therefore \qquad \lambda_{in} = \int_0^r \frac{\mu I x^3}{2\pi r^4}\, dx = \frac{\mu I}{2\pi r^4}\left[\frac{x^4}{4}\right]_0^r = \frac{\mu I r^4}{8\pi r^4} = \frac{\mu I}{8\pi} \text{ WbT/m}$$

But $\qquad \mu = \mu_0\mu_{r\,in}$

where μ_0 is the permeability of free space and it is equal to $4\pi \times 10^{-7}$ H/m, and $\mu_{r\,in}$ is the relative permeability of the conductor material.

The total internal linkage of the conductor is given by

$$\lambda_{in} = \frac{\mu_0\mu_{r\,in}}{8\pi} I = \frac{4\pi \times 10^{-7}\,\mu_{r\,in}}{8\pi} I = \frac{1}{2}\times 10^{-7}\,\mu_{r\,in}\,I \text{ WbT/m} \qquad \qquad ...(7.3.10)$$

For conductors of non-magnetic material $\mu_{r\,in} = 1$

$$\therefore \qquad \lambda_{in} = \frac{1}{2}\times 10^{-7}\,I \text{ WbT/m} \qquad \qquad ...(7.3.11)$$

The inductance due to internal flux

$$L_{in} = \text{internal flux linkage per ampere} = \frac{\lambda_{in}}{I} = \frac{1}{2}\times 10^{-7} \text{ H/m} \qquad \qquad ...(7.3.12)$$

This relation indicates that the internal inductance is independent of the conductor dimensions if the current distribution is uniform.

7.3.2 External Inductance

The cross-section of a conductor and its magnetic field are shown in Fig. 7.2.

The flux lines surrounding the conductor are in the form of concentric circles. The flux density within a cylindrical shell at a distance x metres from the centre and with radial thickness dx outside the conductor is given by

$$B_x = \frac{\mu_0 I}{2\pi x} \text{ Wb/m}^2 \qquad \qquad ...(7.3.13)$$

The flux in the cylindrical shell is

$$d\varphi_x = \frac{\mu_0 I}{2\pi x}\, dx \text{ Wb/m} \qquad \qquad ...(7.3.14)$$

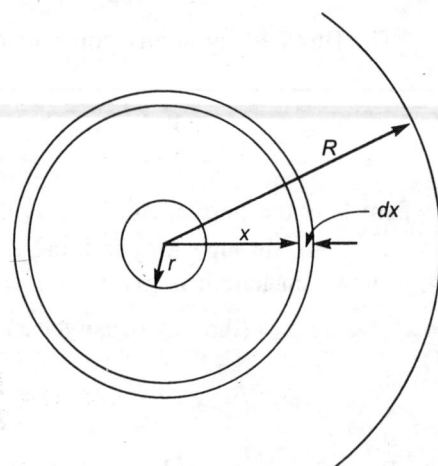

This flux surrounds the whole cross-section of conductor. The flux linkage $d\lambda_x$ due to the flux $d\varphi_x$ are the same as the flux, i.e.,

Fig. 7.2. Conductor and its magnetic field.

$$d\lambda_x = d\varphi_x = \frac{\mu_0 I}{2\pi x}\, dx \qquad \qquad ...(7.3.15)$$

The total flux linkages λ_{ex} between the conductor periphery and a cylinder of very large but finite radius R are obtained by integrating $d\lambda_x$ from $x = r$ to $x = R$.

$$\lambda_{ex} = \int_r^R \frac{\mu_0 I}{2\pi x}\, dx$$

$$= \frac{\mu_0 I}{2\pi} \ln \frac{R}{r} \quad \text{WbT/m} \qquad\qquad \text{...(7.3.16)}$$

$$= 2 \times 10^{-7} I \ln \frac{R}{r} \quad \text{WbT/m} \qquad\qquad \text{...(7.3.17)}$$

$$L_{ex} = \frac{\lambda_{ex}}{I} = 2 \times 10^{-7} \ln \frac{R}{r} \quad \text{H/m} \qquad\qquad \text{...(7.3.18)}$$

The total linkages per metre length of the conductor produced by the entire flux within a cylinder of radius R, is the sum of the internal and external linkages.

$$\lambda = \lambda_{in} + \lambda_{ex} = \frac{1}{2} \times 10^{-7} I + 2 \times 10^{-7} I \ln \frac{R}{r} = 2 \times 10^{-7} I \left(\frac{1}{4} + \ln \frac{R}{r} \right)$$

$$= 2 \times 10^{-7} I \left(\ln e^{1/4} + \ln \frac{R}{r} \right) = 2 \times 10^{-7} I \left(\ln \frac{1}{e^{-1/4}} + \ln \frac{R}{r} \right)$$

$$= 2 \times 10^{-7} I \ln \frac{R}{re^{-1/4}} = 2 \times 10^{-7} I \ln \frac{R}{r'} \quad \text{WbT/m} \qquad\qquad \text{...(7.3.19)}$$

where $r' = re^{-1/4} = 0.7788r$

The total inductance, $L = \frac{\lambda}{I} = 2 \times 10^{-7} \ln \frac{R}{r'}$ H/m $\qquad\qquad$ ...(7.3.20)

where $\ln N = \log_e N = 2.303 \log_{10} N$.

7.4 FLUX LINKAGES IN A GROUP OF CONDUCTORS

Flux linkage relations are required to evaluate the inductance of a transmission line consisting of many parallel conductors.

Consider Fig. 7.3 which illustrates a cross-section of a group of n long, parallel and round conductors. Let the conductors a, b, c, ..., n carrying the currents I_a, I_b, I_c, ..., I_n respectively form a complete circuit. D_{ab}, D_{bc}, D_{ca}, ..., D_{an} represent the distances between the centres of a and b, b and c, c and a, etc.

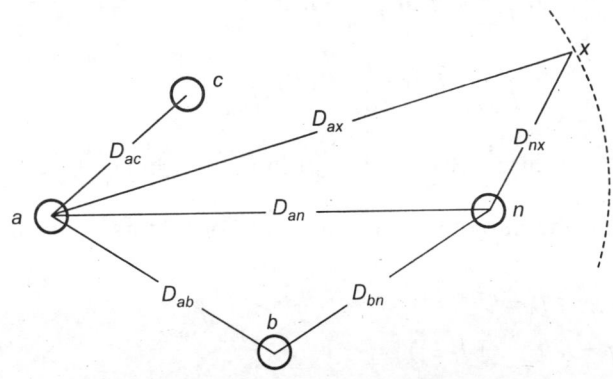

Fig. 7.3. n parallel conductors.

The following assumptions, which are justified for most of the over head lines, are made:

(i) The distances between the conductors are large compared to their radii r_a, r_b, r_c, ..., r_n.

(ii) The current distribution is uniform over the conductors sections.

(iii) The system is unaffected by the external fields.

With these assumptions, the principle of superposition is applicable. The current in each conductor produces a magnetic field within itself and in the region surrounding it. The total flux of the system is the sum of these fluxes. The flux linking any conductor may be obtained by considering the sum of its linkages with all individual fluxes produced by the conductors of the system.

Take a point X at a very large but infinite distance D_{ax} from conductor a. Its distances from other conductors are denoted by D_{bx}, D_{cx}, ..., D_{nx}.

The flux linkages of conductor a with flux which is produced by I_a and which passes between conductor a and point X

$$\lambda_{axa} = 2 \times 10^{-7} I_a \ln \frac{D_{ax}}{r_a} \text{ WbT/m} \qquad \qquad ...(7.4.1)$$

where $\quad r'_a = r_a e^{-1/4} = D_{aa}$ (say) $\qquad \qquad ...(7.4.2)$

The flux linkage of conductor a with flux which is produced by I_b and which passes between the conductor a and point X

$$\lambda_{axb} = 2 \times 10^{-7} I_b \ln \frac{D_{bx}}{D_{ab}} \qquad \qquad ...(7.4.3)$$

Similarly, the flux linkages of conductor a with flux which is produced by I_n and which passes between the conductor a and point X

$$\lambda_{axn} = 2 \times 10^{-7} I_n \ln \frac{D_{nx}}{D_{an}} \qquad \qquad ...(7.4.4)$$

The total flux linkages of a conductor a, due to flux not having a radius greater than D_{ax} will be the sum of a component linkages and is given by

$$\lambda_{ax} = \lambda_{axa} + \lambda_{axb} + \lambda_{axc} + ... + \lambda_{axn}$$

$$= 2 \times 10^{-7} \left[I_a \ln \frac{D_{ax}}{D_{aa}} + I_b \ln \frac{D_{bx}}{D_{ab}} + I_c \ln \frac{D_{cx}}{D_{ac}} + ... + I_n \ln \frac{D_{nx}}{D_{an}} \right] \qquad ...(7.4.5)$$

$$= 2 \times 10^{-7} \left[I_a \ln \frac{1}{D_{aa}} + I_b \ln \frac{1}{D_{ab}} + I_c \ln \frac{1}{D_{ac}} + ... \right.$$

$$\left. I_n \ln \frac{1}{D_{an}} + I_a \ln D_{ax} + I_b \ln D_{bx} + I_c \ln D_{cx} + ... + I_n \ln D_{nx} \right] \qquad ...(7.4.6)$$

Since the sum of all the currents in a system of conductors constituting a complete circuit is zero,

$$I_a + I_b + I_c + ... + I_n = 0 \qquad \qquad ...(7.4.7)$$

and $\quad I_n = -(I_a + I_b + I_c + ... + I_{n-1}) \qquad \qquad ...(7.4.8)$

Substituting the value of I_n in Eq. (7.4.6), we get

$$\lambda_{ax} = 2 \times 10^{-7}\left[I_a \ln \frac{1}{D_{aa}} + I_b \ln \frac{1}{D_{ab}} + \dots + I_n \ln \frac{1}{D_{an}} + I_a \ln D_{ax} + I_b \ln D_{bx} + \dots\right.$$

$$\left. + (-I_a - I_b - \dots - I_{n-1}) \ln D_{nx}\right]$$

$$= 2 \times 10^{-7}\left[I_a \ln \frac{1}{D_{aa}} + I_b \ln \frac{1}{D_{ab}} + \dots + I_n \ln \frac{1}{D_{an}} + I_a \ln \frac{D_{ax}}{D_{nx}} + I_b \ln \frac{D_{bx}}{D_{nx}} + \dots\right.$$

$$\left. + I_{n-1} \ln \frac{D_{(n-1)x}}{D_{nx}}\right] \qquad \dots(7.4.9)$$

If the distances from all the conductors to point X be assumed to be very large, all the ratios such as $\dfrac{D_{ax}}{D_{nx}}$ approach unity in limiting case. The logarithm of such a ratio, i.e., $\ln \dfrac{D_{ax}}{D_{nx}}$ is thus zero in the limit.

The flux linkages of conductor a

$$\lambda_a = 2 \times 10^{-7}\left[I_a \ln \frac{1}{D_{aa}} + I_b \ln \frac{1}{D_{ab}} + I_c \ln \frac{1}{D_{ac}} + \dots + I_n \ln \frac{1}{D_{an}}\right] \qquad \dots(7.4.10)$$

Equation (7.4.10) may be written in a more compact form as

$$\lambda_a = 2 \times 10^{-7}\left[\sum_{x-a}^{x=n} I_x \ln \frac{1}{D_{ax}}\right] \text{WbT/m} \qquad \dots(7.4.11)$$

The linkages about other conductors may be written in a similar manner. Equation (7.4.10) shows that the flux linkages are finite and independent of the distance to the point X, provided that the sum of the currents flowing in all the conductors of the system is zero.

To calculate instantaneous or rms flux linkages, the currents in Eq. (7.4.10) should be expressed as instantaneous or rms currents respectively. The effective inductance of the line in henrys per metre can be obtained by expressing the rms values of I_b, I_c, ..., I_n in terms of I_a and dividing the result by I_a.

7.5 INDUCTANCE OF A TWO-WIRE LINE

Consider a single-phase line consisting of two conductors a and b, of equal radius r. They are situated at a distance D metres. The cross-section of such a line is shown in Fig. 7.4. Let these conductors carry the same current in opposite directions, so that one serves return path for the other.

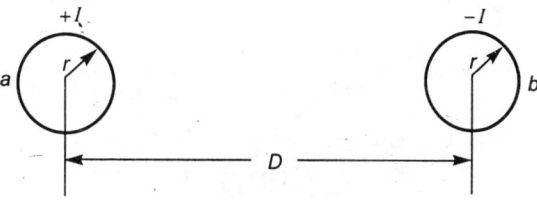

Fig. 7.4. Two-wire line.

The flux linkages of conductor a may be written from Eq. (7.4.10) as

$$\lambda_a = 2 \times 10^{-7}\left[I_a \ln \frac{1}{D_{aa}} + I_b \ln \frac{1}{D_{ab}}\right] \qquad \dots(7.5.1)$$

Here,
$$\left.\begin{array}{ll} I_a = +I, & I_b = -I \\ D_{aa} = re^{-1/4} = r' & D_{ab} = D \end{array}\right\} \qquad \ldots(7.5.2)$$

Substituting these values in Eq. (7.5.1)

$$\lambda_a = 2 \times 10^{-7} \left(I \ln \frac{1}{r'} - I \ln \frac{1}{D} \right) = 2 \times 10^{-7} I \ln \frac{D}{r'} \qquad \ldots(7.5.3)$$

Similarly, the flux linkages of conductor b will be

$$\lambda_b = 2 \times 10^{-7} I \ln \frac{D}{r'} \qquad \ldots(7.5.4)$$

The inductance of conductor a

$$L_a = \frac{\lambda_a}{I} = 2 \times 10^{-7} \ln \frac{D}{r'} \text{ H/m} \qquad \ldots(7.5.5)$$

Similarly, the inductance of conductor b

$$L_b = 2 \times 10^{-7} \ln \frac{D}{r'} \text{ H/m} \qquad \ldots(7.5.6)$$

Inductance per conductor

$$L = L_a = L_b = 2 \times 10^{-7} \ln \frac{D}{r'} \text{ H/m} \qquad \ldots(7.5.7)$$

Loop inductance = inductance of both conductors = $L_a + L_b = 2 \times 2 \times 10^{-7} \ln \frac{D}{r'}$

$$= 4 \times 10^{-7} \ln \frac{D}{r'} \text{ H/m} \qquad \ldots(7.5.8)$$

The inductance of individual conductor is one-half the total inductance of a two-wire line.

Example 7.1 A two-conductor single-phase line operates at 50 Hz. The diameter of each conductor is 20 mm and the spacing between the conductors is 3 m. Calculate (a) the inductance of each conductor per km, (b) the loop inductance of the line per km, (c) the inductive reactance per km, and (d) the loop inductance per km of the line when the conductor material is steel of relative permeability 50.

Solution

$$D = 3 \text{ m}, \quad r = 0.01 \text{ m}$$

$$r' = 0.7788r = 0.7788 \times 0.01 = 7.788 \times 10^{-3} \text{ m}$$

$$\ln \frac{D}{r'} = \ln \frac{3}{7.788 \times 10^{-3}} = 5.953$$

(a) Inductance of each conductor

$$L = 2 \times 10^{-7} \ln \frac{D}{r'} \text{ H/m} = 2 \times 10^{-7} \times 10^3 \times 10^3 \ln \frac{D}{r'} \text{ mH/km}$$

$$= 0.2 \ln \frac{D}{r'} = 0.2 \times 5.953 = 1.19 \text{ mH/km}$$

(b) Loop inductance $= 2 \times$ inductance of each conductor $= 2 \times 1.19 = 2.38$ mH/km

(c) Inductance reactance $= 2\pi f L = 2\pi \times 50 \times 2.38 \times 10^{-3} = 0.7477$ Ω/km

(d) Loop inductance with steel conductors $= 2\ (L_{in} + L_{ex})$

$$= 2\left(\frac{1}{2} \times 10^{-7}\ \mu_{r\,in} + 2 \times 10^{-7} \ln\frac{D}{r}\right) = 10^{-7}\left(50 + 4 \ln\frac{3}{0.01}\right) \text{ H/m}$$

$$= 10^{-7} \times 10^{6}\ (50 + 4 \times 5.703) \text{ mH/km} = 7.281 \text{ mH/km}.$$

7.6 INDUCTANCE OF SYMMETRICAL THREE-PHASE LINE

A three-phase line is said to be symmetrical when its conductors are situated at the corners of an equilateral triangle. Such an arrangement of conductors is also sometimes referred to as *equilateral spacing* and is shown in Fig. 7.5.

Let the spacing between the conductors be D, and the radius of each conductor, r.

Then, by Eq. (7.4.10), the flux linkages of conductor a are

$$\lambda_a = 2 \times 10^{-7}\left(I_a \ln\frac{1}{D_{aa}} + I_b \ln\frac{1}{D_{ab}} + I_c \ln\frac{1}{D_{ac}}\right)$$

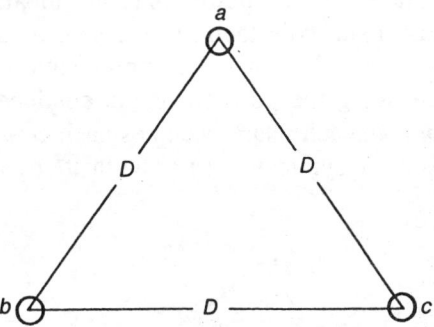

Fig. 7.5. Symmetrical three-phase line.

In this case

$$D_{ab} = D_{bc} = D_{ca} = D$$

and $D_{aa} = r' = re^{-1/4}$

$$\lambda_a = 2 \times 10^{-7}\left(I_a \ln\frac{I}{r'} + I_b \ln\frac{1}{D} + I_c \ln\frac{1}{D}\right) \qquad \text{...(7.6.1)}$$

For a three-wire system, the algebraic sum of the currents in the conductors is zero :

$$I_a + I_b + I_c = 0 \qquad \text{...(7.6.2)}$$

and $I_a = -I_b - I_c$ ...(7.6.3)

so that the Eq. (7.6.1) becomes

$$\lambda_a = 2 \times 10^{-7} \times I_a\left(\ln\frac{1}{r'} - \ln\frac{1}{D}\right) = 2 \times 10^{-7} \times I_a \ln\frac{D}{r'} \qquad \text{...(7.6.4)}$$

The inductance of conductor a is

$$L_a = \frac{\lambda}{I_a} = 2 \times 10^{-7} \ln\frac{D}{r'} \text{ H/m} \qquad \text{...(7.6.5)}$$

$$L_a = 0.2 \ln\frac{D}{r'} \text{ mH/km}$$

The inductance of conductors b and c will also be the same as that of a.

By comparing Eq. (7.6.5) and (7.5.7), it is found that the inductance per conductor of a three-phase symmetrically spaced line is equal to the inductance per conductor of a single-phase line of equal length and with equal spacing between conductors.

7.7 INDUCTANCE OF UNSYMMETRICAL THREE-PHASE LINE

Transmission lines with unsymmetrical conductor arrangements are most commonly used in practice because of their cheapness and convenience in design and construction. For an unsymmetrically spaced three-phase line, the inductances and, therefore, the voltage drops will be different for all the phases even under balanced current conditions. This leads to unbalanced voltages at the receiving end of the line. Moreover, if communication lines are also running adjacent to the power line, an unbalance of voltage is also produced in them. This results in disturbances in them. In order to reduce the inequality of inductance and inductive interference with parallel running communication lines, the line is transposed. The *transposition* is done by changing the position of the conductors so that with a certain length of line, usually called a *barrel*, each phase occupies each conductor position for approximately the same length. Fig. 7.6 is an unsymmetrical line with its transposition cycle.

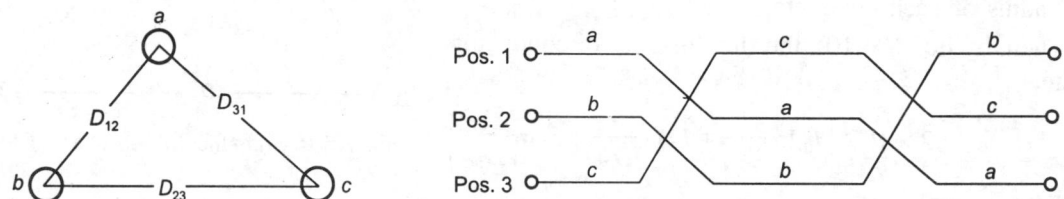

Fig. 7.6. Transposition cycle of unequally spaced three-phase line conductors.

When a non-symmetrically spaced line is properly transposed, the average value of flux linkages of a conductor may be found by adding up its flux linkages for each of its three positions in the transposition cycle and dividing the sum by 3. We shall find the flux linkages of conductor a.

When a is in position 1, b in position 2, and c in position 3, the flux linkages of a are

$$\lambda_{a1} = 2 \times 10^{-7} \left(I_a \ln \frac{1}{r'} + I_b \ln \frac{1}{D_{12}} + I_c \ln \frac{1}{D_{31}} \right) \qquad \text{...(7.7.1)}$$

When a is in position 2, b is in position 3, and c is in position 1, the flux linkages of a are

$$\lambda_{a2} = 2 \times 10^{-7} \left(I_a \ln \frac{1}{r'} + I_b \ln \frac{1}{D_{23}} + I_c \ln \frac{1}{D_{12}} \right) \qquad \text{...(7.7.2)}$$

The flux linkages of conductor a when a is in position 3, b is in position 1 and c is in position 2,

$$\lambda_{a3} = 2 \times 10^{-7} \left(I_a \ln \frac{1}{r'} + I_b \ln \frac{1}{D_{31}} + I_c \ln \frac{1}{D_{23}} \right) \qquad \text{...(7.7.3)}$$

The average value of flux linkages of a is

$$\lambda_a = \frac{1}{3} (\lambda_{a1} + \lambda_{a2} + \lambda_{a3}) \qquad \text{...(7.7.4)}$$

$$= \frac{2}{3} \times 10^{-7} \left[\left(I_a \ln \frac{1}{r'} + I_b \ln \frac{1}{D_{12}} + I_c \ln \frac{1}{D_{31}} \right) + \left(I_a \ln \frac{1}{r'} + I_b \ln \frac{1}{D_{23}} + I_c \ln \frac{1}{D_{12}} \right) \right.$$

$$\left. + \left(I_a \ln \frac{1}{r'} + I_b \ln \frac{1}{D_{31}} + I_c \ln \frac{1}{D_{23}} \right) \right]$$

$$= \frac{2}{3} \times 10^{-7} \left[3 I_a \ln \frac{1}{r'} + I_b \ln \frac{1}{D_{12} D_{23} D_{31}} + I_c \ln \frac{1}{D_{12} D_{23} D_{31}} \right]$$

$$= \frac{2}{3} \times 10^{-7} \left(3 I_a \ln \frac{1}{r'} - I_a \ln \frac{1}{D_{12} D_{23} D_{31}} \right)$$

Since for balanced conditions

$$I_a + I_b + I_c = 0 \; ; \; I_b + I_c = - I_a \qquad \qquad \qquad \ldots (7.7.5)$$

$$\therefore \qquad \lambda_a = 2 \times 10^{-7} \left[I_a \ln \frac{1}{r'} + \frac{1}{3} I_a \ln (D_{12} D_{23} D_{31}) \right]$$

$$= 2 \times 10^{-7} \times I_a \ln \frac{(D_{12} D_{23} D_{31})^{1/3}}{r'} \qquad \qquad \ldots (7.7.6)$$

The average inductance of phase a is

$$L_a = \frac{\lambda_a}{I_a} = 2 \times 10^{-7} \times \ln \frac{(D_{12} D_{23} D_{31})^{1/3}}{r'} \text{ H/m} \qquad \qquad \ldots (7.7.7)$$

Similarly,

$$L_b = L_c = 2 \times 10^{-7} \times \ln \frac{(D_{12} D_{23} D_{31})^{1/3}}{r'} \text{ H/m} \qquad \qquad \ldots (7.7.8)$$

Thus, it is found that the values of the inductance for the three phases are equalized by transposition. The average inductance per phase of a transposed line is

$$L = 2 \times 10^{-7} \times \ln \frac{(D_{12} D_{23} D_{31})^{1/3}}{r'} \text{ H/m} \qquad \qquad \ldots (7.7.9)$$

If the Eqs. (7.7.9) and (7.6.5) are compared, it is seen that the average inductance per metre of any phase of a transposed three-phase line with unequal spacings D_{12}, D_{23}, and D_{31} is equal to the inductance of a three-phase line with similar conductors and symmetrical spacing D provided that

$$D = (D_{12} D_{23} D_{31})^{1/3} \qquad \qquad \ldots (7.7.10)$$

The quantity $(D_{12} D_{23} D_{31})^{1/3}$ is the geometric mean of the three unequal spacings and is called the *equivalent delta spacing* or the *equivalent equilateral spacing* between the conductors of the line. The corresponding value of inductance found will be called the *equivalent inductance*, which may be considered as a series inductance in each phase.

If the conductors are in the same horizontal or vertical plane, the arrangement is known as *flat spacing*. Such configurations of lines are shown in Fig. 7.7.

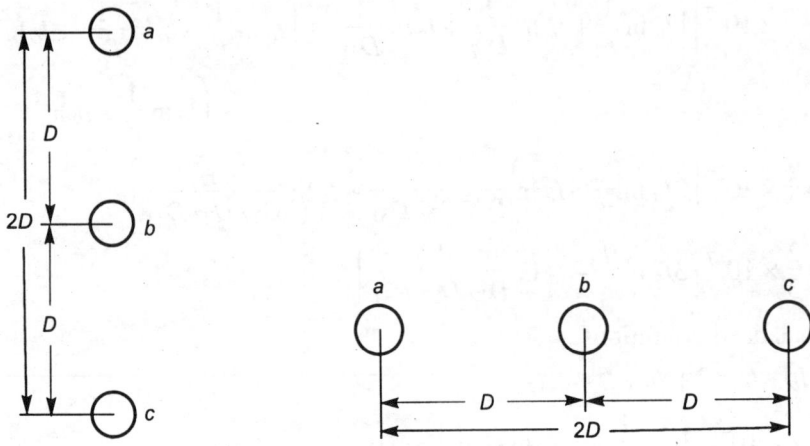

Fig. 7.7. Horizontal and vertical flat spacings of three-phase lines.

For flat spacing

$$D_{12} = D_{23} = \frac{1}{2} D_{31} = D \text{ (say)}$$

∴   the equivalent spacing, $D_{eq} = (D \cdot D \cdot 2D)^{1/3} = 1.26D$ ...(7.7.11)

Although the transposition of unsymmetrical lines helps in equalizing the inductances and reducing the inductive interference with the nearby communication lines, it suffers from a serious drawback that faults due to lightning have got a tendency to get concentrated in transposition structures. At the same time transposition is a costly affair. With transposition, it is found that the risk of outages is increased. For these reasons as far as possible the transposition is avoided in practice, and that if at all it is done, the line is transposed in the switching stations or their neighbourhood for a very small distance.

Example 7.2 A three-phase 50 Hz line consists of three conductors each of diameter 21 mm. The spacing between the conductors is as follows :

A-B = 3 m, B-C = 5m, C-A = 3.6 m.

Find the inductance and inductive reactance per phase per km of the line.

Solution

$$r = \frac{1}{2} \times 21 \times 10^{-3} = 10.5 \times 10^{-3} \text{ m}$$

$$D_{eq} = (3 \times 5 \times 3.6)^{1/3} = 3.78 \text{ m}$$

$$r' = 0.7788r = 0.7788 \times 10.5 \times 10^{-3} = 8.177 \times 10^{-3} \text{ m}$$

Inductance per phase

$$L = 2 \times 10^{-7} \ln \frac{D_{eq}}{r'} = 2 \times 10^{-7} \ln \frac{3.78}{8.177 \times 10^{-3}} = 2 \times 10^{-7} \ln 462.2 = 2 \times 10^{-7} \times 6.136$$

$$= 12.272 \times 10^{-7} \text{ H/m} = 12.272 \times 10^{-7} \times 10^3 \text{ H/km} = 12.272 \times 10^{-4} \text{ H/km}$$

Inductive reactance per phase

$$X_L = 2\pi f L = 2\pi \times 50 \times 12.272 \times 10^{-4} = 0.386 \ \Omega/\text{km}.$$

7.8 METHOD OF GEOMETRIC MEAN DISTANCES

The method of *geometric mean distance* is very convenient and useful to calculate the inductance and capacitance of a line having several conductors connected in parallel for each phase. Thus, it is applicable to all the cases of multi-strand conductor or bundled conductor lines. It is a mathematical concept.

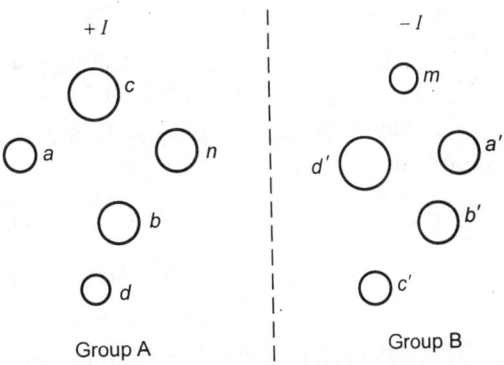

Fig. 7.8. Single-phase line with composite conductors.

Consider a single-phase line consisting of two groups of conductors as shown in Fig. 7.8.

Group A consists of n parallel, round and very long conductors which are connected in parallel, I is the total current carried by group A. Each conductor of the group is assumed to carry a current $+\dfrac{I}{n}$.

Group B consists of m parallel, round and very long conductors which are connected in parallel. Each conductor carries a current $-\dfrac{I}{m}$ where $-I$ is the total current of the return group B.

By Eq. (7.4.10) the flux linkages about any conductor, say a of group A are

$$\lambda_a = 2 \times 10^{-7} \times \frac{I}{n}\left(\ln \frac{1}{D_{aa}} + \ln \frac{1}{D_{ab}} + \ln \frac{1}{D_{ac}} + \dots + \ln \frac{1}{D_{an}} \right)$$

$$- 2 \times 10^{-7} \times \frac{I}{m}\left(\ln \frac{1}{D_{aa'}} + \ln \frac{1}{D_{ab'}} + \ln \frac{1}{D_{ac'}} + \dots + \ln \frac{1}{D_{am}} \right)$$

$$= 2 \times 10^{-7} \, I \ln \left[\frac{(D_{aa'} D_{ab'} D_{ac'} \dots D_{am})^{1/m}}{(D_{aa} D_{ab} D_{ac} \dots D_{an})^{1/n}} \right] \qquad \qquad \text{...(7.8.1)}$$

Since conductor a carries I/n amperes, the inductance of conductor a :

$$L_a = \frac{\lambda_a}{I/n} = 2n \times 10^{-7} \ln \frac{(D_{aa'} D_{ab'} D_{ac'} \dots D_{am})^{1/m}}{(D_{aa} D_{ab} D_{ac} \dots D_{an})^{1/n}} \ \text{H/m} \qquad \qquad \text{...(7.8.2)}$$

Similarly, the inductance of conductor b :

$$L_b = 2n \times 10^{-7} \ln \frac{(D_{ba'} D_{bb'} D_{bc'} \dots D_{bm})^{1/m}}{(D_{ba} D_{bb} D_{bc} \dots D_{bn})^{1/n}} \ \text{H/m} \qquad \qquad \text{...(7.8.3)}$$

The expressions for L_c, L_d, $\dots$, L_n may be written in similar manner. It is found that the conductors of the group have different inductances. The average inductance of a conductor of group A is

$$L_{av} = \frac{1}{n}(L_a + L_b + L_c + \ldots + L_n) \qquad \ldots(7.8.4)$$

The inductance of group A consisting of n conductors connected in parallel electrically will be $\frac{1}{n}$ times the average inductance, i.e.,

$$L_A = \frac{L_{av}}{n} = \frac{1}{n^2}(L_a + L_b + L_c + \ldots + L_n) \qquad \ldots(7.8.5)$$

The values of L_a, L_b, L_c, ..., L_n are substituted in Eq. (7.8.5) and the logarithmic terms are combined. The inductance of group A is then given by

$$L_A = 2 \times 10^{-7} \ln \frac{[(D_{aa'} D_{ab'} D_{ac'} \ldots D_{am})(D_{ba'} D_{bb'} D_{bc'} \ldots D_{bm}) \ldots (D_{na'} D_{nb'} D_{nc'} \ldots D_{nm})]^{1/mn}}{[(D_{aa} D_{ab} D_{ac} \ldots D_{an})(D_{ba} D_{bb} D_{bc} \ldots D_{bn}) \ldots (D_{na} D_{nb} D_{nc} \ldots D_{nn})]^{1/n^2}}$$

$$\ldots(7.8.6)$$

The numerator of the logarithmic term of Eq. (7.8.6) is a product of $(m \times n)$ distances for which the $(m \times n)$th root is taken. It is, therefore, the geometric mean of all the possible $(m \times n)$ mutual distances from conductors of group A to the conductors of group B. This geometric mean is called the *mutual geometric mean distance* D_m between the two conductor groups A and B. Thus in order to determine D_m we use the following procedure :

1. Determine the product of the sequence of m distances from first conductor a of group A to each conductor of group B. In our case this product is

    $(D_{aa'} D_{ab'} D_{ac'} \ldots D_{an}) = D_1$ (say)

2. Determine the product of the sequence of m distances from the second conductor b of group A to each conductor of group B. In our case this product is

    $(D_{ba'} D_{bb'} D_{bc'} \ldots D_{bm}) = D_2$ (say)

3. This process is continued to the last conductor n in group A. The product of the sequence of m distances from the last conductor n of group A to each conductor of group B is given by

    $(D_{na'} D_{nb'} D_{nc'} \ldots D_{nm}) = D_n$ (say)

The mutual GMD between A and B is given by

$D_m = (D_1 D_2 D_3 \ldots D_n)^{1/mn}$

It is to be noted that all the products D_1, D_2, ..., D_n contain m terms.

$D_m = [(D_{aa'} D_{ab'} D_{ac'} \ldots D_{am})(D_{ba'} D_{bb'} D_{bc'} \ldots D_{bm}) \ldots (D_{na'} D_{nb'} D_{nc'} \ldots D_{nm})]^{1/mn}$

$$\ldots(7.8.7)$$

The denominator of the logarithmic term of Eq. (7.8.6) shows that the distances involved here are all within group A, and the denominator represents the produce of $(n \times n)$ distances for which (n^2)th root is taken. It is, therefore, the geometric mean of all the possible n^2 distances of conductors of group A including the distances from the conductors to themselves. The distances from the conductors to themselves have been taken as D_{aa}, D_{bb}, D_{cc}, ..., D_{nn}. The denominator of the logarithmic term of Eq. (7.8.6) is therefore, called the *self geometric mean distance* (self GMD) or sometimes, *geometric mean radius* (GMR) of group A. It is represented by D_{SA}.

$$D_{SA} = [(D_{aa} D_{ab} D_{ac} \ldots D_{an}) (D_{ba} D_{bb} D_{bc} \ldots D_{bn}) \ldots (D_{na} D_{nb} D_{nc} \ldots D_{nn})]^{1/n^2} \quad \ldots(7.8.8)$$

In terms of D_m and DSA, the inductance of conductor A can be written from Eq. (7.8.6) as

$$L_A = 2 \times 10^{-7} \ln \frac{D_m}{D_{SA}} \text{ H/m} \qquad \ldots(7.8.9)$$

Group B consists of m conductors. Its GMR is different from that of group A. However, the GMD of group B is the same as that of group A. The GMR of group B is given by

$$D_{SB} = [(D_{a'a'} D_{a'b'} D_{a'c'} \ldots D_{a'm}) \times (D_{b'a'} D_{b'b'} D_{b'c'} \ldots D_{b'm}) \times (D_{c'a'} D_{c'b'} D_{c'c'} \ldots D_{c'm}) \times \ldots$$

$$\times (D_{ma'} D_{mb'} D_{mc'} \ldots D_{nm})]^{1/m^2}$$

The inductor of group B is then found from

$$L_B = 2 \times 10^{-7} \ln \frac{D_m}{D_{SB}} \text{ H/m}$$

The total inductance of the line is given by

$$L = L_A + L_B = 2 \times 10^{-7} \ln \frac{D_m}{D_{SA}} + 2 \times 10^{-7} \ln \frac{D_m}{D_{SB}} \text{ H/m} \qquad \ldots(7.8.10)$$

7.9 TWO-WIRE LINE

A two-wire line is shown in Fig. 7.4.

$$D_m = D, \ D_{SL} = D_{aa} = r'$$

The inductance of conductor A

$$L_A = 2 \times 10^{-7} \ln \frac{D_m}{D_{SLA}} = 2 \times 10^{-7} \ln \frac{D}{r'} \text{ H/m}$$

Similarly, the inductance of conductor B

$$L_B = 2 \times 10^{-7} \ln \frac{D_m}{D_{SLB}} = 2 \times 10^{-7} \ln \frac{D}{r'} \text{ H/m}$$

Therefore, the total inductance of the line is given by

$$L = L_A + L_B = 4 \times 10^{-7} \ln \frac{D}{r'} \text{ H/m}.$$

7.10 SYMMETRICAL THREE-PHASE LINE

A symmetrical 3-phase line is shown in Fig. 7.5. The mutual geometric mean distance between conductor a, and conductors b and c is

$$D_m = (D \cdot D)^{1/2} = D$$

The self geometric mean distance

$$D_s = D_{aa} = r'$$

$$L_a = 2 \times 10^{-7} \ln \frac{D}{r'} \text{ H/m}$$

Also, by symmetry, $L_a = L_b = L_c$

$\therefore$ inductance per phase, $L = 2 \times 10^{-7} \ln \frac{D}{r'}$ H/m   ...(7.10.1)

7.11 INDUCTANCE OF UNSYMMETRICAL THREE-PHASE LINE

The unsymmetrical three-phase line is shown in Fig. 7.6. When phase a is in position 1, the value of D_m is the geometric mean of all the distances from a to other conductors. In position 1 the GMD is

$$D_{m1} = (D_{12} D_{31})^{1/2}$$ ...(7.11.1)

When phase a is in position 2, the GMD is

$$D_{m2} = (D_{23} D_{31})^{1/2}$$ ...(7.11.2)

When phase a is in position 3, the GMD is

$$D_{m3} = (D_{12} D_{23})^{1/2}$$ ...(7.11.3)

Hence, the equivalent GMD, which is the geometric mean of the three values of the GMD of phase a in the three sections of the transposition cycle, is given by

$$D_m = (D_{m1} D_{m2} D_{m3})^{1/3} = (D_{12} D_{31} D_{23} D_{31} D_{12} D_{23})^{1/6}$$

$$= (D_{12} D_{23} D_{31})^{1/3}$$ ...(7.11.4)

Self GMD of phase a in position 1 of the transposition cycle

$$D_{s1} = D_{aa} = re^{-1/4} = r'$$ ...(7.11.5)

Self GMD of phase a in position 2 of the transposition cycle

$$D_{s2} = D_{aa} = re^{-1/4} = r'$$ ...(7.11.6)

Self GMD of phase a in position 3 of the transposition cycle

$$D_{s3} = D_{aa} = re^{-1/4} = r'$$ ...(7.11.7)

Equivalent self GMD of phase a over a complete transposition cycle is equal to the geometric mean of D_{s1}, D_{s2}, D_{s3}.

Geometric mean of D_{s1}, D_{s2} and D_{s3}

$$D_{sL} = (D_{s1} D_{s2} D_{s3})^{1/3} = (r' \, r' \, r')^{1/3} = r'$$ ...(7.11.8)

The inductance of phase a over a transposition cycle is

$$L_a = 2 \times 10^{-7} \ln \frac{D_m}{D_{sL}}$$

$$= 2 \times 10^{-7} \ln \frac{(D_{12} D_{23} D_{31})^{1/3}}{r'}$$ ...(7.11.9)

Also, $L_a = L_b = L_c$

The average inductance of any phase is, thus

$$L = 2 \times 10^{-7} \ln \frac{(D_{12} D_{23} D_{31})^{1/3}}{r'} \text{ H/m} \qquad \qquad ...(7.11.10)$$

The same result has been deduced in Section 7.7. When $D_{12} = D_{23} = D_{31} = D$, the line becomes symmetrical and the result of Eq. (7.10.1) is obtained.

$$L = 2 \times 10^{-7} \ln \frac{D}{r'} \text{ H/m}.$$

Example 7.3 A three-phase 50 Hz transmission line consists of three equal conductors of radii r, placed in a horizontal plane, with a spacing of 6 m between the middle and each outer conductor, as shown in Fig. 7.9. Determine the inductive reactance per phase per km of the transposed line if the radius of each conductor is 12.5 mm.

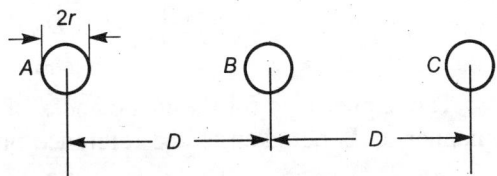

Fig. 7.9. Illustrating Example 7.3.

Solution

$$L = 2 \times 10^{-7} \ln \frac{D_m}{D_{sL}}$$

Here $D_m = (D \cdot D \cdot 2D)^{1/3} = 1.26D = 1.26 \times 6 = 7.56$ m

$D_{sL} = r' = 0.7788r = 0.7788 \times 12.5 \times 10^{-3} = 9.735 \times 10^{-3}$ m

$$L = 2 \times 10^{-7} \ln \frac{7.56}{9.735 \times 10^{-3}} = 13.31 \times 10^{-7} \text{ H/m} = 1.331 \text{ mH/km}$$

Inductive reactance per phase

$$X_L = 2\pi f L = 2\pi \times 50 \times 13.31 \times 10^{-7} \text{ } \Omega/\text{m}$$

$$= 4.181 \times 10^{-4} \text{ } \Omega/\text{m} = 4.181 \times 10^{-4} \times 10^3 \text{ } \Omega/\text{km} = 0.4181 \text{ } \Omega/\text{km}$$

Example 7.4 Determine the effective inductance of each conductor per km of a line consisting of three conductors each of diameter 3 cm placed at the corners of a triangle with sides of 3, 4 and 5 metres. Explain the significance of the complex number in the expression. Also, calculate the inductance of the line if the line is regularly transposed throughout its length.

Solution

The expression of flux linkages of conductor A is given by

$$\lambda_a = 2 \times 10^{-7} \left[\mathbf{I}_a \ln \frac{1}{D_{aa}} + \mathbf{I}_b \ln \frac{1}{D_{ab}} + \mathbf{I}_c \ln \frac{1}{D_{ac}} \right]$$

Let the phase sequence be A-B-C. Take $\mathbf{I}_a$ as the reference phasor. The expression for the currents in the three conductors are given by

$$\mathbf{I}_a = I + j\,0$$

$$\mathbf{I}_b = I\,(-0.5 - j\,0.866)$$

$$\mathbf{I}_c = I\,(-0.5 + j\,0.866)$$

On substituting the values of $\mathbf{I}_a$, $\mathbf{I}_b$, and $\mathbf{I}_c$ in the expression λ_a, it becomes

$$\lambda_a = 2\mathbf{I} \times 10^{-7} \left(\ln \frac{1}{D_{aa}} - 0.5 \ln \frac{1}{D_{ab}} - j\,0.866 \ln \frac{1}{D_{ab}} - 0.5 \ln \frac{1}{D_{ac}} + j\,0.866 \ln \frac{1}{D_{ab}} \right)$$

$$= 2\mathbf{I} \times 10^{-7} \left[\ln \frac{(D_{ab}\,D_{ac})^{1/2}}{D_{aa}} + j\,0.866 \ln \frac{D_{ab}}{D_{ac}} \right]$$

The inductance of conductor A is

$$\mathbf{L}_a = \frac{\lambda_a}{\mathbf{I}} = 2 \times 10^{-7} \left[\ln \frac{(D_{ab}\,D_{ac})^{1/2}}{D_{aa}} + j\,0.866 \ln \frac{D_{ab}}{D_{ac}} \right] \text{ H/m}$$

The expressions for the inductances of conductors of B and C can also be found in a similar manner. If $\mathbf{I}_b$ be taken as the reference phasor, the inductance of conductor B may be given as

$$\mathbf{L}_b = 2 \times 10^{-7} \left[\ln \frac{(D_{bc}\,D_{ab})^{1/2}}{D_{bb}} + j\,0.866 \ln \frac{D_{bc}}{D_{ab}} \right] \text{ H/m}$$

If $\mathbf{I}_c$ be taken as the reference phasor, the inductance of conductor C will be

$$\mathbf{L}_c = 2 \times 10^{-7} \left[\ln \frac{(D_{ca}\,D_{bc})^{1/2}}{D_{cc}} + j\,0.866 \ln \frac{D_{ca}}{D_{bc}} \right] \text{ H/m}$$

In this example,

$$D_{ab} = 3 \text{ m}, \quad D_{bc} = 4 \text{ m}, \quad D_{ca} = 5 \text{ m}, \quad r = 1.5 \text{ cm} = 0.015 \text{ m}$$

$$D_{aa} = D_{bb} = D_{cc} = re^{-1/4} = 0.7788r = 0.7788 \times 0.015 = 0.011682 \text{ m}$$

$$\mathbf{L}_a = 2 \times 10^{-7} \left[\ln \frac{(3 \times 5)^{1/2}}{0.011682} + j\,0.866 \ln \frac{3}{5} \right] \text{ H/m of conductor}$$

$$= 1.1608 - j\,0.0885 \text{ mH/km of conductor}$$

$$\mathbf{L}_b = 2 \times 10^{-7} \left[\ln \frac{(4 \times 3)^{1/2}}{0.011682} + j\,0.866 \ln \frac{4}{3} \right] \text{ H/m of conductor}$$

$$= 1.13843 + j\,0.0498 \text{ mH/km of conductor}$$

$$\mathbf{L}_c = 2 \times 10^{-7} \left[\ln \frac{(5 \times 4)^{1/2}}{0.011682} + j\,0.866 \ln \frac{5}{4} \right] \text{ H/m of conductor}$$

$$= 1.1895 + j\,0.03865 \text{ mH/km of conductor}$$

From the above expression it is observed that λ_a is not in phase with $\mathbf{I}_a$, λ_b is not in phase with $\mathbf{I}_b$, and λ_c is not in phase with $\mathbf{I}_c$. The inductance $\mathbf{L}_a$, $\mathbf{L}_b$, and $\mathbf{L}_c$ are, therefore, complex numbers. The imaginary components represent the power transferred between the phases by mutual induction. The negative imaginary component in the expression for inductance shows that the power is supplied by that phase to other phases. The positive imaginary component shows that the power is received by that phase from other phases. However, the total power transferred in any case is zero. In other words, the mutual power transfer does not affect the power dissipated in various conductors comprising the system.

Inductance of transposed line

$$L_{av} = \frac{1}{3}(L_a + L_b + L_c)$$

$$= \frac{1}{3}(1.1608 - j\,0.0885 + 1.13843 + j\,0.0494 + 1.1895 + j\,0.03865) = 1.163 \ \text{mH/km}$$

7.12 INDUCTANCE OF DOUBLE CIRCUIT SINGLE-PHASE LINE

A double-circuit single-phase line is shown in Fig. 7.10. It consists of four conductors. Conductors a_1 and a_2 are connected in parallel and carry the current in one direction. In effect, they form one conductor A. The conductors b_1 and b_2 are connected in parallel and carry the current in the return direction. These return conductors form the other conductor called phase B. The system is thus similar to the two-wire line.

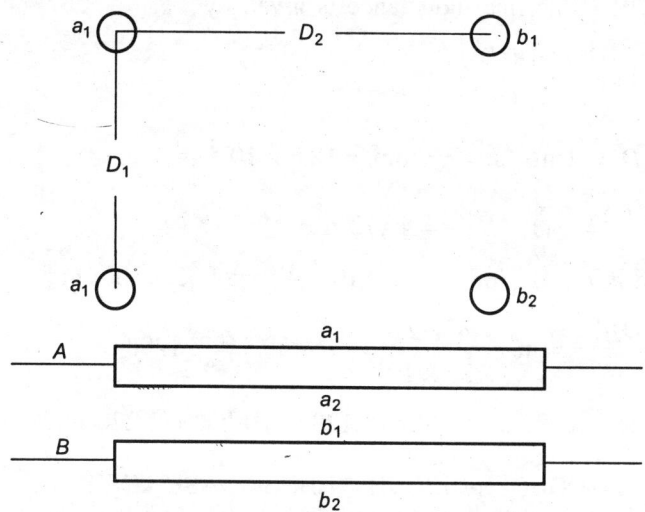

Fig. 7.10. Double-circuit single-phase line.

Let all the conductors be of the same radius r and of the same material. The arrangement is symmetrical and there is no need of transposition. However, if communication circuits are in the vicinity, the transposition may be required to reduce interference. The method of GMD may be utilized to calculate the inductance.

Here, $m = n = 2$.

The distances between the conductors are shown in Fig. 7.10. The self GMD for the pair of wires a_1 and a_2 is given by

$$D_{sL} = (D_{a_1 a_1} D_{a_2 a_2} D_{a_2 a_1} D_{a_1 a_2})^{1/4} = r' \cdot r' \cdot D_1 \cdot D_1)^{1/4} \qquad \ldots (7.12.1)$$

The geometric mean distance (GMD) from A to B is given by

$$D_m = (D_{a_1 b_1} D_{a_1 b_2} D_{a_2 b_1} D_{a_2 b_2})^{1/4} = (D_2 \sqrt{D_1^2 + D_2^2} \cdot \sqrt{D_1^2 + D_2^2} \cdot D_2)^{1/4}$$

$$D_m = (D_2 \sqrt{D_1^2 + D_2^2})^{1/2} \qquad \qquad ...(7.12.2)$$

The inductance of *both conductors*, or of the given line is

$$L = 2 \left(2 \times 10^{-7} \ln \frac{D_m}{D_{sL}} \right) = 4 \times 10^{-7} \ln \frac{(D_2 \sqrt{D_1^2 + D_2^2})^{1/2}}{(r' D_1)^{1/2}}$$

$$= 2 \times 10^{-7} \ln \frac{D_2 (D_1^2 + D_2^2)^{1/2}}{r' D_1} \quad \text{H/m} \qquad \qquad ...(7.12.3)$$

Example 7.5 A double-circuit single-phase line is shown in Fig. 7.10. Conductors a_1 and a_2 forming one path are connected in parallel and carry the current in one direction. Conductors b_1 and b_2 forming the return path are connected in parallel and carry current in the return direction. The diameter of each conductor is 25 mm. Calculate the inductance of the line per kilometre if $D_1 = 1$ m, and $D_2 = 2$ m.

Solution

From Equation (7.12.3) the line inductance is given by

$$L = 2 \times 10^{-7} \ln \frac{D_2 (D_1^2 + D_2^2)^{1/2}}{r' D_1}$$

$$D_1 = 1 \text{ m}, \quad D_2 = 2 \text{ m}, \quad r = \frac{25}{2} \text{ mm} = 12.5 \times 10^{-3} \text{ m}$$

$$D_2 (D_1^2 + D_2^2)^{1/2} = 2 (1 + 4)^{1/2} = 4.472 \text{ m}$$

$$r' D_1 = 0.7788 r D_1 = 0.7788 \times 12.5 \times 10^{-3} \times 1 = 9.735 \times 10^{-3} \text{ m}$$

$$\ln \frac{D_2 (D_1^2 + D_2^2)^{1/2}}{r' D_1} = \ln \frac{4.472}{9.735 \times 10^{-3}} = \ln 459.3 = 6.1298$$

$$L = 2 \times 10^{-7} \times 6.1298 \text{ H/m} = 0.2 \times 6.1298 \text{ mH/km} = 1.226 \text{ mH/km}.$$

7.13 INDUCTANCE OF DOUBLE-CIRCUIT THREE-PHASE LINES

Sometimes it is necessary to carry two circuits on the same tower for greater reliability of power supply. In case one of the circuits goes out of order due to some accident, the other is ready to supply the power requirements. These two three-phase circuits are connected in parallel electrically and are kept on either side of the tower.

The inductive reactance will be the same for these circuits provided that they are identical in construction and are operated in parallel. The effect of mutual inductance between conductors is negligible if large space is provided between them.

Consider the two circuits shown in Fig. 7.11 (a). One circuit consists of conductors a, b and c. The second circuit consists of conductors a', b' and c'. Conductors a and a' are connected in parallel to form phase A. Conductors b and b' are connected in parallel to form phase B. Phase C is composed of conductors c and c' connected in parallel. The spacing between the conductors given in Fig. 7.11 (a) refers to the first section of the transposition cycle. In the second section of transposition cycle conductor a occupies the position of conductor b and in the third section a takes the position of c. Fig. 7.11 (b) shows the transposition cycle of the double-circuit line.

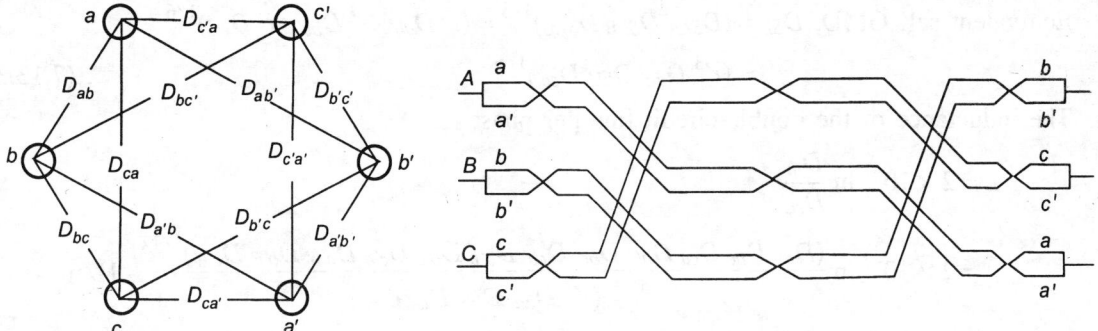

Fig. 7.11. (a) Double-circuit three-phase line. (b) Transposition cycle of a double-circuit three-phase line.

The formula for inductance of a three-phase unsymmetrical line has been found as

$$L = 2 \times 10^{-7} \ln \frac{D_m}{D_{sL}} \qquad \qquad ...(7.13.1)$$

where $D_m = (D_{AB} D_{BC} D_{CA})^{1/3}$

and $\quad D_{sL}$ = equivalent self GMD of a phase conductor over a complete transposition cycle.

The same formula may be extended for double-circuit three-phase line provided that the corresponding geometric mean values are substituted.

For a double-circuit line

$\qquad D_{AB}$ = mutual GMD between phases A and B

$\qquad \qquad$ = mutual GMD between groups a, a' and b, b'

$\qquad \qquad$ = $(D_{ab} D_{ab'} D_{a'b} D_{a'b'})^{1/4}$

$\qquad D_{BC}$ = mutual GMD between phases B and C

$\qquad \qquad$ = mutual GMD between groups b, b' and c, c'

$\qquad \qquad$ = $(D_{bc} D_{bc'} D_{b'c} D_{b'c'})^{1/4}$

$\qquad D_{CA}$ = mutual GMD between phases C and A

$\qquad \qquad$ = mutual GMD between groups c, c' and a, a'

$\qquad \qquad$ = $(D_{ca} D_{ca'} D_{c'a} D_{c'a'})^{1/4}$

$\qquad D_m$ = $(D_{AB} D_{BC} D_{CA})^{1/3}$

$\qquad \qquad$ = $(D_{ab} D_{bc} D_{ca} D_{ab'} D_{bc'} D_{ca'} D_{a'b} D_{b'c} D_{c'a} D_{a'b'} D_{b'c'} D_{c'a'})^{1/12}$ $\qquad$ ...(7.13.2)

$\qquad D_{SLA}$ = self geometric mean distance of phase A, that is, group a, a'

$\qquad \qquad$ = $(D_{aa} D_{a'a'} D_{a'a} D_{aa'})^{1/4} = (r' \, r' \, D_{aa'}^2)^{1/4} = (r' \, D_{aa'})^{1/2}$

$\qquad D_{SLB}$ = self geometric mean distance of phase B, that is, group b, b'

$\qquad \qquad$ = $(D_{bb} D_{b'b'} D_{bb'} D_{b'b})^{1/4} = (r' \, r' \, D_{bb'}^2)^{1/4} = (r' \, D_{bb'})^{1/2}$

$\qquad D_{SLC}$ = self GMD of phase C, that is, group c, c'

$\qquad \qquad$ = $(D_{cc} D_{c'c'} D_{cc'} D_{c'c})^{1/4} = (r' \, r' \, D_{cc'}^2)^{1/4} = (r' \, D_{cc'})^{1/2}$

Equivalent self GMD, $D_{SL} = (D_{SLA} D_{SLB} D_{SLC})^{1/3} = (r' D_{aa'} \cdot r' D_{bb'} \cdot r' D_{cc'})^{1/6}$

$$= (r'^3 D_{aa'} D_{bb'} D_{cc'})^{1/6} \qquad \qquad \text{...(7.13.3)}$$

The inductance of the double-circuit line per phase

$$L = 2 \times 10^{-7} \ln \frac{D_m}{D_{sL}}$$

$$= 2 \times 10^{-7} \ln \frac{(D_{ab} D_{bc} D_{ca} D_{ab'} D_{bc'} D_{ca'} D_{a'b} D_{b'c} D_{c'a} D_{a'b'} D_{b'c'} D_{c'a'})^{1/12}}{(r'^3 D_{aa'} D_{bb'} D_{cc'})^{1/6}} \quad \text{H/m}$$

$$\text{...(7.13.4)}$$

Equation (7.13.11) gives the inductance per phase. In this case a phase is composed of two identical conductors connected in parallel. Therefore, the inductance of *each conductor* is two times the inductance per phase.

The considerations of good voltage regulation, greater power limit, and high power factor require that the inductance should have a low value. The general formula for the line inductance indicates that it will be low if D_{SL} is large and D_m small. It is, therefore, necessary that the conductors constituting same phase should be spaced at as greater distance as possible, and the phases should be spaced as close as permissible. It is indicated that the best arrangement which gives the minimum inductance is that in which the two conductors of each phase are situated diagonally opposite to each other.

7.14 SPECIAL CASES OF DOUBLE-CIRCUIT LINES

Consider first the case when the conductors are situated at the corners of a regular hexagon. The distances are shown in the Fig. 7.12.

Here, $D_{ab} = D_{bc} = D_{ca'} = D_{c'a} = D_{a'b'} = D_{b'c'} = D$

$\qquad D_{ca} = D_{ab'} = D_{bc'} = D_{a'b} = D_{b'c} = D_{c'a'} = \sqrt{3} D$

$\qquad D_{aa'} = D_{bb'} = D_{cc'} = 2D$

The inductance per phase from Eq. (7.13.4)

$$L = 2 \times 10^{-7} \ln \frac{[D^6 \cdot (\sqrt{3}D)^6]^{1/12}}{[2r' (2D)^3]^{1/6}} = 2 \times 10^{-7} \ln \frac{(\sqrt{3}D^2)^{1/2}}{[r' 2D]^{1/2}}$$

$$= 10^{-7} \ln \frac{\sqrt{3} D}{2r'} \quad \text{H/m} \qquad \qquad \text{...(7.14.1)}$$

Now consider a double-circuit three-phase line with flat, vertical spacing (Fig. 7.13). The distance between the conductors are shown in Fig. 7.13.

By Eq. (7.13.4) the inductance of the above line per phase

$$L = 2 \times 10^{-7} \ln \frac{(v \cdot v \cdot 2v \cdot y \cdot y \cdot x \cdot y \cdot y \cdot x \cdot v \cdot v \cdot 2v)^{1/12}}{[r'^3 z \cdot x \cdot z]^{1/6}}$$

$$= 2 \times 10^{-7} \ln \left[2^{1/6} \left(\frac{v}{r'}\right)^{1/2} \left(\frac{y}{z}\right)^{1/3} \right] \quad \text{H/m} \qquad \text{...(7.14.2)}$$

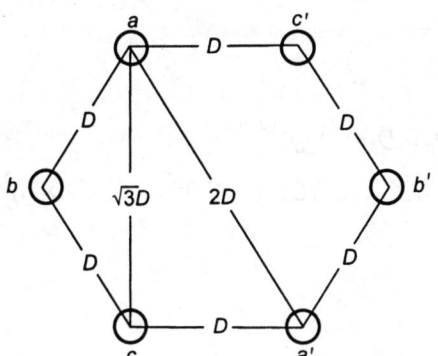

Fig. 7.12. Double-circuit three-phase line with hexagonal spacing.

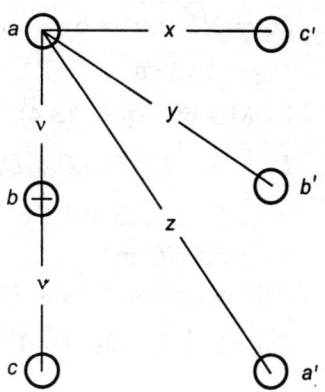

Fig. 7.13. Double-circuit three-phase line with flat vertical spacing.

Example 7.6 A three-phase circuit line consists of 7/4.75 mm hard drawn copper conductors. The arrangement of the conductors is shown in Fig. 7.14. The line is completely transposed Calculate the inductive reactance per phase per km of the system.

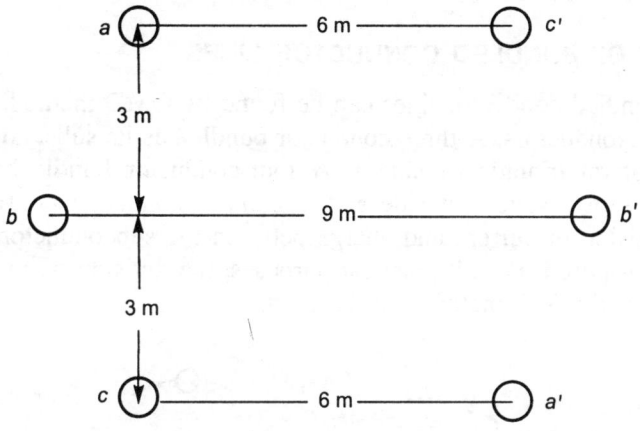

Fig. 7.14. Double-circuit line.

Solution

Diameter of each conductor = $3 \times 4.75 = 14.25$ mm

Radius of each conductor = $\frac{1}{2} \times 14.25 \times 10^{-3}$ m = 7.125×10^{-3} m

$$D_{ab} = \sqrt{3^2 + (1.5)^2} = 3.354 \text{ m}$$

$$D_{bc} = D_{a'b'} = D_{b'c'} = D_{ab} = 3.354 \text{ m}$$

$$D_{ab'} = \sqrt{3^2 + (7.5)^2} = 8.077 \text{ m}$$

$$D_{bc'} = D_{ba'} = D_{b'c} = D_{ab'} = 8.077 \text{ m}$$

$$D_{aa'} = \sqrt{6^2 + 6^2} = 6\sqrt{2} \text{ m}$$

$$D_{cc'} = 6\sqrt{2} \text{ m}$$

Mutual GMD by Eq. (7.13.4)

$$D_m = (D_{ab} D_{bc} D_{ca} D_{ab'} D_{bc'} D_{ca'} D_{a'b} D_{b'c} D_{c'a} D_{a'b'} D_{b'c'} D_{c'a'})^{1/12}$$

$$= (3.354 \times 3.354 \times 6 \times 8.076 \times 8.076 \times 6 \times 8.076 \times 8.076 \times 6 \times 3.354 \times 3.354 \times 6)^{1/12}$$

$$= 5.4576 \text{ m}$$

Self GMD per phase by Eq. (7.13.3)

$$D_{sL} = (r'^3 D_{aa'} D_{bb'} D_{cc'})^{1/6}$$

$$= [(0.7788 \times 7.125 \times 10^{-3})^3 \times 6\sqrt{2} \times 9 \times 6\sqrt{2}]^{1/6} = 0.21913 \text{ m}$$

Inductance per phase

$$L = 2 \times 10^{-7} \ln \frac{D_m}{D_{sL}} = 2 \times 10^{-7} \ln \frac{5.4576}{0.21913} = 6.43 \times 10^{-7} \text{ H/m} = 6.43 \times 10^{-4} \text{ H/km}$$

Inductive reactance per phase

$$X_L = 2\pi f L = 2\pi \times 50 \times 6.43 \times 10^{-7} = 20.2 \times 10^{-5} \text{ }\Omega/\text{m} = 20.2 \times 10^{-2} \text{ }\Omega/\text{km}.$$

7.15   INDUCTANCE OF BUNDLED CONDUCTOR LINES

The inductance of bundled conductor lines can be found by GMD method. The bundle consists of two, three, or four conductors. A three-conductor bundle has its subconductors situated at the vertices of an equilateral triangle of side s. A four-conductor bundle has its subconductors situated at the corners of a square of side s. The bundled-conductor configurations are shown in Fig. 7.15. The division of current and charge between the subconductors is not equal unless subconductors are transposed. For all practical purposes, the division of current and charge may be assumed uniform and GMD method can be used.

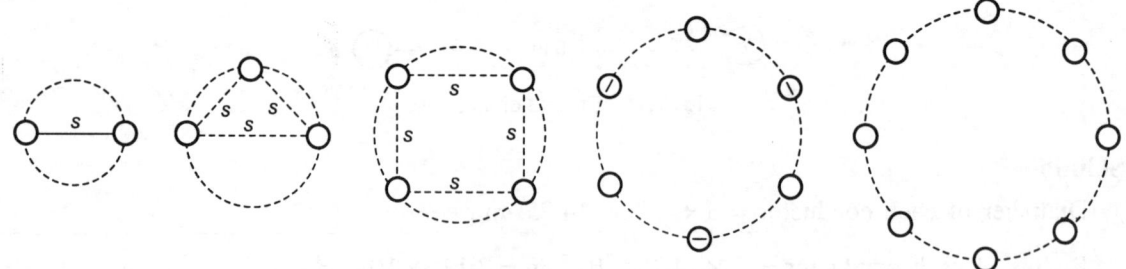

Fig. 7.15. Bundled conductors with 2, 3, 4, 6, 8 subconductors.

Let r = radius of each subconductor

s = spacing between the subconductors

r' = GMR of any one of the subconductors

D_{SL}^b = GMR of the bundled conductor

$$r' = re^{-1/4} = 0.7788r$$

GMR of a bundle consisting of two subconductors

D_{SL}^b = geometric mean of all the possible $(2)^2$ distances of the bundle including

the distances from the subconductors to themselves

$$= [r' \cdot r' \cdot s \cdot s]^{1/4} = (r' \, s)^{1/2} \qquad \qquad \qquad ...(7.15.1)$$

GMR of a bundle consisting of three subconductors

D_{SL}^b = geometric mean of all the possible $(3)^2$ distances of the bundle including

the distances from the subconductors to themselves

$$= [(r' \cdot s \cdot s)^3]^{1/9} = (r' \, s^2)^{1/3} \qquad \qquad \qquad ...(7.15.2)$$

GMR of a bundle consisting of four subconductors

D_{SL}^b = geometric mean of all the possible $(4)^2$ distances of the bundle including

the distances from the subconductors to themselves

$$= [(r' \cdot s \cdot s \sqrt{2} \, s)^4]^{1/16} = (\sqrt{2} \, r' \, s^3)^{1/4} \qquad \qquad ...(7.15.3)$$

GMR of a bundle consisting of n similar subconductors

Let there be n similar subconductors carrying equal currents and symmetrically placed around a ring of radius R (Fig. 7.16). The radius of each subconductor is r.

$$D_{SL}^b = (D_{s1} \cdot D_{s2} \ldots D_{sn})^{1/n}$$

$$= [(D_{11} D_{12} D_{13} \ldots D_{1n})^{1/n} (D_{21} D_{22} D_{23} \ldots D_{2n})^{1/n} \ldots (D_{n1} D_{n2} D_{n3} \ldots D_{nn})^{1/n}]^{1/n}$$

$$...(7.15.4)$$

As the position of each subconductor is identical in the group, the GMR of each subconductor is the same, i.e.,

$$D_{s1} = D_{s2} = \ldots = D_{sn}$$

$$D_{SL}^b = [(D_{s1})^n]^{1/n} = D_{s1}$$

From Fig. 7.16, if $\theta = \pi/n$

$$D_{11} = r'$$

$$D_{12} = 2R \sin \theta$$

$$D_{13} = 2R \sin 2\theta$$

$$D_{14} = 2R \sin 3\theta$$

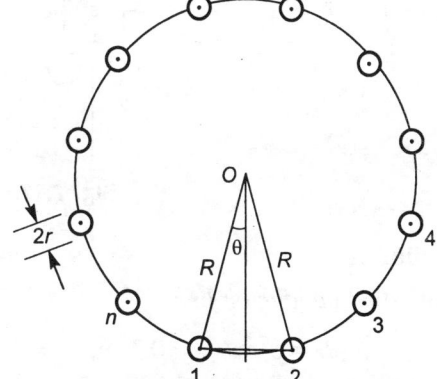

Fig. 7.16. Bundle of n similar conductors.

.

.

.

$$D_{1n} = 2R \sin (n-1) \, \theta$$

$\therefore$ $\quad D_{SL}^b = D_{e1} = [r' \, (2R \sin \theta) (2R \sin 2\theta) \ldots 2R \sin (n-1) \, \theta]^{1/n}$

$$= [r' (2R)^{n-1} \prod_{k=1}^{n-1} \sin k\theta]^{1/n} \qquad \qquad ...(7.15.5)$$

GMD of bundled conductor lines

Let D_m^b = GMD between conductors of one bundle and those of another.

The distance D_m^b is approximately equal to the GMD between the centres of the phase groups

$$D_m^b = (D_{AB} D_{BC} D_{CA})^{1/3} \qquad \qquad ...(7.15.6)$$

where D_{AB}, D_{BC} and D_{CA} are the distances between the centres of the bundles of the phases A, B and C.

Inductance of bundled conductor lines

The inductance of a bundled conductor line is given by

$$L = 2 \times 10^{-7} \ln \frac{D_m^b}{D_{SL}^b} \text{ H/m} \qquad \qquad ...(7.15.7)$$

Example 7.7 Fig. 7.17 shows a twin-conductor circuit of a three-phase 50 Hz line with horizontal spacing of 6 m. Each subconductor of the bundle has a diameter of 25 mm and spacing between the subconductors is 0.3 m. Each phase group shares the total current and charge equally and the line is completely transposed. Determine the line inductance per kilometre and the line inductive reactance per phase per kilometre.

If the subconductors of each phase are replaced by single conductors calculate the percentage decrease in inductive reactance due to bundling. Assume that the area of cross-section of each single conductor is equal to the total area of the two subconductors of a phase.

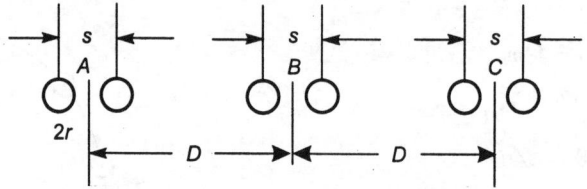

Fig. 7.17. Twin-conductor three-phase line.

Solution

Bundled conductor line

$$D = 6 \text{ m}, \quad s = 0.3 \text{ m}, \quad r = 12.5 \text{ mm} = 12.5 \times 10^{-3} \text{ m}$$

GMR of each group

$$D_{SL}^b = (r' s)^{1/2} = (0.7788 \, rs)^{1/2} = (0.7788 \times 12.5 \times 10^{-3} \times 0.3)^{1/2} = 5.404 \times 10^{-2} \text{ m}$$

GMD between conductors of one bundle and those of another

$$D_m^b = (D_{AB} D_{BC} D_{CA})^{1/3} = (D \cdot D \cdot 2D)^{1/3} = 1.26D = 1.26 \times 6 = 7.56 \text{ m}$$

The inductance of the bundled-conductor line

$$L_b = 2 \times 10^{-7} \ln \frac{D_m^b}{D_{SL}^b} = 2 \times 10^{-7} \ln \frac{7.56}{5.404 \times 10^{-2}} = 9.882 \times 10^{-7} \text{ H/m}$$

The inductive reactance of the bundled-conductor line per phase

$$X_{Lb} = 2\pi f L_b = 2\pi \times 50 \times 9.882 \times 10^{-7} = 31.05 \times 10^{-5} \text{ } \Omega/\text{m}$$

$$= 31.05 \times 10^{-5} \times 10^3 \text{ } \Omega/\text{km} = 0.3105 \text{ } \Omega/\text{km}$$

Equivalent line with single conductor per phase

Let r_1 be the radius of conductor in the equivalent line with single conductor per phase. For equal areas in both systems

$$n\pi r^2 = \pi r_1^2$$

where n is the number of subconductors in a bundle. Here $n = 2$.

$$\therefore \qquad r_1^2 = 2r^2$$

$$r_1 = \sqrt{2} \, r = \sqrt{2} \times 12.5 \times 10^{-3}$$

GMR of each conductor

$$D_{s1} = 0.7788 \, r_1 = 0.7788 \times \sqrt{2} \times 12.5 \times 10^{-3} = 13.767 \times 10^{-3} \text{ m}$$

$$\text{GMD, } D_{m1} = (D_{AB} D_{BC} D_{CA})^{1/3} = (D \cdot D \cdot 2D)^{1/3} = 1.26D = 1.26 \times 6 = 7.56 \text{ m}$$

The inductance of the line with one conductor per phase

$$L_1 = 2 \times 10^{-7} \ln \frac{D_{m1}}{D_{s1}} = 2 \times 10^{-7} \ln \frac{7.56}{13.767 \times 10^{-3}} = 12.616 \times 10^{-7} \text{ H/m}$$

The inductive reactance of line with one conductor per phase

$$X_{L1} = 2\pi f L_1 = 2\pi \times 50 \times 12.616 \times 10^{-7} = 39.63 \times 10^{-5} \text{ } \Omega/\text{m}$$

$$= 39.63 \times 10^{-5} \times 10^3 \text{ } \Omega/\text{km} = 0.3963 \text{ } \Omega/\text{km}$$

Reduction in inductive reactance due to bundling

$$= X_{L1} - X_{Lb}$$

Per unit reduction in inductive reactance

$$= \frac{X_{L1} - X_{Lb}}{X_{L1}} = \frac{2\pi f L_1 - 2\pi f L_b}{2\pi f L_1} = \frac{L_1 - L_b}{L_1}$$

$$= \frac{12.616 \times 10^{-7} - 9.882 \times 10^{-7}}{12.616 \times 10^{-7}} = \frac{12.616 - 9.882}{12.616} = 0.2167$$

Per cent reduction $= 0.2167 \times 100 = 21.67$

Alternatively, the per unit reduction in inductive reactance can be found as follows :

$$\text{Per unit reduction in reactance} = \frac{L_1 - L_b}{L_b} = \frac{\ln \dfrac{D_m}{0.7788 \, r \sqrt{2}} - \ln \dfrac{D_m}{(r' s)^{1/2}}}{\ln \dfrac{D_m}{0.7788 \, r \sqrt{2}}} = \frac{\ln \dfrac{(r' s)^{1/2}}{0.7788 \, r \sqrt{2}}}{\ln \dfrac{D_m}{0.7788 \, r \sqrt{2}}}$$

Now $\ln \dfrac{(r's)^{1/2}}{0.7788 \, r \sqrt{2}} = \ln \dfrac{5.404 \times 10^{-2}}{13.767 \times 10^{-3}} = 1.3674$

and $\ln \dfrac{D_m}{0.7788 \, r \sqrt{2}} = \ln 549.1 = 6.3038$

Per unit reduction $= \dfrac{1.3674}{6.3083} = 0.2167$

Per cent reduction $= 0.2167 \times 100 = 21.67$

Thus, it is seen that there is a 21.67 per cent reduction in the inductive reactance of the line due to bundling (twin-bundle). When GMR of each group is increased, by increasing the number or spacing of the subconductors the reduction in the line inductive reactance is greater.

7.16 LINE CAPACITANCE

Two conductors separated by a dielectric constitute a capacitor. Such a condition is fulfilled by an electric line. The conductors are supported on insulators at the supporting structures and at other places they are separated by air dielectric. The capacitance between the conductors is the charge per unit of potential difference. It is uniformly distributed along the line. The capacitance and conductance form the shunt or parallel admittance of the line. Capacitance effect is negligible small on the performance of short and low voltage lines. In case of high voltage and long lines, the capacitance assumes considerable importance. It affects the regulation, power factor, efficiency of the line and the stability of the system.

When an alternating voltage is applied to the line, a current flows in it. This current is called the *charging current*. It is independent of the load and flows in the line even at no load to supply the charge on the conductors. The charging current depends upon the capacitance of the line, the applied voltage and the frequency.

7.17 ELECTRIC FIELD OF A LONG STRAIGHT CONDUCTOR

Consider a long, straight, isolated conductor carrying a charge $+q$ coulombs per metre. The charge will be uniformly distributed over the surface of the conductor. The electric lines of flux will be straight, radial and uniformly spaced. The points equidistant from the conductor will be at the same potential and have the same flux density. All the cylinders concentric with the conductor will be equipotential surfaces. The electric flux density D_x at a point x metres from the axis of the conductor is the quotient of flux leaving the conductor per metre length and the curved surface of a cylinder 1 metre long and having a radius x metres.

$$D_x = \frac{q}{2\pi x \times 1} \ \text{C/m}^2 \qquad \qquad \qquad ...(7.17.1)$$

The electric field intensity or voltage gradient at the point considered is

$$G_x = \frac{D_x}{\varepsilon} = \frac{q}{(2\pi\varepsilon_0\varepsilon_r) \, x} \ \text{V/m} \qquad \qquad ...(7.17.2)$$

where ε_r is the permittivity of the medium.

The potential difference between two points A and B (Fig. 7.18) situated at distances D_1 and D_2 metres from the conductor is equal to the integral of the electric field intensity over a radial path between the equipotential surfaces passing through A and B. It does not matter whether or not A and B lie on the same radial line.

$$V_{AB} = -\int_{D_1}^{D_2} G_x \, dx = \int_{D_1}^{D_2} \frac{q}{2\pi x \varepsilon} \, dx$$

$$= \frac{q}{2\pi\varepsilon} \ln \frac{D_2}{D_1} \text{ volts} \qquad \qquad \qquad ...(7.17.3)$$

Equation (7.17.3) is very useful specially in determining charges and capacitances of a system of conductors.

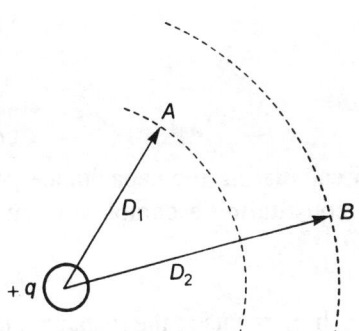

Fig. 7.18.

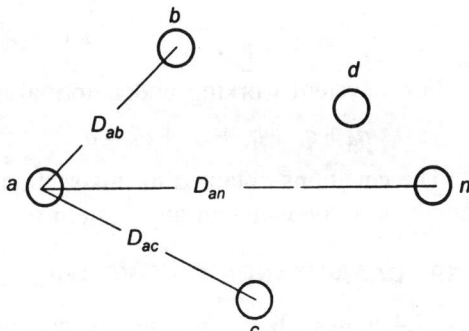

Fig. 7.19. System of n conductors.

7.18 SYSTEM OF CONDUCTORS

Consider a system of n conductors (Fig. 7.19) each of radius r forming a circuit.

Let the respective charges in coulombs per metre be q_a, q_b, q_c, ..., q_n. The spacing between the conductors are denoted by D_{ab}, D_{bc}, D_{cd}, ..., etc. The spacing are assumed to be so large in comparison to their radii that the distribution of charge is uniform around the periphery of each conductor.

The *principle of superposition* will be applied here to find out the potential difference between any two conductors. According to this principle, the difference of potential between to charged conductors is equal to the potential difference due to charge on first conductor alone, plus the potential difference due to the charge on second conductor alone, plus the potential difference due to other charged conductors in the field.

Using the result obtained in Eq. (7.17.3), the potential difference between two conductors a and b is given by

V_{ab} = potential difference between a and b due to charge q_a on a +

potential difference between a and b due to charge q_b on b + ... +

potential difference between a and b due to charge q_n on n

$$= \frac{q_a}{2\pi\varepsilon} \ln \frac{D_{ab}}{r_a} + \frac{q_b}{2\pi\varepsilon} \ln \frac{r_b}{D_{ba}} + \frac{q_c}{2\pi\varepsilon} \ln \frac{D_{cb}}{D_{ca}} + ... + \frac{q_n}{2\pi\varepsilon} \ln \frac{D_{nb}}{D_{na}}$$

If r_a, r_b, r_c, ... are replaced by D_{aa}, D_{bb}, D_{cc}, ... respectively for the sake of symmetrical result

$$V_{ab} = \frac{1}{2\pi\varepsilon}\left[q_a \ln \frac{D_{ab}}{D_{aa}} + q_b \ln \frac{D_{bb}}{D_{ba}} + q_c \ln \frac{D_{cb}}{D_{ca}} + ... + q_n \ln \frac{D_{nb}}{D_{na}}\right] \text{ volts} \qquad ...(7.18.1)$$

Similar expressions for the voltages between a and other conductors can be written as

$$V_{ac} = \frac{1}{2\pi\varepsilon}\left[q_a \ln \frac{D_{ac}}{D_{aa}} + q_b \ln \frac{D_{bc}}{D_{ba}} + q_c \ln \frac{D_{cc}}{D_{aa}} + ... + q_n \ln \frac{D_{nc}}{D_{na}}\right] \text{ volts} \qquad ...(7.18.2)$$

$$V_{an} = \frac{1}{2\pi\varepsilon}\left[q_a \ln \frac{D_{an}}{D_{aa}} + q_b \ln \frac{D_{bn}}{D_{ba}} + q_c \ln \frac{D_{cn}}{D_{ca}} + ... + q_n \ln \frac{D_{nn}}{D_{na}}\right] \text{ volts} \qquad ...(7.18.3)$$

$$V_{an} = \frac{1}{2\pi\varepsilon}\left[\sum_{x=a}^{x=n} q_x \ln \frac{D_{xn}}{D_{xa}}\right] \text{ volts} \qquad ...(7.18.4)$$

For a system working under normal conditions.

$$q_a + q_b + q_c + ... + q_n = 0 \qquad ...(7.18.5)$$

The equations obtained in this section will be utilized in calculating the capacitance per unit length of a conductor in any system of parallel conductors constituting a complete circuit.

7.19 CAPACITANCE OF TWO WIRE LINE

Fig. 7.4 shows a line consisting of two conductors a and b, each of radius r; the distance between the conductors being D.

From Eq. (7.18.1), the potential difference between a and b is

$$V_{ab} = \frac{1}{2\pi\varepsilon}\left[q_a \ln \frac{D_{ab}}{D_{aa}} + q_b \ln \frac{D_{bb}}{D_{ba}}\right] \qquad ...(7.19.1)$$

Here, $\quad q_a + q_b = 0$

so that $\quad q_b = -q_a \qquad ...(7.19.2)$

$$D_{ab} = D_{ba} = D \qquad ...(7.19.3)$$

$$D_{aa} = D_{bb} = r \qquad ...(7.19.4)$$

Substituting these values in Eq. (7.19.1)

$$V_{ab} = \frac{1}{2\pi\varepsilon}\left(q_a \ln \frac{D}{r} - q_a \ln \frac{r}{D}\right) = \frac{1}{2\pi\varepsilon} q_a \ln \left(\frac{D}{r}\right)^2 = \frac{1}{\pi\varepsilon} q_a \ln \frac{D}{r} \qquad ...(7.19.5)$$

The capacitance between the conductors

$$C_{ab} = \frac{q_a}{V_{ab}} = \frac{\pi\varepsilon}{\ln \frac{D}{r}} \text{ F/m} \qquad ...(7.19.6)$$

C_{ab} is referred to as *line-to-line capacitance*. It is shown in Fig. 7.20.

Since the two conductors a and b are oppositely charged, the potential of the points mid-way between the conductors is zero, that is, there is zero potential plane mid-way between a and b. The potential of each conductor is, therefore, $\frac{1}{2}V_{ab}$ with respect to neutral.

The capacitance between each conductor and point of zero potential n is

$$C_n = \frac{q_a}{\frac{1}{2} V_{ab}} = \frac{2\pi\varepsilon}{\ln \frac{D}{r}} \text{ F/m} \qquad \ldots(7.19.7)$$

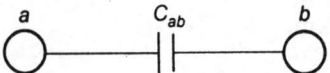

Fig. 7.20. Line-to-line capacitance.

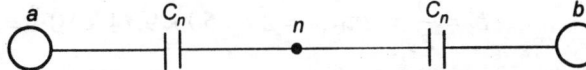

Fig. 7.21. Line-to-neutral capacitances.

C_n is called the *capacitance to neutral* or *capacitance to ground*. The term capacitance to neutral is more common in transmission calculations. C_n is shown in Fig. 7.21.

This is also to be noted that the capacitance C_{ab} between conductors a and b is a combination of two equal capacitances in series. Thus, the capacitance to neutral is twice the capacitance between the conductors, i.e.,

$$C_n = 2C_{ab} \qquad \ldots(7.19.8)$$

The absolute permittivity ε is given by

$$\varepsilon = \varepsilon_0 \, \varepsilon_r$$

where ε_0 is the permittivity of the free space and ε_r is the relative permittivity of the medium.

$$\varepsilon_0 = \frac{1}{4\pi \times 9 \times 10^9} = 8.85 \times 10^{-12} \text{ F/m}$$

For air $\varepsilon_r = 1$

$$C_n = \frac{2\pi\varepsilon}{\ln \frac{D}{r}} = \frac{1}{18 \times 10^9 \ln \frac{D}{r}} \text{ F/m} \qquad \ldots(7.19.9)$$

Capacitive reactance between one conductor and neutral

$$X_c = \frac{1}{2\pi f C_n}$$

Line-to-neutral susceptance, $b_c = \dfrac{1}{X_c} = 2\pi f C_n$

Example 7.8 A two-conductor, single-phase line operates at 50 Hz. The diameter of each conductor is 2 cm and are spaced 3 m apart. Calculate : (a) the capacitance of each conductor to neutral per km, (b) line-to-line capacitance, (c) capacitive susceptance to neutral per km.

Solution

$$D = 3 \text{ m}, \quad r = 0.01 \text{ m}$$

(a) The capacitance of each conductor to neutral

$$C_n = \frac{2\pi\varepsilon_0}{\ln \frac{D}{r}} = \frac{1}{18 \times 10^9 \ln \frac{3}{0.01}} = 9.74 \times 10^{-12} \text{ F/m} = 9.74 \times 10^{-9} \text{ F/km}$$

(b) Line-to-line capacitance

$$C_l = \frac{1}{2} C_n = \frac{1}{2} \times 9.74 \times 10^{-9} = 4.870 \times 10^{-9} \text{ F/km}$$

(c) Capacitive susceptance to neutral

$$b_c = \frac{1}{X_c} = 2\pi f C_n = 2\pi \times 50 \times 9.74 \times 10^{-9} = 3.06 \times 10^{-6} \text{ S/km}$$

7.20 CAPACITANCE OF THE SYMMETRICAL THREE-PHASE LINE

Let a balanced system of voltage be applied to a symmetrical three-phase line shown in Fig. 7.22 (a). The charges on conductors a, b and c are q_a, q_b and q_c respectively.

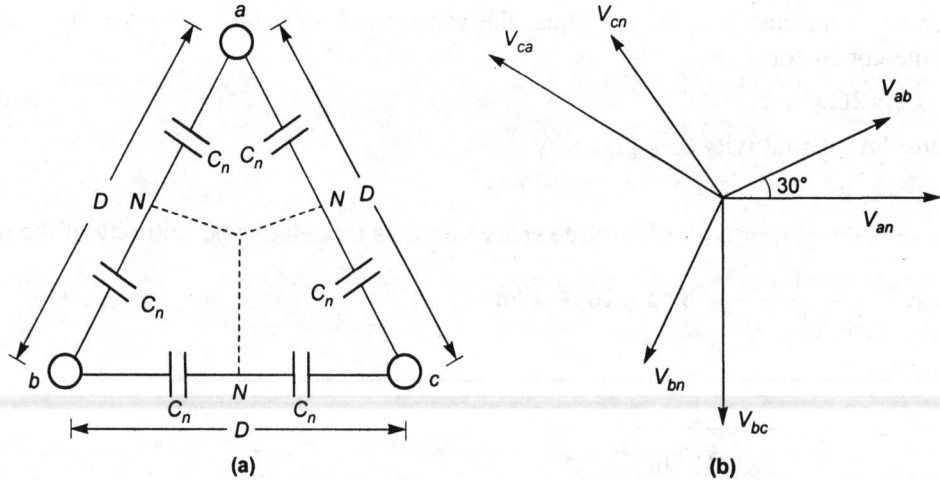

Fig. 7.22. (a) Three-phase line with equilateral spacing. (b) Phasor diagram of voltages.

Take the voltage of conductor a to neutral as a reference phasor [Fig. 7.22 (b)].

$$\mathbf{V}_{an} = V_{an} + j\,0$$

The potential difference between conductors a and b can be written from Eq. (7.18.1) as

$$\mathbf{V}_{ab} = \frac{1}{2\pi\varepsilon}\left(\mathbf{q}_a \ln \frac{D}{r} + \mathbf{q}_b \ln \frac{r}{D} + \mathbf{q}_c \ln \frac{D}{D}\right) \text{ volts} \qquad \qquad …(7.20.1)$$

Similarly, potential difference between a and c is

$$\mathbf{V}_{ac} = \frac{1}{2\pi\varepsilon}\left(\mathbf{q}_a \ln \frac{D}{r} + \mathbf{q}_b \ln \frac{D}{D} + \mathbf{q}_c \ln \frac{r}{D}\right) \text{ volts} \qquad \qquad …(7.20.2)$$

Adding Eqs. (7.20.1) and (7.20.2)

$$\mathbf{V}_{ab} + \mathbf{V}_{ac} = \frac{1}{2\pi\varepsilon}\left[2\mathbf{q}_a \ln \frac{D}{r} + (\mathbf{q}_b + \mathbf{q}_c) \ln \frac{r}{D}\right] \text{ volts} \qquad …(7.20.3)$$

Also, $\qquad \mathbf{q}_a + \mathbf{q}_b + \mathbf{q}_c = 0$ $\qquad \qquad \qquad \qquad \qquad \qquad …(7.20.4)$

$$\mathbf{q}_b + \mathbf{q}_c = -\mathbf{q}_a \qquad \qquad \qquad \qquad \qquad \qquad …(7.20.5)$$

Combining Eqs. (7.20.3) and (7.20.5)

$$\mathbf{V}_{ab} + \mathbf{V}_{ac} = 12\pi\varepsilon \left(2\mathbf{q}_a \ln \frac{D}{r} - \mathbf{q}_a \ln \frac{r}{D} \right) = \frac{1}{2\pi\varepsilon} \left(2\mathbf{q}_a \ln \frac{D}{r} + \mathbf{q}_a \ln \frac{D}{r} \right)$$

$$= \frac{3\mathbf{q}_a}{2\pi\varepsilon} \ln \frac{D}{r} \qquad\qquad\qquad ...(7.20.6)$$

$$\mathbf{V}_{ab} = \sqrt{3}\ V_{an} \underline{/30^\circ}$$

$$\mathbf{V}_{ac} = -\mathbf{V}_{ca} = \sqrt{3}\ V_{an} \underline{/-30^\circ}$$

$$\therefore \qquad \mathbf{V}_{ab} + \mathbf{V}_{ac} = 3V_{an} \qquad\qquad\qquad ...(7.20.7)$$

From Eqs (7.20.6) and (7.20.7)

$$3V_{an} = \frac{3q_a}{2\pi\varepsilon} \ln \frac{D}{r}$$

$$V_{an} = \frac{q_a}{2\pi\varepsilon} \ln \frac{D}{r} \qquad\qquad\qquad ...(7.20.8)$$

The line-to-neutral capacitance

$$C_n = \frac{q_a}{V_{an}} = \frac{2\pi\varepsilon_0}{\ln \dfrac{D}{r}} = \frac{1}{18 \times 10^9 \ln \dfrac{D}{r}} \ \ \text{F/m} = \frac{1}{18 \ln \dfrac{D}{r}} \ \ \mu\text{F/km} \qquad ...(7.20.9)$$

The comparison of Eqs. (7.20.9) and (7.19.7) shows that the capacitance of each conductor to neutral of a three-phase, three-wire, balanced system, with symmetrically spaced conductors, is the same as that of each conductor to neutral a single-phase line of the same size and spacing between the conductors.

7.21 CAPACITANCE OF AN UNSYMMETRICAL THREE-PHASE TRANSPOSED LINE

An exact method of calculation of capacitance for an unsymmetrical three-phase circuit is very complicated. A fair degree of accuracy may be achieved if it is assumed that the charge per unit length of the conductor remains the same in different positions of the transposition cycle. With this assumption $\mathbf{q}_a$, $\mathbf{q}_b$ and $\mathbf{q}_c$ have the same magnitude but 120 degrees apart in phase.

$$\mathbf{q}_a = q_a \underline{/0^\circ} \ ; \qquad \mathbf{q}_b = q_a \underline{/-120^\circ} \ ; \qquad \mathbf{q}_c = q_a \underline{/120^\circ}$$

Referring to Fig. 7.6, the equations for potential difference between conductors a and b can be written for three different positions of the transposition cycle.

The potential difference between a and b when a is in position 1, b in 2 and c in 3,

$$(\mathbf{V}_{ab})_1 = \frac{1}{2\pi\varepsilon} \left(\mathbf{q}_a \ln \frac{D_{12}}{r} + \mathbf{q}_b \ln \frac{r}{D_{12}} + \mathbf{q}_c \ln \frac{D_{23}}{D_{31}} \right) \qquad ...(7.21.1)$$

When a is in position 2, b in position 3, and c in position 1,

$$(\mathbf{V}_{ab})_2 = \frac{1}{2\pi\varepsilon} \left(\mathbf{q}_a \ln \frac{D_{23}}{r} + \mathbf{q}_b \ln \frac{r}{D_{23}} + \mathbf{q}_c \ln \frac{D_{31}}{D_{12}} \right) \qquad ...(7.21.2)$$

When a is in position 3, b in position 1, and c in position 2,

$$(V_{ab})_3 = \frac{1}{2\pi\varepsilon}\left(q_a \ln \frac{D_{31}}{r} + q_b \ln \frac{r}{D_{31}} + q_c \ln \frac{D_{12}}{D_{23}}\right) \qquad ...(7.21.3)$$

The average value of potential difference between conductors a and b in three positions of the transposition cycle,

$$V_{ab} = \frac{1}{3}[(V_{ab})_1 + (V_{ab})_2 + (V_{ab})_3]$$

$$= \frac{1}{6\pi\varepsilon}\left(q_a \ln \frac{D_{12}\,D_{23}\,D_{31}}{r^3} + q_b \ln \frac{r^3}{D_{12}\,D_{23}\,D_{31}} + q_c \ln \frac{D_{12}\,D_{23}\,D_{31}}{D_{12}\,D_{23}\,D_{31}}\right)$$

$$= \frac{1}{2\pi\varepsilon}\left(q_a \ln \frac{D_m}{r} + q_b \ln \frac{r}{D_m}\right) \qquad ...(7.21.4)$$

where $D_m = (D_{12}\,D_{23}\,D_{31})^{1/3}$, is the mutual geometric mean distance between the phases.

Similarly, the average potential different between a and c,

$$V_{ac} = \frac{1}{2\pi\varepsilon}\left(q_a \ln \frac{D_m}{r} + q_c \ln \frac{r}{D_m}\right) \qquad ...(7.21.5)$$

But $\qquad V_{ab} + V_{ac} = 3V_{an}$

Adding Eqs. (7.21.4) and (7.21.5)

$$3V_{an} = \frac{1}{2\pi\varepsilon}\left[2q_a \ln \frac{D_m}{r} + (q_b + q_c) \ln \frac{r}{D_m}\right]$$

Since $q_a + q_b + q_c = 0$ in a three-phase three-wire circuit,

$$q_b + q_c = -q_a$$

$$\therefore \qquad 3V_{an} = \frac{1}{2\pi\varepsilon}\left(2q_a \ln \frac{D_m}{r} - q_a \ln \frac{r}{D_m}\right) = \frac{1}{2\pi\varepsilon}\left(2q_a \ln \frac{D_m}{r} + q_a \ln \frac{D_m}{r}\right) = \frac{3q_a}{2\pi\varepsilon} \ln \frac{D_m}{r}$$

$$V_{an} = \frac{q_a}{2\pi\varepsilon} \ln \frac{D_m}{r} \qquad ...(7.21.6)$$

The capacitance of line-to neutral is thus,

$$C_n = \frac{q_a}{V_{an}} = \frac{2\pi\varepsilon}{\ln \dfrac{D_m}{r}} \text{ F/m} \qquad ...(7.21.7)$$

The same result may be derived by the following *alternative method* :

We have,

$$V_{ab} = \frac{1}{2\pi\varepsilon}\left(q_a \ln \frac{D_m}{r} + q_b \ln \frac{r}{D_m}\right)$$

$$= \frac{1}{2\pi\varepsilon}(q_a - q_b) \ln \frac{D_m}{r} \qquad ...(7.21.8)$$

$$\mathbf{q}_b = \mathbf{q}_a \,\underline{/-120°} = q_a \left(-\frac{1}{2} - j\frac{\sqrt{3}}{2} \right)$$

$$\mathbf{q}_a - \mathbf{q}_b = q_a \left(\frac{3}{2} + j\frac{\sqrt{3}}{2} \right) = \sqrt{3}\, q_a \,\underline{/30°} \qquad \qquad \text{...(7.21.9)}$$

Also, $\mathbf{V}_{ab} = \sqrt{3}\, V_{an} \,\underline{/30°}$ ...(7.21.10)

Combining the Eqs. (7.21.8), (7.21.9) and (7.21.10)

$$\sqrt{3}\, V_{an} \,\underline{/30°} = \frac{3}{2\pi\varepsilon}\, q_a \,\underline{/30°}\, \ln\frac{D_m}{r}$$

$$V_{an} = \frac{q}{2\pi\varepsilon}\, \ln\frac{D_m}{r}$$

$$C_n = \frac{q_a}{V_{an}} = \frac{2\pi\varepsilon_0}{\ln\dfrac{D_m}{r}}$$

$$C_n = \frac{2\pi\varepsilon_0}{\ln\dfrac{D_m}{D_{sc}}} \;\text{F/m} \qquad\qquad\qquad \text{...(7.21.11a)}$$

$$C_n = \frac{1}{18 \times 10^9 \ln\dfrac{D_m}{D_{sc}}} \;\text{F/m} \qquad\qquad\qquad \text{...(7.21.11b)}$$

$$C_n = \frac{1}{18 \ln\dfrac{D_m}{D_{sc}}} \;\mu\text{F/km} \qquad\qquad\qquad \text{...(7.21.11c)}$$

We have replaced r by D_{sc} since it is the self GMD of the surface of a conductor. Equation (7.21.7) may be used to calculate the line-to-neutral capacitance of each conductor when they are at the corners of an equilateral triangle.

In this case

$$D_{12} = D_{23} = D_{31} = D, \quad D_m = (D_{12}\, D_{23}\, D_{31})^{1/3} = D$$

$$C_n = \frac{2\pi\varepsilon_0}{\ln\dfrac{D_m}{r}} = \frac{2\pi\varepsilon_0}{\ln\dfrac{D}{r}} = \frac{1}{18 \times 10^9 \ln\dfrac{D}{r}} \;\text{F/m} \qquad\qquad \text{...(7.12.12)}$$

as obtained before.

For flat spacing (Fig. 7.7)

$$D_{12} = D_{23} = \frac{1}{2} D_{31} = D$$

$$D_m = (D \cdot D \cdot 2D)^{1/3} = 1.26D$$

$$C_n = \frac{2\pi\varepsilon_0}{\ln\dfrac{D_m}{r}} = \frac{1}{18 \times 10^9 \ln\dfrac{1.26D}{r}} \;\text{F/m} \qquad\qquad \text{...(7.21.13)}$$

7.22 CAPACITANCE OF A THREE-PHASE SINGLE-CIRCUIT UNTRANSPOSED LINE

A three-phase single circuit untransposed line is shown in Fig. 7.23.

The potential difference between a and b is given by

$$\mathbf{V}_{ab} = \frac{1}{2\pi\varepsilon_0}\left(\mathbf{q}_a \ln \frac{D_{12}}{r} + \mathbf{q}_b \ln \frac{r}{D_{12}} + \mathbf{q}_c \ln \frac{D_{23}}{D_{31}}\right)$$

The potential difference between a and c is given by

$$\mathbf{V}_{ac} = \frac{1}{2\pi\varepsilon_0}\left(\mathbf{q}_a \ln \frac{D_{31}}{r} + \mathbf{q}_b \ln \frac{D_{23}}{D_{12}} + \mathbf{q}_c \ln \frac{r}{D_{31}}\right)$$

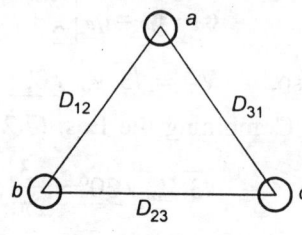

Fig. 7.23. Three-phase single-circuit untransposed line.

But $\qquad \mathbf{V}_{ab} + \mathbf{V}_{ac} = 3\mathbf{V}_{an}$

Therefore,

$$3\mathbf{V}_{an} = \frac{1}{2\pi\varepsilon_0}\left[\mathbf{q}_a\left(\ln \frac{D_{12}}{r} + \ln \frac{D_{31}}{r}\right) + \mathbf{q}_b\left(\ln \frac{r}{D_{12}} + \ln \frac{D_{23}}{D_{31}}\right) + \mathbf{q}_c\left(\ln \frac{D_{23}}{D_{31}} + \ln \frac{r}{D_{31}}\right)\right]$$

Since for a three-phase three-wire circuit

$$\mathbf{q}_a + \mathbf{q}_b + \mathbf{q}_c = 0$$

$$\mathbf{q}_c = -\mathbf{q}_a - \mathbf{q}_b$$

$$\therefore \quad 3\mathbf{V}_{an} = \frac{1}{2\pi\varepsilon_0}\left[\mathbf{q}_a\left(\ln \frac{D_{12}}{r} + \ln \frac{D_{31}}{r}\right) + \mathbf{q}_b\left(\ln \frac{r}{D_{12}} + \ln \frac{D_{23}}{D_{12}}\right) - \mathbf{q}_a \ln \frac{D_{23}\,r}{D_{31}^2} - \mathbf{q}_b \ln \frac{r\,D_{23}}{D_{31}^2}\right]$$

$$= \frac{1}{2\pi\varepsilon_0}\left(\mathbf{q}_a \ln \frac{D_{12}\,D_{31}\,D_{31}^2}{r^2\,r\,D_{23}} + \mathbf{q}_b \ln \frac{r\,D_{23}\,D_{31}^2}{D_{12}^2\,r\,D_{23}}\right)$$

But $\qquad \mathbf{q}_b = q_a\,\underline{/-120°} = q_a\left(-\frac{1}{2} - j\frac{\sqrt{3}}{2}\right)$

$$\therefore \quad 3V_{an} = \frac{1}{2\pi\varepsilon_0}\left(q_a \ln \frac{D_{12}\,D_{31}^3}{r^3\,D_{23}} - \frac{1}{2}\,q_a \ln \frac{D_{31}^2}{D_{12}^2} - j\frac{\sqrt{3}}{2}\,q_a \ln \frac{D_{31}^2}{D_{12}^2}\right)$$

$$= \frac{1}{2\pi\varepsilon_0}\,q_a\left(\ln \frac{D_{12}\,D_{31}^2\,D_{12}}{r^3\,D_{23}\,D_{31}} + j\sqrt{3}\ln \frac{D_{12}}{D_{31}}\right)$$

$$= \frac{1}{2\pi\varepsilon_0}\,q_a\left[\ln \left(\frac{D_{31}}{r}\right)^3 \frac{D_{12}^2}{D_{23}\,D_{31}} + j\sqrt{3}\ln \frac{D_{12}}{D_{31}}\right]$$

$$C_{an} = \frac{q_a}{V_{an}} = \frac{2\pi\varepsilon_0}{\frac{1}{3}\left[\ln \left(\frac{D_{31}}{r}\right)^3 \left(\frac{D_{12}^2}{D_{23}\,D_{31}}\right) + j\sqrt{3}\ln \frac{D_{12}}{D_{31}}\right]}$$

$$= \frac{2\pi\varepsilon_0}{\frac{1}{3}\left[\ln \left(\frac{D_{31}}{r}\right)^3 \left(\frac{D_{12}^2\,D_{12}}{D_{12}\,D_{23}\,D_{31}}\right) + j\sqrt{3}\ln \frac{D_{12}}{D_{31}}\right]} = \frac{2\pi\varepsilon_0}{\frac{1}{3}\left[\ln \left(\frac{D_{31}}{r}\right)^3 \cdot \frac{D_{12}^3}{D_m^3} + j\sqrt{3}\ln \frac{D_{12}}{D_{31}}\right]}$$

where $D_m = (D_{12}\,D_{23}\,D_{31})^{1/3}$

$\therefore$ $\qquad$ $C_{an} = \dfrac{2\pi\varepsilon_0}{\left(\ln\dfrac{D_{31}\,D_{12}}{r\,D_m} + j\dfrac{1}{\sqrt{3}}\ln\dfrac{D_{12}}{D_{31}}\right)}$ F/m

Similarly,

$$C_{bn} = \dfrac{2\pi\varepsilon_0}{\left(\ln\dfrac{D_{12}\,D_{23}}{r\,D_m} + j\dfrac{1}{\sqrt{3}}\ln\dfrac{D_{23}}{D_{12}}\right)}\ \text{F/m}$$

$$C_{cn} = \dfrac{2\pi\varepsilon_0}{\left(\ln\dfrac{D_{31}\,D_{23}}{r\,D_m} + j\dfrac{1}{\sqrt{3}}\ln\dfrac{D_{31}}{D_{23}}\right)}\ \text{F/m}$$

It is found that the phase capacitances are complex numbers.

Special case

For a line with equilateral spacing

$D_{12} = D_{23} = D_{31} = D$ (say)

$$C_{an} = C_{bn} = C_{cn} = C_n = \dfrac{2\pi\varepsilon_0}{\ln\dfrac{D\,D}{r\,D}} = \dfrac{2\pi\varepsilon_0}{\ln\dfrac{D}{r}} = \dfrac{1}{18\times10^9\ln\dfrac{D}{r}}\ \text{F/m}.$$

7.23 CHARGING CURRENT

The current associated with the capacitance of a line is called the *charging current*.

For a single-phase line, the charging current

$$\mathbf{I}_c = \dfrac{V_n}{-j\,X_c} = \dfrac{V}{-j/(\omega C)} = j\omega CV = j\,2\pi f CV\ \text{A} \qquad\qquad ...(7.23.1)$$

where C = line–to–line capacitance in farads

$\qquad X_c$ = capacitive reactance in ohms

$\qquad V$ = line voltage in volts

$$\text{Charging voltamperes} = VI_c = \dfrac{V\cdot V}{X_c} = \dfrac{V^2}{X_c}\ \text{VAr} \qquad\qquad ...(7.23.2)$$

Also, reactive voltamperes generated by the line = charging voltamperes of the line

$$Q = VI_c = \dfrac{V^2}{X_c}\ \text{VAr} \qquad\qquad ...(7.23.3)$$

For a three-phase line, charging current phase

$$\mathbf{I} = \dfrac{V_n}{-j\,X_c} = \dfrac{V_n}{-j/(\omega C_n)} = j\omega C_n V_n\ \text{A} \qquad\qquad ...(7.23.4)$$

where V_n = voltage to neutral in volts = phase voltage in volts

$\qquad C_n$ = capacitance to neutral in farads

Charging voltamperes per phase $= V_n I_c = V_n \cdot \dfrac{V_n}{X_c} = \dfrac{V_n^2}{X_c}$ VAr

Total three-phase charging voltamperes $= 3V_n I_c = \dfrac{3V_n^2}{X_c}$ VAr

Reactive voltamperes generated by the line = charging voltamperes of the line

$$Q_c = 3V_n I_c = \frac{3V_n^2}{X_c} = \frac{3}{X_c}\left(\frac{V_l}{\sqrt{3}}\right)^2 = \frac{V_l^2}{X_c} \text{ VAr} \qquad\qquad ...(7.23.5)$$

where V_l = line-to-line voltage in volts.

Example 7.9 Find the capacitance and capacitive reactance per km of the line described in Example 7.2. If the line operates at 132 kV, find the charging current per km, and the reactive voltamperes generated by the line per km.

Solution

$$C_n = \frac{1}{18\times 10^9 \ln \dfrac{D_m}{D_{sc}}} = \frac{1}{18\times 10^9 \ln \dfrac{(3\times 5\times 3.6)^{1/3}}{10.5\times 10^{-3}}}$$

$$= 9.4385\times 10^{-12} \text{ F/m} = 9.4385\times 10^{-9} \text{ F/km}$$

Capacitive reactance to neutral,

$$X_c = \frac{1}{2\pi f C_n} = \frac{1}{(2\pi\times 50\times 9.4385\times 10^{-9})} = 0.337\times 10^6 \text{ } \Omega/\text{km}$$

Charging current,

$$I_c = \frac{V_n}{X_c} = \frac{(132\times 10^3)/\sqrt{3}}{0.337\times 10^6} = 0.226 \text{ A/km}$$

Reactive voltamperes generated by the line

$$\frac{V_l^2}{X_c} = \frac{(132\times 10^3)^2}{0.336\times 10^6} = 51703 \text{ VAr} = 51.703 \text{ kVAr}$$

7.24 CAPACITANCE BY GMD METHOD

The GMD method of calculation gives the average values of charge or capacitance for the line. The method is particularly useful to find the charge or capacitance for asymmetrical lines. The results so obtained are only one or two per cent lesser than those obtained by rigorous method. The GMD method saves considerable labour. It is an approximate method but is easier to apply.

Equation (7.21.11) is applicable to three-phase double-circuit and bundled conductor lines also, provided that proper values of D_m and D_{sc} are taken. It should be noted clearly that since the charge is always concentrated on the surface of the conductor, the value of D_{sc} in Eq. (7.21.11) is the self-geometric mean distance of the surface of the conductor. But the self GMD of the surface of the conductor is equal to radius r, and *not* to GMR. To find D_{sc} for capacitance calculations r' used in the expression for D_{sL} in inductance calculation is replaced by r.

Example 7.10 Find the capacitance per kilometre to neutral and the capacitive reactance per phase per kilometre described in Example 7.3.

Solution

$$D_m = 7.56 \text{ m}, \quad D_{sc} = r = 12.5 \times 10^{-3} \text{ m}$$

$$C_n = \frac{1}{18 \times 10^9 \ln \dfrac{D_m}{r}} = \frac{1}{18 \times 10^9 \times \ln \dfrac{7.56}{12.5 \times 10^{-3}}}$$

$$= 8.764 \times 10^{-12} \text{ F/m} = 8.764 \times 10^{-12} \times 10^3 \text{ F/km} = 8.674 \times 10^{-9} \text{ F/km}$$

Capacitance reactance per phase per kilometre

$$X_c = \frac{1}{2\pi f C} = \frac{1}{2\pi \times 50 \times 8.674 \times 10^{-9}} = 3.669 \times 10^5 \ \Omega.$$

7.25 CAPACITANCE OF DOUBLE-CIRCUIT THREE-PHASE LINE

Let us first consider a double-circuit three-phase line in which the conductors are situated at the corners of a regular hexagon (Fig. 7.12). We shall use the relation,

$$C_n = \frac{2\pi\varepsilon}{\ln \dfrac{D_m}{D_{sc}}}$$

In this case

$$D_m = 3^{1/4} D \qquad \qquad \qquad ...(7.25.1)$$

$$D_{sc} = (2r D)^{1/2} \qquad \qquad \qquad ...(7.25.2)$$

$$C_n = \frac{2\pi\varepsilon}{\ln \dfrac{3^{1/4} D}{(2r D)^{1/2}}} = \frac{2\pi\varepsilon}{\dfrac{1}{2} \ln \dfrac{\sqrt{3} D}{2r}}$$

$$= \frac{4\pi\varepsilon}{\ln \dfrac{\sqrt{3} D}{2r}} \text{ F/m per phase of two conductors} \qquad \qquad ...(7.25.3)$$

Capacitance to neutral per conductor $= \dfrac{1}{2} \times$ capacitance to neutral per phase.

In a double-circuit line with flat, vertical spacing (Fig. 7.13)

$$D_m = 2^{1/6} v^{1/2} y^{1/3} x^{1/6} \qquad \qquad ...(7.25.4)$$

and $\qquad D_{sc} = r^{1/2} z^{1/3} x^{1/6} \qquad \qquad ...(7.25.5)$

$$\therefore \qquad C_n = \frac{2\pi\varepsilon}{\ln \dfrac{D_m}{D_{sc}}} = \frac{2\pi\varepsilon}{\ln \dfrac{2^{1/6} v^{1/2} y^{1/3} x^{1/6}}{r^{1/2} z^{1/3} x^{1/6}}} = \frac{2\pi\varepsilon}{\ln \left[\dfrac{2^{1/3} v}{r} \left(\dfrac{y}{z} \right)^{2/3} \right]^{1/2}}$$

$$= \frac{4\pi\varepsilon}{\ln\left[\frac{2^{1/3}v}{r}\left(\frac{y}{z}\right)^{2/3}\right]} \text{ F/m/phase} \qquad \qquad ...(7.25.6)$$

It is to be noted that GMD method gives the capacitance per phase rather than the capacitance per conductor.

Example 7.11 Calculate the capacitive reactance per phase per km of the line described in Example 7.6.

Solution

From Example 7.6,

$$D_m = 5.4576 \text{ m}$$

$$D_{sc} = (r^3 D_{aa'} D_{bb'} D_{cc'})^{1/6} = [(7.125 \times 10^{-3})^3 \times 6\sqrt{2} \times 9 \times 6\sqrt{2}]^{1/6} = 0.2483 \text{ m}$$

$$C_n = \frac{1}{18 \times 10^9 \ln \dfrac{D_m}{D_{sc}}} = \frac{1}{18 \times 10^9 \ln \dfrac{5.4576}{0.2483}} = 0.01798 \times 10^{-9} \text{ F/m} = 0.01798 \text{ }\mu\text{F/km}$$

Capacitive reactance to neutral

$$X_c = \frac{1}{2\pi f C_n} = \frac{1}{(2\pi \times 50 \times 0.01798 \times 10^{-6})} = 177051 \text{ }\Omega/\text{km}$$

7.26   EFFECT OF EARTH ON THE LINE CAPACITANCE

In our previous discussion of calculating capacitances, we have neglected the effect of earth. It was assumed there that the conductors are situated in free space. Actually, the conductors run parallel to the ground. The earth is assumed to behave like an infinite, perfectly conducting plane. Its presence, therefore, causes a change in the electric field of the line. The *method of images* as suggested by Lord Kelvin is used to take into account the effect of earth on capacitance of the line.

We know that there is a zero potential plane half-way between two conductors carrying equal and opposite charges. With this fact, it is logical to assume that the field between a single overhead charged conductor and the perfectly conducting plane the earth is identical with the field which would be produced by replacing the conducting plane by a hypothetical conductor similar to original conductor. This hypothetical conductor is located below the surface the earth at a distance equal to the height of the original conductor above the earth. It carries a charge equal and opposite to that of original conductor. Such a hypothetical conductor is called the *image conductor*.

For a system of charged conductors running parallel to the conducting plane, it may be assumed that each has got its own image located directly below it (Fig. 7.24). Each of them has a charge equal and opposite to the original conductor.

We shall utilize the method of images to illustrate the effect of earth on the capacitance of a two-wire line

Fig. 7.25 shows a two-wire single-phase line having conductors a and b. The spacing between

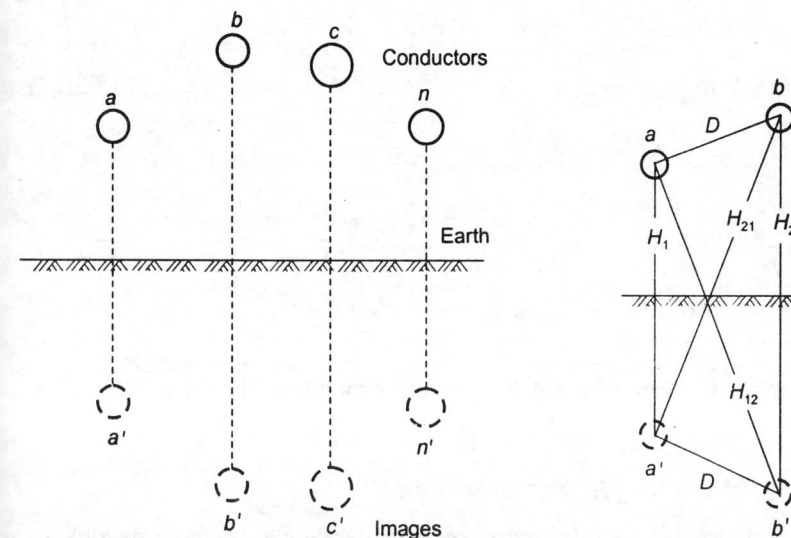

Fig. 7.24. System of conductors and their images.

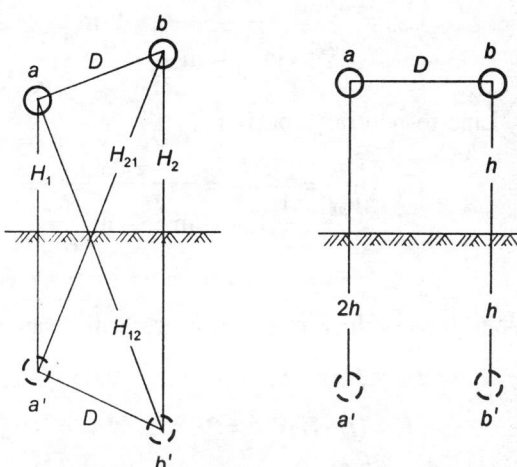

Fig. 7.25. Single-phase single-circuit line.

the conductors is D. a' and b' are images of a and b respectively. The charges on a and b are $+q$ and $-q$ respectively. The potential difference between a and b can be written as

$$V_{ab} = \frac{1}{2\pi\varepsilon_0}\left[q_a \ln \frac{D_{ab}}{D_{aa}} + q_b \ln \frac{D_{bb}}{D_{ba}} + q_{a'} \ln \frac{D_{a'b}}{D_{a'a}} + q_{b'} \ln \frac{D_{b'b}}{D_{b'a}} \right] \qquad \dots(7.26.1)$$

But, $q_a = q,\ q_b = -q_a = -q$

$\qquad q_{a'} = -q,\ q_{b'} = q$

$\qquad D_{aa} = D_{bb} = r$

$\qquad D_{ab} = D_{a'b'} = D,\ D_{aa'} = H_1,\ D_{bb'} = H_2$

$\qquad D_{ab'} = H_{12},\ D_{ba'} = H_{21}$

Substituting these values in Eq. (7.26.1) we get

$$V_{ab} = \frac{1}{2\pi\varepsilon_0}\left(q \ln \frac{D}{r} - q \ln \frac{r}{D} - q \ln \frac{H_{21}}{H_1} + q \ln \frac{H_2}{H_{12}} \right) = \frac{1}{2\pi\varepsilon_0}\left(2q \ln \frac{D}{r} - q \ln \frac{H_{12} H_{21}}{H_1 H_2} \right)$$

$$= \frac{2q}{2\pi\varepsilon_0}\left(\ln \frac{D}{r} - \frac{1}{2} \ln \frac{H_{12} H_{21}}{H_1 H_2} \right) = \frac{q}{\pi\varepsilon_0}\left[\ln \frac{D}{r} - \ln \frac{(H_{12} H_{21})^{1/2}}{(H_1 H_2)^{1/2}} \right]$$

We define the mean distances

$$H_s \overset{\Delta}{=} (H_1 H_2)^{1/2}, \qquad H_m \overset{\Delta}{=} (H_{12} H_{21})^{1/2}$$

Then the expression for V_{ab} can be written as

$$V_{ab} = \frac{q}{\pi\varepsilon_0}\left(\ln \frac{D}{r} - \ln \frac{H_m}{H_s} \right)$$

The line-to-line capacitance

$$C_{ab} = \frac{q}{V_{ab}} = \frac{\pi \varepsilon_0}{\ln \dfrac{D}{r} - \ln \dfrac{H_m}{H_s}} \text{ F/m} \qquad \qquad ...(7.26.2)$$

Line-to-neutral capacitance

$$C_n = \frac{q}{V_{an}} = \frac{q}{\frac{1}{2} V_{ab}} = \frac{2\pi \varepsilon_0}{\ln \dfrac{D}{r} - \ln \dfrac{H_m}{H_s}} = \frac{1}{18 \times 10^9 \left(\ln \dfrac{D}{r} - \ln \dfrac{H_m}{H_s} \right)} \text{ F/m} \qquad ...(7.26.3)$$

Special case

When the conductors a and b are at the same height h from the ground as shown in Fig. 7.25,

$$H_1 = H_2 = 2h, \qquad \qquad H_{12} = H_{21} = (D^2 + 4h^2)^{1/2}$$

$$H_s = (H_1 H_2)^{1/2} = 2h, \qquad H_m = (H_{12} H_{21})^{1/2} = (D^2 + 4h^2)^{1/2}$$

Equation (7.26.3) shows that there is a slight *increase* in capacitance of the line due to presence of earth. The effect diminishes as the height of the conductor above the earth is increased. It is not possible to calculate the capacitance accurately. Capacitance is affected by many variable factors such as the temperature, contour of the ground, vegetation in the vicinity of the line, presence of towers and other objects, etc. Because of these uncertainties, the effect of earth is negligible. The capacitance values obtained with an accuracy of about 0.5 to 1 per cent are regarded satisfactory.

Example 7.12 Determine the capacitance of the line described in Example 7.1 taking into account the effect of ground. The height of conductors above ground is 6 m.

Solution

When the effect of ground is taken into account, the line-to-neutral capacitance is given by

$$C_n = \frac{1}{18 \times 10^9 \left(\ln \dfrac{D}{r} - \ln \dfrac{H_m}{H_s} \right)}$$

$r = 10$ mm $= 10 \times 10^{-3}$, $\quad D = 3$ m

$h = 6$ m, $\quad H_1 = H_2 = 2h = 12$ m

$H_{12} = H_{21} = (D^2 + 4h^2)^{1/2} = (3^2 + 4 \times 6^2)^{1/2} = \sqrt{153}$

$H_m = (H_{12} H_{21})^{1/2} = H_{12} = \sqrt{153}$ m

$H_s = (H_1 H_2)^{1/2} = (2h \cdot 2h)^{1/2} = 2h = 12$ m

$$C_n = \frac{1}{18 \times 10^9 \left(\ln \dfrac{3}{10 \times 10^{-3}} - \ln \dfrac{\sqrt{153}}{12} \right)}$$

$$= \frac{1}{18 \times 10^9 \ln \dfrac{3 \times 12}{10 \times 10^{-3} \times \sqrt{153}}} = 9.792 \times 10^{-12} \text{ F/m}$$

Line-to-line capacitance

$$C_L = \frac{1}{2}\, C_n = 4.896 \times 10^{-12} \text{ F/m} = 4.896 \times 10^{-9} \text{ F/km}$$

7.27 EFFECT OF EARTH ON CAPACITANCE OF SINGLE-CIRCUIT THREE-PHASE LINE WITH TRANSPOSITION

Let us consider a three-phase line with general spacing. The method of images can be applied for determining the capacitance of this line. We shall assume that the line is fully transposed. The conductors, a, b and c carry the charges q_a, q_b and q_c and occupy positions, 1, 2 and 3, respectively, in the first portion of the transposition cycle. The effect of earth is considered by image conductors with charges $-q_a$, $-q_b$, and $-q_c$ respectively as shown in Fig. 7.26.

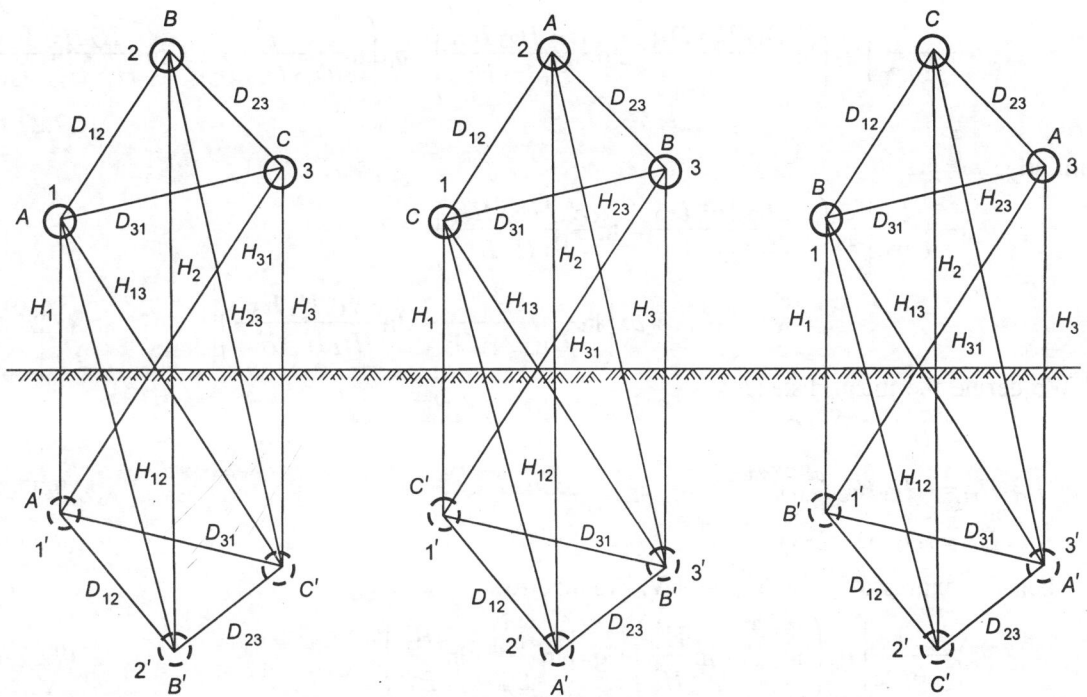

Fig. 7.26. Transposition cycle of a 3-phase line.

The equations for the three sections of the transpositions cycle can be written for the voltage drop V_{ab} as determined by three charged conductors and their images.

For the first portion of the transposition cycle shown in Fig. 7.26 (a),

$$(V_{ab})_1 = \frac{1}{2\pi\varepsilon_0}\left[\left(q_a \ln \frac{D_{12}}{r} + q_b \ln \frac{r}{D_{12}} + q_c \ln \frac{D_{23}}{D_{31}}\right) - \left(q_a \ln \frac{H_{12}}{H_1} + q_b \ln \frac{H_2}{H_{12}} + q_c \ln \frac{H_{23}}{H_{31}}\right)\right]$$

For the second portion of the transposition cycle shown in Fig. 7.26 (b),

$$(V_{ab})_2 = \frac{1}{2\pi\varepsilon_0}\left[\left(q_a \ln \frac{D_{23}}{r} + q_b \ln \frac{r}{D_{23}} + q_c \ln \frac{D_{31}}{D_{12}}\right) - \left(q_a \ln \frac{H_{23}}{H_2} + q_b \ln \frac{H_3}{H_{23}} + q_c \ln \frac{H_{31}}{H_{12}}\right)\right]$$

For the third portion of the transposition cycle shown in Fig. 7.26 (c),

$$(V_{ab})_3 = \frac{1}{2\pi\varepsilon_0}\left[\left(q_a \ln \frac{D_{31}}{r} + q_b \ln \frac{r}{D_{31}} + q_c \ln \frac{D_{12}}{D_{23}}\right) - \left(q_a \ln \frac{H_{31}}{H_3} + q_b \ln \frac{H_1}{H_{31}} + q_c \ln \frac{H_{12}}{H_{23}}\right)\right]$$

The average value of V_{ab} is given by

$$V_{ab} = \frac{1}{3}\left[(V_{ab})_1 + (V_{ab})_2 + (V_{ab})_3\right]$$

$$= \frac{1}{6\pi\varepsilon_0}\left[q_a\left(\ln \frac{D_{12} D_{23} D_{31}}{r^3} - \ln \frac{H_{12} H_{23} H_{31}}{H_1 H_2 H_3}\right) + q_b\left(\ln \frac{r^3}{D_{12} D_{23} D_{31}} - \ln \frac{H_2 H_3 H_1}{H_{12} H_{23} H_{31}}\right)\right.$$

$$\left. + q_c\left(\ln \frac{D_{23} D_{31} D_{12}}{D_{31} D_{12} D_{23}} - \ln \frac{H_{23} H_{31} H_{12}}{H_{31} H_{12} H_{23}}\right)\right]$$

$$= \frac{1}{6\pi\varepsilon_0}\left[q_a\left(\ln \frac{D_{12} D_{23} D_{31}}{r^3} - \ln \frac{H_{12} H_{23} H_{31}}{H_1 H_2 H_3}\right) + q_b\left(\ln \frac{r^3}{D_{12} D_{23} D_{31}} - \ln \frac{H_2 H_3 H_1}{H_{12} H_{23} H_{31}}\right)\right]$$

$$\qquad\qquad\qquad\qquad\qquad\qquad\qquad\qquad\qquad ...(7.27.1)$$

Similarly,

$$V_{ac} = \frac{1}{6\pi\varepsilon_0}\left[q_a\left(\ln \frac{D_{12} D_{23} D_{31}}{r^3} - \ln \frac{H_{12} H_{23} H_{31}}{H_1 H_2 H_3}\right)\right.$$

$$\left. + q_c\left(\ln \frac{r^3}{D_{12} D_{23} D_{31}} - \ln \frac{H_1 H_2 H_3}{H_{12} H_{23} H_{31}}\right)\right] \qquad ...(7.27.2)$$

We define the mean distances

$$H_s = (H_1 H_2 H_3)^{1/3}$$

$$H_m = (H_{12} H_{23} H_{31})^{1/3} \qquad\qquad\qquad\qquad ...(7.27.3)$$

$$D_m = (D_{12} D_{23} D_{31})^{1/3} \qquad\qquad\qquad\qquad ...(7.27.5)$$

Then the expressions for V_{ab} and V_{ac} reduce to

$$V_{ab} = \frac{1}{6\pi\varepsilon_0}\left[q_a\left(\ln \frac{D_m^3}{r^3} - \ln \frac{H_m^3}{H_s^3}\right) + q_b\left(\ln \frac{r^3}{D_m^3} - \ln \frac{H_s^3}{H_m^3}\right)\right] \qquad ...(7.27.6)$$

$$V_{ac} = \frac{1}{6\pi\varepsilon_0}\left[q_a\left(\ln \frac{D_m^3}{r^3} - \ln \frac{H_m^3}{H_s^3}\right) + q_c\left(\ln \frac{r^3}{D_m^3} - \ln \frac{H_s^3}{H_m^3}\right)\right] \qquad ...(7.27.7)$$

But $\qquad V_{ab} + V_{ac} = 3V_{an}$ $\qquad\qquad\qquad\qquad\qquad\qquad\qquad ...(7.27.8)$

and $\qquad q_b + q_c = -q_a$ $\qquad\qquad\qquad\qquad\qquad\qquad\qquad\qquad ...(7.27.9)$

Combination of Eqs. (7.27.7) to (7.27.9) gives

$$3V_{an} = \frac{3q_a}{6\pi\varepsilon_0}\left(\ln \frac{D_m^3}{r^3} - \ln \frac{H_m^3}{H_s^3}\right)$$

$$V_{an} = \frac{q_a}{6\pi\varepsilon_0}\left(3\ln\frac{D_m}{r} - 3\ln\frac{H_m}{H_s}\right)$$

$$V_{an} = \frac{q_a}{2\pi\varepsilon_0}\left(\ln\frac{D_m}{r} - \ln\frac{H_m}{H_s}\right) \qquad \text{...(7.27.10)}$$

$$C_{an} = \frac{q_a}{V_{an}} = \frac{2\pi\varepsilon_0}{\ln\dfrac{D_m}{r} - \ln\dfrac{H_m}{H_s}} \qquad \text{...(7.27.11)}$$

Equation (7.27.11) shows that the effect of ground gives a higher value for the capacitance than that the obtained by neglecting the ground effect.

7.28 CAPACITANCE OF BUNDLED CONDUCTOR LINES

As pointed out in Section 7.24 that the charge is always concentrated on the surface of the conductor, the value of D_{sc} should be taken as the self GMD of the surface of the conductor. But the self GMD of the surface of conductor is equal to its radius, and *not* to GMR. Therefore, to find D_{sc} for capacitance calculation r' used in the Eqs. (7.15.1) to (7.15.5) should be replaced by r.

If D_{sc}^b denotes the modified GMR to be used in calculation of capacitance, we have

$$C_n = \frac{2\pi\varepsilon}{\ln\dfrac{D_m^b}{D_{sc}^b}} \text{ F/m} \qquad \text{...(7.28.1)}$$

Then for a bundle consisting of two subconductors

$$D_{sc}^b = (rs)^{1/2} \qquad \text{...(7.28.2)}$$

For a bundle consisting of three subconductors at the corners of an equilateral triangle of side s

$$D_{sc}^b = (rs^2)^{1/3} \qquad \text{...(7.28.3)}$$

For a bundle consisting of four subconductors in a square formation of side s

$$D_{sc}^b = (\sqrt{2}\, rs^3)^{1/4} \qquad \text{...(7.28.4)}$$

Example 7.13 Find the capacitance per kilometre to neutral and the capacitive reactance to neutral per kilometre of the line described in Example 7.7.

If the subconductors of each phase are replaced by single conductors, calculate the percentage increase in capacitance, due to bundling. Assume that the area of cross-section of each single conductor is equal to the total area of the two subconductors of a phase.

Solution

Bundle Conductor Line

$$D = 6 \text{ m}, \quad s = 0.3 \text{ m}, \quad r = 1.25 \times 10^{-3} \text{ m}$$

GMR of each group

$$D_{sc}^b = (rs)^{1/2} = (12.5 \times 10^{-3} \times 0.3)^{1/2} = (37.5 \times 10^{-4})^{1/2} = 6.123 \times 10^{-2}$$

GMD between conductors of one bundle and those of another

$$D_m^b = 7.56 \text{ m}$$

$$\ln \frac{D_m^b}{D_{sc}^b} = \ln \frac{7.56}{6.123 \times 10^{-2}} = \ln 123.46 = 4.816$$

Line capacitance to neutral

$$C_n = \frac{1}{18 \times 10^9 \ln \dfrac{D_m^b}{D_{sc}^b}} = \frac{1}{18 \times 10^9 \times 4.816} \text{ F/m}$$

$$= \frac{1000}{18 \times 10^9 \times 4.816} \text{ F/km} = 11.535 \times 10^{-9} \text{ F/km}$$

Capacitive reactance to neutral per kilometre

$$X_{cb} = \frac{1}{2\pi f C_n} = \frac{1}{2\pi \times 50 \times 11.535 \times 10^{-9}} = 2.759 \times 10^5 \ \Omega$$

Equivalent Single-Conductor System

GMR of each conductor

$$D_{s1} = r_1 = \sqrt{2} \times r = \sqrt{2} \times 12.5 \times 10^{-3} = 17.67 \times 10^{-3} \text{ m}$$

GMD, $D_{m1} = 7.56$ m

$$\ln \frac{D_{m1}}{D_{s1}} = \ln \frac{7.56}{17.67 \times 10^{-3}} = \ln 427.8 = 6.058$$

Line capacitance to neutral

$$C_1 = \frac{1}{18 \times 10^9 \ln \dfrac{D_{m1}}{D_{s1}}} = \frac{1}{18 \times 10^9 \times 6.085} \text{ F/km}$$

$$= \frac{1000}{18 \times 10^9 \times 6.058} \text{ F/km} = 9.17 \times 10^{-9} \text{ F/km}$$

Capacitive reactance to neutral per kilometre

$$X_{c1} = \frac{1}{2\pi f C_1} = \frac{1}{2\pi \times 50 \times 9.17 \times 10^{-9}} = 3.47 \times 10^4 \ \Omega$$

Increase in Capacitance due to Bundling

It is observed that the capacitance of the line is increased due to bundling.

Increase in capacitance = $C_b - C_1$

Per unit increase in capacitance $= \dfrac{C_b - C_1}{C_1} = \dfrac{13.721 \times 10^{-9} - 9.17 \times 10^{-9}}{9.17 \times 10^{-9}}$

$$= \frac{13.271 - 9.17}{9.17} = 0.4472$$

Per cent increase in capacitance = 44.72

When GMR of each group is increased, by increasing the number or spacing of the subconductors the increase in the capacitance is greater.

Example 7.14 Fig. 7.27 shows a quadruple-conductor circuit of a single-circuit, three-phase, 50 Hz line with a horizontal spacing of 20 m. Each subconductor of the bundle has a diameter of 40 mm and spacing between the subconductors is 0.5 m. Each phase group shares the total current and charge equally and the line is completely transposed. Determine the inductive reactance and capacitive reactance per phase per km of the line.

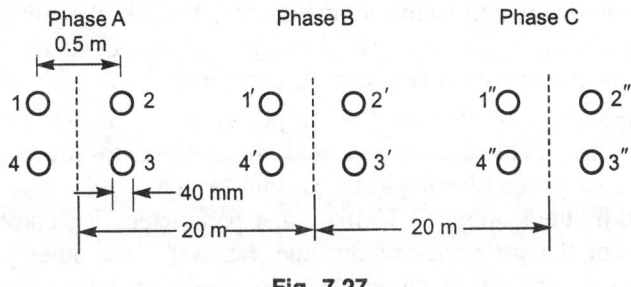

Fig. 7.27.

Solution

$$D = 20 \text{ m}, \quad s = 0.5 \text{ m}, \quad r = 20 \text{ mm} = 20 \times 10^{-3} \text{ m}$$

From Eq. (7.15.3), GMR of a bundle consisting of four subconductors is

$$D_{SL}^b = (\sqrt{2}\, r' s^3)^{1/4} = (\sqrt{2} \times 0.7788 \times 20 \times 10^{-3} \times 0.5^3)^{1/4} = 0.229071 \text{ m}$$

GMD between conductors of one bundle and those of another

$$D_m^b = (D_{AB} D_{BC} D_{CA})^{1/3} = (20 \times 20 \times 40)^{1/3} = 25.198421 \text{ m}$$

The inductance of the bundled conductor line

$$L_b = 2 \times 10^{-7} \ln \frac{D_m^b}{D_{SL}^b} \text{ H/m}$$

$$= 2 \times 10^{-7} \ln \frac{25.198421}{0.229071} = 9.401 \times 10^{-7} \text{ H/m} = 9.401 \times 10^{-4} \text{ H/km}$$

Inductive reactance of the bundled conductor line is given by

$$X_{Lb} = 2\pi f L_b = 2\pi \times 50 \times 9.401 \times 10^{-4} = 0.29534 \text{ }\Omega/\text{km}$$

From Eq. (7.28.4), the modified GMR of a bundle consisting of four subconductors is given by

$$D_{sc}^b = (\sqrt{2}\, rs^3)^{1/4} = (\sqrt{2} \times 20 \times 10^{-3} \times 0.5^3)^{1/4} = 0.2438449 \text{ m}$$

$$C_n = \frac{1}{18 \times 10^9 \ln \dfrac{D_m^b}{D_{sc}^b}} = \frac{1}{18 \times 10^9 \ln \dfrac{25.19842}{0.2438449}}$$

$$= 0.0119783 \times 10^{-9} \text{ F/m} = 11.9783 \times 10^{-9} \text{ F/km}$$

$$X_c = \frac{1}{2\pi fC} = \frac{1}{2\pi \times 50 \times 11.9783 \times 10^{-9}} = 2.65738 \times 10^5 \text{ } \Omega/\text{km}$$

7.29 SHUNT CONDUCTANCE

The line conductors are always separated by a dielectric. In overhead lines, the dielectric is air. The insulators on overhead lines may not be perfect. At high voltages there is a leakage of current from conductor to conductor along the surface or through insulators. There will also be some leakage of current from conductor to conductor through the air between them. Such currents are known as *leakage currents*. The leakage currents depend upon the atmospheric conditions and pollution like moisture and surface deposits.

By applying an alternating electric field some power is lost in the dielectrics due to their imperfections. Such a loss is called *dielectric loss*. In addition to above, there is corona loss also. The *shunt* or *leakage conductance* takes into account all these effects. The shunt conductance is denoted by the symbol G. Unlike other parameters, it cannot be easily calculated from the configuration or the properties of the line material. Like other parameters it may be assumed to be distributed uniformly along the line. Under ordinary conditions the leakage currents are small and, therefore, shunt conductance is usually neglected in calculations.

7.30 INTERFERENCE BETWEEN POWER AND COMMUNICATION LINES

When power lines and communication lines run in close proximity, there may be interference in the communication lines. The interference is mainly due to electromagnetic and electrostatic effects. Currents are induced in the communication lines due to electromagnetic induction. These currents produce distortion in speech signals. The electrostatic effect raises the potential of the communication conductors. The overvoltages may be dangerous for the equipment and personnel.

The interference between power and communication lines may be reduced by increasing the distance between them. In case they are running in proximity, interference may be reduced by transposing the conductors of power and communication lines.

7.31 ELECTROMAGNETIC EFFECT

Consider the three phase conductors, a, b and c of a three-phase power line on a transmission tower as shown in Fig. 7.28.

Let 1 and 2 be the conductors of a neighbouring communication line. These conductors may run on the same transmission towers or they may run on separate communication line poles. Let the currents through power conductors be I_a, I_b and I_c. Suppose that the distances between power conductors and communication conductors be D_{a1}, D_{a2}, D_{b1}, D_{b2}, D_{c1} and D_{c2} as shown in Fig. 7.28. Using Eq. (7.4.10), the flux linkages of conductors 1 and 2 due to currents in the power line, can be written as

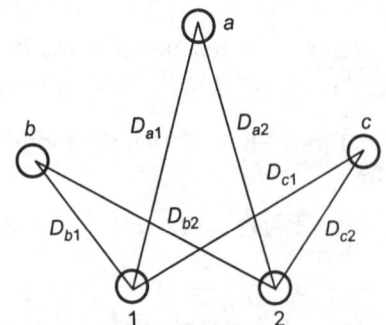

Fig. 7.28. Three-phase power line and communication line.

$$\lambda_1 = 2 \times 10^{-7} \left(I_a \ln \frac{1}{D_{a1}} + I_b \ln \frac{1}{D_{b1}} + I_c \ln \frac{1}{D_{c1}} \right) \text{Wb/m} \qquad \qquad ...(7.31.1)$$

$$\lambda_2 = 2 \times 10^{-7} \left(I_a \ln \frac{1}{D_{a2}} + I_b \ln \frac{1}{D_{b2}} + I_c \ln \frac{1}{D_{c2}} \right) \text{Wb/m} \qquad \qquad ...(7.31.2)$$

The total flux linkages of the communication line are given by

$$\lambda = \lambda_1 - \lambda_2 = 2 \times 10^{-7} \left(I_a \ln \frac{D_{a2}}{D_{a1}} + I_b \ln \frac{D_{b2}}{D_{b1}} + I_c \ln \frac{D_{c2}}{D_{c1}} \right) \text{Wb/m} \qquad \qquad ...(7.31.3)$$

The voltage induced in the communication line loop is given by

$$V = 2\pi f \lambda \ \text{V/m} \qquad \qquad ...(7.31.4)$$

where f is the supply frequency.

The flux linkages and induced voltage depend on currents I_a, I_b and I_c. We shall consider the following cases :

(a) *Balanced currents I_a, I_b and I_c.*

When the power line is transposed and the currents, I_a, I_b and I_c are balanced, the total flux linkages are zero. In practice, the power line is not transposed. In case of untransposed line a negligibly small voltage is induced in the communication line even when the power line currents are balanced.

(b) *Unbalanced currents I_a, I_b and I_c.*

When the power line currents I_a, I_b and I_c are unbalanced, voltages are induced in communication line. It is seen from Eq. (7.31.4) that the induced voltage is proportional to frequency. The third harmonic currents in power conductors are in phase. Therefore, if third harmonic currents are present, a voltage equal to three times the voltage due to fundamental frequency will be induced in the communication line. Also the higher frequencies may lie within audible range producing distortion.

The induced voltage due to electromagnetic induction can be reduced by increasing the distance between the power and communication conductors. Transposition of both the power and communication lines reduces the induced voltage.

7.32 ELECTROSTATIC EFFECT

Fig. 7.29 shows a three-phase power line with conductors a, b and c and a communication line with conductors 1 and 2. The images of a, b and c are a', b' and c' respectively. Similarly, the images of communication line conductors 1 and 2 are 1' and 2'.

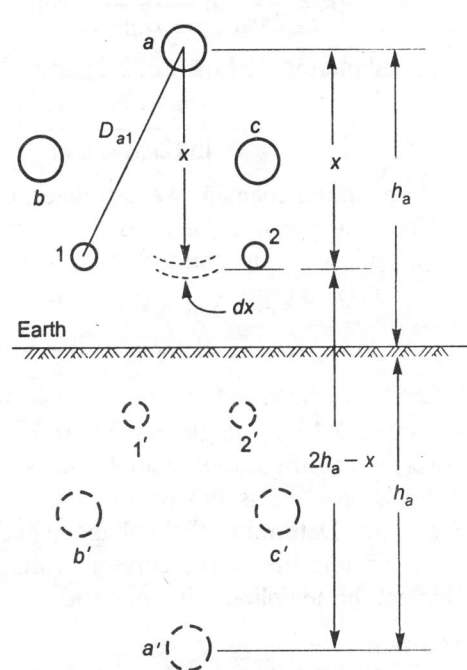

Fig. 7.29. Three-phase power line, communication line and their images.

Consider conductor a of the three-phase line. Let h_a be the height of a from the ground. Let the charge on conductor a be q coulombs per metre length. Then the charge on its image conductor a' will be $-q$ coulombs per metre length. The electric field intensity at a point situated at a distance x from the centre of conductor a is given by

$$G_x = \frac{q}{2\pi\varepsilon_0 \, x} + \frac{q}{2\pi\varepsilon_0 \, (2h_a - x)} \qquad \qquad ...(7.32.1)$$

The potential of conductor a with respect to earth is

$$V_a = \int_r^{h_a} G_x \, dx = \frac{q}{2\pi\varepsilon_0} \int_r^{h_a} \left[\frac{1}{x} + \frac{1}{(2h_a - x)} \right] dx$$

$$V_a = \frac{q}{2\pi\varepsilon_0} \ln \frac{2h_a - r}{r} \text{ volts} \qquad \qquad ...(7.32.2)$$

where r is the radius of conductor a. Suppose that conductor 1 is situated at a distance of D_{a1} from conductor a. The potential of conductor 1 with respect to ground can be found be considering the charge on conductor a and its image conductor a'. Therefore the potential of conductor 1 with respect to ground is given by

$$V_{1a} = \int_{D_{a1}}^{h_a} G_x \, dx = \frac{q}{2\pi\varepsilon_0} \int_{D_{a1}}^{h_a} \left(\frac{1}{x} + \frac{1}{2h_a - x} \right) dx$$

$$V_{1a} = \frac{q}{2\pi\varepsilon_0} \ln \frac{2h_a - D_{a1}}{D_{a1}} \text{ volts} \qquad \qquad ...(7.32.3)$$

Combination of Eqs. (7.32.2) and (7.32.3) gives

$$\mathbf{V}_{1a} = \mathbf{V}_a \frac{\ln \left[(2h_a - D_{a1})/D_{a1} \right]}{\ln (2h_a - r)/r]} \qquad \qquad ...(7.32.4)$$

In a similar manner, we can determine the potential of conductor 1 due to conductors b and c. The total potential of conductor 1 due to conductors a, b and c is given by the phasor sum of potentials due to conductors a, b and c. That is,

$$\mathbf{V}_1 = \mathbf{V}_{a1} + \mathbf{V}_{b1} + \mathbf{V}_{c1} \qquad \qquad ...(7.32.5)$$

Similarly, we can determine the potential of the second conductor 2 of the communication line.

Example 7.15 A single phase 50 Hz power line has the horizontal configuration with 1.1 m between the conductors. A telephone line is run on the same supports as shown in Fig. 7.30. Determine the voltage induced per kilometre in the telephone line if the current in the power line is 60 A. Neglect the telephone line current.

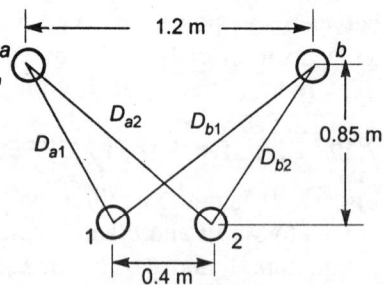

Fig. 7.30.

Solution

Flux linkages of telephone line conductor 1

$$\lambda_1 = 2 \times 10^{-7} \left(I_a \ln \frac{1}{D_{a1}} + I_b \ln \frac{1}{D_{b1}} \right)$$

$$I_a + I_b = 0, \quad I_a = -I_b = I \text{ (say)}$$

$$\lambda_1 = 2 \times 10^{-7} \left(I \ln \frac{1}{D_{a1}} - I \ln \frac{1}{D_{b1}} \right) = 2 \times 10^{-7} I \ln \frac{D_{b1}}{D_{a1}}$$

Flux linkages of telephone line conductor 2

$$\lambda_2 = 2 \times 10^{-7} \left(I_a \ln \frac{1}{D_{a2}} + I_b \ln \frac{1}{D_{b2}} \right)$$

$$\lambda_2 = 2 \times 10^{-7} \left(I \ln \frac{1}{D_{a2}} - I \ln \frac{1}{D_{b2}} \right) = 2 \times 10^{-7} I \ln \frac{D_{b2}}{D_{a2}}$$

Total flux linkages of the telephone circuit

$$\lambda = \lambda_1 - \lambda_2 = 2 \times 10^{-7} I \left(\ln \frac{D_{b1}}{D_{a1}} - \ln \frac{D_{b2}}{D_{a2}} \right) = 2 \times 10^{-7} I \ln \frac{D_{b1} D_{a2}}{D_{b2} D_{a1}} \text{ Wb/m}$$

The voltage induced in the telephone circuit

$$V = 2\pi f \lambda = 2\pi \times 50 \times 2 \times 10^{-7} \times 60 \ln \frac{D_{b1} D_{a2}}{D_{b2} D_{a1}} \text{ V/m}$$

$$D_{a2} = D_{b1} = (0.8^2 + 0.85^2)^{1/2} = (1.3625)^{1/2}$$

$$D_{a1} = D_{b2} = (0.4^2 + 0.85^2)^{1/2} = (0.8825)^{1/2}$$

$$\therefore \quad V = 2\pi \times 50 \times 2 \times 10^{-7} \times 60 \ln \frac{1.3625}{0.8825} = 16373.39 \times 10^{-7} \text{ V/m} = 1.6373 \text{ V/km}$$

Example 7.16 A three phase, 132 kV, 50 Hz, 70 km long transmission line has conductors of 19.53 mm diameter each and the mean equilateral spacing of 6 m. The height of the lowest conductor is 15 m above ground. If a telephone line is run on the same supports as shown in Fig. 7.31, calculate the voltage induced in the telephone circuit due to electromagnetic effect of the power line. The power line carries a balanced current of 200 A. Also calculate the voltage induced in telephone conductor 1 due to electrostatic effect.

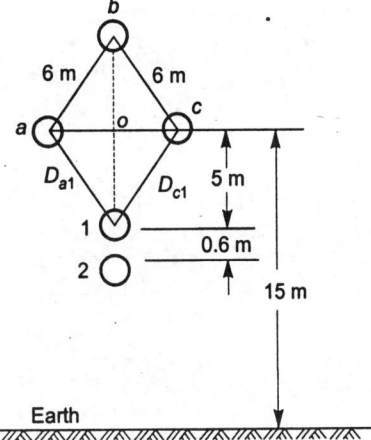

Fig. 7.31.

Solution

$$D_{b0} = 6 \cos 30° = 5.196 \text{ m}$$

$$D_{b1} = D_{b0} + D_{01} = 5.196 + 5 = 10.196 \text{ m}$$

$$D_{b2} = D_{b1} + D_{12} = 10.196 + 0.6 = 10.796 \text{ m}$$

$$D_{a1} = D_{c1} = \sqrt{D_{a0}^2 + D_{01}^2} = \sqrt{3^2 + 5^2} = 5.83 \text{ m}$$

$$D_{a2} = D_{c2} = \sqrt{3^2 + 5.6^2} = 6.353 \text{ m}$$

$$I_a = 200 \underline{/0°} \text{ A}, \qquad I_b = 200 \underline{/-120°} \text{ A}, \qquad I_c = 200 \underline{/120°} \text{ A}$$

The total flux linkages of the telephone circuit are given by Eq. (7.31.3) as

$$\lambda = 2 \times 10^{-7} \left(\mathbf{I}_a \ln \frac{D_{a2}}{D_{a1}} + \mathbf{I}_b \ln \frac{D_{b2}}{D_{b1}} + \mathbf{I}_c \ln \frac{D_{c2}}{D_{c1}} \right)$$

$$= 2 \times 10^{-7} \left(200 \underline{/0^\circ} \ln \frac{6.353}{5.830} + 200 \underline{/-120^\circ} \ln \frac{10.796}{10.196} + 200 \underline{/120^\circ} \ln \frac{6.353}{5.830} \right)$$

$$= 400 \times 10^{-7} \, (0.0859 \underline{/0^\circ} + 0.0572 \underline{/-120^\circ} + 0.0859 \underline{/120^\circ})$$

$$= 400 \times 10^{-7} \, (0.0859 - 0.0286 - j\,0.0495 - 0.0429 + j\,0.0744)$$

$$= 400 \times 10^{-7} \, (0.144 + j\,0.0249)$$

$$= 400 \times 10^{-7} \times 0.02876 \underline{/60^\circ}$$

$$= 1.15 \times 10^{-6} \text{ Wb/m}$$

Voltage induced in the telephone circuit

$$|V| = |2\pi f\lambda|$$

$$= 2\pi \times 50 \times 1.15 \times 10^{-6} \text{ V/m}$$

$$= 0.361 \text{ V /km}$$

In Δ *boc*,

$$(bc)^2 = (bo)^2 + (oc)^2$$

$$(bo)^2 = (bc)^2 - (oc)^2 = 6^2 - 3^2 = 27$$

$\therefore$ $\qquad bo = 5.196$ m

$$h_b = 5.196 + 15 = 20.196 \text{ m}$$

$$h_a = h_c = 15 \text{ m}$$

$$D_{a1} = 5.83 \text{ m}, \quad D_{b1} = 10.196 \text{ m},$$

$$r = \frac{19.53}{2} \text{ mm} = 9.765 \times 10^{-3} \text{ m}$$

$$\mathbf{V}_{1a} = \mathbf{V}_a \frac{\ln\left[\dfrac{2h_a - D_{a1}}{D_{a1}}\right]}{\ln\left[\dfrac{2h_a - r}{r}\right]} = \mathbf{V}_a \frac{\ln\left[\dfrac{2 \times 15 - 5.83}{5.83}\right]}{\ln\left[\dfrac{2 \times 15 - 9.765 \times 10^{-3}}{9.765 \times 10^{-3}}\right]}$$

$$= 0.1771 \, \mathbf{V}_a$$

$$\mathbf{V}_{1b} = \mathbf{V}_b \frac{\ln\left[\dfrac{2h_b - D_{b1}}{D_{b1}}\right]}{\ln\left[\dfrac{2h_b - r}{r}\right]} = \mathbf{V}_b \frac{\ln\left[\dfrac{2 \times 20.196 - 10.196}{10.196}\right]}{\ln\left[\dfrac{2 \times 20.196 - 9.765 \times 10^{-3}}{9.765 \times 10^{-3}}\right]}$$

$$= 0.1304 \, \mathbf{V}_b$$

$$V_{1c} = V_c \frac{\ln\left[\dfrac{2h_c - D_{c1}}{D_{c1}}\right]}{\ln\left[\dfrac{2h_c - r}{r}\right]} = V_c \frac{\ln\left[\dfrac{2\times 15 - 5.83}{5.83}\right]}{\ln\left[\dfrac{2\times 15 - 9.765\times 10^{-3}}{9.765\times 10^{-3}}\right]}$$

$$= 0.1771\, V_a$$

$$V_a = \left(\frac{132}{\sqrt{3}}\right)\times 10^3\, \underline{/0°}$$

$$V_b = \left(\frac{132}{\sqrt{3}}\right)\times 10^3\, \underline{/-120°}$$

$$V_c = \left(\frac{132}{\sqrt{3}}\right)\times 10^3\, \underline{/120°}$$

$$V_1 = V_{1a} + V_{1b} + V_{1c}$$

$$= \frac{132\times 10^3}{\sqrt{3}}\,[0.1771\,\underline{/0°} + 0.1304\,\underline{/-120°} + 0.1771\,\underline{/120°}]$$

$$= \frac{132\times 10^3}{\sqrt{3}}\,(0.1771 - 0.0652 - j\,0.1129 - 0.0885 + j\,0.1534)$$

$$= \frac{132\times 10^3}{\sqrt{3}}\,(0.0234 + j\,0.0405)$$

$$= \frac{132\times 10^3}{\sqrt{3}}\,(0.04677\,\underline{/60°})$$

$$= 3564\ \text{V} = 3.564\ \text{kV}$$

Example 7.17 A multi-conductor single-phase line has three conductors a, b and c each of diameter 40 mm for lead and two conductors d and e of diameter 80 mm for return circuit as shown in Fig. 7.32. Find the inductance per unit length on each side of the line and total inductance of the line.

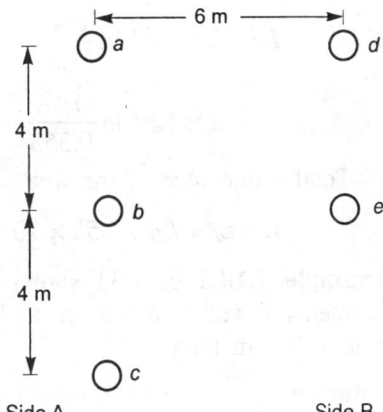

Fig. 7.32.

Solution

The mutual geometric mean distance between two conductor groups A and B

$$D_m = (D_{ad}\, D_{ae}\, D_{bd}\, D_{be}\, D_{cd}\, D_{ce})^{1/6}$$

where $\quad D_{ad} = D_{be} = 6$ m

$$D_{ae} = D_{bd} = D_{ce} = \sqrt{4^2 + 6^2} = 7.21\ \text{m}$$

$$D_{ed} = \sqrt{6^2 + 8^2} = 10\ \text{m}$$

$\therefore \qquad D_m = (6 \times 7.21 \times 7.21 \times 6 \times 10 \times 7.21)^{1/6} = 7.16\ \text{m}$

Self geometric mean distance of side A,

$$D_{SA} = (D_{aa}\, D_{ab}\, D_{ac}\, D_{ba}\, D_{bb}\, D_{bc}\, D_{ca}\, D_{cb}\, D_{cc})^{1/9}$$

Also, $D_{kk} = r'_k = r_k e^{-1/4} = 0.7788 r_k$

$D_{aa} = D_{bb} = D_{cc} = 0.7788 \times 20 \times 10^{-3}$ m

$D_{ab} = D_{ba} = 4$ m, $D_{bc} = D_{cb} = 4$ m

$D_{ac} = D_{ca} = 8$ m

Hence, $D_{SA} = [(D_{aa} D_{bb} D_{cc}) \times (D_{ab} D_{ba}) \times (D_{ac} D_{ca}) \times (D_{bc} D_{cb})^{1/9}$

$= [(0.7788 \times 20 \times 10^{-3})^3 (4 \times 4)(8 \times 8)(4 \times 4)]^{1/9}$

$= 0.734$ m

Self geometric mean distance of side B,

$D_{SB} = (D_{dd} D_{de} D_{ee} D_{ed})^{1/4}$

$D_{dd} = 0.7788 \times 40 \times 10^{-3}$

$D_{ee} = 0.7788 \times 40 \times 10^{-3}$

$D_{de} = D_{ed} = 4$ m

Hence, $D_{SB} = [(0.7788 \times 40 \times 10^{-3})^2 \times 4 \times 4]^{1/4}$

$= 0.353$ m

Inductance of side A,

$$L_A = 2 \times 10^{-7} \ln \frac{D_m}{D_{SA}}$$

$$= 2 \times 10^{-7} \ln \frac{7.16}{0.734} = 4.55 \times 10^{-7} \text{ H/m}$$

Inductance of side B,

$$L_B = 2 \times 10^{-7} \ln \frac{D_m}{D_{SB}}$$

$$= 2 \times 10^{-7} \ln \frac{7.16}{0.353} = 6.02 \times 10^{-7} \text{ H/m}$$

Total inductance of the line

$$L = L_A + L_B = 4.55 \times 10^{-7} + 6.02 \times 10^{-7} = 10.57 \times 10^{-7} \text{ H/m}$$

Example 7.18 Fig. 7.33 shows two conductors A and B made of filaments. The radii of filaments A and B are 3 cm and 4 cm respectively. Calculate the loop inductance of the line if it is 100 m long.

Solution

The mutual geometric mean distance between two conductor groups A and B

$$D_m = (D_{ad} D_{ae} D_{bd} D_{be} D_{cd} D_{ce})^{1/6}$$

$$= (4 \times 4.3 \times 3.5 \times 3.8 \times 2 \times 2.3)^{1/6} = 3.189 \text{ m}$$

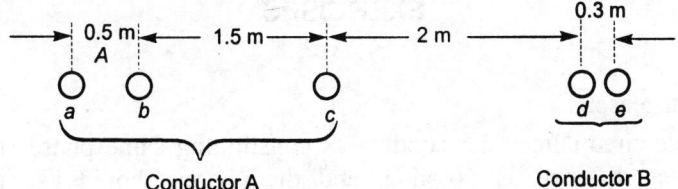

Fig. 7.33.

Self geometric mean distance of side A,

$$D_{SA} = (D_{aa} D_{ab} D_{ac} D_{ba} D_{bb} D_{bc} D_{ca} D_{cb} D_{cc})^{1/9}$$

Also, $D_{kk} = r'_k = r_k e^{-1/4} = 0.7788 r_k$

$$D_{aa} = D_{bb} = D_{cc} = 0.7788 \times 3 \times 10^{-2} \text{ m}$$

$$D_{ab} = D_{ba} = 0.5 \text{ m}, \qquad D_{bc} = D_{cb} = 1.5 \text{ m}, \qquad D_{ca} = D_{ac} = 2 \text{ m}$$

Hence, $D_{SA} = [D_{aa} D_{bb} D_{cc})(D_{ab} D_{ba})(D_{ac} D_{ca})(D_{bc} D_{cb})]^{1/9}$

$$= [(0.7788 \times 3 \times 10^{-2})^3 (0.5)^2 (2)^2 (1.5)^2]^{1/9} = 0.3128 \text{ m}$$

Inductance of side A,

$$L_A = 2 \times 10^{-7} \ln \frac{D_m}{D_{SA}}$$

$$= 2 \times 10^{-7} \ln \frac{3.189}{0.3128} = 4.6438 \times 10^{-7} \text{ H/m}$$

Self geometric mean distance of side B,

$$D_{SB} = (D_{dd} D_{de} D_{ee} D_{ed})^{1/4}$$

$$D_{dd} = 0.7788 \times 4 \times 10^{-2} \text{ m}$$

$$D_{ee} = 0.7788 \times 4 \times 10^{-2} \text{ m}$$

$$D_{de} = D_{ed} = 0.3 \text{ m}$$

$$D_{SB} = [(0.7788 \times 4 \times 10^{-2})^2 (0.3)^2]^{1/4} = 0.09667 \text{ m}$$

Inductance of side B,

$$L_B = 2 \times 10^{-7} \ln \frac{D_m}{D_{SB}}$$

$$= 2 \times 10^{-7} \ln \frac{3.189}{0.09667} = 6.9923 \times 10^{-7} \text{ H/m}$$

Total inductance of the 100 m line

$$L = 100(L_A + L_B)$$

$$= 100(4.6438 + 6.9923) \times 10^{-7} \text{ H}$$

$$= 0.1164 \times 10^{-3} \text{ H} = 0.1164 \text{ mH}$$

EXERCISES

1. Explain with reasons :

   (a) In double-circuit lines the conductors constituting same phase should be spaced at as greater distance as possible, and the phases should be spaced as close as permissible.

   (b) For an unsymmetrically spaced three-phase line, the inductances will be different for all the phases even under balanced current conditions.

   (c) Bundled-conductor lines have lower inductance than single-conductor lines of the same area of cross-section.

   (d) The geometric mean radius (GMR) of a stranded conductor is less than that of a solid conductor of the same overall diameter.

   (e) The formula for the inductance of a single conductor uses geometric mean radius while the capacitance formula uses the actual radius.

   (f) There is an increase in capacitance due to bundling.

2. Derive the expressions for calculating the internal and external flux linkages of conductor carrying current. Hence deduce an expression for the total inductance of a single-phase line.

3. Explain the method of geometric mean distances to calculate the inductance and capacitance of a line having several conductors connected in parallel for each phase.

4. Give the method of calculating the inductance of a bundled conductor line. Derive expressions for geometric mean radii of a bundle consisting of (a) two subconductors, (b) three subconductors, and (c) four subconductors.

5. Derive an expression for the inductance of a symmetrical three-phase line. What is meant by the term equivalent spacing? State its significance.

6. What is meant by the term transposition? Mention its advantages and drawbacks.

7. Derive an expression for the inductance of a double circuit, three-phase line whose conductors are situated at the corners of a regular hexagon.

8. Derive expressions for the line-to-line capacitance and line-to-neutral capacitance of a single-phase line.

9. Derive an expression for the capacitance of a symmetrical three-phase line.

10. What is the effect of earth on line capacitance? Explain the method of images to calculate the capacitance of (a) two-wire single-phase line (b) single-circuit three-phase transposed line.

11. A single-phase line has conductors each of 15 mm diameter and spaced 2 m apart. Calculate the loop inductance per kilometre of the line if the material of the conductor is (a) copper and (b) steel of relative permeability 200.

12. A single-phase line has two conductors separated by a distance of 3 m. Each conductor has a diameter of 25 mm. If the line operates at 10 kV 50 Hz, calculate

    (a) loop inductance per km,

    (b) line capacitance,

(c) capacitive shunt reactance,

(d) charging current per km, and

(e) reactive voltamperes generated per km.

13. A single-phase line consists of two circuits in parallel as shown in Fig. 7.34. Conductors *a* and *a'* in parallel form the lead while conductors *b* and *b'* in parallel form the return circuit. Calculate the total inductance of the line per km assuming that the current is equally shared by the two parallel conductors. The diameter of each conductor is 20 mm.

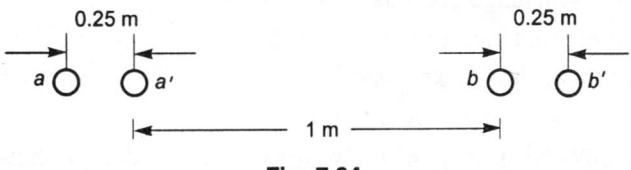

Fig. 7.34.

14. The conductors in a single-phase transmission line are 6 m above the ground. Each conductor is 15 mm diameter and spacing between them is 2.5 m. Calculate

(a) the capacitance per km of the line neglecting the effect of ground

(b) the capacitance per km of the line taking into account the effect of ground, and

(c) percentage increase in capacitance due to the presence of ground.

15. A three-phase, 132 kV, 50 Hz, 80 km transmission line has ACSR conductors of 21 mm diameter each and the mean equilateral spacing is 4.6 m. Calculate

(a) the inductance of the line,

(b) the line capacitance, and

(c) the charging current per km per phase of the line.

16. Calculate the charging current and reactive power generated in a 50 km length of a three-phase, 132 kV, 50 Hz overhead line in which the conductors are 18 mm diameter and spaced (triangularly) at 3 m between centres.

17. The conductors of a three-phase line are placed at the corners of an equilateral triangle. The conductor spacing is 3 m and each copper conductor has a diameter of 12.5 mm. Calculate resistance and series and shunt reactance per metre per phase of the line. The frequency is 50 Hz.

18. A three-phase line has conductors of 5 mm diameter placed at the corners of an equilateral triangle of 1.5 m side. Calculate

(a) the inductive reactance per phase per km and

(b) the capacitive reactance and capacitive susceptance to neutral per phase per km.

19. A single-circuit three-phase, 50 Hz transmission line 10 km long has conductors 30 mm in diameter with spacings as follows:

   A-B = 2 m, B-C = 3 m, C-A = 5 m

Assuming adequate transposition and neglecting the effect of flux linkages within the conductors, determine from first principles

(a) the effective inductive reactance per phase per km, and

(b) the capacitive reactance per phase per km to neutral,

20. Calculate the inductance and capacitance per km of a line consisting of solid conductors of 30 mm diameter placed at the corners of a triangle with sides, 3, 4 and 5 metres. The conductors are adequately transposed.

21. A three-phase 110 kV, 50 Hz transmission line has its conductors arranged in a horizontal plane with 3.5 m between middle conductor and each outside conductor. Each conductor has a diameter of 17.8 mm. The line is completely transposed. Determine

(a) the inductive reactance per phase per km,

(b) the capacitive reactance to neutral per km

(c) the charging current per km, and

(d) the reactive power generated per km.

22. A horizontally spaced three-phase line has ACSR conductors having spacing of 4.5 m between adjacent conductors. Each conductor is 19/5 mm. Assume proper transposition. Find

(a) the inductance per phase per km of the line, and

(b) the capacitance to neutral per km of the line

23. A three-phase line consists of three equal conductors of radius 12.5 mm, placed in a horizontal plane, with a spacing of 6 m between middle and each outer conductor. Calculate

(a) the effective inductance of each conductor in complex form for untransposed line,

(b) the mean inductance of each conductor if the line is transposed,

(c) the effective capacitance of each phase in complex form for untransposed line, and

(d) the mean capacitance of each phase for transposed line. Explain the presence of imaginary terms in (a) and (c).

24. The conductors in the line of Exercise 13 are 10 m above ground. Calculate the line capacitance if the line is properly transposed.

25. Calculate the capacitance per km of a single-phase line in which the conductors are at an average height of 6.5 m above ground. Take the diameter of each conductor as 15.84 mm and the spacing between them 3.5 m.

26. A multi-conductor single-phase line has three conductors each of diameter 20 mm for lead and two conductors of diameter 40 mm for return circuit as shown in Fig. 7.35. Find the inductance on each side of the line and the total inductance of the line.

27. A three-phase double circuit has six conductors located at the corners of a regular hexagon of side 3 m. Each conductor has a diameter of 31.77 mm. The line is completely transposed. Find

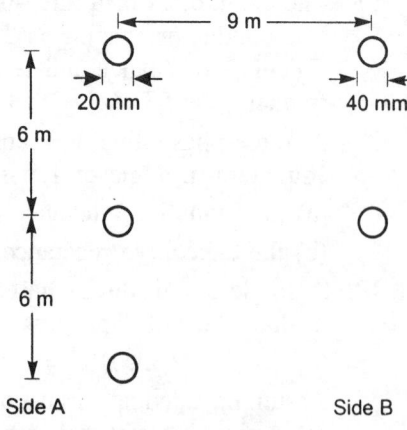

Fig. 7.35.

(a) the inductive reactance per phase per km of the line, and

(b) the capacitive reactance to neutral per km of the line.

If one of the circuits is removed calculate the per unit changes in inductive and capacitive reactances.

28. Fig. 7.36 shows a twin-conductor circuit of a three-phase line with horizontal spacing. The radius of each subconductor is 10 mm. The spacing between the subconductors is 0.5 m. If each phase group shares the total current and charge equally and the line is adequately transposed, determine

(a) the line inductance per km,

(b) the line capacitance per km,

(c) the line inductance of the equivalent single-conductor system,

(d) the line capacitance of the equivalent single-conductor system,

(e) the percentage decrease in the inductance due to bundling, and

(f) the percentage increase in capacitance due to bundling.

Assume that the area of cross-section of each single-conductor is equal to the total area of two subconductors of a phase.

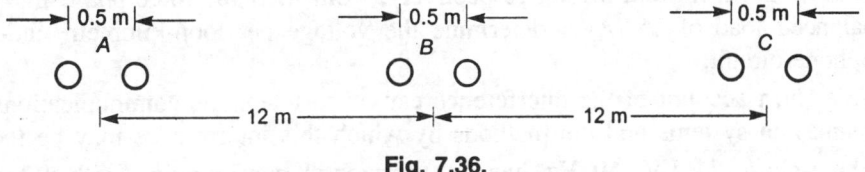

Fig. 7.36.

29. A 750 kV line has a quadruple-conductor circuit of a three-phase line with horizontal spacing as shown in Fig. 7.37.

(a) Calculate the inductive reactance per phase per km at 50 Hz. Each conductor carries 25 per cent of the phase current and the line is properly transposed.

(b) Find the size of a hypothetical single conductor line that would have the same inductance as the given line.

(c) If the line charge per phase divides equally between four subconductors determine the shunt capacitance per phase of the line.

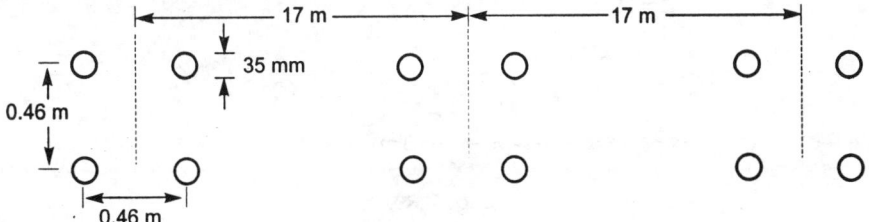

Fig. 7.37.

30. Fig. 7.38 shows a double-circuit three-phase overhead line. If the supply currents and charges are equally divided, calculate the effective inductance and effective capacitance of each phase. The phase sequence is A-B-C and diameter of each conductor is 21 mm.

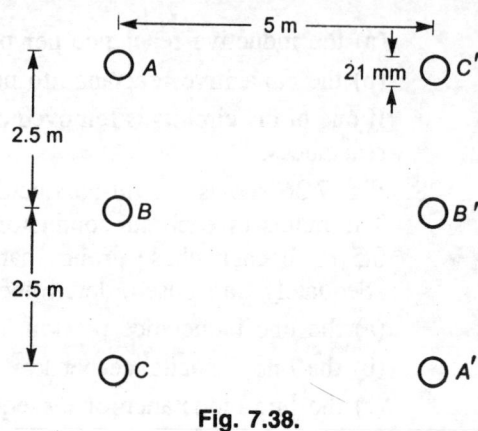

Fig. 7.38.

31. Derive an expression for the voltage induced electromagnetically in a communication circuit, having conductors x and y, by a parallel three-phase 50 Hz power line having conductors a, b and c and carrying a balanced current per phase of 150 A. Show, with an illustration, how transposition may be used to reduce such electromagnetically induced voltages.

32. The three conductors of a three-phase 66 kV transmission line are in the same horizontal plane with 2.70 m between the centre lines of adjacent conductors. Running parallel to them is a 2-wire telephone circuit with its two wires vertically below the centre conductor at 4.5 m and 5.1 m, respectively, from it. If the three-phase line is carrying a balanced load of 25 MVA determine the voltage per loop-kilometre induced to the telephone circuit.

33. Give a short account of the interference caused to telephone communications by power transmission systems and the methods by which this interference may be reduced.

34. A three-phase 11 kV, 50 Hz line has horizontal configuration with 1.3 m between adjacent conductors. The height of the conductors above ground is 12 m. A telephone line is run on the same supports. The horizontal distance between the telephone wires is 0.65 m. The vertical distance between the power line and the telephone line is 1.1 m. Determine the magnitude of the voltage induced in the telephone circuit due to electromagnetic induction if the power line carries a balanced current of 250 A at 50 Hz.

35. A three-phase line is in proximity to a 2-wire telephone line as shown in Fig. 7.39. Determine the voltage, in V/m, induced in the telephone line by the power line carrying balanced currents having magnitudes of 75 A at 50 Hz.

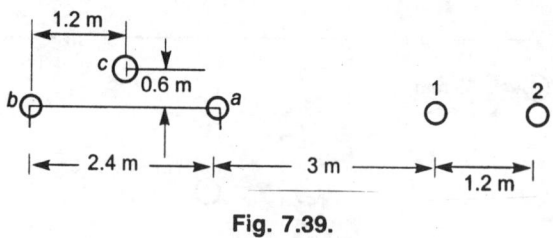

Fig. 7.39.

36. A three-phase transmission line and a telephone line are supported on the same tower as shown in Fig. 7.40. Determine the electromagnetically induced voltage in the telephone line per kilometre. The power line is delivering 40 MW at 0.9 power factor lagging at 132 kV, 50 Hz.

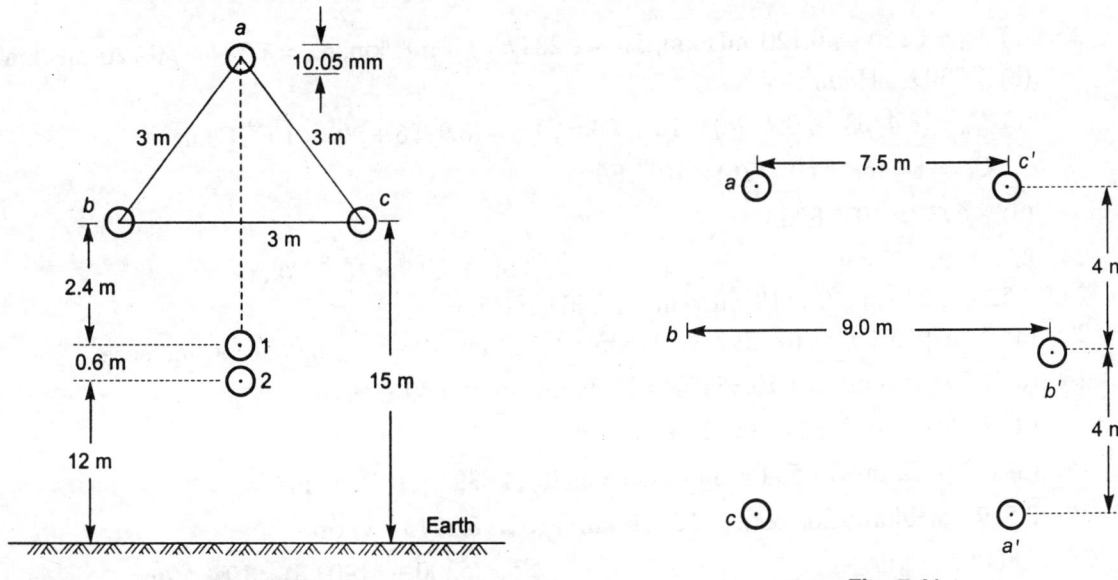

Fig. 7.40.   **Fig. 7.41.**

37. Calculate (a) the inductance per km per phase, and (b) the capacitance to neutral of the double-circuit, three-phase transposed transmission line shown in Fig. 7.41. The radius of each conductor is 1.25 cm.

ANSWERS

11. (a) 2.334 mH/km (b) 22.234 mH/km

12. (a) 2.292 mH/km (b) 0.00507 μF/km (c) 31.39×10^4 Ω/km (d) 31.86 mA/km
(e) 318.57 VAr

13. 1.333 mH/km

14. (a) 9.563×10^{-3} μF/km (b) 9.599×10^{-3} μF/km (c) 0.376%

15. (a) 0.1013 H (b) 0.731 μF (c) 218.77 mA/km

16. 11.45 A; 2617.8 kVAr

17. 1.446×10^{-4} Ω ; 4.036 Ω/m ; 3.537×10^8 Ω

18. (a) 0.4176 Ω/km (b) 3.665×10^5 Ω/km ; 2.7284×10^{-6} S/km

19. (a) 0.3351 Ω (b) 305583 Ω   **20.** 1.11289 mH/km ; 9.984×10^{-9} F/km

21. (a) 0.4056 Ω (b) 355553 Ω (c) 0.1786 A (d) 34.03 kVAr

22. (a) 1.2734 mH (b) 9.0818×10^{-9} F

23. (a) $L_a = 1.354 - j\,0.120$ mH/km, $L_b = 1.2847 + j\,0$ mH/km, $L_c = 1.354 + j\,0.120$ mH/km

(b) 1.3309 mH/km

(c) $C_a = (8.4468 + j\,0.7776) \times 10^{-9}$ F/km, $C_b = (8.9986 + j\,0) \times 10^{-9}$ F/km,

$C_c = (8.4468 + j\,0.7776) \times 10^{-9}$ F/km

(d) 8.6739×10^{-9} F/km

24. 8.78×10^{-9} F/km

25. 4.5867×10^{-9} F/km

26. 0.5288 mH/km ; 0.7119 mH/km ; 1.2407 mH/km

27. (a) 0.167985 Ω/km (b) 292193 Ω/km ; 1.122 pu increase ; 0.0636 pu increase

28. (a) 1.098 mH/km (b) 10.3548×10^{-9} F/km (c) 1.4449 mH/km

(d) 7.9654×10^{-9} F/km (e) 24% (f) 30%

29. (a) 0.291 Ω/km (b) 534.4 mm diameter (c) 4.8397×10^{-9} F/km

30. 0.5797 mH/km ; 20.03066×10^{-9} F/km

32. 8.27134 V/km

34. 7.86061 V/km

35. $(5300 + j\,1903.8) \times 10^{-7}$ V/m

36. 0.6902 V/km

37. 0.607 mH/km, 0.019 μF/km

8

Per Unit Representation

8.1 INTRODUCTION

In the process of computation of power system problems sometimes it is more convenient to express impedances, currents, voltages, and powers in terms of per unit values rather than in ohms, amperes, volts, and watts or vars. *The numerical per unit (pu) value of any quantity is defined as the ratio of its actual value to another arbitrarily chosen value of quantity of the same dimensions, assumed as the base or reference.*

$$\text{Per unit value} \triangleq \frac{\text{actual value}}{\text{base value of the same dimension}} \qquad \qquad ...(8.1.1)$$

$$\text{For any quantity } A, \quad A_{pu} = \frac{A}{A_{base}} \qquad \qquad ...(8.1.2)$$

Thus, any quantity can be converted to a per unit quantity by dividing the numerical value by a chosen base value of the same dimensions. The per unit values are dimensionless. Per cent quantities differ from per unit quantities by a factor of 100. The ratio in per cent is 100 times the value in per unit.

For example, if a base voltage of 10 kV is selected, voltages of 8, 10, and 12 kV would be specified as 0.80, 1.00 and 1.20 per unit respectively or 80, 100, and 120 per cent. A base quantity is designated by the subscript b. The product of two quantities expressed in per unit is expressed in per unit itself, but the product of two quantities expressed in per cent is to be divided by 100 to obtain the result in per cent. Hence the per unit method is preferred.

Let I_A = actual current in amperes

I_b = base current in amperes

V_v = actual voltage in volts

V_b = base voltage in volts

Z_Ω = actual impedance in ohms

Z_b = base impedance in ohms

S_{VA} = actual voltamperes

S_b = base voltamperes.

Per unit current, $I_{pu} \triangleq \dfrac{I_A}{I_b}$ ...(8.1.3)

Per unit voltage, $V_{pu} \triangleq \dfrac{V_v}{V_b}$ ...(8.1.4)

Per unit impedance, $Z_{pu} \triangleq \dfrac{Z_\Omega}{Z_b}$ ...(8.1.5)

$$Z_\Omega = R_\Omega + j\,X_\Omega \; ; \qquad Z_{pu} = \dfrac{Z_\Omega}{Z_b} = \dfrac{R_\Omega}{Z_b} + j\dfrac{X_\Omega}{Z_b}$$

or $\quad Z_{pu} = R_{pu} + j\,X_{pu}$

$\therefore \qquad R_{pu} = \dfrac{R_\Omega}{Z_b}$ ...(8.1.6)

and $\qquad X_{pu} = \dfrac{X_\Omega}{Z_b}$ ...(8.1.7)

Per unit voltamperes, $\quad S_{pu} \triangleq \dfrac{S_{VA}}{S_b}$ ...(8.1.8)

$$S = P + j\,Q = VI* \; ; \qquad S_{pu} = \dfrac{S_{VA}}{S_b} = \dfrac{P}{S_b} + j\dfrac{Q}{S_b}$$

or $\quad S_{pu} = P_{pu} + j\,Q_{pu}$

$\therefore \qquad P_{pu} = \dfrac{P_{watts}}{S_b}$ ...(8.1.9)

and $\qquad Q_{pu} = \dfrac{Q_{vars}}{S_b}$ ...(8.1.10)

Also, for single-phase circuit

$$Z_b = \dfrac{V_b}{I_b}$$ ...(8.1.11)

and $\qquad S_b = V_b\,I_b$ ...(8.1.12)

The values of the base quantities are selected according to convenience. If any two of the four quantities in Eqs. (8.1.11) and (8.1.12) are specified the remaining two are fixed automatically. In power-system calculations, usually voltamperes (S) voltage (V) are specified. Hence S_b and V_b are chosen as base values. The base current I_b and base impedance Z_b are expressed in terms of S_b and V_b.

$$I_b \triangleq \dfrac{S_b}{V_b}$$ ...(8.1.13)

The base impedance is that impedance which has a voltage drop across it equal to the base voltage if the current through it is equal to the base current.

$$Z_b \triangleq \frac{V_b}{I_b} = \frac{V_b\, V_b}{V_b\, I_b} = \frac{V_b^2}{S_b} \text{ ohms} \qquad \qquad ...(8.1.14)$$

$$Y_b = \frac{1}{Z_b} = \frac{S_b}{V_b^2} \text{ siemens}$$

If proper selection of bases is made, the basic circuit relations can be applied to the per unit quantities. For example,

$$V_{pu} = Z_{pu}\, I_{pu} \qquad \qquad ...(8.1.15)$$

$$S_{pu} = V_{pu}\, I_{pu}^* \qquad \qquad ...(8.1.16)$$

Equations (8.1.5) and (8.1.14) can be combined to give

$$Z_{pu} = Z_\Omega \frac{S_b}{V_b^2} \qquad \qquad ...(8.1.17)$$

Also,   base current, $I_b = \dfrac{S_b}{V_b} = \dfrac{\text{base kVA}}{\text{base kV}}$

base kW = numerical value of base kVA

base MW = numerical value of base MVA

The admittance is the reciprocal of the impedance.

Let Y_s = actual admittance in siemens

Y_{pu} = per unit admittance.

$$Y_{pu} \triangleq \frac{1}{Z_{pu}} = \frac{V_b^2}{Z_\Omega\, S_b} = Y_s \frac{V_b^2}{S_b} \qquad \qquad ...(8.1.18)$$

Example 8.1 A 230 kV transmission line has a series impedance of $(4 + j\,60)\ \Omega$ and a shunt admittance of $j\,2 \times 10^{-3}$ S. Using 100 MVA and the line voltage as base values, calculate per unit impedance and per unit admittance of the line.

Solution

$$Z_{pu} = Z_\Omega \frac{(\text{MVA})_b}{(\text{kV})_b^2} = (4 + j\,60) \times \frac{100}{(230)^2}$$

$$= (7.56 + j\,113.4) \times 10^{-3} \text{ pu}$$

$$Y_{pu} = Y_s \frac{V_b^2}{S_b}$$

$$= j\,2 \times 10^{-3} \times \frac{(230)^2}{100}$$

$$= j\,1.058 \text{ pu}$$

8.2 CHANGE OF BASE

It is sometimes necessary to convert per unit quantities from one base to another. Let the base voltamperes and base voltage in system 1 be represented by S_{b1} and V_{b1} respectively. The corresponding values in system 2 are represented by S_{b2} and V_{b2}.

$$\text{Base current in base system 1,} \quad I_{b1} = \frac{S_{b1}}{V_{b1}} \qquad \text{...(8.2.1)}$$

$$\text{Base current in base system 2,} \quad I_{b2} = \frac{S_{b2}}{V_{b2}} \qquad \text{...(8.2.2)}$$

$$\text{The per unit value of current } I \text{ in base system 1,} \quad I_{1pu} = \frac{I}{I_{b1}} \qquad \text{...(8.2.3)}$$

$$\text{The per unit value of current in base system 2,} \quad I_{2pu} = \frac{I}{I_{b2}} \qquad \text{...(8.2.4)}$$

Combining Eqs. (8.2.1) to (8.2.4)

$$I_{2pu} = I_{1pu} \frac{I_{b1}}{I_{b2}} = I_{1pu} \frac{S_{b1}}{V_{b1}} \cdot \frac{V_{b2}}{S_{b2}} = I_{1pu} \frac{S_{b1}}{S_{b2}} \cdot \frac{V_{b2}}{V_{b1}} \qquad \text{...(8.2.5)}$$

The conversion of per unit impedance from one base to another can be made with the help of Eq. (8.1.17).

$$\text{The per unit value of impedance } Z_\Omega \text{ in base system 1,} \quad Z_{1pu} = Z_\Omega \frac{S_{b1}}{V_{b1}^2} \qquad \text{...(8.2.6)}$$

$$\text{The per unit value of impedance } Z_\Omega \text{ in base system 2,} \quad Z_{2pu} = Z_\Omega \frac{S_{b2}}{V_{b2}^2} \qquad \text{...(8.2.7)}$$

Elimination of Z_Ω from Eqs. (8.2.6) and (8.2.7) gives

$$Z_{2pu} = Z_{1pu} \frac{S_{b2}}{S_{b1}} \left(\frac{V_{b1}}{V_{b2}} \right)^2 \qquad \text{...(8.2.8)}$$

Equation (8.2.8) is used for changing the per unit impedance from one set of V and S bases to any other set of V and S bases. It is not to be used for transferring the ohmic value of impedance from one side of transformer to another.

The conversion formula for the admittance can be written as

$$Y_{2pu} = Y_{1pu} \frac{S_{b1}}{S_{b2}} \left(\frac{V_{b2}}{V_{b1}} \right)^2 \qquad \text{...(8.2.9)}$$

8.3 PER UNIT IMPEDANCE OF A TRANSFORMER

Consider a single-phase transformer in which the total series impedance of the two windings referred to the primary is Z_{1e}. Suppose that the rated values are taken as the base quantities.

Base current in the primary = I_1

Base voltage in the primary = V_1

Base impedance in the primary, $Z_{b1} = \dfrac{V_1}{I_1}$

Per unit impedance of the transformer referred to the primary

$$Z_{1epu} = \frac{Z_{1e}}{Z_{b1}} = \frac{Z_{1e}}{V_1/I_1} = \frac{Z_{e1} I_1}{V_1} \qquad \qquad ...(8.3.1)$$

The total series impedance of the two windings referred to the secondary

$$Z_{2e} = Z_{1e} \left(\frac{N_2}{N_1}\right)^2 \qquad \qquad ...(8.3.2)$$

where N_1 and N_2 represent primary and secondary turns respectively. On the secondary side the base quantities are as follows :

Base current $= I_2$

Base voltage $= V_2$

Base impedance, $Z_{b2} = \dfrac{V_2}{I_2}$

Per unit impedance of the transformer referred to the secondary

$$Z_{2epu} = \frac{Z_{2e}}{V_2/I_2} = \frac{Z_{2e} I_2}{V_2} \qquad \qquad ...(8.3.3)$$

Now, $I_2 N_2 = I_1 N_1 \; ; \; I_2 = I_1 \dfrac{N_1}{N_2}$ $\qquad \qquad ...(8.3.4)$

and $\dfrac{V_2}{V_1} = \dfrac{N_2}{N_1} \; ; \; V_2 = \dfrac{N_2}{N_1} V_1$ $\qquad \qquad ...(8.3.5)$

From Eqs. (8.3.2) to (8.3.5)

$$Z_{2epu} = Z_{1e} \left(\frac{N_2}{N_1}\right)^2 \cdot \frac{I_1 N_1}{N_2} \cdot \frac{N_1}{N_2 V_1} = \frac{Z_{1e} I_1}{V_1} \qquad \qquad ...(8.3.6)$$

From Eqs. (8.3.1) and (8.3.6)

$$Z_{2epu} = Z_{1epu}$$

The result of Eq. (8.3.7) may alternatively be deduced as follows :

$$Z_{1epu} = \frac{Z_{1e}}{Z_{b1}} = \frac{Z_{2e} (N_1/N_2)^2}{V_{B1}/I_{b1}} = \frac{Z_{2e} (N_1/N_2)^2}{V_{b2} (N_1/N_2)} \cdot I_{b2} \left(\frac{N_2}{N_1}\right) = \frac{Z_{2e}}{Z_{b2}} = Z_{2epu}$$

Thus, if the rated primary voltage is used as base with primary referred impedance and the rated secondary voltage with secondary referred impedance, the same per unit value of impedance is obtained. Consequently, the per unit impedance of a 2-winding transformer referred to either side is the same. In other words, *the per unit equivalent impedance of a 2-winding transformer is the same whether the calculation is made from the high voltage side or the low voltage side.*

Example 8.2 A three-phase, star-connected system is rated at 50 MVA and 110 kV. Express 40,000 kVA of three-phase apparent power as a per-unit value referred to (a) the three-phase system kVA as base, and (b) the per phase kVA as base.

Solution

(a) For the three-phase base,

base kVA = 40,000 kVA = 1 pu

and   base kV = 110 kV (line-to-line) = 1 pu

$\therefore$ per unit kVA $= \dfrac{40,000}{50,000} = 0.8$ pu

(b) For the per phase base,

base kVA $= \dfrac{1}{3} \times 50,000 = 16.667 = 1$ pu

and   base kV $= \dfrac{110}{\sqrt{3}} = 63.5 = 1$ pu

$\therefore$ per unit kVA $= \dfrac{1}{3} \times \dfrac{40,000}{16.667 \times 10^3} = 0.8$ pu

Example 8.3 A 5 kVA 400/200 V, 50 Hz single-phase transformer has primary and secondary leakage reactance each of 2.5 Ω. Determine the total reactance in per unit.

Solution

The VA on both sides of the transformer is 5000 VA, that is,

$S_b = 5000$ VA.

(a) Total reactance of the transformer referred to primary

$$X_{1e} = X_1 + X_2 \left(\frac{N_1}{N_2}\right)^2 = 2.5 + 2.5 \left(\frac{400}{200}\right)^2 = 2.5 + 10 = 12.5 \ \Omega$$

Base voltage on primary side, $V_{b1} = 400$ V

Now, $X_{pu} = X_\Omega \dfrac{S_b}{V_{b2}}$

Total per unit reactance referred to primary

$$X_{1epu} = X_{1e} \frac{S_b}{V_{b1}^2} = 12.5 \times \frac{5000}{(400)^2} = 0.390625 \text{ pu}$$

(b) Total reactance of the transformer referred to secondary

$$X_{2e} = X_2 + X_1 \left(\frac{N_2}{N_1}\right)^2 = 2.5 + 2.5 \left(\frac{200}{400}\right)^2 = 3.125 \ \Omega$$

Base voltage on secondary side, $V_{b2} = 200$ V

Total per unit reactance referred to secondary

$$X_{2epu} = X_{2e} \frac{S_b}{V_{b2}^2} = 3.125 \times \frac{5000}{(200)^2} = 0.390625 \text{ pu}$$

(c) If the rated primary voltage is used as base, the primary per unit reactance is given by

$$X_{1pu} = X_1 \frac{S_b}{V_{b1}^2} = 2.5 \times \frac{5000}{(400)^2} = 0.078125$$

If the rated secondary voltage is used as base, the secondary per unit reactance is given by

$$X_{2pu} = X_2 \frac{S_b}{V_{b2}^2} = 2.5 \times \frac{5000}{(200)^2} = 0.3125 \ \Omega$$

Sum of per unit reactance on both sides

$$X_{pu} = X_{1pu} + X_{2pu} = 0.078125 + 0.3125 = 0.390625$$

$\therefore \qquad X_{1epu} = X_{2epu} = X_{epu}$

Thus, the same value of per unit reactance of a transformer is obtained regardless of the following methods of determination :

1. Total per unit reactance referred to primary on the primary base voltage.
2. Total per unit reactance referred to secondary on the secondary base voltage.
3. Sum of per unit reactances on both sides. The primary per unit reactance is found on the primary base voltage and the secondary per unit reactance is found on the secondary base voltage. The secondary base voltage is found by multiplying the primary base voltage by the transformer turns ratio.

The calculations for per unit impedance can also be made in a similar manner.

8.4 PER UNIT QUANTITIES IN THREE-PHASE SYSTEMS

Suppose that the suffixes l and p denote the line and phase values in a balanced three-phase system.

In a star connection

$$V_l = \sqrt{3} \ V_p \ ; \quad V_{lb} = \sqrt{3} \ V_{pb}$$

$$I_l = I_p \ ; \quad I_{lb} = I_{pb}$$

$$(V_l)_{pu} = \frac{V_l}{V_{lb}} = \frac{\sqrt{3} \ V_p}{\sqrt{3} \ V_{pb}} = (V_p)_{pu} \qquad\qquad \text{...(8.4.1)}$$

$$(I_l)_{pu} = \frac{I_l}{I_{lb}} = \frac{I_p}{I_{pb}} = (I_p)_{pu} \qquad\qquad \text{...(8.4.2)}$$

Thus, it is seen that in a star connections a per unit phase voltage has the same numerical value as the corresponding per unit line voltage. Also, the per unit phase current has the same numerical value as the corresponding per unit line current.

In a delta connection

$$V_l = V_p \ ; \quad V_{lb} = V_{pb}$$

$$I_l = \sqrt{3} \ I_p \ ; \quad I_{lb} = \sqrt{3} \ I_{pb}$$

$$(V_l)_{pu} = \frac{V_l}{V_{lb}} = \frac{V_p}{V_{pb}} = (V_p)_{pu} \qquad\qquad \text{...(8.4.3)}$$

$$(I_l)_{pu} = \frac{I_l}{I_{lb}} = \frac{\sqrt{3}\, I_p}{\sqrt{3}\, I_{pb}} = (I_p)_{pu} \qquad \qquad ...(8.4.4)$$

These relations show that in the delta connection also a per unit phase voltage has the same numerical value as the corresponding per unit line voltage. Similarly the per unit phase current has the same numerical value as the corresponding per unit line current.

Suppose that the subscript 1ϕ and 3ϕ denote per phase and three-phase values. With both star and delta connections

$$S_{3\phi} = \sqrt{3}\, V_l I_l = 3 V_p I_p$$

$$(S_b)_{3\phi} = \sqrt{3}\, V_{lb} I_{lb} = 3 V_{pb} I_{pb}$$

$$S_{pu} = \frac{S_{3\phi}}{(S_b)_{3\phi}} = \frac{\sqrt{3}\, V_l I_l}{\sqrt{3}\, V_{lb} I_{lb}} = (V_l)_{pu}\, (I_l)_{pu} \qquad \qquad ...(8.4.5)$$

Again, $\quad S_{pu} = \dfrac{3 V_p I_p}{3 V_{pb} I_{pb}} = (V_p)_{pu}\, (I_p)_{pu}$

$\therefore \qquad S_{pu} = (V_l)_{pu}\, (I_l)_{pu} = (V_p)_{pu}\, (I_p)_{pu} \qquad \qquad ...(8.4.6)$

Hence, a per unit phase voltampere has the same numerical value as the corresponding per unit line voltampere irrespective of three-phase connection whether star or delta.

Unless, otherwise specified, a given value of base voltage in a three-phase system is a line-to-line voltage, and a given value of base VA is the total three-phase base VA.

Base impedance using per phase values

$$Z_{pb} = \frac{(V_{pb})^2}{(S_b)_{1\phi}} \qquad \qquad ...(8.4.7)$$

Also, the base impedance using line value of base voltage and total three-phase value of base VA,

$$Z_{lb} = \frac{(V_{lb})^2}{(S_b)_{3\phi}} = \frac{(\sqrt{3}\, V_{pb})^2}{3\,(S_b)_{1\phi}} = \frac{(V_{pb})^2}{(S_b)_{1\phi}} \qquad \qquad ...(8.4.7)$$

These relations show that the base impedance is the same whether (i) per phase values of V_b and S_b are used or (ii) line value of V_b and total three-phase value of S_b are used. But the same basis should be adopted for selecting V_b and S_b to calculate base impedance. In this way the subscripts are eliminated, i.e.,

$$\text{Base impedance, } Z_b = \frac{V_b^2}{S_b} \qquad \qquad ...(8.4.8)$$

$$Z_{pu} = Z_\Omega \frac{(S_b)_{3\phi}}{V_{lb}^2} \qquad \qquad ...(8.4.9)$$

For a three-phase system

(a) $\qquad \text{Base current, } I_b = \dfrac{(S_b)_{1\phi}}{V_{pb}} \qquad \qquad ...(8.4.10)$

(b) $\qquad \text{Base current, } I_b = \dfrac{3\,(S_b)_{1\phi}}{3 V_{pb}} = \dfrac{(S_b)_{3\phi}}{\sqrt{3}\, V_{lb}} \qquad \qquad ...(8.4.11)$

The results derived in this section can be utilised to specify per unit impedance of a three-phase transformer. *The per unit impedance referred to either side of a three-phase transformer is the same irrespective of the three-phase connections whether they are delta/delta, star/star, or delta/star.*

8.5 SELECTION OF BASE VALUES

Generally the per unit values of devices are given in terms of their own VA and voltage ratings. These devices may be included in a power system network. A power system has different voltage levels at different points. It is necessary to refer all of the given per unit values to the system base values. To apply the per unit method to a given problem first a convenient value of MVA is chosen. The same MVA base is used in all parts of the system. It may be the total MVA of the system, the largest MVA of a section, or any round figure such 10, 100, 1000 MVA, etc. After the selection of the base MVA the base voltages for each section are to be chosen. The rated voltage of the largest section may be taken as the base voltage for that section. The base voltages for other sections are then assigned according to the turns ratios of the transformers. When the selection of a common base MVA and base voltages of different sections is made the per unit impedances of various sections can be calculated to draw the single-line diagram giving impedances in per unit values.

8.6 BASE QUANTITIES IN TERMS OF kV AND MVA

In power systems it is a common practice to specify voltage rating in kV and the voltampere rating in MVA. The results already derived in terms of V and VA can be modified.

Base MVA, $S_b = (MVA)_b = V_b I_b \times 10^{-6}$

Base voltage in kV $= (kV)_b = V_b \times 10^{-3}$

$$Z_b = \frac{(kV_l)_b^2}{[(MVA)_b]_{3\varphi}}$$

$$Z_{pu} = Z_\Omega \frac{S_b}{V_b^2} = Z_\Omega \frac{V_b I_b}{V_b^2} = Z_\Omega \frac{V_b I_b \times 10^{-6}}{(V_b \times 10^{-3})^2}$$

$$Z_{pu} = Z_\Omega \frac{(MVA)_b}{(kV)_b^2} \qquad\qquad\qquad ...(8.6.1)$$

$$Z_{2pu} = Z_{1pu} \frac{(MVA)_{b2}}{(MVA)_{b1}} \left[\frac{(kV)_{b1}}{(kV)_{b2}}\right]^2 \qquad ...(8.6.2)$$

$$Z_{pu} = Z_\Omega \frac{[(MVA)_b]_{3\varphi}}{[(kV_l)_b]^2} \qquad\qquad ...(8.6.3)$$

Example 8.4 A portion of a power system consists of two generators in parallel, connected to a step up transformer that links them with a 230 kV transmission line. The ratings of these components are:

Generator G_1: 100 MVA, 12 percent reactance
Generator G_2: 5 MVA, 8 percent reactance

Transformer: 15 MVA, 6 percent reactance

Transmission line: $(4 + j\,60)$ Ω, 230 kV

where the percent reactances are computed on the basis of individual component ratings. Express the reactances and the impedance with 15 MVA as the base value.

Solution

$$Z_{2pu} = Z_{1pu} \frac{S_{b2}}{S_{b1}} \left(\frac{V_{b1}}{V_{b2}} \right)^2$$

For generator G_1,

$$X_{2pu} = X_{1pu} \frac{S_{b2}}{S_{b1}} \left(\frac{V_{b1}}{V_{b2}} \right)^2$$

Since $V_{b2} = V_{b1}$

$$X_{2pu} = 0.12 \left(\frac{15}{10} \right) = 0.18 \text{ pu}$$

Percent reactance of generator $G_1 = 0.18 \times 100 = 18$

Percent reactance of generator $G_2 = 8 \left(\frac{15}{5} \right) = 24$

For the transformer, percent reactance $= 6 \left(\frac{15}{15} \right) = 6$

For the transmission line,

$$Z_{pu} = Z_\Omega \frac{[(MVA)_b]_{3\,\phi}}{[(kV_l)_b]^2}$$

$$= (4 + j\,60) \times \frac{15}{(230)^2} = (1.13 + j + 17) \times 10^{-3} \text{ pu}$$

Percent impedance of the transmission line $= Z_{pu} \times 100$

$$= (1.13 + j\,17) \times 10^{-3} \times 100$$

$$= (0.113 + j\,1.7) \text{ percent}$$

8.7 PER UNIT LOAD IMPEDANCE

In power system analysis loads are represented by constant impedances. Load impedances are generally expressed in terms of active power P and reactive voltamperes Q. Two representations, namely, parallel and series are possible.

Parallel Representation of Load Impedance

Parallel representation of load impedance is shown in Fig. 8.1.

Let P = load power in W

Q = reactive voltamperes in VAr

R_p = load impedance in Ω

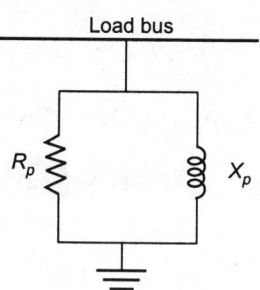

Fig. 8.1. Parallel representation of load impedance.

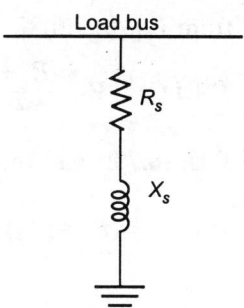

Fig. 8.2. Series representation of load impedance.

X_p = load reactance in Ω

V = load voltage in V

$$P = \frac{V^2}{R_p}$$ ...(8.7.1)

$$R_{pu} = R_p \frac{S_b}{V_b^2} = \left(\frac{V}{V_b}\right)^2 \frac{S_b}{P} = \frac{V_{pu}^2}{P_{pu}}$$ ...(8.7.2)

Similarly,

$$X_{pu} = \left(\frac{V}{V_b}\right)^2 \left(\frac{S_b}{Q}\right) = \frac{V_{pu}^2}{Q_{pu}}$$ ...(8.7.3)

Series Representation of Load Impedance

Series representation of load impedance is shown in Fig. 8.2.

Let R_s = load resistance in Ω

X_s = load reactance in Ω

Z_s = load impedance in Ω

Load current, $$I = \frac{V}{Z_s} = \frac{V}{R_s + j X_s}$$ ...(8.7.4)

$$P + j Q = VI^* = \frac{VV^*}{R_s - j X_s}$$ ...(8.7.5)

$$= \frac{|V|^2}{R_s - j X_s}$$ ...(8.7.6)

$$P - j Q = \frac{|V|^2}{R_s + j X_s}$$ ...(8.7.7)

Multiplying Eqs. (8.7.6) and (8.7.7)

$$P^2 + Q^2 = \frac{|V|^4}{R_s^2 + X_s^2}$$ ...(8.7.8)

Also, from Eq. (8.7.6)

$$P + jQ = |V|^2 \frac{(R_s + jX_s)}{R_s^2 + X_s^2} \qquad \qquad ...(8.7.9)$$

From Eqs. (8.7.8) and (8.7.9)

$$P + jQ = (R_s + jX_s) \frac{P^2 + Q^2}{|V|^2}$$

$$\therefore \qquad Z_s = R_s + jX_s = \frac{|V|^2}{P^2 + Q^2}(P + jQ) \qquad \qquad ...(8.7.10)$$

$$= (kV)^2 \frac{(MW + jMVAr)}{(MW)^2 + (MVAr)^2} \qquad \qquad ...(8.7.11)$$

But $\qquad Z_{pu} = Z_\Omega \dfrac{S_b}{V_b^2}$

$$R_{pu} + jX_{pu} = (R_s + jX_s)\frac{S_b}{V_b^2} \qquad \qquad ...(8.7.12)$$

From Eqs. (8.7.10) and (8.7.12)

$$R_{pu} + jX_{pu} = \frac{|V|^2}{V_b^2} \frac{V_b}{P^2 + Q^2}(P + jQ)$$

$$= V_{pu}^2 \frac{S_b}{P^2 + Q^2}(P + jQ) \qquad \qquad ...(8.7.13)$$

Equating real and imaginary parts

$$R_{pu} = V_{pu}^2 S_b \frac{P}{P^2 + Q^2} \text{ pu} \qquad \qquad ...(8.7.14)$$

$$X_{pu} = V_{pu}^2 S_b \frac{Q}{P^2 + Q^2} \text{ pu} \qquad \qquad ...(8.7.15)$$

8.8   ADVANTAGES OF PER UNIT REPRESENTATION

Per unit system of computation has the following advantages :

1. The ordinary parameters (current, impedance, losses, etc.) vary considerably with the variation of physical size, terminal voltage, power rating, etc., while the per unit parameters are independent of these quantities over a wide range of the same type of apparatus. In other words, the per unit impedance values for apparatus of like ratings lie within a narrow range.

2. Per unit values provide more meaningful information.

3. The chance of confusion between line and phase values in a three-phase balanced system is reduced. A per unit phase quantity has the same numerical value as the corresponding per unit line quantity regardless of the three-phase connection whether star or delta.

4. The impedances of machines are specified by the manufacturers in terms of per unit values.

5. The per unit impedance referred to either side of a single-phase transformer is the same.

6. The per unit impedance referred to either side of a three-phase transformer is the same regardless of the three-phase connections whether they are Δ - Δ, Y - Y or Δ - Y.

7. The computational effort in power systems is very much reduced with the use of per unit quantities. Usually, the per unit quantities being of the order of unity or less can easily be handled with a digital computer. Manual calculations are also simplified. Per unit quantities simplify theoretical deductions and give them more generalised forms.

8.9 ONE-LINE DIAGRAMS

A single-line or one-line diagram is a representation of the essentials of a system in a most simplified form. As already discussed, power system are basically composed of a set of generating plants, a transmission network, and a combination of industrial, commercial and residential loads. Now-a-days, mostly power is generated by three-phase synchronous generators, transmitted by three-phase transmission lines and distributed through three-phase networks.

A balanced three-phase system is studied on a per phase basis. A single-phase circuit consists of one of the three-phase conductors and a neutral return conductor. The loop impedance of a single-phase circuit may be supposed to be concentrated in one conductor only with the impedance of the return conductor assumed to be zero. Thus, a three-phase balanced system is effectively replaced by a single-line diagram. Fig. 8.3 shows a one-line diagram of a power system. G_1, G_2 and G_3 are three synchronous generators. G_1 is grounded through a resistor, G_2 and G_3 through reactors. T_1 and T_2 are two transformers. Circuit breakers are numbered 1-8.

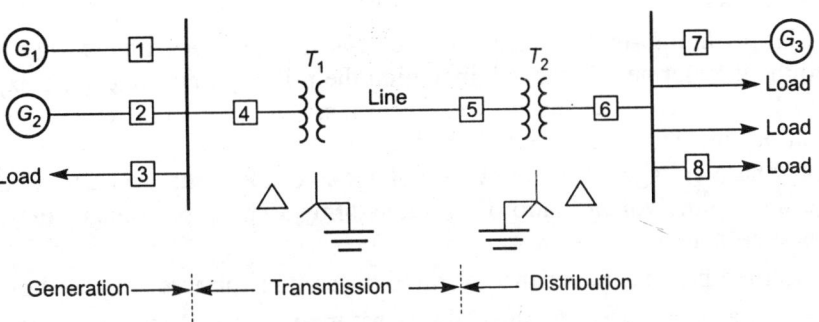

Fig. 8.3. Single-line diagram of a power system.

In the *impedance diagram*, the different components of the power system are replaced by their equivalent circuits. The synchronous generator is replaced by a constant voltage source behind proper impedance. The transformer is replaced by its equivalent circuit. The transmission line is replaced by nominal π-equivalent circuit. The above is only true for normal operation (i.e., for steady-state analysis) when both excitation and three-phase network are balanced and they can be analysed on per phase basis. In many power system studies the synchronous generator resistance, resistances of transformer windings, resistances of transmission lines, line charging and magnetizing circuits of transformers are neglected. The impedance diagram then becomes the *reactance diagram* (Fig. 8.4).

The assumptions are true in many power system studies. The ratings of line, machines and other equipment are given on the one-line diagram in either actual or per unit values.

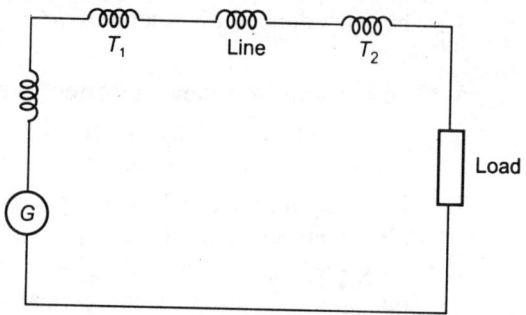

The information supplied by a single-line diagram varies according to the requirement. For example, the one-line diagram for load flow studies may not include circuit breakers. For stability studies circuit breakers and relay positions are shown in the one-line diagram. In short-circuit studies three-separate diagrams to represent positive, negative and zero sequence

Fig. 8.4. Reactance diagram.

networks are shown. The one-line diagram may be converted to an impedance diagram and *vice versa*.

8.10 PREPARATION OF IMPEDANCE DIAGRAMS

In order to prepare a single-line impedance diagram of a power system one-line diagram and specifications of the generators, transformers lines, motors and loads are required. The following considerations are made while preparing impedance diagram :

1. Base values of voltage and VA are selected in part of the system. Unless otherwise mentioned, a given value of base voltage in a three-phase system is the line-to-line voltage and given value of base VA is the total three-phase VA.

2. The base VA is the same in all parts of the system. After selecting a convenient value of base VA for the whole system the rated value of voltage of a particular section is chosen as the base voltage. The base voltages of other sections are then assigned according to line-to-line voltage ratios of transformers. All impedances in the system before the transformer is reached, including the primary impedance, are expressed in per unit to the voltage base chosen for the primary side. All the impedances beyond the transformer including secondary impedance of the transformer are specified in per unit to a new base voltage. This new base voltage is equal to the primary voltage multiplied by the line-to-line voltage ratio of the transformer. The base voltages may be shown on the one-line diagram.

3. The specified per unit impedance of three-phase transformers are based on their ratings.

4. Sometimes a three-phase rating of the transformer is equal to three times the rating of each single-phase transformer. The per unit impedance of the three-phase transformer is equal to the per unit impedance of the single-phase transformer.

5. The per unit impedances of all sections should be calculated on proper base with the help of Eq. (8.2.8).

Example 8.5 A three-phase synchronous generator delivers 10 MVA at a voltage of 10.5 kV. The line impedance is 5 Ω. Determine the voltage drop in the line in per unit and in volts. Use the reference base as 12 MVA at 11 kV.

Solution

(a) We shall first solve this problem using actual voltage and MVA.

We have $S_{3\varphi} = \sqrt{3}\ V_l I_l$ VA ; $I_l = \dfrac{S_{3\varphi}}{\sqrt{3}\ V_l}$

Here $S_{3\varphi} = 10$ MVA $= 10 \times 10^6$ VA

$V_l = 10.5$ kV $= 10.5 \times 10^3$ V

∴ $I_l = \dfrac{S_{3\varphi}}{\sqrt{3}\ V_l} = \dfrac{10 \times 10^6}{\sqrt{3} \times 10.5 \times 10^3} = 549.86$ A

$I_{ph} = I_l = 549.86$ A

Voltage drop in the line per phase $= Z_{ph}\ I_{ph} = 5 \times 549.86 = 2749.3$ V

(b) Now we shall use the per unit method to solve this problem.

We have $S_b = 10$ MVA ; $V_{lb} = 11$ kV

$$V_{pu} = (V_l)_{pu} = \dfrac{V_l}{V_{lb}} = \dfrac{10.5}{11}$$

$$S_{pu} = \dfrac{S_{3\varphi}}{(S_b)_{3\varphi}} = \dfrac{10}{12}$$

Current in per unit, $I_{pu} = \dfrac{S_{pu}}{V_{pu}} = \dfrac{10}{12} \times \dfrac{11}{10.5} = 0.8730158$

Per unit impedance of the line, $Z_{pu} = Z_\Omega \dfrac{[(MVA)_b]_{3\varphi}}{(kV_l)_b^2} = Z_\Omega \dfrac{(S_b)_{3\varphi}}{V_b^2} = 5 \times \dfrac{12}{(11)^2} = 0.4958677$

Voltage drop in the line per phase in per unit

$\Delta V_{pu} = Z_{pu}\ I_{pu} = 0.4958677 \times 0.8730158 = 0.4329$ pu

Voltage drop in the line per phase in volts $= (\Delta V_{pu}) \times$ base voltage per phase

$$= 0.4329 \times \dfrac{11000}{\sqrt{3}} = 2749.36 \text{ V}$$

It is to be noted that the voltage drop in the line impedance is the voltage per phase, and hence the per unit voltage should be multiplied by the phase value of the base voltage.

Example 8.6 Assuming S bases of 25 and 60 MVA calculate the through impedance in ohms between the generator and the output terminals of the transformer for the system shown in Fig. 8.5. The specifications of the generators and transformer are given in Table 8.1.

Table 8.1

Generator 1	Generator 2	Transformer
30 MVA	25 MVA	60 MVA
11 kV	11 kV	11 kV (delta)
		66 kV (star)
$X'' = 0.20$ pu	$X'' = 0.25$ pu	$X = 0.10$ pu

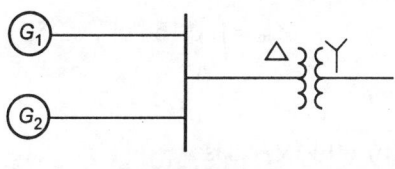

Fig. 8.5.

Solution

We have $Z_{2pu} = Z_{1pu} \left(\dfrac{S_{b2}}{S_{b1}}\right)\left(\dfrac{V_{b1}}{V_{b2}}\right)^2$

On 25 MVA base, for generator 1, $X'' = 0.20 \times \dfrac{25}{30}\left(\dfrac{11}{11}\right)^2 = 0.1667$ pu

On 25 MVA base, for generator 2, $X'' = 0.25 \times \dfrac{25}{25}\left(\dfrac{11}{11}\right)^2 = 0.25$ pu

On 60 MVA base, for generator 1, $X'' = 0.20 \times \dfrac{60}{30}\left(\dfrac{11}{11}\right)^2 = 0.40$ pu

On 60 MVA base, for generator 2, $X'' = 0.25 \times \dfrac{60}{25}\left(\dfrac{11}{11}\right)^2 = 0.60$ pu

Calculation of per unit impedance of the transformer

On 25 MVA base, $X = 0.10 \times \dfrac{25}{60}\left(\dfrac{11}{11}\right)^2 = 0.041667$ pu

On 60 MAV base, $X = 0.1$ pu

The reactance diagrams for S bases of 25 and 60 MVA are given in Fig. 8.6.

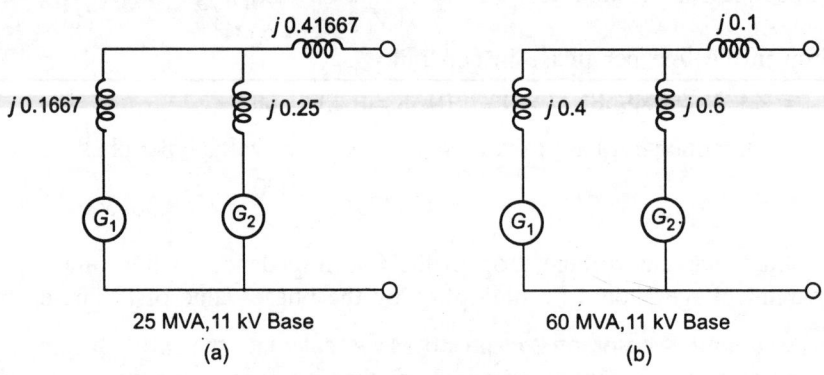

25 MVA, 11 kV Base   60 MVA, 11 kV Base

(a)   (b)

Fig. 8.6. Reactance diagrams.

Calculation of through impedance (reactance) per unit

(a) 25 MVA base

$Z_{pu} = [j\,0.1667 \,\|\, j\,0.25] + j\,0.041667$

$= j\dfrac{0.1667 \times 0.25}{0.1667 + 0.25} + j\,0.041667 = j\,0.100012 + j\,0.041667 = j\,0.141679$ pu

(b) 60 MVA base

$Z_{pu} = [j\,0.4 \,\|\, j\,0.6] + j\,0.1 = j\dfrac{0.4 \times 0.6}{0.4 + 0.6} + j\,0.1 = j\,0.34$ pu

Calculation of impedances (reactances) in ohms

The base impedance is given by

$$Z_b = \frac{(kV_l)_b^2}{(MVA)_b}$$

For 25 MVA base, $Z_b = \frac{(11)^2}{25} = 4.84\ \Omega$

For 60 MVA base, $Z_b = \frac{(11)^2}{60} = 2.01667\ \Omega$

Actual impedance in ohms = per unit impedance × base impedance

Actual impedance in ohms on 25 MVA base $= j\,0.141679 \times 4.84 = j\,0.6857263\ \Omega$

Actual impedance in ohms on 60 MVA base $= j\,0.34 \times 2.01667 = j\,0.6856678\ \Omega$

This problem illustrates the arbitrary selection of the S base, provided the value selected is used consistently throughout the network.

Example 8.7 For the system shown in Fig. 8.7, determine the generator voltage.

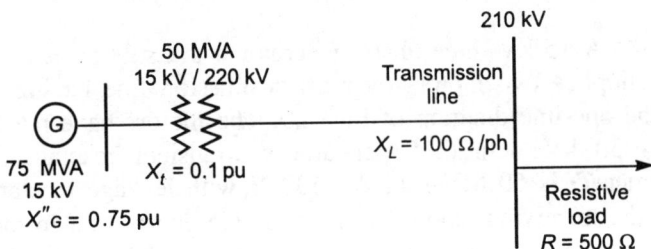

Fig. 8.7.

Solution

Let $S_b = 100$ MVA and $V_b = 15$ kV at the generator terminals.

For generator, $X_{Gpu} = 0.75 \times \frac{100}{75} \left(\frac{15}{15}\right)^2 = 1.0$ pu

For transformer, $X_{tpu} = 0.1 \times \frac{100}{50} \left(\frac{220}{220}\right)^2 = 0.20$ pu

For transmission line, $X_{Lpu} = Z_\Omega \frac{(MVA)_b}{(kV_l)_b^2} = 100 \times \frac{100}{(220)^2} = 0.2066$ pu

For the resistive load

$$R_{pu} = Z_\Omega \frac{(MVA)_b}{(kV_l)_b^2} = 500 \times \frac{100}{(220)^2} = 1.033\ \text{pu}$$

The system current in per unit is given by

$$I_{pu} = \frac{V_{pu}}{R_{pu}} = \frac{210/220}{1.033} = 0.924\ \text{pu}$$

Let $\mathbf{I}_{pu}$ be taken as the reference phasor.

$$\mathbf{I}_{pu} = 0.924 + j\,0 = 0.924\,\underline{/0°}$$

The per unit impedance diagram is shown in Fig. 8.8.

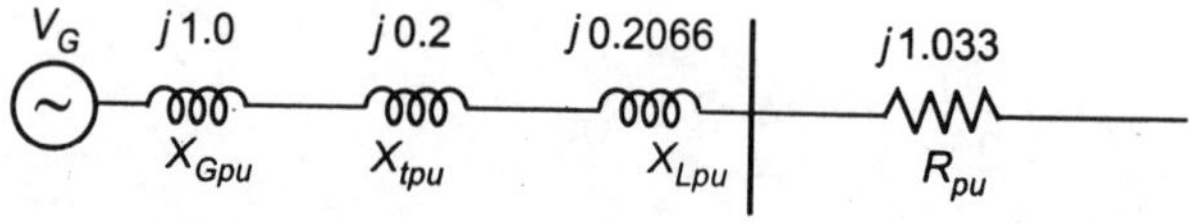

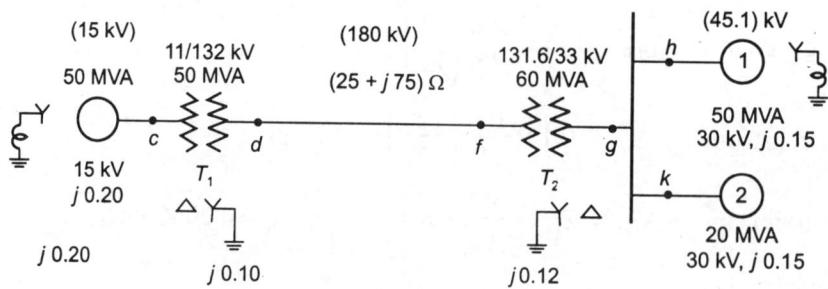

Fig. 8.8.

By KVL,

voltage drop in the network $= I_{pu}\,[R_{pu} + j\,(X_{Gpu} + X_{tpu} + X_{Lpu})]$

$$= 0.924\,[1.033 + j\,(1.0 + 0.2 + 0.2066)]$$

$$= 0.924\,(1.033 + j\,1.4066) = 1.6125355 \ \text{pu}$$

Actual generator terminal voltage, $V_G = 1.6125355 \times 15 = 24.188$ kV line-to-line

Generator terminal voltage per phase $= \dfrac{24.188}{\sqrt{3}} = 13.965$ kV

Example 8.8 A 50 MVA 15 kV three-phase generator has a subtransient reactance of 0.20 per unit. The generator supplies two motors over a transmission line having transformers at both ends, as shown on the one-line diagram of Fig. 8.9. The motors have rated inputs of 30 MVA and 20 MVA, both 30 kV with 0.15 per unit subtransient reactance. The rating of the sending-end transformer T_1 is 50 MVA 11 Δ - 132 Y with leakage reactance of 0.10 per unit. Transformer T_2 at the receiving end has three single-phase transformers connected as a three-phase unit. The rating of each individual transformer is 20 MVA, 33/76 kV with leakage reactance of 0.12 per unit. Series impedance of the line is $(25 + j\,75)$ ohms. Draw the impedance diagram with all impedances marked in per unit. Select the generator rating as the base in the generator circuit.

Fig. 8.9. One-line diagram of Example 8.5.

Solution

Step 1   Selection of base MVA and base voltage in kV.

Let us take a base MVA of 50. Therefore the selected value $S_b = 50$ MVA shall remain the same in all parts of the system.

Let us select the base voltage in the generator circuit as $V_b = 15$ kV. The base voltage on the primary side of the transformer T_1 is 15 kV. We know that for each circuit separated by transformers the base voltage changes in proportions to the turns ratio. The turns ratio is equal to the voltage ratio of the transformer. In order to determine the base kV for various components of the system we use the following formula :

$$\text{Base kV on hv side} = \text{base kV on lv side} \times \frac{(kV)_{hv}}{(kV)_{lv}}$$

or $$\text{base kV on lv side} = \text{base kV on hv side} \times \frac{(kV)_{lv}}{(kV)_{hv}}$$

Base voltage of the transmission line = base voltage on the secondary side of T_1

$$= 15 \times \frac{132}{11} = 180 \text{ kV}$$

The base voltage in the motor circuit is determined with 180 kV applied to the primary of transformer T_2.

Base voltage in the motor circuit = base voltage on the secondary side of the transformer T_2

$$= (\text{base kV on primary of } T_2) \times (\text{turns ratio of } T_2)$$

$$= 180 \times \frac{33}{\sqrt{3} \times 76} = 45.1 \text{ kV}$$

It is to be noted that for a phase voltage of 76 kV, the line voltage is $(\sqrt{3} \times 76)$ kV.

The base voltages are shown in parentheses on the line diagram of Fig. 8.5.

Step 2   Calculation of the per unit reactance of the generator

No calculation is necessary for correcting the value of the generator reactance because it is given as 0.20 pu based on 50 MVA and 15 kV. If a different value of S_b were used in this problem, then correction would be necessary as shown for the transformers, transmission line, and motors.

Step 3   Calculation of the per unit reactance of transformers

The per unit reactances of the transformers are found from the formula

$$Z_{2pu} = Z_{1pu} \left(\frac{S_{b2}}{S_{b1}}\right)\left(\frac{V_{b1}}{V_{b2}}\right)^2 = Z_{1pu} \frac{(MVA)_{b2}}{(MVA)_{b1}} \left[\frac{(kV_l)_{b1}}{(kV_l)_{b2}}\right]^2$$

Per unit reactance of transformer T_1

$$= j\, 0.1 \times \frac{50}{50} \times \left(\frac{11}{15}\right)^2 = j\, 0.0537 \text{ pu}$$

Base voltage on the lv side of the transformer T_2 is 45.1 kV. Total MVA of transformer T_2

$$= 3 \times 20 = 60 \text{ MVA}$$

Per-unit reactance of transformer T_2

$$= j\, 0.12 \times \frac{50}{60} \times \left(\frac{33}{45.1}\right)^2 = j\, 0.0535 \text{ pu}$$

Step 4   Calculation of the per unit impedance of the transmission line

Base kV of the transmission line = 180 kV

$$Z_{pu} = Z_\Omega \frac{(MVA)_b}{(kV_l)_b^2} = (25 + j\,75)\frac{50}{(180)^2} = 0.0385 + j\,0.1157 \text{ pu}$$

Step 5   Calculation of the per unit reactance of motors

The per unit reactances of the motors is found from the formula given in step 3.

$$\text{Per unit reactance of motor 1,} \quad = j\,0.15 \times \frac{50}{30} \times \left(\frac{30}{45.1}\right)^2 = j\,0.1106 \text{ pu}$$

$$\text{Per unit reactance of motor 2,} \quad = j\,0.15 \times \frac{50}{20} \times \left(\frac{30}{45.1}\right)^2 = j\,0.1659 \text{ pu}$$

Step 6   Drawing of per unit impedance diagram

The required per unit impedance diagram is shown in Fig. 8.10.

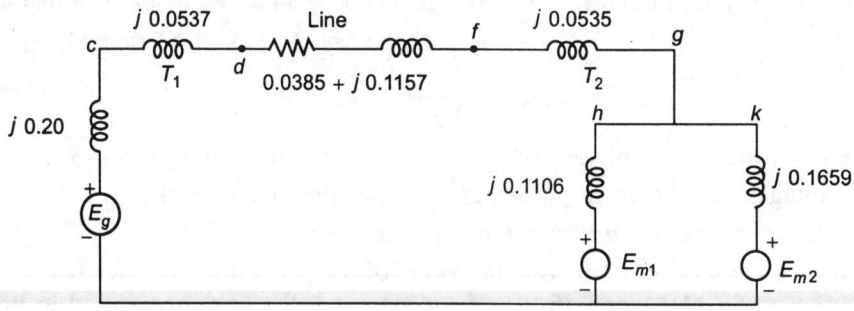

Fig. 8.10. Per-unit impedance diagram.

Example 8.9 If the motors of Example 8.8 have inputs of 24 MW and 16 MW respectively at 30 kV and both operate at unity power factor, find the voltage at the terminals of the generator.

Solution

Total input to motors = 24 + 16 = 40 MW

Base MW = numerical value of base MVA = 50

$$\text{Per unit input} = \frac{\text{actual input}}{\text{base input}} = \frac{40}{50} = 0.8$$

If V_{pu} and I_{pu} denote per unit values of voltage and current at the motor terminals

$$V_{pu}\,I_{pu} = P_{pu}\,; \quad V_{pu} = \frac{30\,\underline{/0^\circ}}{45.1} = 0.6652\,\underline{/0^\circ}\,; \quad I_{pu} = \frac{P_{pu}}{V_{pu}} = \frac{0.8}{0.6652\,\underline{/0^\circ}} = 1.2027\,\underline{/0^\circ} \text{ pu}$$

The voltage at the generator terminals

$$V_G = V_{pu} + (Z_{T1pu} + Z_{Lpu} + Z_{T2pu})\,I_{pu}$$

$$= 0.6652\,\underline{/0^\circ} + [(j\,0.0537 + 0.0385 + j\,0.1157 + j\,0.0535) \times (1.2027\,\underline{/0^\circ})]$$

$$= 0.7115 + j\,0.2681 = 0.76\,\underline{/20.65^\circ} \text{ pu}$$

Generator terminal voltage = 0.76 × 15 = 11.4 kV

Example 8.10 Fig. 8.11 shows a two-machine system. The ratings are as follows :

Synchronous generator 20 MVA, 11 kV, $X'' = 0.15$ pu

Synchronous motor 15 MVA, 11 kV, $X'' = 0.15$ pu

Transformer T_1 25 MVA, 12.5 Δ/132 Y kV

Transformer T_2 20 MVA, 132 Y/11 Δ kV

Line $(200 + j\,500)$ Ω

Static load 5 MVA 0.8 power factor lagging.

Draw the impedance diagram for the system. Choose a base voltage of 132 kV for the transmission line and a base voltampere of 20 MVA.

If the motor is a synchronous machine drawing 15 MVA at 0.9 power factor (leading) and the terminal voltage 1.1 pu find the generator bus voltage.

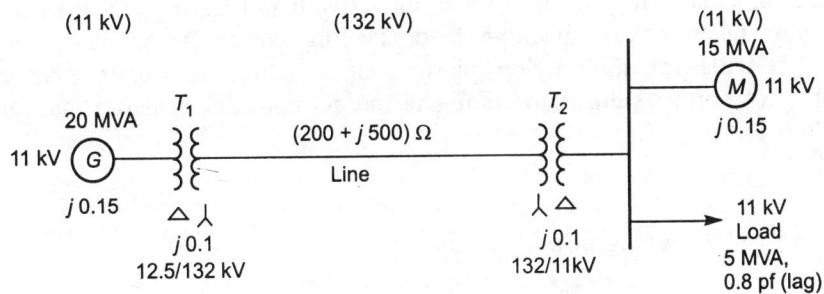

Fig. 8.11. Two-machine system.

Solution

We have $Z_{2pu} = Z_{1pu} \dfrac{(\text{MVA})_{b2}}{(\text{MVA})_{b1}} \left[\dfrac{(\text{kV})_{b1}}{(\text{kV})_{b2}} \right]^2$

Per unit reactance of generator referred to given bases $= j\,0.15 \left(\dfrac{20}{20}\right)\left(\dfrac{11}{12.5}\right)^2 = j\,0.116$

Per unit reactance of motor referred to given bases $= j\,0.15 \left(\dfrac{20}{15}\right)\left(\dfrac{11}{11}\right)^5 = j\,0.200$

Per unit reactance of transformer T_1 referred to given bases $= j\,(0.10)\left(\dfrac{20}{25}\right)\left(\dfrac{132}{132}\right)^2 = j\,0.08$

Per unit reactance of transformer T_2 referred to given bases $= j\,(0.10)\left(\dfrac{20}{20}\right)\left(\dfrac{132}{132}\right)^2 = j\,0.10$

Per unit impedance of transmission line referred to given bases $= (200 + j\,500)\dfrac{20}{(132)^2}$

$$= 0.229 + j\,0.574$$

For a load of 5 MVA 0.8 power factor lagging

$P = 5 \cos \varphi = 5 \times 0.8 = 4$ MW

$$Q = 5 \sin \varphi = 5 \times 0.6 = 3 \text{ MVAr}$$

If load is represented as series constant impedance

$$R_{pu} = V_{pu}^2 \, S_b \, \frac{P}{P^2 + Q^2} \; ; \qquad X_{pu} = V_{pu}^2 \, S_b \, \frac{Q}{P^2 + Q^2}$$

Here $\quad V_{pu} = \dfrac{11}{11} = 1.0 \text{ pu}, \quad S_b = 20 \text{ MVA}$

$$\therefore \qquad R_{pu} = (1.0)^2 \times 20 \times \frac{4}{4^2 + 3^2} = 3.2 \text{ pu}$$

and $\qquad X_{pu} = (1.0)^2 \times 20 \times \dfrac{3}{4^2 + 3^2} = 2.4 \text{ pu.}$

It is a general practice that VAr taken by an inductive load (lagging power factor) is assumed positive and VAr taken by a capacitive load (leading power factor) is assumed negative. Consequently, the rule for multiplying phasors of V and I to obtain phasor voltampere $(P + jQ)$, with Q of correct sign, is to conjugate the current phasor before multiplication.

$$I = \frac{S}{V}$$

$$\therefore \qquad I = \frac{P - jQ}{V} \text{ for lagging power factor}$$

and $\qquad I = \dfrac{P + jQ}{V} \text{ for leading power factor}$

$$I_b = \frac{S_b}{V_b} \; ; \qquad I_{pu} = \frac{I}{I_b} = \frac{S}{V} \cdot \frac{V_b}{S_b} = \frac{S}{S_b} \cdot \frac{1}{V_{pu}}$$

Current taken by the motor which is operating at leading power factor 0.9

$$I_M = \frac{P_M + jQ_M}{S_b} \cdot \frac{1}{V_{pu}} = \frac{15 \, (\cos \varphi_M + j \sin \varphi_M)}{20 \times 1.1}$$

$$\cos \varphi_M = 0.9 \; ; \qquad \sin \varphi_M = 0.436$$

$$I_M = \frac{15 \, (0.9 + j \, 0.436)}{20 \times 1.1} = (0.613 + j \, 0.297) \text{ pu}$$

Current taken by static load which is operating at 0.8 power factor lagging

$$I_L = \frac{P_L - jQ_L}{S_b} \cdot \frac{1}{V_{pu}} = \frac{4 - j3}{20 \times 1.1} = (0.1818 - j \, 0.1364) \text{ pu}$$

Total load current, $I = I_M + I_L = 0.613 + j \, 0.297 + 0.1818 - j \, 0.1364$

$$= 0.7948 + j \, 0.1606 \text{ pu} = 0.81086 \, \underline{/11.42°} \text{ pu}$$

Total per unit impedance between generator and motor bases

Z_T = per unit impedance of T_1 + per unit impedance of the line + per unit impedance of T_2

$$= j \, 0.080 + 0.229 + j \, 0.574 + j \, 0.100 = 0.229 + j \, 0.754 \text{ pu} = 0.788 \, \underline{/73.1°} \text{ pu}$$

Generator bus voltage

$$V_g = V_{pu} + Z_T I$$
$$= 1.1 \underline{/0°} + (0.788 \underline{/73.1°})(0.81086 \underline{/11.42°}) = 1.1 + 0.63895 \underline{/84.53°}$$
$$= 1.1 + 0.061 + j\,0.636 = 1.161 + j\,0.636 = 1.3238 \underline{/28.71°} \text{ pu on 11 kV base}$$
$$|V_g| = 11 \times 1.3238 = 14.56 \text{ kV}$$

EXERCISES

1. Define the terms per unit voltage, per unit impedance and per unit voltamperes. Express per unit impedance in terms of base MVA and base kV for a three-phase system.
2. Derive an expression for per unit impedance of a given base MVA and base kV in terms of new base MVA and new base kV.
3. Show that the per-unit equivalent impedance of a two-winding transformer is the same whether the calculation is made from the high-voltage side or the low-voltage side.
4. A 25 kVA, 220/110 V, three-phase transformer has a 0.04 Ω leakage reactance referred to the low-voltage side. Calculate the per unit leakage impedance referred to the low-voltage and high-voltage sides of the transformer. Use the transformer ratings as the base values.
5. What are the advantages of per-unit representation?

    Draw the reactance diagram for the system shown in Fig. 8.12. The specifications of the components are given in Table 8.2.

Table 8.2

Generator G	Transformer T_1	Transformer T_2	Motor 1	Motor 2	Line
13.8 kV	25 MVA	25 MVA	15 MVA	10 MVA	$X = 65\ \Omega$
25 MVA	13.2/69 kV	69/13.2 kV	13.0 kV	13.0 kV	
$X'' = 0.15$ pu	$X_L = 0.11$ pu	$X_L = 0.11$ pu	$X'' = 0.15$ pu	0.15 pu	

Determine the generator terminal voltage assuming both motors operating at 12 kV, 75% full load, and unity power factor.

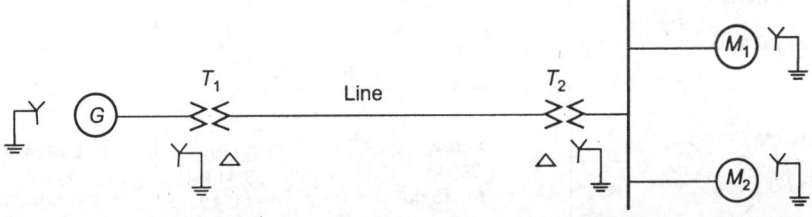

Fig. 8.12.

6. A one-line diagram of a three-phase power system is shown in Fig. 8.13. Choose 13.8 kV, the generator voltage, as the base voltage and 25 MVA as the base MVA. Draw a per unit reactance diagram.

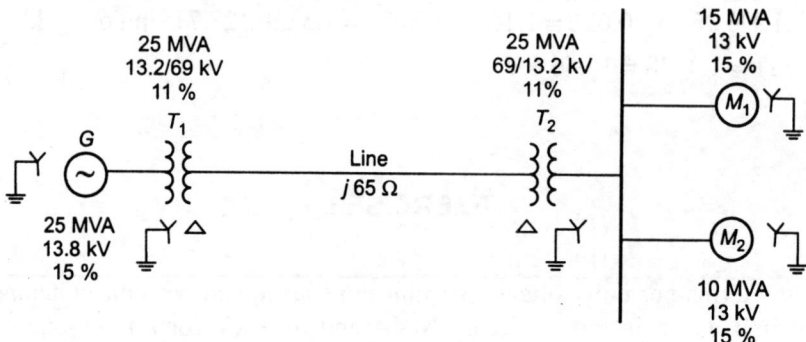

Fig. 8.13.

7. Draw a per unit reactance diagram for the three-phase system shown in Fig. 8.14. Choose 20 MVA and 66 kV as base values.

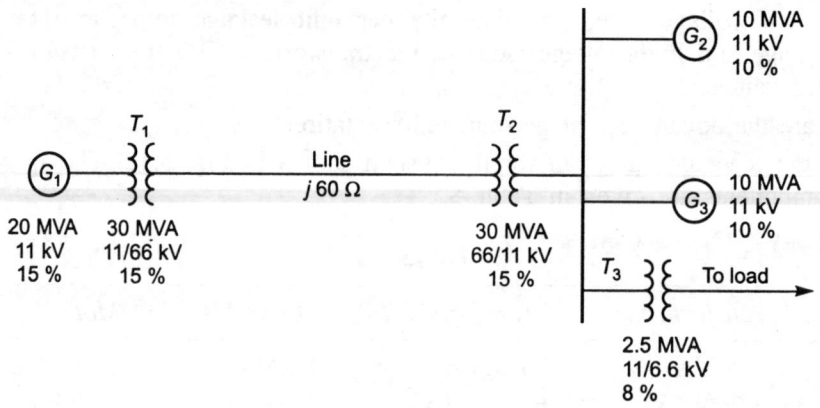

Fig. 8.14.

8. The one-line diagram for a two-generator system is shown in Fig. 8.15. Redraw the diagram to show all values in per unit on a 7000 kVA base.

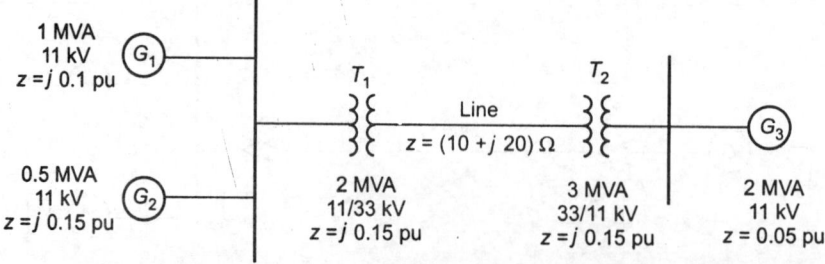

Fig. 8.15.

ANSWERS

4. $j\,0.0826$ pu   **5.** $13.467\,\underline{/26.99°}$ kV
6. Fig. 8.16   **7.** Fig. 8.17
8. Fig. 8.18

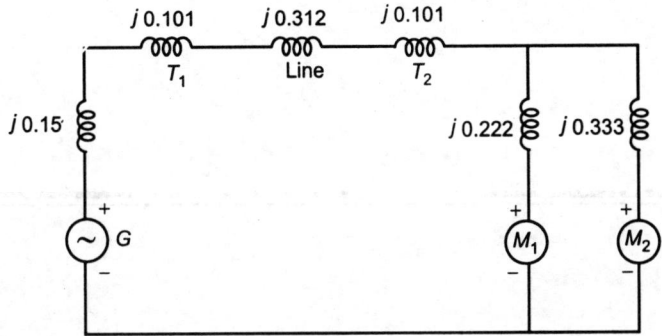

Fig. 8.16.

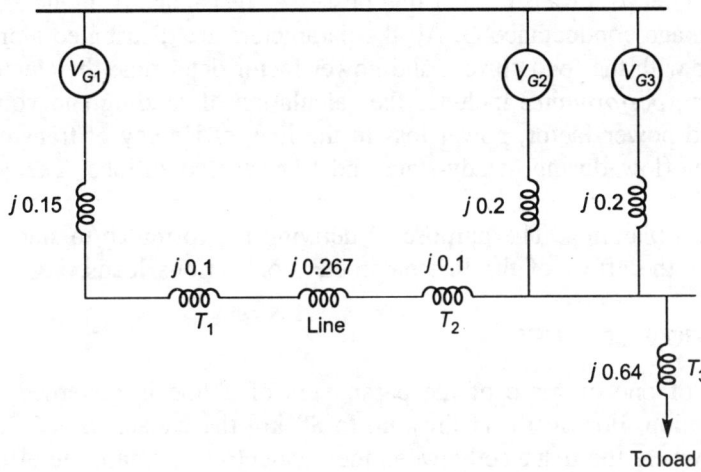

Fig. 8.17.

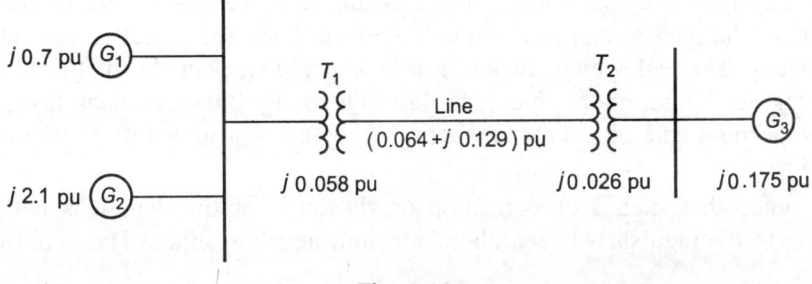

Fig. 8.18.

Short and Medium Lines

9.1 INTRODUCTION

As discussed in Chapter 7, a transmission line possesses resistance R, inductance L, capacitance C, and shunt or leakage conductance G. All the parameters are distributed along the line. These parameters together with the load current and power factor determine the electrical performance of the line. The term *performance* includes the calculation of sending-end voltage, sending-end current, sending-end power factor, power loss in the line, efficiency of transmission, regulation and limits of power flow during steady-state and transient conditions. The values of voltage, current and power factor at the receiving end are usually known. Prior performance calculations are helpful in system planning. The purpose of deriving the formulae to study the performance of a line is to know the effect of the line parameters on various loads.

9.2 CLASSIFICATION OF LINES

The predominance of one or more of the parameters of a line is governed by its length and conductor configuration. For overhead lines up to 80 km the capacitance C is negligibly small but for cable lines where the distance between the conductors is small, the effect of capacitance cannot be ignored. All low-voltage overhead lines having lengths up to 80 km are generally categorised as *short lines*. The lines ranging in length from 80 to 240 km are termed *medium* or *moderately long lines*. For such lines, the capacitance of the line cannot be neglected and it is considered to be lumped at one or more points of the line. The effect of capacitance is more at higher frequency. The leakage conductance or leakance is neglected. The term *long line* refers to a line having its length more than 240 km. The long line treatment takes all the four parameters into account and allows for the fact that they are distributed uniformly over the entire length of the line.

It is to be noted that such a classification on the basis of line length is not necessarily a perfect criterion to distinguish between short, medium and long lines. The classification of the lines in three categories is done according to the accuracy desired. The accuracy is affected by the operating frequency also. The methods used for short and medium lines are approximate while those adopted for long lines are rigorous. The approximate methods are simple without much appreciable error. This chapter is devoted to the performance calculations of a short and medium lines.

9.3 SHORT SINGLE-PHASE LINE

As pointed out earlier, in a short line, the shunt capacitance C and shunt conductance G are neglected. The series resistance R and the series inductance L for the total length of the line is considered. Single phase supply line is usually short in length and operates at relatively low voltages. A single-phase line has two conductors. Each conductor has a resistance R_1 and an inductance L_1. The inductance L_1 is in effect equivalent to an inductive reactance X_1 ($= 2\pi f L_1$). Such a line is shown in Fig. 9.1. For convenience, the resistance and inductive reactance are considered to be lumped in one conductor only and the return conductor is supposed to possess no resistance or inductance. Fig. 9.2 shows the simplified form of Fig. 9.1. It is the equivalent circuit model of a short line. In Fig. 9.2, R and X represent the loop resistance and loop reactance of the line respectively. Thus,

R = loop resistance of the line = resistance of both outgoing and return conductors

$\quad$ = 2 × resistance of one conductor = $2R_1$

and $\quad X$ = loop reactance of the line = reactance of both lead and return conductors

$\quad$ = 2 × inductive reactance of one conductor to neutral = $2X_1$

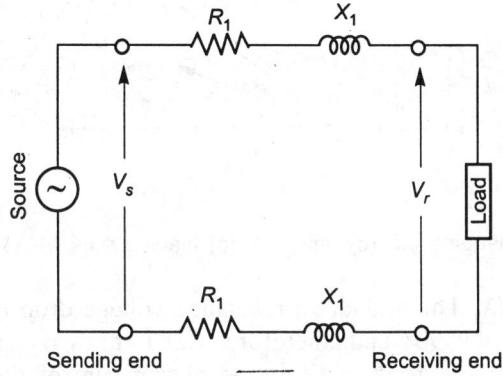

Fig. 9.1. Single-phase line.

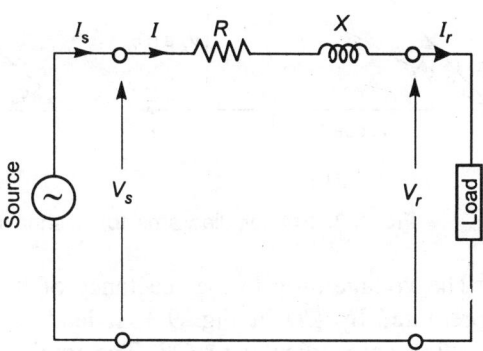

Fig. 9.2. Equivalent-circuit model of a short line.

The end of the line where load is connected is called the *receiving end*. The end where the source of supply is connected is called the *sending end*.

Let $\quad V_r$ = voltage at the receiving end

$\quad V_s$ = voltage at the sending end

$\quad I_r$ = current at the receiving end

$\quad I_s$ = current at the sending end

$\cos \varphi_r$ = power factor of the load

$\cos \varphi_s$ = power factor at the sending end.

Series impedance of the line, $\mathbf{Z} = R + jX$

Since the shunt capacitance and shunt conductance are neglected in a short line, the load current practically remains the same at all points along the length of the line, so that

$$I_s = I_r = I \text{ (say)}$$

9.4 PHASOR DIAGRAM

The phasor diagram for a short line for a load of lagging power factor is shown in Fig. 9.3 (a). Let the receiving-end voltage V_r, which is maintained constant, be taken as reference phasor along OA such that $OA = \mathbf{V}_r$. For lagging power factor $\cos \varphi_r$, the direction of the load current $\mathbf{I}$ lags behind $\mathbf{V}_r$ by an angle φ_r along OB, where $OB = \mathbf{I}$.

The voltage drop in the resistance of the line = IR.

IR is represented by the phasor AC. It is in phase with $\mathbf{I}$ and, therefore, drawn parallel to current phasor OB.

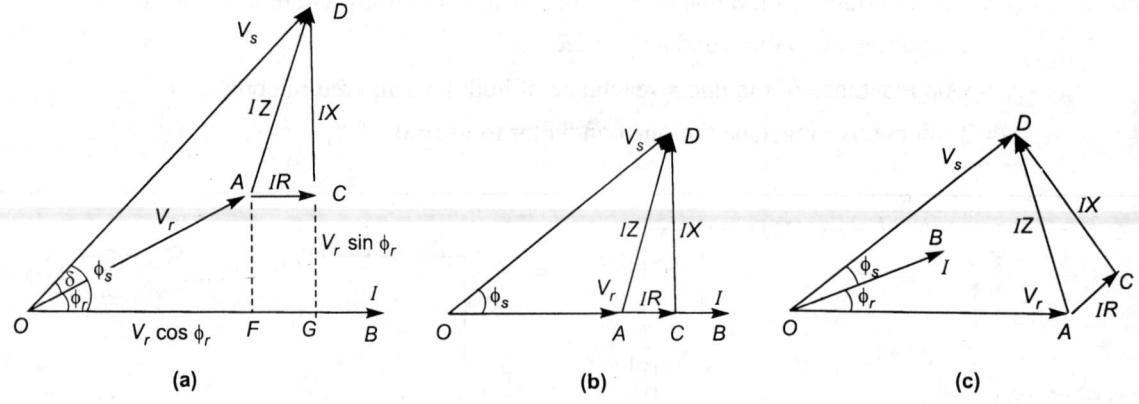

Fig. 9.3. Phasor diagrams for a short line : (a) lagging p.f. (b) unity p.f. (c) leading p.f.

The voltage drop in the reactance of the line = IX. This inductive reactance voltage drop is represented by CD in Fig. 9.3. It leads the current by 90° and, therefore, CD is drawn in a direction perpendicular to OB. The total impedance voltage drop $\mathbf{IZ}$ is the phasor sum of the resistive and reactive voltage drops. It is given by AD in Fig. 9.3.

The sending-end voltage will maintain the voltage $\mathbf{V}_r$ at the receiving end constant and will also supply the total impedance voltage drop in the line. OD represents the sending-end voltage $\mathbf{V}_s$ in both magnitude and direction. The angle by which the current lags behind the voltage at the sending end is φ_s. Therefore, $\cos \varphi_s$ is the power factor of the load measured at the sending end. In Fig. 9.3, δ is the phase displacement between the voltages at the two ends.

The magnitude of V_s can be found from the right angled triangle OGD.

$$OD^2 = OG^2 + GD^2 = (OF + FG)^2 + (GC + CD)^2$$
$$V_s^2 = (V_r \cos \varphi_r + IR)^2 + (V_r \sin \varphi_r + IX)^2$$

$\therefore$ sending-end voltage, $V_s = [(V_r \cos \varphi_r + IR)^2 + (V_r \sin \varphi_r + IX)^2]^{1/2}$ ...(9.4.1)

The power factor of the load measured at the sending end is

$$\cos \varphi_s = \frac{OG}{OD} = \frac{OF + FG}{OD} = \frac{V_r \cos \varphi_r + IR}{V_s} \qquad ...(9.4.2)$$

The phasor diagrams for unity and leading power factors for the same values of V_r and I are shown in Figs. 9.3 (b) and 9.3 (c) respectively.

An alternative expression for V_s can be found by using complex algebra. If $\mathbf{V}_r$ be the reference phasor,

$$\mathbf{V}_r = V_r \underline{/0°} = V_r + j\,0$$

For lagging power factor $\cos \varphi_r$, $\mathbf{I} = I \underline{/- \varphi_r} = I \cos \varphi_r - j\,I \sin \varphi_r$

For leading power factor $\cos \varphi_r$, $\mathbf{I} = I \underline{/+ \varphi_r} = I \cos \varphi_r + j\,I \sin \varphi_r$

For unity power factor, $\mathbf{I} = I \underline{/0°} = I + j\,0$

The line impedance is given by

$$\mathbf{Z} = R + j\,X$$

The sending-end voltage is

$$\mathbf{V}_s = \mathbf{V}_r + \mathbf{ZI}$$

For lagging power factor

$$\mathbf{V}_s = (V_r + j\,0) + (R + j\,X)\,(I \cos \varphi_r - j\,I \sin \varphi_r)$$

$$= (V_r + IR \cos \varphi_r + IX \sin \varphi_r) + (IX \cos \varphi_r - IR \sin \varphi_r)$$

$\therefore$ $V_s = [(V_r + IR \cos \varphi_r + IX \sin \varphi_r)^2 + (IX \cos \varphi_r - IR \sin \varphi_r)^2]^{1/2}$

$$\tan \delta = \frac{IX \cos \varphi_r - IR \sin \varphi_r}{V_r + IR \cos \varphi_r + IX \sin \varphi_r}$$

9.5 SHORT THREE-PHASE LINE

A balanced three-phase circuit may be considered as consisting of three separate identical single-phase circuits. Therefore, the calculations for a balanced three-phase line are carried out in a similar manner as explained for single-phase line, the difference being that *per phase basis* is adopted. When working with balanced three phase lines it is usual to assume that all the given voltages are line-to-line values, that all currents are line currents. Similarly, the given power is the total power for all the three phases and the given reactive voltamperes represent the total reactive voltamperes for all the three phases. Thus, for three-phase line calculations,

power per phase = $(1/3) \times$ (total power)

reactive voltamperes per phase = $(1/3) \times$ (total reactive voltamperes)

Also, in Figs. 9.2 and 9.3,

I = phase current

R = line resistance per phase

X = line reactance per phase

Z = line series impedance per phase

V_s = sending–end phase voltage

V_r = receiving–end phase voltage.

For a balanced 3-phase, star connected line,

$$\text{phase voltage} = \frac{1}{\sqrt{3}} \times \text{line voltage}$$

Fig. 9.2 shows the per-phase model of a short-phase line. The phasor diagram for lagging power factor is shown in Fig. 9.3 (a).

9.6 TRANSMISSION LINE AS A TWO-PORT NETWORK

A transmission line may be viewed as a two-port network, as shown in Fig. 9.4. The voltages and currents at the input and output terminals are expressed in the form of general equations given by

$$\mathbf{V}_s = \mathbf{A}\mathbf{V}_r + \mathbf{B}\mathbf{I}_r \qquad \qquad ...(9.6.1)$$

$$\mathbf{I}_s = \mathbf{C}\mathbf{V}_r + \mathbf{D}\mathbf{I}_r \qquad \qquad ...(9.6.2)$$

where $\mathbf{V}_s$ = sending–end voltage

$\mathbf{I}_s$ = sending–end current

$\mathbf{V}_r$ = receiving–end voltage

$\mathbf{I}_r$ = receiving–end current.

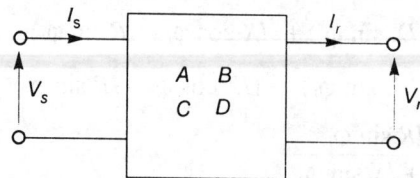

Fig. 9.4. A transmission line as a two-port network.

The **A**, **B**, **C**, **D** constants are called the *general network constants*. They depend on the line parameters and in general are complex. Equations (9.6.1) and (9.6.2) can be put in the matrix form as

$$\begin{bmatrix} \mathbf{V}_s \\ \mathbf{I}_s \end{bmatrix} = \begin{bmatrix} \mathbf{A} & \mathbf{B} \\ \mathbf{C} & \mathbf{D} \end{bmatrix} \begin{bmatrix} \mathbf{V}_r \\ \mathbf{I}_r \end{bmatrix} \qquad \qquad ...(9.6.3)$$

The validity of Eqs. (9.6.1) and (9.6.2) is based on the fact that a transmission line can be represented by a linear, passive, and bilateral network. By virtue of reciprocity, the generalised constants are related to each other by following equation:

$$\mathbf{AD} - \mathbf{BC} = 1 \qquad \qquad ...(9.6.4)$$

Transmission lines of various lengths can be represented by equivalent 2-port networks. We shall determine the **ABCD** constants for various lines. A detailed treatment of general network constants is given in Chapter 11.

9.6.1 ABCD Constants of a Short Line

The sending-end voltage and current can be written from the short-line equivalent network (Fig. 9.2) as

$$\mathbf{V}_s = \mathbf{V}_r + \mathbf{Z}\mathbf{I}_r \qquad\qquad\qquad \ldots(9.6.5)$$

$$\mathbf{I}_s = \mathbf{I}_r$$

Comparing the coefficients of Eqs. (9.6.5) and (9.6.6) with the general equations (9.6.1) and (9.6.2), the **ABCD** constants for a short line are given by

$$\mathbf{A} = 1, \quad \mathbf{B} = \mathbf{Z}, \quad \mathbf{C} = 0, \quad \mathbf{D} = 1.$$

9.7 LINE REGULATION

Voltage regulation of a line is defined as the change in voltage at the receiving end when full load at a given power factor is removed, the voltage at the sending end being kept constant. It is expressed as a fraction or a percentage of the receiving-end voltage at full load. It can be written as :

$$\text{Per unit regulation} \overset{\Delta}{=} \frac{|\mathbf{V}_{rnl}| - |\mathbf{V}_{rfl}|}{|\mathbf{V}_{rfl}|} \qquad\qquad \ldots(9.7.1)$$

$$\text{Per cent regulation} \overset{\Delta}{=} \frac{|\mathbf{V}_{rnl}| - |\mathbf{V}_{rfl}|}{|\mathbf{V}_{rfl}|} \times 100$$

where $|\mathbf{V}_{rnl}|$ = magnitude of receiving–end voltage at no load

$|\mathbf{V}_{rfl}|$ = magnitude of receiving–end voltage at full load.

The voltage $\mathbf{V}_s$ at the sending end is kept constant. It is given by

$$\mathbf{V}_s = \mathbf{A}\mathbf{V}_r + \mathbf{B}\mathbf{I}_r$$

When the load is removed,

$$\mathbf{I}_r = 0, \quad \mathbf{V}_r = \mathbf{V}_{r0}$$

Therefore, $\mathbf{V}_s = \mathbf{A}\mathbf{V}_{r0}$; $\qquad \mathbf{V}_{r0} = \mathbf{V}_s/\mathbf{A}$

where $\mathbf{V}_{r0}$ is the receiving-end voltage at no lead.

$$\text{Line regulation} = \frac{|\mathbf{V}_s|/|\mathbf{A}| - |\mathbf{V}_{rfl}|}{|\mathbf{V}_{rfl}|} \text{ pu} \qquad\qquad \ldots(9.7.2)$$

9.7.1 Line Regulation for Short Lines

In case of a short line, when the load is removed, the voltage at the receiving end is equal to the voltage at the sending end. At full load,

$$|\mathbf{V}_{rfl}| = |\mathbf{V}_r|$$

At no load, $|\mathbf{V}_{rnl}| = |\mathbf{V}_s|$

Therefore, for a short line

$$\text{line regulation} = \frac{|\mathbf{V}_{rnl}| - |\mathbf{V}_{rfl}|}{|\mathbf{V}_{rfl}|} = \frac{|\mathbf{V}_s| - |\mathbf{V}_r|}{|\mathbf{V}_r|} \text{ pu}$$

The regulation depends upon the power factor of the load. If the power factor is lagging, the voltage at the sending end is more than that at the receiving end. For leading power factor, the receiving-end voltage may become greater than at the sending end. The regulation becomes negative in this case.

9.8 LINE EFFICIENCY OR TRANSMISSION EFFICIENCY

The *line efficiency* or *efficiency of transmission* η_T is the ratio of power output of the line to the power input to the line.

$$\eta_T = \frac{\text{power output}}{\text{power input}} \text{ pu} = \frac{\text{power delivered at the receiving end}}{\text{power sent at the sending end}} \text{ pu}$$

$$= \frac{\text{power delivered at the receiving end}}{\text{power delivered at the receiving end} + \text{losses}} \text{ pu}$$

Example 9.1 A single-phase load of 200 kVA is delivered at 2500 V over a transmission line having $R = 1.4$ Ω, $X = 0.8$ Ω. Calculate the current, voltage and power factor at the sending end when the power factor of the load is (a) unity, (b) 0.8 lagging, and (c) 0.8 leading.

Solution

For a single-phase system,

$$\text{output kVA} = \frac{V_r I}{1000}; \qquad 200 = \frac{2500 \times I}{1000}; \qquad I = 80 \text{ A}$$

$$\mathbf{Z} = R + jX = 1.4 + j0.8 = 1.6 \underline{/29.745°} \text{ Ω}$$

Taking $\mathbf{V}_r$ as the reference phasor

$$\mathbf{V}_r = V_r + j0 = 2500 + j0$$

(a)*Unity power factor*

$$\cos \varphi_r = 1, \quad \varphi_r = 0°$$

$$\mathbf{I}_r = I_r \underline{/0°} = 80 \underline{/0°} \text{ A}$$

Current at the receiving end, $|\mathbf{I}_s| = |\mathbf{I}_r| = 80$ A

$$\mathbf{V}_s = \mathbf{V}_r + \mathbf{Z}\mathbf{I}_r = 2500 + (1.6 \underline{/29.745°}) (80 \underline{/0°})$$

$$= 2500 + 128 \underline{/29.745°} = 2500 + 128 (\cos 29.745° + j \sin 29.745°)$$

$$= 2500 + 111.13 + j63.50 = 2611.13 + j63.50 = \sqrt{(2611.13^2 + 63.5^2)} \quad \tan^{-1} \frac{63.50}{2611.13}$$

$$= 2611.9 \underline{/1.393°} \text{ V}$$

Phase difference between $\mathbf{V}_s$ and $\mathbf{I}_s$

$$\varphi_s = 1.393 - 0 = 1.393°$$

Sending-end power factor, $\cos \varphi_s = \cos 1.393° = 0.9997$ (lagging)

(b) *Load power factor equal to 0.8 lagging*

$$\cos \varphi_r = 0.8, \quad \varphi_r = \cos^{-1} 0.8 = 36.8699°$$

$\mathbf{I}_r = I_r \underline{/-\varphi_r} = 80 \underline{/-36.8699°}$ A

$\mathbf{V}_s = \mathbf{V}_r + \mathbf{Z}\mathbf{I}_r = 2500 + (1.6 \underline{/29.747°})\,(80 \underline{/-36.8699°}) = 2500 + 128 \underline{/-7.125°}$

$\quad = 2500 + 128\,(\cos 7.125° - j \sin 7.125°) = 2500 + 127 - j\,15.87$

$\quad = \sqrt{(2627^2 + 15.87^2)}\;\; -\tan^{-1}\dfrac{15.87}{2627} = 2627.06 \underline{/-0.3463°}$ V

Phase difference between $\mathbf{V}_s$ and $\mathbf{I}_s$

$\quad \varphi_s = 0.3463° - (-36.8699°) = 36.5236°$

$\quad \cos \varphi_s = \cos 36.5236° = 0.8036$ (lagging)

(c) *Load power factor equal to 0.8 leading*

Here $\quad \mathbf{I}_r = I_r \underline{/+\varphi_r°} = 80 \underline{/36.8699°}$ A

$\quad\quad \mathbf{V}_s = \mathbf{V}_r + \mathbf{Z}\mathbf{I}_r$

$\quad\quad\quad = 2500 + (1.6 \underline{/29.745°})\,(80 \underline{/36.8699°}) = 2500 + 128\,(\cos 66.615° + j \sin 66.615°)$

$\quad\quad\quad = 2500 + 50.8 + j\,117.486 = \sqrt{(2550.8^2 + 117.486^2)}\;\;\tan^{-1}\dfrac{117.486}{2550.8}$

$\quad\quad\quad = 2553.5 \underline{/2.6371°}$ V

$\quad\quad \varphi_s = -2.6371° + 36.8699° = 34.2328°$

$\quad\quad \cos \varphi_s = \cos(34.2328°) = 0.82676$ (leading).

Example 9.2 15000 kVA is received at 33 kV at 0.85 power factor lagging over an 8 km three-phase overhead line. Each line has $R = 0.29\ \Omega$ per km, and $X = 0.65\ \Omega$ per km. Calculate (a) the voltage at the sending end (b) the power factor at the sending end, (c) the regulation, and (d) the efficiency of the transmission line.

Solution

The receiving-end kVA per phase $= \dfrac{15000}{3} = 5000$

Phase voltage at the receiving end, $V_{rp} = \dfrac{1}{\sqrt{3}} \times V_{rl} = \dfrac{1}{\sqrt{3}} \times 33 \times 10^3 = 19052$ V

Let I_{rp} = current per phase at the receiving end

$\dfrac{V_{rp}\,I_{rp}}{1000}$ = receiving-end kVA per phase

$I_{rp} = \dfrac{1000 \times 5000}{19502} = 262.43$ A

Let $\mathbf{V}_{rp}$ be taken as reference phasor.

$\quad \mathbf{V}_{rp} = V_{rp} \underline{/0°} = V_{rp} + j\,0 = 19052 + j\,0$

Power factor at the receiving end, $\cos \varphi_r = 0.85$ lagging, $\varphi_r = 31.7883°$

Therefore, $\mathbf{I}_{rp} = I_{rp} \underline{/-\varphi_r} = 262.43 \underline{/-31.7883°}$ A

Line constants

Resistance per phase, $R_p = 0.29 \times 8 = 2.32\ \Omega$

Inductive reactance per phase, $X_p = 0.65 \times 8 = 5.20\ \Omega$

Series impedance per phase, $\mathbf{Z}_p = R_p + j\, X_p = 2.32 + j\, 5.20$

$$= \sqrt{(2.32^2 + 5.20^2)}\quad \tan^{-1}\frac{5.20}{2.32} = 5.694\ \underline{/65.9558°}\ \Omega$$

(a) Let $\mathbf{V}_{sp}$ = phase voltage at the sending end

$\mathbf{V}_{sp} = \mathbf{V}_{rp} + \mathbf{Z}_p\mathbf{I}_{rp} = 19051 + j\,0 + (5.694\ \underline{/65.9558°}) \times (262.43\ \underline{/-31.7883°})$

$\quad = 19052 + 1494.3\ \underline{/34.1675°} = 19052 + 1236.4 + j\,839.2$

$\quad = 20288.4 + j\,839.2 = 20306.2\ \underline{/2.3686°}\ $ V

Line voltage at the sending end, $|\mathbf{V}_{sl}| = \sqrt{3}\ |\mathbf{V}_{sp}| = \sqrt{3} \times 20306.2 = 35171$ V = 35.171 kV.

(b) Phase difference between $\mathbf{V}_s$ and $\mathbf{I}_s$

$\quad \varphi_s = 2.3686° - (-31.7883°) = 34.1569°$

Power factor at the sending end, $\cos \varphi_s = \cos 34.1569° = 0.8275$ (lagging)

(c) $\qquad$ Line regulation $= \dfrac{|\mathbf{V}_{sp}|/|\mathbf{A}| - |\mathbf{V}_{rp}|}{|\mathbf{V}_{rp}|} = \dfrac{|\mathbf{V}_{sp}| - |\mathbf{V}_{rp}|}{|\mathbf{V}_{rp}|}\qquad (\because\ \mathbf{A} = 1)$

$$= \frac{20306.2 - 19052}{19052} = 0.0658\ \text{pu or 6.58 per cent}$$

(d) Transmission efficiency, $\eta_T = \dfrac{\text{power output}}{\text{power output} + \text{losses}}$

Power output $= 3V_{rp}\, I_{rp} \cos \varphi_r = 15000 \times 1000 \times 0.85 = 1275 \times 10^4$ W

Power loss in the line $= 3I_{rp}^2\, R_p = 3 \times (262.43)^2 \times 2.32 = 478857$ W

$$\eta_T = \frac{1275 \times 10^4}{1275 \times 10^4 + 478857} = 0.9638\ \text{pu} = 96.38\ \text{per cent.}$$

Example 9.3 A three-phase line, 10 km long, delivers 5 MW at 11 kV, 50 Hz, 0.8 power factor lagging. The power loss in the line is 10 per cent of the power delivered. The line conductors are situated at the corners of an equilateral triangle of 2 m side. Calculate the voltage and power factor at the sending end.

Solution

Phase voltage at the receiving end, $V_{rp} = \dfrac{11 \times 10^3}{\sqrt{3}} = 6350$ V

Receiving-end current, $I_{rp} = \dfrac{P}{3V_{rp}\cos \varphi_r} = \dfrac{5 \times 10^6}{3 \times 6350 \times 0.8} = 328$ A

Since the line is short, this current will be assumed to be the same at all points along the line.

If R_p be the resistance of each conductor, the power loss in the line is $3I_{rp}^2\, R_p$.

But,   power loss = 10 per cent of power delivered

$$3I_{rp}^2 R_p = \frac{10}{100} \times 5 \times 10^6 \; ; \qquad R_p = \frac{5 \times 10^5}{3 \times (328)^2} = 1.549 \; \Omega$$

Also, $R_p = \rho \dfrac{l}{a}$

For hard-drawn copper,

$$\rho = 1.7774 \; \mu\Omega \; \text{cm} = 1.7774 \times 10^{-6} \; \Omega \; \text{cm}$$

$$l = 10 \; \text{km} = 10 \times 10^3 \times 10^2 = 10 \times 10^5 \; \text{cm}$$

$\therefore$ $1.549 = \dfrac{1}{a} \times 1.7774 \times 10^{-6} \times 10 \times 10^5$

$$a = 1.147 \; \text{cm}^2$$

If r be the radius of each conductor

$$a = \pi r^2$$

$$r = \sqrt{\frac{a}{\pi}} = \sqrt{\frac{1.147}{\pi}} = 0.6042 \; \text{cm}, \qquad r' = 0.7788 \; r$$

Spacing between the conductors, $D = 2 \; \text{m} = 200 \; \text{cm}$

The inductance of each conductor

$$L = 2 \times 10^{-7} \ln \frac{D}{r'} \; \text{H/m} = 2 \times 10^{-7} \ln \frac{200}{0.6042 \times 0.7788} = 1.21 \times 10^{-6} \; \text{H/m}$$

$$= 1.21 \times 10^{-6} \times 10 \times 10^3 \; \text{H} = 1.21 \times 10^{-2} \; \text{H}$$

Reactance per phase, $X_{Lp} = 2\pi f L = 2\pi \times 50 \times 1.21 \times 10^{-2} = 3.8 \; \Omega$

Taking $\mathbf{V}_r$ as a reference phasor,

$$\mathbf{V}_r = 6350 \underline{/0^\circ}$$

$$\varphi_r = \cos^{-1}(0.8) = 36.8698^\circ$$

$$\mathbf{Z}_p = 1.549 + j\,3.8 = 4.1 \underline{/67.8226^\circ} \; \Omega$$

Voltage at the sending end, $\mathbf{V}_{sp} = \mathbf{V}_{rp} + \mathbf{I}_{rp}\mathbf{Z}$

$$\mathbf{V}_{sp} = 6350 \underline{/0^\circ} + (328 \underline{/-36.87^\circ})(4.1 \underline{/67.82^\circ}) = 6350 \underline{/0^\circ} + 1344.8 \underline{/30.95^\circ}$$

$$= 6350 + j\,0 + 1153.32 + j\,691.62 = 7503.32 + j\,691.62 = 7535.13 \underline{/5.266^\circ} \; \text{V}$$

The line voltage at the sending end, $V_{sl} = \sqrt{3} \times 7535.13 = 13501.228 \; \text{V} = 13.051 \; \text{kV}$

The phase difference between $\mathbf{V}_s$ and $\mathbf{I}_s$

$$\varphi_s = 5.266^\circ - (-36.87^\circ) = 42.136^\circ$$

Sending-end power factor, $\cos \varphi_s = \cos 42.136^\circ = 0.7416$ (lagging)

9.9 LINE WITH TRANSFORMERS

Transformers are installed at both the ends of the line. Step-up transformers at the sending end raise the generator voltage to transmission level, and the step-down transformers lower the transmission voltage at the load-end to a value suitable for distribution purposes (Fig. 9.5). Thus, the two transformers, T_1 and T_2 at the ends are included in the transmission system. T_1 and T_2 are the step-up and step-down transformers respectively. It is, therefore, necessary that the resistances and leakage reactances of the transformers should also be taken into account in the calculation of the line performance. Fig. 9.5 is the single line diagram of the system showing one phase only.

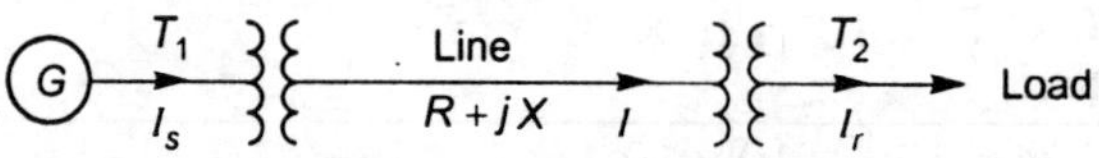

Fig. 9.5. Single-line diagram of a system.

The no-load current of a power transformer is negligibly small in comparison with its load current. The shunt branch in its equivalent circuit is, therefore, omitted for calculation work. The approximate equivalent circuit of a transformer is shown in Fig. 9.6.

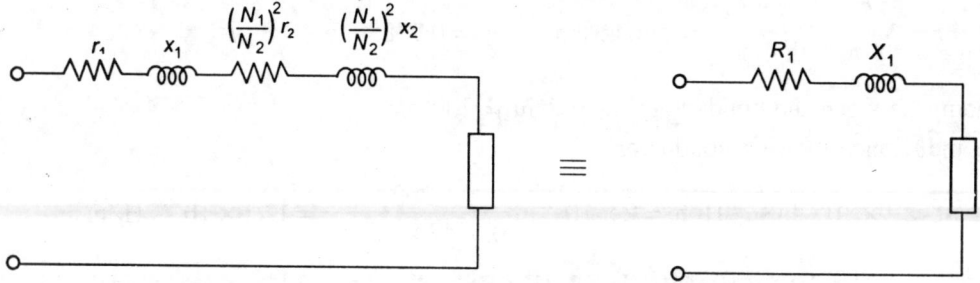

Fig. 9.6. Approximate equivalent circuit of a transformer.

Here, r_1, r_2 = resistances of hv and lv windings to neutral respectively

x_1, x_2 = leakage reactances of hv and lv windings to neutral respectively

N_1, N_2 = number of turns on hv and lv windings respectively.

The equivalent resistance to neutral of transformer referred to high voltage side is R_1. It is given by

$$R_1 = r_1 + \left(\frac{N_1}{N_2}\right)^2 r_2$$

Similarly, the equivalent leakage reactances to neutral of transformer referred to hv side is X_1 and is given by

$$X_1 = x_1 + \left(\frac{N_1}{N_2}\right)^2 x_2$$

The equivalent resistances and reactances of both the transformers are found in terms of hv sides. Let the resistance and reactance of the load-end transformer be R_2 and X_2 respectively referred to hv winding. The equivalent circuit of the transmission system having two transformers at the line ends takes the form shown in Fig. 9.7.

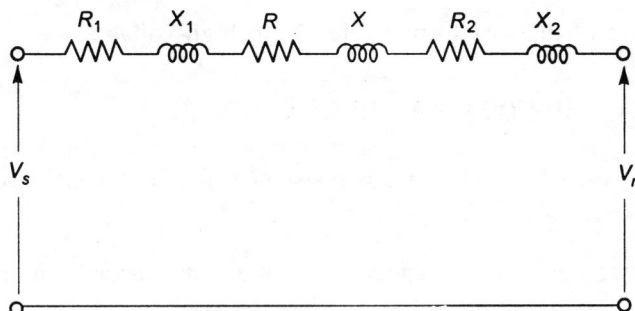

Fig. 9.7. Equivalent circuit of transmission circuit.

The resistances of both the transformers are added to the line resistance to get the total resistance of the transmission system. The total resistance of the line and the transformer is R_t. It is given by

$$R_t = R_1 + R + R_2$$

Similarly, if X_t denotes the total reactance of the line and the transformers, then

$$X_t = X_1 + X + X_2$$

After calculating R_t and X_t the line calculations are made in the usual manner as already given.

Example 9.4 A three-phase load of 3000 kVA, 0.8 power factor, is supplied at 11 kV from a step-down transformer having a ratio 3 : 1. The primary side of the transformer is connected to a transmission line, the constants of which are: resistance per conductor, 2 ohms; reactance per conductor, 3 ohms. The resistance and reactance per phase of the primary windings of the transformer (which are star-connected) are 5 and 10 ohms respectively, and the corresponding values for the secondary winding (which are delta-connected) are 1.5 ohms and 3 ohms respectively. Determine the voltage and power factor at the sending end of the transmission line.

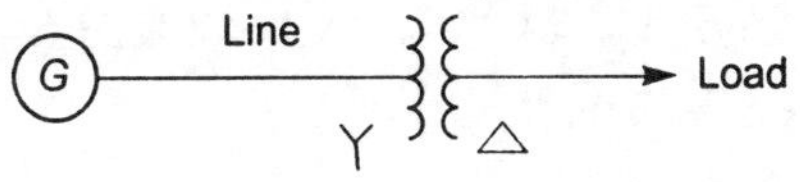

$$G \quad\text{——————}\underset{\text{Y}}{\overset{\text{Line}}{\text{))}}\underset{\triangle}{\text{((}}\text{——→ Load}$$

Fig. 9.8. Single-line diagram of the system of Example 9.4.

Solution

$$\frac{\text{Line voltage on primary of transformer}}{\text{Line voltage on secondary of transformer}} = \frac{3}{1}$$

$$\frac{\sqrt{3}\,V_{p1}}{V_{p2}} = \frac{3}{1}, \ \frac{V_{p1}}{V_{p2}} = \sqrt{3} = \frac{N_1}{N_2}$$

where, N_1 and N_2 are the number of turns on primary and secondary sides of the transformer.

Equivalent resistance of the transformer referred to high-voltage side

$$= r_1 + \left(\frac{N_1}{N_2}\right)^2 r_2 = 5 + (\sqrt{3})^2 \times 1.5 = 9.5 \ \Omega/\text{ph}$$

Equivalent reactance of the transformer referred to high-voltage side

$$= x_1 + \left(\frac{N_1}{N_2}\right)^2 x_2 = 10 + (\sqrt{3})^2 \times 3 = 19 \ \Omega/\text{ph}$$

Therefore, the system resistance per phase = resistance of line + resistance of transformer

$$R = 2 + 9.5 = 11.5 \ \Omega$$

The system reactance per phase = reactance of line + reactance of transformer.

$$X = 3 + 19 = 22 \ \Omega$$

Load voltage in terms of high-voltage side, $V_{rp} = 11\sqrt{3} = 19.05$ kV = 19050 V

$$\text{kVA} = \frac{\sqrt{3} \ V_l \ I_l}{1000}$$

Line current, $I_l = \dfrac{\text{kVA} \times 1000}{\sqrt{3} \ V_l} = \dfrac{3000 \times 1000}{\sqrt{3} \times 33 \times 10^3} = 52.486$ A

Taking $\mathbf{V}_{rp}$ as the reference phasor so that

$$\mathbf{V}_{rp} = 19050 \underline{/0^\circ} = 19050 + j\,0$$

$$\varphi_r = \cos^{-1} 0.8 = 36.87^\circ$$

$$\mathbf{Z} = R + j\,X = 11.5 + j\,22 = 24.824 \underline{/62.4^\circ} \ \Omega$$

$$\mathbf{I}_r = 52.486 \underline{/-36.87^\circ} \ \text{A}$$

Sending-end voltage per phase

$$\mathbf{V}_{sp} = \mathbf{V}_{rp} + \mathbf{Z}\mathbf{I}_r$$

$$= 19050 \underline{/0^\circ} + (24.824 \underline{/62.4^\circ})(52.486 \underline{/-36.87^\circ}) = 19050 + j\,0 + 1302.9 \underline{/25.53^\circ}$$

$$= 19050 + j\,0 + 1175.7 + j\,561.5 = 20225.7 + j\,561.5 = 20233.5 \underline{/1.59^\circ} \ \text{V}$$

Line voltage at the sending end, $V_{sl} = \sqrt{3} \ V_{sp} = \sqrt{3} \times 20233.5 = 35045$ V = 35.045 kV

The phase difference between $\mathbf{V}_s$ and $\mathbf{I}_s$

$$\varphi_s = 1.59^\circ - (-36.87^\circ) = 38.46^\circ$$

Sending-end power factor, $\cos \varphi_s = \cos 38.46^\circ = 0.783$ (lagging)

Example 9.5 A substation receives 6000 kVA at 6 kV, 0.8 power factor lagging, on the low voltage side of a transformer from a generation station through a three phase cable system having resistance of 7 Ω and reactance of 2 Ω per phase. Identical 6600/33000 V transformers are installed at each end, 6600 V side being delta and 33000 V side star connected. The resistance and reactance of each transformer are 1 Ω and 9 Ω respectively, referred to hv side. Calculate the voltage at the generating station busbars.

Solution

Consider the equivalent resistances and reactances in terms of high voltage side. Then, we have

Resistance of cable = 7 Ω per phase

Reactance of cable = 2 Ω per phase

Resistance of both the transformers = $2 \times 1 = 2$ Ω/ph

Resistance of both the transformers = $2 \times 9 = 18$ Ω/ph

Total reactance of the system referred to hv side = resistance of transformers + resistance of the cable.

$$R = 2 + 7 = 9 \text{ } \Omega/\text{ph}$$

Similarly, total reactance of the system = 18 + 2 = 20 Ω/ph

The voltage at the receiving end = 6000 V

The receiving-end voltage referred to hv side $= \dfrac{6000 \times 33000}{6600} = 30000$ V

The line current referred to hv side $I = \dfrac{6000 \times 10^3}{\sqrt{3} \times 30000} = 115$ A

Phase voltage at the receiving end, $V_r = \dfrac{1}{\sqrt{3}} \times 30000 = 13725$ V

$$V_s^2 = (IR + V_r \cos \varphi_r)^2 + (IX + V_r \sin \varphi_r)^2$$
$$= (115 \times 9 + 17325 \times 0.8)^2 + (115 \times 20 + 17325 \times 0.6)^2$$
$$= (1035 + 13860)^2 + (2300 + 10395.5)^2 = (14895)^2 + (12695)^2$$
$$= 221.9 \times 10^6 + 161.2 \times 10^6 = 383.1 \times 10^6$$

$V_s = 19.57$ kV phase to neutral = 33.89 kV line to line.

This is the voltage on the hv side of the sending-end step-up transformer. The corresponding voltage on its lv side

$$= 33.89 \times \frac{6600}{33000} \text{ kV} = 6778 \text{ V}$$

The voltage at the generator busbars = 6778 V

9.10   MEDIUM LINES

It has been mentioned in section 9.2 that the capacitance of medium length lines is significant. When the effect of capacitance is not negligible, it may be assumed to be concentrated at one or more definite points along the line. A number of localized capacitance models have been used to make approximate line performance calculations. The following models are commonly used :

(a) Nominal T model

(b) Nominal Π model

It should be noted that the nominal T and Π models are not equivalent representations. They are different representations for an actual line.

9.11 NOMINAL T MODEL OF A MEDIUM LINE

In a nominal T model of a medium line, the total line capacitance is assumed to be concentrated at the middle point of the line, while the series impedance in split into two equal parts. The nominal T model of a medium line is shown in Fig. 9.9.

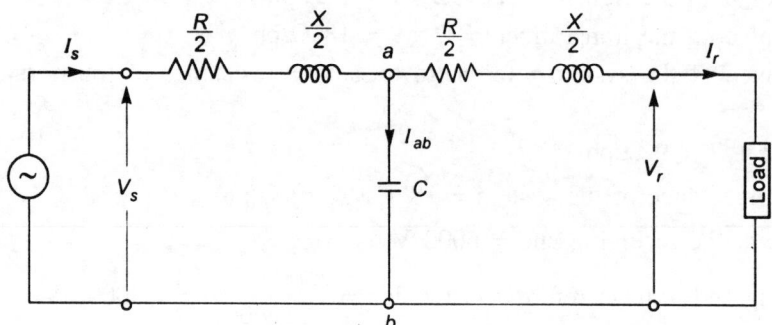

Fig. 9.9. Nominal T model of a medium line.

Series impedance of the line $\mathbf{Z} = R + jX$

Shunt admittance of the line $\mathbf{Y} = j\omega C$

Based on the assumption that V_r and I_r are known, the corresponding sending-end quantities can be obtained by application of KVL and KCL to the circuit shown in Fig. 9.9. By KVL

$$\mathbf{V}_{ab} = V_r + \frac{\mathbf{Z}}{2}\mathbf{I}_r$$

Current in the capacitor, $\mathbf{I}_{ab} = \dfrac{\mathbf{V}_{ab}}{\mathbf{Z}_{ab}} = \mathbf{Y}\mathbf{V}_{ab}$

By KCL at node a,

$$\mathbf{I}_s = \mathbf{I}_r + \mathbf{I}_{ab} = \mathbf{I}_r + \mathbf{Y}\mathbf{V}_{ab} = \mathbf{I}_r + \mathbf{Y}\left(\mathbf{V}_r + \frac{\mathbf{Z}}{2}\mathbf{I}_r\right)$$

or
$$\mathbf{I}_s = \mathbf{Y}\mathbf{V}_r + \left(1 + \frac{\mathbf{ZY}}{2}\right)\mathbf{I}_r \qquad\qquad …(9.11.1)$$

By KVL

$$\mathbf{V}_s = \mathbf{V}_{ab} + \frac{\mathbf{Z}}{2}\mathbf{I}_s = \mathbf{V}_r + \frac{\mathbf{Z}}{2}\mathbf{I}_r + \frac{\mathbf{Z}}{2}\left[\mathbf{Y}\mathbf{V}_r + \left(1 + \frac{\mathbf{ZY}}{2}\right)\mathbf{I}_r\right]$$

or
$$\mathbf{V}_s = \left(1 + \frac{\mathbf{ZY}}{2}\right)\mathbf{V}_r + \mathbf{Z}\left(1 + \frac{\mathbf{ZY}}{4}\right)\mathbf{I}_r \qquad\qquad …(9.11.2)$$

Equations (9.11.1) and (9.11.2) give the sending-end current and sending-end voltage respectively. These equations can be written in the matrix form as

$$\begin{bmatrix} \mathbf{V}_s \\ \mathbf{I}_s \end{bmatrix} = \begin{bmatrix} 1 + \dfrac{\mathbf{ZY}}{2} & \mathbf{Z}\left(1 + \dfrac{\mathbf{ZY}}{4}\right) \\ \mathbf{Y} & 1 + \dfrac{\mathbf{ZY}}{2} \end{bmatrix} \begin{bmatrix} \mathbf{V}_r \\ \mathbf{I}_r \end{bmatrix} \qquad\qquad …(9.11.3)$$

Also,
$$\begin{bmatrix} \mathbf{V}_s \\ \mathbf{I}_s \end{bmatrix} = \begin{bmatrix} \mathbf{A} & \mathbf{B} \\ \mathbf{C} & \mathbf{D} \end{bmatrix} \begin{bmatrix} \mathbf{V}_r \\ \mathbf{I}_r \end{bmatrix}$$

Hence the ABCD constants for nominal T-circuit model of a medium line are

$$\mathbf{A} = \mathbf{D} = 1 + \frac{\mathbf{ZY}}{2} ; \qquad \mathbf{B} = \mathbf{Z}\left(1 + \frac{\mathbf{ZY}}{4}\right) ; \qquad \mathbf{C} = \mathbf{Y} \qquad \qquad ...(9.11.4)$$

9.11.1 Phasor Diagram

The phasor diagram of the nominal T circuit of Fig. 9.9 is shown in Fig. 9.10. It is drawn for a lagging power factor $\cos \varphi_r$.

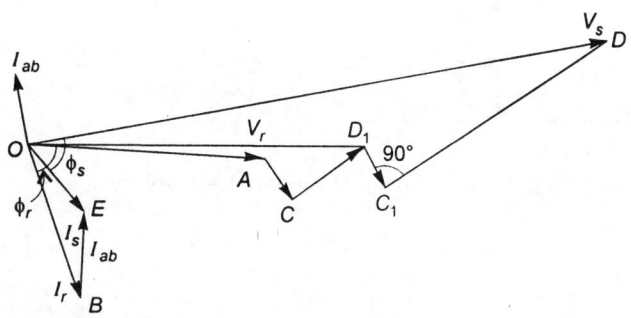

Fig. 9.10. Phasor diagram of a nominal T network.

In the phasor diagram :

$OA = \mathbf{V}_r$ = receiving–end voltage to neutral. It is taken as a reference phasor.

$OB = \mathbf{I}_r$ = load current lagging behind $\mathbf{V}_r$ by an angle φ_r . $\cos \varphi_r$ is the power factor of the load.

$AC = I_r \dfrac{R}{2}$ = voltage drop in the resistance of the right hand half of the line. It is parallel to $\mathbf{I}_r$.

$CD_1 = I_r \dfrac{X}{2}$ = voltage drop in the reactance of the right hand half of the line. It is perpendicular

        to OB, i.e., $\mathbf{I}_r$.

$OD_1 = \mathbf{V}_{ab}$ = voltage at the mid–point of the line across the capacitance C.

$BE = \mathbf{I}_{ab}$ = current in the capacitor. It leads the voltage $\mathbf{V}_{ab}$ by 90°.

$OE = \mathbf{I}_s$ = sending–end current, the phasor sum of load current and capacitor current.

$D_1C_1 = I_s \dfrac{R}{2}$ = voltage drop in the resistance in the left hand side of the line. It is parallel to $\mathbf{I}_s$.

$C_1D = I_s \dfrac{X}{2}$ = voltage drop in the reactance in the left half of the line. It is perpendicular to $\mathbf{I}_s$.

$OD = \mathbf{V}_s$ = sending–end voltage. It is the phasor sum of $\mathbf{V}_{ab}$ and the impedance voltage drop in

        the left-hand half of the line.

    φ_s = phase angle at the sending end. $\cos \varphi_s$ is the power factor at the sending end.

9.12 NOMINAL Π MODEL OF A MEDIUM LINE

The nominal Π (*pi*) model of a medium line assumes that one-half of the total line capacitance is concentrated at each end of the line and the total resistance and inductive reactance are concentrated at the centre of the line. Fig. 9.11 shows the nominal Π model of the line. In this circuit,

$$\mathbf{V}_{ab} = \mathbf{V}_r, \quad \mathbf{Z}_{ab} = \frac{1}{\mathbf{Y}_{ab}}$$

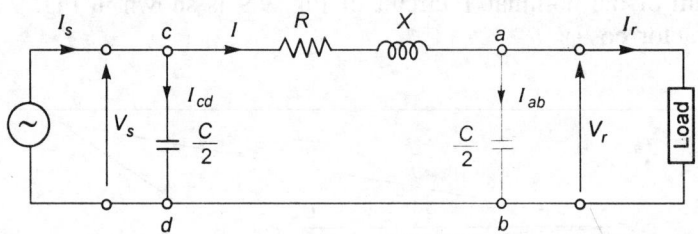

Fig. 9.11. Nominal Π model of a medium line.

By Ohm's law

$$\mathbf{I}_{ab} = \frac{\mathbf{V}_{ab}}{\mathbf{Z}_{ab}} = \frac{\mathbf{Y}}{2}\mathbf{V}_r$$

By KCL at node *a*,

$$\mathbf{I} = \mathbf{I}_r + \mathbf{I}_{ab} = \mathbf{I}_r + \frac{\mathbf{Y}}{2}\mathbf{V}_r$$

Voltage at the sending end

$$\mathbf{V}_s = \mathbf{V}_{cd} = \mathbf{V}_{ab} + \mathbf{ZI} = \mathbf{V}_r + \mathbf{Z}\left(\mathbf{I}_r + \frac{\mathbf{Y}}{2}\mathbf{V}_r\right)$$

or $$\mathbf{V}_s = \left(1 + \frac{\mathbf{ZY}}{2}\right)\mathbf{V}_r + \mathbf{ZI}_r \qquad\qquad …(9.12.1)$$

By Ohm's law

$$\mathbf{I}_{cd} = \frac{\mathbf{V}_{cd}}{\mathbf{Z}_{cd}} = \frac{\mathbf{Y}}{2}\mathbf{V}_s = \frac{\mathbf{Y}}{2}\left[\left(1 + \frac{\mathbf{ZY}}{2}\right)\mathbf{V}_r + \mathbf{ZI}_r\right]$$

Sending-end current is found by applying KCL at node *c*

$$\mathbf{I}_s = \mathbf{I} + \mathbf{I}_{cd} = \mathbf{I}_r + \frac{\mathbf{Y}}{2}\mathbf{V}_r + \frac{\mathbf{Y}}{2}\left[\left(1 + \frac{\mathbf{ZY}}{2}\right)\mathbf{V}_r + \mathbf{ZI}_r\right]$$

or $$\mathbf{I}_s = \mathbf{Y}\left(1 + \frac{\mathbf{ZY}}{4}\right)\mathbf{V}_r + \left(1 + \frac{\mathbf{ZY}}{2}\right)\mathbf{I}_r \qquad\qquad …(9.12.2)$$

Equations (9.12.1) and (9.12.2) can be written in matrix form as

$$\begin{bmatrix} \mathbf{V}_s \\ \mathbf{I}_s \end{bmatrix} = \begin{bmatrix} \left(1 + \dfrac{\mathbf{ZY}}{2}\right) & \mathbf{Z} \\ \mathbf{Y}\left(1 + \dfrac{\mathbf{ZY}}{4}\right) & \left(1 + \dfrac{\mathbf{ZY}}{2}\right) \end{bmatrix} \begin{bmatrix} \mathbf{V}_r \\ \mathbf{I}_r \end{bmatrix}$$

Also, $$\begin{bmatrix} \mathbf{V}_s \\ \mathbf{I}_s \end{bmatrix} = \begin{bmatrix} \mathbf{A} & \mathbf{B} \\ \mathbf{C} & \mathbf{D} \end{bmatrix} \begin{bmatrix} \mathbf{V}_r \\ \mathbf{I}_r \end{bmatrix}$$ ...(9.12.3)

Hence the ABCD constants for nominal Π-circuit model of a medium line are

$$\mathbf{A} = \mathbf{D} = 1 + \frac{\mathbf{ZY}}{2} ; \qquad \mathbf{B} = \mathbf{Z} ; \qquad \mathbf{C} = \mathbf{Y}\left(1 + \frac{\mathbf{ZY}}{4}\right)$$ ...(9.12.4)

9.12.1 Phasor Diagram

The phasor diagram of a nominal Π-circuit is shown in Fig. 9.12. It is also drawn for a lagging power factor of the load. In the phasor diagram the quantities shown are as follows :

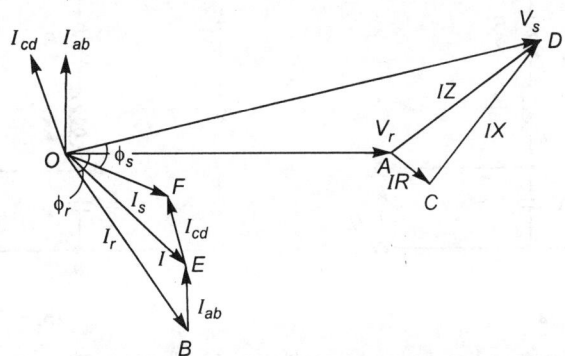

Fig. 9.12. Phasor diagram a nominal Π network.

$OA = \mathbf{V}_r$ = receiving-end voltage to neutral. It is taken as a reference phasor.

$OB = \mathbf{I}_r$ = load current lagging $\mathbf{V}_r$ by an angle φ_r.

$BE = \mathbf{I}_{ab}$ = current in receiving-end capacitance. It leads $\mathbf{V}_r$ by 90°.

The line current $\mathbf{I}$ is the phasor sum of $\mathbf{I}_r$ and $\mathbf{I}_{ab}$. It is shown by OE in the diagram.

$AC = IR$ = voltage drop in the resistance of the line. It is parallel to $\mathbf{I}$.

$CD = IX$ = inductive voltage drop in the line. It is perpendicular to $\mathbf{I}$.

$AD = \mathbf{IZ}$ = voltage drop in the line impedance.

$OD = \mathbf{V}_s$ = sending-end voltage to neutral. It is the phasor sum of $\mathbf{V}_r$ and $\mathbf{IZ}$.

The current taken by the capacitance at the sending end is $\mathbf{I}_{cd}$. It leads the sending-end voltage $\mathbf{V}_s$ by 90°.

$OF = \mathbf{I}_s$ = the sending-end current. It is the phasor sum of $\mathbf{I}$ and $\mathbf{I}_{cd}$.

φ_s = phase angle between $\mathbf{V}_s$ and $\mathbf{I}_s$ at the sending end, cos φ_s will give the sending-end power factor.

9.13 CALCULATION OF TRANSMISSION EFFICIENCY AND REGULATION OF MEDIUM LINES

If $\mathbf{V}_r$ denotes the line-to-neutral (phase) voltage at the receiving end in volts, $\mathbf{S}_s$ denotes the sending-end voltamperes, and $\mathbf{I}_s^*$ is the complex conjugate of $\mathbf{I}_s$, then

$$\mathbf{S}_s = 3\mathbf{V}_{sp}\,\mathbf{I}_{sp}^* = P_s + j\,Q_s$$

where P_s = active power in watts at the sending end

Q_s = reactive voltamperes at the sending end in VAr

The transmission efficiency can be calculated as follows :

$$\eta_T = \frac{P_r}{P_s}$$

In order to calculate regulation we have to calculate the receiving end voltage at no load V_{rnl}. The nominal T circuit for this condition reduces to the circuit shown in Fig. 9.13 (a).

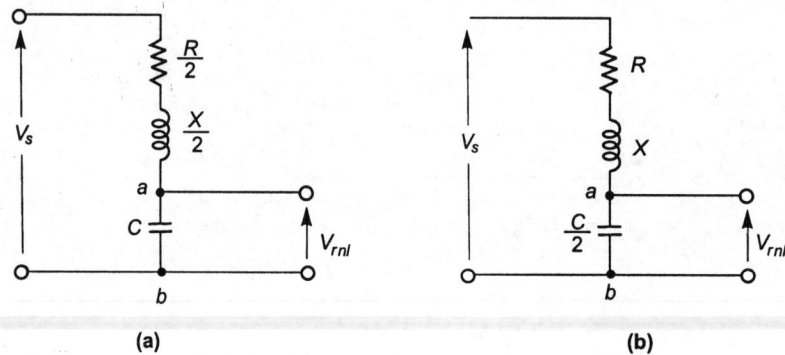

Fig. 9.13. (a) Nominal T circuit under no-load conditions. (b) Nominal Π circuit under no-load conditions.

Here $\quad \mathbf{V}_{rnl} = \mathbf{V}_{ab} = \dfrac{|\mathbf{V}_s|\left(-\dfrac{j}{\omega C}\right)}{\left(\dfrac{R}{2} + j\dfrac{X}{2} - \dfrac{j}{\omega C}\right)}$

The nominal Π-circuit under no-load conditions is shown in Fig. 9.13 (b).

Here $\quad \mathbf{V}_{rnl} = \mathbf{V}_{ab} = \dfrac{|\mathbf{V}_s|\left[-\dfrac{j}{(\omega C/2)}\right]}{\left[R + jX - \dfrac{j}{(\omega C/2)}\right]}$

Alternative method of calculation of regulation

We have $\mathbf{V}_s = \mathbf{A}\mathbf{V}_r + \mathbf{B}\mathbf{I}_r$

When the load is removed,

$$\mathbf{I}_r = 0, \qquad \mathbf{V}_r = \mathbf{V}_{r0} = \mathbf{V}_{rnl}$$

$\therefore \qquad \mathbf{V}_s = \mathbf{A}\mathbf{V}_{rnl}$

$$\mathbf{V}_{rnl} = \frac{\mathbf{V}_s}{\mathbf{A}}$$

$$|\mathbf{V}_{rnl}| = \left|\frac{\mathbf{V}_s}{\mathbf{A}}\right|$$

Per unit regulation $= \dfrac{|\mathbf{V}_{rnl}| - |\mathbf{V}_{rfl}|}{|\mathbf{V}_{rfl}|} = \dfrac{\left|\dfrac{\mathbf{V}_s}{\mathbf{A}}\right| - |\mathbf{V}_{rfl}|}{|\mathbf{V}_{rfl}|}$

For nominal T model of the line

$$A = 1 + \frac{ZY}{2}$$

For nominal Π-model of the line

$$A = 1 + \frac{ZY}{2}$$

Example 9.6 A three-phase, 50 Hz, transmission line, 40 km long delivers 36 MW at 0.8 power factor lagging at 60 kV (phase). The line constants per conductor are, $R = 2.5\ \Omega$, $L = 0.1$ H, $C = 0.25\ \mu$F. Shunt leakage may be neglected. Determine the voltage, current, power factor, active power and reactive voltamperes at the sending end. Also, determine the efficiency and regulation of the line. Use (a) nominal T method, (b) nominal Π method.

Solution

Phase voltage at the receiving end, $V_r = 60$ kV $= 60 \times 10^3$ V

Power per phase $= \dfrac{1}{3} \times 36$ MW $= 12 \times 10^6$ W

Therefore, the receiving-end current, $I_r = \dfrac{12 \times 10^6}{60 \times 10^3 \times 0.8} = 250$ A

Taking $\mathbf{V}_r$ as the reference phasor,

$\qquad \mathbf{V}_r = V_r + j\,0$

$\qquad \cos \varphi_r = 0.8,\ \varphi_r = 36.8699°$

Resistance per phase $R = 2.5\ \Omega$

Inductive reactance per phase, $X_L = 2\pi f L = 2\pi \times 50 \times 0.1 = 31.416\ \Omega$

Series impedance per phase, $\mathbf{Z} = R + j\,X = 2.5 + j\,31.4 = 31.499\ \underline{/85.448°}\ \Omega$

Shunt admittance per phase, $Y = 2\pi f C = 2 \times \pi \times 50 \times 0.25 \times 10^{-6} = 7.854 \times 10^{-5}$ S

$\qquad \mathbf{Y} = 0 + j\,7.854 \times 10^{-5} = 7.854 \times 10^{-5}\ \underline{/90°}\ $ S

Calculations by nominal T method

The nominal T circuit for the line is shown in Fig. 9.9.

$$V_{ab} = V_r + \frac{Z}{2} I_r = 60 \times 10^3 \underline{/0°} + \frac{1}{2}(250 \underline{/-36.869°})(31.499 \underline{/85.448°})$$

$$= 60 \times 10^3 \underline{/0°} + 3937.375 \underline{/48.579°} = 60 \times 10^3 + j\,0 + 2604.915 + j\,2952.514$$
$$= 62604.915 + j\,2952.514 = 62674.498 \underline{/2.7°} \text{ V}$$

The current in the capacitor

$$I_{ab} = YV_{ab} = (7.854 \times 10^{-5} \underline{/90°})(62674.498 \underline{/2.7°}) = 4.922 \underline{/92.7°} = -0.2319 + j\,4.917$$
$$I_s = I_r + I_{ab} = 200 - j\,150 - 0.2319 + j\,4.917 = 199.768 - j\,145.083 = 246.893 \underline{/-35.989°}$$

Voltage drop in the left-hand half of the line

$$= I_s \frac{Z}{2} = \frac{1}{2}(246.893 \underline{/-35.989°})(31.499 \underline{/85.448°})$$

$$= 3888.441 \underline{/49.459°} = 2527.456 + j\,2954.986$$

Voltage at the sending end,

$$V_s = V_{ab} + I_s \frac{Z}{2} = 62604.915 + j\,2952.514 + 2527.456 + j\,2954.986$$

$$= 65132.371 + j\,5907.5 = 65399.727 \underline{/5.183°} \text{ V}$$

Line voltage at the sending end $= \sqrt{3} \times$ phase voltage at the sending end
$$= \sqrt{3} \times 65400 = 113276 \text{ V} = 113.276 \text{ kV}$$

Phase difference between V_s and I_s

$$\varphi_s = 5.183° - (-35.989°) = 41.172°$$

Sending-end power factor, $\cos \varphi_s = \cos 41.172° = 0.7527$ (lagging)

Total sending-end voltamperes, $S_{s3\varphi} = 3V_{sp} I_{sp}^* = 3 \times 65400 \underline{/5.183°} \times 246.893 \underline{/+35.989°}$
$$= 48440407 \underline{/41.172°} = 36462873 + j\,31889370$$

Also, $\quad S_{s3\varphi} = P_s + j\,Q_s$

$\therefore \quad P_s = 36462873$ watts

$\quad Q_s = 31889370$ vars

Transmission efficiency, $\eta_T = \dfrac{P_r}{P_s} = \dfrac{36 \times 10^6}{36462873} = 0.9873$ pu $= 98.73$ per cent

Alternative method of determining transmission efficiency

Power loss in the line $= 3I_r^2 \dfrac{R}{2} + 3I_s^2 \dfrac{R}{2}$

$$= 3 \times 250^2 \times 1.25 + 3 \times (246.893)^2 \times 1.25 = 462.961 \times 10^3 \text{ W}$$

Transmission efficiency $= \dfrac{\text{power output}}{\text{power output} + \text{power loss}}$

$$= \dfrac{36 \times 10^6}{36 \times 10^6 + 462.961 \times 10^3} = 0.9873 \text{ pu} = 98.73 \text{ per cent}$$

$$A = 1 + \frac{ZY}{2} = 1 + \frac{31.499}{2} \underline{/85.448°} \times 7.854 \times 10^{-5} \underline{/90°}$$

$$= 1 + 1.23696 \times 10^{-3} \underline{/175.448°} = 0.9987669 + j\,0.0982 \times 10^{-3}$$

$$|A| = 0.998767$$

$$|\mathbf{V}_{rnl}| = \left|\frac{\mathbf{V}_s}{A}\right| = \frac{65400}{0.998767} = 65481$$

$$\text{Regulation} = \frac{|\mathbf{V}_{rnl}| - |\mathbf{V}_{rfl}|}{|\mathbf{V}_{rfl}|} = \frac{65481 - 60000}{60000} = 0.0913456 \text{ pu} = 9.13456 \text{ per cent}$$

Calculation by nominal Π method

The nominal Π-circuit for the line is shown in Fig. 9.11.

$$\mathbf{I}_{ab} = \frac{Y}{2}\mathbf{V}_r = \left(\frac{1}{2} \times 7.854 \times 10^{-5} \underline{/90°}\right)(60 \times 10^3 \underline{/0°}) = 2.356 \underline{/90°} = j\,2.356$$

$$\mathbf{I} = \mathbf{I}_r + \mathbf{I}_{ab} = 200 - j\,150 + j\,2.356 = 200 - j\,147.644 = 248.594 \underline{/-36.435°}$$

Voltage drop per phase = $\mathbf{IZ}$ = $(248.594 \underline{/-36.435°})(31.499 \underline{/85.448°})$
$$= 7830.462 \underline{/49.013°} = 5135.905 + j\,5910.890$$

Voltage at the sending end per phase, $\mathbf{V}_s = \mathbf{V}_r + \mathbf{IZ} = 60 \times 10^3 + j\,0 + 5135.905 + j\,5910.890$
$$= 65135.905 + j\,5910.890 = 65403.553 \underline{/5.185°}$$

Sending-end line voltage = $65403.553 \times \sqrt{3}$ V = 113.28 kV

$$\mathbf{I}_{cd} = \frac{Y}{2}\mathbf{V}_s = \left(\frac{1}{2} \times 7.854 \times 10^{-5} \underline{/90°}\right)(65403.553 \underline{/5.185°})$$

$$= 2.568 \underline{/95.185°} = -0.232 + j\,2.558$$

Sending-end current,

$$\mathbf{I}_s = \mathbf{I} + \mathbf{I}_{cd} = 200 - j\,147.644 - 0.232 + j\,2.558$$

$$= 199.768 - j\,145.086 = 246.895 \underline{/-35.989°}$$

Phase difference between $\mathbf{V}_s$ and $\mathbf{I}_s$

$$\varphi_s = 5.185° - (-35.989°) = 41.174°$$

Sending-end power factor, $\cos\varphi_s = \cos 41.174° = 0.7527$ (lagging)

Total sending-end voltamperes, $\mathbf{S}_{s3\varphi} = 3\mathbf{V}_{sp}\mathbf{I}_{sp}^* = 3 \times 65403.5 \underline{/5.185°} \times 246.895 \underline{/+35.989°}$
$$= 48443391 \underline{/41.174°} = 36464006 + j\,31892608$$

Also, $\mathbf{S}_{s3\varphi} = P_s + j\,Q_s$

$\therefore$ $P_s = 36464006$ watts

     $Q_s = 31892608$ vars

Transmission efficiency, $\eta_T = \dfrac{P_r}{P_s} = \dfrac{36 \times 10^6}{3646006} = 0.987275$ pu = 98.7275 per cent

Alternative method of determining transmission efficiency

Power loss in the line, $p_L = 3I^2R = 3 \times (248.594)^2 \times 2.5 = 463.49 \times 10^3$ W

Transmission efficiency, $\eta_T = \dfrac{P_r}{P_r + p_L}$

$$= \frac{36 \times 10^6}{36 \times 10^6 + 463.69 \times 10^3} = 0.9873 \text{ pu} = 98.73 \text{ per cent}$$

$$|\mathbf{A}| = \left| 1 + \frac{\mathbf{ZY}}{2} \right| = 0.998767$$

$$|\mathbf{V}_{rnl}| = \left| \frac{\mathbf{V}_s}{\mathbf{A}} \right| = \frac{65403.5}{0.998767} = 65484 \text{ V}$$

Regulation $= \dfrac{|\mathbf{V}_{rnl}| - |\mathbf{V}_{rfl}|}{|\mathbf{V}_{rfl}|} = \dfrac{65484 - 60000}{60000} = 0.0914 \text{ pu} = 9.14 \text{ per cent}$

Example 9.7 A three-phase, 50 Hz, 150 km line operates at 110 kV between the lines at the sending end. The total inductance and capacitance per phase are 0.2 H ad 1.5 µF. Neglecting losses calculate the value of receiving-end load having a power factor of unity for which the voltage at the receiving end will be the same as that at the sending end. Assume one-half of the total capacitance of the line to be concentrated at each end.

Solution

The circuit for the given line is shown in Fig. 9.14.

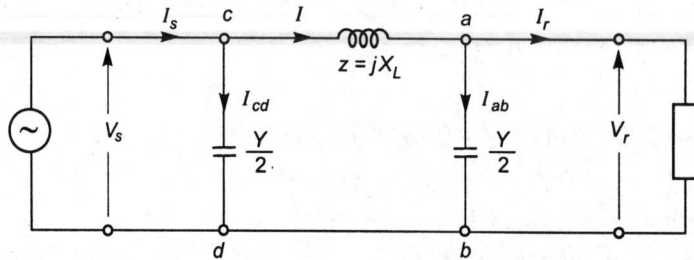

Fig. 9.14. Illustrating Example 9.7.

Taking $\mathbf{V}_r$ as the reference phasor, i.e., $\mathbf{V}_r = V_r + j\,0$

$$V_r = V_s = \frac{110 \times 10^3}{\sqrt{3}} = 63508.53 \text{ V/phase}$$

Inductive reactance per phase, $X_L = 2\pi f L = 2 \times \pi \times 50 \times 0.2 = 62.832 \ \Omega$

Series impedance per phase, $\mathbf{Z} = j\,X_L = j\,62.832 \ \Omega$

Shunt admittance per phase, $Y = 2\pi f C = 2 \times \pi \times 50 \times 1.5 \times 10^{-6} = 4.712 \times 10^{-4}$ S

$$\mathbf{Y} = j\,4.712 \times 10^{-4} = 4.712 \times 10^{-4} \underline{/90^\circ} \text{ S}$$

Current in the load-end capacitor,

$$\mathbf{I}_{ab} = \frac{\mathbf{Y}}{2} \mathbf{V}_r = \left(\frac{1}{2} \times 4.712 \times 10^{-4} \underline{/90°}\right) (63508.53 \underline{/0°}) = 14.963 \underline{/90°} = j\,14.963 \text{ A}$$

Let the load current be I_r. Since the load power factor is unity,

$$\mathbf{I}_r = I_r \underline{/0°} = I_r + j\,0$$

Current through the inductive reactance, $\mathbf{I} = \mathbf{I}_r + \mathbf{I}_{ab} = I_r + j\,14.963$

Sending-end voltage,

$$\mathbf{V}_s = \mathbf{V}_r + \mathbf{IZ} = V_r \underline{/0°} + (I_r + j\,14.963)\,(j\,62.832) = (V_r - 940.155) + j\,62.832\,I_r$$

$$V_s^2 = (V_r - 940.155)^2 + (62.832\,I_r)^2$$

$$(63508.53)^2 = (63508.53 - 940.155)^2 + (62.832\,I_r)^2$$

$$I_r = 173.275 \text{ A}$$

Example 9.8 A three-phase 132 kV transmission line is connected to a 50 MW load at a power factor of 0.85 lagging. The line constants of the 80 km long line are $\mathbf{Z} = 96 \underline{/78°}$ Ω and $\mathbf{Y} = 0.001 \underline{/90°}$ S. Using nominal T-circuit representation, calculate : (a) the $\mathbf{A}, \mathbf{B}, \mathbf{C}$, and $\mathbf{D}$ constants of the line; (b) sending-end voltage; (c) sending-end current; (d) sending-end power factor; (e) efficiency of transmission.

Solution

$$V_{rp} = \frac{132 \times 10^3}{\sqrt{3}} = 76210 \text{ V}$$

$$I_{rp} = \frac{P}{\sqrt{3}\,V_L \cos \varphi_r} = \frac{50 \times 10^6}{\sqrt{3} \times 132 \times 10^3 \times 0.85} = 257.29 \text{ A}$$

Taking $\mathbf{V}_{rp}$ as the reference phasor,

$$\mathbf{V}_{rp} = V_{rp} \underline{/0°} = 76210 \underline{/0°} \text{ V}$$

$$\mathbf{I}_{rp} = 257.29 \underline{/- \cos^{-1} 0.85°} = 257.29 \underline{/- 31.79°} \text{ A}$$

(a) For the nominal T-circuit representation, the $\mathbf{A}, \mathbf{B}, \mathbf{C}$, and $\mathbf{D}$ constants are

$$\mathbf{A} = 1 + \frac{1}{2}\mathbf{ZY} = 1 + \frac{1}{2}\,(96 \underline{/78°})\,(0.001 \underline{/90°}) = 1 + 0.048 \underline{/168°}$$

$$= 1 + (-0.04695 + j\,0.00998) = (0.95305 + j\,0.00998 = 0.9531 \underline{/0.6°}$$

$$\mathbf{B} = \mathbf{Z}\left(1 + \frac{1}{4}\mathbf{ZY}\right) = (96 \underline{/78°})\left(1 + \frac{0.048}{2} \underline{/168°}\right) = (96 \underline{/78°})\,(1 - 0.02348 + j\,0.00499)$$

$$= 96 \underline{/78°}\,(0.97652 + j\,0.00499) = (96 \underline{/78°}\,(0.97653 \underline{/0.29°} = 93.75 \underline{/78.29°} \text{ Ω}$$

$$\mathbf{C} = \mathbf{Y} = 0.001 \underline{/90°} \text{ S}$$

$$\mathbf{D} = \mathbf{A} = 1 + \frac{1}{2}\mathbf{ZY} = 0.9531 \underline{/0.6°}$$

(b) Phase voltage at the sending end,

$$\mathbf{V}_{sp} = \mathbf{AV}_{rp} + \mathbf{BI}_{rp} = (0.9531 \underline{/0.6°})\ (76210 \underline{/0°}) + (93.75 \underline{/78.29°})\ (257.29 \underline{/-31.79°})$$
$$= 72635.7 \underline{/0.6°} + 24121 \underline{/46.5°} = 72631.7 + j\ 760.6 + 16603.8 + j\ 17496.7$$
$$= 89235.5 + j\ 18257.3 = 91084 \underline{/11.56°}\ \text{V}$$

Line voltage at the sending end,

$$\mathbf{V}_{sl} = \sqrt{3}\ \mathbf{V}_{sp} = \sqrt{3} \times 91084 \underline{/11.56°} = 157762 \underline{/11.56°}\ \text{V} = 157.762 \underline{/11.56°}\ \text{kV}$$

(c) Sending-end current

$$\mathbf{I}_s = \mathbf{CV}_{rp} + \mathbf{DI}_{rp} = (0.001 \underline{/90°})\ (76210 \underline{/0°}) + (0.9531 \underline{/0.6°})\ (257.29 \underline{/-31.79°})$$
$$= 76.21 \underline{/90°} + 245.2 \underline{/31.19°} = j\ 76.21 + 209.76 - j\ 126.98$$
$$= 209.76 - j\ 50.77 = 215.82 \underline{/-13.6°}\ \text{A}$$

(d) Angle between $\mathbf{V}_{sp}$ and $\mathbf{I}_{sp}$

$$\varphi_s = 11.56 - (-13.6) = 25.16°$$

Power factor at the sending end, $\cos \varphi_s = \cos 25.16° = 0.9051$ (lagging)

(e) Efficiency of transmission,

$$\eta_t = \frac{\text{output}}{\text{input}} = \frac{50 \times 10^6}{\sqrt{3}\ V_{sl}\ I_s\ \cos \varphi_s} = \frac{50 \times 10^6}{(\sqrt{3} \times 157762 \times 215.82 \times 0.9051)} = 0.9367\ \text{pu} = 93.67\%$$

EXERCISES

1. Explain with the aid of a phasor diagram, the effect of power factor upon the voltage drop in a transmission line.

   A line, having $R = 0.05\ \Omega$, $X = 0.07\ \Omega$, delivers 10 kW at 230 V, 0.8 power factor (lagging). Calculate the voltage regulation and efficiency of transmission line. (L.U.)

2. A single phase line supplied a load of 100 kVA at a power factor of 0.75 (lagging) and a voltage of 2200 V. The line has a total resistance of 1.5 ohms and a total inductive reactance of 2.5 ohms. Calculate the voltage at the sending end and the input power in kW. (I.E.E.)

3. Deduce an expression for calculating the approximate voltage regulation of a short transmission line.

   A single-phase line has an impedance of $5 \underline{/60°}$ ohms and supplies a load of 120 amperes, 3300 volts at 0.8 lagging power factor. Calculate the sending-end voltage and draw a phasor diagram approximately to scale. (C & G)

4. Show that when the voltage drop due to resistance and reactance is small compared with the line voltage, the voltage drop along a three-phase transmission line per ampere km is given by the expression $\sqrt{3}\ (R \cos \varphi + X \sin \varphi)$, where R is the resistance per km of conductor, X the reactance per km of conductor, and $\cos \varphi$ the power factor of the load.

   Find the voltage drop along a three-phase transmission line, the line voltage at the load being 3000 V and the length of the line being 30 km, when 5000 kVA are delivered at a power factor 0f 0.8, the current lagging. The resistance and reactance per km are 0.72 Ω and 0.6 Ω respectively. (L.U.)

5. Each line of a short three-phase transmission line, has an impedance of $1 + j\,1.5$ ohms. The line delivers 500 kVA at 3000 V at a power factor of 0.75 (lagging). Calculate the efficiency of transmission and the sending-end voltage. Draw the relevant phasor diagram. (I.E.E.)

6. Deduce an approximate expression for the voltage drop in a short transmission line.

A three-phase line, 3 km long, delivers 3000 kW at power factor 0.8 (lagging) to a load. If the voltage at the supply end is 11 kV, determine the voltage at the load and the efficiency of transmission. The resistance per km of each conductor is 0.4 ohm and the reactance (line to neutral) per km of each conductor is 0.3 ohm. (I.E.E.)

7. What load at 0.8 power factor, lagging, can be delivered by a three-phase line 5 km long with a voltage drop of 10 per cent? The station voltage is 11000 V, and the resistance and the reactance per km of line are 0.9 Ω and 0.08 Ω respectively.

(C & G)

8. A three-phase transmission line, 20 km long, delivers 1000 kW at 30 kV, 0.8 power factor (lagging). Calculate the voltage at the generator end if the resistance and reactance per km of conductor are 1.25 Ω and 0.6 Ω respectively.

If a 30/10 kV transformer is connected at the end of this line, calculate the voltage at the generator end of the line if the load of 1000 kW at 0.8 power factor is delivered at the secondary side of the transformer at 10 kV. Referred to the neutral on the secondary side the equivalent resistance and reactance of the transformer are 0.8 Ω and 2.5 Ω respectively. (L.U.)

9. Show that a three-phase system, having a line voltage V, total power P, and impedance per line Z, has the same regulation and efficiency as a single-phase system with a voltage V, power P, and loop impedance Z.

10. A three-phase overhead transmission line 100 km long has the following constants per km; resistance and reactance per phase 0.15 Ω and 0.4 Ω respectively, and line to neutral admittance due to capacitance 0.5×10^{-6} S. Find the sending-end voltage when the load delivered is 30 MVA, power factor 0.8 lagging, at 66 kV. (C & G)

11. A three-phase transmission line delivers 20000 kVA, power factor 0.8 lagging at 66 kV at the receiving end. Each conductor has a resistance of 10 ohms, and an admittance due to capacitance between conductors (line to neutral) of 4×10^{-4} S. Calculate the voltage, current, and power factor at the transmitting end, and the efficiency of transmission, neglecting line leakance. (L.U.)

12. A three-phase transmission line, 100 km long, has the following constants: Resistances per line, 12 ohms; inductance per line, 0.155 henry; capacitance between each line and neutral, 1.2 μF. A load of 30 MW at 0.8 power factor (lagging) is connected to the distant end and is to be supplied at 120 kV, 50 Hz. Determine the voltage at the sending end of the line; using the nominal T method. (I.E.E.)

13. A three-phase, 50 Hz overhead transmission line, 60 km long with 132 kV between the lines at the receiving end, has the following constants :

Resistance per km per phase = 0.25 ohm

Inductance per km per phase – 2.0 millihenrys

Capacitance per km per phase = 0.014 microfarad.

Determine, using an approximate method of allowing for the capacitance, the voltage, current power factor at the sending end regulation and efficiency when the load at the receiving end is 70000 kW at 0.8 power factor lagging.

Draw a phasor diagram for the circuit assumed. (L.U.)

14. Deduce an expression for the approximate voltage drop in a short transmission line. Explain how the effect of capacitance is taken into account in calculations of the approximate voltage drop in long lines, and draw phasor diagram for such a line.

A three-phase line, 100 km long, has constants per km per conductor as follows: Resistance, 0.5 ohm; inductance, 2 milli-henrys, capacitance to neutral, 0.015 microfarad. Calculate the voltage required at the generating end in order that a load of 10 MVA at 0.8 power factor (lagging) may be supplied at 120 kV. The frequency is 50 Hz.

(I.E.E.)

15. A three-phase, 50 Hz transmission line, 50 km long, supplies at the receiving end, a load of 60 MVA, 0.8 power factor (lagging), at 120 kV. Determine the voltage, current and power factor at the sending end, using the nominal Π-method. The resistance per km of conductor is 0.3 Ω. the inductance per km per phase is 2 mH and the capacitance to neutral per km is 0.015 μF. (C & G)

16. Draw and explain the phasor diagram for a transmission line assuming that half the line capacitance is concentrated at each end of the line.

A 50 Hz, three-phase line 100 km long, delivers a load of 40000 kVA at 110 kV and a lagging power factor of 0.7. The line constants (line-to-neutral values) are resistance 11 ohms, inductive reactance 38 ohms, capacitive susceptance 3×10^{-4} S leakage negligible. Find the sending-end voltage, current, power factor and power input.

(C & G)

17. A three-phase load of 9000 kVA at 0.9 power factor (lagging) is received at 60 kV and 50 Hz from an overhead transmission feeder 50 km long, for which $R = 0.67$ Ω, $X = 0.67$ Ω per conductor per km, and the capacitance to neutral is 0.014 μF per km. Calculate the voltage and the power factor at the sending end of the line. (L.U.)

18. A three-phase overhead transmission line, 80 km long, delivers 24 MVA at 66 kV, 50 Hz, 0.8 power factor (lagging). The line conductors have a diameter of 1.5 cm and are symmetrically spaced at a distance of 2.5 m. Determine the regulation and efficiency of the line, using the nominal Π method. The line resistance is 8.72 Ω/phase.

ANSWERS

1. 1.93%; 97.1%

2. 2.326 kV; 78.102 kW

3. 3.637 kV

4. 4686 V

5. 93.1%; 3291 V   6. 10.5 kV; 95.6%

7. 14.4 kW   8. 31.14 kV; 31.9 kV

10. 82.96 kV

11. 73.7 kV; 165.26 A; 0.7998 (lag); 94.83%

12. 131.347 kV

13. 154.838 kV; 369.1 A; 0.7714 (lag) 17.84%, 91.6%

14. 124.679 kV   15. 135.23 kV; 278 A; 0.7923 (lag)

16. 122.3 kV ; 195.63 A ; 0.710 (lag) 29363 kW

17. 66.54 kV ; 0.9196 (lag)   18. 14.49% ; 94.47%

10

Long Transmission Lines

10.1 INTRODUCTION

The line parameters R, L, C, and G are distributed uniformly along the whole length of the line. It may be assumed that a line consists of a large number of short sections connected together. Fig. 10.1 shows an approximation of this distributed nature. The short sections are considered individually. If the size of these sections be made infinitely small by considering the line to be made up of infinite number of sections, these parameters may be assumed to be lumped in the individual sections. Another assumption made is to consider the series elements R and L to be lumped in one conductor. On a per phase basis, a long transmission line may, therefore, be represented as shown in Fig. 10.2.

10.2 EXACT SOLUTION OF A LONG LINE

Let the following notations be used in the treatment to follow :

r = resistance per unit length, per phase

l = inductance per unit length, per phase

c = capacitance per unit length, per phase

x = inductive reactance per unit length, per phase

$\mathbf{z}$ = series impedance per unit length, per phase

g = shunt leakage conductance, per phase to neutral per unit length

b = shunt leakage susceptance, per phase to neutral per unit length

$\mathbf{y}$ = shunt admittance per unit length, per phase to neutral.

$\therefore \qquad \mathbf{z} = r + jx$

$\qquad \mathbf{y} = g + jb$

Consider a short section of line length ds situated at a distance s from the receiving end of

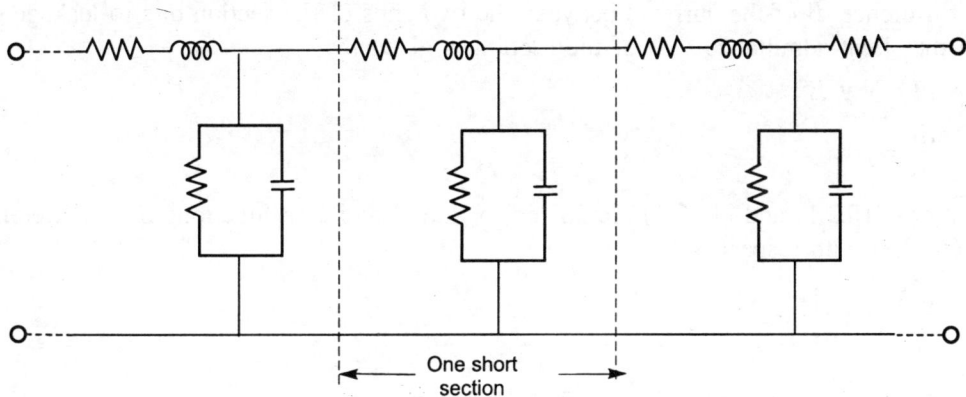

Fig. 10.1. Representation of a transmission line showing the distributed nature of parameters.

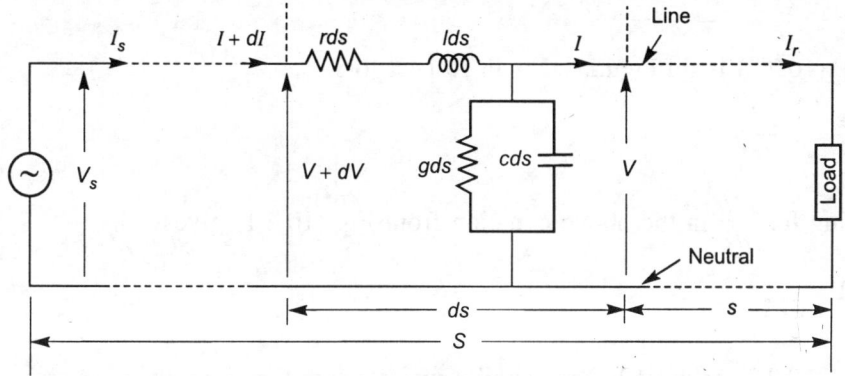

Fig. 10.2. Incremental length of the transmission line.

the line. The series impedance of this section = $z\,ds$ and the shunt admittance of this section = $y\,ds$.

The current is not uniform along the line due to the fact that a part of it leaks as charging and corona-loss currents through b and g respectively. The voltage is also different at different points because of voltage drop in line reactance and inductive reactance.

For steady-state sinusoidal supply system, let

$\mathbf{V}$ = voltage at a distance s from the load end

$\mathbf{V} + d\mathbf{V}$ = voltage at distance $(s + ds)$ from the load end

$\mathbf{I}$ = current at a distance s from the load end

$\mathbf{I} + d\mathbf{I}$ = current at a distance $(s + ds)$ from the load end.

The difference in voltages between the ends of the assumed section of length ds is $d\mathbf{V}$. The differences, as already mentioned, is caused by the voltage drop in the series impedance $z\,ds$ of the section :

$\therefore$ $d\mathbf{V} = \mathbf{I}\,z\,ds$

$$\frac{d\mathbf{V}}{ds} = \mathbf{I}\,z \qquad \qquad \text{...(10.2.1)}$$

The difference $d\mathbf{I}$ of the currents between the two ends of the section due to leakage current through the shunt admittance $\mathbf{y}\,ds$ of the section is given by

$$d\mathbf{I} = \mathbf{V}\,\mathbf{y}\,ds$$

$$\frac{d\mathbf{I}}{ds} = \mathbf{y}\,\mathbf{V} \qquad\qquad \text{...(10.2.2)}$$

Equations (10.2.1) and (10.2.2) are solved to find $\mathbf{V}$ and $\mathbf{I}$ as functions of s. Differentiation of Eq. (10.2.1) with respect to s gives

$$\frac{d^2\mathbf{V}}{ds^2} = \mathbf{z}\,\frac{d\mathbf{I}}{ds}$$

The value of $\dfrac{d\mathbf{I}}{ds}$ is substituted from Eq. (10.2.2) to give

$$\frac{d^2\mathbf{V}}{ds^2} = \mathbf{zy}\,\mathbf{V} \qquad\qquad \text{...(10.2.3)}$$

Now, Eq. (10.2.2) is differentiated with respect to s.

$$\frac{d^2\mathbf{I}}{ds^2} = \mathbf{y}\,\frac{d\mathbf{V}}{ds}$$

Substituting for $\dfrac{d\mathbf{V}}{ds}$ in the above equation from Eq. (10.2.1) gives

$$\frac{d^2\mathbf{I}}{ds^2} = \mathbf{zy}\,\mathbf{I} \qquad\qquad \text{...(10.2.4)}$$

Equations (10.2.3) and (10.2.4) are similar in form and, therefore, their general solutions will also be similar. Taking Eq. (10.2.3) and putting $\mathbf{zy} = \gamma^2$, we obtain

$$\frac{d^2\mathbf{V}}{ds^2} = \gamma^2\,\mathbf{V}$$

This is a linear differential equation of second order with constant coefficients. The general solution of this equation is of the form

$$\mathbf{V} = \mathbf{C}_1 e^{\gamma s} + \mathbf{C}_2 e^{-\gamma s} \qquad\qquad \text{...(10.2.5)}$$

where $\mathbf{C}_1$ and $\mathbf{C}_2$ are arbitrary constants whose values depend upon the boundary conditions, that is, $\mathbf{C}_1$ and $\mathbf{C}_2$ are found from the known values of $\mathbf{V}$ and $\mathbf{I}$ at some point of the line.

For finding the value of $\mathbf{I}$ we differentiate Eq. (10.2.5) with respect to s to get

$$\frac{d\mathbf{V}}{ds} = \mathbf{C}_1\,\gamma e^{\gamma s} - \mathbf{C}_2\,\gamma e^{-\gamma s}$$

Combining this equation with Eq. (10.2.1) gives

$$\mathbf{zI} = (\mathbf{C}_1\,\gamma e^{\gamma s} - \mathbf{C}_2\,\gamma e^{-\gamma s})$$

$$\mathbf{I} = \frac{\sqrt{\mathbf{zy}}}{\mathbf{z}}\,(\mathbf{C}_1\,e^{\gamma s} - \mathbf{C}_2\,e^{-\gamma s}) = \sqrt{\frac{\mathbf{y}}{\mathbf{z}}}\,(\mathbf{C}_1\,e^{\gamma s} - \mathbf{C}_2\,e^{-\gamma s}) \qquad \text{...(10.2.6)}$$

The values of $\mathbf{V}$ and $\mathbf{I}$ at the receiving end where $s = 0$ are $\mathbf{V}_r$ and $\mathbf{I}_r$ respectively. Substituting these values in Eqs. (10.2.5) and (10.2.6), we obtain

$$\mathbf{V}_r = \mathbf{C}_1 + \mathbf{C}_2 \qquad\qquad (\because\ e^0 = 1)$$

$$\mathbf{I}_r = \sqrt{\frac{\mathbf{y}}{\mathbf{z}}}\,(\mathbf{C}_1 - \mathbf{C}_2)$$

The values of arbitrary constants are found from these simultaneous equations as

$$\mathbf{C}_1 = \frac{1}{2}\left(\mathbf{V}_r + \sqrt{\frac{\mathbf{z}}{\mathbf{y}}}\,\mathbf{I}_r\right) = \frac{1}{2}(\mathbf{V}_r + \mathbf{Z}_o\,\mathbf{I}_r) \quad \text{and} \quad \mathbf{C}_2 = \frac{1}{2}\left(\mathbf{V}_r - \sqrt{\frac{\mathbf{z}}{\mathbf{y}}}\,\mathbf{I}_r\right) = \frac{1}{2}(\mathbf{V}_r - \mathbf{Z}_o\,\mathbf{I}_r)$$

where $\mathbf{Z}_o \triangleq \sqrt{\dfrac{\mathbf{z}}{\mathbf{y}}}$

The values of $\mathbf{C}_1$ and $\mathbf{C}_2$ are substituted in Eqs. (10.2.5) and (10.2.6) to obtain the steady-state values of $\mathbf{V}$ and $\mathbf{I}$ at any intermediate point distant s from the receiving end.

$$\mathbf{V} = \frac{1}{2}(\mathbf{V}_r + \mathbf{Z}_o\,\mathbf{I}_r)\,e^{\gamma s} + \frac{1}{2}(\mathbf{V}_r - \mathbf{Z}_o\,\mathbf{I}_r)\,e^{-\gamma s} \qquad\qquad \text{...(10.2.7)}$$

$$\mathbf{I} = \frac{1}{2}\left(\mathbf{I}_r + \frac{\mathbf{V}_r}{\mathbf{Z}_o}\right)e^{\gamma s} + \frac{1}{2}\left(\mathbf{I}_r - \frac{\mathbf{V}_r}{\mathbf{Z}_o}\right)e^{-\gamma s} \qquad\qquad \text{...(10.2.8)}$$

10.3 PHYSICAL INTERPRETATION OF THE LONG LINE EQUATIONS

Here we shall discuss the physical interpretation of Eqs. (10.2.7) and (10.2.8). Since both the quantities $\mathbf{z}$ and $\mathbf{y}$ are complex, $\gamma\ (= \sqrt{\mathbf{zy}})$ will also be a complex quantity. It may, therefore, be written in the form

$$\gamma = \alpha + j\,\beta$$

where α and β are real and positive quantities.

$$e^{\gamma s} = e^{(\alpha + j\beta)s} = e^{\alpha s}\,e^{j\beta s}$$

The term $e^{\alpha s}$ is real. It increases exponentially at the rate of e^{α} per unit length of line.

$$e^{j\beta s} = \cos\beta s + j\sin\beta s = 1\,\underline{/\beta s}$$

The operator $e^{j\beta s}$ has a magnitude 1 for every value of s but it has an angle which increases with s. For unit distance along the line the angle changes by β radians. It is thus observed that the exponential factor $e^{\alpha s}$ when operated upon a phasor quantity will change its magnitude only at the rate of e^{α} per unit length of line. The second exponential factor $e^{j\beta s}$ when operated upon a phasor will keep its magnitude constant but will change its phase by β radians per unit length of line. The factor $e^{\alpha s}$ may be considered as *magnitude operator* and the factor $e^{j\beta s}$ is the *rotational operator*. Thus, if a phasor be operated by a factor $e^{\gamma s}$ both magnitude and phase of the phasor will be changed. As we proceed from the receiving end to the sending end of the line, that is in the direction of s increasing, the amplitude of the voltage component,

$\frac{1}{2} (\mathbf{V}_r + \mathbf{Z}_a \mathbf{I}_r) e^{\gamma s}$ increases exponentially at the rate of e^{α} per unit length of line and its phase is advanced by β radians per unit length of line. On the other hand, if we move from the sending end to the receiving end, this component of voltage decreases in amplitude and retards in phase. This is the characteristic of a travelling wave which decreases in amplitude and retards in phase along the direction of propagation. This term, therefore, represents a voltage wave originated at the sending end and travelling towards the receiving end of the line. This component of voltage is called the *incident* voltage or *forward* voltage wave.

With a similar argument it is clear that the current component $\frac{1}{2} (\mathbf{I}_r + \mathbf{V}_r / \mathbf{Z}_o) e^{\gamma s}$ is the incident current wave travelling from sending end towards the receiving end of the line.

Now, $\qquad e^{-\gamma s} = e^{-(\alpha + j\beta) s} = e^{-\alpha s} \cdot e^{-j\beta s}$

As s increases, the magnitude of $e^{-\alpha s}$ decreases,

$$e^{-j\beta s} = \cos \beta s - j \sin \beta s = 1 \underline{/-\beta s}$$

The operator $e^{-j\beta s}$ has always a magnitude equal to 1 but produces a rotation opposite to $e^{j\beta s}$. It retards the phase of a phasor upon which it is operated by β radians per unit distance. If the factor $e^{-\alpha s}$ operates upon the voltage component $\frac{1}{2} (\mathbf{V}_r - \mathbf{Z}_o \mathbf{I}_r)$, it decreases its magnitude and if this component is operated by $e^{-j\beta s}$ its magnitude remains the same but its phase goes on retarding by β radians per unit length of the line as we move from the receiving end towards the sending end. Thus, the voltage component $\frac{1}{2} (\mathbf{V}_r - \mathbf{Z}_o \mathbf{I}_r) e^{-\gamma s}$ represents a wave which decreases in magnitude exponentially with the distance and is retarded in phase as it moves from the receiving end to the sending end. This component of voltage is, therefore, termed *reflected* or *backward wave*. Similarly, $\frac{1}{2} (\mathbf{I}_r - \mathbf{V}_r / \mathbf{Z}_o) e^{-\gamma s}$ is the reflected or backward current curve.

The voltage and current at a point along the line consists of two components; namely, the incident and the reflected waves travelling in opposite directions. It should be noted that although although the components of voltage and current at any point along the line are travelling waves, the resultant instantaneous voltage and current, which are obtained by adding the incident and reflected components at that point, are not travelling waves, but, *standing* or *stationary waves*.

10.4 PROPAGATION CONSTANT

We have seen earlier that the magnitude and the phase of a travelling wave is governed by the complex quality γ. In other words, γ governs the propagation of component waves. It is, therefore, called the *propagation constant*. The real part of propagation constant is α. It determines the change in magnitude per unit length of the line of the wave, and is termed *attenuation constant*. It is expressed in nepers per unit length. β is the imaginary part of the propagation constant. It determines the change in phase of the wave per unit length of the line. It is, therefore, called *phase constant* or *wave length constant*. β is expressed in radians per unit length.

The numerical value of propagation can be determined by the relation

$$\gamma = \sqrt{(\mathbf{z} \cdot \mathbf{y})} = \sqrt{(r + j\omega l)(g + j\omega c)} \qquad \qquad \text{...(10.4.1)}$$

10.5 WAVELENGTH AND VELOCITY OF PROPAGATION

The particular value of the length of the line for which the voltage or current undergoes a phase shift of 2π radians is called a *wave length*. It is represented by λ. Since the phase shift per unit distance is β radians, the phase shift for a distance λ will be $\lambda\beta$.

$\therefore$ $\beta\lambda = 2\pi$ radians

and $\lambda = \dfrac{2\pi}{\beta}$ metres   ...(10.5.1)

The velocity of propagation of the wave v_p is given by

$$v_p = \lambda f \text{ m/s}$$

v_p is also called the *phase velocity*. It can also be written as

$$v_p = \frac{2\pi}{\beta} f = \frac{\omega}{\beta} \qquad \qquad \qquad ...(10.5.2)$$

10.6 CHARACTERISTIC IMPEDANCE Z_o

The quantity $\sqrt{\dfrac{\mathbf{z}}{\mathbf{y}}}$ is a complex number as $\mathbf{y}$ and $\mathbf{z}$ are complex. It is denoted by $\mathbf{Z}_o$. It has the dimension of an impedance since

$$\frac{\mathbf{z}}{\mathbf{y}} = \sqrt{\frac{\text{ohms/unit length}}{\text{siemens/unit length}}} = \text{ohms}$$

$$\mathbf{Z}_o = \sqrt{\frac{\mathbf{z}}{\mathbf{y}}} = \sqrt{\frac{\mathbf{z}S}{\mathbf{y}S}} = \sqrt{\frac{\mathbf{Z}}{\mathbf{Y}}} = \sqrt{\frac{r + j\omega l}{g + j\omega c}} \qquad \qquad ...(10.6.1)$$

This quantity depends upon the characteristic of the line per unit length. It is, therefore called *characteristic impedance* of the line. It is independent of the length of the line. It depends upon the radius and spacing between the conductors. The reciprocal of $\mathbf{Z}_o$ is called the *characteristic admittance* of the line. It is denoted by $\mathbf{Y}_o$.

$\therefore$ $\mathbf{Y}_o = \dfrac{1}{\mathbf{Z}_o} = \sqrt{\dfrac{\mathbf{y}}{\mathbf{z}}} = \sqrt{\dfrac{\mathbf{y}S}{\mathbf{z}S}} = \sqrt{\dfrac{\mathbf{Y}}{\mathbf{Z}}}$ ...(10.6.2)

$\mathbf{Z}_o$ is the impedance offered to the propagation of a voltage or current wave during its travel along the line.

For a lossless line, $r = 0$, $g = 0$, and the characteristic impedance becomes

$$\mathbf{Z}_o = \sqrt{\frac{l}{c}} = \sqrt{\frac{lS}{cS}} = \sqrt{\frac{L}{C}} \qquad \qquad \qquad ...(10.6.3)$$

The characteristic impedance in this special case is called the *surge impedance* or *natural impedance* of the line.

10.7 HYPERBOLIC FORM OF LINE EQUATIONS

In previous section, the equations governing the behaviour of transmission lines in steady-state were expressed in exponential form. Hyperbolic functions are used to reduce these equations to more compact forms. Use is made of the following relations between hyperbolic and exponential functions :

$$\sinh \theta = \frac{e^\theta - e^{-\theta}}{2} \qquad \qquad ...(10.7.1)$$

$$\cosh \theta = \frac{e^\theta + e^{-\theta}}{2} \qquad \qquad ...(10.7.2)$$

Rearranging the terms of Eq. (10.2.7) we get,

$$\mathbf{V} = \mathbf{V}_r \left(\frac{e^{\gamma s} + e^{-\gamma s}}{2} \right) + \mathbf{Z}_0 \, \mathbf{I}_r \left(\frac{e^{\gamma s} - e^{-\gamma s}}{2} \right)$$

$$\mathbf{V} = \mathbf{V}_r \cosh \gamma s + \mathbf{Z}_0 \, \mathbf{I}_r \sinh \gamma s \qquad \qquad ...(10.7.3)$$

Similarly, the rearrangement of the terms of Eq. (10.2.8) gives

$$\mathbf{I} = \mathbf{I}_r \left(\frac{e^{\gamma s} + e^{-\gamma s}}{2} \right) + \mathbf{V}_r \, \mathbf{Z}_0 \left(\frac{e^{\gamma s} - e^{-\gamma s}}{2} \right)$$

$$\mathbf{I} = \mathbf{I}_r \cosh \gamma s + \frac{\mathbf{V}_r}{\mathbf{Z}_0} \sinh \gamma s \qquad \qquad ...(10.7.4)$$

Equations (10.7.3) and (10.7.4) are the general equations of voltage and current at any distance s from the receiving end of the line. They are in hyperbolic form. The values of voltage and current at the sending end can be found by substituting $s = S$ in Eqs. (10.7.3) and (10.7.4). At $s = S$, $\mathbf{V} = \mathbf{V}_s$, and $\mathbf{I} = \mathbf{I}_s$. Thus we have

$$\mathbf{V}_s = \mathbf{V}_r \cosh \gamma S + \mathbf{Z}_0 \, \mathbf{I}_r \sinh \gamma S \qquad \qquad ...(10.7.5)$$

$$\mathbf{I}_s = \mathbf{I}_r \cosh \gamma S + \frac{\mathbf{V}_r}{\mathbf{Z}_0} \sinh \gamma S \qquad \qquad ...(10.7.6)$$

We define the following **ABCD** parameters :

$$
\left.
\begin{aligned}
\mathbf{A} &\triangleq \cosh \gamma S \\
\mathbf{B} &\triangleq \mathbf{Z}_0 \sinh \gamma S \\
\mathbf{C} &\triangleq \frac{1}{\mathbf{Z}_0} \sinh \gamma S \\
\mathbf{D} &\triangleq \cosh \gamma S
\end{aligned}
\right\} \qquad ...(10.7.7)
$$

Equations (10.7.5) and (10.7.6) may, therefore, be written as

$$\mathbf{V}_s = \mathbf{A}\mathbf{V}_r + \mathbf{B}\mathbf{I}_r \qquad \qquad ...(10.7.8)$$

$$\mathbf{I}_s = \mathbf{C}\mathbf{V}_r + \mathbf{D}\mathbf{I}_r \qquad \qquad ...(10.7.9)$$

In matrix form

$$
\begin{bmatrix} \mathbf{V}_s \\ \mathbf{I}_s \end{bmatrix} = \begin{bmatrix} \mathbf{A} & \mathbf{B} \\ \mathbf{C} & \mathbf{D} \end{bmatrix} \begin{bmatrix} \mathbf{V}_r \\ \mathbf{I}_r \end{bmatrix} \qquad \qquad ...(10.7.10)
$$

These equations are used in evaluating the performance of long lines. The constants **A**, **B**, **C** and **D** are called *transmission parameters* or *chain parameters*. It is to be noted clearly that in Eqs. (10.7.8) and (10.7.9) the value of voltage must be expressed in volts and must be the line-to-neutral voltage.

10.8 EVALUATION OF ABCD PARAMETERS

The hyperbolic functions involved in the transmission line equations are not easily evaluated with the help of ordinary tables of hyperbolic functions. This is due to the fact that they are functions of complex arguments. The following methods are used to calculate the hyperbolic functions $\cosh \gamma S$ and $\sinh \gamma S$ for determining the **ABCD** parameters of transmission line :

We have $\gamma = \alpha + j\beta$

(1) *Use of complex exponentials*

$$\cosh \gamma S = \cosh (\alpha S + j\beta S) = \frac{1}{2} [e^{(\alpha S + j\beta S)} + e^{-(\alpha S + j\beta S)}]$$

$$= \frac{1}{2} (e^{\alpha S} \cdot e^{j\beta S} + e^{-\alpha S} \cdot e^{-j\beta S}) = \frac{1}{2} (e^{\alpha S} \underline{/\beta S} + e^{-\alpha S} \underline{/-\beta S}) \qquad \ldots(10.8.1)$$

and $\quad \sinh \gamma S = \sinh (\alpha S + j\beta S) = \frac{1}{2} [e^{(\alpha S + j\beta S)} - e^{-(\alpha S + j\beta S)}]$

$$= \frac{1}{2} (e^{\alpha S} \cdot e^{j\beta S} - e^{-\alpha S} \cdot e^{-j\beta S}) = \frac{1}{2} (e^{\alpha S} \underline{/\beta S} - e^{-\alpha S} \underline{/-\beta S}) \qquad \ldots(10.8.2)$$

Since $\quad \gamma = \sqrt{zy}$

$$\gamma S = \sqrt{(zS)(yS)} = \sqrt{ZY} \qquad \ldots(10.8.3)$$

Also, $\quad Z_o = \sqrt{\dfrac{z}{y}} = \sqrt{\dfrac{zS}{yS}} = \sqrt{\dfrac{Z}{Y}} \qquad \ldots(10.8.4)$

The **ABCD** parameters can, therefore, be written as

$$A = \cosh \gamma S = \cosh \sqrt{ZY} \qquad \ldots(10.8.5)$$

$$B = Z_o \sinh \gamma S = \sqrt{\frac{Z}{Y}} \sinh \sqrt{ZY} \qquad \ldots(10.8.6)$$

$$C = \frac{1}{Z_o} \sinh \gamma S = \sqrt{\frac{Y}{Z}} \sinh \sqrt{ZY} \qquad \ldots(10.8.7)$$

$$D = \cosh \gamma S = \cosh \sqrt{ZY} = A \qquad \ldots(10.8.8)$$

(2) *Use of identities*

The hyperbolic sines and cosines of the complex argument γS can be separated into real and imaginary parts of the use of following identities :

$$\cosh \gamma S = \cosh (\alpha S + j\beta S) = \cosh \alpha S \cos \beta S + j \sinh \alpha S \sin \beta S$$

$$\sinh \gamma S = \sinh (\alpha S + j\beta S) = \sinh \alpha S \cos \beta S + j \cosh \alpha S \sin \beta S$$

It should be noted that the unit of βS is the radian.

(3) *Use of power series*

$$\cosh \gamma S = 1 + \frac{(\gamma S)^2}{2!} + \frac{(\gamma S)^4}{4!} + \ldots \qquad \ldots(10.8.9)$$

$$\sinh \gamma S = \gamma S + \frac{(\gamma S)^3}{3!} + \frac{(\gamma S)^5}{5!} + \ldots \qquad \ldots(10.8.10)$$

The above series converge rapidly for the values of γS usually found for power lines. Sufficient accuracy can be obtained by taking only the first two terms.

Thus $\quad \cosh \gamma S \simeq 1 + \frac{(\gamma S)^2}{2!} = 1 + \frac{ZY}{2} \qquad \ldots(10.8.11)$

$$\sinh \gamma S \simeq \gamma S + \frac{(\gamma S)^3}{3!} = \sqrt{ZY}\left(1 + \frac{1}{6}ZY\right) \qquad \ldots(10.8.12)$$

Usually the above approximations are satisfactory for overhead lines upto 500 km. Therefore,

$$A = D = \cosh \gamma S = 1 + \frac{1}{2}ZY \qquad \ldots(10.8.13)$$

$$B = Z_o \sinh \gamma S = Z\left(1 + \frac{ZY}{6}\right) \qquad \ldots(10.8.14)$$

$$C = \frac{1}{Z_o} \sinh \gamma S = Y\left(1 + \frac{ZY}{6}\right) \qquad \ldots(10.8.15)$$

Example 10.1 A 275 kV overhead transmission line has the following characteristics :

$Z = 12.5 + j\,66\ \Omega,\ Y = 4.4 \times 10^{-4}\,\underline{/90°}\ S.$

Calculate the **ABCD** constants and the surge impedance of the line.

Solution

$\quad Z = 12.5 + j\,66 = 67.17\,\underline{/79.276°}\ \Omega$

$\quad Y = 4.4 \times 10^{-4}\,\underline{/90°}\ S$

$\quad \gamma S = \sqrt{ZY} = [(67.17\,\underline{/79.2756°}) \times (4.4 \times 10^{-4}\,\underline{/90°})]^{1/2}$

$\qquad = [0.02956\,\underline{/169.2756°})]^{1/2} = \sqrt{0.02956}\,\underline{/(1/2)(169.2756°)}$

$\qquad = 0.1719\,\underline{/84.6378°} = 0.0161 + j\,0.1712$

$\qquad = \alpha S + j\beta S$

$\therefore \qquad \alpha S = 0.0161,\ \beta S = 0.1712\ \text{rad}$

We convert βS to degrees. Therefore,

$$\beta S = 0.172 \times \frac{180}{\pi}$$

$$= 9.809°$$

Method of complex exponentials

Using Eq. (10.8.1)

$$\cosh \gamma S = \frac{1}{2}(e^{\alpha S} \underline{/\beta S} + e^{-\alpha S} \underline{/-\beta S}) = \frac{1}{2}(e^{0.0161} \underline{/9.809°} + e^{-0.0161} \underline{/-9.809°})$$

$$= \frac{1}{2}(1.0162 \underline{/9.809°} + 0.984 \underline{/-9.809°})$$

$$= \frac{1}{2}(1.0013 + j\,01731 + 0.9696 - j\,0.1676)$$

$$= \frac{1}{2}(1.9709 + j\,5.5 \times 10^{-3}) = 0.9855 \underline{/0.1599°}$$

$$\therefore \qquad A = D = \cosh \gamma S = 0.9855 \underline{/0.1599°}$$

We now calculate $\sinh \gamma S$ as

$$\sinh \gamma S = \frac{1}{2}(e^{\alpha S} \underline{/\beta S} - e^{-\alpha S} \underline{/-\beta S}) = \frac{1}{2}(e^{0.0161} \underline{/9.809°} - e^{-0.0161} \underline{/-9.809°}$$

$$= \frac{1}{2}(1.0013 + j\,0.1731 - 0.9696 + j\,0.1676)$$

$$= 0.01585 + j\,0.17035 = 0.1711 \underline{/84.68°}$$

We have

$$Z_o = \sqrt{\frac{Z}{Y}} = \sqrt{\frac{67.17 \underline{/79.2756°}}{4.4 \times 10^{-4} \underline{/90°}}}$$

$$= \sqrt{152659} \underline{/(\frac{1}{2})(79.2756 - 90°)} = 390.7 \underline{/-5.36°}\ \Omega$$

Therefore,

$$B = Z_o \sinh \gamma S = (390.7 \underline{/-5.36°})(0.1711 \underline{/84.68°}) = 66.85 \underline{/79.32°}\ \Omega$$

Also $\qquad C = \frac{1}{Z_o} \sinh \gamma S = \frac{0.1711 \underline{/84.68°}}{390.7 \underline{/-5.36°}} = 4.3793 \times 10^{-4} \underline{/90.04°}\ S$

Method of identities

$$\cosh \gamma S = \cosh \alpha S \cos \beta S + j \sinh \alpha S \sin \beta S$$

$$\cosh \alpha S = \cosh (0.0161) = \frac{1}{2}(e^{0.0161} + e^{-0.0161}) = \frac{1}{2}(1.0162 + 0.9840) = 1.00$$

$$\sinh \alpha S = \sinh (0.0161) = \frac{1}{2}(e^{0.0161} - e^{-0.0161}) = \frac{1}{2}(1.0162 - 0.9840) = 0.0161$$

If the electronic calculator has built-in hyperbolic functions the intermediate steps can be omitted.

We have

$$\cos \beta S = \cos 9.809° = 0.98538$$
$$\sin \beta S = \sin 9.809° = 0.17036$$

Therefore,

$$\cosh \gamma S = \cosh \alpha S \cos \beta S + j \sinh \alpha S \sin \beta S = 1.00 \times 0.98538 + j\, 0.0161 \times 0.17036$$

$$= 0.9538 + j\, 2.743 \times 10^{-3} = 0.9854 \underline{/0.1595°}$$

$$\sinh \gamma S = \sinh \alpha S \cos \beta S + j \cosh \alpha S \sin \beta S$$

$$= 0.0161 \times 0.98538 + j\, 1.0 \times 0.17036 = 0.01586 + j\, 0.17036 = 0.17109 \underline{/84.68°}$$

Therefore we have

$$\mathbf{A} = \mathbf{D} = \cosh \gamma S = 0.9854 \underline{/0.1595°}$$

$$\mathbf{B} = \mathbf{Z}_o \sinh \gamma S = (390.7 \underline{/-5.36°})\,(0.17109 \underline{/86.68°}) = 66.847 \underline{/79.32°}\ \Omega$$

$$\mathbf{C} = \frac{1}{\mathbf{Z}_o} \sinh \gamma S = \frac{0.17109 \underline{/86.68°}}{390.7 \underline{/-5.36°}} = 4.3790 \times 10^{-4} \underline{/90.04°}\ S$$

Power series method

$$\mathbf{A} = \mathbf{D} = \cosh \gamma S \simeq 1 + \frac{1}{2} \mathbf{ZY}$$

$$= 1 + \frac{1}{2}\,(67.17 \underline{/79.2756°})\,(4.4 \times 10^{-4} \underline{/90°}) = 1 + 0.014777 \underline{/169.2756°}$$

$$= 1 - 0.01451 + j\, 2.74993 \times 10^{-3} = 0.98549 + j\, 2.74993 \times 10^{-3} = 0.985493 \underline{/0.15988°}$$

Also, $\sinh \gamma S \simeq \mathbf{ZY}\left(1 + \dfrac{\mathbf{ZY}}{6}\right)$

$$\mathbf{B} = \mathbf{Z}_o \sinh \gamma S \simeq \mathbf{Z}\left(1 + \frac{\mathbf{ZY}}{6}\right)$$

$$\mathbf{C} = \frac{1}{\mathbf{Z}_o} \sinh \gamma S \simeq \mathbf{Y}\left(1 + \frac{\mathbf{ZY}}{6}\right)$$

$$1 + \frac{\mathbf{ZY}}{6} = 1 + \frac{0.02956}{6} \underline{/169.2756°} = 0.9951 + j\, 1677 \times 10^{-4} = 0.9951 \underline{/0.052786°}$$

$$\mathbf{B} = \mathbf{Z}\left(1 + \frac{\mathbf{ZY}}{6}\right) = (67.17 \underline{/79.2756°}) \times (0.9951 \underline{/0.052786°})$$

$$= 66.8408 \underline{/79.328°}\ \Omega$$

$$\mathbf{C} = \mathbf{Y}\left(1 + \frac{\mathbf{ZY}}{6}\right) = (4.4 \times 10^{-4} \underline{/90°})\,(0.9951 \underline{/0.052786°})$$

$$= 4.3784 \times 10^{-4} \underline{/90.052786°}\ S$$

It is seen that the results obtained by the three methods are the same.

Example 10.2 If the line given in Example 10.1 delivers 250 MW at a lagging power factor of 0.9, determine the sending-end voltage, sending-end current, line charging current, efficiency of transmission, and voltage regulation.

Solution

Phase voltage at the receiving end, $V_{rp} = \dfrac{275 \times 10^3}{\sqrt{3}} = 158.77 \times 10^3$ V

Taking $\mathbf{V}_{rp}$ as the reference phasor.

$\qquad \mathbf{V}_{rp} = V_{rp} \underline{/0°} = 158.77 \times 10^3 \underline{/0°}$ V

Receiving-end current, $I_r = \dfrac{P}{\sqrt{3}\ V_l \cos \varphi_r} = \dfrac{250 \times 10^6}{\sqrt{3} \times 275 \times 10^3 \times 0.9} = 583.2$ A

$\qquad \mathbf{I}_r = I_r \underline{/- \varphi_r°} = 583.2 \underline{/- \cos^{-1} 0.9°} = 583.2 \underline{/- 25.84°}$ A

From Example 10.1

$\qquad \mathbf{A} = 0.9854 \underline{/0.1595°} = \mathbf{D}\ ; \quad \mathbf{B} = 66.85 \underline{/79.32°}\ \Omega\ ; \quad \mathbf{C} = 4.3793 \times 10^{-4} \underline{/90.04°}$ S

Phase voltage at the sending end,

$\qquad \mathbf{V}_{sp} = \mathbf{AV}_{rp} + \mathbf{BI}_r$

$\qquad\qquad = (0.9854 \underline{/0.1595°})\,(158.77 \times 10^3 \underline{/0°}) + (66.85 \underline{/79.32°})\,(583.2 \underline{/- 25.84°})$

$\qquad\qquad = 156452 \underline{/0.1595°} + 38987 \underline{/53.48°} = 156451 + j\,435.5 + 23201 + j\,31331$

$\qquad\qquad = 179652 + j\,31766.5 = 182439 \underline{/10.03°}$ V

Line voltage at the sending end, $V_{sl} = \sqrt{3}\ V_{sp} = \sqrt{3} \times 182439 = 315994$ V

Sending-end current,

$\qquad \mathbf{I}_s = \mathbf{CV}_{rp} + \mathbf{DI}_r$

$\qquad\qquad = (4.3793 \times 10^{-4} \underline{/90.04°})\,(158.77 \times 10^3 \underline{/0°}) + (0.9854 \underline{/0.1595°})\,(538.2 \underline{/- 25.84°})$

$\qquad\qquad = 69.53 \underline{/90.04°} + 574.68 \underline{/- 25.68°} = - 0.0485 + j\,69.53 + 517.92 - j\,249.03$

$\qquad\qquad = 517.87 - j\,179.5 = 548.1 \underline{/- 19.1°}$ A

Charging current = current at the sending end at no load.

At no load,

$\qquad \mathbf{I}_r = 0, \quad \mathbf{I}_s = \mathbf{I}_{so} =$ charging current

$\qquad \mathbf{I}_{so} = \mathbf{CV}_r = 69.53 \underline{/90.04°}$ V

Sending-end power factor, $\cos \varphi_s = \cos [10.03° - (- 19.1°)] = \cos 29.13° = 0.8735$ (lagging)

Sending-end power, $P_s = 3V_{sp} I_s \cos \varphi_s = 3 \times 182439 \times 548.1 \times 0.8735$

$\qquad\qquad\qquad\qquad = 262.036 \times 10^6$ W $= 262.036$ MW

Alternatively, the power at the sending end can be calculated as follows :

$\qquad P_s = \mathrm{Re}\,[3\mathbf{V}_{sp}\,\mathbf{I}^*] = \mathrm{Re}\,[3 \times (182439 \underline{/10.03°})\,(548.1 \underline{/+ 19.1°})]$

$\qquad\qquad = \mathrm{Re}\,(299.984 \times 10^6 \underline{/29.13°}) = \mathrm{Re}\,(262.041 \times 10^6 + j\,146.03 \times 10^6)$

$\qquad\qquad = 262.041 \times 10^6$ W $= 262.041$ MW

Transmission efficiency, $\eta_T = \dfrac{P_r}{P_s} = \dfrac{250}{262.041} = 0.954$ pu $= 95.4$ per cent

Again, $\mathbf{V}_s = \mathbf{AV}_r + \mathbf{BI}_r$

At no load,

$\mathbf{I}_r = 0, \quad \mathbf{V}_r = \mathbf{V}_{rnl}$

$\therefore \qquad \mathbf{V}_s = \mathbf{AV}_{rnl}, \ \mathbf{V}_{rnl} = \dfrac{\mathbf{V}_s}{\mathbf{A}}$

$|\mathbf{V}_{rnl}| = \dfrac{|\mathbf{V}_s|}{|\mathbf{A}|} = \dfrac{182439}{0.9854} = 185142 \text{ V}$

$|\mathbf{V}_{rfl}| = 158.77 \times 10^3 \text{ V}$

$\text{Regulation} = \dfrac{|\mathbf{V}_{rnl}| - |\mathbf{V}_{rfl}|}{|\mathbf{V}_{rfl}|} = \dfrac{185142 - 158.77 \times 10^3}{158.77 \times 10^3} = 0.1661 \text{ pu} = 16.61 \text{ per cent.}$

10.9 FERRANTI EFFECT

A long transmission line has a large capacitance. If such a line is open-circuited or very lightly loaded at the receiving end, the magnitude of the voltage at the receiving end becomes higher than the voltage at the sending end. This phenomenon is called *Ferranti effect*. It was first noticed by Ferranti on overhead lines supplying a lightly loaded network. Ferranti effect is due to charging current of the line. The value of current at the sending end at no load and normal operating voltage applied at the sending end is called the *charging current*.

Ferranti effect can be explained by considering a nominal π model of the line. Fig. 10.3 (b) shows the phasor diagram of Fig. 10.3 (a). Here *OE* represents the receiving end voltage V_r. *OH* represents the current I_{c1} through the capacitor $C/2$ at the receiving end. The voltage drop $I_{c1} R$ across the resistance R is shown by *EF*. It is in phase with I_{c1}. The voltage drop across X is $I_{c1} X$. It is represented by the phasor *FG* which leads the phasor $I_{c1} R$ by 90°. The phasor *OG* represents the sending-end voltage V_s under no-load condition. It is seen from the phasor diagram that $V_s < V_r$. In other words, the voltage at the receiving end is greater than the voltage at the sending end when the line is at no load.

In practice, the capacitance of the line is not concentrated at some definite points. It is distributed uniformly along the whole length of the line. Therefore, as we move from the sending

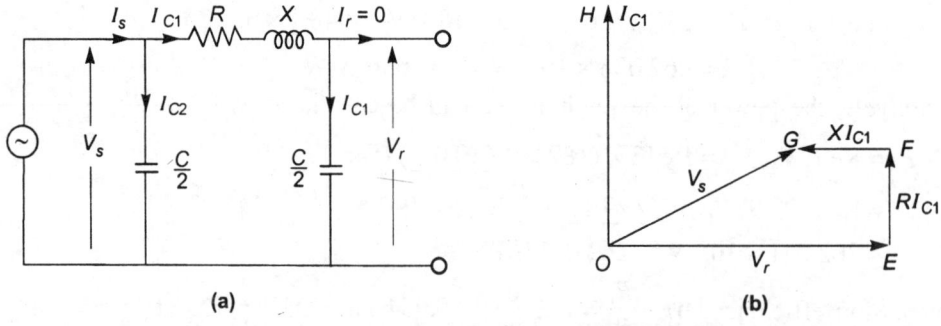

Fig. 10.3. (a) Nominal π-model of the line at no load. (b) Phasor diagram.

end to the receiving end of the line, the voltage goes on increasing from point to point. At no load or light load the voltage at the receiving end is quite large as compared to the constant voltage at the sending end.

For a nominal π-model of a line

$$V_s = \left(1 + \frac{ZY}{2}\right) V_r + ZI_r \qquad \qquad ...(10.9.1)$$

At no load, $I_r = 0$

$$\therefore \qquad V_s = \left(1 + \frac{ZY}{2}\right) V_r$$

$$V_s - V_r = \frac{ZY}{2} V_r \qquad \qquad ...(10.9.2)$$

$$Z = (r + j\omega l)\, S, \ \ Y = (j\omega c)\, S$$

If the resistance of the line is neglected,

$$Z = j\omega l S$$

and $\qquad V_s - V_r = \frac{1}{2}\,(j\omega l S)\,(j\omega c S)\, V_r = -\frac{1}{2}\,(\omega^2 S^2)\, lc V_r \qquad \qquad ...(10.9.3)$

For overhead lines

$$\frac{1}{\sqrt{lc}} = \text{velocity of propagation of electromagnetic waves on the line} = 3 \times 10^8 \text{ m/s}$$

$$V_s - V_r = -\frac{1}{2}\,(2\pi f)^2\, S^2] \cdot \frac{1}{(3 \times 10^8)^2}\, V_r = -\left(\frac{4\pi^2}{18} \times 10^{-6}\right) f^2\, S^2\, V_r \qquad ...(10.9.4)$$

Equation (10.9.4) shows that $(V_s - V_r)$ is negative. That is, $V_r > V_s$. This equation also shows that Ferranti effect depends on frequency and electrical length of the line. The conductor diameter and spacing have no bearing on Ferranti effect.

In general, for any line

$$\mathbf{V}_s = \mathbf{A V}_r + \mathbf{B I}_r$$

At no load,

$$\mathbf{I}_r = 0, \quad \mathbf{V}_r = \mathbf{V}_{rnl}$$

$$\therefore \qquad \mathbf{V}_s = \mathbf{A V}_{rnl}, \ |\mathbf{V}_{rnl}| = \frac{|\mathbf{V}_s|}{|\mathbf{A}|}$$

For a long line $\mathbf{A}$ is less than unity and it decreases with the increase in length of line. Hence $V_{rnl} > V_s$. As the line length increases the rise in the voltage at the receiving end at no load becomes more predominant.

10.10 SURGE IMPEDANCE LOADING (SIL)

A transmission line may be considered as generating capacitive reactive voltamperes in shunt capacitance, and consuming (absorbing) inductive reactive voltamperes in its series inductance.

The load at which the inductive and capacitive reactive voltamperes are equal and opposite is called the *surge impedance load* (SIL) or *natural load* of the line.

Let V = phase voltage at the receiving end

I = phase current

L = series inductance per phase

X_L = series inductive reactance per phase

X_C = shunt capacitive reactance per phase

Z_o = surge impedance per phase.

Capacitive voltamperes (VAr) generated in the line

$$= \frac{V^2}{X_C} = V^2 \omega C \text{ per phase} \qquad \qquad ...(10.10.1)$$

Inductive reactive voltamperes (VAr) absorbed by the line

$$= I^2 X_L = I^2 \omega L \text{ per phase} \qquad \qquad ...(10.10.2)$$

Under natural load conditions

$$V^2 \omega C = I^2 \omega L \qquad \qquad ...(10.10.3)$$

$$\therefore \qquad \frac{V}{I} = \sqrt{\frac{L}{C}} = Z_o \qquad \qquad ...(10.10.4)$$

At this load the voltage V and current I are in the same phase at all points along the line. Under such conditions the line is terminated in its surge impedance Z_o which is purely resistive. When a line is terminated in its surge impedance Z_o the power delivered by it is called the *natural load* or *surge impedance load* (SIL). Thus, *surge impedance load of a line may be defined as the power delivered by it to a purely resistive load equal to its surge impedance.*

A *lossless line operating at its nominal voltage terminating in its surge impedance Z_o is said to be surge impedance loaded.*

If P_o is its natural load, $(SIL)_{1\,\varphi}$ of the line per phase

$$(SIL)_{1\,\varphi} = P_o = V_p I_p \cos \varphi$$

Since the load is purely resistive,

$$\cos \varphi = 1$$

and $\qquad P_o = V_p I_p = V_p \dfrac{V_p}{Z_o}$

or $\qquad P_o = \dfrac{V_p^2}{Z_o} \text{ W/phase} \qquad \qquad ...(10.10.5)$

Thus, per phase power transmitted under surge impedance loading is V_p^2 / Z_o watts, where V_p is the phase voltage.

Line voltage $V_L = \sqrt{3}\, V_p$

Total natural load of the line $(SIL)_{3\,\varphi} = 3P_o = \dfrac{3V_p^2}{Z_o} = \dfrac{V_L^2}{Z_o} \text{ W}$

If kV_L is the receiving-end voltage in kV, then

$$(SIL)_{3\,\varphi} = \frac{(kV_L)^2}{Z_o} \text{ MW} \qquad\qquad ...(10.10.6)$$

It is seen that SIL is independent of the distance and depends mainly on the voltage.

Thus, SIL may be used to find the permissible power transfer capability of any line for various transmission distances. In practice, SIL is always less than the rated capacity of a line. If the load is less than SIL, reactive voltamperes are generated and the receiving-end voltage will be greater than the sending-end voltage. On the other hand, if the load is greater than SIL, reactive voltamperes are absorbed by the line. In this case the receiving-end voltage is less than the sending-end voltage. The compensation requirements and power loss increase when the load on the system is greater than the natural load. If the resistance and shunt conductance are neglected and the load is equal to the natural load the receiving-end voltage will be equal to the sending-end voltage.

Typical surge impedance values for different lines are given as follows :

> For single-conductor line $Z_o = 400$ Ω/phase
> For 2-conductor bundle line $Z_o = 320$ Ω/phase
> For 3-conductor bundle line $Z_o = 280$ Ω/phase
> For 4-conductor bundle line $Z_o = 260$ Ω/phase

It is seen that multiple conductor lines have lower surge impedance than single-conductor lines. Consequently, the natural load of multiple conductor lines is greater than that of single-conductor lines. Surge impedance values for cables are about one-tenth of those for overhead lines.

Surge impedance load (SIL) can be found with the help of the relation (10.10.6).

For 132 kV, single-conductor line, $Z_o = 400$ Ω, SIL $= \dfrac{(132)^2}{400} = 43.5$ MW

For 230 kV, 2-conductor bundle line, $Z_o = 320$ Ω, SIL $= \dfrac{(230)^2}{320} = 165$ MW

For 400 kV, 4-conductor bundle line, $Z_o = 260$ Ω, SIL $= \dfrac{(400)^2}{260} = 615$ MW

For 500 kV, SIL = 830 MW, for 700 kV, SIL = 1600 MW, and for 1100 kV, SIL = 5000 MW.

EHV lines can be loaded upto $P = k P_o$, where the factor k varies from 0.85 (for long lines) to 1.3 (for short lines).

Surge impedance loading is, in many respects, the ideal loading. Not only is reactive power production equal to reactive power consumption, but the voltage and current profiles are uniform along the line. The uniform (flat) voltage profile is especially desirable since voltage can be held near the maximum value. The voltages and currents are also in phase at every point along the line.

Long lines cannot be loaded much above the uncompensated surge impedance loading.

EXERCISES

1. In a three-phase line with 132 kV at the receiving end the following are the transmission constants :

   $A = D = 0.98 \underline{/3°}$, $B = 110 \underline{/75°}$ Ω, $C = 0.0005 \underline{/88°}$ S.

   If the load at the receiving end is 50 MVA at 0.8 lagging power factor, determine the value of the sending-end voltage.

2. If, for a given three-phase line, $A = D = 0.95 \underline{/3°}$, $B = 100 \underline{/60°}$ Ω and $C = 0.001 \underline{/91°}$ S, calculate the current and voltage at the sending end of the line when the power delivered at the receiving end is 100 MVA at 132 kV and a lagging power factor of 0.8. (Assume capacitance concentrated at the centre of the line).   (L.U.)

3. The following data refer to 220 kV, 200 km long overhead line:

   Resistance per km per phase = 0.18 Ω

   Inductive reactance per km per phase = 0.68 Ω

   Shunt leakage susceptance per phase to neutral per km = 4.4×10^{-6} S

   Shunt leakage conductance = 0.

   Calculate the general network constants, and the surge impedance.

4. A three-phase transmission line has the following line parameters :

   $A = D = 0.96 \underline{/2°}$, $B = 55 \underline{/65°}$ Ω/phase, $C = 0.00005 \underline{/80°}$ S/phase.

   Determine the sending-end voltage and power factor when the line supplies a load of 45 MW at 132 kV and 0.8 power factor lagging.

5. What is Ferranti effect? Deduce an expression for the voltage rise of an unloaded line.

ANSWERS

1. 165 kV   2. 350 A ; 198 kV

3. $A = D = 0.941 \underline{/0.95°}$, $B = 137.9 \underline{/75.47°}$ Ω, $C = 0.8627 \times 10^{-3} \underline{/90.3°}$ S,

   $Z_0 = 400 \underline{/-7.5°}$ Ω

4. 148 kV ; 0.84 (lag)

General Network Constants

11.1 INTRODUCTION

A network having two input and two output terminals is known as a *two-port network*. It may also be called a *two-terminal-pair network*. In Fig. 11.1, *a*, *b* represent the input pair terminals and *c*, *d* the output pair terminals. The two pairs of terminals are usually shown to be enclosed in a box. A circuit consisting of any arrangement of its components is connected to these terminals.

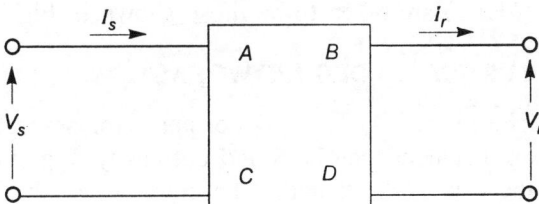

Fig. 11.1. Two-port network.

An electric network containing one or more sources of energy is called an *active network* while a network consisting only of passive components, such as resistances, inductances and capacitances and no source of electrical energy is termed *passive network*. A network is said to be *linear* if the impedances of its elements are independent of the amount of current passing through them or voltage across them. A *bilateral network* passes current equally in both directions.

On a phase-neutral basis a symmetrically loaded transmission line may be considered to be a passive, linear and bilateral two-terminal pair network. One pair of terminals constitutes the sending end while the other pair represents the receiving end. The two pairs of input and output terminals are widely separated.

The sinusoidal voltages and currents in steady-state at the input and output terminals of a two-port network are expressed in the form of the general equations given by

$$\left.\begin{array}{l} V_s = AV_r + BI_r \\ I_s = CV_r + DI_r \end{array}\right\} \qquad\qquad\qquad ...(11.1.1)$$

where, A, B, C, D are called the *general network constants* of the system. These constants are known by other names like *transmission parameters, chain parameters* and *auxiliary network constants*. In the above equations V_s, I_s are respectively the voltage and current at the sending or input end, and V_r, I_r represent the voltage and current respectively at the receiving or output end. The constants A, B, C, D are all complex numbers. The constants A and D are dimensionless, the constants B has the dimension of an impedance (ohms) and the constant C has the dimension of an admittance (siemens).

Eq. (11.1.1) can be put in the matrix form as :

$$\begin{bmatrix} V_s \\ I_s \end{bmatrix} = \begin{bmatrix} A & B \\ C & D \end{bmatrix} \begin{bmatrix} V_r \\ I_r \end{bmatrix} \qquad \qquad ...(11.1.2)$$

The matrix $\begin{bmatrix} A & B \\ C & D \end{bmatrix}$ is called the *transfer matrix* or *transmission matrix* of the network.

The relations of a two-port network expressed in terms of general network constants A, B, C, D are particularly useful in the analysis of line networks. They are utilized to draw circle diagrams. The overall constants of a combination of networks can be found easily provided that the A, B, C, D constants of the various networks are known. This makes their use very significant in the power system analysis. These constants are indicated in the two-terminal-pair network box as shown in Fig. 11.1. The directions of the currents and polarities of voltages in Eq. (11.1.1) are taken to be those shown in Fig. 11.1.

11.2 CASCADED NETWORKS

The overall A, B, C, D constants for several 2-port networks connected in cascade (or chain arrangement) can be found out easily. Fig. 11.2 shows two cascaded networks, and one that is the equivalent of both. The constants of the two component networks are A_1, B_1, C_1, D_1 and A_2, B_2, C_2, D_2. Let the constants for the equivalent network be A_0, B_0, C_0, D_0.

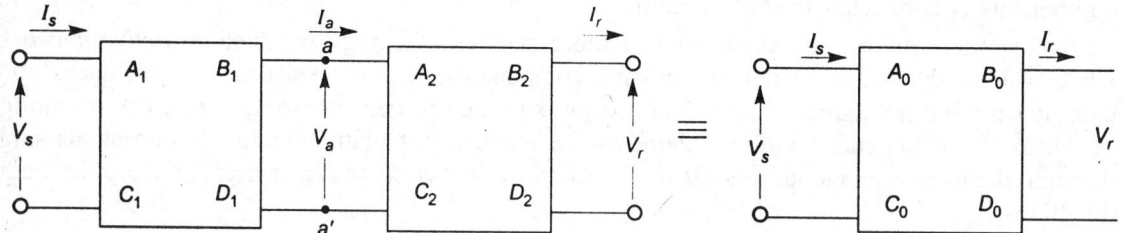

Fig. 11.2. The cascaded networks and their equivalents.

Let V_a and I_a be the voltage and current respectively at the junction a of the two networks,

$$\left. \begin{array}{l} V_a = A_2 V_r + B_2 I_r \\ I_a = C_2 V_r + D_2 I_r \end{array} \right\} \qquad \qquad ...(11.2.1)$$

For the network 1,

$$\left. \begin{array}{l} V_s = A_1 V_a + B_1 I_a \\ I_s = C_1 V_a + D_1 I_a \end{array} \right\} \qquad \qquad ...(11.2.2)$$

Substituting the values of V_a and I_a from the first of equations in the second set, we have

$$V_s = A_1 (A_2 V_r + B_2 I_r) + B_1 (C_2 V_r + D_2 I_r)$$

$$= (A_1 A_2 + B_1 C_2) V_r + (A_1 B_2 + B_1 D_2) I_r \qquad \text{...(11.2.3)}$$

and $\quad I_s = C_1 (A_2 V_r + B_2 I_r) + D_1 (C_2 V_r + D_2 I_r)$

$$= (C_1 A_2 + D_1 C_2) V_r + (C_1 B_2 + D_1 D_2) I_r \qquad \text{...(11.2.4)}$$

The sending-end voltage and current for the equivalent network with constants A_0, B_0, C_0, D_0 are given by

$$\left. \begin{array}{l} V_s = A_0 V_r + B_0 I_r \\ I_s = C_0 V_r + D_0 I_r \end{array} \right\} \qquad \text{...(11.2.5)}$$

Equating the constants of V_r and I_s the overall constants for the two networks in cascade are :

$$\left. \begin{array}{l} A_0 = A_1 A_2 + B_1 C_2 \\ B_0 = A_1 B_2 + B_1 D_2 \\ C_0 = C_1 A_2 + D_1 C_2 \\ D_0 = C_1 B_2 + D_1 D_2 \end{array} \right\} \qquad \text{...(11.2.6)}$$

Matrix method

For the first network

$$\begin{bmatrix} V_s \\ I_s \end{bmatrix} = \begin{bmatrix} A_1 & B_1 \\ C_1 & D_1 \end{bmatrix} \begin{bmatrix} V_a \\ I_a \end{bmatrix} \qquad \text{...(11.2.7)}$$

But V_a and I_a are the input voltage and current respectively of the second network, so that

$$\begin{bmatrix} V_a \\ I_a \end{bmatrix} = \begin{bmatrix} A_2 & B_2 \\ C_2 & D_2 \end{bmatrix} \begin{bmatrix} V_r \\ I_r \end{bmatrix} \qquad \text{...(11.2.8)}$$

Combining these equations

$$\begin{bmatrix} V_s \\ I_s \end{bmatrix} = \begin{bmatrix} A_1 & B_1 \\ C_1 & D_1 \end{bmatrix} \begin{bmatrix} A_2 & B_2 \\ C_2 & D_2 \end{bmatrix} \begin{bmatrix} V_r \\ I_r \end{bmatrix} \qquad \text{...(11.2.9)}$$

For the equivalent network

$$\begin{bmatrix} V_s \\ I_s \end{bmatrix} = \begin{bmatrix} A_0 & B_0 \\ C_0 & D_0 \end{bmatrix} \begin{bmatrix} V_r \\ I_r \end{bmatrix} \qquad \text{...(11.2.10)}$$

Comparing Eqs. (11.2.9) and (11.2.1) we get

$$\begin{bmatrix} A_0 & B_0 \\ C_0 & D_0 \end{bmatrix} = \begin{bmatrix} A_1 & B_1 \\ C_1 & D_1 \end{bmatrix} \begin{bmatrix} A_2 & B_2 \\ C_2 & D_2 \end{bmatrix} \qquad \text{...(11.2.11)}$$

Thus, the transmission matrix of a cascade two 2-port networks is equal to the product of the transmission matrices of the individual 2-port networks taken in order. It is this property which makes the transmission matrix so useful. This result may be generalized for any number of 2-port networks connected in cascade. The overall transmission matrix is equal to the matrix product of the transmission matrices of the individual networks, taken in order.

It should be noted clearly that in finding network overall constants the order of the networks should not be changed. If the networks are interchanged in position, that is, network 1 is at the receiving end and network 2 at the sending end the constants A_0, B_0, C_0, D_0 are obtained by interchanging the subscripts 1 and 2. For the network combinations shown in Fig. 11.3, the general constants may be written as :

$$\begin{bmatrix} A_0 & B_0 \\ C_0 & D_0 \end{bmatrix} = \begin{bmatrix} A_2 & B_2 \\ C_2 & D_2 \end{bmatrix} \begin{bmatrix} A_1 & B_1 \\ C_1 & D_1 \end{bmatrix} \qquad \text{...(11.2.12)}$$

$$\left. \begin{aligned} A_0 &= A_1A_2 + B_2C_1 \\ B_0 &= A_2B_1 + B_2D_1 \\ C_0 &= A_1C_2 + C_1D_2 \\ D_0 &= B_1C_2 + D_1D_2 \end{aligned} \right\} \qquad \text{...(11.2.13)}$$

The importance of the order of matrices of multiplication lies in the fact that $[A][B] \neq [B][A]$.

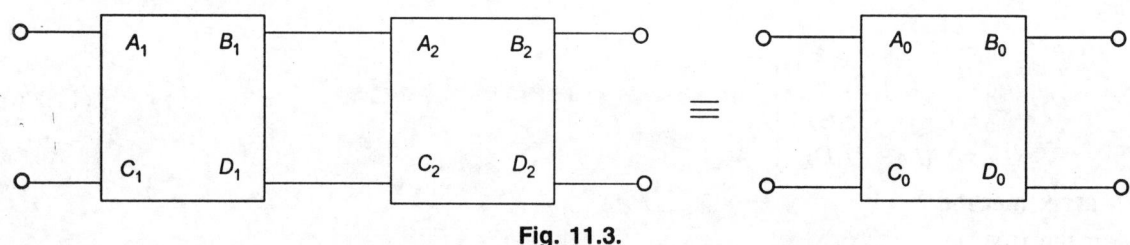

Fig. 11.3.

11.3 RELATIONS BETWEEN ABCD CONSTANTS

The relations between *ABCD* constants of a passive, linear and bilateral network can be found with the help of reciprocity theorem. First a voltage V is applied to the input terminals keeping the output terminals short-circuited [Fig. 11.4 (a)]. Since under short circuit $V_r = 0$, Eq. (11.1.1) gives

$$V = BI_{rs} \qquad \text{...(11.3.1)}$$

and $\qquad I_{ss} = DI_{rs} \qquad \text{...(11.3.2)}$

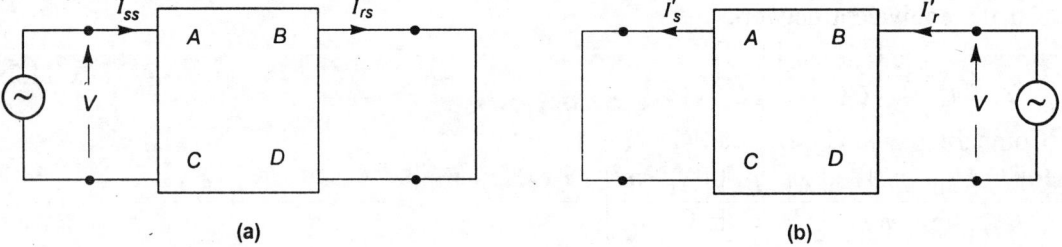

(a)   (b)

Fig. 11.4.

Now, the voltage V is applied to the output terminals and the input terminals are short circuited [Fig. 11.4 (b)]. The direction of flow of currents at the input and output terminals are reversed and the sending-end voltage V_s becomes zero. Equation (11.1.1) becomes

$$0 = AV - BI'_r$$

or $\qquad I'_r = \dfrac{AV}{B}$ $\qquad\qquad\qquad\qquad\qquad\qquad\qquad\qquad\qquad\qquad$ ...(11.3.3)

and $\qquad -I'_s = CV - DI'_r$ $\qquad\qquad\qquad\qquad\qquad\qquad\qquad\qquad\qquad$ ...(11.3.4)

Since the network is passive, by the reciprocity theorem

$$I'_s = I_{rs} \qquad\qquad\qquad\qquad\qquad\qquad ...(11.3.5)$$

Combining Eqs. (11.3.1), (11.3.3), (11.3.4) and (11.3.5) we get

$$-I_{rs} = CV - \frac{DAV}{B}$$

$$-\frac{V}{B} = CV - \frac{DAV}{B}$$

Dividing both the sides of the above equation by $-\dfrac{V}{B}$ we get

$$AD - BC = 1 \qquad\qquad\qquad\qquad\qquad\qquad ...(11.3.6)$$

Equation (11.3.6) is one of the required relations between the network constants. This relation may also be put in the determinant form as :

$$\begin{vmatrix} A & B \\ C & D \end{vmatrix} = 1$$

This relation permits finding the fourth constant if any three are known.

Another relation of importance is that $A = D$ for symmetrical networks. A network is symmetrical if it is electrically identical from either end. For a symmetrical network the input and output terminals may be interchanged without affecting the network behaviour.

If the network is supplied at the input terminals and short-circuited at the output terminals as in Fig. 11.4 (a), Eqs. (11.3.1) and (11.3.2) give input impedance seen from sending end.

$$Z_{st} = \frac{V}{I_{ss}} = \frac{B}{D} \qquad\qquad\qquad\qquad\qquad ...(11.3.7)$$

If the network is supplied at the output terminals and short-circuited at the input terminals, as in Fig. 11.4 (b), Eq. (11.3.3) gives input impedances seen from the receiving end.

$$Z_{rl} = \frac{V}{I'_r} = \frac{B}{A} \qquad\qquad\qquad\qquad\qquad ...(11.3.8)$$

But for a symmetrical network,

$$Z_{sl} = Z_{rl}$$

$$\frac{B}{D} = \frac{B}{A}$$

or $\qquad A = D$ $\qquad\qquad\qquad\qquad\qquad\qquad\qquad\qquad\qquad$ ...(11.3.9)

11.4 OUTPUT IN TERMS OF INPUT

The general equations giving sending-end voltage and current in terms of receiving-end voltage and current are

$$V_s = AV_r + BI_r$$

$$I_s = CV_r + DI_r$$

$\therefore \quad DV_s - BI_s = ADV_r + BDI_r - BCV_r - BDI_r = (AD - BC)\, V_r = V_r \text{ (since } AD - BC = 1)$

or $\quad V_r = DV_s - BI_s \hspace{6cm} \text{...(11.4.1)}$

Also, $\quad -CV_s + AI_s = -ACV_r - BCI_r + ACV_r + ADI_r = (AD - BC)\, I_r = I_r$

or $\quad I_r = -CV_s + AI_s \hspace{6cm} \text{...(11.4.2)}$

Equations (11.4.1) and (11.4.2) give output voltage and current in terms of input voltage and current and the A, B, C, D, constants. In matrix from they may be expressed as

$$\begin{bmatrix} V_r \\ I_r \end{bmatrix} = \begin{bmatrix} D & -B \\ -C & A \end{bmatrix} \begin{bmatrix} V_s \\ I_s \end{bmatrix} \hspace{5cm} \text{...(11.4.3)}$$

Matrix method

We have $\begin{bmatrix} V_s \\ I_s \end{bmatrix} = \begin{bmatrix} A & B \\ C & D \end{bmatrix} \begin{bmatrix} V_r \\ I_r \end{bmatrix}$

Premultiplying each side of the above relation with $\begin{bmatrix} A & B \\ C & D \end{bmatrix}^{-1}$

$$\begin{bmatrix} A & B \\ C & D \end{bmatrix}^{-1} \begin{bmatrix} V_s \\ I_s \end{bmatrix} = \begin{bmatrix} A & B \\ C & D \end{bmatrix}^{-1} \begin{bmatrix} A & B \\ C & D \end{bmatrix} \begin{bmatrix} V_r \\ I_r \end{bmatrix}$$

But $\quad \begin{bmatrix} A & B \\ C & D \end{bmatrix}^{-1} = \dfrac{1}{(AD - BC)} \begin{bmatrix} D & -B \\ -C & A \end{bmatrix} = \begin{bmatrix} D & -B \\ -C & A \end{bmatrix}$

Since $\quad AD - BC = 1$

$\therefore \quad \begin{bmatrix} D & -B \\ -C & A \end{bmatrix} \begin{bmatrix} V_s \\ I_s \end{bmatrix} = \begin{bmatrix} V_r \\ I_r \end{bmatrix}$

or $\quad \begin{bmatrix} V_r \\ I_r \end{bmatrix} = \begin{bmatrix} D & -B \\ -C & A \end{bmatrix} \begin{bmatrix} V_s \\ I_s \end{bmatrix}$

$\therefore \quad V_r = DV_s - BI_s$

$\quad I_r = -CV_s + AI_s$

11.5 TYPICAL TRANSMISSION NETWORKS

In many cases a complicated 2-port network may be found to be an assembly of simpler networks. For example, a transmission network is made up of cascaded series and shunt elements. We shall, therefore, determine the constants of several of the simple basic networks. The A, B, C, D constants for a complicated network can be found in terms of the constants of these basic component networks. For the given network, equations are written for the input voltage and current (V_s, I_s), and the output voltage and current (V_r, I_r) either by Kirchhoff's voltage and current laws (KVL and KCL) or by matrix method depending upon the convenience. The A, B, C, D constants are then determined by comparison of the equations so obtained with general Eq. (11.1.1) or Eq. (11.1.2).

11.5.1   Series Impedance Circuit

A circuit having series impedance Z is shown in Fig. 11.5. Such a case is found in a short transmission line where the line capacitance is negligible and the shunt admittance Y is zero. For a transformer with magnetizing current shown in Fig. 11.5 we may write :

$$\left.\begin{array}{l} V_s = V_r + ZI_r \\ I_s = I_r \end{array}\right\} \qquad \text{...(11.5.1)}$$

or $$\begin{bmatrix} V_s \\ I_s \end{bmatrix} = \begin{bmatrix} 1 & Z \\ 0 & 1 \end{bmatrix} \begin{bmatrix} V_r \\ I_r \end{bmatrix} \qquad \text{...(11.5.2)}$$

By comparing these equations with the general Eqs. (11.1.1) and (11.1.2) the general constants for the series impedance network can be written as :

$$\left.\begin{array}{ll} A = 1 & B = Z \\ C = 0 & D = 1 \end{array}\right\} \qquad \text{...(11.5.3)}$$

The transfer matrix for the network is $\begin{bmatrix} 1 & Z \\ 0 & 1 \end{bmatrix}$.

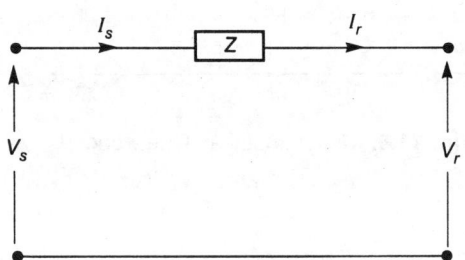

Fig. 11.5. Series impedance circuit.

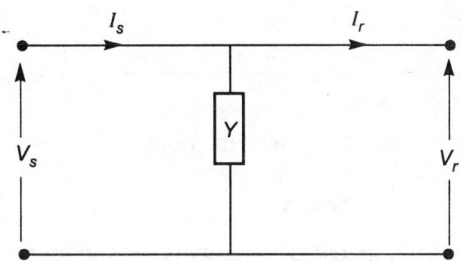

Fig. 11.6. Shunt admittance circuit.

11.5.2   Shunt Admittance Circuit

Fig. 11.6 shows a transmission network with a shunt admittance Y. Such a network may represent the magnetizing current circuit of a transformer or a shunt capacitor. For the network shown in Fig. 11.6 we may write :

$$\left.\begin{array}{l} V_s = V_r \\ I_s = YV_r + I_r \end{array}\right\} \qquad \text{...(11.5.4)}$$

or $$\begin{bmatrix} V_s \\ I_s \end{bmatrix} = \begin{bmatrix} 1 & 0 \\ Y & 1 \end{bmatrix} \begin{bmatrix} V_r \\ I_r \end{bmatrix} \qquad \text{...(11.5.5)}$$

Hence, $\quad A = 1, \quad B = 0, \quad C = Y, \quad D = 1$ ...(11.5.6)

11.5.3   Half T-Network

A half T-network is shown in Fig. 11.7.

$$I_s = V_r Y + I_r$$

$$V_s = V_r + I_s Z = V_r + (V_r Y + I_r) Z = (1 + ZY) V_r + ZI_r$$

or
$$\begin{bmatrix} V_s \\ I_s \end{bmatrix} = \begin{bmatrix} (1 + ZY) & Z \\ Y & 1 \end{bmatrix} \begin{bmatrix} V_r \\ I_r \end{bmatrix}$$ ...(11.5.7)

Hence, $A = 1 + ZY$, $B = Z$, $C = Y$, $D = 1$ ...(11.5.8)

Matrix method

The half T-network can be considered as the cascade connection of two sections. One section is a series impedance Z and the other a shunt admittance Y. The overall constants are obtained from the matrix product of the transfer matrices of each section in the correct order.

$$\begin{bmatrix} V_s \\ I_s \end{bmatrix} = \begin{bmatrix} 1 & Z \\ 0 & 1 \end{bmatrix} \begin{bmatrix} 1 & 0 \\ Y & 1 \end{bmatrix} \begin{bmatrix} V_r \\ I_r \end{bmatrix} = \begin{bmatrix} 1 + ZY & Z \\ Y & 1 \end{bmatrix} \begin{bmatrix} V_r \\ I_r \end{bmatrix}$$

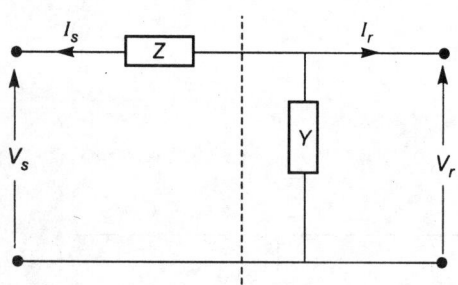
Fig. 11.7. Half T-network.

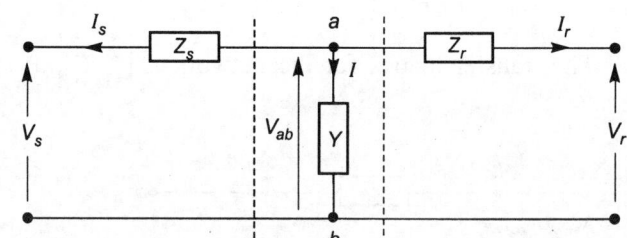
Fig. 11.8. Unsymmetrical T-network.

11.5.4 T-Network

An unsymmetrical T-network is shown in Fig. 11.8.

From the network,

$$V_{ab} = V_r + Z_r I_r$$

$$I = Y V_{ab} = Y (V_r + Z_r I_r)$$

$$I_s = I + I_r = Y (V_r + Z_r I_r) + I_r = Y V_r + (1 + Z_r Y) I_r$$ ...(11.5.9)

$$V_s = V_{ab} + Z_s I_s = (V_r + Z_r I) + Z_s (Y V_r + I_r + Z_r Y I_r)$$

$$= (1 + Z_s Y) V_r + (Z_s + Z_r + Z_s Z_r Y) I_s$$ ...(11.5.10)

or
$$\begin{bmatrix} V_s \\ I_s \end{bmatrix} = \begin{bmatrix} (1 + Z_s Y) & (Z_s + Z_r + Z_s Z_r Y) \\ Y & 1 + Z_r Y \end{bmatrix} \begin{bmatrix} V_r \\ I_r \end{bmatrix}$$ ...(11.5.11)

Hence, $A = 1 + Z_s Y$, $B = Z_s + Z_r + Z_s Z_r Y$, $C = Y$, $D = I + Z_r Y$ ...(11.5.12)

Matrix method

The T-network can be regarded as the cascade connection of three sections. The overall constants are obtained by the matrix product of transmission matrices of each of the sections in correct order.

$$\begin{bmatrix} V_s \\ I_s \end{bmatrix} = \begin{bmatrix} 1 & Z_s \\ 0 & 1 \end{bmatrix} \begin{bmatrix} 1 & 0 \\ Y & 1 \end{bmatrix} \begin{bmatrix} 1 & Z_r \\ 0 & 1 \end{bmatrix} \begin{bmatrix} V_r \\ I_r \end{bmatrix} = \begin{bmatrix} (1 + Z_s Y) & Z_s \\ Y & 1 \end{bmatrix} \begin{bmatrix} 1 & Z_r \\ 0 & 1 \end{bmatrix} \begin{bmatrix} V_r \\ I_r \end{bmatrix}$$

$$= \begin{bmatrix} (1 + Z_s\, Y) & (Z_s + Z_r + Z_s\, Z_r\, Y) \\ Y & 1 + Z_r\, Y \end{bmatrix} \begin{bmatrix} V_r \\ I_r \end{bmatrix}$$

For a symmetrical T-network (Fig. 11.9)

$$Z_s = Z_r = \frac{1}{2} Z_T \text{ (say)}$$

so that the series impedance of the network is $Z_s + Z_r = Z_T$. Also assume that $Y = Y_T$. The general network constants for a symmetrical T-network are

$$\left.\begin{array}{l} A = 1 + \dfrac{1}{2} Z_T\, Y_T, \quad B = Z_T\left(1 + \dfrac{1}{4} Z_T\, Y_T\right) \\[2mm] C = Y_T, \quad D = 1 + \dfrac{1}{2} Z_T\, Y_T \end{array}\right\} \qquad \text{...(11.5.13)}$$

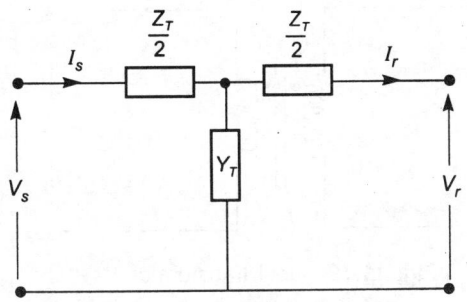

Fig. 11.9. Symmetrical T-circuit.

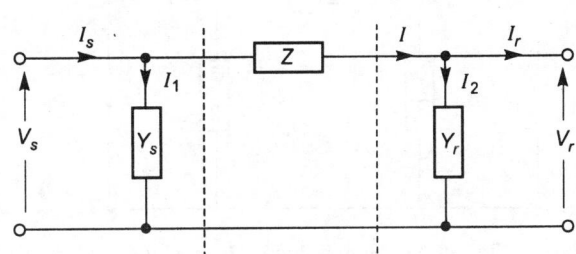

Fig. 11.10. Unsymmetrical π-network.

11.5.5 Π-Network

Fig. 11.10 shows an unsymmetrical π-network.

$$I_2 = Y_r\, V_r \; ; \; I = I_r + I_2 = I_r + Y_r\, V_r$$

$$V_s = V_r + IZ = V_r + (I_r + Y_r\, V_r)\, Z = (1 + ZY_r)\, V_r + ZI_r \qquad \text{...(11.5.14)}$$

$$I_s = I + I_1 = I_r + Y_r\, V_r + V_s\, Y_s = I_r + Y_r\, V_r + (V_r + ZY_r\, V_r + ZI_r)\, Y_s$$

$$\quad = (Y_s + Y_r + ZY_s\, Y_r)\, V_r + (1 + ZY_s)\, I_r \qquad \text{...(11.5.15)}$$

$$\begin{bmatrix} V_s \\ I_s \end{bmatrix} = \begin{bmatrix} (1 + ZY_r) & Z \\ (Y_s + Y_r + ZY_s\, Y_r) & (1 + ZY_s) \end{bmatrix} \begin{bmatrix} V_r \\ I_r \end{bmatrix} \qquad \text{...(11.5.16)}$$

$$\therefore \quad \left.\begin{array}{l} A = 1 + ZY_r, \quad B = Z \\ C = Y_s + Y_r + ZY_s\, Y_r, \quad D = 1 + ZY_s \end{array}\right\} \qquad \text{...(11.5.17)}$$

Matrix method

The Π-network is made up of three cascaded single-element sections, and hence the overall constants may be obtained by the matrix product of the transmission matrices of each section in correct order.

$$\begin{bmatrix} V_s \\ I_s \end{bmatrix} = \begin{bmatrix} 1 & 0 \\ Y_s & 1 \end{bmatrix} \begin{bmatrix} 1 & Z \\ 0 & 1 \end{bmatrix} \begin{bmatrix} 1 & 0 \\ Y_r & 1 \end{bmatrix} \begin{bmatrix} V_r \\ I_r \end{bmatrix}$$

$$= \begin{bmatrix} 1 & Z \\ Y_s & (1+ZY_s) \end{bmatrix} \begin{bmatrix} 1 & 0 \\ Y_r & 1 \end{bmatrix} \begin{bmatrix} V_r \\ I_s \end{bmatrix} = \begin{bmatrix} (1+ZY_r) & Z \\ (Y_s+Y_r+ZY_s\,Y_r) & (1+ZY_s) \end{bmatrix} \begin{bmatrix} V_r \\ I_r \end{bmatrix}$$

For a symmetrical Π-network (Fig. 11.11)

$$Y_s = Y_r = \frac{1}{2} Y_\pi \text{ (say) and } Z = Z_\pi$$

The general constants for a symmetrical Π network are

$$\left. \begin{aligned} & A = 1 + \frac{1}{2} Z_\pi\, Y_\pi, \quad B = Z_\pi \\ & C = Y_\pi \left(1 + \frac{1}{4} Z_\pi\, Y_\pi\right), \quad D = 1 + \frac{1}{2} Z_\pi\, Y_\pi \end{aligned} \right\} \qquad \text{...(11.5.18)}$$

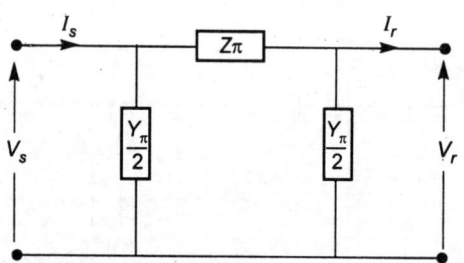

Fig. 11.11. Symmetrical Π-network.

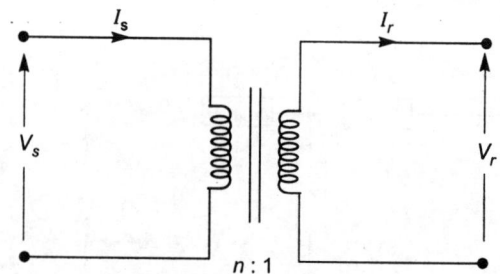

Fig. 11.12. Ideal transformer.

11.5.6 Ideal Transformer

Fig. 11.12 shows an ideal transformer which is assumed to be loss free. Let the turn ratio N_p/N_s be n.

$$V_s = n\, V_r \qquad \text{...(11.5.19)}$$

$$I_s = \frac{1}{n}\, I_r \qquad \text{...(11.5.20)}$$

$$\begin{bmatrix} V_s \\ I_s \end{bmatrix} = \begin{bmatrix} n & 0 \\ 0 & \dfrac{1}{n} \end{bmatrix} \begin{bmatrix} V_r \\ I_r \end{bmatrix} \qquad \text{...(11.5.21)}$$

$$\therefore \qquad A = n, \quad B = 0, \quad C = 0, \quad D = \frac{1}{n} \qquad \text{...(11.5.22)}$$

11.5.7 Actual Transformer

An actual transformer (Fig. 11.13) may be represented by a series impedance Z, a shunt admittance Y (referred to the primary) and an ideal transformer. The three sections are connected in cascade. The overall constants are obtained by the product of the transmission matrices of each section taken in order.

$$\begin{bmatrix} V_s \\ I_s \end{bmatrix} = \begin{bmatrix} 1 & Z \\ 0 & 1 \end{bmatrix} \begin{bmatrix} 1 & 0 \\ Y & 1 \end{bmatrix} \begin{bmatrix} n & 0 \\ 0 & \dfrac{1}{n} \end{bmatrix} \begin{bmatrix} V_r \\ I_r \end{bmatrix} = \begin{bmatrix} (1+ZY) & Z \\ Z & 1 \end{bmatrix} \begin{bmatrix} n & 0 \\ 0 & \dfrac{1}{n} \end{bmatrix} \begin{bmatrix} V_r \\ I_r \end{bmatrix}$$

$$= \begin{bmatrix} n\,(1+ZY) & \dfrac{Z}{n} \\[2mm] n\,Y & \dfrac{1}{n} \end{bmatrix} \begin{bmatrix} V_r \\ I_r \end{bmatrix} \qquad \qquad \dots(11.5.23)$$

$\therefore \qquad A = n\,(1+ZY), \quad B = \dfrac{1}{n}\,Z, \quad C = n\,Y, \quad D = \dfrac{1}{n} \qquad \qquad \dots(11.5.24)$

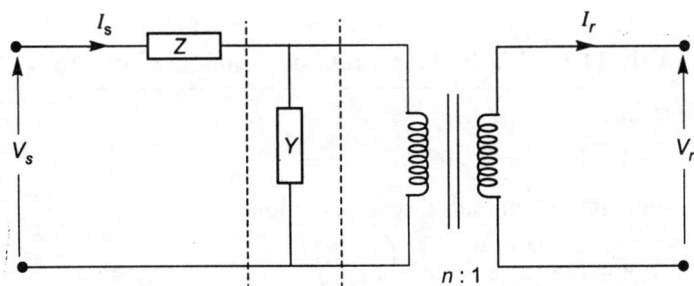

Fig. 11.13. Actual transformer.

11.5.8 Transmission Line

The general network constants of a transmission line with distributed parameters are derived in Section 10.7, and are given below :

$$\left. \begin{aligned} A &= \cosh \gamma S \\ B &= Z_0 \sinh \gamma S \\ C &= \dfrac{1}{Z_0} \sinh \gamma S \\ D &= \cosh \gamma S \end{aligned} \right\} \qquad \qquad \dots(11.5.25)$$

These constants may also be put in terms of Z and Y as follows :

$$\gamma S = (z \cdot y)^{1/2}\, S = (zSyS)^{1/2} = (ZY)^{1/2}$$

$$Z_0 = \left(\frac{Z}{Y}\right)^{1/2} = \left(\frac{Z \cdot Z}{Z \cdot Y}\right)^{1/2} = \frac{Z}{(ZY)^{1/2}}$$

$$\frac{1}{Z_0} = \left(\frac{Y}{Z}\right)^{1/2} = \left(\frac{Y \cdot Y}{Z \cdot Y}\right)^{1/2} = \frac{Y}{(ZY)^{1/2}}$$

$$\left. \begin{aligned} A &= D = \cosh (ZY)^{1/2} \\ B &= Z_0 \sinh (ZY)^{1/2} = Z\,\frac{\sinh (ZY)^{1/2}}{(ZY)^{1/2}} \\ C &= \frac{1}{Z_0} \sinh (ZY)^{1/2} = Y\,\frac{\sinh (ZY)^{1/2}}{(ZY)^{1/2}} \end{aligned} \right\} \qquad \dots(11.5.26)$$

$\therefore$

The A, B, C, D constants of a transmission line can also be expressed in the form of convergent series.

$$A = D = \cosh (ZY)^{1/2} = \left[1 + \frac{1}{2} (ZY) + \frac{1}{24} (ZY)^2 + \frac{1}{720} (ZY)^3 + ... \right] \quad\quad ...(11.5.27)$$

$$B = Z \left[1 + \frac{1}{6} (ZY) + \frac{1}{120} (ZY)^2 + \frac{1}{5040} (ZY)^3 + ... \right] \quad\quad ...(11.5.28)$$

$$C = Y \left[1 + \frac{1}{6} (ZY) + \frac{1}{120} (ZY)^2 + \frac{1}{5040} (ZY)^3 + ... \right] \quad\quad ...(11.5.29)$$

The accuracy of the result is not affected much if the terms other than first degree within the brackets are omitted in the calculation work. Table 11.1 shows the $ABCD$ constants of transmission lines.

Table 11.1   ABCD Constants of Transmission Lines

1. Short line

$$A = 1, \quad B = Z, \quad C = 0, \quad D = A$$

2. Medium line (nominal T representation)

$$A = 1 + \frac{ZY}{2}, \quad B = Z\left(1 + \frac{ZY}{4}\right), \quad C = Y, \quad D = A$$

3. Medium line (nominal π representation)

$$A = 1 + \frac{ZY}{2}, \quad B = Z, \quad C = Y\left(1 + \frac{ZY}{4}\right), \quad D = A$$

4. Long line

$$A = \cosh \gamma S = \cosh \sqrt{ZY} = 1 + \frac{ZY}{2} + \frac{(ZY)^2}{4!} + ...$$

$$B = Z_0 \sinh \gamma S = Z \frac{\sinh \sqrt{ZY}}{\sqrt{ZY}} = Z\left[1 + \frac{ZY}{3!} + \frac{(ZY)^2}{5!} + ...\right]$$

$$C = \frac{1}{Z_0} \sinh \gamma S = Y \frac{\sinh\sqrt{ZY}}{\sqrt{ZY}} = Y\left[1 + \frac{ZY}{3!} + \frac{(ZY)^2}{5!} + ...\right]$$

$$D = A$$

11.5.9   Transmission Line with Transformers at Both Ends

Fig. 11.14 shows a transmission line with transformers at each of its ends.

The following equations can be written by inspection :

$$V_2 = V_r + Z_2 I_3 ; \quad I_3 = I_r + I_4 = I_r + V_r Y_2$$

$$V_1 = AV_2 + BI_3 ; \quad I_2 = CV_2 + DI_3$$

$$V_s = V_1 + Z_1 I_2 ; \quad I_s = I_1 + I_2 = V_s Y_1 + I_2$$

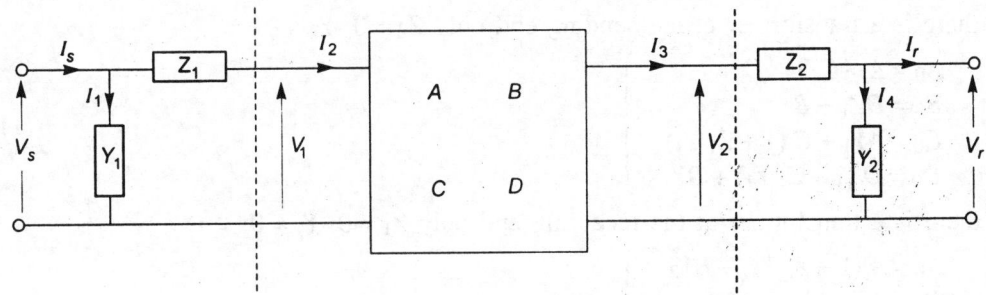

Fig. 11.14. Transmission line with end transformers.

Now $I_2 = C (V_r + Z_2 I_3) + DI_3 = CV_r + (CZ_2 + D) I_3$

$\quad = CV_r + (CZ_2 + D) (I_r + V_r Y_2) = [C (1 + Z_2 Y_2) + DY_2] V_r + (CZ_2 + D) I_r$

$V_1 = A (V_r + Z_2 I_3) + BI_3 = AV_r + (AZ_2 + B) I_3$

$\quad = [AV_r + (AZ_2 + B) (I_r + V_r Y_2) = [A (1 + Z_2 Y_2) + BY_2] V_r + (AZ_2 + B) I_r$

$V_s = V_1 + Z_1 I_2$

$\quad = [A (1 + Z_2 Y_2) + BY_2] V_r + (AZ_2 + B) I_r + Z_1 [C (1 + Z_2 Y_2) + DY_2] V_r + Z_1 (CZ_2 + D) I_r$

$\quad = [A (1 + Z_2 Y_2) + BY_2 + CZ_1 (1 + Z_2 Y_2) + DZ_1 Y_2] V_r + (AZ_2 + B + CZ_1 Z_2 + DZ_1) I_r$

$$\dots(11.5.30)$$

$I_s = V_s Y_1 + I_2$

$\quad = [AY_1 (1 + Z_2 Y_2) + (BY_1 Y_2 CZ_1 Y_1 (1 + Z_2 Y_2) + DZ_1 Y_1 Y_2] V_r$

$\quad\quad + (AZ_2 Y_1 + BY_1 + CZ_1 Z_2 Y_1 + DZ_1 Y_1) I_r + [C (1 + Z_2 Y_2) + DY_2] V_r + (CZ_2 + D) I_r$

$\quad = [AY_1 (1 + Z_2 Y_2) + BY_1 Y_2 + C (1 + Z_1 Y_1) (1 + Z_2 Y_2) + DY_2 (1 + Z_1 Y_1)] V_r$

$\quad\quad + [AZ_2 Y_1 + BY_1 + CZ_2 (1 + Z_1 Y_1) + D (1 + Z_1 Y_1)] I_r \quad\dots(11.5.31)$

If the general constants for the whole are A_0, B_0, C_0, D_0, we can write

$$V_r = A_0 V_r + B_0 I_r \qquad\qquad\qquad\qquad\qquad\dots(11.5.32)$$

$$I_s = C_0 V_r + D_0 I_r \qquad\qquad\qquad\qquad\qquad\dots(11.5.33)$$

Comparison of Eq. (11.5.30) with Eq. (11.5.31) and Eq. (11.5.31) with Eq. (11.5.33) gives

$$\left.\begin{array}{l} A_0 = A (1 + Z_2 Y_2) + BY_2 + CZ_1 (1 + Z_2 Y_2) + DZ_1 Y_2 \\ B_0 = AZ_2 + B + CZ_1 Z_2 + DZ_1 \\ C_0 = AY_1 (1 + Z_2 Y_2) + BY_1 Y_2 + C (1 + Z_1 Y_1) (1 + Z_2 Y_2 + DY_2 (1 + Z_1 Y_1) \\ D_0 = AZ_2 Y_1 + BY_1 + CZ_2 (1 + Z_1 Y_1) + D (1 + Z_1 Y_1) \end{array}\right\} \dots(11.5.34)$$

Special cases

(a) If the line is symmetrical $A = D$. Also, if the terminal transformers are identical $Z_1 = Z_2 = Z$ (say) and $Y_1 = Y_2 = Y$ (say), the general constant are

$$\left.\begin{array}{l} A_0 = D_0 = (1 + 2YZ) + BY + CZ (1 + ZY) \\ B_0 = 2AZ + B + CZ^2 \\ C_0 = 2AY (1 + ZY) + BY^2 + C (1 + ZY)^2 \end{array}\right\} \dots(11.5.35)$$

(b) If there is a transformer at the sending end only $Z_2 = 0$, $Y_2 = 0$

$$\left.\begin{aligned}
A_0 &= A + CZ_1 \\
B_0 &= DZ_1 + B \\
C_0 &= AY_1 + C(1 + Z_1 Y_1) \\
D_0 &= D(1 + Z_1 Y_1) + BY_1
\end{aligned}\right\} \qquad \text{...(11.5.36)}$$

(c) If there is a transformer at the receiving end only $Z_1 = 0$, $Y_1 = 0$

$$\left.\begin{aligned}
A_0 &= A(1 + Z_2 Y_2) + BY_2 \\
B_0 &= AZ_2 + B \\
C_0 &= DY_2 + C(1 + Z_2 Y_2) \\
D_0 &= D + CZ_2
\end{aligned}\right\} \qquad \text{...(11.5.37)}$$

11.6 NETWORKS IN PARALLEL

Two networks connected in parallel are shown in Fig. 11.15. A new network may be determined which will be equivalent in performance to the two networks in parallel.

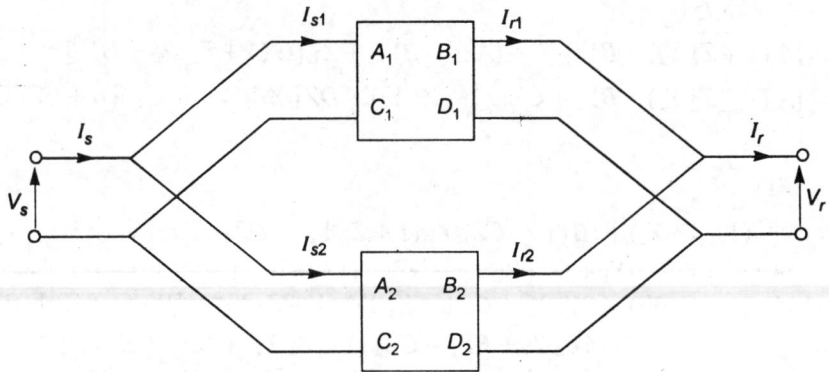

Fig. 11.15. Two networks in parallel.

Let I_{s1}, I_{s2} = sending–end currents in the networks 1 and 2

I_{r1}, I_{r2} = receiving–end currents in the networks 1 and 2.

For the network 1,

$$V_s = A_1 V_r + B_1 I_{r1} \qquad \text{...(11.6.1)}$$

$$I_{s1} = C_1 V_r + D_1 I_{r1} \qquad \text{...(11.6.2)}$$

For the network 2,

$$V_s = A_2 V_r + B_2 I_{r2} \qquad \text{...(11.6.3)}$$

$$I_{s2} = C_2 V_r + D_2 I_{r2} \qquad \text{...(11.6.4)}$$

Also, $\quad I_r = I_{r1} + I_{r2} \qquad \text{...(11.6.5)}$

and $\quad I_s = I_{s1} + I_{s2} \qquad \text{...(11.6.6)}$

Multiplying Eq. (11.6.1) by B_2 and Eq. (11.6.3) by B_1 and adding, we get

$$V_s = (B_1 + B_2) = (A_1 B_2 + A_2 B_1) V_r + B_1 B_2 (I_{r1} + I_{r2})$$

$$\therefore \qquad V_s = \frac{A_1 B_2 + A_2 B_1}{B_1 + B_2} V_r + \frac{B_1 B_2}{B_1 + B_2} I_r \qquad \qquad ...(11.6.7)$$

Adding Eqs. (11.6.2) and (11.6.4) and using Eqs. (11.6.5) and (11.6.8) :

$$I_{s1} + I_{s2} = (C_1 + C_2) V_r + D_1 I_{r1} + D_2 I_{r2}$$

$$I_s = (C_1 + C_2) V_r + D_1 (I_r - I_{r2}) + D_2 I_{r2} = (C_1 + C_2) V_r + (D_2 - D_1) I_{r2} + D_1 I_r$$

Substituting the value of I_{r2} from Eq. (11.6.3), we have

$$I_s = (C_1 + C_2) V_r + (D_2 - D_1) \left(\frac{V_s - A_2 V_r}{B_2} \right) + D_1 I_r$$

$$= (C_1 + C_2) V_r + \frac{D_2 - D_1}{B_2} V_r - \frac{A_2 (D_2 - D_1)}{B_2} V_r + D_1 I_r$$

Putting the value of V_s from Eq. (11.6.7) we obtain

$$I_s = (C_1 + C_2) V_r + \frac{D_2 - D_1}{B_2} \left[\frac{A_1 B_2 + A_2 B_1}{B_1 + B_2} V_r + \frac{B_1 B_2}{B_1 + B_2} I_r \right] - \frac{A_2 (D_2 - D_1)}{B_2} V_r + D_1 I_r$$

$$= \left[C_1 + C_2 + \frac{D_2 - D_1}{B_2} \left(\frac{A_1 B_2 + A_2 B_1}{B_1 + B_2} - A_2 \right) \right] V_r + \left(\frac{D_2 - D_1}{B_2} \cdot \frac{B_1 B_2}{B_1 + B_2} + D_1 \right) I_r$$

$$= \left[C_1 + C_2 \frac{(A_1 - A_2)(D_2 - D_1)}{B_1 + B_2} \right] V_r + \left(\frac{B_1 D_2 + D_1 B_2}{B_1 + B_2} \right) I_r \qquad ...(11.6.8)$$

Therefore, from Eqs. (11.6.7) and (11.6.8) the general circuit constants of a network equivalent to two networks in parallel are

$$\left. \begin{array}{l} A_0 = \dfrac{A_1 B_2 + B_1 A_2}{B_1 + B_2} \\[2mm] B_0 = \dfrac{B_1 B_2}{B_1 + B_2} \\[2mm] C_0 = C_1 + C_2 + \dfrac{(A_1 - A_1)(D_1 - D_2)}{B_1 + B_2} \\[2mm] D_0 = \dfrac{B_1 D_2 + B_2 D_1}{B_1 + B_2} \end{array} \right\} \qquad \qquad ...(11.6.9)$$

For two identical networks operating in parallel, and each having constants $ABCD$,

$$A_0 = A \; ; \quad B_0 = \frac{B}{2} \; ; \quad C_0 = 2C \; ; \quad D_0 = D \qquad \qquad ...(11.6.10)$$

Similarly, for n identical networks in parallel

$$A_0 = A \; ; \quad B_0 = \frac{B}{n} \; ; \quad C_0 = n\,C \; ; \quad D_0 = D \qquad \qquad ...(11.6.11)$$

11.7 EQUIVALENT NETWORKS

A power system includes generators, transformers and loads in addition to transmission line. To study the behaviour of the power system as a whole it is convenient to replace the system by an equivalent network. In the steady-state, the equivalent network of a power line will represent the line exactly at one definite frequency only at which the elements of the equivalent network are calculated. A network is said to be equivalent to a system if the terminal voltages and the currents of the circuit bear the same relation to the terminal voltages and currents of the system. Thus, an equivalent network of a line is a circuit in which voltage and current at each end are given by the equations of the forms similar to those of transmission line. The constants, *ABCD*, of the network will also be of the same magnitude as those of transmission line. Although the equivalent network at its terminals has the same behaviour as at the sending and receiving ends of the actual line, it gives no information regarding the conditions along the line.

An equivalent network simplifies the calculations. The utility of equivalent networks is particularly significant to the systems containing transmission lines with terminal transformers, the lines having different *ABCD* constants running in parallel, and the artificial model networks of the long lines built to study their behaviour, or other complicated transmission systems.

A transmission line has parameters distributed throughout its length. Like other two-port networks, it can be represented by an equivalent *T* or π network. In an equivalent *T* or π network these parameters are lumped in three branches.

11.8 EQUIVALENT T-NETWORK OF A LINE

The *T* equivalent network of a transmission line consists of three elements $\frac{1}{2}Z_T$, $\frac{1}{2}Z_T$, and Y_T connected to form *T* (Fig. 11.16). Z_T and Y_T represent the total series impedance and shunt admittance of the line respectively.

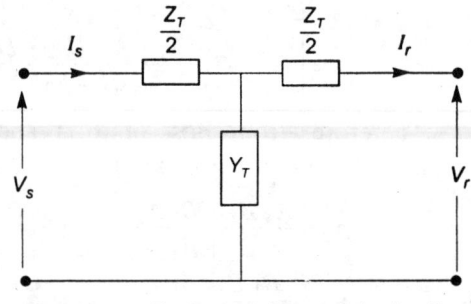

Fig. 11.16. Equivalent *T*-network of a line.

The *ABCD* constants of a symmetrical *T* network have already been derived in Section 11.5.4 as

$$A = D = 1 + \frac{1}{2}Z_T Y_T \; ; \quad B = Z_T\left(1 + \frac{1}{4}Z_T Y_T\right) \; ;$$

$$C = Y_T$$

From the known values of the *ABCD* constants of the line, the values of the elements $\frac{1}{2}Z_T$ and Y_T of its equivalent circuit are given from the above relations as :

$$\frac{1}{2}Z_T = \frac{A-1}{C} = \frac{\cosh \gamma S - 1}{\frac{1}{Z_0}\sinh \gamma S} = Z_0 \frac{2\sinh^2 \frac{\gamma S}{2}}{2\sinh \frac{\gamma S}{2} \cdot \cosh \frac{\gamma S}{2}} = Z_0 \tanh \frac{\gamma S}{2}$$

$$Z_0 = \left(\frac{z}{y}\right)^{1/2} = \left(\frac{z \cdot z}{z \cdot y}\right)^{1/2} \frac{S}{S} = \frac{z\,S}{\gamma S} = \frac{Z}{\gamma S} = \frac{\frac{Z}{2}}{\frac{\gamma S}{2}}$$

$$\therefore \qquad \frac{1}{2} Z_T = \frac{Z}{2} \cdot \frac{\tanh \dfrac{\gamma S}{2}}{\dfrac{\gamma S}{2}} \qquad \qquad \dots(11.8.1)$$

and

$$Y_T = C = \frac{1}{Z_0} \sinh \gamma S$$

$$\frac{1}{Z_0} = \left(\frac{y}{z}\right)^{1/2} = \left(\frac{y \cdot y}{z \cdot y}\right)\frac{S}{S} = \frac{y \, S}{\gamma S} = \frac{Y}{\gamma S}$$

$$\therefore \qquad Y_T = Y \frac{\sinh \gamma S}{\gamma S} \qquad \qquad \dots(11.8.2)$$

11.9 EQUIVALENT Π NETWORKS OF A LINE

The equivalent π network of a line consists of a series element having an impedance Z_π equal to the total impedance of the line and two shunt elements each having an admittance equal to half of the total shunt admittance of the line. Fig. 11.17 shows the equivalent π network of a line.

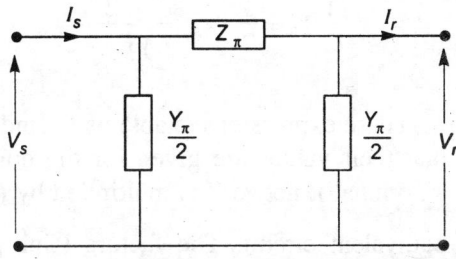

Fig. 11.17. Equivalent Π-network of a line.

The constants may be written from Eq. (11.5.17) as

$$A = D = 1 + \frac{1}{2} Z_\pi Y_\pi \; ; \quad B = Z_\pi \; ;$$

$$C = Y_\pi \left(1 + \frac{1}{4} Z_\pi Y_\pi\right)$$

The elements of the equivalent π network of the line are, therefore, given by

$$Z_\pi = B = Z_0 \sinh \gamma S = Z \frac{\sinh \gamma S}{\gamma S} \qquad \qquad \dots(11.9.1)$$

and

$$\frac{1}{2} Y_\pi = \frac{A-1}{B} = \frac{\cosh \gamma S - 1}{Z_0 \sinh \gamma S} = \left(\frac{y}{z}\right)^{1/2} \tanh \frac{\gamma S}{2} = \frac{Y}{2} \frac{\tanh \dfrac{\gamma S}{2}}{\dfrac{\gamma S}{2}} \qquad \dots(11.9.2)$$

11.10 THE NOMINAL T- AND Π-EQUIVALENT NETWORK OF A LINE

For electrically short lines γS is small so that $\sinh \gamma S = \gamma S$ and $\tanh \gamma S = \gamma S$,

then, $\qquad \dfrac{1}{2} Z_{Tn} = \dfrac{1}{2} Z, \qquad Y_{Tn} = Y$

$Z_{\pi n} = Z, \qquad \dfrac{1}{2} Y_{\pi n} = \dfrac{1}{2} Y$

Such approximate equivalent networks are termed *nominal equivalent networks*. The nominal equivalent networks do not take into account the uniform distribution of the parameters along the line. The parameters are assumed to be lumped or concentrated. To distinguish between the elements of a nominal T of π networks and those of their corresponding exact networks, the elements of the nominal networks are designated by using suffix n as given above.

$$\frac{1}{2} Z_T = \frac{1}{2} Z_{Tn} \frac{\tanh \dfrac{\gamma S}{2}}{\dfrac{\gamma S}{2}} \qquad \ldots(11.10.1)$$

$$Y_T = Y_{Tn} \frac{\sinh \gamma S}{\gamma S} \qquad \ldots(11.10.2)$$

$$Z_\pi = Z_{\pi n} \frac{\sinh \gamma S}{\gamma S} \qquad \ldots(11.10.3)$$

$$\frac{1}{2} Y_\pi = \frac{1}{2} Y_{\pi n} \frac{\tanh \dfrac{\gamma S}{2}}{\dfrac{\gamma S}{2}} \qquad \ldots(11.10.4)$$

These expressions enable us to find out the elements of an exact equivalent network provided that their values are given for the nominal circuits. For example, the series impedance $Z_{\pi n}$ of a nominal π network is multiplied by $(\sinh \gamma S)/(\gamma S)$ to give its corresponding value for the exact equivalent circuit. The factors $(\sinh \gamma S)/(\gamma S)$ and $\left(\tanh \dfrac{\gamma S}{2}\right)\left(\dfrac{\gamma S}{2}\right)$ are, therefore, sometimes referred to as *correction factors*. Their values approach unity for electrically short lines.

11.11 EXPERIMENTAL DETERMINATION OF ABCD CONSTANTS

The general network constants can be determined experimentally by open- and short-circuit tests.

Open-Circuit Test

In open-circuit or no-load test (Fig. 11.18), the output terminals of the network are open. The voltage measured across them is V_r. Since the circuit is open at the receiving end, the receiving-end current I_r is zero. The sending-end current and voltage are I_{s0} and V_{s0} respectively.

Substituting these values in Eq. (11.1.1)

$$V_{s0} = A V_r$$

$\therefore \qquad A = \dfrac{V_{s0}}{V_r} \qquad \ldots(11.11.1)$

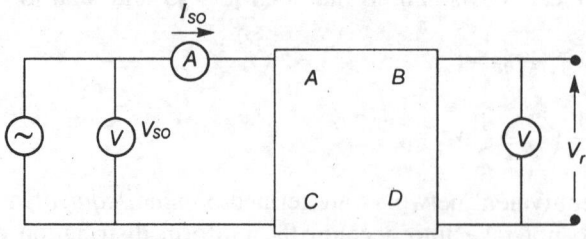

Fig. 11.18. Open-circuit test.

and $\qquad I_{s0} = CV_r$

$\therefore \qquad C = \dfrac{I_{s0}}{V_r}$ $\qquad\qquad\qquad\qquad\qquad\qquad\qquad\qquad$ ...(11.11.2)

Sending-end impedance with receiving end open-circuited

$$Z_{so} = \frac{V_{so}}{I_{so}} = \frac{A}{C} \qquad\qquad\qquad\qquad\qquad\qquad ...(11.11.3)$$

Thus, the constant A is the ratio of the sending end and receiving voltages if the receiving end is open-circuited. The constant C is the ratio of sending-end current to the receiving-end voltage when the receiving end is open.

Short-Circuit Test

In the short-circuit test (Fig. 11.19), the output terminals are short-circuited so that full load current I_r, flows in the short-circuit. The receiving-end voltage V_r is zero. The sending-end current and voltage are I_{ss} and V_{ss} respectively. The substitution of these values in Eq. (11.1.1) gives

$\qquad V_{ss} = BI_r$

$\therefore \qquad B = \dfrac{V_{ss}}{I_r}$ $\qquad\qquad\qquad\qquad\qquad\qquad\qquad\qquad$ ...(11.11.4)

and $\qquad I_{ss} = DI_r$

$\therefore \qquad D = \dfrac{I_{ss}}{I_r}$ $\qquad\qquad\qquad\qquad\qquad\qquad\qquad\qquad$ ...(11.11.5)

Sending-end impedance with receiving end short-circuited

$$Z_{ss} = \frac{V_{ss}}{I_{ss}} = \frac{B}{D} \qquad\qquad\qquad\qquad\qquad\qquad ...(11.11.6)$$

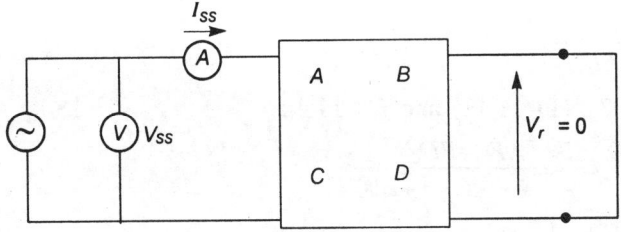

Fig. 11.19. Short-circuit test.

Thus, the constant B is given by the ratio of sending-end voltage to the receiving-end current when the receiving end is short-circuited. The constant D is the ratio of the sending-end current to the receiving-end current at short circuit.

The quantities V_{ss}, I_{ss}, V_{so}, I_{so}, V_r and I_r are in complex form. Their phase relations should also be measured to calculate the $ABCD$ constants. The measurement of phase angles is easy for a compact two port network and, therefore, the computation of $ABCD$ constant is also easy.

For a transmission line, the *ABCD* constants are difficult to determine as the sending and receiving ends are at a certain distance. The magnitudes of the voltages and currents at the two ends can be measured easily, but the phase relationship between these quantities cannot found by a simple method.

To compute the *ABCD* constants of transmission line, it is reduced to an equivalent *T* or *Π* circuit. The impedances at each end of the line are then measured by open-and short-circuit tests,

The voltage V_r is applied at the receiving end and the sending end is open-circuited. The direction of flow of current reverses. The Eqs. (11.4.1) and (11.4.2) then assume the forms

$$V_r = DV_s + BI_s \qquad \qquad \text{...(11.11.7)}$$

$$-I_r = -CV_s - AI_s$$

$$I_r = CV_s + AI_s \qquad \qquad \text{...(11.11.8)}$$

With sending end open, $I_s = 0$

Therefore, from Eqs. (11.11.7) and (11.11.8)

$$V_{ro} = DV_s \qquad \qquad \text{...(11.11.9)}$$

and $\qquad I_{so} = CV_s \qquad \qquad \text{...(11.11.10)}$

The receiving-end impedance with sending end open is given by

$$Z_{ro} = \frac{V_{ro}}{I_{ro}} = \frac{D}{C} \qquad \qquad \text{...(11.11.11)}$$

Now, the sending end is short-circuited and the measurements are made at the receiving end. Substituting $V_s = 0$ in Eqs. (11.11.7) and (11.11.8), we get

$$V_{rs} = BI_{ss} \qquad \qquad \text{...(11.11.12)}$$

$$I_{rs} = AI_{ss} \qquad \qquad \text{...(11.11.13)}$$

The receiving-end impedance with the sending end short-circuited is

$$Z_{rs} = \frac{V_{ro}}{I_{rs}} = \frac{BI_{ss}}{AI_{ss}} = \frac{B}{A} \qquad \qquad \text{...(11.11.14)}$$

From Eqs. (11.11.3), (11.11.11) and (11.11.14)

$$Z_{ro} - Z_{rs} = \frac{D}{C} - \frac{B}{A} = \frac{AD - BC}{AC} = \frac{1}{AC}$$

$$\frac{Z_{ro} - Z_{rs}}{Z_{so}} = \frac{C}{A} \times \frac{1}{AC} = \frac{1}{A^2}$$

$$A = \left(\frac{Z_{so}}{Z_{ro} - Z_{rs}} \right)^{1/2} \qquad \qquad \text{...(11.11.15)}$$

From the measured values of the open-end short-circuit impedances at the ends the constant *A* can be calculated. The other constants are found by Eqs. (11.11.3), (11.11.11) and (11.11.14).

Example 11.1 A 132 kV transmission line, having constants A, B, C, D is preceded and terminated by 132/33 kV transformers. The short-circuit impedances of the transformers measured on their 33 kV sides are Z_1 for the input transformers, and Z_2 for the output transformer. Neglecting the magnetizing currents of the transformers, calculate the constants, A, B, C, D representing the complete system.

Solution

Fig. 11.20 shows the given system.

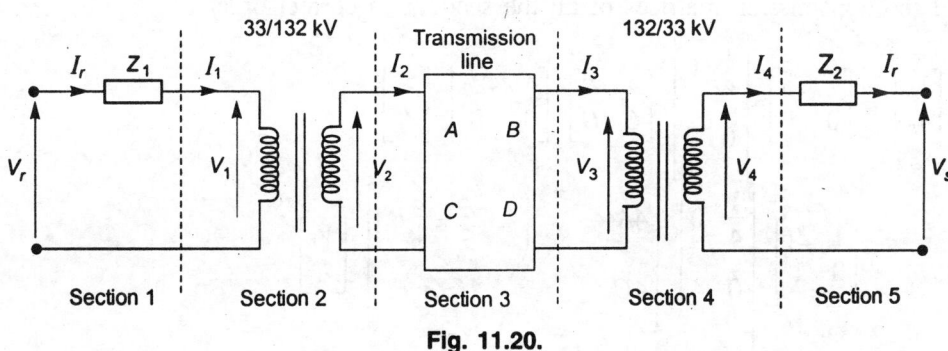

Fig. 11.20.

$$V_4 = V_r + Z_2 I_r \; ; \qquad I_4 = I_r$$

$$\begin{bmatrix} V_4 \\ I_4 \end{bmatrix} = \begin{bmatrix} 1 & Z_2 \\ 0 & 1 \end{bmatrix} \begin{bmatrix} V_r \\ I_r \end{bmatrix} \qquad \qquad \ldots(E11.1.1)$$

$$V_3 = 4V_4 \; ; \qquad I_3 = \frac{1}{4} I_4$$

$$\begin{bmatrix} V_3 \\ I_3 \end{bmatrix} = \begin{bmatrix} 4 & 0 \\ 0 & \dfrac{1}{4} \end{bmatrix} \begin{bmatrix} V_4 \\ I_4 \end{bmatrix} \qquad \qquad \ldots(E11.1.2)$$

$$V_2 = AV_3 + BI_3 \; ; \qquad I_2 = CV_3 + DI_3$$

or $$\begin{bmatrix} V_2 \\ I_2 \end{bmatrix} = \begin{bmatrix} A & B \\ C & D \end{bmatrix} \begin{bmatrix} V_3 \\ I_3 \end{bmatrix} \qquad \qquad \ldots(E11.1.3)$$

$$V_1 = \frac{1}{4} V_2 \; ; \qquad I_1 = 4I_2$$

or $$\begin{bmatrix} V_1 \\ I_1 \end{bmatrix} = \begin{bmatrix} \dfrac{1}{4} & 0 \\ 0 & 4 \end{bmatrix} \begin{bmatrix} V_2 \\ I_2 \end{bmatrix} \qquad \qquad \ldots(E11.1.4)$$

$$V_s = V_1 + Z_1 I_1 \; ; \qquad I_s = I_1$$

or $$\begin{bmatrix} V_s \\ I_s \end{bmatrix} = \begin{bmatrix} 1 & Z_1 \\ 0 & 1 \end{bmatrix} \begin{bmatrix} V_1 \\ I_1 \end{bmatrix} \qquad \qquad \ldots(E11.1.5)$$

Combining the matrix Eqs. (E11.1.1) to (E11.1.5)

$$\begin{bmatrix} V_s \\ I_s \end{bmatrix} = \begin{bmatrix} 1 & Z_1 \\ 0 & 1 \end{bmatrix} \begin{bmatrix} \frac{1}{4} & 0 \\ 0 & 4 \end{bmatrix} \begin{bmatrix} A & B \\ C & D \end{bmatrix} \begin{bmatrix} 4 & 0 \\ 0 & \frac{1}{4} \end{bmatrix} \begin{bmatrix} 1 & Z_2 \\ 0 & 1 \end{bmatrix} \begin{bmatrix} V_r \\ I_r \end{bmatrix} \qquad ...(E11.1.6)$$

Eq. (E11.1.6) can be written directly by considering the given system as the cascade connection of five angle element sections and the overall constants are determined by the matrix product of the transmission matrices of all the sections in correct order.

$$\therefore \quad \begin{bmatrix} V_s \\ I_s \end{bmatrix} = \begin{bmatrix} 1 & Z_1 \\ 0 & 1 \end{bmatrix} \begin{bmatrix} \frac{1}{4} & 0 \\ 0 & 4 \end{bmatrix} \begin{bmatrix} A & B \\ C & D \end{bmatrix} \begin{bmatrix} 4 & 4Z_2 \\ 0 & 1 \end{bmatrix} \begin{bmatrix} V_r \\ I_r \end{bmatrix}$$

$$= \begin{bmatrix} 1 & Z_1 \\ 0 & 1 \end{bmatrix} \begin{bmatrix} \frac{1}{4} & 0 \\ 0 & 4 \end{bmatrix} \begin{bmatrix} 4A & \left(4AZ_2 + \frac{1}{4}B\right) \\ 4C & 4CZ_2 \end{bmatrix} \begin{bmatrix} V_r \\ I_r \end{bmatrix}$$

$$= \begin{bmatrix} 1 & Z_1 \\ 0 & 1 \end{bmatrix} \begin{bmatrix} A & \left(AZ_2 + \frac{1}{16}B\right) \\ 16C & (16CZ_2 + D) \end{bmatrix} \begin{bmatrix} V_r \\ I_r \end{bmatrix}$$

$$= \begin{bmatrix} (A + 16Z_1 C) & \left(AZ_2 + \frac{1}{16}B + 16CZ_1 Z_2 + DZ_1\right) \\ 16C & (16CZ_2 + D) \end{bmatrix} \begin{bmatrix} V_r \\ I_r \end{bmatrix} \quad ...(E11.1.7)$$

If the overall constants for the whole system are A_o, B_o, C_o, D_o

$$V_s = A_o V_r + B_o I_r \; ; \qquad I_s = C_o V_r + D_o I_r$$

or

$$\begin{bmatrix} V_s \\ I_s \end{bmatrix} = \begin{bmatrix} A_o & B_o \\ C_o & D_o \end{bmatrix} \begin{bmatrix} V_r \\ I_r \end{bmatrix} \qquad ...(E11.1.8)$$

By comparing Eqs. (E11.1.7) and (E11.1.8) we get

$$A_o = A + 16CZ_1$$

$$B_o = AZ_2 + \frac{1}{16}B + 16CZ_1 Z_2 + DZ_1$$

$$C_o = 16C$$

$$D_o = 16CZ_2 + D$$

The constants may also be deduced without the use of matrices by process of elimination of quantities I_1, I_2, I_3, I_4, V_1, V_2, V_3, and V_4 from the Eqs. (E11.1.1) to (E11.1.5) as follows :

$$I_s = I_1 = 4I_2 = 4\,(CV_3 + DI_3) = 4\left(4CV_4 + \frac{1}{4}DI_4\right)$$

$$= 16CV_4 + DI_4 = 16C \, (V_r + Z_2 \, I_r) + DI_r$$

$$= 16CV_r + (16CZ_2 + D) \, I_r \qquad\qquad \text{...(E11.1.9)}$$

$$V_s = V_1 + I_s \, Z_1$$

$$V_1 = \frac{1}{4} V_2 = \frac{1}{4} (AV_3 + BI_3) = \frac{1}{4}\left(4AV_4 + \frac{1}{4} BI_4\right) = A \, (V_r + Z_2 \, I_r) + \frac{1}{16} BI_r$$

$$V_s = A \, (V_r + Z_2 \, I_r) + \frac{1}{16} BI_r + 16CZ_1 \, V_r + 16CZ_1 \, Z_2 \, I_r + DZ_1 \, I_r$$

$$= (A + 16CZ_1) \, V_r + \left(AZ_2 + \frac{1}{16} B + 16CZ_1 \, Z_2 + DZ_1\right) I_r \qquad \text{...(E11.1.10)}$$

$$\therefore \qquad A_o = A + 16CZ_1$$

$$B_o = AZ_2 + \frac{1}{16} B + 16CZ_1 \, Z_2 + DZ_1$$

$$C_o = 16C$$

$$D_o = 16CZ_2 + D$$

Example 11.2 In a three-phase transmission line with 132 kV at the receiving end the following are the transmission constants : $\mathbf{A} = \mathbf{D} = 0.98\,\underline{/3°}$, $\mathbf{B} = 110\,\underline{/75°}\ \Omega$, $\mathbf{C} = 0.0005\,\underline{/88°}$ S. If the load at the receiving end is 50 MVA at 0.8 lagging power factor determine the voltage, current and power factor at the sending end.

Solution

The phase voltage at the receiving end, $V_{rp} = \dfrac{132 \times 10^3}{\sqrt{3}} = 76210$ V

Taking $\mathbf{V}_{rp}$ as the reference phasor,

$$\mathbf{V}_{rp} = 76210\,\underline{/0°} = 76210 + j\,0$$

$$\text{MVA} = \sqrt{3} \, V_{rl} \, I_{rl} \times 10^{-6}$$

$$50 = \sqrt{3} \times 132 \times 10^3 \times I_r \times 10^{-6}$$

$$I_r = \frac{50 \times 10^6}{\sqrt{3} \times 132 \times 10^3} = 218.7 \text{ A}$$

$$\mathbf{I}_r = I_r\,\underline{/-\varphi_r} = 218.7\,\underline{/-\cos^{-1} 0.8} = 218.7\,\underline{/-36.87°} \text{ A}$$

$$\mathbf{V}_{sp} = \mathbf{A}\mathbf{V}_{rp} + \mathbf{B}\mathbf{I}_r$$

$$= (0.98\,\underline{/3°}) \, (76210\,\underline{/0°}) + (110\,\underline{/75°}) \, (218.7\,\underline{/36.87°})$$

$$= 74685.8\,\underline{/3°} + 24057\,\underline{/38.13°} = 74583.5 + j\,3908.8 + 18923.5 + j\,14854$$

$$= 93507 + j\,18763 = 95371\,\underline{/11.346°} \text{ V}$$

Line voltage at the sending end, $|\,\mathbf{V}_{sl}\,| = \sqrt{3}\ |\,\mathbf{V}_{sp}\,| = \sqrt{3} \times 95371 = 165187$ V $= 165.187$ kV

$$\mathbf{I}_s = \mathbf{C}\mathbf{V}_{rp} + \mathbf{D}\mathbf{I}_r$$

$$= (0.0005\,\underline{/88°}) \, (76210\,\underline{/0°}) + (0.98\,\underline{/3°}) \, (218.7\,\underline{/-36.87°})$$

$$= 38.105 \underline{/88°} + 214.326 \underline{/-33.87°} = 1.33 + j\,38.08 + 177.95 - j\,119.45$$
$$= 179.28 - j\,81.37 = 196.88 \underline{/-24.41°} \text{ A}$$

The phase angle between sending-end voltage and sending-end current

$$\varphi_s = 11.346 - (-24.41) = 35.756°$$

The power factor at the sending end, $\cos \varphi_s = \cos 35.756° = 0.8115$ (lagging)

Example 11.3 A three-phase transmission line has the following circuit constants : $\mathbf{A}_1 = 0.97 \underline{/0.6°}$, $\mathbf{B}_1 = 60 \underline{/70°}$ Ω/phase. If a second line having the constants $\mathbf{A}_2 = 0.97 \underline{/0.4°}$, $\mathbf{B}_2 = 50 \underline{/76°}$ Ω/phase is connected in parallel with the first line, determine the sending-end voltage when delivering 50 MW at 132 kV and 0.8 lagging power factor at the receiving end.

Solution

The phase voltage at the receiving end, $V_{rp} = \dfrac{132 \times 10^3}{\sqrt{3}} = 76210$ V

Taking $\mathbf{V}_{rp}$ as the reference phasor,

$$\mathbf{V}_{rp} = 76210 \underline{/0°} = 76210 + j\,0$$
$$P_r = \sqrt{3}\; V_{rl}\, I_{rl} \cos \varphi_r$$
$$50 \times 10^6 = \sqrt{3} \times 132 \times 10^3 \times I_r \times 0.8$$
$$I_r = \frac{50 \times 10^6}{\sqrt{3} \times 132 \times 10^3 \times 0.8} = 273.3 \text{ A}$$
$$\mathbf{I}_r = I_r \underline{/-\varphi_r} = 273.3 \underline{/-\cos^{-1} 0.8} = 273.3 \underline{/-36.87°} \text{ A}$$

When the two transmission lines are connected in parallel

$$\mathbf{A} = \frac{\mathbf{A}_1\,\mathbf{B}_2 + \mathbf{A}_2\,\mathbf{B}_1}{\mathbf{B}_1 + \mathbf{B}_2}\;;\qquad \mathbf{B} = \frac{\mathbf{B}_1\,\mathbf{B}_2}{\mathbf{B}_1 + \mathbf{B}_2}$$

$$\mathbf{A}_1\,\mathbf{B}_2 + \mathbf{A}_2\,\mathbf{B}_1 = (0.97 \underline{/0.6°})\,(50 \underline{/76°}) + (0.97 \underline{/0.4°})\,(60 \underline{/70°})$$
$$= 48.5 \underline{/76.6°} + 58.2 \underline{/70.4°} = 11.24 + j\,47.18 + 19.52 + j\,54.82$$
$$= 30.76 + j\,102 = 106.54 \underline{/73.22°}$$

$$\mathbf{B}_1 + \mathbf{B}_2 = 60 \underline{/70°} + 50 \underline{/76°} = 20.52 + j\,56.38 + 12.096 + j\,48.51$$
$$= 32.616 + j\,104.89 = 109.84 \underline{/72.727°}$$

$$\mathbf{B}_1\,\mathbf{B}_2 = (60 \underline{/70°})\,(50 \underline{/76°}) = 3000 \underline{/146°}$$

$$\mathbf{A} = \frac{106.54 \underline{/73.22°}}{109.84 \underline{/72.727°}} = 0.9699 \underline{/0.493°}$$

$$\mathbf{B} = \frac{3000 \underline{/146°}}{109.84 \underline{/72.727°}} = 27.31 \underline{/73.273°}$$

$$\mathbf{V}_{sp} = \mathbf{A}\mathbf{V}_{rp} + \mathbf{B}\mathbf{I}_r$$
$$= (0.9699 \underline{/0.493°})\,(76210 \underline{/0°}) + (27.31 \underline{/73.273°}) \times (273.3 \underline{/-36.87°})$$
$$= 73916 \underline{/0.493°} + 7463.8 \underline{/36.4°} = 73913 + j\,636 + 6007.5 + j\,4429$$

$$= 79920.5 + j\,5065 = 80081\,\underline{/3.626°}\ \text{V}$$

Line voltage at the sending end, $|\mathbf{V}_{sl}| = \sqrt{3}\,|\mathbf{V}_{sp}| = \sqrt{3} \times 80081 = 138704\ \text{V} = 138.704\ \text{kV}$

Example 11.4 A transmission circuit is represented by a symmetrical π network in which the series impedance is $120\,\underline{/60°}\ \Omega$, and each shunt admittance is $2.5 \times 10^{-3}\,\underline{/90°}$ S. Calculate (a) the value of the general circuit constants **ABCD**, and (b) the characteristic impedance of the circuit.

Solution

The general circuit constants of a symmetrical Π network are given by Eq. (11.5.18) as

$$\mathbf{A} = \mathbf{D} = 1 + \frac{1}{2}\,\mathbf{Z}_\pi\,\mathbf{Y}_\pi\,, \qquad \mathbf{B} = \mathbf{Z}_\pi\,, \qquad \mathbf{C} = \mathbf{Y}_\pi\left(1 + \frac{1}{4}\,\mathbf{Z}_\pi\,\mathbf{Y}_\pi\right)$$

In this problem,

$$\mathbf{B} = \mathbf{Z}_\pi = 120\,\underline{/60°}\ \Omega\,, \qquad \frac{1}{2}\,\mathbf{Y}_\pi = 2.5 \times 10^{-3}\,\underline{/90°}\ \text{S}$$

(a) $\quad \mathbf{A} = \mathbf{D} = 1 + (120\,\underline{/60°})\,(2.5 \times 10^{-3}\,\underline{/90°})$

$\qquad = 1 + 0.3\,\underline{/150°} = 1 - 0.2598 + j\,0.15 = 0.7402 + j\,0.15 = 0.7552\,\underline{/11.456°}$

$\quad \mathbf{C} = (2 \times 2.5 \times 10^{-3}\,\underline{/90°})\,(1 + 0.15\,\underline{/150°})$

$\qquad = (5 \times 10^{-3}\,\underline{/90°})\,(1 - 0.1299 + j\,0.075) = (5 \times 10^{-3}\,\underline{/90°})\,(0.8701 + j\,0.075)$

$\qquad = (5 \times 10^{-3}\,\underline{/90°})\,(0.8733\,\underline{/4.927°}) = (4.3665 \times 10^{-3}\,\underline{/94.927°}$ S

(b) Characteristic impedance, $\mathbf{Z}_0 = \sqrt{\dfrac{\mathbf{B}}{\mathbf{C}}} = \left(\dfrac{120\,/60°}{4.3665 \times 10^{-3}\,/99.927°}\right)^{1/2} = 165.78\,\underline{/-17.46°}\ \Omega$

Example 11.5 A 2-port resistive network has input terminals A and B, and output terminals C and D. The resistances measured across AB when terminals CD are first short-circuited and then open-circuited are, respectively, $720\ \Omega$ and $1240\ \Omega$. The resistances measured across CD with AB open-circuited is $910\ \Omega$. Determine the equivalent T-network.

Solution

Fig. 11.21 shows the equivalent T-network.

Resistance across AB with CD short-circuited, $R_1 + \dfrac{R_2\,R_3}{R_2 + R_3} = 720$ ...(E11.5.1)

Resistance across AB with CD open-circuited, $R_1 + R_3 = 1240$ ...(E11.5.2)

Resistance across CD with AB open-circuited, $R_2 + R_3 = 910$ ...(E11.5.3)

From Eqs. (E11.5.1) and (E11.5.2)

$$R_3 - \frac{R_2\,R_3}{R_2 + R_3} = 1240 - 720$$

$$\frac{R_2\,R_3 + R_3^2 - R_2\,R_3}{R_2 + R_3} = 520$$

$$R_3^2 = 520\,(R_2 + R_3) = 520 \times 910$$

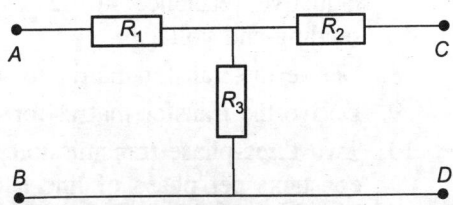

Fig. 11.21.

$\therefore \qquad R_3 = 687.9 \ \Omega$

$R_2 = 910 - 687.9 = 222.1 \ \Omega$

$R_1 = 1240 - 687.9 = 552.1 \ \Omega$

EXERCISES

1. Determine the **A**, **B**, **C** and **D** constants for a three-phase 50 Hz, 100 km long transmission line having the following uniformly distributed parameters per km per phase: resistance = 0.25 Ω, inductance = 2.0 mH, capacitance = 0.014 μF. (I.E.E.)

2. A transmission circuit is represented by a T-network in which each series impedance is $90 \underline{/60°}$ ohms, and the shunt admittance is $1.5 \times 10^{-3} \underline{/90°}$ S. Determine (a) the general circuit constants A, B, C, D; and (b) the open-circuit driving point impedance of the network. (I.E.E.)

3. Explain what is meant by the terms of *nominal-Π* and *equivalent-Π* when applied to a uniform transmission line.

   Determine from first principles the values of the elements of an equivalent-Π section for a transmission line having a characteristic impedance Z_o and a propagation constant γ. (I.E.E.)

4. A three-phase, 160 km long transmission the delivers 350 A at a power factor 0.9 (lagging) at 132 kV. The transmission constants for the line are as follows :

   $\mathbf{A} = \mathbf{D} = 0.9844 \underline{/0° \ 19'}, \ \mathbf{B} = 69.9646 \underline{/70° \ 9'}, \ \mathbf{C} = 0.000469 \underline{/90° \ 6'}.$

   Calculate the voltage, current and power factor at the sending end.

5. Show that in a symmetrical single-phase 2-port network $A = (1 + BC)^{1/2}$. Prove the relation $AD - BC = 1$ for the following cases; (a) long transmission line, (b) *T*-network, (c) Π-network.

6. A three-phase transmission line delivers 20 MW at a power factor of 0.8 (lagging), 32 kV. The transmission constants for the line, considered as a π-network, are as follows :

   $\mathbf{A} = \mathbf{D} = 1 \ ; \ \mathbf{B} = (1 + j \ 3) \ \Omega \ ; \ \mathbf{C} = 2 \times 10^{-4} \ \text{S}.$

   Determine the sending-end current and voltage of the line, and an approximate value for the reactive MVA taken by the line.

7. A 50 Hz, three-phase line, 100 km long, delivers a load of 30 MVA at 120 kV, 0.8 lagging power factor. The line constants for each conductor are : resistance 15 Ω, inductive reactance 40 Ω, capacitive susceptance to neutral 2.5×10^{-4} S. Find the sending-end voltage.

8. Derive the transfer matrix for the network shown in Fig. 11.22.

9. Derive the transfer matrix for the network shown in Fig. 11.23.

10. Two three-phase transmission lines P and Q are connected in series. The transmission constants per phase of line P are : $\mathbf{A}_1 = \mathbf{D}_1 = 0.98 \underline{/2°}, \ \mathbf{B}_1 = 28 \underline{/69°}, \ \mathbf{C}_1 = 0.0002 \underline{/88°}$

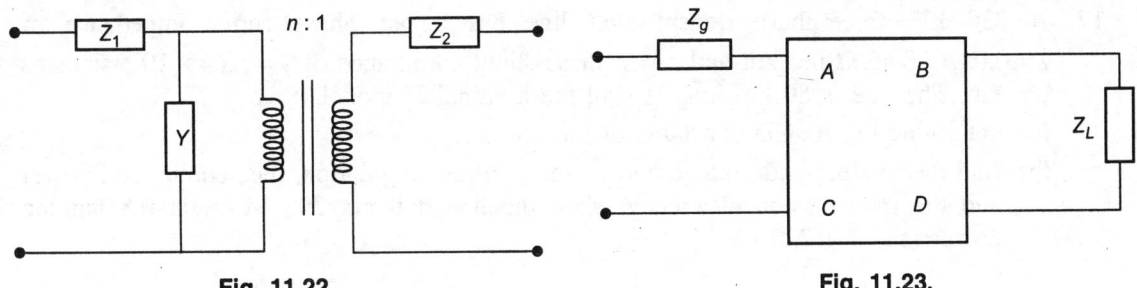

Fig. 11.22.   **Fig. 11.23.**

and those of line Q are : $\mathbf{A}_2 = \mathbf{D}_2 = 0.95\,\underline{/3°}$, $\mathbf{B}_2 = 40\,\underline{/85°}$, $\mathbf{C}_2 = 0.0004\,\underline{/90°}$. Calculate the constants **A**, **B**, **C** and **D** for a single three-phase line equivalent to the two lines in series.

(L.U.)

11. Show that in a symmetrical single-phase 2-port transmission network $\mathbf{A} = \sqrt{(1 + \mathbf{BC})}$. Determine by reasonable approximations, the constants **A**, **B**, **C**, **D** for 1 km of line connected to 100 m of cable each having the inductance and capacitance as follows : overhead line, inductance, 2.4 mH/km, capacitance 0.01 μF/km, cable inductance 1.1 mH/km, capacitance 0.5 μF/km. The line and the cable may each be regarded as a T-network operating at 50 Hz, the cable termination being regarded as the receiving end.

(I.E.E.)

12. A transmission line consists of two circuits P and Q connected in series, the circuit P being at the sending end of the line. The circuits have the following auxiliary constants :

Circuit P :
$\quad \mathbf{A}_p = 0.982\,\underline{/1.2°}$, $\mathbf{B}_p = 77.3\,\underline{/80.0°}$, $\mathbf{C}_p = 0.000452\,\underline{/91.0°}$, $\mathbf{D}_p = 0.982\,\underline{/1.2°}$

Circuit Q :
$\quad \mathbf{A}_q = 0.808\,\underline{/2.0°}$, $\mathbf{B}_q = 30\,\underline{/45.0°}$, $\mathbf{C}_q = 0.001\,\underline{/92.0°}$, $\mathbf{D}_q = 0.808\,\underline{/2.0°}$.

Develop expressions for each of the four auxiliary constants **A**, **B**, **C** and **D** of the whole line and calculate the numerical value of the constant **A**. (L.U.)

13. A 100 km, 50 Hz 275 kV, three-phase line has the following constants per km : resistance = 0.05 Ω; inductance = 1.2 mH; capacitance to neutral = 0.025 μF.

Using a Π equivalent circuit, determine the sending-end voltage and current when the line provides 330 MVA at 275 kV with a power factor of 0.8 lagging.

14. Show that for a transmission line

$$A = \frac{V_s I_s + V_r I_r}{V_r I_s + V_s I_r} \; ; \qquad B = \frac{V_s^2 - V_r^2}{V_r I_s + V_s I_r}$$

15. The **ABCD** constants of a lossless three-phase, 500 kV transmission line are:
$\quad \mathbf{A} = \mathbf{D} = 0.86 + j\,0$, $\mathbf{B} = 0 + j\,130.2$, $\mathbf{C} = j\,0.002$

Determine the sending-end quantities and the voltage regulation when the line delivers 1000 MVA at 0.8 lagging power factor at 500 kV.

16. The **ABCD** constants of a 3-phase, 345 kV transmission line are:
$\quad \mathbf{A} = \mathbf{D} = 0.98182 + j\,0.0012447$, $\mathbf{B} = 4.035 + j\,58.947$, $\mathbf{C} = j\,0.00061137$

The line delivers 400 MVA at 0.8 lagging power factor of 345 kV. Determine the sending-end quantities, voltage regulation, and transmission efficiency.

17. A 230 kV, three-phase transmission line has a per phase series impedance of $z = 0.05 + j\,0.45\ \Omega$ per km and a per phase shunt admittance of $y = j\,3.4 \times 10^{-6}$ siemens per km. The line is 80 km long. Using the nominal Π model,

(a) determine the ABCD constants of the line;

(b) find the sending-end voltage and current, voltage regulation, the sending-end power and the transmission efficiency when the line delivers 200 MVA at 0.8 lagging power factor at 220 kV.

ANSWERS

1. $\mathbf{A} = \mathbf{D} = 0.986\,\underline{/0.316°}\ \ \mathbf{B} = 67\,\underline{/68.4°}\ \Omega\ ;\ \mathbf{C} = 0.00044\,\underline{/90°}$ S

2. (a) $\mathbf{A} = \mathbf{D} = 0.885\,\underline{/4°\,20'}\ ;\ \mathbf{B} = 170\,\underline{/62°\,10'}\ \Omega\ ;\ \mathbf{C} = 1.5 \times 10^{-3}\,\underline{/90°}$ S.

 (b) $590\,\underline{/-85°\,40'}\ \Omega$

4. 163.14 kV ; 330.64 A ; 0.862 (lag)   6. 34.1 kV ; 44.8 A ; 16.6 MVAr

7. 125.9 kV

8.
$$\begin{bmatrix} n\,(1 + Z_1\,Y)\,n\,Z_2 & (1 + Z_1\,Y) + \dfrac{Z_1}{n} \\[2mm] n\,Y & n\,Z_2 + \dfrac{1}{n} \end{bmatrix}$$

9.
$$\begin{bmatrix} A + \dfrac{B}{Z_L} + Z_g\left(C + \dfrac{D}{Z_L}\right) & B + DZ_g \\[2mm] C + \dfrac{D}{Z_L} & D \end{bmatrix}$$

10. $\mathbf{A} = 0.914\,\underline{/5.35°}\ ;\ \mathbf{B} = 65.5\,\underline{/81°}\ \Omega\ ;\ \mathbf{C} = 0.00041\,\underline{/87.6°}$ S ; $\mathbf{D} = 0.924\,\underline{/5°}$

11. $\mathbf{A} = \mathbf{D} = 1\ ;\ \mathbf{B} = 3.787\,\underline{/90°}\ \Omega\ ;\ \mathbf{C} = 1.885 \times 10^{-5}\,\underline{/90°}$ S

12. 0.7171   13. 304 kV ; 617 A

15. 622.153 kV, $794.649\,\underline{/-1.33°}$ A, 800 MW $+ j\,305.408$ MVAr, 44.687%

16. 387.025 kV, $592.29\,\underline{/-27.325°}$ A, 324.87 MW $+ j\,228.25$ MVAr, 14.259%, 98.5%

17. (a) $0.9951 + j\,0.000544,\ 4 + j\,36\ \Omega,\ j\,0.0002713$ S

 (b) 242.67 kV, $502.38\,\underline{/-33.69°}$ A, 10.847%, 163.18 MW $+ j\,134.02$ MVAr, 98.052%

Power Circle Diagrams

12.1 INTRODUCTION

The performance of a transmission line can be studied either by analytical methods or by graphical methods. Purely analytical methods sometimes involve tedious calculations. They take considerable time and labour. Power circle diagrams provide a convenient graphical method for studying the performance of a transmission line under the conditions of the varying load. Circle diagrams are specially useful in system-stability studies, reactive volt-ampere calculations, and in the system design and operation. Circle diagrams are easy to construct and use. Their accuracy in fairly high.

12.2 RECEIVING-END VOLTAGE PHASOR DIAGRAM

The receiving-end phasor diagram for voltage is drawn with the help of the general equation

$$\mathbf{V}_s = \mathbf{A}\mathbf{V}_r + \mathbf{B}\mathbf{I}_r$$

The use of the above equation gives generalized result which may be applied to any specific case. The complex constants **A** and **B** may be written in the form of their magnitudes and arguments as

$$\mathbf{A} = A\underline{/\alpha}, \qquad \mathbf{B} = B\underline{/\beta}$$

Let $\mathbf{V}_r$ be taken as the reference phasor. The receiving-end current $\mathbf{I}_r$ is assumed to have a phase difference of φ_r lagging. The constant **A** ($= \cosh \gamma S$) has a magnitude slightly less than unity. The argument α is a small positive angle. Angle β is nearly equal to 90°.

$$\mathbf{A}\mathbf{V}_r = A\underline{/\alpha} \cdot V_r\underline{/0°} = AV_r\underline{/\alpha}$$

It indicates that the phasor $\mathbf{A}\mathbf{V}_r$ is leading $\mathbf{V}_r$ by angle α. Similarly, expressing $\mathbf{B}\mathbf{I}_r$ in the polar form as

$$\mathbf{B}\mathbf{I}_r = B\underline{/\beta} \cdot I_r\underline{/-\varphi_r} = BI_r\underline{/\beta - \varphi_r}$$

Thus, the phasor $\mathbf{B}I_r$ leads $\mathbf{V}_r$ by an angle $(\beta - \varphi_r)$. $\mathbf{V}_s$ will be obtained by adding phasors $\mathbf{A}\mathbf{V}_r$ and $\mathbf{B}I_r$. $\mathbf{V}_s$ leads $\mathbf{V}_r$ by an angle δ called the *torque angle*, *load angle* or *transmission angle*. The phasor diagram is shown in Fig. 12.1. $\mathbf{V}_s$ may be written as

$$\mathbf{V}_s = V_s \underline{/\delta}$$

δ changes with the change in load. Draw a line xx through O_r making an angle φ_r with phasor $\mathbf{B}I_r$. The line $x_1O_1x_1$ is drawn parallel to xO_rx and the line yO_ry is drawn perpendicular to xO_rx. It is observed that the line xO_rx makes an angle β with the reference phasor $\mathbf{V}_r$.

12.3 PHASOR VOLT-AMPERES

The phasor volt-amperes $\mathbf{S}$ is the product of phasor voltage and the conjugate of phasor current. The conjugate phasor has the same modulus but its angle is of the opposite sign.

Let the voltage at the receiving-end be taken as reference phasor. Then

$$\mathbf{V}_r = V_r \underline{/0°}$$

For an inductive load the current lags the voltage. If φ_r be the phase angle of the current

$$\mathbf{I}_r = I_r \underline{/-\varphi_r}$$

The conjugate of $\mathbf{I}_r$ will be denoted by $\mathbf{I}^*$. It is given by

$$\mathbf{I}_r^* = I_r \underline{/+\varphi_r}$$

Phasor volt-amperes at the receiving-end of the line

$$\mathbf{S}_r = \mathbf{V}_r \cdot \mathbf{I}_r^* = V_r \underline{/0°} \cdot I_r \underline{/+\varphi_r} = V_r I_r \underline{/\varphi_r}$$

$$= V_r I_r \cos \varphi_r + j \, V_r I_r \sin \varphi_r = P_r + j \, Q_r \quad \text{(say)} \qquad \qquad ...(12.3.1)$$

The phasor product $\mathbf{S}_r$ so obtained is the sum of two components. The real part, denoted by P_r is called the *active* or *real power*. The imaginary part Q_r is known as *reactive volt-amperes*. P_r is expressed in watts, kilowatts, or megawatts. The quadrature component Q_r is expressed in vars (from the initials of volt amperes reactive), kilovars or megavars.

For a capacitive load for which the power factor is leading, the phasor volt amperes $\mathbf{S}_r$ is given by

$$\mathbf{S}_r = \mathbf{V}_r \cdot \mathbf{I}_r^* = V_r \underline{/0°} \cdot I_r \underline{/-\varphi_r} = V_r I_r \underline{/-\varphi_r}$$

$$= V_r I_r \cos \varphi_r - j \, V_r I_r \sin \varphi_r = P_r - j \, Q_r \qquad \qquad ...(12.3.2)$$

Q_r obtained in this case is negative. We shall adopt the sign convention that the volt amperes reactive taken by an inductive load are positive and by a capacitive load negative. The signs of Q_r obtained in Eqs. (12.3.1) and (12.3.2) agree with the convention of sign adopted by us.

The phasor volt-amperes $\mathbf{S}$ may also be found by multiplying the phasor current and the conjugate of the phasor voltage, but the convention adopted by us is more generally accepted.

12.4 RECEIVING-END POWER CIRCLE DIAGRAM

Receiving-end power circle diagram may be derived from graphical or analytical methods. First we shall draw the circle diagram based upon the graphical considerations. The receiving-and power circle diagram may be developed from the voltage phasor diagram of Fig. 12.1. Each phasor on this diagram is multiplied by $\mathbf{V}_r/\mathbf{B}$. The results of the multiplication are shown as follows :

$$\mathbf{V}_r \cdot \frac{\mathbf{V}_r}{\mathbf{B}} = \frac{V_r^2}{B} \underline{/-\beta}$$

$$A\mathbf{V}_r \cdot \frac{\mathbf{V}_r}{\mathbf{B}} = \frac{AV_r^2}{B} \underline{/\alpha - \beta}$$

$$\mathbf{B}\mathbf{I}_r \cdot \frac{\mathbf{V}_r}{\mathbf{B}} = V_r I_r \underline{/-\varphi_r}$$

$$\mathbf{V}_s \cdot \frac{\mathbf{V}_r}{\mathbf{B}} = \frac{V_s V_r}{B} \underline{/\delta - \beta}$$

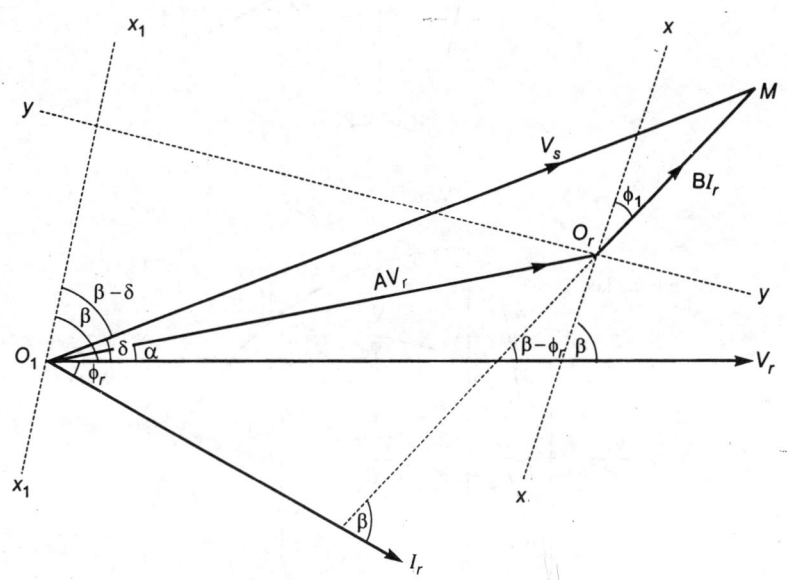

Fig. 12.1. Receiving-end voltage phasor diagram.

By such a multiplication all the voltage phasors of Fig. 12.1 are converted into volt-amperes and are rotated through an angle $-\beta$. The diagram so obtained is shown in Fig. 12.2.

Let the line xO_rx be taken as horizontal axis, yO_ry the vertical axis, and O_r the origin of coordinates. In Fig. 12.2 the product $\mathbf{V}_r\mathbf{I}_r$ gives the total volt-amperes at the receiving-end. The horizontal projection of $\mathbf{V}_r\mathbf{I}_r$ is $V_r I_r \cos \varphi_r$ which is the real power. The vertical projection of $\mathbf{V}_r\mathbf{I}_r$ is $V_r I_r \sin \varphi_r$ which is the volt-amperes reactive. For this reason, the xO_rx and yO_ry may be marked as real power and reactive volt-amperes axes respectively.

Under the sign convention adopted that an inductive load draws positive reactive volt-amperes, the diagram shown in Fig. 12.2 should be rotated about the horizontal axis. The gives the diagram shown in Fig. 12.3.

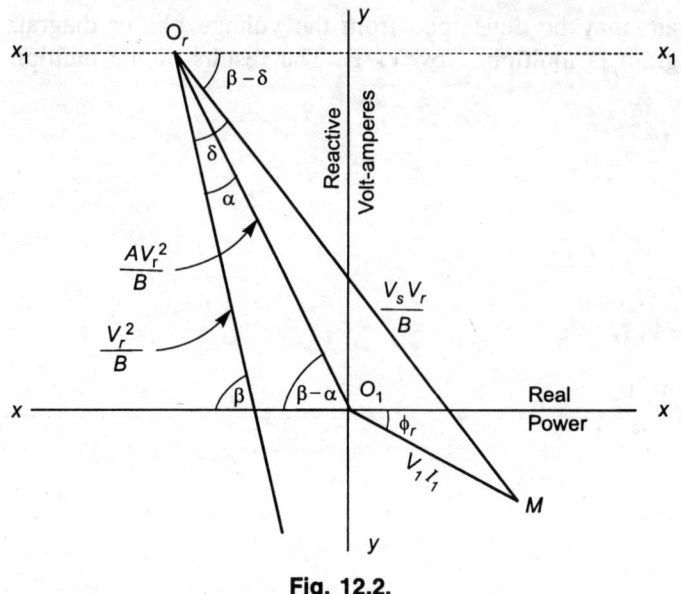

Fig. 12.2.

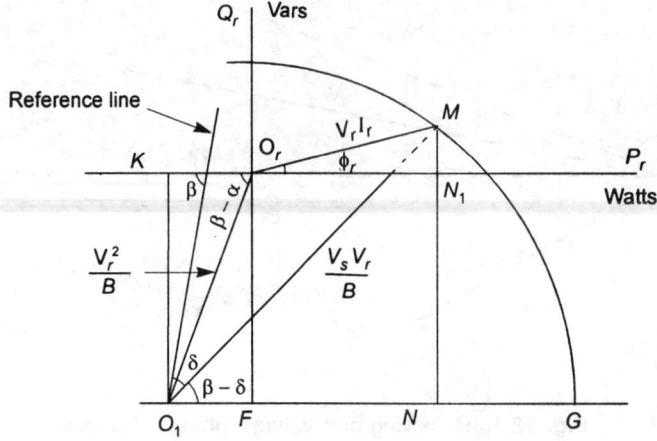

Fig. 12.3. Receiving-end power circle diagram.

The following facts are derived from Fig. 12.3. If the sending-end and receiving-end voltages V_s and V_r are kept constant the position of the point O_1 does not change as the distance $O_1O_r \left(= \dfrac{AV_r^2}{B} \right)$ is a constant and is independent of current I_r. Also the distance of the operating point M from O_1, i.e., $\dfrac{V_s V_r}{B}$ is constant for fixed values of V_s and V_r. With the variation of the load I_r and φ_r, the position of the point M and therefore, the distance O_rM ($= V_r I_r$) will change, but the point M will remain at a constant distance $O_1M \left(= \dfrac{V_s V_r}{B} \right)$ from the fixed point O_1. Thus,

M will move along a circle of radius $\dfrac{V_s V_r}{B}$ with its centre at O_1. Since the locus of the operating point M is a circle with the axis of reference as real power, and reactive voltamperes at the receiving-end, the diagram so obtained is called the *receiving-end power circle diagram.*

The coordinates of the centre of the receiving-end power circle diagram referred to a system of rectangular coordinates in which the abscissae represent the real power P_r, and the ordinates represent the reactive volt-amperes Q_r are $(- O_r K, - KO_1)$ in Fig. 12.3.

$$\left.\begin{aligned} - O_r K &= -\frac{A}{B} V_r^2 \cos (\beta - \alpha) \text{ watts} \\[2mm] - KO_1 &= -\frac{A}{B} V_r^2 \sin (\beta - \alpha) \text{ vars} \\[2mm] \text{Radius of a receiving–end power circle, } \rho_r = O_1 M &= \frac{V_s V_r}{B} \text{ volt–amperes} \end{aligned}\right\} \quad \ldots (12.4.1)$$

The distance of the centre O_1 from the origin O_r is

$$O_r O_1 = \frac{A}{B} V_r^2$$

The equation of the power circle referred to P_r and Q_r as reference axes, centre $(- O_r K, - KO_1)$ and radius $O_1 M$ may be written as

$$(P_r + O_r K)^2 + (Q_r + KO_1)^2 = (O_1 M)^2$$

Substituting the values obtained from Eq. (12.4.1) in the above equation, we get

$$\left[P_r + \frac{A}{B} V_r^2 \cos (\beta - \alpha) \right]^2 + \left[Q_r + \frac{A}{B} V_r^2 \sin (\beta - \alpha) \right]^2 = \left(\frac{V_s V_r}{B} \right)^2 \qquad \ldots (12.4.2)$$

It is to be noted that line $\dfrac{V_r^2}{B}$ in Fig. 12.3 is marked as *reference line* from which torque angle δ is measured. This line in Fig. 12.3 corresponds to phasor V_r, which was originally taken as reference phasor in Fig. 12.1. In Fig. 12.3, δ is the angle between the line $\dfrac{V_r^2}{B}$ and the line $O_1 M$.

We shall consider two cases. In the first case, suppose that V_r is fixed and it is desired to draw receiving-end power circle diagrams for different values of V_s. The position of the centre O_1, which is independent of V_s is unchanged while the radii of the circles will be different. In other words, we get the circles having the same centre O_1, but different radii corresponding to different values of V_s. Such concentric circles for a fixed value of V_r and different value of V_s are shown in Fig. 12.4.

Secondly, let V_s be kept constant and it is desired to draw receiving-end power circles for different values of V_r. The circles will not be concentric. The positions of the centres will be different for different values of V_r, but they shall be on the line through the origin and making an angle $(\beta - \alpha)$ with the horizontal (Fig. 12.5).

If V_s and V_r are expressed in phase volts, i.e., line to neutral volts, the coordinates of the diagram given in Fig. 12.3 will represent watts per phase and vars per phase respectively. Each

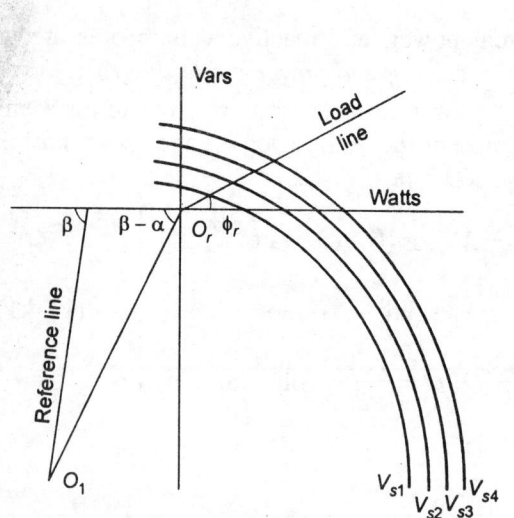

Fig. 12.4. Receiving-end power circles for different values of V_s and a fixed value of V_r.

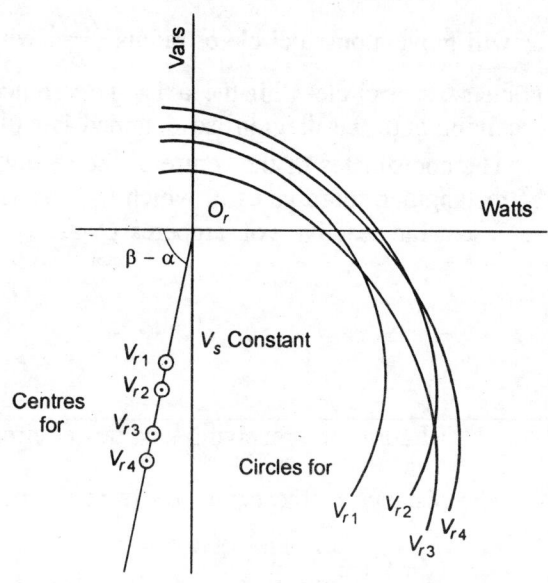

Fig. 12.5.

length on the diagram is determined by the product of two voltages or the square of the voltage. For a balanced three-phase circuit the line-to-line voltage is $\sqrt{3}$ times the phase voltage. Thus, if V_s and V_r are expressed in line-to-line volts, each length on the diagram becomes three times its value when V_s and V_r are expressed in phase volts. The coordinates in this case will, therefore, represent the total three-phase watts and vars respectively. If the voltages, V_s and V_r are expressed in line-to-line kilovolts the coordinates of the diagram will represent megawatts and megavars respectively for all the three phases.

Thus, if in Eq. (12.4.1) the voltages are in line-to-line kilovolts,

$$\left.\begin{array}{l} \text{abscissa of the centre of the receiving–end circle} = -\dfrac{A}{B}\,V_r^2\cos(\beta-\alpha)\ \text{MW} \\[2mm] \text{ordinate of the centre of the receiving–end circle} = -\dfrac{A}{B}\,V_r^2\sin(\beta-\alpha)\ \text{MVAr} \\[2mm] \text{radius of the receiving–end circle} = \dfrac{V_s V_r}{B}\ \text{MVA} \end{array}\right\} \quad \ldots(12.4.3)$$

12.5 ANALYTICAL METHOD

In the analytical method the power-circle diagram for the receiving-end of a line is drawn from the consideration of the phasor volt-amperes at that end.

The general transmission equation is given by

$$\mathbf{V}_s = \mathbf{A}\mathbf{V}_r + \mathbf{B}\mathbf{I}_r$$

$$\mathbf{I}_r = \frac{\mathbf{V}_s}{\mathbf{B}} - \frac{\mathbf{A}}{\mathbf{B}}\,\mathbf{V}_r = \frac{V_s\,\underline{/\delta}}{B\,\underline{/\beta}} - \frac{A\,\underline{/\alpha}}{B\,\underline{/\beta}}\,V_r\,\underline{/0°} = \frac{V_s}{B}\,\underline{/\delta-\beta} - \frac{AV_r}{B}\,\underline{/\alpha-\beta}$$

$$\mathbf{I}_r^* = \frac{V_s}{B}\,\underline{/\beta.-\delta} - \frac{AV_r}{B}\,\underline{/\beta-\alpha} \qquad\qquad \ldots(12.5.1)$$

Phasor volt-amperes at the receiving end of the line

$$\mathbf{S}_r = \mathbf{V}_r \, \mathbf{I}_r^*$$

$$P_r + j\,Q_r = \frac{V_s V_r}{B} \underline{/\beta - \delta} - \frac{AV_r^2}{B} \underline{/\beta - \alpha}$$

$$= \frac{V_s V_r}{B} [\cos(\beta - \delta) + j \sin(\beta - \delta)] - \frac{AV_r^2}{B} [\cos(\beta - \alpha) + j \sin(\beta - \alpha)]$$

$$= \left[\frac{V_s V_r}{B} \cos(\beta - \delta) - \frac{AV_r^2}{B} \cos(\beta - \alpha) \right] + j \left[\frac{V_s V_r}{B} \sin(\beta - \delta) - \frac{AV_r^2}{B} \sin(\beta - \alpha) \right]$$

$$...(12.5.2)$$

By separating real and imaginary parts, we get

$$P_r = \frac{V_s V_r}{B} \cos(\beta - \delta) - \frac{AV_r^2}{B} \cos(\beta - \alpha) \qquad\qquad ...(12.5.3)$$

$$Q_r = \frac{V_s V_r}{B} \sin(\beta - \delta) - \frac{AV_r^2}{B} \sin(\beta - \alpha) \qquad\qquad ...(12.5.4)$$

$$P_r + \frac{AV_r^2}{B} \cos(\beta - \alpha) = \frac{V_s V_r}{B} \cos(\beta - \delta) \qquad\qquad ...(12.5.5)$$

$$Q_r + \frac{AV_r^2}{B} \sin(\beta - \alpha) = \frac{V_s V_r}{B} \sin(\beta - \delta) \qquad\qquad ...(12.5.6)$$

Squaring and adding Eqs. (12.5.5) and (12.5.6), we get

$$\left[P_r + \frac{AV_r^2}{B} \cos(\beta - \alpha) \right]^2 + \left[Q_r + \frac{AV_r^2}{B} \sin(\beta - \alpha) \right]^2 = \left(\frac{V_s V_r}{B} \right)^2 [\cos^2(\beta - \delta) + \sin^2(\beta - \delta)]$$

$$= \left(\frac{V_s V_r}{B} \right)^2 \qquad\qquad ...(12.5.7)$$

If V_s and V_r are kept constant, Eq. (12.5.7) represents a circle referred to a system of rectangular coordinates in which the abscissae represent real power P_r and the ordinates represent reactive voltamperes Q_r. Equation (12.5.7) is, therefore, called the *power-circle equation at the receiving-end* of a transmission network.

If (P_{ro}, Q_{ro}) are the coordinates of the centre and ρ_r the radius of the circle, then from Eq. (12.5.7), we have

$$\left.\begin{array}{l} P_{ro} = -\dfrac{AV_r^2}{B} \cos(\beta - \alpha) \text{ watts} \\[2ex] Q_{ro} = -\dfrac{AV_r^2}{B} \sin(\beta - \alpha) \text{ vars} \\[2ex] \rho_r = \dfrac{V_s V_r}{B} \text{ volt–amperes} \end{array}\right\} \qquad\qquad ...(12.5.8)$$

Equation (12.5.7) may be written in the form

$$(P_r - P_{ro})^2 + (Q_r - Q_{ro})^2 = \rho_r^2 \qquad \qquad \text{...(12.5.9)}$$

It is observed that Eq. (12.5.8) obtained analytically from the considerations of the phasor volt-amperes at the receiving-end are identical with Eq. (12.4.1), which were derived graphically from the receiving-end voltage phasor diagram.

Thus, both the analytical and graphical methods give similar results.

12.6 SENDING-END VOLTAGE PHASOR DIAGRAM

The basic voltage phasor diagram may be constructed from the equation

$$\mathbf{V}_r = \mathbf{DV}_s - \mathbf{BI}_s$$

Let us take $\mathbf{V}_s$ as the reference phasor so that

$$\mathbf{V}_s = V_s\underline{/0°}$$

$$\mathbf{DV}_s = D\underline{/\Delta} \cdot V_s\underline{/0°} = DV_s\underline{/\Delta}$$

Thus, the phasor $\mathbf{DV}_s$ leads the phasor $\mathbf{V}_s$ by an angle Δ.

$$\mathbf{BI}_s = B\underline{/\beta}\, I_s\underline{/-\varphi_s} = BI_s\underline{/\beta - \varphi_s}$$

The phasor $\mathbf{BI}_s$ leads the phasor $\mathbf{V}_s$ by an angle $(\beta - \varphi_s)$. The receiving-end voltage $\mathbf{V}_r$ is obtained by subtracting the phasor $\mathbf{BI}_s$ from the phasor $\mathbf{DV}_s$.

$$\mathbf{V}_r = V_r\underline{/-\delta}$$

The sending-end voltage phasor diagram is shown in Fig. 12.6.

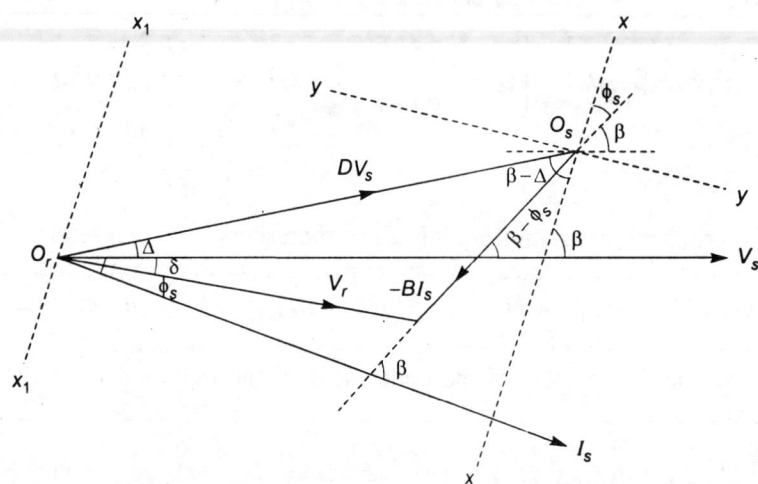

Fig. 12.6. Sending-end voltage phasor diagram.

12.7 SENDING-END POWER CIRCLE DIAGRAM

The sending-end power circle is drawn graphically from the voltage phasor diagram at the sending-end in exactly the same manner in which the receiving-end power circle is drawn.

Each phasor in Fig. 12.6 is multiplied by $\dfrac{-\mathbf{V}_s}{\mathbf{B}}$

$$\dfrac{-\mathbf{V}_s}{\mathbf{B}} = \dfrac{V_s / 180°}{B / \underline{\beta}} = \dfrac{V_s}{B} / \underline{180° - \beta}$$

$$\dfrac{\mathbf{V}_r(-\mathbf{V}_s)}{\mathbf{B}} = V_r / \underline{-\delta} \cdot \dfrac{V_s / 180°}{B / \underline{\beta}} = \dfrac{V_s V_r}{B} / \underline{180° - \beta - \delta}$$

$$\dfrac{-\mathbf{V}_s}{\mathbf{B}} \cdot \mathbf{DV}_s = \dfrac{DV_s^2}{B} / \underline{\Delta + 180° - \beta}$$

$$(-\mathbf{BI}_s)\left(\dfrac{-\mathbf{V}_s}{\mathbf{B}}\right) = \mathbf{V}_s \mathbf{I}_s$$

By such a multiplication it is found that all the voltage phasors are converted into volt-amperes and Fig. 12.6 is rotated through the angle $(180° - \beta)$. The modified diagram is shown in Fig. 12.7.

In Fig. 12.7, the phasor $\mathbf{V}_s \mathbf{I}_s$ representing the volt-amperes at the sending-end is inclined at an angle φ_s with the horizontal axes x-x. The projection of $\mathbf{V}_s \mathbf{I}_s$ on horizontal axis x-x gives the real power, and its projection on the vertical axis y-y gives the reactive voltamperes. The horizontal axis is, therefore, marked in watts and the vertical axis is marked in reactive volt-amperes (vars).

We have assumed that an inductive load draws positive reactive volt-amperes. We have also assumed that the sending-end power factor cos φ_s is lagging.

In order to plot the reactive volt-amperes of an inductive load in the positive, i.e., upward direction the diagram of Fig. 12.7 is rotated about the horizontal axis. This gives Fig. 12.8. The distances in Fig. 12.8 are marked as magnitudes only.

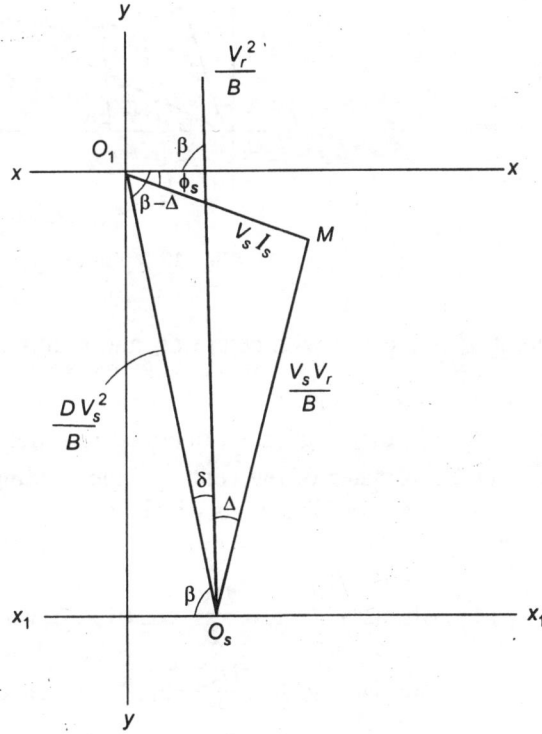

Fig. 12.7.

If V_s and V_r are kept constant the position of point O_1 becomes fixed. The distance of the operating point M from O_1, i.e., $O_1 M \left(= \dfrac{V_s V_r}{B}\right)$ is also a constant for the fixed values of V_s and V_r. Any variation in load will change the position of operating point M, but it shall remain fixed at a distance $O_1 M \left(= \dfrac{V_s V_r}{B}\right)$ from the fixed point O_1. Thus, the locus of the operating

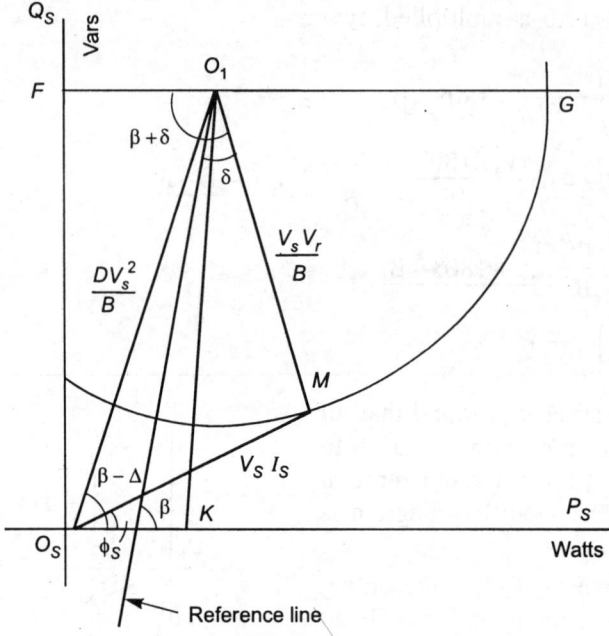

Fig. 12.8. Sending-end power circle diagram.

point M is a circle with centre O_1 and radius $O_1 M \left(= \dfrac{V_s V_r}{B} \right)$. Since the locus of the operating point is a circle with the axes of reference as active power and reactive volt-amperes at the sending-end, the diagram so obtained is called the *sending-end power circle diagram.*

The coordinates of the centre of the sending-end power circle diagram referred to a system of rectangular coordinates in which the abscissae represent the real power P_s and the ordinates represent the reactive volt-amperes Q_s at the sending-end are $(O_s K, KO_1)$ in Fig. 12.8.

$$\left. \begin{array}{l} \text{Abscissa, } O_s K = \dfrac{DV_s^2}{B} \cos (\beta - \Delta) \text{ watts} \\[3mm] \text{Ordinate, } KO_1 = \dfrac{DV_s^2}{B} \sin (\beta - \Delta) \text{ vars} \\[3mm] \text{Radius of the sending–end power circle, } \rho_s = O_1 M = \dfrac{V_s V_r}{B} \text{ volt–amperes} \end{array} \right\} \quad ...(12.7.1)$$

The distance of the centre O_1 from origin O_s is

$$O_s O_1 = \dfrac{DV_s^2}{B}$$

The equation of the power circle referred to P_s and Q_s as reference axes, centre $(O_s K, KO_1)$ and radius $O_1 M$ may be written as

$$(P_s - O_s K)^2 + (Q_s - KO_1)^2 = (O_1 M)^2$$

Substituting the values obtained from Eq. (12.7.1) in the above equation, we obtain

$$\left[P_s - \frac{DV_s^2}{B} \cos (\beta - \Delta) \right]^2 + \left[Q_s - \frac{DV_s^2}{B} \sin (\beta - \Delta) \right]^2 = \left(\frac{V_s \cdot V_r}{B} \right)^2 \qquad \dots (12.7.2)$$

$\dfrac{V_s^2}{B}$ is the *reference line* from which torque angle δ is measured on the sending-end power circle diagram.

If V_s be maintained constant then we get a family of concentric circles for various values of V_r. If V_r be kept constant then again we get a family of circles for various values of V_s, but these circles are not concentric. The centres of the circle all lie on the same line making an angle $(\beta - \Delta)$ with the horizontal.

12.8 ANALYTICAL METHOD

The sending-end power circle diagram for a transmission network may also be drawn from the consideration of phasor volt-amperes at the sending-end.

Consider the general transmission equations

$$\mathbf{V}_r = \mathbf{DV}_s - \mathbf{BI}_s$$

$$\mathbf{I}_s = \frac{\mathbf{DV}_s}{\mathbf{B}} - \frac{\mathbf{V}_r}{\mathbf{B}}$$

Let $\mathbf{V}_s$ be taken as phasor of reference. Then we have

$$\mathbf{V}_s = V_s \underline{/0^\circ}$$

$$\mathbf{V}_r = V_r \underline{/-\delta}$$

$$\mathbf{D} = D \underline{/\Delta}$$

$$\mathbf{I}_s = \frac{D \underline{/\Delta}}{B \underline{/\beta}} V_s \underline{/0^\circ} - \frac{V_r \underline{/-\delta}}{B \underline{/\beta}} = \frac{DV_s}{B} \underline{/(\Delta - \beta)} - \frac{V_r}{B} \underline{/-(\beta + \delta)}$$

The conjugate of $\mathbf{I}_s$ is given by

$$\mathbf{I}_s^* = \frac{DV_s}{B} \underline{/(\beta - \Delta)} - \frac{V_r}{B} \underline{/(\beta + \delta)}$$

Phasor volt-amperes at the sending-end is given by

$$\mathbf{S}_s = \mathbf{V}_s \mathbf{I}_s^*$$

$$P_s + j\,Q_s = \frac{DV_s^2}{B} \underline{/(\beta - \Delta)} - \frac{V_s V_r}{B} \underline{/(\beta + \delta)}$$

$$= \frac{DV_s^2}{B} [\cos (\beta - \Delta) + j \sin (\beta - \Delta)] - \frac{V_s V_r}{B} [\cos (\beta + \delta) + j \sin (\beta + \delta)]$$

$$= \left[\frac{DV_s^2}{B} \cos (\beta - \Delta) - \frac{V_s V_r}{B} \cos (\beta + \delta) \right] + j \left[\frac{DV_s^2}{B} \sin (\beta - \Delta) - \frac{V_s V_r}{B} \sin (\beta + \delta) \right]$$

$$\dots (12.8.1)$$

Separating the real and imaginary parts, we get

$$P_s = -\frac{V_s V_r}{B} \cos(\beta + \delta) + \frac{DV_s^2}{B} \cos(\beta - \Delta) \qquad \text{...(12.8.2)}$$

$$Q_s = -\frac{V_s V_r}{B} \sin(\beta + \delta) + \frac{DV_s^2}{B} \sin(\beta - \Delta) \qquad \text{...(12.8.3)}$$

From Eqs. (12.8.2) and (12.8.3) we obtain

$$\left[P_s - \frac{DV_s^2}{B} \cos(\beta - \Delta)\right]^2 + \left[Q_s - \frac{DV_s^2}{B} \sin(\beta - \Delta)\right]^2$$

$$= \left(\frac{V_s V_r}{B}\right)^2 [\cos^2(\beta + \delta) + \sin^2(\beta + \delta)]$$

$$= \left(\frac{V_s V_r}{B}\right)^2 \qquad \text{...(12.8.4)}$$

If V_s and V_r are kept constant, Eq. (12.8.4) represents a circle referred to a system of rectangular coordinates in which the abscissae represent the real power P_s at the sending-end, and the ordinates represent reactive volt-amperes Q_s at the sending-end. Eq. (12.8.4) is, therefore, called the *power-circle equation at the sending-end* of a transmission network.

Coordinates of the centre of power circle :

$$\left.\begin{array}{l} \text{Abscissa, } P_{so} = \dfrac{DV_s^2}{B} \cos(\beta - \Delta) \text{ watts} \\[3mm] \text{Ordinate, } Q_{so} = \dfrac{DV_s^2}{B} \sin(\beta - \Delta) \text{ vars} \\[3mm] \text{Radius of the circle, } \rho_s = \dfrac{V_s V_r}{B} \text{ volt--amperes} \end{array}\right\} \qquad \text{...(12.8.5)}$$

Eq. (12.8.4) can be written in the form

$$(P_s - P_{so}) + (Q_s - Q_{so})^2 = \rho_s^2 \qquad \text{...(12.8.4a)}$$

When V_s and V_r are expressed in line-to-line kilovolts, coordinates of the centre of the power circle :

$$\left.\begin{array}{l} \text{Abscissa, } P_{so} = \dfrac{DV_s^2}{B} \cos(\beta - \Delta) \text{ MW} \\[3mm] \text{Ordinate, } Q_{so} = \dfrac{DV_s^2}{B} \sin(\beta - \Delta) \text{ MWAr} \\[3mm] \text{Radius of the circle, } \rho_s = \dfrac{V_s V_r}{B} \text{ MVA} \end{array}\right\} \qquad \text{...(12.8.6)}$$

It is observed that Eqs. (12.8.5), obtained analytically from the considerations of phasor volt-amperes at the sending-end, are identical with Eqs. (12.7.1), which were derived graphically from the sending-end voltage phasor diagram. Thus, both graphical and analytical methods give similar results.

12.9 POWER TRANSMITTED OVER A LINE

The power received for a torque angle δ is given by Eq. (12.5.2) as

$$P_r = \frac{V_s V_r}{B} \cos(\beta - \delta) - \frac{A V_r^2}{B} \cos(\beta - \alpha)$$

The same result may be derived by referring to receiving-end power circle diagram given in Fig. 12.3.

$$P_r = V_r I_r \cos \varphi_r = O_r N_1 = FN = O_1 N - O_1 F$$

$$P_r = O_1 M \cos(\beta - \delta) - O_1 O_r \cos(\beta - \alpha)$$

$$= \frac{V_s V_r}{B} \cos(\beta - \delta) - \frac{A V_r^2}{B} \cos(\beta - \alpha) \qquad \qquad ...(12.9.1)$$

For given values of V_s and V_r there is a maximum amount of power which can be received by the load. The power received is a maximum when $\delta = \beta$. Thus, the maximum power received by the load is

$$P_{rm} = \frac{V_s V_r}{B} - \frac{A V_r^2}{B} \cos(\beta - \alpha) \qquad \qquad ...(12.9.2)$$

The maximum power is called the *power limit* or the *load limit*.

The maximum power is given by the maximum horizontal coordinate of the power circle. In Fig. 12.3,

$$P_{rm} = FG = O_1 G - O_1 F$$

$$= \frac{V_s V_r}{B} - \frac{A V_r^2}{B} \cos(\beta - \alpha) \qquad \qquad ...(12.9.2)$$

Since $O_1 F = \rho_r$ = radius of the circle

$O_1 F = P_{ro}$ = horizontal coordinate of the centre.

Therefore, the maximum power P_{rm} is equal to the difference between the numerical values of ρ_r and P_{ro}, i.e., $P_{rm} = |\rho_r| - |P_{ro}|$.

The corresponding reactive volt-amperes for maximum power received at the load can be found from Eq. (12.5.4) by substituting $\beta = \delta$ in it. It is given by

$$Q_{rm} = -\frac{A V_r^2}{B} \sin(\beta - \alpha) \qquad \qquad ...(12.9.3)$$

In Fig. 12.3, Q_{rm} is given by $O_r F$, i.e.,

$$Q_{rm} = O_r F = -\frac{A V_r^2}{B} \sin(\beta - \alpha) \qquad \qquad ...(12.9.3)$$

The power delivered to the line at the sending-end is written from Eq. (12.8.2) as

$$P_s = -\frac{V_s V_r}{B} \cos(\beta + \delta) + \frac{D V_s^2}{B} \cos(\beta - \Delta) \qquad \qquad ...(12.9.4)$$

$$= \frac{V_r V_s}{B} \cos[180 - (\beta + \delta)] + \frac{D V_s^2}{B} \cos(\beta - \Delta)$$

The power delivered is a maximum when

$$\beta + \delta = 180°$$

or $\qquad \delta = 180° - \beta$

$\therefore \qquad P_{sm} = \dfrac{V_s V_r}{B} + \dfrac{DV_s^2}{B} \cos(\beta - \Delta)$

From Fig. 12.8, the maximum power delivered to the line is given by FG, which is the maximum horizontal projection of $V_s I_s$.

$$P_{sm} = FG = O_1 G + FO_1$$

$$= \dfrac{V_s V_r}{B} + \dfrac{DV_s^2}{B} \cos(\beta - \Delta) \qquad \qquad ...(12.9.5)$$

The corresponding reactive volt-amperes at the sending-end are obtained from Eq. (12.8.3) by substituting $\beta + \delta = 180°$ in it.

$\therefore \qquad Q_{sm} = \dfrac{DV_s^2}{B} \sin(\beta - \Delta) \qquad \qquad ...(12.9.6)$

The same result may be obtained from the sending-end power circle given by Fig. 12.8.

$$Q_{sm} = O_s F = \dfrac{DV_s^2}{B} \sin(\beta - \Delta) \qquad \qquad ...(12.9.6)$$

12.10 PER UNIT VALUE CIRCLE DIAGRAMS

For the receiving-end power circle diagram, with constant V_r, the coordinates of the centre of the circle remain the same while the radius of the circle depends and increases with V_s. With different values of V_s the power diagrams are concentric circles. However, if V_s is kept constant and V_r is varied, both the coordinates of the circle and the radii vary and the circle diagrams are no more concentric. Hence the receiving-end power circle chart is valid only for a given receiving-end voltage for which it is constructed. For each receiving end voltage a new chart is to be prepared. Nevertheless, it can be proved that all the centres of the circles lie on the same line through the origin. Similarly, for the sending-end power diagrams, the circles are concentric only for constant V_s, and they are not concentric for different values of V_s.

If the sending-end and receiving-end circle diagrams are drawn on the same chart, the chart is required to be prepared for one receiving end voltage, or one sending-end voltage, and one set of circles will not be concentric.

In power system analysis, many quantities are given in power unit values, and so it is advantageous to solve the problems in terms of per unit values. Per unit values are dimensionless and hence, independent of the actual values of the quantities. Charts constructed in per unit values will be useful for different sets of sending-end and receiving-end voltages.

12.11 RECEIVING-END PER UNIT CIRCLE DIAGRAMS

Let V_b be the scalar value base voltage at both the ends of the line.

Then, $\quad\dfrac{\text{sending–end actual voltage}}{\text{base voltage}} = \dfrac{V_s}{V_b} \overset{\Delta}{=} v_s =$ per unit voltage at the sending end

and $\quad\dfrac{\text{receiving–end actual voltage}}{\text{base voltage}} = \dfrac{V_r}{V_b} \overset{\Delta}{=} v_r =$ per unit voltage at the receiving end.

The quantities on the previous diagrams are divided by $\dfrac{V_b^2}{B}$. Such a division results in the following results :

$$\text{Radius of the receiving–end circle} = \frac{V_s}{V_b} \cdot \frac{V_r}{V_b} = v_s\, v_r \text{ pu}$$

Coordinates of the centre of the receiving–end circle :

$$\left. \begin{aligned} \text{Abscissa} &= -A \left(\frac{V_r}{V_b}\right)^2 \cos(\beta - \alpha) = -A\, v_r^2 \cos(\beta - \alpha) \text{ pu} \\[2mm] \text{Ordinate} &= -A \left(\frac{V_r}{V_b}\right)^2 \sin(\beta - \alpha) = -A\, v_r^2 \sin(\beta - \alpha) \text{ pu} \end{aligned} \right\} \quad \ldots(12.11.1)$$

The receiving-end power circles are drawn for fixed values of v_r and various values of v_s with the help of Equation (12.11.1). We get concentric circles with their centres at n_r and origin O_r. For a different value of v_r, but a fixed value of $v_s\, v_r$, we get another set of concentric circles with their centres at n'_r and origin O_r (Fig. 12.9).

The coordinates of centre n_r are found by Eq. (12.11.1). The point n'_r lies on the line $n_s\, O_r$. If we shift the point n'_r so that it coincides with n_r both the families of the circles coincide.

It is observed that

$$n_r\, O_r = [(\text{abscissa})^2 + (\text{ordinate})^2]^{1/2}$$
$$= A v_r^2 \text{ pu}$$

For each value of v_r there is a new origin on the line of origins $n_r\, O_r$. The reference line is drawn n_r making an angle β with the horizontal. The line of origin $n_r\, O_r$ which makes an angle $(\beta - \alpha)$ with the horizontal is also drawn through n_r. The torque angles are measured from the reference line. Concentric circles for convenient per unit values $v_s\, v_r$ are drawn with centre n_r.

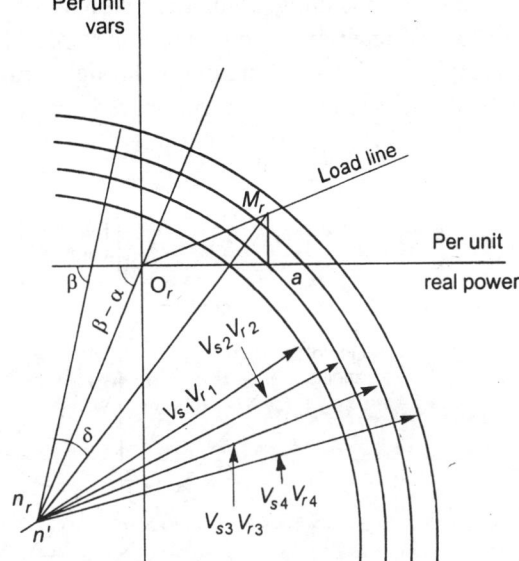

Fig. 12.9. Per unit receiving-end circle diagram.

For a given value of v_r let O_r (Fig. 12.9) be the origin. A line $O_r\, M_r$ is drawn through O_r making an angle φ_r (phase angle of the load) with the horizontal through O_r. $O_r\, M_r$ is then the *load line* for the inductive load. Real power is read to the right of O_r and positive vars and read upwards from O_r. The point M_r, which is the intersection of the load line and the power circle for given v_s, is called the *operating point*. It corresponds to an inductive load in Fig. 12.9. The horizontal and vertical components of $O_r M_r$ determine the per unit real power and vars respectively. These per unit values when multiplied by $\dfrac{V_b^2}{B}$ give the three-phase real power and the vars respectively, provided that V_b is expressed as line-to-line voltage in determining the per unit quantities. The torque angle δ measured from the reference line is shown in Fig. 12.9.

12.12 SENDING-END PER UNIT CIRCLE DIAGRAMS

If V_b is the same base voltage employed for the receiving-end diagrams, then radius of the sending-end circle is $v_s v_r$ pu, and the centre of the circle has the following coordinates :

$$\text{Abscissa} = - D\, v_s^2 \cos (\beta - \Delta) \text{ pu}$$

$$\text{Ordinate} = - D\, v_s^2 \sin (\beta - \Delta) \text{ pu}$$

The power diagrams can now be drawn for different values of v_s, but for a constant value of the product $v_s v_r$, as in the case of the power diagrams at the receiving-end. The centres of these circles will be on the same line through the origin. If such of centres n'_s, n''_s are shifted to coincide with n_s, all the circles coincide and the origin O_s in Fig. 12.10 is shifted to the same extent. The operating point for given v_s, v_r and P_s can be obtained in the same manner as in the case of receiving-end power diagrams. The torque angle can be measured from a line making an angle β with the horizontal through origin O_s. The line of origins makes an angle $(\beta - \Delta)$ with the horizontal. It can be proved that the distance

$$n_s\, O_s = D v_s^2 \text{ pu}$$

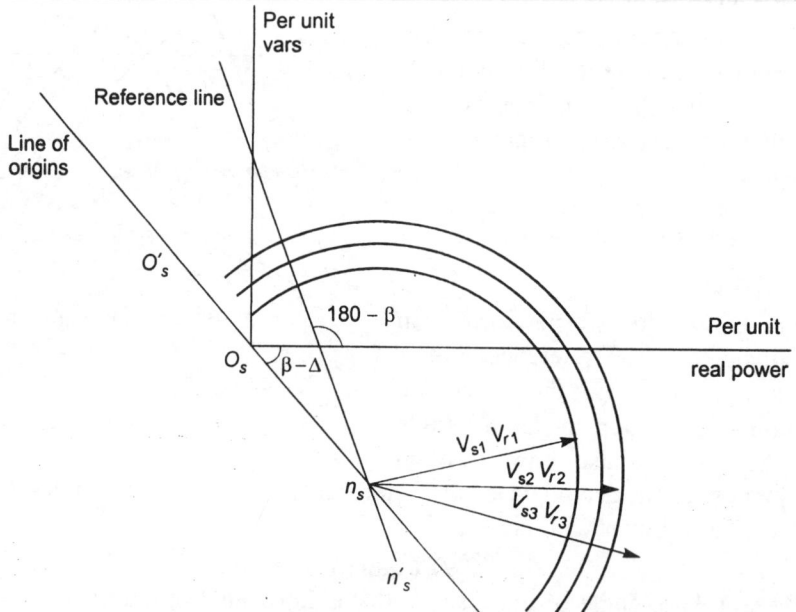

Fig. 12.10. Per unit sending-end circle diagram.

12.13 UNIVERSAL POWER CIRCLE DIAGRAMS

The same set of per unit value power circle diagrams, which can be used to study the sending-end and receiving-end conditions, are called *universal diagrams*. Only one set of circle is adequate for these purposes if certain modifications are made in the origins of per unit value circle diagrams.

The radii of the circles, i.e., $v_s\, v_r$ are the same in both the sets of circles. But the centres of the circles are at different points on the same respective lines or alternatively, the origins are at different points on the respective lines from the corresponding centres. Both the set of circle diagrams are normally drawn such that the inductive vars are read positive upwards from the origin O_r or O_s. If this convention is to be maintained, it can be seen that it is not possible to use only one set of circles for both the terminal conditions as they cannot be made to coincide by rotating them suitably. To make possible the same set of circles for both ends, the circle diagrams for sending-end power are constructed as shown in Fig. 12.11, by which they can be made to coincide with receiving-end circles. But this necessitates the reading of inductive vars vertically downwards positive from O_s against the usual convention. With this altered convention for the vars for the sending end both sets of circle can be made to coincide at one point n.

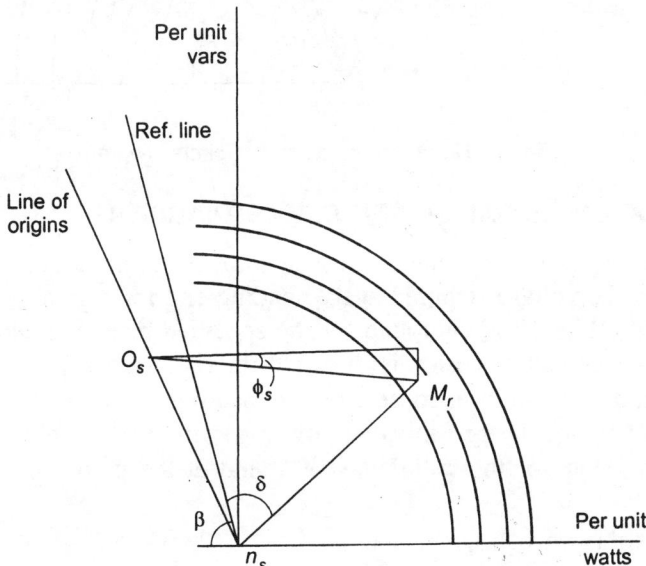

Fig. 12.11. Per unit sending-end circle diagram.

Fig. 12.11 shows the circles having the radii equal to the same per unit values of $v_s\, v_r$ as the radii of the receiving-end circles. In Fig. 12.11 the line of origins is drawn at an angle of $180° - (\beta - \Delta)$ with the horizontal. The distance of origin O_s corresponding to a given value of v_s is equal to Dv_s^2 per unit. Sending-end reference line makes an angle $(180° - \beta)$ with the horizontal. The real power is read to the right from the point O_s and positive vars are read downwards from O_s. The point M_s in Fig. 12.11 corresponds to an inductive load. The torque angle δ for the load is also shown in Fig. 12.11. It is measured from the sending-end reference line.

When the points n_r and n_s are made to coincide at n the two sets of the circles also coincide. The combined diagram of both the sets is shown in Fig. 12.12.

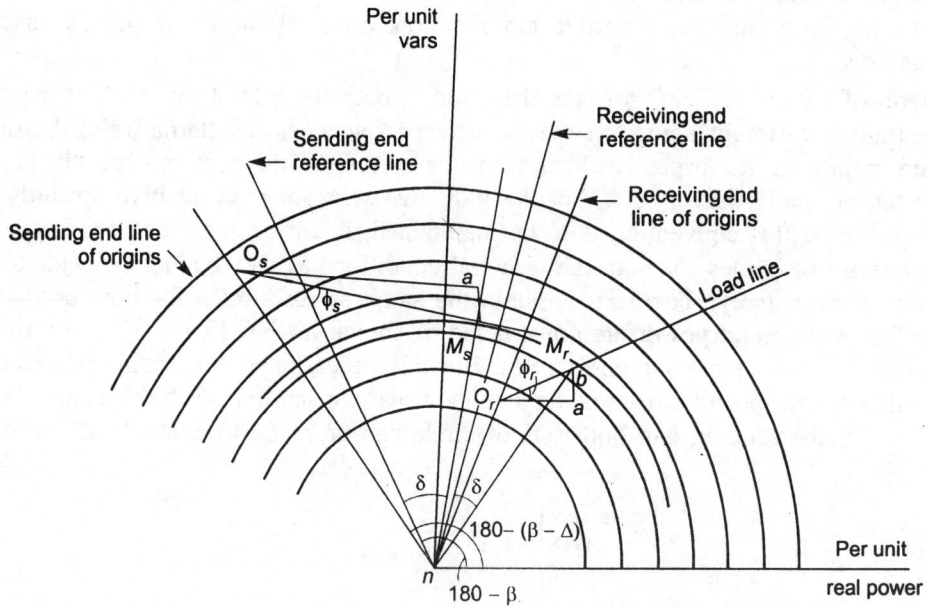

Fig. 12.12. A universal power circle diagram.

12.14 THE USE OF UNIVERSAL POWER CIRCLE DIAGRAMS

Circle diagrams may be used conveniently to give a lot of information. For example, they can be used to determine the voltage required at the sending-end for a given load and receiving-end voltage. The origin O_r (Fig. 12.12) is found for the specified receiving-end voltage v_r. The load line for the given power factor, $\cos \varphi_r$ is drawn from O_r at the angle φ_r from the horizontal through O_r. The operating point M_r is located on the load line such that the horizontal component $O_r a$ of the line $O_r M_r$ gives the per unit real power and the vertical component $a M_r$ gives the vars at the load. The value of the radius $v_s v_r$ is found at the point M_r. Since v_r is known, v_s can be calculated.

When the load is thrown off, I_r becomes zero.

$$\therefore \qquad V_s = A V_{ro}$$

and $\qquad v_s = A v_{ro}$

where v_{ro} is per unit receiving-end voltage at no load.

$$v_{ro} = \frac{v_s}{A}$$

The regulation $= \dfrac{v_{nl} - v_{fl}}{v_{fl}} = \dfrac{v_{ro} - v_r}{v_r}$ pu.

The circle diagrams can also be employed to determine the vars that must be supplied by the synchronous or static capacitors at the receiving-end to improve the power factor and reduce voltage regulation.

In order to keep a constant voltage v_{rl} at the receiving-end for a given sending-end voltage and power, the amount of vars to be supplied by synchronous or static capacitors may be determined.

The origin O_r and the point M_r corresponding to load are located in the usual manner. If a vertical through a intersects the circle of radius $v_{s1}\,v_{r1}$ at b, then $a\,M_r$ represents the total vars required by the load. The length ab represents the vars supplied by the line for the specified voltages v_{rl} and v_{s1} and load $O_r\,a$. The length $b\,M_r$ determines the amount of vars to be supplied by the synchronous or static capacitors.

The improved power factor at the receiving-end $= \cos\underline{/b\,O_r\,a} = \cos\left(\tan^{-1}\dfrac{ab}{O_r\,a}\right)$

In order to have unity power factor at the load all the power $a\,M_r$ must be supplied by the synchronous capacitor. The voltage at the sending-end is then found by determining the value of $v_s\,v_r$ at a.

The sending-end quantities namely real power and vars can be determined if the voltage, power, and power factor at the receiving-end and torque angle are known. For the given load and receiving-end origin O_r and the load point M_r are determined. The torque angle δ is then measured from the receiving-end reference line. It is the angle $O_r\,n\,M_r$. The value of $v_s\,v_r$ is found at M_r. The sending-end reference line and line of origins are drawn. A line is drawn through n making an angle δ with the sending-end reference line. The point M_s is obtained by the intersection of this line and the circle of radius $v_s\,v_r$ already drawn. The value of v_s being known, the origin O_s may be located at a distance Dv_s^2 per unit on the line of origins. The horizontal and vertical components of $O_s\,M_s$ give the real power and vars at the sending-end.

Power factor at the sending-end,

$\cos\varphi_s = \cos\underline{/M_s\,O_s\,a'}$

$$= \cos\left(\tan^{-1}\frac{\text{sending–end vars}}{\text{sending–end real power}}\right) = \cos\left(\tan^{-1}\frac{M_s\,a'}{O_s\,a'}\right)$$

The power loss in the line is the difference between the sending-end and receiving-end powers.

Power loss $= P_s - P_r = O_s\,a' - O_r\,a$ pu.

EXERCISES

1. Explain the function of a synchronous phase modifier placed at the receiving-end of the transmission line.
2. Derive the equation of a receiving-end power circle diagram in terms of general circuit constants. Show how this diagram can be used to determine (a) the maximum power that can be transmitted, and (b) synchronous phase modifier rating under given operating conditions.

3. What is a universal power circle diagram of a transmission line? Explain its construction. What information can be obtained from it?

4. In a three-phase line with 132 kV at the receiving-end the following are the constants :

$$A = D = 0.98 \underline{/3°}, \ B = 110 \underline{/75°} \ \Omega, \ C = 0.0005 \underline{/88°} \ S.$$

If the load at the receiving-end is 50 MVA at 0.8 lagging power factor, determine graphically or otherwise (a) the sending-end voltage, and (b) the leading reactive MVA at the receiving-end if the sending-end voltage is 140 kV.

5. A three-phase transmission line has a resistance of 6 Ω/phase and a reactance of 20 Ω/phase. The sending-end voltage is 66 kV and the voltage at the receiving-end is maintained constant by a synchronous phase modifier. Determine the MVAr of the synchronous phase modifier when the load at the receiving-end is 75 MW at 0.8 power factor lagging, and also the maximum load which can be transmitted over the line, the voltage being 66 kV at both ends.

6. A three-phase transmission line has a resistance per phase of 5 Ω and an inductive reactance per phase of 12 Ω and the line voltage at the receiving-end is 33 kV. Determine the voltage at the sending-end when the load at the receiving-end is 20 MVA at 0.8 power factor lagging.

   The voltage at the sending-end is maintained constant at 36 kV by means of a synchronous phase modifier at the receiving-end, which has the same rating at zero load at the receiving-end as for the full load of 16 MW. Determine the power factor of the full load output and the rating of the synchronous phase modifier. (L.U.)

7. Explain how a receiving-end power circle is drawn. Derive an expression for the maximum received power with the sending-end receiving-end voltages kept constant.

8. A 132 kV three phase line has the following line constants :

$$A = 0.9 \underline{/2.5°}, \ B = 100 \underline{/70°} \ \Omega, \ C = 0.0006 \underline{/80°} \ S.$$

Draw the receiving-end power circle for a load of 40 MW at 0.8 power factor lagging at the receiving-end and determine the sending-end voltage.

ANSWERS

4. (a) V_{sl} = 165 kV (b) Q_{pm} = 34.5 MVAr   **5.** 96.8 MVAr ; 148 MW

6. 40.1 kV ; 0.9 (lag) 7.92 MVAr   **8.** V_{sl} = 151 kV

13

Control of Voltage and Reactive Power

13.1 INTRODUCTION

For a short line, the magnitude of voltage drop in the line is

$$|\Delta V| = |V_s| - |V_r|$$

$$|\Delta V| = IR \cos \varphi_r + IX \sin \varphi_r = \frac{1}{V_r}[(V_r I \cos \varphi_r) R + (V_r I \sin \varphi_r) X] = \frac{RP_r + XQ_r}{V_r}$$

For a transmission line, $X \gg R$ and R is negligibly small. Therefore

$$|\Delta V| = \frac{XQ_r}{V_r} \qquad Q_r = \frac{V_r}{X}|\Delta V|$$

This relation shows that the reactive power Q_r is proportional to the magnitude of the voltage drop in the line. Thus voltage control and reactive power control are interrelated. The voltage at the consumer terminals must be maintained constant within prescribed limits irrespective of the type and magnitude of the load. The maintenance of constant voltage is a complicated problem due to fact that the system is supplied from many sources and it supplies many loads at several voltage levels.

In order to maintain the voltages at their prescribed values at all times, it is necessary to maintain the balance of reactive power in the system. In other words, the reactive power generation should be exactly equal to the reactive power consumed (absorbed). Any mismatch in the reactive power balance affects the bus voltage magnitudes without much affecting the system frequency.

In practice, it is difficult to maintain the balance of reactive powers because of the varied unpredictable demands of the consumers. Therefore an unbalance always exists between the supply and demand conditions of reactive power. Hence there is a variation of voltages from their respective rated values.

Whereas active power is only supplied by generators, reactive power may be supplied from several sources. A list of generators and consumers of reactive power is as follows:

1. *Generators of Q*
   (a) Synchronous generators and synchronous motors
   (b) Static capacitors
   (c) Distributed capacitance of overhead lines, and cables.

2. *Consumers of Q*
   (a) Synchronous machines
   (b) Inductive static loads
   (c) Induction motors
   (d) Distributed inductance of overhead lines and cables
   (e) Transformer inductance.

We know that an overexcited synchronous machine (generator or motor) produces reactive power and acts as a shunt capacitor. Similarly, an underexcited synchronous machine consumes reactive power and acts as a shunt inductor.

Lightly loaded transmission lines or lines under no-load conditions supply reactive power because of their shunt capacitance. When the lines are fully loaded, they operate beyond their 'natural load' and absorb reactive power. Underground cables supply reactive power under all conditions of loading. Transformers absorb low reactive power under no-load conditions. When the transformers are fully loaded, they absorb high values of reactive power.

Inductive reactors absorb reactive power. They may be connected in series or parallel with the networks. While series reactors are used for limiting the fault currents, shunt reactors are used for reactive power control. Shunt reactors are connected in the transmission circuit during light-load conditions.

Capacitors generate reactive power and may be connected in series or parallel with the system. Series capacitors compensate the line reactance in long overhead lines and thus improve the stability limit. Shunt capacitors are used for reactive compensation.

The voltage at the receiving end of short lines is maintained nearly constant by adjusting the field excitation of the alternators at the sending end. For long lines this method is not suitable as it results in greater line losses and voltage drop.

13.2 METHODS OF VOLTAGE CONTROL

The voltages at different buses of the power system vary with the changes in load. The voltage is normally high at light-load conditions and low at heavy-load conditions. To keep network voltages within permissible limits, means must be provided to control the voltage, that is, to increase the circuit voltage when it is too low and to reduce it when it is too high.

The following methods are used for voltage control in a power system :

1. Tap-changing transformers
2. Shunt reactors
3. Synchronous phase modifiers
4. Shunt capacitors

5. Series capacitors

6. Static VAR systems (SVS).

Adjusting the system voltage by means of shunt reactive elements is known as *shunt compensation*. The shunt compensation consists of (a) *static shunt compensation*, and (b) *synchronous compensation*.

The static shunt compensation uses shunt reactors, shunt capacitors, and static var system (SVS). In synchronous compensation, synchronous phase modifiers are used.

Adjusting the system voltage by connecting capacitors in series with the line is called *static series compensation*.

13.3 TAP-CHANGING TRANSFORMERS

The change of voltage is affected by changing the number of turns of the transformer provided with taps. For sufficiently close control of voltage, taps are usually provided on the high voltage winding of the transformer. There are two types of tap-changing transformers :

(a) Off-load tap-changing transformers

(b) On-load tap-changing transformers.

Tap-changing is the most widely used method of controlling voltage at all levels.

13.3.1 Off-Load Tap-Changing Transformer

Fig. 13.1 shows the off-load changing transformer. In this method, the transformer is disconnected from the main supply when the tap setting is to be changed. The tap setting is usually done manually.

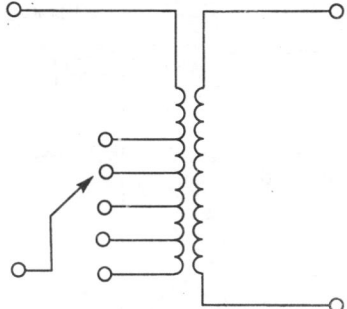

13.3.2 On-Load Tap-Changing Transformer

In order that the supply may not be interrupted, on-load tap changing transformers are used. Such a transformer is known as tap-changing under-load (TUCL) transformer. While tap changing, two essential conditions are to be fulfilled, namely :

Fig. 13.1. Off-load tap-changing transformer.

1. The load circuit should not be broken to avoid arcing and prevent the damage of contacts.

2. No part of the winding should be short circuited while adjusting the taps, as the contacts and the short-circuited part of the winding may be damaged.

Fig. 13.2 shows the tap changing employing a centre-tapped reactor R. Here S is the diverter switch and 1, 2, 3, 4, 5 are selector switches. The transformer is in operation with switches 1 and S closed. To change to tap 2, switch S is opened and 2 closed. Switch 1 is then opened and S closed to complete the tap change. It is to be noted that diverter switch operates on load, and no current flows in the selector switches during tap changing. During normal operation the current flows in the reactor in both the halves in opposite directions, and therefore, the reactor cores is unmagnetised and the reactance voltage drop is eliminated. During the tap change only half of the reactance which limits the current, is connected in the circuit.

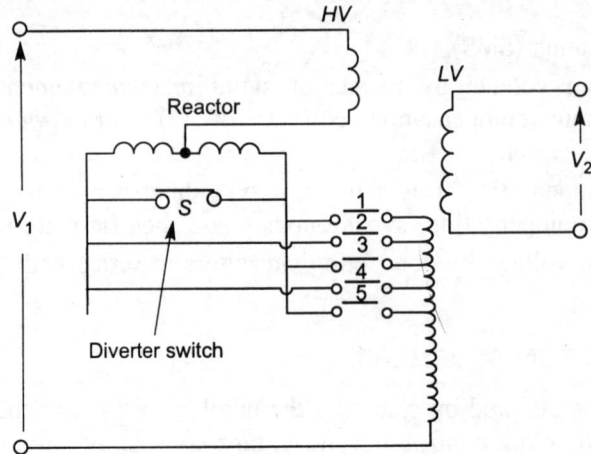

Fig. 13.2. On-load tap changing using a reactor.

A schematic block diagram of a control system for an on-load tap changing transformer is shown in Fig. 13.3. The voltage regulating equipment for automatic control on load tap changer comprises a line drop compensator, voltage sensitive regulating relay, time delay relay, etc.

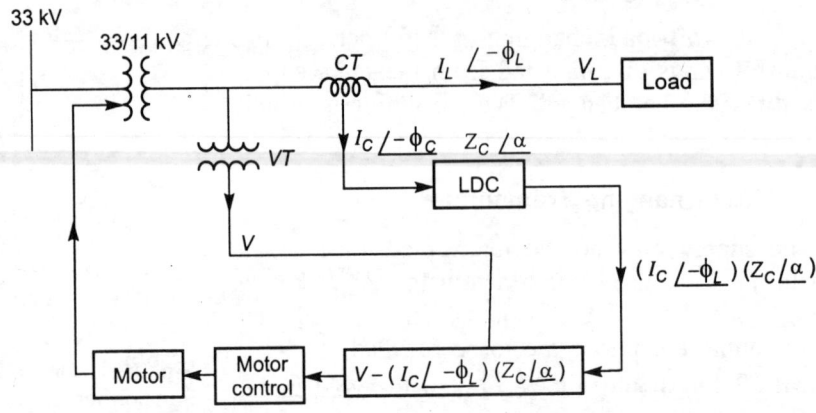

Fig. 13.3. Block diagram of a control system for an on-load tap-changing transformer.

The line drop compensator (LDC) is a replica of transmission line. It consists of adjustable resistor and reactor elements. A time delay relay is provided with the voltage sensitive relay. The time delay in the voltage sensing circuit prevents the tap-changing operation during transient voltages or short-time disturbances.

The LDC is connected to the secondary of the current transformer (CT). The line voltage is sensed from secondary of a voltage transformer (VT) by voltage sensitive relay. The variation of load current producing voltage drop in the transmission line is simulated in LDC. The voltage drop across LDC is injected to the voltage regulator circuit.

When the line voltage varies beyond certain set value, the voltage sensitive relay sends command to on-load tap changer through motor driven mechanism. There are high and low limits of travel for tap-adjusting mechanism. The motor rotates in the required direction to adjust the tap.

13.4 SHUNT REACTORS

Shunt reactor is an inductive current element connected between line and neutral to compensate for capacitive current from transmission lines or underground cables.

Shunt reactors are used in long distance EHV and UHV transmission lines. They are needed when the line is to be charged or when the line is lightly loaded. Under these conditions, the shunt capacitance of the line predominates and the receiving-end voltage is higher than the sending-end voltage. This is known as Ferranti effect. Shunt reactors are used to compensate the capacitive VAr of the line and therefore, the voltage is regulated within the prescribed limits. During normal loaded conditions, the series inductive reactance of the line produces a voltage drop of IX_L and the voltage at the receiving end of the line is reduced. Therefore, shunt reactors are switched off under normal loaded conditions of the line.

Shunt reactors are installed in sending-end substations, receiving-end substations and intermediate substations of long EHV and UHV AC lines. For very long lines, shunt reactors are installed at an interval of about 300 km in intermediate substations to limit the voltage at the intermediate points during low loads.

Shunt reactors are connected to the tertiary windings of power transformers via circuit breakers. EHV shunt reactors may be connected to the transmission line directly without any switching device.

In construction the shunt reactors are identical with power transformers, except for their cores. They have similar windings, tanks, bushings, radiators, etc. An air gap is provided within the reactor core to prevent magnetic saturation.

Shunt reactors are subjected to overvoltages like power transformers. To avoid insulation failures, reactors are to be removed at such occasions when they are most needed. It is difficult to switch off large shunt reactors as the switching overvoltages are of high magnitudes.

13.5 SHUNT CAPACITORS

Shunt capacitors are the capacitors connected in parallel with the lines. They are installed near the load terminals, in receiving-end substations, distribution substations and in switching substations. Shunt capacitors inject leading reactive voltamperes (VAr) to counteract some or all of lagging inductive VAr at the point of installation. They are arranged in three-phase banks.

A serious disadvantage with shunt capacitors is that at no load or light load, the receiving-end voltage may considerably exceed the sending-end voltage (Ferranti effect). To overcome this difficulty the capacitor bank is provided with fixed and variable elements. When the voltage rises the variable elements may be removed (switched off) from the bank to decrease the capacitance. With the fall of voltage the variable elements are added to (switched in) the bank to increase the capacitance. Thus, shunt capacitors are switched in during heavy, low power factor loads.

In transmission lines shunt capacitors are connected either to the tertiary winding of the power transformers or to the busbars. Most of the industrial loads (induction motors, welding equipment, furnace transformers, etc.) draw inductive currents of poor power factor. Shunt capacitors are used with individual equipment to improve power factor. The improved power factor reduces the kVA drawn from the supply. The power losses are reduced and the efficiency is increased. The smaller voltage drop in the line results in good voltage regulation. Thus, shunt capacitors regulate the voltage and reactive power flows at the points where they are installed.

13.6 SERIES COMPENSATION

Capacitors are connected in series with the line at suitable location. Series capacitors increase transmission capacity, improve system stability, control voltage regulation and ensure proper load division among parallel feeders. These advantages are discussed below :

(a) Increase in Power Transfer Capability

The power transfer over a line is given by

$$P_1 = \frac{V_s V_r}{X_L} \sin \delta \qquad \qquad ...(13.6.1)$$

where P_1 = power transferred per phase (W)

$\quad\quad V_s$ = sending–end phase voltage (V)

$\quad\quad V_r$ = receiving–end phase voltage (V)

$\quad\quad X_L$ = series inductive reactance of the line per phase (Ω)

$\quad\quad \delta$ = phase angle between V_s and V_r .

If a capacitor having capacitive reactance X_C is connected in series with the line, the reactance of the line is reduced from X_L to $(X_L - X_C)$. The power transfer is given by

$$P_2 = \frac{V_s V_r}{X_L - X_C} \sin \delta \qquad \qquad ...(13.6.2)$$

$$\therefore \qquad \frac{P_2}{P_1} = \frac{X_L}{X_L - X_C} = \frac{1}{1 - \dfrac{X_C}{X_L}} = \frac{1}{1 - k}$$

where $k \triangleq \dfrac{X_C}{X_L}$

The factor k is known as *degree of compensation* or *compensation factor*. Thus,

$$\text{per unit compensation, } k \triangleq \frac{X_C}{X_L} \text{ pu} \qquad \qquad ...(13.6.3)$$

$$\text{percentage compensation} \triangleq \frac{X_C}{X_L} \times 100 \text{ \%} \qquad \qquad ...(13.6.4)$$

where X_L = total series inductive reactance of the line per phase

$\quad\quad X_C$ = capacitive reactance of the capacitor bank per phase.

In practice, k lies between 0.4 and 0.7. For $k = 0.5$,

$$\frac{P_2}{P_1} = \frac{1}{1 - k} = \frac{1}{1 - 0.5} = 2$$

Thus, the power transfer is doubled by 50% compensation.

(b) Improvement in System Stability

From Eqs. (13.6.1) and (13.6.2), it is seen that for the same power transfer and for the same values of V_s and V_r, the phase angle δ in case of series compensated line is less than that for the uncompensated line. The reduced value of δ gives higher stability.

(c) Load Division among Parallel Lines

Series capacitors are used in transmission systems to modify the load division between parallel lines. Suppose that a new line with large power transfer capability is to be paralleled with an already existing line. It may be difficult to load the new line without overloading the old line. In such a case the reduction of series reactance by series compensation ensures proper load division among parallel circuits for maximum power transfer and reduced losses.

(d) Control of Voltage

An advantage of using series capacitors is that there is an automatic change in VAr with the change in load current and thus drops in voltage levels due to sudden load variations are corrected instantly.

Series capacitors are very commonly used with EHV and UHV lines.

13.7 LOCATION OF SERIES CAPACITORS

Many technical and economic considerations decide the location of series capacitors. The series capacitors may be located at the sending end, receiving end, or at the centre of the line. Sometimes they are distributed at two or more points along the line. The degree of compensation and the characteristics of the line decide the location of the capacitors. Their installation at the terminals provides the facility of maintenance but the overvoltages appearing across the terminals of the capacitors under fault conditions will overstress the capacitors. The capacitors are installed in the intermediate switching stations of comparatively long lines. The location at the centre of the line also reduces the rating of the capacitor. The *rating* of the series capacitor is given by

$$Q_C = 3I^2 X_C \times 10^{-6} \text{ MVAr}$$

where I is the line current. Capacitor banks consist of small units connected in series, parallel, or both to get the desired voltage and VAr rating.

13.8 PROTECTIVE SCHEMES FOR SERIES CAPACITORS

As the capacitors are in series with the line, sometimes they have to carry large currents due to overloads or sustained faults. The overcurrents will cause excessive voltage drop across the capacitors. The design, based on these voltage drops, is, of course, not economical. Also, it is not desirable to remove them from the circuit at such instants to maintain stability. To protect the capacitors from such abnormal voltages, spark gaps and surge diverters are connected across the capacitor terminals. A circuit breaker is also connected in parallel with it. The protective gap is usually designed to spark at two or four times the rated normal voltage. Normally, the breaker is open, but when the gap breaks down, the breaker closes. When the line current comes to normal, the breaker opens nd thereby, the capacitor in the line is reconnected. Some of the methods, to protect series capacitors are shown in Fig. 13.4.

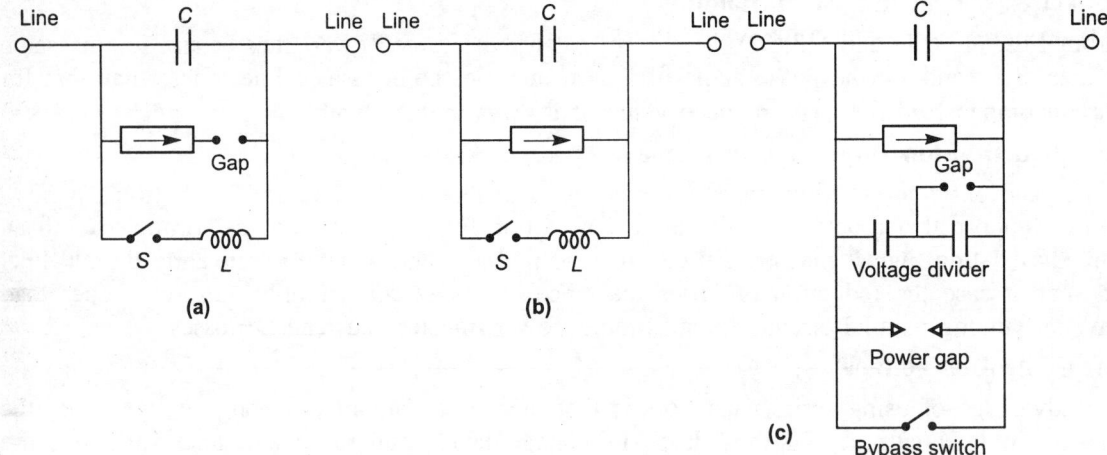

Fig. 13.4. Protective schemes for series capacitors during abnormal conditions.

13.9 PROBLEMS ASSOCIATED WITH SERIES CAPACITORS

Some of the major problems associated with the series-capacitor applications are given below :

(a) Series compensated lines have a tendency to produce series resonance at frequencies lower than power frequencies. This is known as *subsynchronous resonance* (SSR). The subsynchronous resonance currents produce mechanical resonance in turbogenerator shafts which may result in high torsional stresses in the rotor shafts. Thus turbogenerator shafts may be completely sheared and cause major damage to generating units.

Let f = normal (synchronous) frequency

f_r = subsynchronous resonance frequency of series compensated line

$X_L = 2\pi fL$ = series inductive reactance of the line at normal frequency.

At subsynchronous resonance

$$2\pi f_r L = \frac{1}{2\pi f_r C}$$

$$f_r^2 = \frac{1}{2\pi L} \cdot \frac{1}{2\pi C}$$

Dividing both sides by f^2 we get

$$\frac{f_r^2}{f^2} = \frac{1}{2\pi fL} \cdot \frac{1}{2\pi fC} = \frac{X_C}{X_L} = k$$

$$\therefore \qquad f_r = f\sqrt{k} \qquad\qquad\qquad \dots(13.9.1)$$

This relation shows that subsynchronous resonance occurs at frequency f_r, equal to the normal frequency multiplied by the square root of degree of compensation.

The condition of subsynchronous resonance may occur during faults, or switching operations. The problem of subsynchronous resonance with series compensated lines are overcome by the following methods :

(1) Use of filters, (2) bypassing the series capacitor bank under resonance conditions, and (3) tripping of generator units under resonance conditions.

(b) Switching in of an unloaded transformer at the end of a series compensated line may produce *ferroresonance*. This may result in sustained oscillations. The frequency of oscillation is an integral multiple of the system frequency. The oscillations may be suppressed by using shunt reactors across the capacitors or short circuiting the capacitors temporarily, or a bypass breaker.

(c) When starting induction motors, resonance conditions may be produced causing the motor to lock in at a fraction of the synchronous speed. A resistance may be connected in parallel at the time of starting to overcome this difficulty.

(d) Lightly loaded synchronous motors have got a tendency to hunt.

(e) Series capacitors produce high recovery voltages across the circuit breaker contacts.

(f) If the degree of compensation and location of capacitors are not proper, the distance relays used for line protection may not function properly.

13.10 SERIES CAPACITOR VERSUS SHUNT CAPACITOR

The basic function of capacitors, whether they are series or shunt, installed as a single unit or as a bank, is to regulate the voltage and reactive power flows at the point where they are installed. The shunt capacitor does it by changing the power factor of the load, whereas the series capacitor does it by directly reducing the inductive reactance of the line. A series capacitor provides for a voltage rise which increases automatically and instantaneously with the increase of the load current. Also, a series capacitor produces more net voltage rise than a shunt capacitor at lower power factors which produces more voltage drop. However, a series capacitor improves the system power factor much less than a shunt capacitor and has a little effect on source current.

13.11 SYNCHRONOUS PHASE MODIFIERS

A synchronous phase modifier is a synchronous motor running without a mechanical load. It is connected in parallel with the load at the receiving end of the line. It can generate or absorb reactive voltamperes (VAr) by varying the excitation of its field winding. It can be made to take leading current with overexcitation of the field winding. In such a case it delivers inductive (or absorbs capacitive) VAr. On the other hand, if the excitation is adjusted to a smaller value it takes a lagging current, and therefore, supplies capacitive (or absorbs inductive) VAr. When the load is high the low voltage at the far end is increased by running the machine overexcited. When the load is low, the high voltage is reduced by running the machine underexcited. Thus, the current drawn by a synchronous phase modifier can be varied from lagging to leading smoothly by varying its excitation.

The other common names given to a synchronous phase modifier are *synchronous condenser*, *synchronous capacitor* and *synchronous compensator*. It is a very convenient device to keep the receiving-end voltage constant under any condition of load. It also improves the power factor. The output can be varied smoothly. Synchronous phase modifier is connected at the receiving end of the line to the tertiary of the power transformers (Fig. 13.5). It has the disadvantage of being relatively costly. Its installation, maintenance and operation are not easy. It is difficult to increase its capacity in order to cope with the increasing demand.

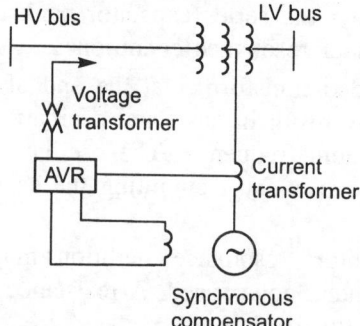

Fig. 13.5. Connection of synchronous phase modifier.

13.11.1 Rating of a Phase Modifier

The rating of a phase modifier can be determined by either analytical method or graphical method. The two methods are discussed separately.

Analytical Method

Equation (12.5.9) can be conveniently used for determining the rating of a phase modifier. It is repeated below for reference.

$$(P_r - P_{ro})^2 + (Q_r - Q_{ro})^2 = \rho_r^2 \qquad \qquad ...(13.11.1)$$

where

$$P_{ro} = -\frac{AV_r^2}{B} \cos(\beta - \alpha) \text{ MW}$$

$$Q_{ro} = -\frac{AV_r^2}{B} \sin(\beta - \alpha) \text{ MVAr} \qquad \qquad ...(13.11.2)$$

$$\rho_r = \frac{V_s V_r}{B} \text{ MVA}$$

P_r = load in MW

Q_r = MVAr supplied by the line

V_s and V_r are line-to-line voltages in kV.

For a short line having resistance R and inductance L only

$$A = 1, \qquad \alpha = 0, \qquad B = Z, \qquad \beta = \tan^{-1}\frac{X}{R}, \qquad \cos\beta = \frac{R}{Z}, \qquad \sin\beta = \frac{X}{Z}$$

$$P_{ro} = -\frac{V_r^2}{Z} \cos\beta = -\frac{V_r^2}{Z} \cdot \frac{R}{Z} = \frac{V_r^2}{Z^2} R \text{ MW}$$

$$Q_{ro} = -\frac{V_r^2}{Z} \sin\beta = -\frac{V_r^2}{Z} \cdot \frac{X}{Z} = \frac{V_r^2}{Z^2} X \text{ MVAr}$$

$$\rho_r = \frac{V_s V_r}{Z} \text{ MVA}$$

If $\cos\varphi_r$ = load power factor, then

MVAr required by the load = MVA $\sin \varphi_r$ = (MVA $\cos \varphi_r$) $\dfrac{\sin \varphi_r}{\cos \varphi_r}$ = $P_r \tan \varphi_r$

Total MVAr supplied by the phase modifiers, $Q_{pm} = P_r \tan \varphi_r - Q_r$ ...(13.11.3)

Sign Convention. It is also to be noted that inductive (lagging) VAr is positive and capacitive (leading) VAr is negative.

Maximum power transmitted. The maximum power transmitted is found from Eq. (12.9.2) as follows :

$$P_{rm} = \frac{V_s V_r}{B} - \frac{A V_r^2}{B} \cos (\beta - \alpha) = |\rho_r| - |P_{ro}|$$

Thus, P_{rm} is also equal to the difference between the numerical values of ρ_r and P_{ro}.

Graphical Methods

(a) *Receiving-end power circle diagram method.* The coordinates of the centre of the receiving-end power circle diagram referred to a system of rectangular coordinates in which the abscissae represent the real power P_r, and the ordinates represent the reactive voltamperes Q_r are given by (P_{ro}, Q_{ro})

where $P_{ro} = -\dfrac{A V_r^2}{B} \cos (\beta - \alpha)$

$Q_{ro} = -\dfrac{A V_r^2}{B} \sin (\beta - \alpha)$

The radius of the circle, $\rho_r = \dfrac{V_s V_r}{B}$

P_{ro}, Q_{ro} and ρ_r are calculated. A suitable scale is chosen. The centre O_1 (Fig. 13.6) is located by taking $O_r K$ equal to $|P_{ro}|$ and $K O_1$ equal to $|Q_{ro}|$. From O_1 an arc of the circle is drawn with calculated value of the radius ρ_r equal to $O_1 G$. The load line $O_r L$ is then drawn from the origin O_r at an angle φ_r with the horizontal. It cuts the circle at M. The receiving-end power and voltamperes are represented by $O_r N$ and MN respectively. For a given load $P_r = O_r A$ at power factor $\cos \varphi_r$ (lagging), the distance AB represents the reactive voltamperes $P_r \tan \varphi_r$ required by the load, and the ordinate AC if the receiving-end power circle represents the reactive voltamperes Q_r supplied by the line. The total VAr supplied by the phase modifier is given by

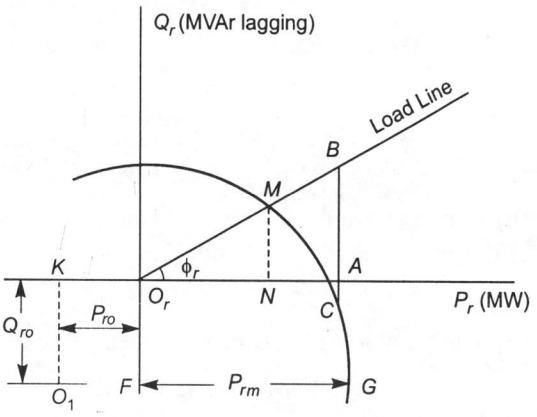

Fig. 13.6. Receiving-end power circle.

$$Q_{pm} = P_r \tan \varphi_r - Q_r$$

The vertical line BC between the load line and the receiving-end power circle represents Q_{pm}.

To find the *maximum power transferred* (P_{rm}) a horizontal line is drawn from O_r cutting the vertical axis at F and the circle at G. Then the distance FG represents P_{rm}. Alternatively, P_{rm} can be found from the difference between the numerical values of ρ_r and P_{ro}.

For minimum synchronous phase modifier capacity V_s must be such that the lagging VAr taken by the phase modifier at no load is equal to the leading VAr taken at full load.

(b) *Universal power circle diagram method.* The rating of a synchronous phase modifier can also be determined by the use of universal power circle diagram. This method is given in Section 12.14.

Example 13.1 A three-phase transmission line has a resistance 10 Ω per phase and a reactance of 30 Ω per phase. (a) Determine, graphically or otherwise, the maximum power which may be transmitted if 132 kV were maintained at each end. (b) What is the phase difference between the receiving-end and sending-end voltages for maximum power transmitted. (c) Also, determine the rating of a synchronous phase modifier required to supply 100 MW at 0.9 power factor lagging at the receiving end.

Solution

Analytical method

(a) The maximum power received by the load is

$$P_{rm} = \frac{V_s V_r}{B} - \frac{A V_r^2}{B} \cos (\beta - \alpha)$$

For a short line

$$A = 1, \qquad \alpha = 0, \qquad \mathbf{B} = \mathbf{Z}, \qquad \beta = \tan^{-1} \frac{X}{R}, \qquad \cos \beta = \frac{R}{Z}, \qquad \sin \beta = \frac{X}{Z}$$

In this problem, $R = 10$ Ω/phase, $X = 30$ Ω/phase, $V_s = V_r = 132$ kV

$$\mathbf{B} = \mathbf{Z} = R + j X = 10 + j\, 30 = 31.62 \underline{/71.56^\circ}$$

$$\mathbf{B} = B \underline{/\beta}, \ B = 31.62 \ \Omega, \ \beta = 71.56^\circ$$

$$P_{rm} = \frac{132 \times 132}{31.62} - \frac{1 \times (132)^2}{31.62} \cos (71.56^\circ - 0^\circ) = 551\, (1 - \cos 71.56^\circ) = 376.7 \ \text{MW}$$

(b) For the given values of V_s and V_r, the power transfer is a maximum when $\delta = \beta$, where δ is the angle between $\mathbf{V}_s$ and $\mathbf{V}_r$.

$$\therefore \qquad \delta = \beta = 71.56^\circ$$

(c) The equation of the receiving-end circle is

$$(P_r - P_{ro})^2 + (Q_r - Q_{ro})^2 = \rho_r^2$$

$$P_{ro} = -\frac{V_r^2}{Z^2} R = -\frac{(132)^2 \times 10}{(10)^2 + (30)^2} = -174.24 \ \text{MW}$$

$$Q_{ro} = -\frac{V_r^2}{Z^2} X = -\frac{(132)^2 \times 30}{(10)^2 + (30)^2} = -522.72 \ \text{MVAr}$$

$$\rho_r = \frac{V_s V_r}{Z} = \frac{132 \times 132}{31.62} = 551 \text{ MVA}$$

Putting $P_r = 100$ MW in the equation of the receiving-end circle we get,

$$(100 + 174.24)^2 + (Q_r + 522.72)^2 = (551)^2$$

$$(Q_r + 522.72)^2 = (551)^2 - (274.24)^2$$

$$Q_r + 522.72 = (303601 - 75207.6)^{1/2}$$

$$Q_r = 477.90 - 522.72 = -44.82 \text{ MVAr}$$

Rating of phase modifier at full load

$$Q_{pm} = P_r \tan \varphi_r - Q_r = 100 \times 0.4843 - (-44.82) = 93.25 \text{ MVAr (leading)}$$

Graphical method

Choose the following scales :

    Horizontal : 1 cm = 100 MW

    Vertical : 1 cm = 100 MVAr

The horizontal coordinate of the circle = – 1.74 cm

The vertical coordinate of the circle = – 5.22 cm

Radius of the circle = 5.51 cm

Take $O_r K$ equal to 1.74 cm and KO_1 equal to 5.22 cm. With centre O_1 and radius $O_1 G$ equal to 5.51 cm draw a circle. The distance FG gives the maximum power transmitted. It is equal to 3.75 cm (375 MW).

The 0.9 power factor lagging load line $O_r L$ is drawn an angle $\varphi_r = \cos^{-1} 0.9 = 25.8°$. For a load of 100 MW take $O_r A$ equal to 100 MW (1 cm). Draw a vertical line BAC through A cutting the circle and the load line. The length BC represents Q_{pm}. Here $BC = 0.95$ cm and hence $Q_{pm} = 95$ MVAr.

Example 13.2 A three-phase overhead line has resistance and reactance per phase of 25 Ω and 90 Ω respectively. The sending-end voltage is 145 kV, while the load-end voltage is maintained at 132 kV for all loads by an automatically controlled synchronous phase modifier. If the MVAr of the modifier has the same value for zero load as for a load of 50 MW, find the rating of the modifier and the power factor of this load.

Solution

Analytical method

$$Z^2 = R^2 + X^2 = (25)^2 + (90)^2 = 8725 \; ; \qquad Z = \sqrt{8725} = 93.4 \ \Omega$$

$$P_{ro} = -\frac{V_r^2}{Z^2} R = -\frac{(132)^2}{8725} \times 25 = -49.92 \text{ MW}$$

$$Q_{ro} = -\frac{V_r^2}{Z^2} X = -\frac{(132)^2}{875} \times 90 = -179.73 \text{ MVAr}$$

$$\rho_r = \frac{V_s V_r}{Z} = \frac{145 \times 132}{93.4} = 204.92 \text{ MVA}$$

$$(P_r - P_{ro})^2 + (Q_r - Q_{ro})^2 = \rho_r^2$$

$$(P_r + 49.92)^2 + (Q_r + 179.73)^2 = (204.92)^2$$

At no load $P_r = 0$

$$(Q_r + 179.73)^2 = (204.92)^2 - (49.92)^2 = 41992 - 2492 = 39500$$

$$Q_r + 179.73 = \sqrt{39500} = 198.74$$

$$Q_r = 198.74 - 179.73 = 19.01 \text{ MVAr}$$

$$Q_{pm} = P_r \tan \varphi_r - Q_r = 0 - 19.01 = 19.01 \text{ MVAr (lagging)}$$

$$\tan \varphi_r = \frac{Q_r}{P_r} = \frac{19.01}{50} = 0.3802 \; ; \qquad \varphi_r = 20.81°$$

Power factor $\cos \varphi_r = \cos 20.81° = 0.9347$ (lagging)

Graphical method

Choose the following scales :

    Horizontal : 1 cm = 20 MW

    Vertical : 1 cm = 20 MVAr

∴ $P_{ro} = -49.92$ MW $= -2.5$ cm

    $Q_{ro} = -179.73$ MVAr $= -9.0$ cm

    $\rho_r = 204.92$ MVA $= 10.25$ cm

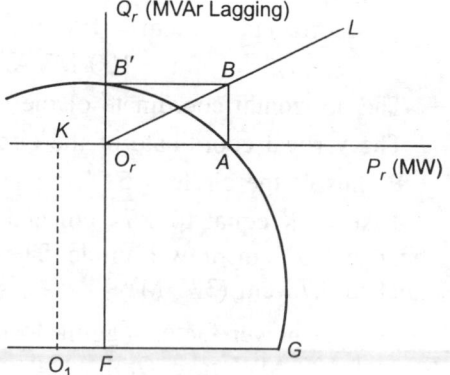

Take $O_r K$ equal to 2.5 cm and KO_1 equal to 9.0 cm. With centre O_1 and radius $O_1 G$ equal to 10.25 cm draw a circle as shown in Fig. 13.7.

Fig. 13.7.

The operating point for zero load is B'. Therefore, $O_r B'$ represents the amount of shunt compensation to maintain the specified voltages under no-load condition. Thus at no load the synchronous phase modifier takes lagging MVAr given by $O_r B' = 1$ cm = 20 MVAr.

For a load of 50 MW, take $OA = 50$ MW = 2.5 cm. Draw a perpendicular $AB = O_r B'$ and a line $O_r BL$ joining O_r and B. Then $O_r L$ is the load line and $\angle LO_r A = \varphi_r$. By measurement $\varphi_r = 22°$. Therefore, the rating of the synchronous phase modifier is 20 MVAr (lagging). The power factor at 50 MW is $\cos \varphi_r$, i.e., $\cos 22°$ or 0.9272 lagging.

Example 13.3 A 320 km 275 kV three-phase line has the following general parameters :

$$\mathbf{A} = 0.94 \underline{/1.0°}, \; \mathbf{B} = 107 \underline{/78°} \; \Omega.$$

If the receiving-end voltage is 275 kV, determine (a) the sending-end voltage necessary if a load of 300 MW at 0.9 lagging power factor is being delivered at the receiving end, (b) the maximum power that can be transmitted if the sending-end voltage is held at 290 kV, (c) the additional reactive MVA that will have to be provided at the receiving end when delivering 450 MVA at 0.9 lagging power factor, the supply voltage being 290 kV.

Solution

Analytical method

(a) Line voltage at the receiving end, $V_{rl} = 275$ kV

Phase voltage at the receiving end, $V_{rp} = \dfrac{1}{\sqrt{3}} \times$ line voltage $= \dfrac{1}{\sqrt{3}} \times 275 \times 10^3 = 158771$ V

$$\mathbf{V}_{rp} = V_{rp} \,\underline{/0^\circ} = 158771 \,\underline{/0^\circ} \text{ V}$$

$$\sqrt{3} \, V_{rl} I_r \cos \varphi_r = P_r$$

$$\sqrt{3} \times 275 \times 10^3 \times I_r \times 0.9 = 300 \times 10^6 \; ; \qquad I_r = 700 \text{ A}$$

$$\mathbf{I}_r = I_r \,\underline{/-\varphi_r} = 700 \,\underline{/-\cos^{-1} 0.9} = 700 \,\underline{/-25.84^\circ} \text{ A}$$

$$\mathbf{V}_{sp} = \mathbf{A V}_{rp} + \mathbf{B I}_r = (0.94 \,\underline{/1.0^\circ})\,(158771 \,\underline{/0^\circ}) + (107 \,\underline{/78^\circ})\,(700 \,\underline{/-25.84^\circ})$$

$$= 149245 \,\underline{/1.0^\circ} + 74900 \,\underline{/52.16^\circ} = 149222 + j\,2604.7 + 45948 + j\,59150.5$$

$$= 195170 + j\,61755 = 204707 \,\underline{/17.56^\circ} \text{ V}$$

Line voltage at the sending end, $\mathbf{V}_{sl} = \sqrt{3} \, \mathbf{V}_{sp} = \sqrt{3} \times 204707 \,\underline{/17.56^\circ}$

$$= 354563 \,\underline{/17.56^\circ} \text{ V} = 354.563 \,\underline{/17.56^\circ} \text{ kV}$$

(b) $\quad P_{rm} = \dfrac{V_s V_r}{B} - \dfrac{A V_r^2}{B} \cos(\beta - \alpha) = \dfrac{290 \times 275}{107} - \dfrac{0.94 \times (275)^2}{107} \cos(78^\circ - 1^\circ)$

$$= 745.3 - 149.45 = 595.85 \text{ MW}$$

(c) The equation of the receiving-end circle is

$$(P_r - P_{ro})^2 + (Q_r - Q_{ro})^2 = \rho_r^2$$

Here, $\quad P_r = \text{MVA} \cos \varphi_r = 450 \times 0.9 = 405$ MW

$$P_{ro} = -\dfrac{A V_r^2}{B} \cos(\beta - \alpha) = -\dfrac{0.94 \times (275)^2}{107} \cos(78^\circ - 1^\circ) = -149.45 \text{ MW}$$

$$Q_{ro} = -\dfrac{A V_r^2}{B} \sin(\beta - \alpha) = -\dfrac{0.94 \times (275)^2}{107} \sin(78^\circ - 1^\circ) = -647.34 \text{ MVAr}$$

$$\rho_r = \dfrac{V_s V_r}{B} = \dfrac{290 \times 275}{107} = 745.3 \text{ MVA}$$

Substituting the values of P_r, P_{ro}, Q_{ro}, and ρ_r in the equation of the receiving-end circle, we get

$$(405 + 149.450)^2 + (Q_r + 647.34)^2 = (745.3)^2$$

$$(Q_r + 647.34)^2 = (745.3)^2 - (554.45)^2$$

$$Q_r + 647.34 = 498 \; ; \qquad Q_r = 498 - 647.34 = -149.34 \text{ MVAr}$$

Rating of the phase modifier, $Q_{pm} = P_r \tan \varphi_r - Q_r = 405 \tan 29.84^\circ - (-149.34) = 345.5$ MVAr

Graphical method

Choose the following scales :

    Horizontal : 1 cm = 50 MW

    Vertical : 1 cm = 50 MVAr

The horizontal coordinate of the circle = -149.45 MW = -2.989 cm

The vertical coordinate of the circle = -647.34 MVAr = -12.95 cm

Radius of the circle = 745.3 MVA = 14.90 cm

Take $O_r K = 2.989$ cm and $KO_1 = 12.95$ cm. With centre O_1 and radius $O_1 G$ equal to 14.9 cm draw the circle as shown in Fig. 13.8. The distance FG gives the maximum power transmitted. It is equal to 1.95 cm (597.5 MW). Draw the load line $O_r L$ at an angle φ_r equal to $\cos^{-1} 0.9 = 25.84°$. For a load of 300 MW take $O_r N$ equal to 6 cm. Draw a vertical line NM through N cutting the load line at M.

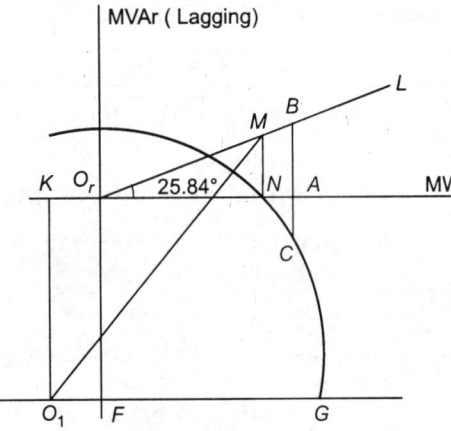

Fig. 13.8.

Then, $O_1 M = 18.2$ cm = 910 MVA

$$\therefore \quad 910 = \frac{V_s V_r}{B} = \frac{V_s \times 275}{107}$$

$$V_s = \frac{910 \times 107}{275} = 354 \text{ kV}$$

For a load of 450 MVA at 0.9 p.f. lagging, locate point B on the load line such that $O_r B = 450$ MVA (= 9 cm). Draw a vertical line BAC cutting the circle at point C. The length BC represents the rating of the phase modifier, Q_{pm}. Here $BC = 6.8$ cm. Hence $Q_{pm} = 6.8 \times 50 = 340$ MVAr (leading).

Example 13.4 A three-phase 50 Hz transmission line has resistance of 14 Ω and an inductive reactive of 48 Ω per phase. The capacitive susceptance to neutral is 4×10^{-4} siemen. Find the MVAr rating on no load and full load of a synchronous phase modifier to maintain the sending-end and receiving-end voltages constant at 70 kV and 66 kV respectively when the line is delivering a load of 24 MVA at 0.8 power factor lagging. Use the nominal π equivalent circuit for solution.

Solution

$$\mathbf{Z} = R + jX = 14 + j\,48 = 50\,\underline{/73.74°}\ \Omega$$

$$\mathbf{Y} = g + jb = j\,4 \times 10^{-4} = 4 \times 10^{-4}\,\underline{/90°}\ \text{S}$$

For a nominal π circuit

$$\mathbf{A} = 1 + \frac{1}{2}\,\mathbf{ZY} = 1 + \frac{1}{2} \times (50\,\underline{/73.74°})\,(4 \times 10^{-4}\,\underline{/90°})$$

$$= 1 + 0.01\,\underline{/163.74°} = 1 - 9.6 \times 10^{-3} + j\,2.8 \times 10^{-3} = 0.9904\,\underline{/0.162°}$$

$$\mathbf{B} = \mathbf{Z} = 50\,\underline{/73.74°}\ \Omega$$

$$\beta = 73.74°, \quad \alpha = 0.162°, \quad \beta - \alpha = 73.578°$$

$$P_{ro} = -\frac{AV_r^2}{B} \cos(\beta - \alpha) = -\frac{0.9904 \times (66)^2}{50} \cos 73.578° = -24.39 \text{ MW}$$

$$Q_{ro} = -\frac{AV_r^2}{B} \sin(\beta - \alpha) = -\frac{0.9904 \times (66)^2}{50} \sin 73.578° = 82.76 \text{ MVAr}$$

$$\rho_r = \frac{V_s V_r}{B} = \frac{70 \times 66}{50} = 92.4 \text{ MVAr}$$

Rating at no load

$$(P_r - P_{ro})^2 + (Q_r - Q_{ro}) = \rho_r^2$$

At no load, $P_r = 0$

$\therefore$
$$P_{ro}^2 + (Q_r - Q_{ro})^2 = \rho_r^2$$

$$(24.39)^2 + (Q_r + 82.76)^2 = (92.4)^2$$

$$(Q_r + 82.76)^2 = 8537.76 - 594.87$$

$$Q_r = 89.12 - 82.76 = 6.36 \text{ MVAr}$$

$$Q_{pm} = P_r \tan \varphi_r - Q_r = 0 - 6.36 = -6.36 = 6.36 \text{ MVAr (lagging)}$$

Rating at 24 MVA at 0.8 power factor

$$P_r = (\text{MVAr}) \cos \varphi_r = 24 \times 0.8 = 19.2 \text{ MW}$$

$$(P_r - P_{ro})^2 + (Q_r - Q_{ro})^2 = \rho_r^2$$

$$(19.2 + 24.39)^2 + (Q_r + 82.76)^2 = (92.4)^2$$

$$(Q_r + 82.76)^2 = (92.4)^2 - (43.59)^2 = 6637.8$$

$$Q_r = \sqrt{6637.8} - 82.76 = 81.47 - 82.76 = -1.287 \text{ MVAr}$$

$$Q_{pm} = P_r \tan \varphi_r - Q_r = 19.2 \times 0.75 - (-1.287) = 15.687 \text{ MVAr (leading)}$$

This problem can also be solved by graphical method.

13.12 STATIC VAR SYSTEMS (SVS)

In EHV transmission practice, when the voltage at a bus falls below the reference value, capacitive vars are to be injected. When the bus voltage becomes higher than the reference value, inductive vars are supplied to lower the bus voltage. In conventional methods of shunt compensation, shunt reactors are connected during low loads, and shunt capacitors are connected during heavy loads or low lagging power factor loads. Such switching operations are very slow because of the greater time (3-4 cycles) required for the operation of the circuit breakers. Moreover, circuit breakers are not suitable for frequent switching during voltage variations. These limitations have been overcome by static var systems (SVS). In a static var system, thyristors are used as switching devices instead of circuit breakers. Thyristor switching is faster than mechanical switching and also it is possible to have transient-free operation by controlling the instant of switching.

The advent of high-speed, high-current switching made possible by thyristors has introduced a new concept is providing reactive compensation for optimum EHV/UHV system performance. The static var compensators (SVC) use shunt reactor and shunt capacitor combinations with high-voltage, high-current thyristor control for obtaining fast and accurate control of reactive power flow. The static var compensation (SVC) is also known as *static var system* (SVS).

The first use of static var compensators in EHV/UHV transmission started in 1960s and was based on saturated reactors. The first thyristor switched compensators were installed in late 1970s. Since that time the use of SVS has become very popular to replace synchronous compensation (using synchronous phase modifiers).

13.13 SVS SCHEMES

Many SVS schemes are in operation. Some of the commonly used schemes are as follows :
1. Thyristor controlled reactor (TCR)
2. Thyristor switched capacitor (TSC)
3. Fixed capacitor (FC), thyristor controlled reactor (TCR) scheme.
4. Thyristor switched capacitor/ (TSC), thyristor controlled reactor (TCR) scheme

13.14 THYRISTOR CONTROLLED REACTOR (TCR)

A single-phase thyristor controlled reactor X is shown in Fig. 13.9. The current through the reactor can be varied by controlling the firing angles of back to back pair of thyristors connected in series with the reactor. The TCR scheme is used in EHV lines for providing lagging vars during low loads or load rejections.

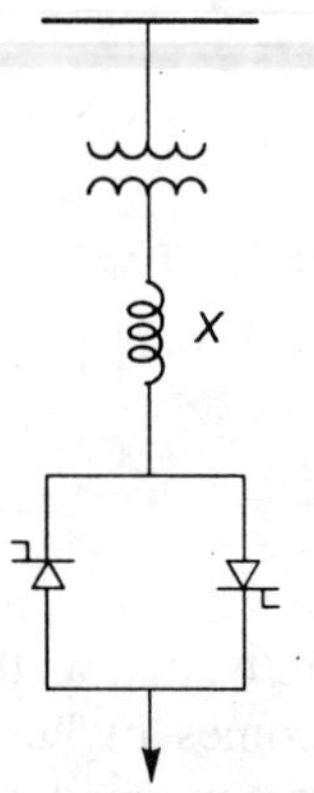

Fig. 13.9. Single-phase thyristor-controlled reactor (TCR).

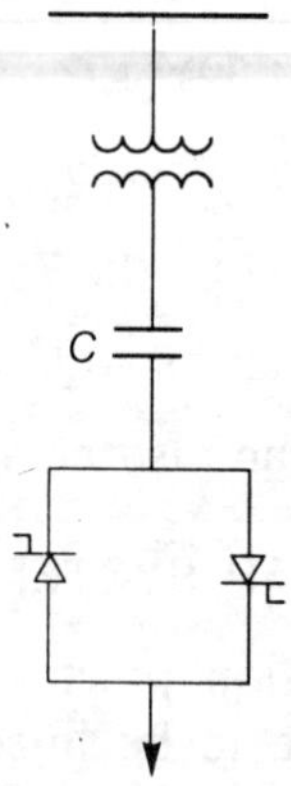

Fig. 13.10. Single-phase thyristor-switched capacitor (TSC).

13.15 THYRISTOR SWITCHED CAPACITOR (TSC)

A single-phase thyristor switched capacitor C is shown in Fig. 13.10. The current through the capacitor can be varied by controlling the firing angles of back to back thyristors connected in series with the capacitor. The TSC scheme is used in EHV lines for providing leading vars during high loads.

13.16 FC-TCR TYPE VAR COMPENSATOR

The fixed capacitor (FC), thyristor-controlled reactor
(TCR) type var compensator is shown in Fig. 13.11.
This arrangement provides discrete leading vars from
the capacitors and continuously lagging vars from the
thyristor-switched reactors. Leading vars are supplied
by two or more capacitor banks. The current through
the reactor can be varied by controlling the firing
angles of back to back pairs of thyristors connected in
series with the reactor. Harmonics are generated
because of switching operations. These harmonics are
injected into the system. There are several problems
associated with the injection of harmonics. Therefore,
these harmonics must be eliminated. Since 6-pulse
thyristors are to be used for 3 phases, fifth and seventh
harmonics are generated in addition to third
harmonics. Small reactors are usually connected in the
fixed capacitor branches, in order to tune these

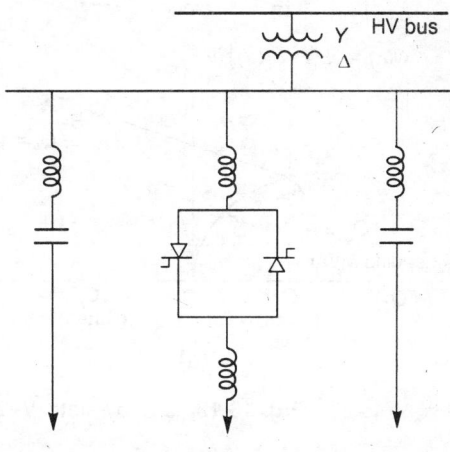

To other phases in delta (Δ)

Fig. 13.11. FC-TCR type var compensator.

branches as filters for the fifth and seventh harmonics. The TCR and the secondary winding of
the coupling transformer in delta are connected in delta. (Fig. 13.12). These delta connections
eliminate third harmonics. The steady-state characteristic of the FC-TCR system is shown in
Fig. 13.13.

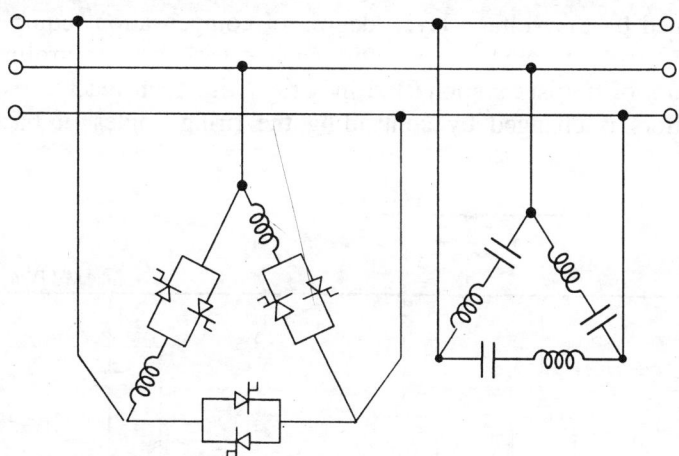

Fig. 13.12. Delta connection of FC-TCR type var compensator.

Here, B_C = susceptance of the capacitor

  $B_{L(\alpha)}$ = susceptance of the reactor at firing angle α.

The control range is AB with a positive slope. The slope is determined by the control of the
firing angle. It is to be noted that the current in the reactor is controlled by the firing angle of
the back-to-back connected thyristors. As the inductance is varied, the susceptance varies over

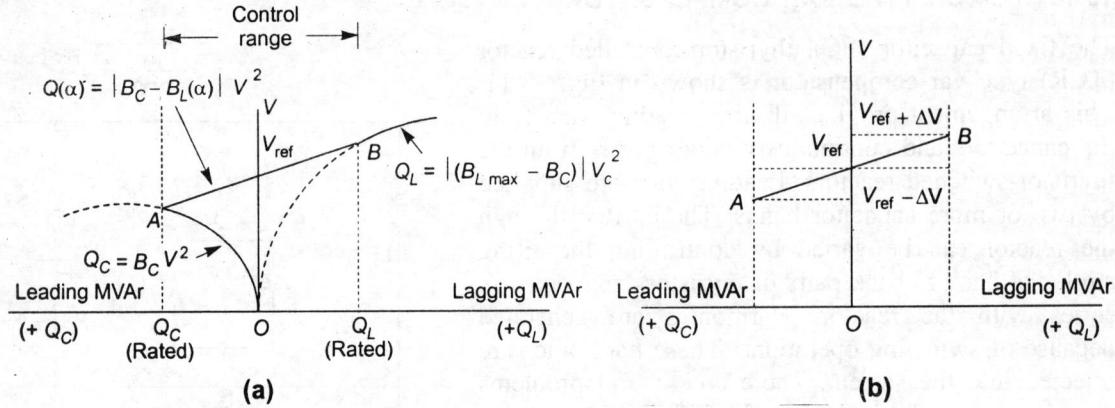

Fig. 13.13. Steady-state V-Q characteristic of FC-TCR type var compensator.

a wide range within limits $B_{L\,min}$ and $B_{L\,max}$ (corresponding to $X_{L\,max}$ and $X_{L\,min}$). The voltage varies within the limits $(V_{ref} - \Delta V)$ and $(V_{ref} + \Delta V)$ as shown in Fig. 13.13 (b). It is seen from Fig. 13.13 (a) that outside the control interval AB, the characteristic is the same as that of a capacitor (low voltage) or an inductor (high voltage).

13.17 TSC-TCR SCHEME

The thyristor switched capacitor (TSC) — thyristor controlled reactor (TCR) type var compensator is shown in Fig. 13.14. It consists of a number of TSC and TCR banks. Their number is determined by the voltage level, degree of compensation required, current rating of thyristors, etc. The current through the reactors can be varied by controlling the firing angles of back to back pairs of thyristors connected in series with each reactor. Similarly, the current through the capacitors is changed by controlling the firing angles of back to back pairs of

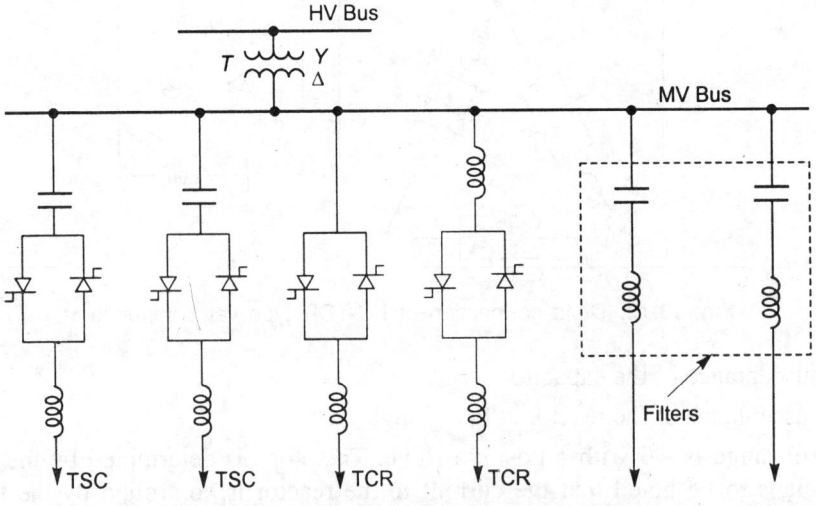

Fig. 13.14. TSC-TCR type var compensator.

thyristors connected in series with each capacitor. During heavy loads, the thyristors of TSC are made to conduct for longer duration in each cycle. Thus, leading, vars are provided by TSC during heavy loads. During low loads, the thyristors of TCR are made to conduct for longer durations in each cycle. Thus, lagging vars are supplied by TCR during low loads. Therefore, a TSC-TCR type var compensator is capable of providing the leading and lagging vars both rapidly continuously, and independently. For continuous control of reactive power flow, it is necessary that the TCR rating should be slightly higher than the rating of individual capacitor bank. Some fixed capacitors are also connected in the circuit in addition to switched capacitors. These capacitors serve as fillers for harmonics when only the reactors are switched. It is usually necessary to connect inductor in series with each switched capacitor bank to limit the current in the thyristor switches and to reduce the risk of resonance with the *AC* line impedance.

The power transformer *T* is used to step down bus voltage (say 400 kV or 220 kV) to a lower voltage level such as 33 kV for economic design of SVC.

The steady-state V-Q characteristic of TSC-TCR type var compensator is similar to that of FC-TCR type var compensator as shown in Fig. 13.13.

The TSC-TCR method of var compensator is better than the FC-TCR type compensator. The following are the advantages of TSC-TCR system over FC-TCR system :

(1) Improved performance during large system disturbances.

(2) Lower power loss.

The TSC-TCR system of var compensation has replaced the earlier method of using synchronous phase modifiers (synchronous capacitors) for var compensation.

13.18 ADVANTAGES OF SVS

Static VAR systems (SVS) offer the following advantages :
1. The power transfer capability of the lines is increased.
2. Transient stability of the system is improved.
3. The dynamic system stability is also improved due to increased damping provided.
4. It is possible to damp subsynchronous resonance frequency oscillations.
5. Steady-state and temporary overvoltages can be controlled.
6. Load power factor is improved. Consequently, line losses are reduced and the system efficiency is improved.
7. The dynamic response of SVS is very fast.
8. Their maintenance is very simple.

13.19 APPLICATIONS OF SVS

The following are the important applications of static var systems in EHV transmission circuits :
1. A static var system provides a fast, smooth and stepless variation of compensation of reactive power injected into the line. Thus, it ensures an accurate voltage control of buses over a wide range of loads.
2. The static var systems are used to control the reactive power demand of large fluctuating loads such as rolling mills, arc furnaces, etc.
3. If a large load is switched off, the voltage at the sending end of the line increases

(Ferranti effect). The static var system provides a fast change in the reactive power compensation to regulate the voltage. In other words, the SVS provides control of dynamic overvoltages caused by load rejection.

4. In HVDC converter stations, the provision of the static var system mainly helps to have fast control of reactive power flow. Thus, voltage fluctuations are controlled and the problem of instability is overcome.

EXERCISES

1. Show that the voltage control and reactive power control are interrelated.
2. Name the generators and consumers of reactive power in a power system.
3. What are the different methods for voltage control in a power system?
4. Explain the terms series and shunt compensations.
5. What are the advantages of series compensation? What are the problems associated with series capacitors?
6. Explain the term subsynchronous resonance. Name the methods to overcome the problem of SSR.
7. What are static var systems? Describe the SVS schemes commonly used in EHV/UHV transmission.
8. State the advantages of SVS over other methods of voltage control.
9. Explain why static compensation is preferred over synchronous compensation in modern power systems.
10. Explain why series compensation leads to improvement in system stability. Compare the performances of series and shunt capacitors in a power system.
11. In a three-phase line with 132 kV at the receiving end the following are the constants :
$$\mathbf{A} = \mathbf{D} = 0.98 \underline{/3°}, \ \mathbf{B} = 110 \underline{/75°} \ \Omega, \ \mathbf{C} = 0.0005 \underline{/88°} \ \text{S}.$$
If the load at the receiving end is 50 MVA at 0.8 lagging power factor, determine graphically or otherwise (a) the sending-end voltage, and (b) the leading reactive MVA at the receiving end if the sending-end voltage is 140 kV.
12. A three-phase transmission line has a resistance per phase of 5 Ω and an inductive reactance per phase of 12 Ω and the line voltage at the receiving end is 33 kV.
    (a) Determine the voltage at the sending end when the load at the receiving end is 20 MVA at 0.8 power factor lagging.
    (b) The voltage at the sending end is maintained constant at 36 kV by means of a synchronous phase modifier at the receiving end, which has the same rating at zero load at the receiving end as for the full load of 16 MW. Determine the power factor of the full-load output and the rating of the synchronous phase modifier.   (L.U.)
13. A three-phase transmission line 25 km in length supplies a load of 10 MW at 0.8 power factor lagging at voltage of 33 kV. The resistance and reactance per km per conductor are 0.35 Ω and 0.6 Ω respectively. Neglecting the capacitance of the line, determine the

rating of synchronous capacitor, operating at zero power factor, connected at the load end of the line such that the sending-end voltage may be 33 kV. (L.U.)

14. A three-phase 50 Hz transmission line has the following values per phase per km : $R = 0.25\ \Omega$, $L = 2$ mH, $C = 0.014\ \mu$F. The line is 50 km long, the voltage at the receiving end is 132 kV and the power delivered is 80 MVA at 0.8 power factor lagging.

    If the voltage at the sending end is maintained at 140 kV by a synchronous phase modifier at the receiving end, determine the kVAr of this machine (i) with no load, (ii) with full load at the receiving end.

15. A three-phase transmission line has a resistance 6 Ω/phase and a reactance of 20 Ω/phase. The sending-end voltage is 66 kV and the voltage at the receiving end is maintained constant by a synchronous phase modifier. Determine the MVAr of the synchronous phase modifier when the load at the receiving end is 75 MW at 0.8 power factor lagging, and also the maximum load which can be transmitted over the line, the voltage being 66 kV at both ends. (L.U.)

16. A three-phase transmission line is to automatically regulate to zero the voltage regulation by means of a synchronous phase modifier at the load end. If the full load output is 50 MW at 0.8 power factor lagging delivered at 200 kV and the line-to-neutral impedance is $(20 + j\,60)\ \Omega$, find the input to synchronous set under these conditions. Deduce any formula employed. (H.N.C.)

17. A three-phase overhead line has the following general parameters :

    $\mathbf{A} = 0.8705\ \underline{/2.3^\circ}$, $\mathbf{B} = 187\ \underline{/75.1^\circ}\ \Omega$.

    Find the MVAr rating on no load and full load of a synchronous phase modifier to maintain the voltages constant at 154 kV at both ends. The load at the receiving end is 50 MVA at 0.85 power factor lagging. Also, determine the maximum load that can be transmitted.

18. A three-phase overhead line has resistance and reactance of 12 Ω and 40 Ω respectively per phase. The supply voltage is 132 kV and the load-end voltage is maintained constant at 132 kV for all loads by an automatically controlled synchronous phase modifier. Determine the kVAr of the modifier when the load at the receiving end is 120 MW at power factor 0.8 lagging. (H.N.C.)

ANSWERS

11. (a) $V_{sl} = 165$ kV (b) $Q_{pm} = 34.5$ MVAr (leading)

12. (a) 40.1 kV (b) 0.9 (lag), 7.92 MVAr   **13.** 14.7 MVAr

14. (i) 30.4 MVAr (lagging) (ii) 42.6 MVAr (leading)

15. 96.8 MVAr, 148 MW   **16.** 60.3 MVAr

17. $Q_{rm} = 17.09$ MVAr (inductive) at no load; $Q_{pm} = 29.63$ MVAr (capacitive) at full load; $P_{rm} = 94.2$ MW

18. 145 MVAr.

14

Load Flow Analysis

14.1 INTRODUCTION

Load flow (or power flow) analysis is the determination of current, voltage, active power, and reactive voltamperes at various points in a power system operating under normal steady-state or static conditions. Load flow studies are made to plan the best operation and control of the existing system as well as to plan the future expansion to keep pace with the load growth. Such studies help in ascertaining the effects of new loads, new generating stations, new lines and new interconnections before they are installed. The prior information serves to minimize the system losses and to provide a check on the system stability.

The mathematical formulation of load flow problem results in a set of *algebraic non-linear* equations. A lot of calculation work is involved in the solution of these equations. Hand computations are very tedious and time consuming. Earlier load flow studies were made by ac network analysers (analog computers). Digital computers, because of greater flexibility, economy accuracy and quicker operation, have practically replaced network analysers for the solution of load flow problems.

14.2 LOAD FLOW PROBLEM

Load flow studies are performed to calculate the magnitude and phase angles of voltages at the buses, and also the active power and reactive voltamperes flow for the given terminal or bus conditions. The following variables are associated with each bus or node :

1. Magnitude of the voltage, $|V_i|$.
2. Phase angle of the voltage, δ_i.
3. Active power P_i.
4. Reactive voltamperes Q_i.

Three types of buses or nodes are identified in a power system network for load flow studies.

In each bus two variables are known (specified) and two are to be determined. The bus classification depends upon the specified variables. The buses are classified as follows :

(a) Swing Bus or Reference Bus or Slack Bus

Voltage magnitude $|V_i|$ and phase angle δ_i are specified for this bus. This bus is first to respond to a changing load condition.

(b) Generator Bus or Voltage-Controlled Bus or PV Bus

Here $|V_i|$ and P_i are specified. Often the upper and lower limits of Q are also specified. The phase angles of the voltages and the reactive powers are to be determined.

(c) Load Bus or P-Q Bus

Here the active power P_i and reactive voltamperes Q_i are specified.

Buses with neither generator nor load may be considered as load buses where $P_i = Q_i = 0$. If any bus in a power system network has both load and generator, then load is generally treated as negative generation.

Table 14.1 shows the types of buses and the associated known and unknown variables.

Table 14.1   Bus Types for Power Flow Analysis

Bus type	Specified variables	Unknown variables
Reference bus or slack bus	$|V_i|$, δ_i	P_i, Q_i
Generator bus or voltage controlled bus or PV bus	P_i, $|V_i|$	Q_i, δ_i
Load bus or PQ bus	P_i, Q_i	$|V_i|$, δ_i

One of the generator buses is selected as the reference bus for the reason given below :

The losses in the system remain unknown until the load flow solution is complete. It is for this reason that one of the generator buses is made to take the additional real and reactive powers to supply the transmission losses P_L and Q_L. This bus is, therefore, known as the *slack* or *swing bus*. Since the voltages throughout the system must be close to 1.0 per unit, the voltage at the slack bus is assigned to be 1.0 per unit. The voltage of the slack bus is taken as reference and therefore its angle δ_i is equal to zero. Generally, the bus connected to the largest generating station is selected as the slack bus. The slack bus is usually numbered as bus 1.

In load flow studies single-phase representation with positive-sequence network is used since power system is usually balanced under normal conditions of operation.

The load flow problem is divided into the following steps :

1. A suitable mathematical network model to give relationships between voltages, powers, and reactive voltamperes is formulated.

2. Powers, vars, and voltages are specified at various buses.

3. Numerical solution of the load flow problem subject to the restraints given in 2 is found to give the bus voltages.

4. Flow of power and vars is found in all the lines of the network.

14.3 BUS ADMITTANCE MATRIX Y_{bus}

Consider a small power system network (Fig. 14.1) consisting of two generating stations, three transmission lines, one load and a static capacitor connected to load bus 3. We shall assume that the network is symmetrical and operating under balanced conditions.

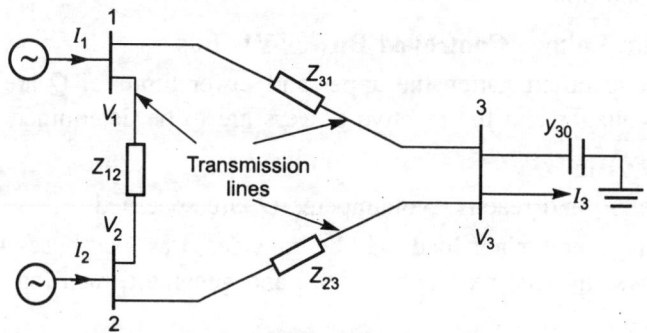

Fig. 14.1. Power system network for power flow.

We are interested in the steady-state solution of the network shown in Fig. 14.1. This is a three-node network. We can write the node voltage equations as

$$I_1 = (y_{12} + y_{31}) V_1 - y_{12} V_2 - y_{31} V_3$$
$$I_2 = - y_{12} V_1 + (y_{12} + y_{23}) V_2 - y_{23} V_3$$
$$- I_3 = - y_{31} V_1 - y_{23} V_2 + (y_{31} + y_{23} + y_{30}) V_3$$

where $y_{12} = \dfrac{1}{z_{12}}$, $y_{23} = \dfrac{1}{z_{23}}$, $y_{31} = \dfrac{1}{z_{31}}$.

In matrix form

$$\begin{bmatrix} I_1 \\ I_2 \\ -I_3 \end{bmatrix} = \begin{bmatrix} y_{12} + y_{31} & - y_{12} & - y_{31} \\ - y_{12} & (y_{12} + y_{23}) & - y_{23} \\ - y_{31} & - y_{23} & (y_{31} + y_{32} + y_{30}) \end{bmatrix} \begin{bmatrix} V_1 \\ V_2 \\ V_3 \end{bmatrix}$$

It is to be noted that all injected currents are positive and extracted currents are negative. The above equations can be written as

$$\begin{bmatrix} I_1 \\ I_2 \\ -I_3 \end{bmatrix} = \begin{bmatrix} Y_{11} & Y_{12} & Y_{13} \\ Y_{21} & Y_{22} & Y_{23} \\ Y_{31} & Y_{32} & Y_{33} \end{bmatrix} \begin{bmatrix} V_1 \\ V_2 \\ V_3 \end{bmatrix}$$

where $Y_{11} = y_{12} + y_{13}$; $Y_{22} = y_{21} + y_{23}$; $Y_{33} = y_{30} + y_{31} + y_{32}$

and $Y_{12} = Y_{21} = - y_{12}$; $Y_{23} = Y_{32} = - y_{23}$; $Y_{31} = Y_{13} = - y_{31}$

The elements Y_{11}, Y_{22}, Y_{33} forming the diagonal terms are called *self admittances*. The self admittance of a node x is equal to the sum of admittances of all the elements connected to node x. In general, the diagonal element Y_{pp} of the bus admittance matrix is equal to the sum of admittances of all the elements connected to bus p.

That is, $Y_{pp} = y_{p1} + y_{p2} + \dots + y_{pn}$

where y_{pq} is the admittance of the element connected between the buses p and q.

The elements Y_{12}, Y_{13}, Y_{21}, Y_{23}, Y_{31}, Y_{32} forming the off-diagonal terms are called *mutual admittances*.

$$Y_{12} = Y_{21} = -y_{12} \; ; \quad Y_{23} = Y_{32} = -y_{23} \; ; \quad Y_{31} = Y_{13} = -y_{31}$$

It is to be noted that all mutual admittance terms have a negative sign. In general, the off-diagonal term of the bus admittance matrix is equal to the negative of the admittance connected between node p and node q.

That is, $Y_{pq} = -y_{pq}$

For a network having n nodes (buses) excluding ground, a set of following equations, one for each node, can be written as

$$I_1 = Y_{11} V_1 + Y_{12} V_2 + \ldots + Y_{1n} V_n$$

$$I_2 = Y_{21} V_1 + Y_{22} V_2 + \ldots + Y_{2n} V_n$$

$$\vdots \qquad \vdots \qquad \vdots \qquad \qquad \vdots$$

$$I_n = Y_{n1} V_1 + Y_{n2} V_2 + \ldots + Y_{nn} V_n$$

These equations can be written in matrix forms as

$$\begin{bmatrix} I_1 \\ I_2 \\ \vdots \\ I_n \end{bmatrix} = \begin{bmatrix} Y_{11} & Y_{12} & \ldots & Y_{1n} \\ Y_{21} & Y_{22} & \ldots & Y_{2n} \\ \vdots & \vdots & & \vdots \\ Y_{n1} & Y_{n2} & \ldots & Y_{nn} \end{bmatrix} \begin{bmatrix} V_1 \\ V_2 \\ \vdots \\ V_n \end{bmatrix} \qquad \ldots(14.3.1)$$

In a more compact form

$$\mathbf{I}_{bus} = \mathbf{Y}_{bus} \mathbf{V}_{bus} \qquad \ldots(14.3.2)$$

where $\quad \mathbf{I}_{bus} = $ bus current vector $= \begin{bmatrix} I_1 \\ I_2 \\ \vdots \\ I_n \end{bmatrix} \qquad \ldots(14.3.3)$

$\mathbf{V}_{bus} = $ bus voltage vector $= \begin{bmatrix} V_1 \\ V_2 \\ \vdots \\ V_n \end{bmatrix} \qquad \ldots(14.3.4)$

$\mathbf{Y}_{bus} = $ bus admittance matrix $= \begin{bmatrix} Y_{11} & Y_{12} & \ldots & Y_{1n} \\ Y_{21} & Y_{22} & \ldots & Y_{2n} \\ \vdots & \vdots & & \vdots \\ Y_{n1} & Y_{n2} & \ldots & Y_{nn} \end{bmatrix} \qquad \ldots(14.3.5)$

Bus voltages are measured with respect to ground. Eq. (14.3.2) is called the *nodal current equation*. It is a vector equation consisting of n scalar equations.

If the power system elements have mutual coupling, the bus admittance matrix cannot be found directly by inspection of the one line diagram. In presence of mutual coupling between power system elements the inspection method fails. In such a case Y_{bus} can be formed from *graph theoretic approach*. However, the mutual coupling between power system elements exist only in case of transmission lines running in parallel for a long distance. But this coupling is also weak. Therefore, for all practical purposes the mutual coupling can be ignored and Y_{bus} is formed by inspection method.

Advantages of bus admittance matrix Y_{bus}

The main advantages of the bus admittance matrix Y_{bus} are as follows :

1. Data preparation is simple.
2. Its formation and modification is easy.
3. Since the bus admittance matrix is a sparse matrix (that is, most of its elements are zero), the computer memory requirements are less. For a large power system more than 90 per cent of its off-diagonal elements are zero. This is due to the fact that in power system networks each node (bus) is connected to not more than three nodes in general and an element Y_{pq} exists only if a transmission line links nodes p and q.

Example 14.1 Determine Y_{bus} for the five-bus system shown in Fig. 14.2. Assume that the lines shown dotted are not connected and the shunt admittances at the buses and mutual couplings between the lines are neglected.

Solution

$$y_{12} = \frac{1}{z_{12}} = \frac{1}{j\,0.2} = -j\,5$$

$$y_{23} = \frac{1}{z_{23}} = \frac{1}{j\,0.25} = -j\,4$$

$$y_{34} = \frac{1}{z_{34}} = \frac{1}{j\,0.3} = -j\,3.33$$

$$y_{45} = \frac{1}{z_{45}} = \frac{1}{j\,0.25} = -j\,4$$

$$y_{14} = \frac{1}{z_{14}} = \frac{1}{j\,0.5} = -j\,2$$

$$y_{15} = \frac{1}{z_{15}} = \frac{1}{j\,0.2} = -j\,5$$

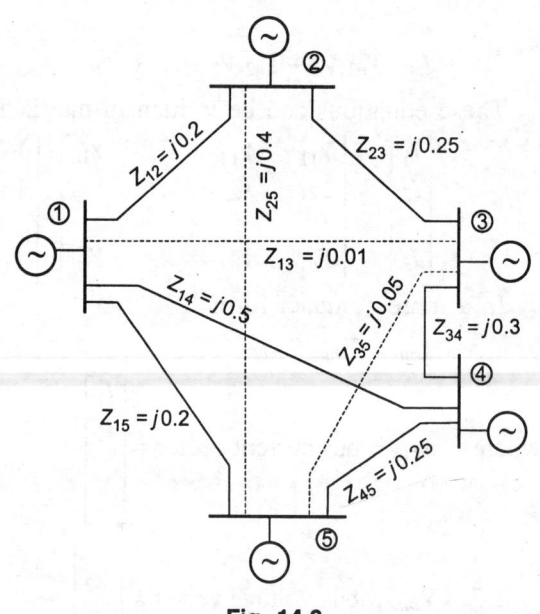

Fig. 14.2.

$$Y_{11} = y_{12} + y_{14} + y_{15} = -j\,5 - j\,2 - j\,5 = -j\,12$$

$$Y_{22} = y_{12} + y_{23} = -j\,5 - j\,4 = -j\,9$$

$$Y_{33} = y_{23} + y_{34} = -j\,4 - j\,3.33 = -j\,7.33$$

$$Y_{44} = y_{34} + y_{14} + y_{45} = -j\,3.33 - j\,2 - j\,4 = -j\,9.33$$

$$Y_{55} = y_{45} + y_{15} = -j\,4 - j\,5 = -j\,9$$

$$Y_{12} = -y_{12} = j\,5$$

$$Y_{23} = -y_{23} = j\,4$$

$$Y_{34} = -y_{34} = j\,3.33$$

$$Y_{45} = -y_{45} = j\,4$$

$$Y_{14} = -y_{14} = j\,2$$

$$Y_{15} = -y_{15} = j\,5$$

The bus admittance matrix is given by

$$
Y_{bus} =
\begin{bmatrix}
Y_{11} & Y_{12} & Y_{13} & Y_{14} & Y_{15} \\
Y_{21} & Y_{22} & Y_{23} & Y_{24} & Y_{25} \\
Y_{31} & Y_{32} & Y_{33} & Y_{34} & Y_{35} \\
Y_{41} & Y_{42} & Y_{43} & Y_{44} & Y_{45} \\
Y_{51} & Y_{52} & Y_{53} & Y_{54} & Y_{55}
\end{bmatrix}
=
\begin{bmatrix}
-j\,12 & j\,5 & 0 & j\,2 & j\,5 \\
j\,5 & -j\,9 & j\,4 & 0 & 0 \\
0 & j\,4 & -j\,7.33 & j\,3.33 & 0 \\
j\,2 & 0 & j\,3.33 & -j\,9.33 & j\,4 \\
j\,5 & 0 & 0 & j\,4 & -j\,9
\end{bmatrix}
$$

Example 14.2 In Example 14.1, if the line between buses 2 and 5 with an impedance $j\,0.4$ is connected, determine the modified bus admittance matrix. The other lines shown dotted are not connected.

Solution

By addition of one new line between buses 2 and 5, only four elements namely, Y_{22}, Y_{55}, Y_{25}, Y_{52} are modified.

$$y_{25} = \frac{1}{z_{25}} = \frac{1}{j\,0.4} = -j\,2.5$$

$$Y_{22\,(new)} = Y_{22\,(old)} + y_{25} = -j\,9 - j\,2.5 = -j\,11.5$$

$$Y_{55\,(new)} = Y_{55\,(old)} + y_{52} = -j\,9 - j\,2.5 = -j\,11.5$$

$$Y_{25} = -y_{25} = j\,2.5 = Y_{52}$$

Therefore, the modified bus admittance matrix is given by

$$
Y_{bus} =
\begin{bmatrix}
-j\,12 & j\,5 & 0 & j\,2 & j\,5 \\
j\,5 & -j\,11.5 & j\,4 & 0 & j\,2.5 \\
0 & j\,4 & -j\,7.33 & j\,3.33 & 0 \\
j\,2 & 0 & j\,3.33 & -j\,9.33 & j\,4 \\
j\,5 & j\,2.5 & 0 & j\,4 & -j\,11.5
\end{bmatrix}
$$

Example 14.3 In Example 14.1, if the lines between buses 3 and 5, and 1 and 3 are connected, determine the modified bus admittance matrix. The line between buses 2 and 5 is not connected.

Solution

By addition of two lines between buses 3 and 5, and 1 and 3 we have to modify only seven elements namely, Y_{11}, Y_{33}, Y_{55}, Y_{31}, Y_{13}, Y_{35}, Y_{53}.

$$y_{13} = \frac{1}{z_{13}} = \frac{1}{j\,0.01} = -j\,100 \; ; \qquad\qquad y_{35} = \frac{1}{z_{35}} = \frac{1}{j\,0.05} = -j\,20$$

$$Y_{11\,(new)} = Y_{11\,(old)} + y_{13} = -j\,12 - j\,100 = -j\,112$$

$$Y_{33\,(new)} = Y_{33\,(old)} + y_{13} + y_{35} = -j\,7.33 - j\,100 - j\,20 = -j\,127.33$$

$$Y_{55\,(new)} = Y_{55\,(old)} + Y_{35} = -j\,9 - j\,20 = -j\,29$$

$$Y_{13} = -y_{13} = j\,100 = Y_{31} \; ; \qquad\qquad Y_{35} = -y_{35} = j\,20 = Y_{53}$$

$$Y_{bus} = \begin{bmatrix} -j\,112 & j\,5 & j\,100 & j\,2 & j\,5 \\ j\,5 & -j\,9 & j\,4 & 0 & 0 \\ j\,100 & j\,4 & -j\,127.33 & j\,3.33 & j\,20 \\ j\,2 & 0 & j\,3.33 & -j\,9.33 & j\,4 \\ j\,5 & 0 & j\,20 & j\,4 & -j\,29 \end{bmatrix}$$

Example 14.4 Determine Y_{bus} for the 3-bus system shown in Fig. 14.3. The line series impedances are as follows :

Line (bus to bus)	Impedance (pu)
1-2	$0.06 + j\,0.18$
1-3	$0.03 + j\,0.09$
2-3	$0.08 + j\,0.24$

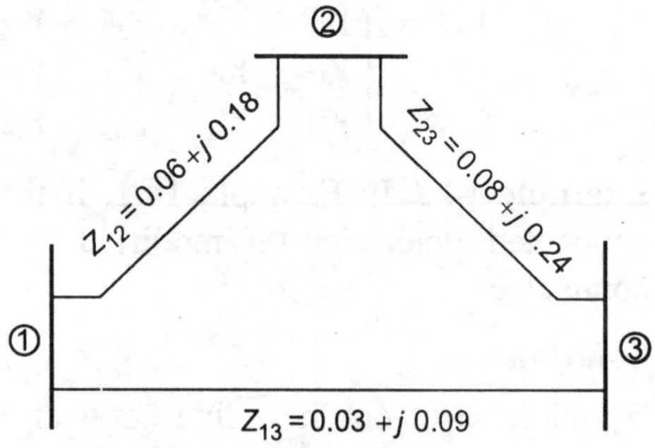

Neglect the shunt capacitances of the lines.

Solution

$$y_{12} = \frac{1}{z_{12}} = \frac{1}{0.06 + j\,0.18} = 1.66 - j\,5$$

$$y_{13} = \frac{1}{z_{13}} = \frac{1}{0.03 + j\,0.09} = 3.33 - j\,10$$

$$y_{23} = \frac{1}{z_{23}} = \frac{1}{0.08 + j\,0.24} = 1.25 - j\,3.75$$

$$Y_{11} = y_{12} + y_{13} = 1.66 - j\,5 + 3.33 - j\,10 = 5 - j\,15$$

$$Y_{22} = y_{12} + y_{23} = 1.66 - j\,5 + 1.25 - j\,3.75 = 2.91 - j\,8.75$$

$$Y_{33} = y_{23} + y_{13} = 1.25 - j\,3.75 + 3.33 - j\,10 = 4.58 - j\,13.75$$

$$Y_{12} = -y_{12} = -1.66 + j\,5 = Y_{21}$$

$$Y_{23} = -y_{23} = -1.25 + j\,3.75 = Y_{32}$$

$$Y_{13} = -y_{13} = -3.33 + j\,10 = Y_{31}$$

$$Y_{bus} = \begin{bmatrix} 5 - j\,15 & -1.66 + j\,5 & -3.33 + j\,10 \\ -1.66 + j\,5 & 2.91 - j\,8.75 & -1.25 + j\,3.75 \\ -3.33 + j\,10 & -1.25 + j\,3.75 & 4.58 - j\,13.75 \end{bmatrix}$$

Fig. 14.3.

Example 14.5 In Example 14.4, each line has a total shunt admittance of $-j\,5.0$ pu. Determine the modified bus admittance matrix.

Solution

Fig. 14.4 shows the 3-bus system where the shunt admittances of the line are also indicated.

For the line between buses 1 and 2, the total shunt admittance $= y_c + y_d = -j\,5.0$ and $y_c = y_d$

For the line between buses 1 and 3, the total shunt admittance $= y_a + y_b = -j\,5.0$ and $y_a = y_b$

For the line between buses 2 and 3, the total shunt admittance $= y_e + y_f = -j\,5.0$ and $y_e = y_f$

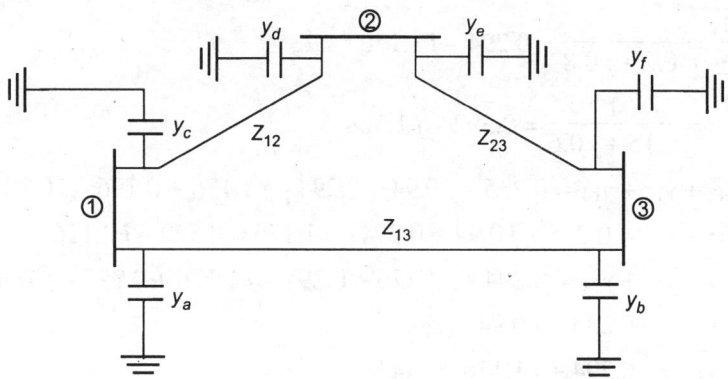

Fig. 14.4.

When we consider the effect of shunt admittances, only the diagonal elements in Y_{bus} are modified. We have

$$y_a = y_b = y_c = y_d = y_e = y_f = -j\,2.5$$

$$Y_{11\,(new)} = Y_{11\,(old)} + y_a + y_c = (5 - j\,15) + (-j\,5) = 5 - j\,20$$

$$Y_{22\,(new)} = Y_{22\,(old)} + y_d + y_e = 2.91 - j\,8.75 - j\,5 = 2.91 - j\,13.5$$

$$Y_{33\,(new)} = Y_{33\,(old)} + y_b + y_f = 4.58 - j\,13.75 - j\,5 = 4.58 - j\,18.75$$

$$\therefore \qquad Y_{bus} = \begin{bmatrix} 5 - j\,20 & -1.66 + j\,5 & -3.33 + j\,10 \\ -1.66 + j\,5 & 2.91 - j\,13.5 & -1.25 + j\,3.75 \\ -3.33 + j\,10 & -1.25 + j\,3.75 & 4.58 - j\,18.75 \end{bmatrix}$$

Example 14.6 Determine Y_{bus} for the 4-bus system shown in Fig. 14.5. The line series impedances are as follows :

Line (bus to bus)	Impedance (pu)
1-2	$0.25 + j\,1.0$
1-3	$0.20 + j\,0.8$
1-4	$0.30 + j\,1.2$
2-3	$0.20 + j\,0.8$
3-4	$0.15 + j\,0.6$

Neglect the shunt capacitances of the lines.

Solution

$$y_{12} = \frac{1}{z_{12}} = \frac{1}{0.25 + j\,1.0} = 0.235 - j\,0.94$$

$$y_{13} = \frac{1}{z_{13}} = \frac{1}{0.2 + j\,0.8} = 0.294 - j\,1.176$$

$$y_{14} = \frac{1}{z_{14}} = \frac{1}{0.3 + j\,1.2} = 0.196 - j\,0.784$$

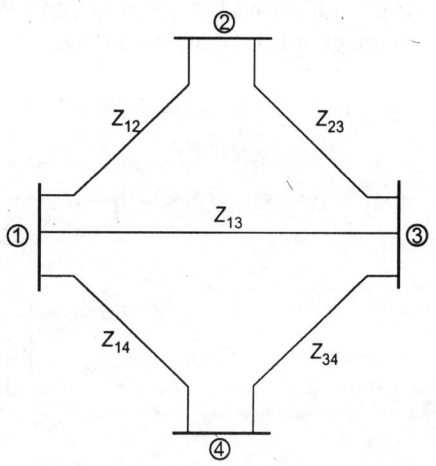

Fig. 14.5.

$$y_{23} = \frac{1}{z_{23}} = \frac{1}{0.2 + j\,0.8} = 0.294 - j\,1.176$$

$$y_{34} = \frac{1}{z_{34}} = \frac{1}{0.15 + j\,0.6} = 0.392 - j\,1.568$$

$Y_{11} = y_{12} + y_{13} + y_{14} = 0.235 - j\,0.94 + 0.294 - j\,1.176 + 0.196 - j\,0.784 = 0.725 - j\,2.9$

$Y_{22} = y_{12} + y_{23} = 0.235 - j\,0.94 + 0.294 - j\,1.176 = 0.529 - j\,2.116$

$Y_{33} = y_{23} + y_{13} + y_{34} = 0.294 - j\,1.176 + 0.294 - j\,1.176 + 0.392 - j\,1.568 = 0.98 - j\,3.92$

$Y_{12} = -y_{12} = -0.235 + j\,0.94 = Y_{21}$

$Y_{13} = -y_{13} = -0.294 + j\,1.176 = Y_{31}$

$Y_{14} = -y_{14} = -0.196 + j\,0.784 = Y_{41}$

$Y_{23} = -y_{23} = -0.294 + j\,1.176 = Y_{32}$

$Y_{34} = -y_{34} = -0.392 + j\,1.568 = Y_{43}$

$Y_{44} = y_{14} + y_{34} = 0.196 - j\,0.784 + 0.392 - j\,1.568 = 0.588 - j\,2.352$

$$Y_{bus} = \begin{bmatrix} 0.725 - j\,2.9 & -0.235 + j\,0.94 & -0.294 + j\,1.176 & -0.196 + j\,0.784 \\ -0.235 + j\,0.94 & 0.529 - j\,2.116 & -0.294 + j\,1.176 & 0 \\ -0.294 + j\,1.17 & -0.294 + j\,1.176 & 0.98 - j\,3.92 & -0.392 + j\,1.568 \\ -0.196 + j\,0.784 & 0 & -0.392 + j\,1.568 & 0.588 + j\,2.352 \end{bmatrix}$$

Example 14.7 In Example 14.6, each line between buses, 1-2, 1-3, and 1-4 has a total shunt admittance of $-j\,0.16$ pu. The shunt admittances of the remaining lines are neglected. Determine the bus admittance matrix if the series impedances are the same as in Example 14.6.

Solution

Fig. 14.6 shows the given system. When shunt admittances are also considered, only the diagonal elements of Y_{bus} are modified.

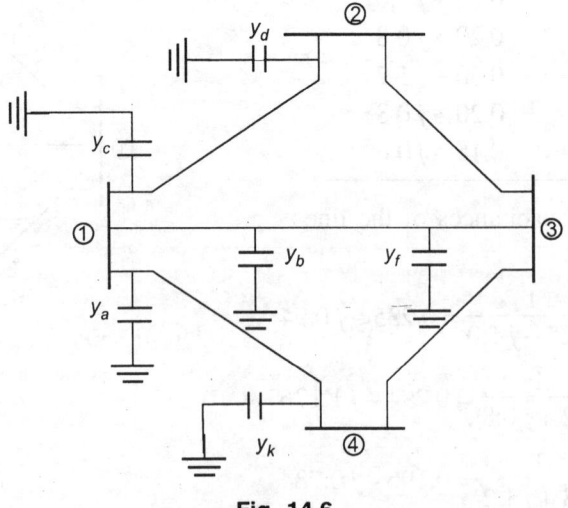

Fig. 14.6.

Total shunt admittance between buses 1-2 $= y_c + y_d = -j\,0.16$

Total shunt admittance between buses 1-4 $= y_a + y_k = -j\,0.16$

Total shunt admittance between buses 1-3 $= y_b + y_f = -j\,0.16$

Also, $y_c = y_d = y_a = y_k = y_b = y_f$

$$Y_{11\,(new)} = Y_{11\,(old)} + y_a + y_b + y_c = Y_{11\,(old)} + 3y_a = (0.725 - j\,2.9) + 3 \times \left(-\frac{1}{2}j\,0.16\right)$$

$$= 0.725 - j\,3.14$$

$$Y_{22\,(new)} = Y_{22\,(old)} + y_d = 0.529 - j\,2.116 + \left(-\frac{1}{2}j\,0.16\right)$$

$$= 0.529 - j\,2.196$$

$$Y_{33\,(new)} = Y_{33\,(old)} + y_f = 0.98 - j\,3.92 - \frac{1}{2}\,(j\,0.16)$$

$$= 0.98 - j\,4$$

$$Y_{44\,(new)} = Y_{44\,(old)} + y_k = 0.588 - j\,2.352 - \frac{1}{2}\,(j\,0.16)$$

$$= 0.588 - j\,2.432$$

$$Y_{bus} = \begin{bmatrix} (0.725 - j\,3.14) & (-0.235 + j\,0.94) & (-0.294 + j\,1.176) & (-0.196 + j\,0.784) \\ (-0.235 + j\,0.94) & (0.529 - j\,2.196) & (-0.294 + j\,1.176) & 0 \\ (-0.294 + j\,1.176) & (-0.294 + j\,1.176) & 0.98 - j\,4 & (-0.392 + j\,1.568) \\ (-0.196 + j\,0.784) & 0 & (0.392 + j\,1.568) & (0.588 - j\,2.432) \end{bmatrix}$$

14.4 STATIC LOAD FLOW EQUATIONS (SLFE)

From the nodal current equations, the total current entering the ith bus of an n-bus system is given by

$$\mathbf{I}_i = \mathbf{Y}_{i1}\,\mathbf{V}_1 + \mathbf{Y}_{i2}\,\mathbf{V}_2 + \ldots + \mathbf{Y}_{ij}\,\mathbf{V}_j + \ldots + \mathbf{Y}_{in}\,\mathbf{V}_n = \sum_{k=1}^{n} \mathbf{Y}_{ik}\,\mathbf{V}_k \qquad \ldots(14.4.1)$$

Let $\mathbf{V}_k = V_k\,\underline{/\delta_k}\,;$ $\mathbf{Y}_{ik} = Y_{ik}\,\underline{/\theta_{ik}}$

$\therefore$ $$\mathbf{I}_i = \sum_{k=1}^{n} Y_{ik}\,V_k\,\underline{/(\delta_k + \theta_{ik})} \qquad \ldots(14.4.2)$$

The complex power injected into the ith bus is

$$\mathbf{S}_i = P_i + j\,Q_i = \mathbf{V}_i\,\mathbf{I}_i^* \qquad \ldots(14.4.3)$$

Since $\mathbf{V}_i = V_i\,\underline{/\delta_i}$

$$\mathbf{S}_i = P_i + j\,Q_i = V_i \sum_{k=1}^{n} Y_{ik}\,V_k\,\underline{/(\delta_i - \delta_k - \theta_{ik})} \qquad \ldots(14.4.4)$$

Separation of real and imaginary parts gives

$$P_i = V_i \sum_{k=1}^{n} Y_{ik} V_k \cos (\delta_i - \delta_k - \theta_{ik}) \qquad i = 1, 2, ..., n \qquad ...(14.4.5)$$

$$Q_i = V_i \sum_{k=1}^{n} Y_{ik} V_k \sin (\delta_i - \delta_k - \theta_{ik}) \qquad i = 1, 2, ..., n \qquad ...(14.4.6)$$

Equations (14.4.5) and (14.4.6) are called *static load flow equations* (SLFE). Equation (14.4.5) gives n real power flow equations. Similarly, Eq. (14.4.6) gives n reactive power flow equations. Thus, Eqs. (14.4.5) and (14.4.6) represent $2n$ power flow equations. At each bus we have four variables P_i, Q_i, V_i and δ_i resulting in total of $4n$ variables. In order to find a solution it is necessary to specify two variables at each bus. Thus the number of unknown variables is reduced to $2n$. The solution of these remaining $2n$ variables is done by numerical methods because Eqs. (14.4.5) and (14.4.6) are nonlinear.

No exact analytical solution of nonlinear equations is possible. These equations may be solved by *iterative* techniques which employ successive approximations eventually converging upon a solution. Before the advent of digital computers these trial and error techniques were tedious and time consuming. However, today these methods find widespread applications for solving load flow problems. The *iteration procedure* involves an initial assumed value for each of the unknown independent variable. These numerical values are substituted in the original equation to obtain a new set of corrected values of these independent variables. The second set is used to find the third corrected set. The process is repeated. Each calculation of a new set of variables is called an *iteration*. The iteration is continued until the unknown values converge within required limits.

14.5 METHODS OF LOAD FLOW SOLUTION

A large number of iterative methods using both Y_{bus} and Z_{bus} are available in literature. The desirable features of an ideal load flow method are as follows :

(1) High speed, that is, fast convergence
(2) Minimal storage
(3) Simplicity and ease of programming
(4) Reliability for ill-conditioned system such as systems having junctions of very high and low series impedances, long EHV lines, large series capacitances, series and shunt compensation. These factors affect the convergence.

No single load flow method satisfies all these requirements. In practice, all these features are not required simultaneously in all situations. A compromise is to be made in the choice of a particular method.

14.6 GAUSS-SEIDEL METHOD USING Y_{bus}

From the nodal current equations, the total current entering the kth bus of an n-bus system is given by

$$I_k = Y_{k1} V_1 + Y_{k2} V_2 + \dots + Y_{kn} V_n = \sum_{i=1}^{n} Y_{ki} V_i \qquad \dots(14.6.1)$$

The complex power injected into the kth bus is

$$S_k = P_k + j Q_k = V_k I_k^* \qquad \dots(14.6.2)$$

The complex conjugate of Eq. (14.6.2) gives

$$S_k^* = P_k - j Q_k = V_k^* I_k \qquad \dots(14.6.3)$$

$$I_k = \frac{1}{V_k^*} (P_k - j Q_k) \qquad \dots(14.6.4)$$

Elimination of I_k from Eqs. (14.6.1) and (14.6.4) gives

$$Y_{k1} V_1 + Y_{k2} V_2 + \dots + Y_{kk} V_k + \dots + Y_{kn} V_n = \frac{1}{V_k^*} (P_k - j Q_k) \qquad \dots(14.6.4a)$$

Therefore, the voltage at any bus k where P_k and Q_k are specified is given as

$$V_k = \frac{1}{Y_{kk}} \left[\frac{P_k - j Q_k}{V_k^*} - \sum_{\substack{i=1 \\ i \neq k}}^{n} Y_{ki} V_i \right] \qquad \dots(14.6.5)$$

Equation (14.6.5) is the heart of the iterative algorithm.
At bus 2,

$$V_2 = \frac{1}{Y_{22}} \left[\frac{P_2 - j Q_2}{V_2^*} - Y_{21} V_1 - Y_{23} V_3 - \dots - Y_{2n} V_n \right] \qquad \dots(14.6.6)$$

At bus 3,

$$V_3 = \frac{1}{Y_{33}} \left[\frac{P_3 - j Q_3}{V_3^*} - Y_{31} V_1 - Y_{32} V_2 - Y_{34} V_4 - \dots - Y_{3n} V_n \right] \qquad \dots(17.6.7)$$

For kth bus the voltage at the $(r+1)$th iteration is given by

$$V_k^{(r+1)} = \frac{1}{Y_{kk}} \left[\frac{P_k - j Q_k}{(V_k^{(r)})^*} - \sum_{i=1}^{k-1} Y_{ki} V_i^{(r+1)} - \sum_{i=k+1}^{n} Y_{ki} V_i^{(r)} \right] \qquad \dots(14.6.8)$$

In the above equations, the quantities P_k, Q_k, Y_{kk} and Y_{ki} are known and do not vary during the iteration cycle. Let us define

$$C_k \overset{\Delta}{=} \frac{P_k - j Q_k}{Y_{kk}} \text{ for } k = 2, 3, \dots, n \qquad \dots(14.6.9)$$

$$D_{ki} \overset{\Delta}{=} \frac{Y_{ki}}{Y_{kk}} \text{ for } k = 2, 3, \dots, n \quad i = 1, 2, \dots, n \quad (\text{except } i \neq k) \qquad \dots(14.6.10)$$

The values of C_k and D_{ki} are computed in the beginning and then used in every iteration. This saves computer time considerably.

For kth bus the voltage at the $(r+1)$th iteration can be written as

$$V_k^{(r+1)} = \frac{C_k}{(V_k^{(r)})^*} - \sum_{i=1}^{k-1} D_{ki} V_i^{(r+1)} - \sum_{i=k+1}^{n} D_{ki} V_i^{(r)} \qquad \ldots(14.6.11)$$

$$V_k^{(r+1)} = \frac{C_k}{(V_k^{(r)})^*} - \sum_{\substack{i=1 \\ i \neq k}}^{n} D_{ki} V_i^{(r)} \quad k = 2, 3, \ldots, n \qquad \ldots(14.6.12)$$

14.6.1 Computation for Load Buses

Let us assume that all the buses are of P-Q type except the slack bus which is of V-δ type.

Let bus 1 be designated the swing bus. Since the voltage at the slack bus is fixed in magnitude and phase, it does not vary during iterative procedure and there is no iteration involved for this bus.

The solution begins with bus 2.

The iteration procedure is as follows :

1. Assume initial values of load bus voltages and angles for generator bus voltage (except for the slack bus). Let the assumed values be

   $$V_2^{(0)}, V_3^{(0)}, \ldots, V_n^{(0)} .$$

   The superscript (0) indicates an initial approximation.

2. Calculate $V_2^{(1)}$ in terms of the initial assumed voltages as follows :

   $$V_2^{(1)} = \frac{1}{Y_{22}} \left[\frac{P_2 - j Q_2}{(V_2^{(0)})^*} - Y_{21} V_1 - Y_{23} V_3^{(0)} - \ldots - Y_{2n} V_n^{(0)} \right] \qquad \ldots(14.6.13)$$

3. When the corrected value of $V_2^{(1)}$ is determined, we determine $V_2^{(1)*}$. The corrected value of $V_2^{(1)*}$ is then substituted back into the Eq. (14.6.13) in place of $V_2^{(0)*}$. Thus a new corrected value of V_2 is obtained. This process is continued for a specified number of iterations. It should be noted that the last iteration may not give the correct value of V since the correction is also dependent upon other assumed voltages.

4. Using the corrected value of the voltage from step 3 and other assumed values of voltages we perform calculations for bus 3. Here also we perform several iterations as we have done for bus 2.

5. We then go to buses 4, 5, 6, ..., etc. and continue the same procedure of iteration until all buses have been considered and we have obtained a new set of the values of all bus voltages in the network.

6. Repeat the iteration process from step 1 to step 5 until the difference ΔV in old and new values of bus voltages for all the buses of the network (except the slack bus) is within a specified limit of tolerance. The iterations process is then said to converge to a solution. Let the iteration count be denoted by r. Then the magnitude of change in voltage at bus k between two consecutive iterations is given by

   $$|\Delta V_k^{(r+1)}| = |V_k^{(r+1)} - V_k^{(r)}| \qquad \ldots(14.6.14)$$

For all *P-Q* buses the criterion for convergence is $|\Delta V_k| < \varepsilon$ where ε is the tolerance level. Typical values of ε range from 0.01 to 0.0001.

14.6.2 Computation for PV Buses

At a *PV* bus, *P* and $|V|$ are specified and *Q* and δ are unknown. At the *k*th bus, the voltages magnitude is to be maintained at a specified value $|V_i|_{sp}$. The values of *Q* and δ are to be updated in every iteration.

From Eq. (14.6.4a),

$$Q_k = -\operatorname{Im}\left[V_k^* \sum_{i=1}^{n} Y_{ki} V_i \right] \qquad \ldots(14.6.15)$$

where the symbol Im means 'the imaginary part of'. If *k*th bus is the *PV*·bus the following procedure is used :

The revised value of Q_k is found from Eq. (14.6.15) by substituting most updated values of voltages on the right hand side. Thus, for $(r+1)$th iteration we have

$$Q_k^{(r+1)} = -\operatorname{Im}\left[(V_k^{(r)})^* \sum_{i=1}^{n} Y_{ki} V_i^{(r+1)} + (V_k^{(r)})^* \sum_{i=1}^{n} Y_{ki} V_i^{(r)} \right] \qquad \ldots(14.6.16)$$

The revised value of δ_k is found from Eq. (14.6.8) as follows :

$$\delta_k^{(r+1)} = \text{angle of } V_k^{(r+1)}$$

$$= \text{angle of }\left[\frac{P_k - j\, Q_k^{(r+1)}}{Y_{kk}\, (V_k^{(r)})^*} - \sum_{i=1}^{k-1} \frac{Y_{ki}}{Y_{kk}} V_i^{(r+1)} - \sum_{i=k+1}^{n} \frac{Y_{ki}}{Y_{kk}} V_i^{(r)} \right] \qquad \ldots(14.6.17)$$

For a *PV* bus the upper limit Q_{max} and lower limit Q_{min} of *Q* to hold the generation vars within limits is also given. That is

$$Q_{k\,(min)} < Q_k < Q_{k\,(max)} \qquad \ldots(14.6.18)$$

Therefore, the calculated $Q_k^{(r+1)}$ is checked for limits of Q_k, that is

$$Q_{k\,(min)} < Q_k^{(r+1)} < Q_{k\,(max)} \qquad \ldots(14.6.19)$$

If $Q_k^{(r+1)}$ is within its limits then calculate new value of C_k and use Eq. (14.6.11) to calculate $V_k^{(r+1)}$ using $|V_k|_{sp}$ and $\delta_k^{(r)}$ for the magnitude and phase angle of $V_k^{(r)}$. Now put $|V_k^{(r+1)}| = |V_k|_{sp}$ and retain the phase angle $\delta_k^{(r+1)}$ and move to the next bus.

If $Q_k^{(r+1)} > Q_{k\,(max)}$, then put $Q_k^{(r+1)} = Q_{k\,(max)}$. Now we find the new value of C_k and treat the *k*th bus as *PQ* bus and continue the computations similar to a load (*PQ*) bus.

If $Q_k^{(r+1)} < Q_{k\,(min)}$, then $Q_{k\,(min)}$ is taken as the reactive power at bus *k*, that is,

$$Q_k^{(r+1)} = Q_{k\,(min)}$$

Now we find the new value of C_k and treat the *k*th bus as *PQ* bus. The computation is then continued as in case of a *PQ* bus.

The flow chart for a load flow study using Gauss-Seidel method is given in Fig. 14.7. Here the buses are numbered as follows :

$i = 1$ slack bus

$i = 2, 3, ..., m$ PV buses

$i = m + 1, m + 2, ..., n$ PQ buses

14.6.3 Acceleration Factors

In the Gauss-Seidel method, a large number of iterations are required to arrive at the specified (desired) convergence. The rate of convergence can be increased by the use of acceleration factor to the solution obtained after each iteration. The acceleration factor is a multiplier that enhances correction between the values of voltage in two successive iterations. For the ith bus,

let $V_i^{(r)}$ = value of voltage at the rth iteration

$V_i^{(r+1)}$ = value of voltage at the $(r + 1)$th iteration

$V_{i\,(accelerated)}^{(r+1)}$ = accelerated new value of the voltage at the $(r + 1)$th iteration

r = iteration count

α = accelerating factor

then $V_{i\,(accelerated)}^{(r+1)} = V_i^{(r)} + \alpha\,[V_i^{(r+1)} - V_i^{(r)}]$

Thus, after calculating $V_i^{(r+1)}$ at $(r + 1)$th iteration we calculate the new estimated bus voltage $V_{i\,(accelerated)}^{(r+1)}$ and this new estimate replaces the calculated value $V_i^{(r+1)}$. Different accelerating factors may be used for real and imaginary components of the voltage. That is, if V_i is resolved into real and imaginary components as

$V_i = a_i + j\,b_i$

and if α and β are the acceleration factors associated with a_i and b_i then

$a_{i\,(accelerated)}^{(r+1)} = a_i^{(r)} + \alpha\,[a_1^{(r+1)} - a_i^{(r)}]$

$b_{i\,(accelerated)}^{(r+1)} = b_i^{(r)} + \beta\,[b_i^{(r+1)} - b_i^{(r)}]$

The choice of a specific value of acceleration factor depends upon the system parameters. The optimum valve of α usually lies in the range 1.2 to 1.6 for most systems, but a value of 1.6 is widely used in practice.

14.7 NEWTON-RAPHSON (NR) METHOD

This is an iterative technique for solving a set of simultaneous nonlinear equations in an equal number of unknowns.

Let the n equations in n unknowns be

$$\left.\begin{array}{l} f_1\,(x_1, x_2, ..., x_n) = y_1 \\ f_2\,(x_1, x_2, ..., x_n) = y_2 \\ ...\quad ...\quad ...\quad ...\quad ... \\ f_n\,(x_1, x_2, ..., x_n) = y_n \end{array}\right\} \qquad ...(14.7.1)$$

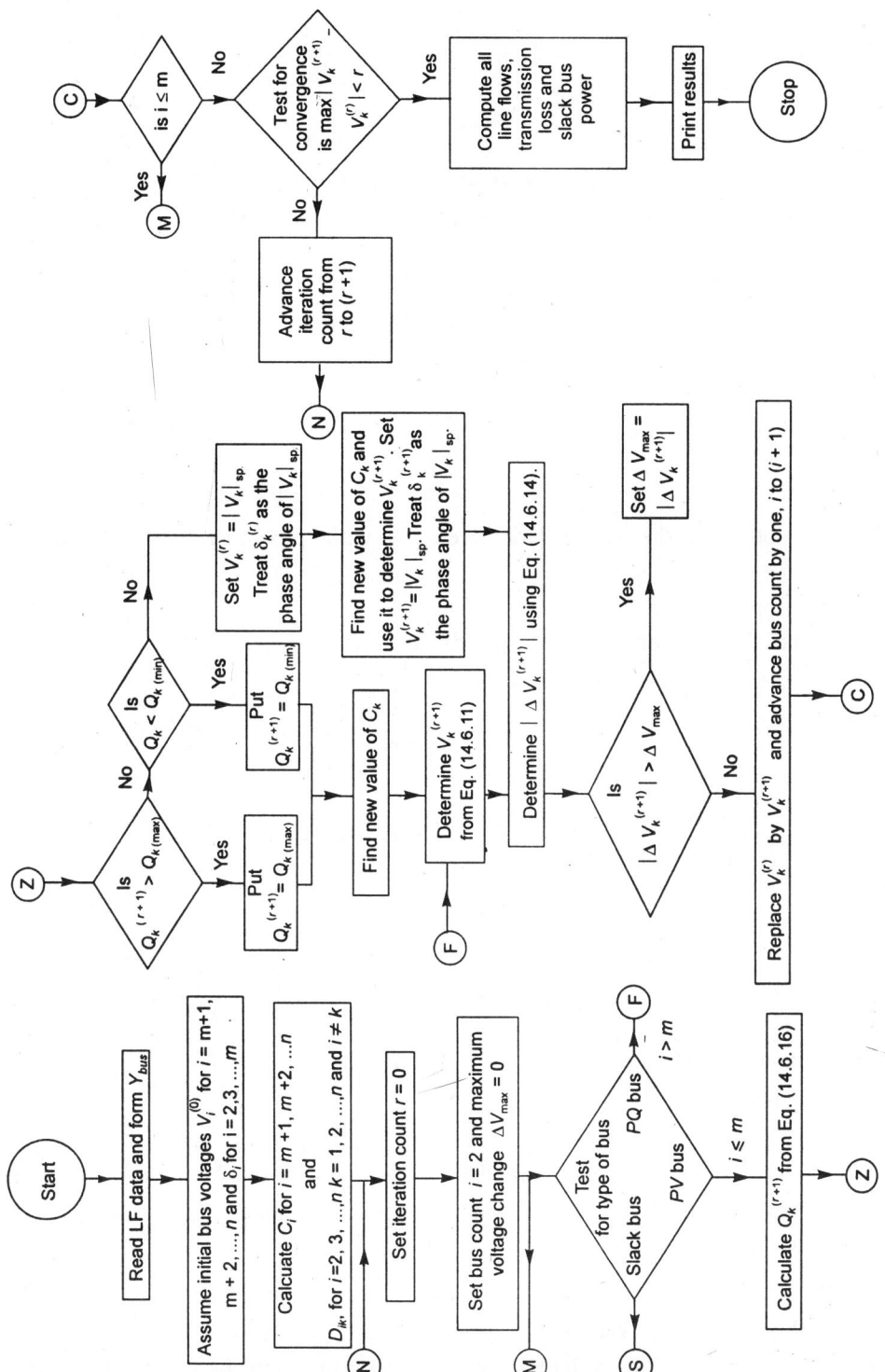

Fig. 14.7. Flowchart for Gauss-Seidel flow algorithm.

In general,

$$f_i(x_1, x_2, \ldots, x_n) = y_i \qquad i = 1, 2, \ldots n \qquad \ldots(14.7.2)$$

Assume initial values of unknowns as

$$\mathbf{X}^0 = x_1^{(0)}, x_2^{(0)}, \ldots, x_n^{(0)}$$

These values are not far from the actual solution. Let

$$\Delta \mathbf{X}^0 = \Delta x_1, \Delta x_2, \ldots, \Delta x_n$$

be the corrections, which being added to the initial guess, give the actual solution.

$$\therefore \qquad f_1(x_1^{(0)} + \Delta x_1, x_2^{(0)} + \Delta x_2, \ldots, X_{n(0)} + \Delta x_n) = y_i \qquad i = 1, 2, \ldots, n \qquad \ldots(14.7.3)$$

Expanding these equations by Taylor's theorem we get

$$f_i(x_1^0, x_2^0, \ldots, x_n^0) + \left(\frac{\partial f_i}{\partial x_i}\right)_{x_1 = x_0} \Delta x_1 + \left(\frac{\partial f_i}{\partial x_2}\right) \Delta x_2 + \ldots + \left(\frac{\partial f_i}{\partial x_n}\right) \Delta x_n + \text{higher order terms} = y_i$$

$$\ldots(14.7.4)$$

It is to be noted that the partial derivatives of f_i with respect to $x_i, x_2, \ldots, x_n$ are evaluated at the solution estimates $x_1^0, x_2^0, \ldots, x_n^0$. Since our initial estimate is close to the true solution we neglect higher order terms.

Equation (14.7.4) can be written in matrix form as

$$\begin{bmatrix} y_1 - f_1(x_1^0, x_2^0, \ldots, x_n^0) \\ y_2 - f_2(x_1^0, x_2^0, \ldots, x_n^0) \\ \ldots \quad \ldots \quad \ldots \quad \ldots \quad \ldots \\ y_n - f_n(x_1^0, x_2^0, \ldots, x_n^0) \end{bmatrix} = \begin{bmatrix} \Delta f_1^0 \\ \Delta f_2^0 \\ \ldots \\ \Delta f_n^0 \end{bmatrix} = \begin{bmatrix} \left(\dfrac{\partial f_1}{\partial x_1}\right)^0 & \left(\dfrac{\partial f_1}{\partial x_2}\right)^0 & \cdots & \left(\dfrac{\partial f_1}{\partial x_n}\right)^0 \\ \left(\dfrac{\partial f_2}{\partial x_1}\right)^0 & \left(\dfrac{\partial f_2}{\partial x_2}\right)^0 & \cdots & \left(\dfrac{\partial f_2}{\partial x_n}\right)^0 \\ \ldots & \ldots & \ldots & \ldots \\ \left(\dfrac{\partial f_n}{\partial x_1}\right)^0 & \left(\dfrac{\partial f_n}{\partial x_2}\right)^0 & \cdots & \left(\dfrac{\partial f_n}{\partial x_n}\right)^0 \end{bmatrix} \begin{bmatrix} \Delta x_1 \\ \Delta x_2 \\ \ldots \\ \Delta x_n \end{bmatrix} \qquad \ldots(14.7.5)$$

In vector matrix form Eq. (14.7.5) can be written as

$$\Delta \mathbf{f} = \mathbf{J} \, \Delta \mathbf{X} \qquad \ldots(14.7.6)$$

where $\mathbf{J}$ is a matrix of partial derivatives known as the Jacobian matrix or simply Jacobian of the simultaneous equations. It is obtained by differentiating the function vector $\mathbf{f}$ with respect to x and evaluating it at x^0. $\Delta \mathbf{f}$ is the column vector of mismatches of the functions with the initial guess for x_i. That is

$$\Delta \mathbf{f} = y_1 - f_1(x_1^{(0)}, x_2^{(0)}, \ldots, x_n^{(0)})$$

$\Delta \mathbf{X}$ is the column vector of connection terms $\Delta x_1, \Delta x_2, \ldots, \Delta x_n$.

Equation (14.7.6) provides a linearized relationship between the errors $\Delta \mathbf{f}$ and the correction $\Delta \mathbf{X}$ through the Jacobian $\mathbf{J}$. A solution for $\Delta \mathbf{X}$ is obtained by using any suitable method of solving a set of linear algebraic equations. Equation (14.7.6) gives ΔX, that is, $\Delta x_1, \Delta x_2, ..., \Delta x_n$. The correction is added to the initial guess to complete the first iteration. The next better solution is obtained as follows :

$$x_1^{(1)} = x_1^{(0)} + \Delta x_1$$

$$x_2^{(1)} = x_2^{(0)} + \Delta x_2$$

$$x_n^{(1)} = x_n^{(0)} + \Delta x_n$$

In general, for the $(r+1)$th iteration

$$x^{(r+1)} = x^{(r)} + \Delta x^{(r)}$$

The iteration process is continued until the errors $\Delta f_1, \Delta f_2, ..., \Delta f_n$ are lower than a specified tolerance. At the end of each iteration, the Jacobian $\mathbf{J}$ is re-evaluated.

14.8 NEWTON-RAPHSON METHOD FOR LOAD FLOW SOLUTION

There are two methods of solution for the load flow using Newton-Raphson method. The first method uses rectangular coordinates for the variables, while the second method uses the polar coordinate form. The polar coordinate method is widely used.

The equation for the complex power at node i in the polar form is given in Eq. (14.4.4). Equations (14.4.5) and (14.4.6) give the active and reactive powers at bus i. For the sake of convenience, let us reproduce these equations.

$$\mathbf{S}_i = P_i + j\, Q_i = \mathbf{V}_i \sum_{k=1}^{n} \mathbf{Y}_{ik}^* \mathbf{V}_k^* \qquad \qquad ...(14.8.1)$$

$$= \sum_{k=1}^{n} (V_i\, V_k\, Y_{ik}) \,\underline{/(\delta_i - \delta_k - \theta_{ik})} \qquad \qquad ...(14.8.2)$$

$$P_i = \sum_{k=1}^{n} (V_i\, V_k\, Y_{ik}) \cos(\delta_i - \delta_k - \theta_{ik}) \qquad \qquad ...(14.8.3)$$

$$Q_i = \sum_{k=1}^{n} (V_i\, V_k\, Y_{ik}) \sin(\delta_i - \delta_k - \theta_{ik}) \qquad \qquad ...(14.8.4)$$

Equations (14.8.3) and (14.8.4) can also be written as

$$P_i = V_i\, V_i\, Y_{ii} \cos \theta_{ii} + \sum_{\substack{k=1 \\ k \neq i}}^{n} (V_i\, V_k\, Y_{ik}) \cos(\delta_i - \delta_k - \theta_{ik}) \qquad \qquad ...(14.8.3a)$$

$$Q_i = -V_i\, V_i\, Y_{ii} \sin \theta_{ii} + \sum_{\substack{k=1 \\ k \neq i}}^{n} (V_i\, V_k\, Y_{ik}) \sin(\delta_i - \delta_k - \theta_{ik}) \qquad \qquad ...(14.8.4a)$$

We have $\Delta f = J \, \Delta X$ ...(14.8.5)

If $\quad \Delta P_i = P_{i\,(sp)} - P_{i\,(cal)}$

then $\quad i = 1, 2, ..., n, \quad i \neq$ slack

and if $\quad \Delta Q_i = Q_{i\,(sp)} - Q_{i\,(cal)}$

then $\quad i = 1, 2, ..., n, \quad i \neq$ slack, $\quad i \neq PV$ bus

where the subscripts sp and cal denote the specified and calculated values respectively then Eq. (14.8.5) can be written as

$$\begin{bmatrix} \Delta P \\ \Delta Q \end{bmatrix} = \begin{bmatrix} H & N \\ M & L \end{bmatrix} \begin{bmatrix} \Delta \delta \\ \Delta V \end{bmatrix} \qquad ...(14.8.6)$$

The off-diagonal and diagonal elements of the sub-matrices H, N, M and L are determined by differentiating Eqs. (14.8.3) and (14.8.4) with respect to δ and $|V|$. Off-diagonal elements of H

$$H_{ik} \triangleq \frac{\partial P_i}{\partial \delta_k} = V_i V_k Y_{ik} \sin (\delta_i - \delta_k - \theta_{ik}), \quad i \neq k \qquad ...(14.8.7)$$

Diagonal elements

$$H_{ii} \triangleq \frac{\partial P_i}{\partial \delta_i} = - V_i \sum_{\substack{k=1 \\ k \neq 1}}^{n} Y_{ik} V_k \sin (\delta_i - \delta_k - \theta_i) \qquad ...(14.8.8)$$

Using Eq. (14.8.4a), we have

$$\frac{\partial P_i}{\partial \delta_i} = - [Q_i + V_i^2 Y_{ii} \sin (- \theta_{ii})]$$

$$H_{ii} = - Q_i - V_i^2 Y_{ii} \sin \theta_{ii} = - Q_i - B_{ii} V_i^2 \qquad ...(14.8.9)$$

The off-diagonal and diagonal elements of N are given by

$$\frac{\partial P_i}{\partial |V_k|} = V_i V_k \cos (\delta_i - \delta_k - \theta_{ik}) \qquad ...(14.8.10)$$

$$\frac{\partial P_i}{\partial |V_i|} = 2V_i Y_{ii} \cos \theta_{ii} + \sum_{\substack{k=1 \\ k \neq i}}^{n} V_k Y_{ik} \cos (\delta_i - \delta_k - \theta_{ik}) \qquad ...(14.8.11)$$

The off-diagonal and diagonal elements of M matrix are

$$\frac{\partial Q_i}{\partial \delta_k} = - V_i V_k Y_{ik} \cos (\delta_i - \delta_k - \theta_{ik}) \text{ for } i \neq k \qquad ...(14.8.12)$$

$$\frac{\partial Q_i}{\partial \delta_i} = \sum_{\substack{k=1 \\ k \neq i}}^{n} V_i V_k Y_{ik} \cos (\delta_i - \delta_k - \theta_{ik}) \qquad ...(14.8.13)$$

The off-diagonal and diagonal elements of L matrix are

$$\frac{\partial Q_i}{\partial V_k} = V_i Y_{ik} \sin(\delta_i - \delta_k - \theta_{ik}) \text{ for } i \neq k \qquad \qquad \ldots(14.8.14)$$

$$\frac{\partial Q_i}{\partial V_i} = -2V_i Y_{ii} \sin \theta_{ii} + \sum_{\substack{k=1 \\ k \neq i}}^{n} V_k Y_{ik} \sin(\delta_i - \delta_k - \theta_{ik}) \qquad \qquad \ldots(14.8.15)$$

It is seen from the elements of the Jacobian that there is no symmetry in the results. Multiplying and dividing by V the voltage magnitude increment ΔV to bring symmetry in the result we have

$$\Delta P = H \Delta \delta + (VN) \frac{\Delta V}{V}$$

$$\Delta Q = M \Delta \delta + (VL) \frac{\Delta V}{V}$$

Let $VN = N'$ and $VL = L'$. We can write

$$\begin{bmatrix} \Delta P \\ \Delta Q \end{bmatrix} = \begin{bmatrix} H & N' \\ M & L' \end{bmatrix} \begin{bmatrix} \Delta \delta \\ \dfrac{\Delta V}{V} \end{bmatrix}$$

In this case it will be seen that

$$H_{ik} = L'_{ik} \quad \text{and} \quad N'_{ik} = -M_{ik}$$

This property of symmetry of elements reduces computer time and storage.

Treatment of generator buses

For a generator bus the reactive power Q_i is not specified but the voltage magnitude $|\Delta V| = \Delta V$ is specified.

If $\qquad \mathbf{V}_i = v'_i + j v''_i$

$\qquad |\mathbf{V}_i|^2 = v'^2_i + v''^2_i$

then at all generator buses, the variable ΔQ_i is to be replaced by $\Delta |\mathbf{V}_i|^2$.

The elements of M are given by

$$M_{ik} = \frac{\partial (|V_i|)^2}{\partial \delta_k} = 0, \qquad i \neq k$$

and $\qquad M_{ii} = \dfrac{\partial (|V_i|)^2}{\partial \delta_i} = 0$

The elements of L are given by

$$L_{ik} = |\mathbf{V}_k| \frac{\partial (|V_i|)^2}{\partial |\mathbf{V}_k|} = 0, \qquad i \neq k$$

$$L_{ii} = |\mathbf{V}_i| \frac{\partial (|V_i|)^2}{\partial |\mathbf{V}_i|} = 2|\mathbf{V}_i|^2$$

14.9 COMPUTATIONAL PROCEDURE FOR NEWTON-RAPHSON METHOD

The computational procedure for Newton-Raphson method using polar coordinate is as follows :

1. Form Y_{bus}.

2. Assume initial values of bus voltages $|V_i|^0$ and phase angles δ_i^0 for $i = 2, 3, ..., n$ for load buses and phase angles for *PV* buses. Normally we set the assumed bus voltage magnitude and its phase angle equal to slack bus quantities $|V_1| = 1.0$, $\delta_1 = 0°$.

3. Compute P_i and Q_i for each load bus from the following equations :

$$P_i = \sum_{k=1}^{n} V_i V_k Y_{ik} \cos(\delta_i - \delta_k - \theta_{ik}) \qquad ...(14.9.1)$$

$$Q_i = \sum_{k=1}^{n} V_i V_k Y_{ik} \sin(\delta_i - \delta_k - \theta_{ik}) \qquad ...(14.9.2)$$

4. Compute the scheduled errors ΔP_i and ΔQ_i for each load bus from the following relations

$$\Delta P_i^{(r)} = P_{i\,sp} - P_{i\,(cal)}^{(r)} \qquad i = 2, 3, ..., n \qquad ...(19.9.3)$$

$$\Delta Q_i^{(r)} = Q_{i\,sp} - Q_{i\,(cal)}^{(r)} \qquad i = 2, 3, ..., n \qquad ...(14.9.4)$$

For *PV* buses, the exact value of Q_i is not specified, but its limits are known. If the calculated value of Q_i is within limits, only ΔP_i is calculated. If the calculated value of Q_i is beyond the limits, then an appropriate limit is imposed and ΔQ_i is also calculated by subtracting the calculated value of Q_i from the appropriate limit. The bus under consideration is now treated as a load (*PQ*) bus.

5. Compute the elements of the Jacobian matrix
$$\begin{bmatrix} H & N' \\ M & L' \end{bmatrix}$$
using the estimated $|V_i|$ and δ_i from step 2.

6. Obtain $\Delta \delta$ and $\Delta|V_i|$ from equation

$$\begin{bmatrix} \Delta P \\ \Delta Q \end{bmatrix} = \begin{bmatrix} H & N' \\ M & L' \end{bmatrix} \begin{bmatrix} \Delta \delta \\ \dfrac{\Delta V}{V} \end{bmatrix} \qquad ...(14.9.5)$$

7. Using the values of $\Delta \delta_i$ and $\Delta|V_i|$ calculated in step 6, modify the voltage magnitude and phase angle at all load buses by the equations

$$|V_i^{(r+1)}| = |V_i^{(r)}| + \Delta|V_i^{(r)}| \qquad ...(14.9.6)$$

$$\delta_i^{(r+1)} = \delta_i^{(r)} + \Delta \delta_i^{(r)} \qquad ...(14.9.7)$$

Start the next iteration cycle at step 2 with these modified $|V_i|$ and δ_i.

8. Continue until scheduled errors $\Delta P_i^{(r)}$ and $\Delta Q_i^{(r)}$ for all load buses are within a specified tolerance, that is,

$$\Delta P_i^{(r)} < \varepsilon, \ \Delta Q_i^{(r)} < \varepsilon$$

where ε denotes the tolerance level for load buses.

9. Calculate line flows and power at the slack bus exactly in the same manner as in the GS method.

The flowchart for Newton-Raphson method using polar coordinates for load flow solution is given in Fig. 14.8.

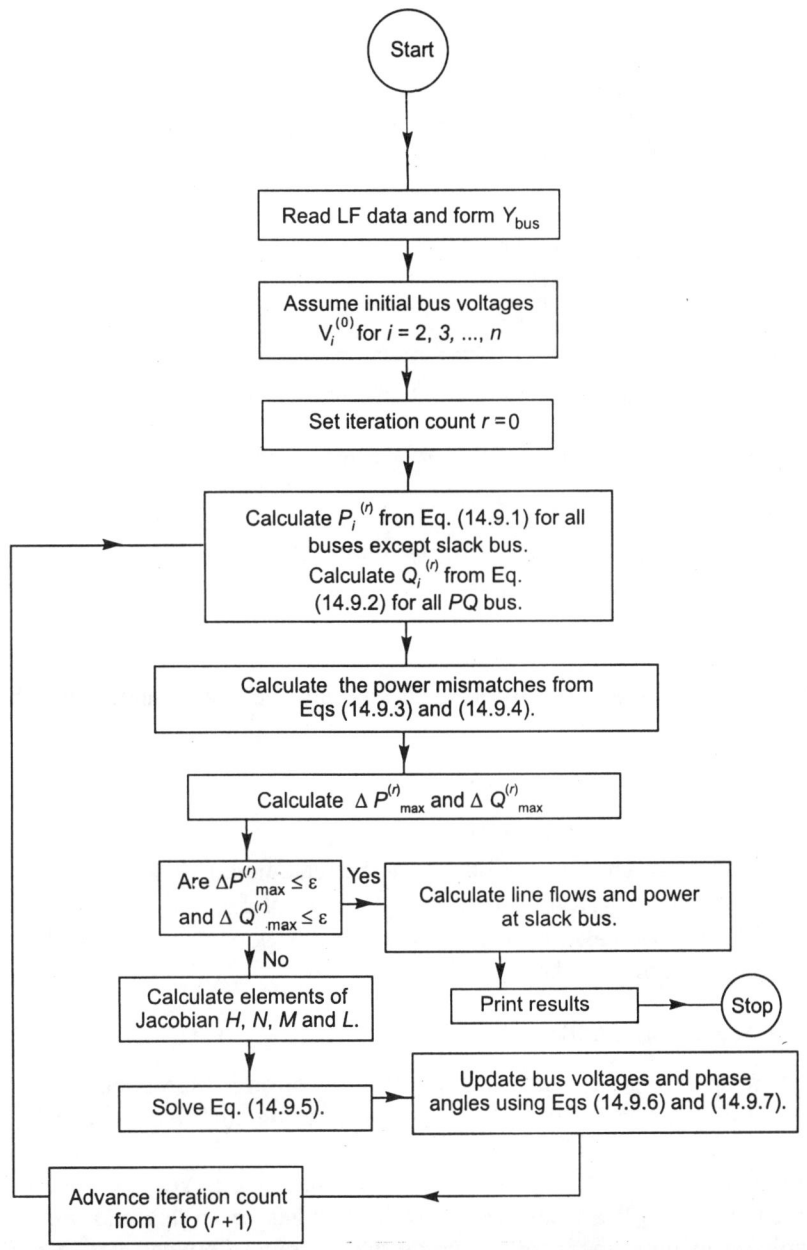

Fig. 14.8. Flowchart for load flow solution using NR method in polar coordinates.

14.10 POWER FLOW THROUGH LINES AND SLACK BUS POWER

The final step in power flow study in the flow of power through the lines. Consider the line connected two buses i and m. The line and transformers at each end can be represented by a nominal π-circuit. This circuit has a series admittance Y_{im} and two shunt admittances y_{im0} and y_{mi0} as shown in Fig. 14.9.

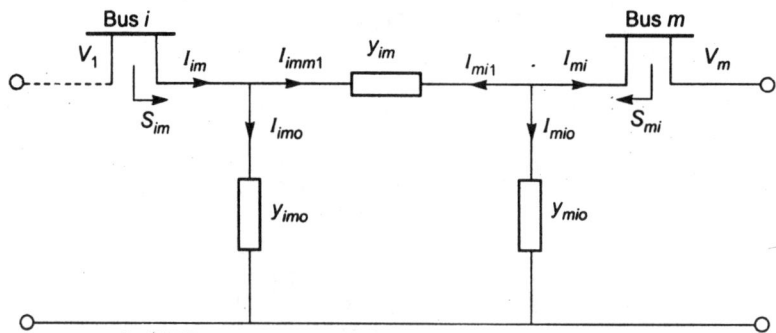

Fig. 14.9. Π-representation of a line and transformers connected between two buses.

The current supplied by bus i to the line is given by

$$I_{im} = \dot{I}_{im1} + I_{im0} = (V_i - V_m)\, y_{im} + V_i\, y_{im0} \qquad \qquad ...(14.10.1)$$

The complex power supplied by bus i to the line is given by

$$S_{im} = P_{im} + j\, Q_{im} = V_i\, \dot{I}_{im}^*$$

$$= V_i^*\, (V_i^* - V_m^*)\, y_{im}^* + V_i\, V_i^*\, y_{im0}^* \qquad \qquad ...(14.10.2)$$

Similarly, the complex power fed by the mth bus into the line connecting mth and ith buses is

$$S_{mi} = V_m\, (V_m^* - V_i^*)\, y_{im}^* + V_m\, V_m^*\, y_{mi0}^* \qquad \qquad ...(14.10.3)$$

$$= V_m^2\, y_{im}^* - V_m\, V_i^*\, y_{im} + V_m^2\, y_{mi0}^*$$

Thus, power flow through all the lines can be calculated. The power loss in the line connecting ith and mth bus is given by the algebraic sum of S_{im} and S_{mi}. The total transmission loss is the sum of losses over all the lines. The slack bus power is calculated by summing the flows on the lines terminating at the slack bus.

14.11 DECOUPLED LOAD FLOW METHODS

In a power system operating in steady-state, there is a strong interdependence between active powers and bus voltage angles. Similarly, there is a strong interdependence between reactive powers and voltage magnitudes. Thus real power changes, (ΔP) are less sensitive to changes in voltage magnitude and are mainly sensitive to change in bus voltage angles. Similarly, the reactive power changes (ΔQ) are less sensitive to changes in angles and are mainly sensitive to change in voltage magnitude. In other words the coupling between active power P and the bus voltage magnitude $|V|$ is relatively weak. Similarly, the coupling between reactive power Q and bus voltage phase angle is also weak. This weak coupling is utilized in the development of decoupled load flow (DLF) method in which P is decoupled from ΔV and Q is decoupled from $\Delta \delta$. With this assumptions Eq. (14.9.5) is reduced to

$$\begin{bmatrix} \Delta P \\ \Delta Q \end{bmatrix} = \begin{bmatrix} H & 0 \\ 0 & L \end{bmatrix} \begin{bmatrix} \Delta \delta \\ \dfrac{\Delta V}{V} \end{bmatrix} \qquad \qquad ...(14.11.1)$$

where the matrices N and M are neglected. Equation (14.11.1) is the decoupled equation which can be expanded as

$$[\Delta P] = [H] [\Delta \delta] \qquad \qquad ...(14.11.2)$$

$$[\Delta Q] = [L] \begin{bmatrix} \dfrac{\Delta |V|}{|V|} \end{bmatrix} \qquad \qquad ...(14.11.3)$$

We have already proved the following relations :

$$L'_{ik} = H_{ik} = V_i V_k Y_{ik} \sin (\delta_i - \delta_k - \theta_{ik}) \qquad \qquad ...(14.11.4)$$

$$H_{ii} = - Q_i - B_{ii} V_i^2 \qquad \qquad ...(14.11.5)$$

$$L'_i = Q_i - B_{ii} V_i^2 \qquad \qquad ...(14.11.6)$$

We can solve Eqs. (14.11.2) and (14.11.3) simultaneously at each iteration. The H and L matrices are updated in each iteration using Eqs. (14.11.4), (14.11.5) and (14.11.6).

In order to have faster convergence, the following method is used :

(i) Perform each iteration by first solving Eq. (14.11.2) for $\Delta \delta$.

(ii) Use the updated value of δ in making and then solving Eq. (14.11.3) for $\Delta |V|$.

The decoupled load flow (DLF) method requires less memory in storing the Jacobian as compared to the Newton-Raphson method. However, the time required per iteration of the DLF method is practically the same as that of NR method. In DLF method more number of iterations are required for convergence because of the approximations made in it.

14.12 FAST DECOUPLED LOAD FLOW (FDLF) METHOD

In the case of fast decoupled load flow (FDLF) method following approximations are further made in evaluating Jacobian element.

$$\left. \begin{array}{l} \cos (\delta_i - \delta_k) \cong 1 \\ \sin (\delta_i - \delta_k) \cong 0 \\ G_{ik} \sin \delta_{ik} \ll B_{ik} \\ Q_i \ll B_{ii} |V_i|^2 \end{array} \right\} \qquad \qquad ...(14.12.1)$$

With these assumptions, the Jacobian elements now become

$$H_{ik} = L_{ik} = - V_i V_k B_{ik} \quad i \neq k \qquad \qquad ...(14.12.2)$$

and $$L_{ii} = H_{ii} = - B_{ii} |V_i|^2 \qquad \qquad ...(14.12.3)$$

Equation (14.11.2) and (14.11.3) can be written as

$$[\Delta P] = [V_i \, V_k \, B'_{ik}] \, [\Delta \delta] \qquad \qquad ...(14.12.4)$$

$$[\Delta Q] = [V_i \, V_k \, B''_{ik}] \left[\frac{\Delta |V|}{|V|} \right] \qquad \qquad ...(14.12.5)$$

where B'_{ik} and B''_{ik} are the elements of $[-B_{ik}]$ matrix.

We can obtain further decoupling and simplification as follows :

(a) Neglect from $[B'']$ the angle shifting effects of phase shifters.

(b) Neglect from $[B']$ the representation of those network elements that affect reactive power flows, that is, shunt reactances and transformers' off-nominal in-phase taps.

(c) Dividing Eqs. (14.12.4) and (14.12.5) by $|V_i|$ and putting $|V_i| = 1.0$ pu in the equations.

(d) Neglecting series resistance in calculating the elements of $[B']$.

With these assumptions Eqs. (14.12.4) and (14.12.5) are simplified as follows :

$$\left[\frac{\Delta P}{|V|} \right] = [B'] \, [\Delta \delta] \qquad \qquad ...(14.12.6)$$

$$\left[\frac{\Delta Q}{|V|} \right] = [B''] \, [\Delta |V|] \qquad \qquad ...(14.12.7)$$

It is to be noted that in Eqs. (14.12.6) and (14.12.7) both $[B']$ and $[B'']$ are real, sparse and have the structures of $[H]$ and $[L]$ respectively. Since they contain only admittances, they are constant and do not change during successive iterations. They are to be evaluated only once and inverted only once during the first iteration. They are then used in all successive iterations. Equations (14.12.6) and (14.12.7) are solved alternately using the most recent voltage values. This method is very fast and converges very reliably in two to five iterations. The accuracy is fairly good even for large systems. It is found that a good approximate solution is obtained after first or second iteration.

14.13 DC LOAD FLOW METHOD

Consider bus i connected to bus j over an impedance of Z_{ij}. The active power flow is given by

$$P_{ij} = \frac{V_i \, V_j}{Z_{ij}} \sin (\delta_i - \delta_j) \qquad \qquad ...(14.13.1)$$

where $V_i = V_i \angle \delta_i$

$V_j = V_j \angle \delta_j$

Let us make the following simplifying assumptions :

$$X_{ij} \simeq Z_{ij} \qquad (\because \; X_{ij} \gg R_{ij})$$

$$V_i \simeq 1.0 \text{ pu}$$

$$V_j \simeq 1.0 \text{ pu}$$

$$\sin (\delta_i - \delta_j) \simeq \delta_i - \delta_j$$

With these assumptions, Eq. (14.13.1) can be written as

$$P_{ij} \simeq \frac{\delta_i - \delta_j}{X_{ij}} \simeq B_{ij} (\delta_i - \delta_j) \qquad \qquad ...(14.13.2)$$

This equation can be written in matrix form as

$$[P] = [B] [\delta] \qquad \qquad ...(14.13.2)$$

$$\therefore \qquad [\delta] = [B]^{-1} [P] \qquad \qquad ...(14.13.3)$$

or $\qquad [\delta] = [X] [P] \qquad \qquad ...(14.13.4)$

where the $[B]$ matrix is an $(n-1) \times (n-1)$ matrix for an n-bus system. The diagonal and off-diagonal elements of the $[B]$ matrix can be obtained by adding the series susceptances of the branches connected to bus i and by setting them equal to the negated series susceptance of branch ij, respectively. The linear Eq. (14.13.4) can be solved by using matrix methods.

This approximate method of calculating the real power flows by solving first for the bus angles is known as the *dc load-flow method*, in contrast with the exact nonlinear solution, which is known as the ac solution. The dc load-flow method is very fast because of the linear approximation made. This method is useful to perform a large number of load-flow runs needed for comprehensive contingency analysis on large scale power systems.

14.14 COMPARISON OF LOAD FLOW ANALYSIS METHODS

The choice of a particular method of load flow analysis depends upon the size of the system, rate of convergence, simplicity, computer memory, etc.

Advantages of Gauss Seidel (GS) Method

1. It can be easily programmed.
2. The solution technique is simple.
3. Computer memory requirements are smaller.
4. It takes less computational time per iteration.

Limitations of Gauss-Seidel Method

1. The rate of convergence is slow and therefore, larger number of iterations are required. The GS method would take hundreds of iterations to converge if a system with several hundred buses were to be analysed.
2. The number of iterations increases directly with the number of buses in the system.
3. This method is sensitive to the choice of reference bus.

The GS method is used only for the system having small number of buses.

Advantages of Newton-Raphson (NR) Method

1. Newton-Raphson method possesses quadratic convergence characteristic. Therefore, the convergence is very fast.
2. The number of iterations are independent of the size of the system. Solution to a high accuracy is obtained nearly always in two to three iterations for both small and large systems.
3. The NR Method convergence is not sensitive to the choice of slack bus.
4. Overall there is a saving in computation time since fewer number of iterations are required for convergence.

Limitations of NR Method

1. The solution technique is difficult.
2. It takes longer time as the elements of the Jacobian are to be computed for each iteration.
3. The computer memory requirement is large.

The NR method is more complicated than the GS method, however, it has advantages that far outweigh its shortcomings of complexity. It is the most reliable and powerful technique for solving load flow problems.

Although a large number of load flow methods are available in literature it has been observed that only the Newton-Raphson and fast-decoupled load-flow methods are most popular. The fast decoupled load flow is definitely superior to the Newton-Raphson method from the point of view of speed and storage.

Example 14.8 A three-bus power system is shown in Fig. 14.10. The relevant per unit line admittances on 100 MVA base are indicated on the diagram and bus data are given in Table 14.2.

Form Y_{bus} and determine the voltages at bus 2 and 3 after the second iteration using Gauss-Seidel method. Take the acceleration factor $\alpha = 1.6$.

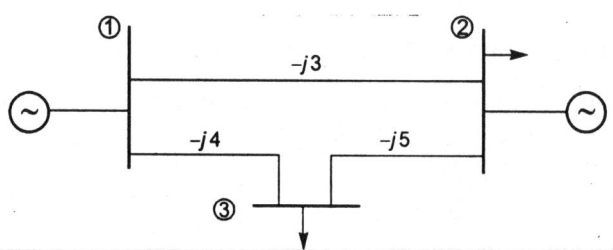

Fig. 14.10. A three-bus power system.

Table 14.2

Bus number	Type	Generation		Load		Bus voltage	
		P_G (MW)	Q_G (MVAr)	P_L (MW)	Q_L (MVAr)	V pu	δ deg.
1	Slack	?	?	0	0	1.02	0°
2	PQ	25	15	50	25	?	?
3	PQ	0	0	60	30	?	?

Solution

We first calculate the elements of the bus admittance matrix Y_{bus}.

Y_{11} = sum of admittances of all the elements connected to node (bus) $1 = -j\,3 - j\,4 = -j\,7$

Y_{22} = sum of all admittances connected to bus $2 = -j\,3 - j\,5 = -j\,8$

Y_{33} = sum of all admittances connected to bus $3 = -j\,4 - j\,5 = -j\,9$

$Y_{12} = Y_{21} = -$ (admittance connected between nodes 1 and 2) $= -(-j\,3) = j\,3$

$Y_{23} = Y_{32} = -$ (admittance connected between nodes 2 and 3) $= -(-j\,5) = j\,5$

$Y_{31} = Y_{13} = -$ (admittance connected between nodes 1 and 3) $= -(-j\,4) = j\,4$

The bus admittance matrix is given by

$$Y_{bus} = \begin{bmatrix} Y_{11} & Y_{12} & Y_{13} \\ Y_{21} & Y_{22} & Y_{23} \\ Y_{31} & Y_{32} & Y_{33} \end{bmatrix} = \begin{bmatrix} -j\,7 & j\,3 & j\,4 \\ j\,3 & -j\,8 & j\,5 \\ j\,4 & j\,5 & -j\,9 \end{bmatrix}$$

Now we shall calculate bus powers. The powers for load buses are taken as negative and those for generator buses as positive.

At bus 2,

$$P_2 = \frac{P_G - P_L}{\text{base MVA}} = \frac{25 - 50}{100} = -0.25 \text{ pu}$$

$$Q_2 = \frac{Q_G - Q_L}{\text{base MVA}} = \frac{15 - 25}{100} = -0.10 \text{ pu}$$

At bus 3,

$$P_3 = \frac{P_G - P_L}{\text{base MVA}} = \frac{0 - 60}{100} = -0.60 \text{ pu}$$

$$Q_3 = \frac{Q_G - Q_L}{\text{base MVA}} = \frac{0 - 30}{100} = -0.30 \text{ pu}$$

The voltage at bus k at the $(r+1)$th iteration is given by

$$\mathbf{V}_k^{(r+1)} = \frac{1}{Y_{kk}} \left[\frac{P_k - j\,Q_k}{(\mathbf{V}_k^{(r)})^*} - Y_{k1}\,\mathbf{V}_1^{(r+1)} - \dots - Y_{kk-1}\,\mathbf{V}_{k-1}^{(r+1)} - \dots\, Y_{kn}\,\mathbf{V}_n^{(r)} \right]$$

First iteration

We are given that

$$\mathbf{V}_1 = 1.02 \underline{/0°}$$

$$\mathbf{V}_k^{(0)} = 1 \underline{/0°} \text{ for } k = 2, 3, \dots$$

The voltage at bus 2 is given by

$$\mathbf{V}_2^{(1)} = \frac{1}{Y_{22}} \left[\frac{P_2 - j\,Q_2}{(\mathbf{V}_2^{(r)})^*} - Y_{21}\,\mathbf{V}_1 - Y_{23}\,\mathbf{V}_3^{(0)} \right]$$

$$= \frac{1}{-j\,8} \left[\frac{-0.25 + j\,0.1}{(1 \underline{/0°})^*} - (j\,3)\,(1.02 \underline{/0°}) - (j\,5)\,(1 + j\,0) \right]$$

$$= \frac{1}{-j\,8} (-0.25 + j\,0.1 - j\,3.06 - j\,5) = \frac{1}{-j\,8} (-0.25 - j\,7.96)$$

$$= \frac{7.9639 \underline{/-91.799°}}{8 \underline{/-90°}} = 0.9955 \underline{/-1.799°} = 0.995 - j\,0.0312$$

$$\Delta \mathbf{V}_2^{(1)} = \mathbf{V}_2^{(1)} - \mathbf{V}_2^{(0)} = 0.995 - j\,0.0312 - (1 + j\,0) = -0.005 - j\,0.0312$$

The accelerated value of $\mathbf{V}_2^{(1)}$ is given by

$$\mathbf{V}_2^{(1)}{}_{(acc)} = \mathbf{V}_2^{(0)} + \alpha\,\Delta \mathbf{V}_2^{(1)} = (1 + j\,0) + 1.6\,(-0.005 - j\,0.0312) = 0.992 - j\,0.0499$$

The voltage at bus 3 is given by

$$\mathbf{V}_3^{(1)} = \frac{1}{Y_{33}} \left[\frac{P_3 - j Q_3}{(\mathbf{V}_3^{(0)})^*} - Y_{31} \mathbf{V}_1 - Y_{32} \mathbf{V}_2^{(1)} \right]$$

$$= \frac{1}{-j\,9} \left[\frac{-0.6 + j\,0.3}{(1 + j\,0)^*} - (j\,4)\,(1.02 + j\,0) - j\,5\,(0.992 - j\,0.0499) \right]$$

$$= \frac{1}{-j\,9} (-6 + j\,0.3 - j\,4.08 - j\,4.96 - 0.2495) = \frac{1}{-j\,9} (-0.8495 - j\,8.74)$$

$$= \frac{8.781 \,/\!-95.55°}{9 \,/\,90°} = 0.9756 \,/\!-5.55° = 0.971 - j\,0.0943$$

$$\Delta \mathbf{V}_3^{(1)} = \mathbf{V}_3^{(1)} - \mathbf{V}_3^{(0)} = 0.971 - j\,0.0943 - (1 + j\,0) = -0.029 - j\,0.0943$$

$$\mathbf{V}_{3\,(acc)}^{(1)} = \mathbf{V}_3^{(0)} + 1.6\,\Delta\,\mathbf{V}_3^{(1)} = (1 + j\,0) + 1.6\,(-0.029 - j\,0.0943) = 0.9536 - j\,0.1509$$

The voltages at the end of first iteration are

$$\mathbf{V}_1^{(1)} = 1.02 + j\,0$$

$$\mathbf{V}_2^{(1)} = 0.992 - 0.0499$$

$$\mathbf{V}_3^{(1)} = 0.9536 - j\,0.1509$$

Second iteration

The voltage at bus 2 is given by

$$\mathbf{V}_2^{(1)} = \frac{1}{Y_{22}} \left[\frac{P_2 - j Q_2}{(\mathbf{V}_2^{(1)})^*} - Y_{21} \mathbf{V}_1 - Y_{23} \mathbf{V}_3^{(1)} \right]$$

$$= \frac{1}{-j\,8} \left[\frac{(-0.25 + j\,0.1)}{(0.992) - j\,0.0499)^*} - (j\,3)\,(1.02 + j\,0) - (j\,5)\,(0.9536 - j\,0.1509) \right]$$

$$= \frac{1}{-j\,8} \left[\frac{0.26926 \,/\,158.2°}{0.9932 \,/\,2.88°} - j\,3.06 - j\,4.768 - 0.7545 \right]$$

$$= 0.125 \,/\,90° \,[-0.2463 + j\,0.1132 - j\,7.868 - 0.7545]$$

$$= 0.125 \,/\,90° \,(-1.0008 - j\,7.7148) = (0.125 \,/\,90°)\,(7.7794 \,/\!-97.39°)$$

$$= 0.9724 \,/\!-7.39° = 0.9643 - j\,0.125$$

$$\Delta \mathbf{V}_2^{(2)} = \mathbf{V}_2^{(2)} - \mathbf{V}_2^{(1)} = (0.9643 - j\,0.125) - (0.992 - j\,0.0499) = -0.0277 - j\,0.0751$$

$$\mathbf{V}_{(2)\,(acc)}^2 = \mathbf{V}_2^{(1)} + \alpha\,(\Delta\,\mathbf{V}_2^{(2)})$$

$$= (0.992 - j\,0.0499) + 1.6\,(-0.0277 - j\,0.0751) = 0.9477 - j\,0.170$$

The voltage at bus 3 is given by

$$\mathbf{V}_3^{(2)} = \frac{1}{Y_{33}} \left[\frac{P_3 - j Q_3}{\mathbf{V}_3^{(1)*}} - Y_{31} \mathbf{V}_1 - Y_{32} \mathbf{V}_2^{(2)} \right]$$

$$= \frac{1}{-j\,9} \left[\frac{(-0.6 + j\,0.3)}{(0.9536 - j\,0.01509)^*} - (j\,4)\,(1.02 + j\,0) - (j\,5)\,(0.9477 - j\,0.17) \right]$$

$$= \frac{1}{9 \angle -90°} \left[\frac{0.67082 \angle 153.43°}{0.96546 \angle 8.99°} - j\,4.08 - j\,4.7385 - 0.85 \right]$$

$$= \frac{1}{9 \angle -90°} \left[-0.5652 + j\,0.4040 - j\,4.08 - j\,4.7385 - 0.85 \right]$$

$$= \frac{1}{9 \angle -90°} (-1.4152 - j\,8.4145) = \frac{8.5327 \angle -99.55°}{9 \angle 90°}$$

$$= 0.9481 \angle -9.55° = 0.93496 - j\,0.1573$$

$$\Delta V_3^{(2)} = V_3^{(2)} - V_3^{(1)} = 0.93496 - j\,0.1573 - (0.9536 - j\,0.1509) = -0.01864 - j\,0.0064$$

$$V_{3(acc)}^{(2)} = V_3^{(1)} + \alpha\,(\Delta V_3^{(2)})$$

$$= 0.9536 - j\,0.1509 + 1.6\,(-0.01864 - 0.0064) = 0.92377 - j\,0.16114$$

The bus voltages at the end of second iterations are given by

$$V_1^{(2)} = 1.02 + j\,0$$

$$V_2^{(2)} = 0.9477 - j\,0.170$$

$$V_2^{(3)} = 0.92377 - j\,0.16114$$

EXERCISES

1. Develop the equations for real and reactive bus powers.

   Show that a diagonal element of a Y_{bus} is equal to the sum of admittances directly connected to that bus and an off-diagonal element is equal to the negative of the sum of admittances directly connected between the buses.

2. What are the advantages of Y_{bus} over Z_{bus}?

   Derive equations for elements of Jacobian using polar coordinates.

3. Compare the performance of Gauss-Seidel and Newton-Raphson methods for load flow solution. Explain the method of formation of Y_{bus}.

4. What is the significance of load flow analysis in a power system?

   Give the classification of various types of buses in a power system for load flow studies. Justify the classification.

5. Explain the computational procedure for load flow solution using Gauss-Seidel method when the system contains all types of buses.

6. Explain the computational procedure for load flow solution using Newton-Raphson method when the system contains all types of buses.

7. Using Gauss-Seidel method give a flow chart for a load flow study on a power system having only P-Q buses. What modification is made in the flow chart to account for PV buses?

8. Give a flow chart for load flow study using Newton-Raphson method. How does the method get modified to account for PV buses?

CHAPTER 15

Economic Operation of Power Systems

15.1 INTRODUCTION

The sizes of electric power systems are increasing rapidly to meet the energy requirements. A number of power plants are connected in parallel to supply the system load by interconnection of power stations. With the development of integrated power systems (that is, grid systems) it becomes necessary to operate the plant units most economically. The *economic scheduling* of generators aims to guarantee at all times the optimum combination of generators connected to the system to supply the load demand. The economic dispatch problem involves two separate steps namely the *unit commitment* and the *on-line economic dispatch*. The unit commitment is the selection of units that will supply the anticipated load of the system over a required period of time at minimum cost as well as provide a specified margin of the operating reserve, known as the *spinning reserve*. The function of the on-line economic dispatch is to distribute the load among the generating units actually paralleled with the system in such a manner as to minimize the total cost of supplying the minute to minute requirements of the system.

The main factor controlling the most desirable load allocation between various generating units is the total running cost. The operating cost of a thermal plant is mainly the cost of the fuel. Fuel supplies for thermal plants can be coal, natural gas, oil, or nuclear fuel. The other costs such as costs of labour, supplies, maintenance, etc., being difficult to be determined and approximate, are assumed to vary as a fixed percentage of the fuel cost. Therefore, these costs are included in the fuel cost. Thus, the operating cost of a thermal plant, which is mainly the fuel cost, is given as a function of generation. This function is defined as a nonlinear function of plant generation.

For optimal operation of thermal system, our problem is to find the generation of different units so that the total fuel cost is a minimum subject to satisfying certain constraints. Basically there are two types of constraints, namely, equality constraints and inequality constraints.

15.2 INCREMENTAL FUEL COST

The input-output curves of generating units of thermal plant are important to describe the efficiency of the plant. A typical input-output curve of a unit is shown in Fig. 15.1. It is an experimental curve plotted with abscissa as output power P_i in MW and ordinate as fuel (heat) input in joules per hour (J/h) of the ith unit. The ordinates of the curves may be converted to fuel cost in C_i Rs/h by multiplying the fuel input by the cost of the fuel in rupees per joule. This curve is shown in Fig. 15.2 where $P_{i\,min}$ is the minimum loading limit below which it is not economical to operate the unit and $P_{i\,max}$ is the maximum active power output of ith unit. The majority of generating units have a nonlinear generation cost function C_i. We shall assume that the variation of fuel cost of each generator (C_i) with the active power output (P_i) is given by a quadratic polynomial. We can write

$$C_i = \alpha_i + \beta_i\,P_i + \gamma_i\,P_i^2 \qquad\qquad ...(15.2.1)$$

where C_i = fuel cost of generator i

P_i = power output of generator i

α_i, β_i, and γ_i are constants, α_i includes salary and wages, interest and depreciation and is independent of generation. β_i represents basically the fuel cost and γ_i is a measure of losses in the system. Usually β_i dominates.

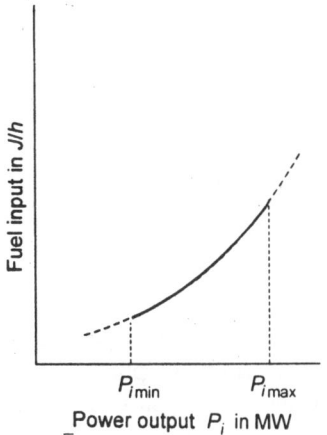

Fig. 15.1. Typical input-output curve of a generating unit.

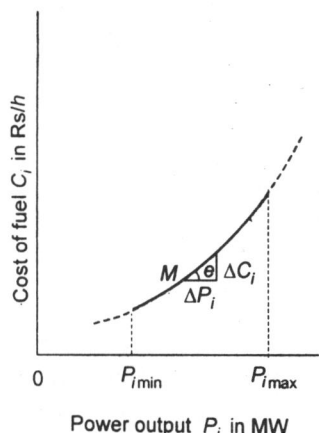

Fig. 15.2. Fuel-cost curve.

The slope of the cost curve at a point M is given by

$$\tan\theta = \frac{\Delta C_i}{\Delta P_i}$$

where ΔC_i is the increase in fuel cost corresponding to an increase of power output ΔP_i. The *incremental fuel cost* for a generator for any given electrical power output is defined as the limiting value of the ratio of the increase in cost of fuel in Rs/h to the corresponding increase in electrical power output in MW when the increase in power output tends to zero.

Incremental fuel cost for the ith generator is given by

$$(IC)_i = \underset{\delta P \to 0}{\text{Lt}} \frac{\partial C_i}{\partial P_i} \bigg|_{\substack{\Delta P_i = 0 \\ \text{for all } j \text{ except } j = 1}} = \frac{dC_i}{dP_i} \qquad \ldots(15.2.2)$$

The incremental cost is equal to the slope of the fuel cost curve. If the cost curve is approximated as a quadratic polynomial (with positive coefficients) as in Eq. (15.2.1), we have

$$(IC)_i = \frac{dC_i}{dP_i} = \frac{d}{dP_i} (\alpha_i + \beta_i P_i + \gamma_i P_i^2)$$

$$(IC)_i = \beta_i + 2\gamma_i P_i \text{ Rs/MWh} \qquad \ldots(15.2.3)$$

Equation (15.2.3) is of the form $y = mx + c$.

Thus the plot of $(IC)_i$ versus P_i is a straight line. In other words, the incremental cost curves are linear (with positive coefficients). A plot of incremental cost versus power output is called *incremental cost curve*. It is shown in Fig. 15.3.

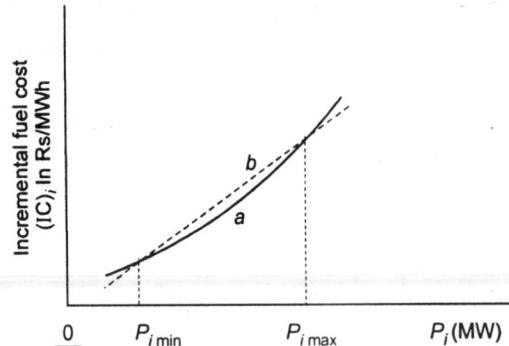

Fig. 15.3. Incremental cost curve for generator i. (a) Actual incremental cost curve; (b) Approximated (linear) incremental cost curve.

For better accuracy when the cost is represented by a nth degree polynomial, incremental fuel cost may be represented by short line segments from point to point. Alternatively, in the inverse form output power P_i may be expressed in terms of polynomials of $(IC)_i$ as given by

$$P_i = \alpha_{i0} + \alpha_{i1} (IC)_i + \alpha_{i2} (IC)_i^2 + \ldots \qquad \ldots(15.2.4)$$

15.3 ECONOMIC DISPATCH NEGLECTING TRANSMISSION LOSSES

Consider n generators in the same plant or close enough electrically so that the line losses may be neglected. Let $C_1, C_2, \ldots, C_n$ be the operating costs of individual units for the corresponding power outputs $P_1, P_2, \ldots, P_n$ respectively. If C is the total operating cost of the entire system and P_R is the total power received by the plant bus and transferred to the load, then

$$C = C_1 + C_2 + \ldots + C_n = \sum_{i=1}^{n} C_i \qquad \ldots(15.3.1)$$

$$P_R = P_1 + P_2 + \ldots + P_n = \sum_{i=1}^{n} P_i \qquad \ldots(15.3.2)$$

Our problem can be stated as follows :

Minimize $\quad C = \sum_{i=1}^{n} C_i \qquad \ldots(15.3.1)$

subject to the constraint

$$P_R - \sum_{i=1}^{n} P_i = 0 \qquad \ldots(15.3.2)$$

Equation (15.3.2) shows that if transmission losses are neglected, the total demand P_R at any instant must be met by the total generation. It is to be noted that Eq. (15.3.2) is the *equality constraint*.

This is a constrained minimization problem. We shall use the classical Lagrangian multiplier technique to solve it. For this purpose first we form the *augmented* or *Lagrange cost function* C^* defined by

$$C^* \triangleq C + \lambda f \qquad \ldots(15.3.3)$$

where f is the equality constraint equation given by

$$P_R = \sum_{i=1}^{n} P_i \triangleq f(P_1, P_2, \ldots, P_n) = 0 \qquad \ldots(15.3.4)$$

and λ is the Lagrange multiplier. Combination of Eqs. (15.3.3) and (15.3.4) gives

$$C^* \triangleq C + \lambda \left(P_R - \sum_{i=1}^{n} P_i \right) \qquad \ldots(15.3.5)$$

Equation (15.3.5) can be solved for minimum by determining the partial derivative of the function C^* with respect to variable P_i (plant generation) and equating it equal to zero.

$$\frac{\partial C^*}{\partial P_i} = \frac{\partial C}{\partial P_i} + \lambda \frac{\partial}{\partial P_i} \left(P_R - \sum_{i=1}^{n} P_i \right) = 0 \qquad i = 1, 2, \ldots, n \qquad \ldots(15.3.6)$$

or $\qquad \dfrac{\partial C^*}{\partial P_i} = \dfrac{\partial C}{\partial P_i} + \lambda (0 - 1) = 0$

$\therefore \qquad \dfrac{\partial C}{\partial P_i} = \lambda \qquad \ldots(15.3.7)$

Since C_i, is a function of P_i only, the partial derivatives become full derivatives, that is,

$$\frac{\partial C_i}{\partial P_i} = \frac{dC_i}{dP_i} \qquad \ldots(15.3.8)$$

Therefore, the condition for optimum operation is

$$\frac{dC_1}{dP_1} = \frac{dC_2}{dP_2} = \ldots = \frac{dC_n}{dP_n} = \lambda \qquad \ldots(15.3.9)$$

Since $\frac{dC_i}{dP_i}$ is the incremental cost generation $(IC)_i$ for generator (i) Eq. (15.3.9) shows that the criterion for most economical division of load between units within a plant (or between units close enough electrically so that the line losses may be neglected) is that all the units must operate at the same incremental fuel cost. This is known as the *principle of equal λ criterion* or the *equal incremental cost-loading principle* for economic operation. It should be noted that n equations of Eq. (15.3.9) and the constraint equation (15.3.4) are sufficient for determining $(n + 1)$ unknowns $P_1, P_2, \ldots, P_n$ and λ.

Example 15.1 The fuel costs of a two-unit plant are given by

$$C_1 = 100 + 2P_1 + 0.005\ P_1^2\ ; \qquad C_2 = 200 + 2P_2 + 0.01\ P_2^2$$

where P_1 and P_2 are in MW. The plant supplies a load of 450 MW. Find economic load scheduling of the two units and the incremental fuel cost. Neglect losses.

Solution

The incremental fuel cost of the first generator is given by

$$(IC)_1 = \frac{dC_1}{dP_1} = 2 + 2 \times 0.005\ P_1 = 2 + 0.01\ P_1\ \text{Rs/MWh}$$

The incremental fuel cost of the second generator

$$(IC)_2 = \frac{dC_2}{dP_2} = 2 + 0.02\ P_2$$

For optimum load division the two incremental costs should be equal, that is,

$$(IC)_1 = (IC)_2$$
$$2 + 0.01\ P_1 = 2 + 0.02\ P_2$$
$$P_1 = 2P_2 \qquad \ldots(E15.1.1)$$

Since the total load is shared by the two generators

$$P_1 + P_2 = 450 \qquad \ldots(E15.1.2)$$

Combination of Eqs. (E15.1.1) and (E15.1.2) gives

$$2P_2 + P_2 = 450,\ P_2 = 150\ \text{MW and}\ P_1 = 300\ \text{MW}$$

The incremental fuel cost of generator 1 corresponding to load $P_1 = 150$ MW is given by

$$(IC)_1 = 2 + 0.01\ P_1 = 2 + 0.01 \times 300 = 5\ \text{Rs/MWh}$$

The incremental fuel cost of generator 2 corresponding to load $P_2 = 150$ MW is given by

$$(IC)_2 = 2 + 0.02\ P_2 = 2 + 0.02 \times 150 = 5\ \text{Rs/MWh}$$

Hence the incremental fuel cost of the plant for most economic operation is Rs 5 per MWh.

Example 15.2 The incremental fuel costs in rupees per MWh for a plant consisting of two units are given by

$$\frac{dC_1}{dP_1} = 0.16\,P_1 + 30$$

$$\frac{dC_2}{dP_2} = 0.20\,P_2 + 25$$

Assume that both units are operating all the time throughout the year. The maximum and minimum loads on each unit are 200 MW and 50 MW respectively. If the total load varies between 100 MW and 400 MW find the load division between two units as the system load varies over the full range.

Solution

For minimum load $P_1 = 50$ MW on unit 1

$$\frac{dC_1}{dP_1} = 0.16 \times 50 + 30 = 38 \text{ Rs/MWh}$$

For minimum load $P_2 = 50$ MW on unit 2

$$\frac{dC_2}{dP_2} = 0.20 \times 50 + 25 = 35 \text{ Rs/MWh}$$

Thus, when the load is 100 MW, it is shared equally by the two units but unit 1 operates at a higher incremental fuel cost than unit 2. The incremental fuel cost of unit 1 is Rs 38 per MWh and that of unit 2 is Rs 35 per MWh. With the increase of the plant output, the more load should be added to unit 2 till IC of unit 2 also becomes Rs 38 per MWh. Until that point is reached, the IC of the plant is determined by unit 2 alone. When the load on the plant is 100 MW, each unit operates at its minimum load of 50 MW with plant $IC = 35$ Rs/MWh.

When $\quad \dfrac{dC_2}{dP_2} = 38$

$$0.20\,P_2 + 25 = 38$$

or $\quad P_2 = 65$ MW

Since the load on unit 1 at $IC = 38$ is 50 MW, the total load delivered at equal incremental costs of Rs 38/MWh is $(50 + 65) = 115$ MW. From this point onwards, increase the load on each unit so that the units operate at the same incremental costs and these operating conditions are found by assuming successively higher values of IC and calculating P_1, P_2 and total load. This process is continued till the load on unit 1 reaches its upper limit of 200 MW. Then the load on unit 2 is increased keeping the load on unit 1 equal to 200 MW corresponding to its upper limit of 200 MW. At this stage the two units will not operate at the same IC, and unit 1 will determine the plant IC. The load on unit 2 is increased till it also reaches its upper limit of 200 MW so that the total load on both the units together becomes 400 MW. The results are shown in Table 15.1.

Table 15.1 Load Division for Example 15.2

Plant IC (λ) Rs/MWh	P_1 MW	P_2 MW	Plant output ($P_1 + P_1$) MW
35	50	50	100
38	50	65	115
40	62.5	75	137.5
45	93.75	100	193.75
50	125	125	250
55	156.25	150	306.25
60	187.5	175	362.5
62	200	185	385
63	200	190	390
64	200	195	395
65	200	200	400

Example 15.3 Determine the saving in fuel cost in Rs/h for the economic distribution of total load of 115 MW between the two units of plant described in Example 15.2 compared with equal distribution of the same total load.

Solution

From Table 15.1 it is found that for equal incremental costs, unit 1 should supply 50 MW and unit 2 should supply 65 MW corresponding to a total of 115 MW. For equal sharing each unit will supply 57.5 MW.

The increase in cost of unit 1 is given by

$$\Delta C_1 = \int_{50}^{57.5} dC_i = \int_{50}^{57.5} (0.16\, P_1 + 30)\, dP_1 = \left| 0.16 \frac{P_1^2}{2} + 30 P_1 \right|_{50}^{57.5}$$

$$= \frac{0.16}{2} (57.5^2 - 50^2) + 30 (57.5 - 50) = 64.5 + 225 = 289.5 \ \text{Rs/h}$$

Change in cost of unit 2 is given by

$$\Delta C_2 = \int_{65}^{57.5} dC_2 = \int_{65}^{57.5} (0.20\, P_2 + 25)\, dP_2 = \left| 0.20 \frac{P_2^2}{2} + 25 P_2 \right|_{65}^{57.5}$$

$$= 0.10 (57.5^2 - 65^2) + 25 (57.5 - 65) = -91.875 - 187.5 = -279.375 \ \text{Rs/h}$$

The negative sign shows that there is a decrease in cost. The net increase in cost due to departure from economic loading is given by

$$\Delta C = \Delta C_1 + \Delta C_2 = 289.5 - 279.375 = 10.125 \ \text{Rs/h}$$

Thus, the saving in fuel cost due to economic loading is Rs 10.125 per hour. If it is assumed that the system load remains constant throughout the year, the total yearly saving

$$= \text{Rs } 10.125 \times 365 \times 24 = \text{Rs } 88695$$

This saving justifies the need of economic loading of units and the necessity of using devices for controlling the loading of each unit automatically.

15.4 TRANSMISSION LOSS AS A FUNCTION OF PLANT GENERATION

Consider a simple power system consisting of two generating plants and one load as shown in Fig. 15.4.

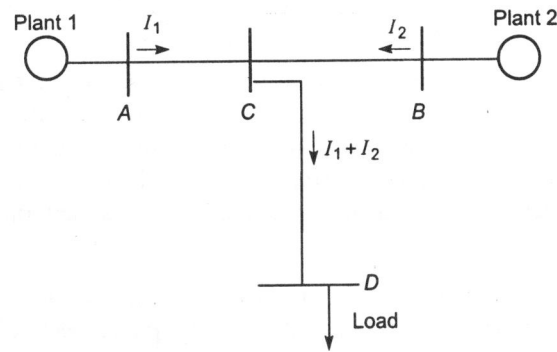

Fig. 15.4. A simple system connecting two generating plants to one load.

Let R_{AC}, R_{BC}, and R_{CD} be the resistance of the lines AC, BC and CD respectively. For the given system we can write the transmission losses as

$$P_L = 3 \mid I_1 \mid^2 R_{AC} + 3 \mid I_2 \mid^2 R_{BC} + 3 \mid I_1 + I_2 \mid^2 R_{CD} \qquad ...(15.4.1)$$

If we assume that I_1 and I_2 are in phase,

$$\mid I_1 + I_2 \mid = \mid I_1 \mid + \mid I_2 \mid \qquad ...(15.4.2)$$

$\therefore \qquad P_L = 3 \mid I_1 \mid^2 (R_{AC} + R_{CD}) + 3 \mid I_2 \mid^2 (R_{BC} + R_{CD}) + 6 \mid I_1 \mid \mid I_2 \mid R_{CD} \qquad ...(15.4.3)$

Let P_1 and P_2 be the three-phase power output of plants 1 and 2 at power factors of $\cos \varphi_1$ and $\cos \varphi_2$, and V_1 and V_2 be the bus voltages at the plants.

$$\mid I_1 \mid = \frac{P_1}{\sqrt{3} \mid V_1 \mid \cos \varphi_1}, \qquad \mid I_2 \mid = \frac{P_2}{\sqrt{3} \mid V_2 \mid \cos \varphi_2}$$

Substituting these values of I_1 and I_2 in Eq. (15.4.3) we get

$$P_L = P_1^2 \frac{R_{AC} + R_{CD}}{\mid V_1 \mid^2 \cos^2 \varphi_1} + 2P_1 P_2 \frac{R_{CD}}{\mid V_1 \mid \mid V_2 \mid \cos \varphi_1 \cos \varphi_2} + P_2^2 \frac{R_{BC} + R_{CD}}{\mid V_2 \mid^2 \cos^2 \varphi_2} \qquad ...(15.4.4)$$

Equation (15.4.4) can be written as

$$P_L = P_1^2 B_{11} + 2P_1 P_2 B_{12} + P_2^2 B_{22} \qquad ...(15.4.5)$$

$$B_{11} = \frac{R_{AC} + R_{CD}}{|V_1|^2 \cos^2 \varphi_1}$$

where
$$B_{22} = \frac{R_{BC} + R_{CD}}{|V_2|^2 \cos^2 \varphi_2}$$

$$\left.\rule{0pt}{60pt}\right\} \quad ...(15.4.6)$$

$$B_{12} = \frac{R_{CD}}{|V_1| |V_2| (\cos \varphi_1)(\cos \varphi_2)}$$

The terms B_{11}, B_{12}, and B_{22} are called *loss coefficients* or *B coefficients*.

If the voltages in Eq. (15.4.6) are line-to-line voltages in kilovolts and line resistances are in ohms, the units for the loss coefficients will be in reciprocal megawatts (MW^{-1}). Then, in Eq. (15.4.5), with three-phase powers in megawatts, P_L will be in megawatts also. If all the quantities are in per unit, the coefficients will also be in per unit.

It is seen that the loss coefficients depend on source voltages and power factors. The source voltages and power factors depend on and vary with system operating conditions. However, *B* coefficients are constants. It is sufficiently accurate to calculate *B* coefficients for some average operating conditions and use these values for economic loading for all the load variations. However, for large load variations or for major system changes, several sets of loss coefficients are used.

Example 15.4 For the system shown in Fig. 15.4, the voltage at bus C is $V_C = 1.0 \underline{/0°}$ pu. The currents in lines AC and BC are $1.05 \underline{/0°}$ pu and $0.9 \underline{/0°}$ pu respectively. The line impedances are as follows :

$$Z_{AC} = (0.05 + j\,0.20) \text{ pu}, \quad Z_{BC} = (0.04 + j\,0.16) \text{ pu}, \quad Z_{CD} = (0.03 + j\,0.12) \text{ pu}.$$

Find the loss coefficients and the transmission loss.

Solution

$$V_A = V_C + Z_{AC}\,I_1 = 1\underline{/0°} + (0.05 + j\,0.2)(1.05\underline{/0°})$$
$$= 1.0525 + j\,0.21 \text{ pu} = V_A \cos \varphi_A + j\,V_A \sin \varphi_A$$

$\therefore \qquad V_A \cos \varphi_A = 1.0525$

$$V_B = V_C + Z_{BC}\,I_2 = 1\underline{/0°} + (0.04 + j\,0.16)(0.9\underline{/0°})$$
$$= 1.036 + j\,0.144 \text{ pu} = V_B \cos \varphi_B + j\,V_B \sin \varphi_B$$

$\therefore \qquad V_B \cos \varphi_B = 1.036$

Power output of plant 1

$$P_1 = \text{Re}\,[V_A\,I_1^*] = \text{Re}\,[(1.0525 + j\,0.21) \times 1.05\underline{/0°}] = 1.1051 \text{ pu}$$

Power output of plant 2

$$P_2 = \text{Re}\,[V_B\,I_2^*] = \text{Re}\,[(1.036 + j\,0.144) \times 0.9\underline{/0°}] = 0.9324 \text{ pu}$$

$$B_{11} = \frac{R_{AC} + R_{CD}}{(V_A \cos \varphi_A)^2} = \frac{0.05 + 0.03}{(1.0525)^2} = 0.072218 \text{ pu}$$

$$B_{12} = \frac{R_{CD}}{(V_A \cos \varphi_A)(V_B \cos \varphi_B)} = \frac{0.03}{(1.0525) \times (1.036)} = 0.027513 \text{ pu}$$

$$B_{22} = \frac{R_{BC} + R_{CD}}{(V_B \cos \varphi_B)^2} = \frac{0.04 + 0.03}{(1.036)^2} = 0.0652196 \text{ pu}$$

Transmission loss

$$P_L = P_1^2 B_{11} + P_2^2 B_{22} + 2P_1 P_2 B_{12}$$

$$= (1.1051)^2 \times 0.072218 + (0.9324)^2 \times 0.0652196 + 2 \times 1.1051 \times 0.9324 \times 0.027513$$

$$= 0.2015944 \text{ pu}$$

Check

Transmission loss = sum of $I^2 R$ losses in lines AC, BC and CD

$$= I_1^2 R_{AC} + I_2^2 R_{BC} + (I_1 + I_2)^2 R_{CD}$$

$$= (1.05)^2 \times 0.05 + 0.9^2 \times 0.04 + (1.05 + 0.9)^2 \times 0.03 = 0.2016 \text{ pu}$$

15.5 GENERAL LOSS FORMULA

Ideally, the exact power flow equations should be used to account for the transmission loss in the system. However, it is a common practice to express the transmission loss in terms of active power generations only. This approach is commonly known as the *loss formula* or *B-coefficient method*. The simplest form of loss equation for a system of k plants is given by

$$P_L = \sum_{m=1}^{k} \sum_{n=1}^{k} (P_m B_{mn} P_n) \qquad \qquad ...(15.5.1)$$

where P_L = transmission loss

P_m = active power generation at mth plant

P_n = active power generation at nth plant

$B_{mn} = B_{nm}$

The coefficient B_{mn} are commonly known as *loss coefficients* or *B-coefficients*. The formula given by Eq. (15.5.1) is called *George's formula*.

For a two-plant system,

$$P_L = \sum_{m=1}^{2} \sum_{n=1}^{2} (P_m B_{mn} P_n) = \sum_{m=1}^{2} (P_m B_{m1} P_1 + P_m B_{m2} P_2)$$

$$= P_1 B_{11} P_1 + P_1 B_{12} P_2 + P_2 B_{21} P_1 + P_2 B_{22} P_2$$

$$= P_1^2 B_{11} + 2P_1 P_2 B_{12} + P_2^2 B_{22} \qquad \qquad ...(15.5.2)$$

For a three-plant system,

$$P_L = P_1^2 B_{11} + P_2^2 B_{22} + P_3^2 B_{33} + 2P_1 P_2 B_{12} + 2P_2 P_3 B_{23} + 2P_3 P_1 B_{13} \qquad ...(15.5.3)$$

The matrix form of the transmission-loss equation is given by

$$P_L = \mathbf{P}^T \mathbf{B} \mathbf{P} \qquad \qquad ...(15.5.4)$$

where $\mathbf{P}^T$ is the transpose of $\mathbf{P}$.

For a total of k sources,

$$\mathbf{P} = \begin{bmatrix} P_1 \\ P_2 \\ \cdot \\ \cdot \\ \cdot \\ P_k \end{bmatrix} \quad \text{and} \quad \mathbf{B} = \begin{bmatrix} B_{11} & B_{12} & B_{1k} \\ B_{21} & B_{22} & B_{2k} \\ \cdot & \cdot & \cdot \\ \cdot & \cdot & \cdot \\ \cdot & \cdot & \cdot \\ B_{k1} & B_{k2} & B_{kk} \end{bmatrix} \qquad \qquad \ldots(15.5.5)$$

15.6 OPTIMUM LOAD DISPATCH CONSIDERING TRANSMISSION LOSSES

When the distances of generating plants from the loads are different, the cost of different transmission losses will affect the economic distribution. Consider n generating plants. Let $C_1, C_2, \ldots, C_n$ be the fuel costs of individual plants for the corresponding electrical power outputs of $P_1, P_2, \ldots, P_n$ respectively. Let P_R be the total power received by the loads (that is, the load demand) and P_L the total transmission losses. For n plants, the total fuel cost is

$$C = C_1 + C_2 + \ldots + C_n = \sum_{i=1}^{n} C_i \qquad \qquad \ldots(15.6.1)$$

The total input to the network from all plants is

$$P = P_1 + P_2 + \ldots + P_n = \sum_{i=1}^{n} P_i \qquad \qquad \ldots(15.6.2)$$

Since the total demand and the transmission losses must be met by the total generation at that instant,

$$P_R + P_L = P_1 + P_2 + \ldots + P_n = \sum_{i=1}^{n} P_i \qquad \qquad \ldots(15.6.3)$$

We shall assume that the total transmission loss P_L is a function of generation. That is,

$$P_L \triangleq \varphi(P_1, P_2, \ldots, P_n) \qquad \qquad \ldots(15.6.4)$$

The optimal load dispatch problem considering transmission losses can be stated as follows:

$$\text{Minimize} \quad C = \sum_{i=1}^{n} C_i \qquad \qquad \ldots(15.6.5)$$

subject to the constraint

$$P_R + P_L - \sum_{i=1}^{n} P_i = 0 \qquad \qquad \ldots(15.6.6)$$

Equation (15.6.6) is the *active power balance equation.*

Making use of Lagrangian multiplier λ, the augmented cost function C^* is defined as

$$C^* \triangleq C + \lambda \left(P_R + P_L - \sum_{i=1}^{n} P_i \right) \qquad \qquad ...(15.6.7)$$

For economical load dispatch,

$$\frac{\partial C^*}{\partial P_i} = 0$$

$\therefore$

$$\frac{\partial C^*}{\partial P_i} = \frac{\partial C}{\partial P_i} + \lambda \frac{\partial P_L}{\partial P_i} - \lambda = 0$$

$$\frac{\partial C}{\partial P_i} = \lambda \left(1 - \frac{\partial P_L}{\partial P_i} \right)$$

Since the cost of a unit depends on the output of that unit only C_i is a function of P_i only, and therefore, the partial derivatives become full derivatives and

$$\frac{\partial C_i}{\partial P_i} = \frac{dC_i}{dP_i}$$

and

$$\frac{dC_i}{dP_i} = \lambda \left(1 - \frac{\partial P_L}{\partial P_i} \right)$$

$$\frac{dC_i}{dP_i} + \lambda \frac{\partial P_L}{\partial P_i} = \lambda \qquad \qquad ...(15.6.9)$$

The partial derivative $(\partial P_L/\partial P_i)$ is known as the *incremental transmission loss* (ITL) associated with generator i. It gives the extra system loss incurred by an increment of active power injection by generator i. The partial derivative $(\partial C_i/\partial P_i)$ is the incremental fuel cost $(IC)_i$ of generator i.

Equation (15.6.9) represents a set of n equations with $(n+1)$ unknowns. Here n generators are unknown as λ is also unknown. These equations are called *coordination equations* because they coordinate the incremental transmission losses with the incremental cost of production.

Equation (15.6.9) can be written as

$$\frac{1}{\left(1 - \dfrac{\partial P_L}{\partial P_i} \right)} \frac{dC_i}{dP_i} = \lambda$$

or

$$L_i \frac{dC_i}{dP_i} = \lambda \qquad i = 1, 2, ..., n \qquad \qquad ...(15.6.10)$$

where

$$L_i \triangleq \frac{1}{1 - \dfrac{\partial P_L}{\partial P_i}} = \frac{1}{1 - (ITL)_i} \qquad \qquad ...(15.6.11)$$

The factor L_i, is called the *penalty factor* for plant i. It depends upon the location of the plant. The larger the incremental transmission loss, the larger is the penalty factor.

Equation (15.6.10) can be written as

$$L_i \frac{dC_1}{dP_1} = L_2 \frac{dC_2}{dP_2} = \ldots = L_n \frac{dC_n}{dP_n} = \lambda \qquad \ldots(15.6.12)$$

Equation (15.6.12) shows that *for minimum fuel cost, the incremental fuel cost of each plant multiplied by its penalty factor is the same for all plants in the system. In other words, the optimal generations are obtained when each plant is operated such that the penalised incremental costs are equal.*

Equation (15.6.9) can be written in the alternative form as

$$(IC)_i = \lambda \, [1 - (ITL)_i] \qquad i = 1, 2, \ldots, n \qquad \ldots(15.6.13)$$

Special case

Equation (15.6.12) gives the criterion for optimum loading of plants when the transmission losses are considered. If the losses are neglected, $P_L = 0$ and

$$\frac{dC_i}{dP_i} = \lambda$$

or $\qquad \dfrac{dC_1}{dP_1} = \dfrac{dC_2}{dP_2} = \ldots = \dfrac{dC_n}{dP_n} = \lambda$

This relationship is already obtained in Eq. (15.3.9).

Example 15.5 On the system consisting of two generating plants the incremental costs in rupees per megawatt hour with P_1 and P_2 in megawatts are

$$\frac{dC_1}{dP_1} = 0.15 \, P_1 + 150 \; ; \qquad \frac{dC_2}{dP_2} = 0.25 \, P_2 + 175$$

The system is operating on economic dispatch with $P_1 = P_2 = 200$ MW and $\dfrac{\partial P_L}{\partial P_2} = 0.2$. Find the penalty factor of plant 1.

Solution

On economic dispatch,

$$L_1 \frac{dC_1}{dP_1} = L_2 \frac{dC_2}{dP_2} \qquad \ldots(E15.5.1)$$

$$L_2 = \frac{1}{1 - \dfrac{\partial P_L}{\partial P_2}} = \frac{1}{1 - 0.2} = 1.25$$

$$P_1 = P_2 = 200 \text{ MW}$$

Substitution of the values of P_1, P_2, L_2, in Eq. (E15.5.1) gives

$$L_1 \, (0.15 \times 200 + 150) = 1.25 \, (0.25 \times 200 + 175)$$

$$180 L_1 = 281.25, \; L_1 = \frac{281.25}{180} = 1.5625$$

Example 15.6 A power system has two generating plants and the power is being dispatched economically with $P_1 = 150$ MW and $P_2 = 275$ MW. The loss coefficients are :

$$B_{11} = 0.10 \times 10^{-2} \text{ MW}^{-1}$$

$$B_{12} = -0.01 \times 10^{-2} \text{ MW}^{-1}$$

$$B_{22} = 0.13 \times 10^{-2} \text{ MW}^{-1}.$$

To raise the total load on the system by 1 MW will cost an additional Rs 200 per hour. Find (a) the penalty factor for plant 1, and (b) the additional cost per hour to increase the output of plant 1 by 1 MW.

Solution

(a) For a system with two plants,

$$P_L = P_1^2 B_{11} + 2P_1 P_2 B_{12} + P_2^2 B_{22}$$

$$\frac{\partial P_L}{\partial P_1} = 2P_1 B_{11} + 2P_2 B_{12} + 0$$

$$= 2 \times 150 \times 0.1 \times 10^{-2} + 2 \times 275 (-0.01 \times 10^{-2}) = 0.3 - 0.055 = 0.245$$

Penalty factor for plant 1

$$L_1 = \frac{1}{1 - \dfrac{\partial P_L}{\partial P_1}} = \frac{1}{1 - 0.245} = 1.3245$$

(b) Incremental cost of the system is given by

$$\lambda = \text{Rs } 200 \text{ per MWh}$$

We have $L_1 \dfrac{dC_1}{dP_1} = \lambda$

$$1.3245 \frac{dC_1}{dP_1} = 200 ; \qquad \frac{dC_1}{dP_1} = \frac{200}{1.3245} = 151 \text{ Rs/MWh}$$

Example 15.7 In a two-plant system, the entire load is located at plant 2, which is connected to plant 1 by a transmission line. Plant 1 supplies 100 MW of power with a corresponding transmission loss of 5 MW. Calculate the penalty factors for the two plants.

Solution

For a two-plant system,

$$P_L = P_1^2 B_{11} + 2P_1 P_2 B_{12} + P_2^2 B_{22} \qquad \qquad \text{...(E15.7.1)}$$

Since all the load is at plant 2, varying P_2 does not affect the transmission loss P_L. Thus, from Eq. (E15.7.1)

$$P_L = P_1^2 B_{11}$$

$$5 = P_1^2 B_{11} = (100)^2 B_{11} = 10^4 B_{11}$$

$$\therefore \qquad B_{11} = \frac{5}{10^4}$$

and

$$\frac{\partial P_L}{\partial P_1} = \frac{\partial}{\partial P_1}(P_1^2 B_{11})$$

$$= 2P_1 B_{11} = 2 \times 100 \times \frac{5}{10^4} = 0.1$$

Penalty factor for plant 1

$$L_1 = \frac{1}{1 - \dfrac{\partial P_L}{\partial P_1}} = \frac{1}{1 - 0.1} = 1.111$$

Now,

$$\frac{\partial P_L}{\partial P_2} = 0$$

$\therefore$ penalty factor for plant 2

$$L_2 = \frac{1}{1 - \dfrac{\partial P_L}{\partial P_2}} = 1$$

15.7 ITERATIVE METHOD OF SOLVING COORDINATION EQUATION

The coordination equation for nth plant is

$$\frac{dC_n}{dP_n} + \lambda \frac{\partial P_L}{\partial P_n} = \lambda \qquad \qquad \qquad \dots(15.7.1)$$

Let us assume that the plants have the quadratic cost functions of the form

$$C_n = \frac{1}{2} \alpha_n P_n^2 + \beta_n P_n + \gamma_n \qquad \qquad \dots(15.7.2)$$

$\therefore$

$$(IC)_n = \frac{dC_n}{dP_n} = \alpha_n P_n + \beta_n \text{ Rs/MWh} \qquad \dots(15.7.3)$$

It is seen that for a quadratic cost function, the incremental cost curve [$(IC)_n$ vs P_n curve] is a straight line given by Eq. (15.7.3). In this equation

α_n = slope of incremental production cost curve

β_n = intercept of incremental production cost curve on $(IC)_n$ axis.

From Eq. (15.5.1)

$$\frac{\partial P_L}{\partial P_n} = 2 \sum_{m=1}^{k} P_m B_{mn} \qquad \qquad \dots(15.7.4)$$

Thus the coordination Eq. (15.7.1) can be written as

$$\alpha_n P_n + \beta_n + 2\lambda \sum_{m=1}^{k} B_{mn} P_m = \lambda$$

Collecting all coefficients of P_n we get

$$P_n (\alpha_n + 2\lambda B_{mn}) = -\lambda \sum_{\substack{m=1 \\ m \neq n}}^{k} 2B_{mn} P_m - \beta_n + \lambda$$

Solving for P_n we get

$$P_n = \frac{1 - \dfrac{\beta_n}{\lambda} - \displaystyle\sum_{\substack{m=1 \\ m \neq n}}^{k} 2B_{mn}P_m}{\dfrac{\alpha_n}{\lambda} + 2B_{nn}} \qquad n = 1, 2, ..., k \qquad ...(15.7.5)$$

Since the incremental transmission loss depends on power outputs from all the plants, the coordination equation (15.7.5) cannot be solved directly. It is well suited for solution by iterative method. The following procedure is used :

1. Assume a suitable value of $\lambda = \lambda_0$. This value should be greater than the largest intercept of the incremental cost of the various units. Calculate $P_1, P_2, ..., P_n$ based on equal incremental costs.

2. Calculate the generation at all buses with the help of Eq. (15.7.5) keeping in mind that the values of powers to be substituted on the right-hand side of Eq. (15.7.5) during zeroth iteration correspond to the values calculated in step 1. For subsequent iterations the values of powers to be substituted correspond to the powers in the previous iteration. However, if any generator violates the limit of generation that generator is fixed at the limit violated.

3. Check if the difference in power at all generator buses between two consecutive iterations is less than a specified value, otherwise go back to step 2.

4. Calculated the losses using the relation

$$P_L = \sum_{m=1}^{k} \sum_{n=1}^{k} P_m B_{mn} P_n$$

5. Calculate

$$\Delta P = \left| \sum_{n=1}^{k} P_n - P_R - P_L \right|$$

6. If ΔP is less than a specified value ε, stop calculations and calculate cost of generation with the values of powers. If $\Delta P < \varepsilon$ is not satisfied go to step 7.

7. Update λ as

$$\lambda^{(k+1)} = \lambda^{(k)} - \Delta \lambda^{(k)}$$

where $\Delta \lambda$ is the step size. The step size $\Delta \lambda$ may be selected on the basis of per unit power mismatch, that is,

$$\Delta \lambda^{(k)} = \frac{\Delta P^{(k)}}{P_R}$$

The initial guess and the step size are the two main considerations in any iterative solution as the convergence and number of iterations depend on these factors. Since the transmission losses are 10 to 15 percent of the total generation, we can neglect losses for initial guess in iteration.

For a system where transmission losses are neglected,

$$\frac{dC_i}{dP_i} = \lambda$$

From Eq. (15.7.3)

$$\frac{dC_i}{dP_i} = \alpha_i P_i + \beta_i$$

$\therefore \qquad \alpha_i P_i + \beta_i = \lambda$

$$P_i = \frac{1}{\alpha_i}(\lambda - \beta_i) \qquad\qquad\qquad …(15.7.6)$$

Taking the sum of all generations

$$\sum_{i=1}^{n} P_i = \lambda \sum_{i=1}^{n} \frac{1}{\alpha_i} - \sum_{i=1}^{n} \frac{\beta_i}{\alpha_i}$$

But $\qquad \displaystyle\sum_{i=1}^{n} P_i = P_R$

$\therefore \qquad \displaystyle P_R = \lambda \sum_{i=1}^{n} \frac{1}{\alpha_i} - \sum_{i=1}^{n} \frac{\beta_i}{\alpha_i}$

or $\qquad \displaystyle \lambda = \frac{\left(P_R + \displaystyle\sum_{i=1}^{n} \frac{\beta_i}{\alpha_i} \right)}{\displaystyle\sum_{i=1}^{n} \frac{1}{\alpha_i}} \qquad\qquad …(15.7.7)$

We calculate λ from Eq. (15.7.7) and substitute in Eq. (15.7.6) to get individual P_i.

This method does not require iterative solution for a system where transmission losses are neglected.

Starting values of λ, that is, $\lambda^{(0)}$ and plant generations $P_i^{(0)}$ to be used in step 1 of iterative procedure may be obtained from Eq. (15.7.7) and

$$P_i^{(0)} = \frac{1}{\alpha_1}(\lambda^{(0)} - \beta_i) \text{ for } i = 1, 2, …, n$$

Example 15.8 A system consists of two plants connected by a transmission line as shown in Fig. 15.5. The load is at plant 2. If a load of 125 MW is transmitted from plant 1 to the load, there is a loss of 12.5 MW. Determine the generation schedule and the load demand if the cost of the received power is Rs 70 per MWh. Assume that the incremental costs of the two plants are given by

$$\frac{dC_1}{dP_1} = 0.25\,P_1 + 40 \text{ Rs/MWh} \;; \qquad \frac{dC_2}{dP_2} = 0.20\,P_2 + 50 \text{ Rs/MWh}.$$

Solve the problem using (a) coordination equations; (b) penalty factor method.

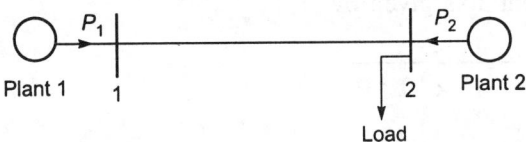

Plant 1   1   2   Plant 2

Load

Fig. 15.5. Illustrating Example 15.8.

Solution

Since the load is at the bus of plant 2, therefore, the line loss will not be affected by variation of P_2. Thus,

$$B_{12} = B_{21} = 0 \text{ and } B_{22} = 0$$

For a 2-plant system

$$P_L = P_1^2\,B_{11} + P_2^2\,B_{22} + 2P_1\,P_2\,B_{12}$$

When $P_1 = 125$ MW, $P_L = 12.5$ MW,

we have $12.5 = (125)^2\,B_{11} + 0 + 0$

or $B_{11} = \dfrac{12.5}{(125)^2} = 8 \times 10^{-4} \text{ MW}^{-1}$

Therefore, $P_L = 8 \times 10^{-4}\,P_1^2\;$; $\dfrac{\partial P_L}{\partial P_1} = 16 \times 10^{-4}\,P_1$

We are also given $\lambda = 70$ Rs/MWh.

(a) *Use of coordinates equation*

The coordination equation for plant 1 is given by

$$\frac{dC_1}{dP_1} + \lambda\,\frac{\partial P_L}{\partial P_1} = \lambda$$

Substituting the values of $\dfrac{dC_1}{dP_1}$, λ and $\dfrac{\partial P_L}{\partial P_1}$ we have

$$0.25\,P_1 + 40 + 70 \times 16 \times 10^{-4}\,P_1 = 70$$

$$(0.25 + 0.112)\,P_1 = 30, \quad P_1 = 82.8729 \text{ MW}$$

The coordination equation for plant 2 is given by

$$\frac{dC_2}{dP_2} + \lambda \frac{\partial P_L}{\partial P_2} = \lambda$$

or $0.20\, P_2 + 50 + 0 = 70,\ P_2 = 100$ MW

The line loss is given by

$$P_L = 8 \times 10^{-4}\, P_1^2 = 8 \times 10^{-4} \times (82.8729)^2 = 5.494 \text{ MW}$$

Therefore, the total load, $P_R = P_1 + P_2 - P_L = 82.8729 + 100 - 5.494 = 177.3789$ MW

(b) *Use of penalty factor*

The penalty factor for plant 1 is given by

$$L_1 = \frac{1}{1 - \dfrac{\partial P_L}{\partial P_1}} = \frac{1}{(1 - 16 \times 10^{-4}\, P_1)}$$

For optimal dispatch

$$L_1 \frac{dC_1}{dP_1} = \lambda \ ; \qquad \frac{1}{(1 - 16 \times 10^{-4}\, P_1)} \times (0.25\, P_1 + 40) = 70$$

or $0.25\, P_1 + 40 = 70 - 70 \times 16 \times 10^{-4}\, P_1$

or $P_1 = \dfrac{30}{0.25 + 0.112} = 82.8729$ MW

Penalty factor for plant 2,

$$L_2 = \frac{1}{1 - \dfrac{\partial P_L}{\partial P_2}} = \frac{1}{1 - 0} = 1$$

For optimal dispatch,

$$L_2 \frac{dC_2}{dP_2} = \lambda \ ; \qquad 1 \times (0.20\, P_2 + 50) = 70$$

$$P_2 = \frac{20}{0.20} = 100 \text{ MW}$$

It is seen that both the methods give identical results. This is due to the fact that the penalty factor has been derived from the coordination equation.

EXERCISES

1. The fuel costs of two generators are given by

$$C_1 = 1.6 + 15 P_1 + 0.1\, P_1^2 \text{ Rs/h} \ ; \qquad C_2 = 1.8 + 25 P_2 + 0.1\, P_2^2 \text{ Rs/h}.$$

   If the total demand on the generators is 250 MW, find the economic loading of two generators.

2. The incremental costs in Rs/MWh of two turboalternators is as follows :

$$\frac{dC_1}{dP_1} = 0.2\,P_1 + 60 \;; \qquad \frac{dC_2}{dP_2} = 0.3\,P_2 + 40$$

The rating of units are 150 MW and 250 MW. How will the load of 200 MW be shared between two units for most economic operation? Determine the saving in cost in Rs/h for economical load allocation compared to the loading in proportion to the rating of the units.

3. Derive coordination equation for economic load scheduling in a large power system. Give the sequence of steps to determine economic generation of various plants for a given load demand.

4. For economic load scheduling in a plant with k units, prove that

$$\lambda = \frac{\left[P_R + \displaystyle\sum_{n=1}^{k} \left(\frac{\beta_n}{\alpha_n} \right) \right]}{\displaystyle\sum_{n=1}^{k} \left(\frac{1}{\alpha_n} \right)}$$

where the symbols have their usual meanings.

5. Show that when a number of generating units are operating in parallel and supplying power into a transmission network, the most economical scheduling of loads is obtained when their incremental costs of received power are equal. Derive an equation coordinating the incremental cost of production, the incremental transmission loss and the incremental cost of received power.

ANSWERS

1. P_1 = 150 MW ; P_2 = 100 MW
2. P_1 = 80 MW ; P_2 = 120 MW, Rs 6.25 per hour.

16

Symmetrical Faults

16.1 INTRODUCTION

Normally, a power system operates under balanced conditions. Under abnormal (that is, fault) conditions, the system may become unbalanced. If the insulation of the system fails at any point or if two or more conductors that normally operate with a potential difference come in contact with each other, a short-circuit, or fault, is said to occur. The contact may be a physical metallic one, or it may occur through an arc.

A fault may occur on a power system due to a number of reasons. Some of the common causes have their origins in natural disturbances like lightning, high-speed winds, earthquakes, earth tremors, snow, frost etc. Generators, transformers, and other protective switchgear may fail due to insulation breakdown. There may be accidental faults such as falling of a tree along a line, vehicles colliding with supporting structures, airplane crashing with the line, birds shorting line. Sometimes sabotage also results in creating faults. Contamination of insulators may also result in a fault. Wind and ice loading may cause insulator strings to fail mechanically. Deterioration of insulation due to ageing and over loading of underground cables results in short circuits. Sometimes small animals like rats, lizards, etc. enter switchgear to create faults.

Faults may occur at different points in a power system. However, here we will be mostly concerned with faults on transmission lines. Faults that occur on a transmission line are broadly classified as follows :

1. Three-phase short circuits
2. Unsymmetrical faults.

Three-Phase Short Circuits

In such types of faults all the three phases are short circuited to each other and often to earth also. Such faults are balanced and symmetrical in the sense that the system remains balanced even after the fault. A three-phase short circuit occurs rarely but it is most severe type of fault involving largest currents. For this reason, the balanced short-circuit calculations are performed to determine these large currents to be used to determine the rating of the circuit breakers.

Unsymmetrical Faults

Unsymmetrical faults involve only one or two phases. For unsymmetrical faults, voltages and currents become unbalanced and each phase is to be treated individually for calculation purpose. This chapter is concerned with symmetrical three-phase short-circuit faults. The analysis of unsymmetrical faults is discussed in Chapter 18.

16.2 EFFECTS OF FAULTS

Faults can damage or disrupt power systems in several ways. Faults give rise to abnormal operating conditions, usually excessive voltages and currents at certain points on the system. Large voltages stress insulation beyond their breakdown value while large currents result in overheating of power system components. Sustained overheating may reduce useful life of the equipment. Sometimes faults lower system voltages below their permissible limits. Faults can cause the three-phase system to become unbalanced with the result that three-phase equipment operates improperly. Sometimes faults block the flow of power. Faults can cause system to become unstable. Hence, it is necessary that, upon the occurence of the fault, the faulty section should be disconnected as rapidly as possible in order that the normal operation of the rest of the system is not affected. If this is not done, the equipment may be damaged and the power supply is disrupted. The relays should immediately detect the existence of the fault and initiate circuit breaker operation to disconnect the faulty section.

16.3 PURPOSE OF FAULT ANALYSIS

Fault analysis is also known as *short-circuit study* or *short-circuit analysis*. A fault analysis includes the following:
  - To determine the values of voltages and currents at different points of the system during the fault.
  - Determination of the ratings of the required circuit breakers.
  - Selection of appropriate schemes of protective relaying.

Thus, the purpose of fault analysis is to save the system from abnormal conditions within minimum time.

16.4 SIMPLIFYING ASSUMPTIONS

Accurate fault calculations involve much labour and time. In practice, the following simplifying assumptions are made in the analysis of faults :
  1. Each synchronous machine model is represented by a constant-voltage source behind (that is, in series with) a proper reactance. This reactance may be subtransient reactance (X''_d), transient reactance (X'_d), or synchronous reactance at steady state (X_d). If we are interested to determine current immediately after the fault we take subtransient reactance X''_d. If we wish to determine the fault current after about 3 to 4 cycles we use transient reactance.

2. In the transformer models, the shunt elements that account for magnetizing current and core losses are neglected.

3. All transformers are considered to be at their normal taps.

4. In the transmission-line models the shunt capacitances are neglected.

5. All series resistances in generators, transmission lines, and transformers are neglected. This assumption is usually made for hand calculations. With this assumption, the power system network will contain only reactances, and therefore, the system can be represented by its most simplified reactance diagram. However, if the calculations are to be made with the help of a digital computer, this assumption is unnecessary.

6. Load impedances are neglected and therefore, the prefault system is *unloaded*. In other words, the power system network becomes open circuited. Hence we neglect the normal load currents (that is, prefault currents). With the system network open, all the prefault bus voltages will have the same magnitude and phase angle. Therefore, in per unit analysis, all prefault voltages in the network including the prefault voltage at the fault point can be taken as $1.0 \underline{/0°}$ pu. This is commonly known as *flat profile*. When the prefault currents are also to be considered, we shall use superposition principle.

These assumptions result in considerable simplification in calculations without seriously affecting accuracy. Actually the results obtained are on the safer side.

16.5 THEVENIN'S EQUIVALENT CIRCUIT

Thevenin's theorem may be stated as follows :

A linear network N consisting of sources and impedances can be replaced at any pair of terminals a-b by a simple equivalent network consisting of a voltage source V_T in series with an impedance Z_T. The source voltage V_T is the voltage across the terminals a-b when they are open circuited. The series impedance Z_T is the network impedance as seen from the terminals a-b when all the sources are replaced by their internal impedances. The voltage V_T is called the *Thevenin voltage* and the impedance Z_T is called the *Thevenin impedance*. The equivalent circuit consisting of V_T in series with Z_T is known as *Thevenin's equivalent circuit*.

Thevenin's theorem is very useful in fault calculations. Usually we have to determine the network impedance as seen from the fault point. In most cases Thevenin voltage V_T is assumed to be 1.0 per unit.

16.6 SHORT-CIRCUIT CAPACITY, SCC

The *short-circuit capacity* (SCC) of a bus of a network is defined as the product of the magnitudes of the prefault voltage and the post fault current. The short-circuit capacity is also known as the *fault level*.

$$|SCC| \stackrel{\Delta}{=} V^0 |I_F| \text{ VA} \qquad \qquad ...(16.6.1)$$

where V_0 = prefault voltage in volts

I_F = post–fault current in amperes.

For a solid fault the fault impedance $Z_f = 0$ and the fault current is given by

$$I_F = \frac{V_T}{Z_T} \qquad \qquad ...(16.6.2)$$

where V_T = Thevenin's voltage per phase in volts

$\quad\;\; Z_T$ = Thevenin's impedance in ohms.

In our case, $V_T = V^0$, therefore

$$|\,SCC\,|_{1\,\varphi} = |\,V_T\,|\,|\,I_F\,| = \frac{|\,V_T\,|^2}{Z_T} \; VA/phase \qquad \qquad ...(16.6.3)$$

We have $\quad Z_{T\,pu} = Z_T \dfrac{S_b}{V_b^2} \qquad \qquad ...(16.6.4)$

where S_b = base voltamperes in VA

$\quad\;\; V_b$ = base voltage in volts

$\quad Z_{T\,pu}$ = Thevenin's impedance in per unit.

If V_T is chosen as base voltage, $V_T = V_b$. Therefore, Eq. (16.6.4) may be written as

$$Z_{T\,pu} = Z_T \frac{S_b}{V_T^2} \qquad \qquad ...(16.6.5)$$

$$\frac{V_T^2}{Z_T} = \frac{S_b}{Z_{T\,pu}} \qquad \qquad ...(16.6.6)$$

Combination of Eqs. (16.6.3) and (16.6.6) gives

$$|\,SCC\,|_{1\,\varphi} = \frac{S_b}{Z_{T\,pu}} \; VA \; per \; phase \qquad \qquad ...(16.6.7)$$

If $(kV_T)_p$ is the phase voltage in kV, then from Eq. (16.6.3)

$$|\,SCC\,|_{1\,\varphi} = \frac{|\,(kV_T)_p\,|^2}{Z_T} \; MVA/phase$$

Total $|\,SCC\,|$ for all the three phases is given by

$$|\,SCC\,|_{3\,\varphi} = 3\,|\,SCC\,|_{1\,\varphi} = \frac{3\,|\,(kV_T)_p\,|^2}{Z_T} \; MVA$$

If $(kV_T)_l$ is the line voltage in kV, then $(kV_T)_l = \sqrt{3} \,(kV_T)_p$

$$\therefore \qquad |\,SCC\,|_{3\,\varphi} = \frac{|\,(kV_T)_l\,|^2}{Z_T} \qquad \qquad ...(16.6.8)$$

$$Z_{T\,pu} = Z_T \frac{[(MVA)_b]_{3\,\varphi}}{[(kV_T)_b]^2} \qquad \qquad ...(16.6.9)$$

Combination of Eqs. (16.6.8) and (16.6.9) gives

$$| \text{SCC} |_{3\,\varphi} = \frac{[(\text{MVA})_b]_{3\,\varphi}}{Z_{T\,pu}} \text{ MVA} \qquad \qquad \ldots (16.6.10)$$

We can also write Eq. (16.6.10) as

$$(\text{SSC})_{3\,\varphi} = \frac{(S_b)_{3\,\varphi}}{Z_{T\,pu}} \text{ MVA} \qquad \qquad \ldots (16.6.11)$$

where $(S_b)_{3\,\varphi}$ is the total three-phase MVA.

Equation (16.6.11) shows that the fault level (SCC) can be found if the total per unit impedance from the voltage source to the fault is known.

The fault current (short-circuit current) can be found as follows :

$$\sqrt{3}\ V_{lb}\, I_F \times 10^{-6} = \text{short-circuit MVA in all the three phases} \qquad \ldots (16.6.12)$$

where V_{lb} is in volts and I_F in amperes.

16.7 PROCEDURE FOR CALCULATING SHORT-CIRCUIT VOLTAMPERES AND SHORT-CIRCUIT CURRENT

For calculating the short-circuit voltamperes S_{sc} and the short-circuit current I_{sc} the following procedure is used :

1. Draw a single-line diagram of the complete network. On this diagram, indicate the rating, voltage, resistance, and reactance of all generators, transformers, transmission lines, loads, etc.

2. Select a common base S_b (kVA or MVA) and convert all impedances to per-unit values on the same voltamperes base S_b.

3. Corresponding to the single-line diagram of the network, draw the reactance (or impedance) diagram showing one phase of the system and the neutral. On this diagram, indicate all the per-unit resistances and per-unit reactances of the components calculated in step 2.

4. Calculate the total per unit impedance from the source to the fault point by circuit analysis. This may involve series-parallel combination, star-delta or delta-star transformations. It should be noted that this total per unit impedance is the Thevenin impedance of the network as seen from the fault point.

5. Determine the fault MVA (short-circuit MVA) and fault current (short-circuit current) from Eqs. (16.6.11) and (16.6.12).

16.8 STAR-DELTA AND DELTA-STAR TRANSFORMATIONS

Star-delta and delta-star transformations are used in network reduction. From circuit theory we know that the equivalent star impedance connected to a given terminal is equal to the product of the two delta impedances connected to the same terminal divided by the sum of the delta impedances. We use the following results with reference to Fig. 16.1 :

$\Delta \to Y$	$Y \to \Delta$
$Z_1 = \dfrac{Z_{12}\,Z_{31}}{Z_{12} + Z_{23} + Z_{31}}$	$Z_{12} = Z_1 + Z_2 + \dfrac{Z_1\,Z_2}{Z_3}$
$Z_2 = \dfrac{Z_{12}\,Z_{23}}{Z_{12} + Z_{23} + Z_{31}}$	$Z_{23} = Z_2 + Z_3 + \dfrac{Z_2\,Z_3}{Z_1}$
$Z_3 = \dfrac{Z_{31}\,Z_{23}}{Z_{12} + Z_{23} + Z_{31}}$	$Z_{31} = Z_3 + Z_1 + \dfrac{Z_3\,Z_1}{Z_2}$

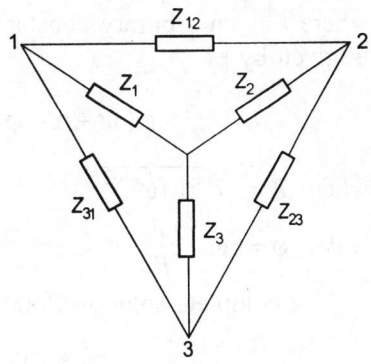

Fig. 16.1. Star-delta transformations.

16.9 TRANSIENT IN A SERIES R-L CIRCUIT

Consider a series R-L circuit to which a sinusoidal voltage is suddenly applied by closing a switch S as shown in Fig. 16.2. Let the applied voltage be given by

$$v = V_m \sin(\omega t + \alpha) \qquad \qquad \dots(16.9.1)$$

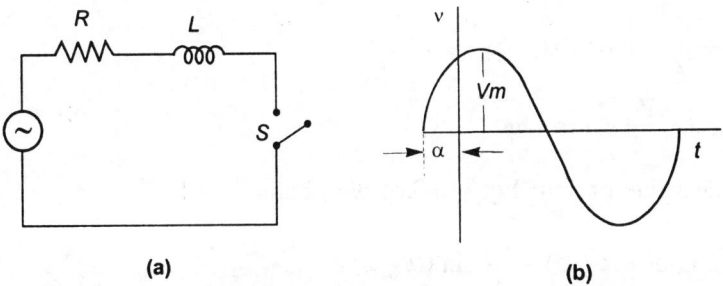

(a)	(b)

Fig. 16.2. (a) Series R-L circuit (b) Voltage waveform.

When the switch S is closed we can write the voltage equation by KVL as

$$v_R + v_L = v$$

$$Ri + \frac{L\,di}{dt} = V_m \sin(\omega t + \alpha) \qquad \qquad \dots(16.9.2)$$

where i is the instantaneous current. Equation (16.9.2) can be solved by a number of methods. The solution consists of two parts, namely,

$$i = i_n + i_f \qquad \qquad \dots(16.9.3)$$

In this equation i_n is called the *natural response* and is the general solution of the homogeneous equation (input set to zero). The component i_f is called the *forced response* and is a particular solution of Eq. (16.9.2).

From circuit theory

$$i_n = k e^{-\frac{R}{L}t} \qquad \qquad \dots(16.9.4)$$

where k is an arbitrary constant to be determined from initial conditions. The forced response is given by

$$i_f = \frac{V_m}{Z} \sin(\omega t + \alpha - \varphi) \qquad \qquad ...(16.9.5)$$

where $Z = \sqrt{R^2 + (\omega L)^2}$

and $\quad \varphi = \tan^{-1} \frac{\omega L}{R}$

The complete solution (total response) is therefore given by

$$i = i_n + i_f = ke^{-\frac{R}{L}t} + \frac{V_m}{Z} \sin(\omega t + \alpha - \varphi) \qquad \qquad ...(16.9.6)$$

The initial conditions are applied to the total response. In our case the initial conditions are those existing at the time of closing the switch. Since the current is zero just before closing the switch, the initial conditions are

$$i = 0 \text{ at } t = 0$$

Putting these values of i and t in Eq. (16.9.6) we get

$$0 = ke^0 + \frac{V_m}{Z} \sin(0 + \alpha - \varphi)$$

which gives $\quad k = -\dfrac{V_m}{Z} \sin(\alpha - \varphi)$

Substituting the value of k in Eq. (16.9.6) we obtain

$$i = \frac{V_m}{Z} \sin(\omega t + \alpha - \varphi) - \frac{V_m}{Z} \sin(\alpha - \varphi) e^{-\frac{R}{L}t} \qquad \qquad ...(16.9.7)$$

Equation (16.9.7) shows that the current i has two components :
1. The steady-state component (forced response) i_f given by

$$i_f = \frac{V_m}{Z} \sin(\omega t + \alpha - \varphi) \qquad \qquad ...(16.9.8)$$

2. The exponentially decaying transient component (natural response) i_n given by

$$i_n = -\frac{V_m}{Z} \sin(\alpha - \varphi) e^{-\frac{R}{L}t} \qquad \qquad ...(16.9.9)$$

The component i_n is also known as *d.c. offset component* since it is unidirectional. The d.c. component decays exponentially with a time constant (L/R). The angle α denotes the instant in the voltage cycle at which the switch is closed. If the switch is closed when $\alpha = \varphi$, the d.c. offset term will be zero and the total current wave will be symmetrical about the time axis. If the switch is closed at random, there is an unpredictable amount of offset and the total current becomes unsymmetrical about the time axis. However, if the switch is closed when $(\alpha - \varphi) = \pm \pi/2$, the d.c. offset component will be a maximum, and the first peak of the resultant current i will become twice the peak value of the final steady-state current. This effect is known as *doubling effect*.

16.10 SUDDEN SHORT CIRCUIT AT THE ARMATURE TERMINALS OF A THREE-PHASE GENERATOR

The current flowing in the armature of a three phase synchronous generator when its terminals are suddenly short circuited is similar to the current flowing in series R-L circuit when a sinusoidal voltage is suddenly applied to it. Fig. 16.3 (a) shows a three-phase synchronous generator subjected to symmetrical short circuit. Fig. 16.3 (b) shows its equivalent single-phase circuit. In a series R-L circuit we assume that reactance X_L $(= 2\pi f L)$ remains constant. However, we shall see that the synchronous generator offers time varying reactance which changes from X''_d to X'_d and finally to X_d.

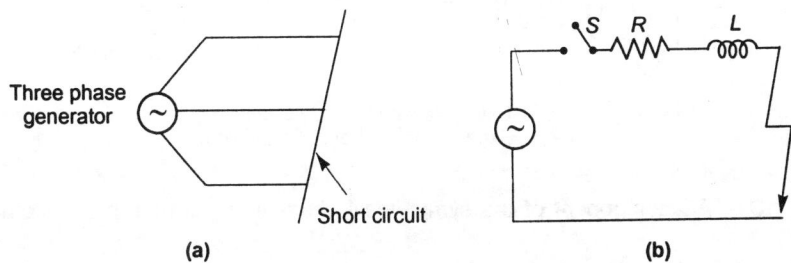

Fig. 16.3. (a) Three-phase synchronous generator subjected to symmetrical short circuit. (b) Equivalent single-phase circuit.

Consider a three-phase generator on no load, running at its synchronous speed and carrying a constant field current. Suddenly, the three phases are short-circuited. The short-circuit is each phase consists of a steady-state a.c. component and a transient d.c. offset. Since the voltages of the three phases have 120° phase shift between them, the angle α will be different in all the three phases. In other words, the short circuit occurs at different points on the voltage wave of each phase. Since in a practical situation, we do not know when in a sinusoidal cycle a fault will occur, we are not able to predict the amount of offset. Because of its unpredictability, let us remove (subtract) the d.c. offset term from the current wave forms. Fig. 16.4 shows the short-circuit current for one phase of a synchronous generator with d.c. offset component neglected. The dashed envelope is called the *symmetrical short-circuit armature current*.

The wave may be divided into three distinct time periods :

(a) **Subtransient Period :** This period lasts for only about 2 cycles. During this period the current decays very rapidly.

(b) **Transient Period :** The transient period lasts for about 30 cycles. During this period the current decreases some what slowly.

(c) **Steady-State Period :** The current reaches its steady-state value.

The r.m.s. value of initial current (that is, the current at the instant of short circuit) is called *subtransient current* **I″**. The corresponding reactance of the winding is called the *direct-axis subtransient reactance* X''_d. This reactance is essentially due to the presence of damper windings. When the transient envelope is extrapolated backwards in time it cuts the vertical axis at point b. The r.m.s. value of current represented by the intercept Ob is called *transient current* I'.

$$I' = \frac{Ob}{\sqrt{2}}$$

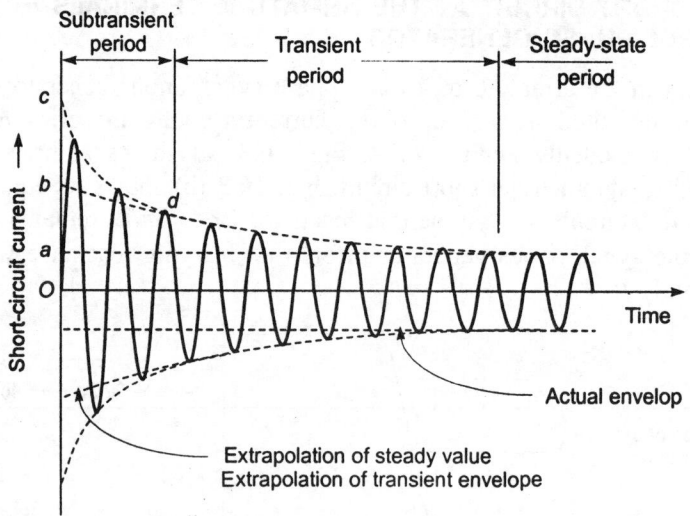

Fig. 16.4. AC component of the symmetrical short-circuit current in an alternator.

The corresponding reactance is called *direct axis transient reactance* X'_d. This reactance is essentially due to the field winding. Similarly, the r.m.s. value of current represented by the intercept *Ob*, that is, $Ob/\sqrt{2}$ is called the *steady-state short-circuit current I*. The corresponding reactance is called *direct-axis synchronous reactance* X_d. If E_0 is the r.m.s. value of open circuit phase voltage, we can write

$$I'' = \frac{Oa}{\sqrt{2}} = \frac{E_0}{X''_d}$$

$$I' = \frac{Ob}{\sqrt{2}} = \frac{E_0}{X'_d}$$

$$I = \frac{Oc}{\sqrt{2}} = \frac{E_0}{X_d}$$

where I'' = r.m.s. value of subtransient current

I' = r.m.s. value of transient current

I = r.m.s. value of steady–state short–circuit current.

Example 16.1 A three-phase transmission line operating at 33 kV and having a resistance and reactance of 5 Ω and 20 Ω respectively is connected to a generating station busbar through a 15 MVA step-up transformer which has a reactance of 0.06 per unit. Connected to the busbars are two generators, one 10 MVA having 0.10 per unit reactance and another 5 MVA having 0.075 per unit reactance. Calculate the short-circuit MVA and the fault current when a three-phase short-circuit occurs (a) at the high-voltage terminals of the transformer (b) at the load end of the transmission line.

Solution

The single-line diagram of the network is shown in Fig. 16.5 (a). Let 15 MVA be taken as the base MVA.

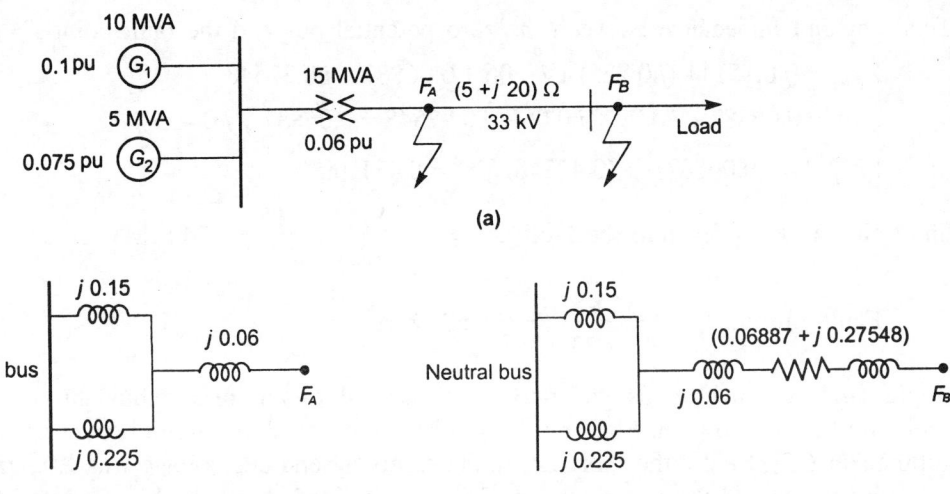

Fig. 16.5. (a) Single line diagram (b) Equivalent single-phase circuit with fault at F_A (c) Equivalent single-phase circuit with fault at F_B.

Per unit impedance of generator 1 on 15 MVA base $= j\,0.10 \times \dfrac{15}{10} = j\,0.15$ pu

Per unit impedance of generator 2 on 15 MVA base $= j\,0.075 \times \dfrac{15}{5} = j\,0.225$ pu

Per unit impedance of transformer on 15 MVA base $= j\,0.06$ pu

Per unit impedance of the transmission line

$$Z_{pu} = Z_\Omega \frac{(MVA)_b}{(kV_l)_b^2} = (5 + j\,20)\frac{15}{(33)^2} = (0.06887 + j\,0.27548)\ \text{pu}$$

(a) Three-Phase Fault at F_A

The equivalent single-phase circuit with fault at point F_A is shown in Fig. 16.5 (b). The total impedance from the generator neutral to the fault at F_A is

$$Z_{e\,pu} = (j\,0.15)\ ||\ (j\,0.225) + j\,0.06 = \frac{j\,0.15 \times j\,0.225}{j\,0.15 + j\,0.0225} + j\,0.06 = j\,0.15\ \text{pu}$$

Short-circuit MVA fed into the fault at F_A

$$S_{sc} = \frac{S_b}{|\,Z_{e\,pu}\,|} = \frac{15}{0.15} = 100\ \text{MVA}$$

Fault-current, $I_F = \dfrac{100 \times 10^6}{\sqrt{3} \times 33 \times 10^3} = 1749.5\ \text{A}$

(b) Three-Phase Fault at F_B

The equivalent single-phase circuit with fault at F_B is shown in Fig. 16.5 (c). This diagram also does not contain any source.

Total per unit impedance between the zero potential bus and the fault point F_B is

$$Z_{e\ pu} = (j\,0.15) \,||\, (j\,0.225) + j\,0.06 + 0.06887 + j\,0.27548$$

$$= 0.06887 + j\,0.09 + j\,0.06 + j\,0.27548 = 0.06887 + j\,0.42548$$

$$|Z_{e\ pu}| = [(0.06887)^2 + (0.42548)^2]^{1/2} = 0.431 \text{ pu}$$

Short-circuit MVA fed into the fault at $F_B = \dfrac{S_b}{|Z_{e\ pu}|} = \dfrac{15}{0.431} = 34.8$ MVA

Fault current, $I_F = \dfrac{34.8 \times 10^6}{\sqrt{3} \times 33 \times 10^3} = 608.8$ A

Example 16.2 A synchronous generator is connected to an infinite bus through a 138 kV transmission line as shown in Fig. 16.6. A solid three phase short-circuit occurs on the line near circuit breaker CB_1. Before the short circuit the receiving-end bus voltage was 1.0 per unit value, unity power factor and the generator was 75 per cent loaded, on the basis of its MVA rating. Determine the subtransient, transient, and synchronous short-circuit currents.

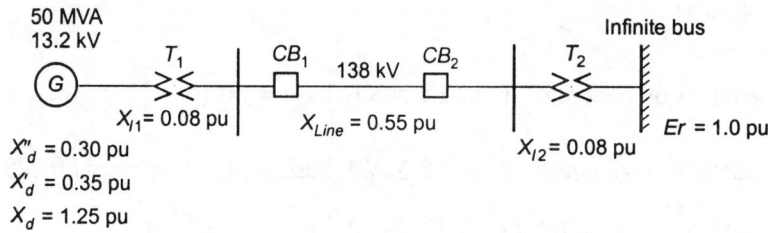

Fig. 16.6. Illustrating Example 16.2.

Solution

Calculation of the Prefault Voltage Behind the Subtransient Reactance

Total impedance from the internal voltage of the generator to the infinite bus under subtransient conditions

$$Z''_T = j\,X''_d + j\,X_{l1} + j\,X_{line} + j\,X_{l2} = j\,0.30 + j\,0.08 + j\,0.55 + j\,0.08 = j\,1.01 \text{ pu}$$

The receiving-end voltage, $E_r = 1\underline{/0^\circ} = (1 + j\,0)$ pu. Since the generator is 75% loaded on the basis of its MVA rating, the current from the generator is given by $I_G = 0.75$ pu.

Let E''_{int} = subtransient internal voltage of the generator.

$\therefore \qquad E''_{int} = E_r + I_G\,Z''_T = 1 + j\,0 + 0.75 \times j\,1.01 = 1 + j\,0.7575 = 1.2545\underline{/37.14^\circ}$ pu

Calculation of the Subtransient Current to the Fault

Let the reactance from the internal voltage of the generator to the fault point be X''_{gF}. Then

$$X''_{gF} = X''_d + X_{l1} = 0.30 + 0.08 = 0.38$$

The subtransient current from the generator to the fault $= \dfrac{E''_{int}}{X''_{gF}} = \dfrac{1.2545}{0.38} = 3.3013$ pu

The reactance from the infinite bus to the fault, $X''_{bF} = 0.08 + 0.55 = 0.63$ pu

The subtransient current from the infinite bus to the fault $= \dfrac{E_r}{X''_{bF}} = \dfrac{1.00}{0.63} = 1.5873$ pu

Total subtransient short-circuit current at the fault = subtransient current from the generator to the fault + subtransient current from the infinite bus to the fault

$= 3.3013 + 1.5873 = 4.8886$ pu

Calculation of the Effect of the Maximum d.c. Component Offset

The maximum possible d.c. offset is taken to be $\sqrt{2}$ times the symmetrical wave, and the wave of the total offset wave is the short-circuit current I_{sc}.

$$I_{sc} = \sqrt{I_n^2 + I_w^2}$$

where I_n = current with d.c. offset neglected

I_w = current with d.c. offset.

Current (with maximum d.c. component) from the generator $= \sqrt{2} \times 3.3013$

Current (with maximum d.c. component) from the infinite bus $= \sqrt{2} \times 1.5873$

Total current (with maximum d.c. component) from the generator and infinite bus

$= \sqrt{2} \times 3.3013 + \sqrt{2} \times 1.5873 = \sqrt{2} \times 4.8886 = 6.91248$ pu

The greatest r.m.s. value of the short circuit current $= \sqrt{(4.8886)^2 + (6.91248)^2} = 8.46645$ pu

Conversion of the Per-Unit Current to Amperes

Let I_{lb} be the base current.

$$\sqrt{3}\, V_{lb}\, I_{lb} \times 10^{-6} = \text{base MVA}$$

$$I_{lb} = \frac{\text{base MVA}}{\sqrt{3} \times V_{lb} \times 10^{-6}} = \frac{50}{\sqrt{3} \times 138 \times 10^3 \times 10^{-6}} = 209.18 \text{ A}$$

Therefore, the short-circuit current is given by

$I_{sc} = $ base current $\times$ pu current $= 209.18 \times 8.46645 = 1771$ A

Calculation of the Prefault Voltage Behind Transient Reactance

Total impedance from the internal voltage of the synchronous generator to the infinite bus under transient conditions

$$\mathbf{Z}'_T = j\,X'_d + j\,X_{l1} + j\,X_{line} + j\,X_{l2} = j\,0.35 + j\,0.08 + j\,0.55 + j\,0.08 = j\,1.06 \text{ pu}$$

Let $\mathbf{E}'_{int}$ = transient internal voltage of the generator.

$$\mathbf{E}'_{int} = \mathbf{E}_r + \mathbf{I}_G\,\mathbf{Z}'_T = 1 + j\,0 + 0.75 \times j\,1.06 = 1 + j\,0.795 = 1.2775\,\underline{/38.485°} \text{ pu}$$

Calculation of the Transient Current to the Fault

Let the total reactance from the internal voltage of the synchronous generator to the fault point be X'_{gF}. Then

$$X'_{gF} = X'_d + X_{l1} = 0.35 + 0.08 = 0.43 \text{ pu}$$

The transient current from the generator to the fault

$$I'_g = \frac{E'_{int}}{X'_{gF}} = \frac{1.2775}{0.43} = 2.9709 \text{ pu}$$

It is to be noted that only the magnitude is considered, not the angle.

The transient current from the infinite bus to the fault

$$I'_b = \frac{E_r}{X'_{bF}} = \frac{1.0}{0.08 + 0.55} = 1.5873 \text{ pu}$$

Total transient short-circuit current at the fault point = transient current from the generator to the fault + transient current from the infinite bus to the fault

$$I'_T = I'_g + I'_b = 2.9709 + 1.5873 = 4.5582 \text{ pu}$$

Conversion of the Per Unit Current to Amperes

The transient short-circuit current in amperes at the fault point is given by

$$I_{sc} = \text{base current} \times \text{pu current} = 209.18 \times 4.5582 = 953.48 \text{ A}$$

Calculation of the Prefault Voltage Behind Synchronous Reactance

Total impedance from the internal voltage of the synchronous generator to the infinite bus under synchronous conditions

$$Z_{total} = j\,X_d + j\,X_{l1} + j\,X_{line} + j\,X_{l2} = j\,1.25 + j\,0.08 + j\,0.55 + j\,0.08 = j\,1.96 \text{ pu}$$

Therefore, $\mathbf{E}_{int} = \mathbf{E}_r + \mathbf{Z}_{total}\,\mathbf{I}_G = 1 + j\,0 + j\,1.96 \times 0.75 = 1 + j\,1.47 = 1.778\,\underline{/55.77°}$ pu

Calculation of Synchronous Current

The synchronous reactance from the internal voltage of the generator to the fault point is given by

$$X_{gF} = X_{d1} + X_{l1} = 1.25 + 0.08 = 1.33 \text{ pu}$$

The synchronous current from the generator to the fault

$$I_g = \frac{E_{int}}{X_{gF}} = \frac{1.778}{1.33} = 1.3368 \text{ pu}$$

The synchronous current from the infinite bus to the fault

$$I_b = \frac{E_r}{X_{bF}} = \frac{1.0}{0.08 + 0.55} = 1.5873 \text{ pu}$$

The total synchronous short-circuit current at the fault point in per unit

$$I_T = 1.3368 + 1.5873 = 2.9241 \text{ pu}$$

The total synchronous short-circuit current at the fault point in amperes

$$I_S = \text{base current} \times \text{pu current} = 209.18 \times 2.9241 = 611.66 \text{ A}$$

16.11 CONSIDERATION OF PREFAULT LOAD CURRENT

In general, the fault currents are much greater than load currents and therefore, the load currents are neglected in fault calculations. However, in some cases it becomes necessary to consider the effect of load current. When this is done, we cannot take a flat voltage profile at all buses. Therefore, all the prefault voltages in the network including the prefault voltage at the fault point depart from the value of $1\,/0°$ pu. In such cases first we calculate the prefault voltage at the fault point. The fault current is then found using this value of prefault voltage. By superposition, the total current is the phasor sum of the load current and the fault current.

16.12 CURRENT-LIMITING REACTORS

Fault currents may be large enough to cause damage to the line and other equipment of a power system network. The interrupting capacities of circuit breakers to handle such currents would also be very large. Fault current is limited by the system reactance, which includes the impedance of the generators, transformers, lines, and other components of the system. Modern generators have reactances large enough to limit fault currents. Old generators have low values of reactance. The fault levels increase with the growth of the interconnected system. Therefore, if the system is large or some of the generators are old, the fault current can be kept within safe limits by increasing the system reactance. This is done by connecting reactors at strategic points in the system. Current-limiting reactors are coils used to limit current during fault conditions. Such rectors have large values of inductive reactances and low ohmic resistances.

16.12.1 Construction of Reactors

For current-limiting reactors, it is important that magnetic saturation at high current does not reduce the coil reactance. Reactors can be either air-cored type or iron-cored type. Air-cored reactors do not have magnetic saturation, and therefore their reactances are independent of current. For this reason, air-cored reactors are most commonly used these days. There are two main types of air-cored reactors : dry-type reactor and oil immersed reactor.

(a) Dry-type air-cored reactors

The windings of a dry-type reactor are embedded in vertical supports of concrete. The winding is bolted to concrete base which is insulated from ground by porcelain post insulators which also support the reactor. Dry type reactors are usually cooled by natural ventilation. Sometimes, forced air cooling is also used. For dry-type reactors located near metal objects, such as I-beams, plates, channels, etc. magnetic shielding is provided to prevent the reactor flux from inducing currents in the surrounding metal objects. Dry type air-cored reactors are used upto 33 kV.

(b) Oil-immersed air-cored reactors

In this type the coil is placed in an oil tank. The oil serves the purpose of insulation as well as cooling. Oil-immersed reactors can be cooled by any of the methods used for cooling power transformers. An oil-immersed reactor possesses the following advantages over dry-type reactor of the same rating :

(a) Higher safety against flashover

(b) Higher thermal capacity.

Oil-immersed reactors are built for both outdoor and indoor installations and for any voltage level.

Reactors are usually built as single-phase units. In general, the ratio of effective resistance of a rector to its inductive reactance is about 0.03.

16.12.2 Location of Reactors

Current-limiting reactors may be connected (a) in series with each generator, (b) in series with each feeder and (c) between busbar sections.

Generator Reactors

When the reactors are connected in series with each generator, they are known as *generator reactors*. Fig. 16.7 shows the location of generator reactors. The feeders are connected directly to the busbars. Modern generators are designed to have sufficiently large reactance to protect them even in dead short-circuits at their terminals. Thus, current-limiting reactors may only be used with old generators having low values of reactance. Generator reactors suffer from the following drawbacks:

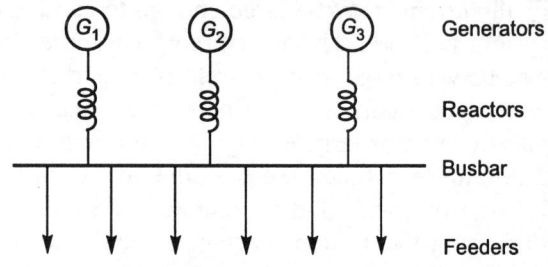

Fig. 16.7. Generator reactors.

1. The full-load current flowing in the reactor under normal operation produces a constant voltage drop and power loss in each reactor.

2. If a busbar or feeder fault occurs close to the busbars, the voltage at the busbars drops to a low value with the result that generators may lose synchronism and the supply may be interrupted.

Feeder Reactors

When the reactors are connected in series with each feeder, they are known as *feeder reactors*. Fig. 16.8 shows feeder reactors. There are two main advantages of feeder reactors: (1) If there is fault on any feeder, the voltage drop in its reactor will not affect the busbar voltage, therefore there is a little chance for the generator to lose synchronism. (2) The fault on a feeder will not affect other feeders.

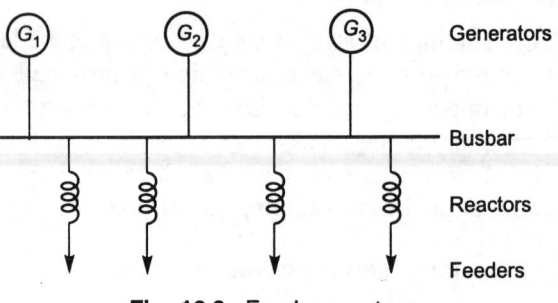

Fig. 16.8. Feeder reactors.

The feeder reactors have the following limitations :

(a) There is a constant voltage drop and power loss in each reactor during normal operating conditions.

(b) If a fault occurs at the busbars, no protection is provided to the generators.

(c) If the number of generators is increased, the sizes of the feeder reactors should also be increased.

Busbar Reactors

Generating station busbars are sectionalized and reactors are connected between sections. This is the most common method of connection of reactors. Two methods of reactors connections are used. These are (a) ring system, and (b) tie-bus system or tie-bar system. The ring system and tie-bus system are shown in Fig. 16.9.

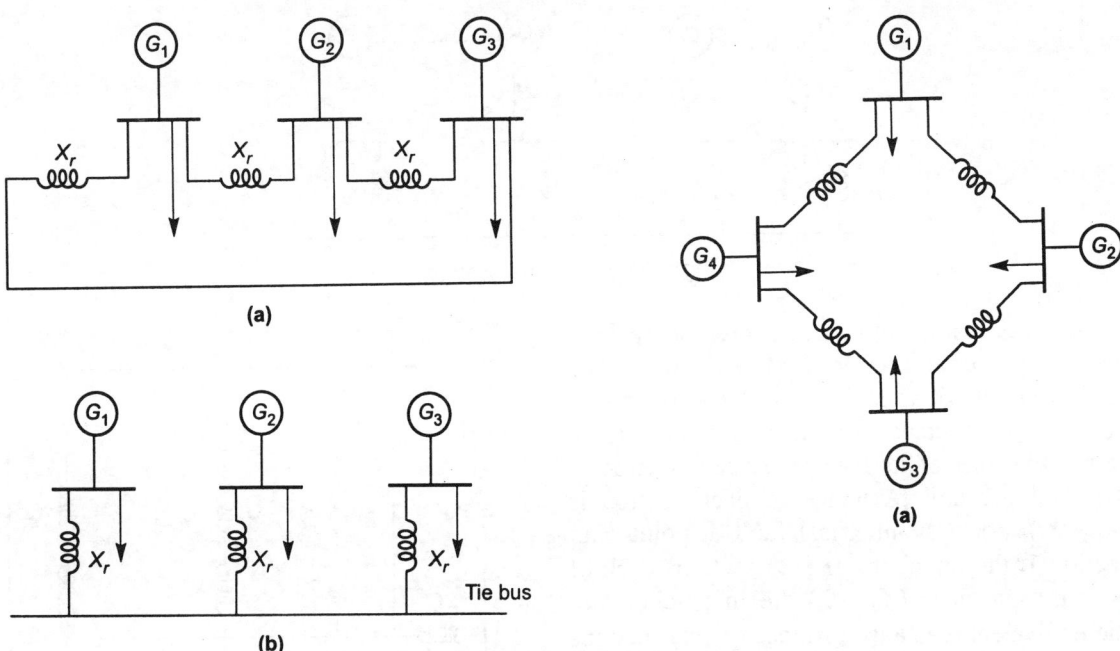

Fig. 16.9. Busbar reactors : (a) Ring system, (b) Tie-bus system.

Under normal operation each generator supplies feeder connected to its own section, and there will be no current through the reactors. Thus, there is no voltage drop or power loss in the reactor during normal operation. The busbar reactors localize the faults. For example, if a fault occurs on a feeder, only that busbar section is affected to which it is connected. The other sections continue to operate normally. In the ring system, the current transferred between two sections flows through two paths in parallel, whereas in the tie-bus system the current flows through two reactors in series. Therefore, the reactors in the tie-bus system have only one-third of the reactance of ring reactors. However, they carry twice as much current as the ring reactors. The tie-bus system is more flexible than the ring system. With the tie-bus arrangement extra generators may be added to the system without addition of extra circuit breakers or increasing the existing reactance. This is explained in the next section.

16.12.3 Rating of Reactors

The rating of reactors is expressed in terms of the MVA that it is designed to carry at rated current and voltage. The reactance is expressed in per unit and is the ratio of voltage drop across the reactor at the rated current to the line-to-neutral voltage of the system.

16.13 SHORT CIRCUIT MVA IN A TIE-BUS SYSTEM

Consider the busbars of a generating station divided into sections, to each of which is connected a generator of per unit reactance X_g. Each sectional busbar is connected to a common tie bus through a reactor of per unit reactance X_r. Let S be the MVA rating of each generator. The per unit values are all on a base of S MVA. The arrangement is shown in Fig. 16.10.

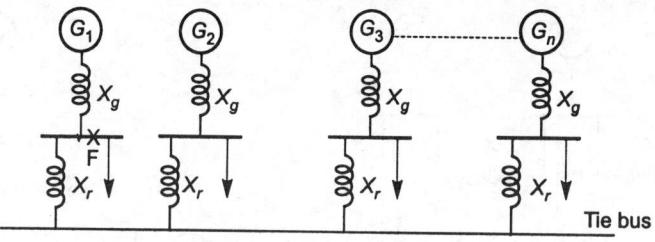

Fig. 16.10. Tie-bus system.

Suppose that a symmetrical three-phase fault occurs on busbar section 1 at a point F. The equivalent circuit after the fault is shown in Fig. 16.11. The reactance of each generator is in series with that of its reactor. Except section 1, the total per unit reactance of each section is $(X_g + X_r)$. For a symmetrical fault at point F in section 1, the remaining $(n-1)$ sections each of per unit reactance $(X_g + X_r)$ are in parallel. Let the equivalent reactance of these $(n-1)$ sections [shown dotted in rectangle of Fig. 16.11] in parallel be $X_{(n-1)}$.

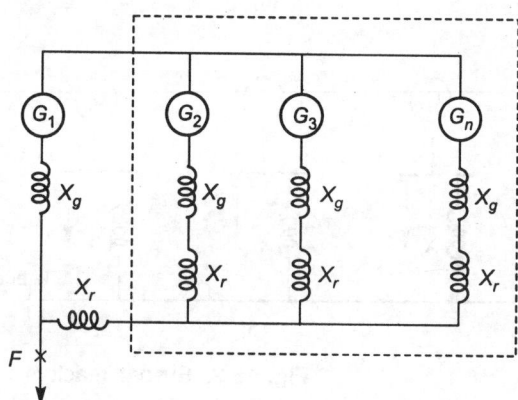

Fig. 16.11. Equivalent circuit after the fault at F.

$$\therefore \qquad X_{n-1} = \frac{1}{n-1}(X_g + X_r) \qquad ...(16.13.1)$$

X_{n-1} is in series with the reactance X_r of the faulty bus section as shown in Fig. 16.11. Let X_T be the Thevenin equivalent reactance at fault point F.

$$\therefore \qquad X_T = X_g \,||\, (X_{n-1} + X_r)$$

$$\frac{1}{X_T} = \frac{1}{X_g} + \frac{1}{X_{n-1} + X_r}$$

$$= \frac{1}{X_g} + \frac{1}{\dfrac{1}{n-1}(X_g + X_r) + X_r} = \frac{1}{X_g} + \frac{n-1}{X_g + n X_r} = \frac{n(X_g + X_r)}{X_g(X_g + n X_r)} \qquad ...(16.13.2)$$

Short circuit MVA

$$S_{sc} = \frac{S_b}{X_T} = \frac{S}{X_g} \times \frac{n(X_g + X_r)}{X_g + n X_r} = \frac{S}{X_g} \times \frac{X_g + X_r}{\dfrac{1}{n} X_g + X_r} \qquad ...(16.13.3)$$

When $n \to \infty$, $\dfrac{1}{n} \to 0$

$$\therefore \qquad \mathrm{Lt}_{n \to \infty} S_{sc} = \frac{S}{X_g X_r}(X_g + X_r) = S\left(\frac{1}{X_g} + \frac{1}{X_r}\right) \qquad ...(16.13.4)$$

Equation (16.13.4) gives the short circuit MVA with very large number of busbar sections.

It shows that the short circuit MVA becomes independent of the number of busbar sections if *n* is very large.

Thus, in a tie-bar system, with large value of busbar sections, extra generators may be added to the system without addition of extra circuit beakers or increasing the existing reactance. This is one of the advantages of the tie-bar system.

Example 16.3 The 33 kV busbars of a station are in two sections *A* and *B* separated by a reactor. *A* is fed from four 10 MVA generators each having 0.20 per unit reactance and *B* is fed from the grid through a 50 MVA transformer of 0.10 per unit reactance. The circuit breakers have each a rupturing capacity of 500 MVA. Find the reactance of the reactor to prevent the circuit breakers being overloaded, if a symmetrical short circuit occurs on an outgoing feeder connected to it.

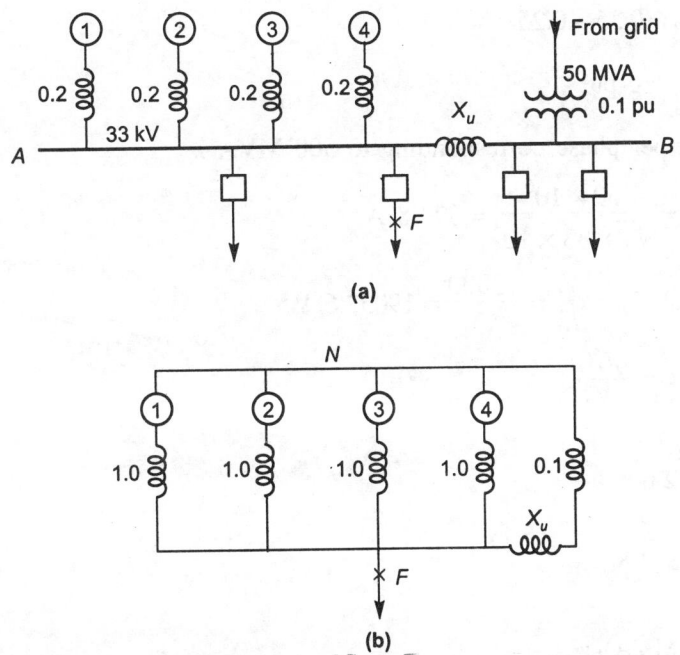

Fig. 16.12. (a) Single-line diagram; (b) Equivalent single phase circuit after fault at *F*.

Solution

The system layout is shown in Fig. 16.12 (a). Let 50 MVA be taken as the base MVA.

Per unit reactance of each generator on 50 MVA base $= \dfrac{50 \times 0.20}{10} = 1.0$ pu

Let the per unit reactance of the reactor be X_u on 50 MVA base. For the symmetrical fault at point *F*, the equivalent single-phase circuit is shown in Fig. 16.12 (b).

The four generators are connected in parallel each having a per unit reactance of 1.0.

The combined per unit reactance of the generators $= \dfrac{1.0}{4} = 0.25$

Combined per unit reactance of transformer and reactor in series $= X_u + 0.1$.

Thevenin equivalent per unit reactance at the fault point F is

$$X_T = \frac{0.25\,(X_u + 0.1)}{0.25 + X_u + 0.1} = \frac{0.25\,(X_u + 0.1)}{X_u + 0.35}$$

Short circuit MVA fed into the fault at F

$$S_{sc} = \frac{S_b}{X_T} = \frac{50\,(X_u + 0.35)}{0.25\,(X_u + 0.1)}$$

If the short circuit MVA is not to exceed 500 MVA, then

$$\frac{50\,(X_u + 0.35)}{0.25\,(X_u + 0.1)} = 500 \;; \qquad X_u + 0.35 = \frac{500 \times 0.25}{50}\,(X_u + 0.1)$$

$$2.5\,X_u - X_u = 0.35 - 0.25$$

$$X_u = \frac{0.10}{1.5} = \frac{1}{15} \text{ pu}$$

Full-load current per phase corresponding to 500 MVA is

$$I_{fl} = \frac{S_b}{\sqrt{3}\,V_l} = \frac{50 \times 10^6}{\sqrt{3} \times 33 \times 10^3} = 874.8 \text{ A}$$

Voltage to neutral, $V_n = \dfrac{V_l}{\sqrt{3}} = \dfrac{33000}{\sqrt{3}} = 19052.6 \text{ V}$

Per unit reactance $= \dfrac{I_{fl}\,X_\Omega}{V_n}$

$$\frac{1}{15} = \frac{874.8\,X_\Omega}{19052.6}$$

$$X_\Omega = \frac{19052.6}{15 \times 874.8} = 1.452 \;\Omega$$

Alternative method

$$Z_{pu} = Z_\Omega \frac{[(MVA)_b]_{3\,\varphi}}{(kV_{lb})^2}$$

$$\frac{1}{15} = X_\Omega \frac{50}{(33)^2}$$

$$X_\Omega = \frac{(33)^2}{15 \times 50} = 1.452 \;\Omega$$

Example 16.4 A small generating station has two alternators A and B of 3 MVA and 4.5 MVA and per unit reactances 0.07 and 0.08. The circuit breakers are rated at 150 MVA. It is intended to extend the system by a supply from the grid through a transformer of 7.5 MVA with 0.075 per unit reactance. The two busbar sections working on 3.3 kV are connected by a reactor. Find the reactance of the reactor necessary to protect the switchgear if a short circuit occurs on an outgoing feeder connected to the alternator busbars.

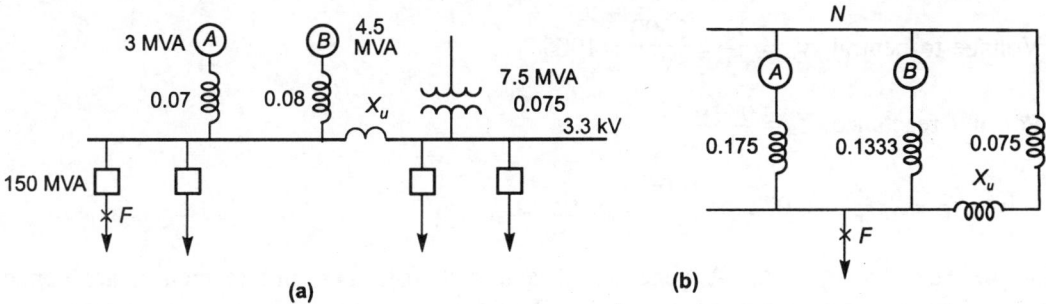

Fig. 16.13. (a) Single-line diagram. (b) Equivalent single-phase circuit after fault at F.

Solution

The system arrangement is shown in Fig. 16.13 (a).

Let 7.5 MVA be taken as the base MVA.

Per unit reactance of generator A on 7.5 MVA base $= \dfrac{7.5}{3} \times 0.07 = 0.175$ pu

Per unit reactance of generator B on 7.5 MVA base $= \dfrac{7.5}{4.5} \times 0.08 = 0.1333$ pu

Per unit reactance of transformer on 7.5 MVA base $= 0.075$ pu. Let the per unit reactance of the reactor be X_u on 7.5 MVA base. For a fault at point F the equivalent single-phase circuit is shown in Fig. 16.13 (b).

The combined per unit reactance of generators A and B in parallel

$$= \frac{0.175 \times 0.1333}{0.175 + 0.1333} = 0.07566 \text{ pu}$$

Thevenin equivalent reactance at the fault point F

$$X_T = \frac{0.07556 \,(X_u + 0.075)}{0.07566 + X_u + 0.075} = \frac{0.07566 \,(X_u + 0.075)}{X_u + 0.15066}$$

Short circuit MVA fed into the fault, $S_{sc} = \dfrac{S_b}{X_T} = \dfrac{7.5 \,(X_u + 0.15066)}{0.07566 \,(X_u + 0.075)}$

If the short circuit is not to exceed 150 MVA, then

$$\frac{7.5 \,(X_u + 0.15066)}{0.07566 \,(X_u + 0.075)} = 150$$

$$X_u + 0.15066 = 1.5132 \, X_u + 0.1135$$

$$0.5132 \, X_u = 0.03716$$

$$X_u = \frac{0.03716}{0.5132} = 0.0724 \text{ pu}$$

Full load current corresponding to 7.5 MVA

$$I_{fl} = \frac{S_b}{\sqrt{3} \, V_l} = \frac{7.5 \times 10^6}{\sqrt{3} \times 3.3 \times 10^3} = 1312 \text{ A}$$

Voltage to neutral, $V_n = \dfrac{V_l}{\sqrt{3}} = \dfrac{3300}{\sqrt{3}} = 1905$ V

Per unit reactance, $X_u = \dfrac{I_{fl} X_\Omega}{V_n}$

$$X_\Omega = X_u \dfrac{V_n}{I_{fl}} = 0.0724 \times \dfrac{1905}{1312} = 0.105 \ \Omega.$$

Example 16.5 Three 20 MVA generators, each with 0.15 per unit reactance, are connected through three reactors to a common busbar. Each feeder connected to the generator side of a reactor has 200 MVA circuit breaker. Determine the minimum value of reactor reactance if the busbar voltage is 11 kV. Find also the reactor reactance if the three reactors are ring connected.

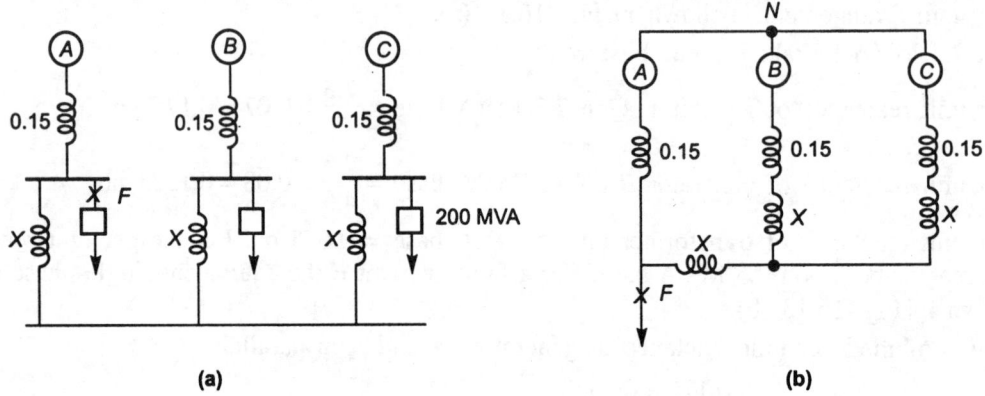

(a)   **(b)**

Fig. 16.14. (a) Single-line diagram of tie-bus system; (b) Equivalent circuit after the fault at F.

Solution

(a) Tie-bus System

Fig. 16.14 (a) shows the tie-bus system. Consider a symmetrical three-phase fault at point F. Let X be the per unit reactance of each reactor on a 20 MVA base. The equivalent circuit after the fault at F is shown in Fig. 16.4 (b).

$\qquad X_T =$ Thevenin equivalent reactance at fault point F

$\qquad\quad =$ total reactance between the zero potential bus and point F.

$\therefore \qquad X_T = [\{(0.15 + X) \ || \ (0.15 + X)\} + X] \ || \ (0.15)$

$$= \left\{ \dfrac{1}{2}(0.15 + X) + X \right\} \ || \ (0.15) = (0.075 + 1.5 \ X) \ || \ (0.15)$$

$$= \dfrac{(0.075 + 1.5 \ X) \ (0.15)}{0.075 + 1.5 \ X + 0.15} = \dfrac{(0.05 + X) \ (0.15)}{0.15 + X} \quad \text{pu}$$

Short-circuit MVA fed into the fault, $S_{sc} = \dfrac{\text{base MVA}}{\text{equivalent per unit reactance}} = \dfrac{20 \ (X + 0.15)}{0.15 \ (X + 0.05)}$

This should not exceed 200 MVA.

$$\therefore \quad \frac{20\,(X+0.15)}{0.15\,(X+0.05)} = 200$$

$$X + 0.15 = 1.5\,(X + 0.05) \; ; \qquad X = 0.15 \text{ pu}$$

Full-load current, $I_{fl} = \dfrac{20 \times 10^6}{\sqrt{3} \times 11000} = 1049.7$ A

Phase voltage, $V_p = \dfrac{11000}{\sqrt{3}} = 6350.8$ V

Ohmic reactance of each reactor, $X_\Omega = X_{pu} \times \dfrac{V_p}{I_{fl}} = 0.15 \times \dfrac{6350.8}{1049.7} = 0.9075\ \Omega$

(b) Ring System

Fig. 16.15 (a) shows the ring system of reactors. In order to determine Thevenin equivalent reactance at F we use the equivalent circuit after the fault at F as shown in Fig. 16.15 (b). The delta-connected system between NEH is replaced by the star-connected system as shown in Fig. 16.15 (c) where

$$X_1 = \frac{0.15 \times 0.15}{0.15 + 0.15 + X} = \frac{0.0225}{0.3 + X}$$

$$X_2 = X_3 = \frac{0.15\,X}{0.15 + 0.15 + X} = \frac{0.15\,X}{0.3 + X}$$

Further simplification of network is shown in Figs. 16.15 (d) to (f). The Thevenin equivalent reactance after fault at F is given from Fig. 16.15 (f) as

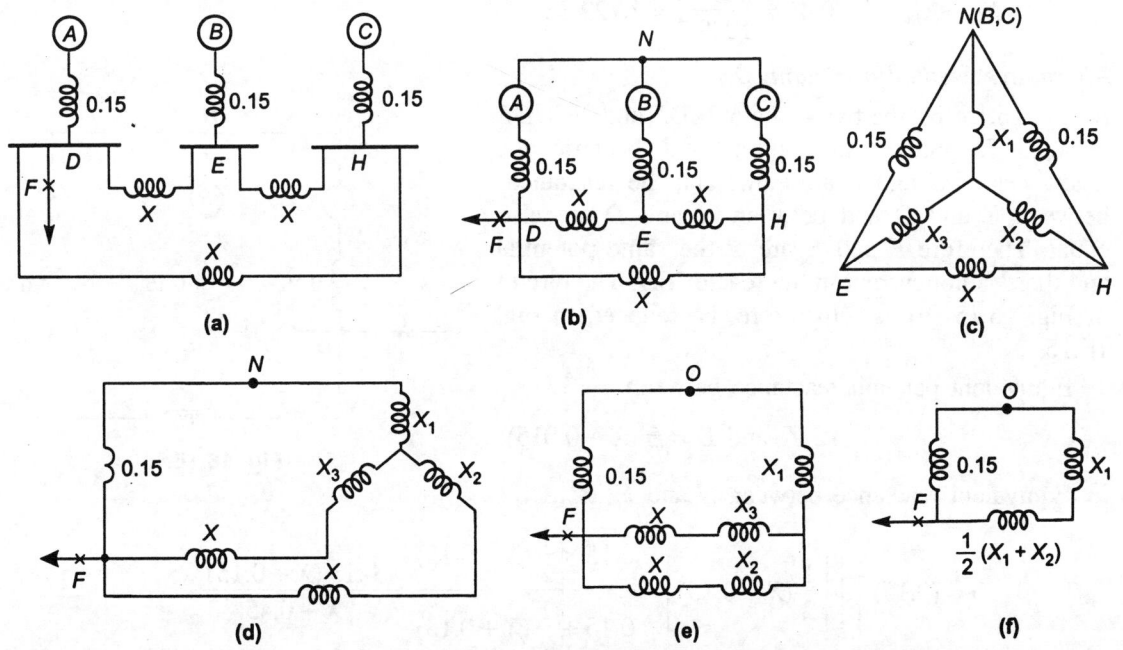

Fig. 16.15.

$$X_T = 0.15 \left|\left| \left[X_1 + \frac{1}{2}(X + X_2) \right] \right. = 0.15 \left|\left| \left[\frac{0.0225}{0.3 + X} + \frac{1}{2}\left(X + \frac{0.15 X}{0.3 + X} \right) \right] \right.\right.\right.$$

$$= 0.15 \left|\left| \left[\frac{X}{2} + \frac{0.0225 + 0.075 X}{0.3 + X} \right] = 0.15 \left|\left| \left[\frac{X}{2} + \frac{0.075 (0.3 + X)}{0.3 + X} \right] \right.\right.\right.\right.$$

$$= 0.15 \left|\left| \left(\frac{X}{2} + 0.075 \right) = \frac{0.15 \left(\dfrac{X}{2} + 0.075 \right)}{0.15 + \dfrac{X}{2} + 0.075} \right.\right.$$

$$= \frac{0.15 (X + 0.15)}{X + 0.45}$$

Short-circuit MVA fed into the fault F

$$S_{sc} = \frac{S_b}{X_T} = \frac{20 (X + 0.45)}{0.15 (X + 0.15)}$$

Since the short-circuit MVA is not to exceed 200 MVA, we have

$$\frac{20 (X + 0.45)}{0.15 (X + 0.15)} = 200$$

$$X + 0.45 = 1.5 (X + 0.15) ; \qquad X = 0.45 \text{ pu}$$

Reactance of each reactor

$$X_\Omega = X_{pu} \frac{V_p}{I_{fl}} = 0.45 \times \frac{6350.8}{1049.7} = 2.722 \ \Omega$$

Alternative Method of Finding X_T

In Fig. 16.15 (b), the two paths *NBEDF* and *NCHDF* between N and F are symmetrical because the reactances of B and C are equal and the reactances between E and D and between H and D are also equal. Therefore E and H are at the same potential and there is not current in the reactor *EH*. The circuit of Fig. 16.15 (b) can, therefore, be reduced to Fig. 16.16.

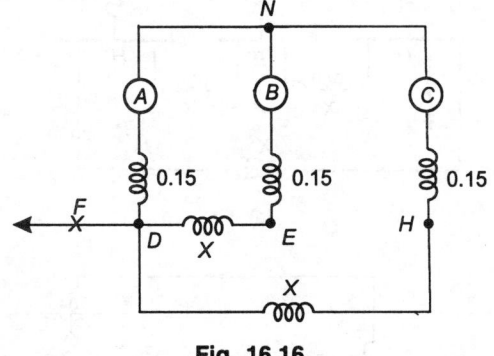

Fig. 16.16.

Equivalent per unit reactance between

$$N \text{ and } D = \frac{1}{2} (X + 0.015)$$

Equivalent reactance between N and F is

$$X_T = (0.15) \left|\left| \left[\frac{1}{2} (X + 0.15) \right] = \frac{0.15 \times \dfrac{1}{2} (X + 0.15)}{0.15 + \dfrac{1}{2} (X + 0.15)} = \frac{0.15 (X + 0.15)}{X + 0.45} \right.\right.$$

as obtained before.

Example 16.6 Three 11 kV solidly grounded generators are connected to three reactors in a tie-bus arrangement as shown in Fig. 16.17 (a). The reactances of the generators and reactors are 0.2 and 0.1 pu, respectively based on 25 MVA base. If there is a symmetrical three-phase fault at point F, determine (a) short-circuit MVA and (b) fault current distribution in the system.

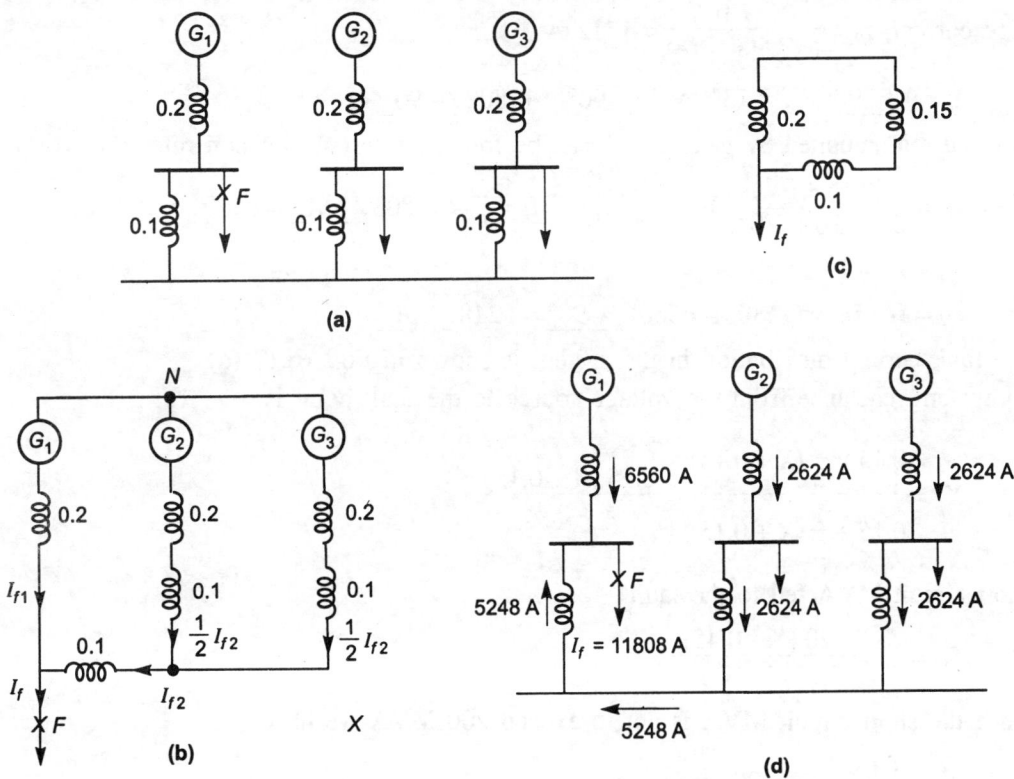

Fig. 16.17. (a) Single-line diagram of the tie-bus system. (b) Equivalent circuit after fault at F. (c) Application of current-division rule. (d) Fault-current distribution in the system.

Solution

Fig. 16.17 (b) shows the equivalent circuit after the fault at point F.

X_T = Thevenin equivalent reactance at fault point F

= total reactance between the zero potential bus N and point F

$$= [\{(0.2+0.1) \parallel (0.2+0.1)\} + 0.1] \parallel (0.2) = \left(\frac{0.3}{2}+0.1\right) \parallel (0.2) = 0.25 \parallel 0.2$$

$$= \frac{0.25 \times 0.2}{0.25+0.2} = \frac{1}{9} \text{ pu}$$

(a) Short-circuit MVA

$$S_{sc} = \frac{\text{base MVA}}{X_T} = \frac{25}{1/9} = 225 \text{ MVA}$$

Fault current at point F is

$$I_f = \frac{V_f}{Z_T} = \frac{1 / 0°}{j\,(1/9)} = 9 / -90° \text{ pu}$$

Base current, $I_b = \dfrac{25 \times 10^6}{\sqrt{3} \times 11 \times 10^3} = 1312 \text{ A}$

$I_f = 9 / -90°$ pu $= 1312 \times 9 / -90° = 11808 / -90°$ A

Fault current supplied by generator 1 can be found by current-division rule [Fig. 16.17 (c)].

$$I_{f1} = \frac{(0.15 + 0.1)}{0.15 + 0.1 + 0.2} I_f = \frac{0.25}{0.45} I_f = \frac{5}{9} I_f = \frac{5}{9} \times 11808 / -90° = 6560 / -90° \text{ A}$$

$I_{f1} + I_{f2} = I_f$

$I_{f2} = I_f - I_{f1} = (11808 - 6560) / -90° = 5248 / -90°$ A

The fault-current distribution in the system is shown in Fig. 16.17 (d).

Equivalent reactance from the voltage source to the fault point is

$$X_e = \frac{0.15 \times \dfrac{1}{2}\,(X + 0.15)}{0.15 + \dfrac{1}{2}\,(X + 0.15)} = \frac{0.15\,(X + 0.15)}{X + 0.45}$$

Short circuit MVA fed to the fault

$$S_{sc} = \frac{S_b}{X_e} = \frac{20\,(X + 0.45)}{0.15\,(X + 0.15)}$$

Since the short circuit MVA is not to exceed 200 MVA we have

$$\frac{20\,(X + 0.45)}{0.15\,(X + 0.15)} = 200$$

$X + 0.45 = 1.5\,(X + 0.15)$

$X = 0.45$ pu

Reactance of each reactor

$$X_\Omega = X_u \frac{V_n}{I_{f1}} = 0.45 \times \frac{6350}{1050} = 2.72 \ \Omega.$$

EXERCISES

1. (a) Enumerate the positions in which current limiting reactors may be connected.
   (b) Explain the object of sectionalizing the busbars in a large power section. (c) What are the advantages of the tie-bar system over the ring system?

2. A three-phase transmission line, operating 33 kV and having a resistance and reactance of 6 Ω and 21 Ω respectively, is connected to the generating station busbars through a 3000 kVA step-up transformer which has a reactance of 5 per cent. Connected to the busbars are two similar 10000 kVA alternators, each with a reactance of 10 per cent. Calculate the kVA at a short circuit fault between phases occurring (a) at the high voltage terminals of the transformer, (b) at the load end of the transmission line.

3. Explain what is meant by the percentage rating of a current limiting reactor.

   A transformer rated at 30000 kVA and having a short circuit reactance of 5 per cent is connected to the busbars of a transformer station which is supplied through two 33 kV feeder cables each having an impedance of $(1 + j\,2)$ ohms. One of the feeder is connected to a generating station with plant rated at 60000 kVA connected to its busbars having a short circuit reactance of 10 per cent and the other feeder to a station with 80000 kVA of generating plant with a reactance of 15 per cent. Calculate the kVA supplied to the fault in the event of a short circuit occurring between the secondary terminals of the transformer.

4. A generating station has three busbar sections which are connected by two reactors each of 20 per cent reactance and rated at 200 MVA. Generators of total capacity 100 MVA, total reactance of 20 per cent and of negligible resistance are connected to each busbar. Calculate the MVA fed to a fault, under short-circuit conditions, on the central busbar station.

5. Three 60 MVA ac generators, each having 15 per cent reactance are connected via three 36 MVA reactors each of 10 per cent reactance to a common busbar. The feeders are each connected to the junction of each alternator and its reactor. What must be the minimum rating of each feeder circuit breaker?

6. Describe the *star* or *tie-bar* method of interconnecting the busbar sections in a generating station and compare it with other busbar arrangements.

   A station contains 4 busbar sections to each of which is connected a generating unit of 30 MVA having 12 per cent leakage reactance, the busbar reactor having a reactance of 10 per cent. Calculate the maximum MVA fed into a fault on any busbar section and also the maximum MVA if the number of similar busbar sections were increased to infinity. Deduce any formula used.

7. Each of the three generators in a central station has a short circuit reactance of 20 per cent based upon the respective ratings of 75 MVA, 90 MVA. Each machine is connected to its own sectional busbar and each busbar is connected to a tie-bar through a reactor of 10 per cent reactance based upon the rating of the alternator connected to it. Calculate the MVA fed into a short circuit occurring between the bars of the section to which the 110 MVA machine is connected.

8. The main busbars in a generating station are divided into three sections, each section being connected to a tie-bar by a similar reactor. One 20000 kVA, three-phase 50 Hz, 11000 V generator, having a short circuit reactance of 15 per cent, is connected to each section busbar. When a short circuit takes place between the phases of one of the section busbars, the voltage on the remaining sections falls to 60 per cent of the normal value. Calculate the reactance in ohms of each reactor.

ANSWERS

2. (a) 45200 kVA (b) 23940 kVA
4. 1160 MVA
6. 422 MVA, 550 MVA
8. 0.454 Ω

3. 286000 kVA
5. 583 MVA
7. 917 MVA

Symmetrical Components

17.1 INTRODUCTION

A three-phase system is said to be *symmetrical* when the system viewed from any phase is similar. Thus, in a three-phase symmetrical system the self impedances of all the three phases are equal and the mutual impedances, if any, between the three phases are the same. Three-phase voltages (or currents) are said to be *balanced* if the magnitudes of the three voltages (or currents) are equal and they are separated from each other by 120° (electrical) in phase. Because of the *symmetry* of the system and the *balanced* nature of voltages (or currents), the analysis is made on a *single-phase basis*. This is due to the fact that a symmetrical system, because of the balanced voltages, gives rise to balanced currents.

When the system is unbalanced the voltages, currents and the phase impedances are in general unequal. Such a system can be solved by a symmetrical per phase technique known as the *method of symmetrical components*. This method was proposed by Fortescue in 1918 and is often called the *three-component method*. It provides a means of extending per phase analysis to systems with unbalanced loads, or with unbalanced termination like short circuit or fault. It is assumed that the system is balanced ($Z_a = Z_b = Z_c$) with unbalanced termination.

17.2 FORTESCUE'S THEOREM

Fortescue's theorem is stated as follows :

An unbalanced set of n phasors may be resolved into $(n - 1)$ balanced n-phase systems of different phase sequence and one zero-phase sequence system. According to Fortescue a *zero-phase sequence system* is one in which all phasors are of equal magnitude and angle or they are all identical. Fortescue's theorem can be applied to a more practical three-phase system.

17.3 PHASE SEQUENCE

Phase-sequence of the phasors is the order in which they pass through a positive maximum.

Thus, a phase sequence *a b c* implies that the maxima occur in the order *a*, *b*, *c*. If *a b c* is taken as the positive phase sequence then *a c b* represents the negative phase sequence. It should be noted that for *both* positive and negative phase sequences, the direction of rotation of phasors is taken to be anticlockwise. The positive sequence voltage is defined to have the phase sequence of the original system. We arbitrarily label this sequence *a b c*. The negative sequence voltage then has the phase sequence *a c b*.

17.4 α-OPERATOR

The phasor α is an operator which when operates upon a phasor rotates it by +120° without changing the magnitude of the phasor upon which it operates. It has a magnitude of unity and angle 120°.

$$\alpha \triangleq 1\underline{/120°} = 1 \cdot e^{j\,2\pi/3} = \cos 120° + j \sin 120° = -0.5 + j\,0.866$$

$$\alpha^2 = 1\underline{/240°} = e^{j\,4\pi/3} = e^{-j\,2\pi/3} = -0.5 - j\,0.866$$

$$\alpha^3 = 1\underline{/0°} = e^{j\,2\pi} = 1$$

$$\alpha^4 = \alpha, \qquad \alpha^5 = \alpha^2$$

$$1 + \alpha + \alpha^2 = 0$$

$$\alpha^* = \alpha^2, \qquad (\alpha^2)^* = \alpha$$

Fig. 17.1 shows phasors representing various powers of α.

It is to be noted that the operator $-\alpha$ does not produce rotation through −120°. This can be seen from Fig. 17.1.

$$-\alpha = 1\underline{/-60°}.$$

The symbols *a*, *h*, or λ are sometimes used instead of α.

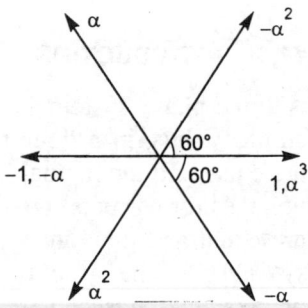

Fig. 17.1. Phasor diagram of the various powers of α.

17.5 SYMMETRICAL COMPONENTS OF AN UNBALANCED THREE-PHASE SYSTEM

According to Fortescue's theorem, three unsymmetrical and unbalanced phasors (voltages or currents) of a three-phase system can be resolved into following three component sets of balanced phasors which possess certain symmetry.

1. A set of three phasors equal in magnitude, displaced from each other by 120° in phase, and having the *same* phase sequence as the original unbalanced phasors. This set of balanced phasors is called *positive phase-sequence components*.
2. A set of three phasors equal in magnitude, displaced from each other by 120° in phase, and having the phase sequence *opposite* to that of the original phasors. This set of balanced phasors is called *negative phase sequence components*.
3. A set of three phasors equal in magnitude with zero phase displacement from each other. This set is called *zero phase sequence components*. The components of this set are all identical.

These three sets of balanced phasors are called the *symmetrical components* of the original unbalanced phasors.

Suppose that the three phasors are represented by *a*, *b*, and *c* such that the phase sequence is *a b c*. Thus, the phase sequence of the positive phase sequence components is *a b c* and the phase sequence of the negative phase sequence components is *a c b*. It is assumed that the subscripts 0, 1, 2 refer to zero sequence, positive sequence, and negative sequence respectively.

If V_a, V_b and V_c, represent an unbalanced set of voltage phasors, the three balanced sets are written as

$$(V_{a0}, V_{b0}, V_{c0}) \qquad \text{zero sequence set}$$

$$(V_{a1}, V_{b1}, V_{c1}) \qquad \text{positive sequence set}$$

$$(V_{a2}, V_{b2}, V_{c2}) \qquad \text{negative sequence set.}$$

Fig. 17.2 shows (V_a, V_b, V_c) and the symmetrical components of these unbalanced voltage phasors.

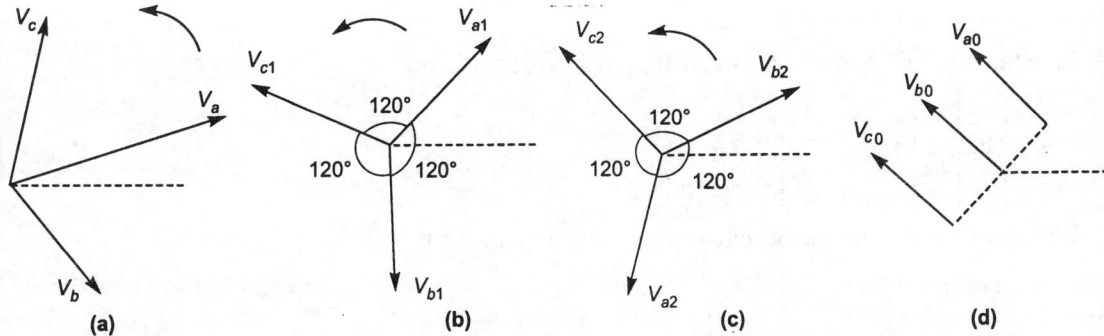

Fig. 17.2. (a) Original system of three unbalanced phasors; (b) Positive-sequence components; (c) Negative-sequence components; (d) Zero-sequence components.

Since each of the original unbalanced phasors is the sum of its components we may write

$$V_a = V_{a0} + V_{a1} + V_{a2} \qquad \qquad ...(17.5.1)$$

$$V_b = V_{b0} + V_{b1} + V_{b2} \qquad \qquad ...(17.5.2)$$

$$V_c = V_{c0} + V_{c1} + V_{c2} \qquad \qquad ...(17.5.3)$$

It is to be noted that the three-phase balanced system is a special case of a general three-phase system in which zero and negative sequence components are zero.

17.6 COMPONENT SYNTHESIS

The determination of original phasors in terms of sequence components is called *component synthesis*. The calculations of symmetrical component phasors are always made in terms of phase *a* as the symmetrical phase. Use is made of the operator α to express each component of V_b and V_c in terms of the components of V_a. Thus, for a balanced position phase sequence *a b c* [Fig. 17.2 (a)] we may write the following relations :

$$V_{b1} = \alpha^2 V_{a1} \qquad \qquad ...(17.6.1)$$

$$V_{c1} = \alpha V_{a1} \qquad \qquad ...(17.6.2)$$

For negative phase sequence [Fig. 17.2 (c)]

$$V_{b2} = \alpha \, V_{a2} \qquad \qquad \ldots(17.6.3)$$

$$V_{c2} = \alpha^2 \, V_{a2} \qquad \qquad \ldots(17.6.4)$$

For zero phase sequence [Fig. 17.2 (d)]

$$V_{b0} = V_{a0} \qquad \qquad \ldots(17.6.5)$$

$$V_{c0} = V_{a0} \qquad \qquad \ldots(17.6.6)$$

On substituting these values in equations (17.5.1) to (17.5.3) V_a, V_b and V_c can be written in terms of phase a components as follows :

$$V_a = V_{a0} + V_{a1} + V_{a2} \qquad \qquad \ldots(17.6.7)$$

$$V_b = V_{a0} + \alpha^2 \, V_{a1} + \alpha \, V_{a2} \qquad \qquad \ldots(17.6.8)$$

$$V_c = V_{a0} + \alpha \, V_{a1} + \alpha^2 \, V_{a2} \qquad \qquad \ldots(17.6.9)$$

Equations (17.6.7) to (17.6.9) can be put in matrix form

$$\begin{bmatrix} V_a \\ V_b \\ V_c \end{bmatrix} = \begin{bmatrix} 1 & 1 & 1 \\ 1 & \alpha^2 & \alpha \\ 1 & \alpha & \alpha^2 \end{bmatrix} \begin{bmatrix} V_{a0} \\ V_{a1} \\ V_{a2} \end{bmatrix} \qquad \qquad \ldots(17.6.10)$$

Equations (17.6.10) can be put in a more compact form

$$\mathbf{V}_{abc} = \mathbf{A} \mathbf{V}_{012} \qquad \qquad \ldots(17.6.11)$$

where
$$\mathbf{V}_{abc} = \begin{bmatrix} V_a \\ V_b \\ V_c \end{bmatrix} \triangleq \text{ phase voltage vector} \qquad \qquad \ldots(17.6.12)$$

$$\mathbf{V}_{012} = \begin{bmatrix} V_{a0} \\ V_{a1} \\ V_{a2} \end{bmatrix} \triangleq \text{ symmetrical sequence voltage vector} \qquad \qquad \ldots(17.6.13)$$

$$\mathbf{A} = \begin{bmatrix} 1 & 1 & 1 \\ 1 & \alpha^2 & \alpha \\ 1 & \alpha & \alpha^2 \end{bmatrix} \qquad \qquad \ldots(17.6.14)$$

Equation (17.6.11) is referred to as the *synthesis equation*. The voltage vectors $\mathbf{V}_{abc}$ and $\mathbf{V}_{012}$ represent the actual phase voltages and the symmetrical voltage components respectively. $\mathbf{A}$ is called the *symmetrical component transform matrix* which transforms the phase voltages $\mathbf{V}_{abc}$ into component voltages $\mathbf{V}_{012}$.

17.7 COMPONENT ANALYSIS

The determination of sequence component in terms of original phasors is called *component analysis*. Premultiplying both the sides of Eq. (17.6.11) by $\mathbf{A}^{-1}$.

$$\mathbf{A}^{-1} \mathbf{V}_{abc} = \mathbf{A}^{-1} \mathbf{A} \mathbf{V}_{012}$$

$$\mathbf{V}_{012} = \mathbf{A}^{-1} \mathbf{V}_{abc} = C \, V_{abc} \qquad \qquad \ldots(17.7.1)$$

where $\mathbf{C} = \mathbf{A}^{-1}$ $\qquad \qquad \ldots(17.7.2)$

But $\qquad A^{-1} = \dfrac{1}{3}\begin{bmatrix} 1 & 1 & 1 \\ 1 & \alpha & \alpha^2 \\ 1 & \alpha^2 & \alpha \end{bmatrix}$ ...(17.7.3)

Equation (17.7.1) is called the *analysis equation*. It gives the symmetrical components of voltages in terms of original voltage phasors.

Equation (17.7.1) may be written in the form

$$\begin{bmatrix} V_{a0} \\ V_{a1} \\ V_{a2} \end{bmatrix} = \frac{1}{3}\begin{bmatrix} 1 & 1 & 1 \\ 1 & \alpha & \alpha^2 \\ 1 & \alpha^2 & \alpha \end{bmatrix}\begin{bmatrix} V_a \\ V_b \\ V_c \end{bmatrix} \qquad \text{...(17.7.4)}$$

These relations are of basic importance. In terms of separate equations we may write

$$V_{a0} = \frac{1}{3}\left(V_a + V_b + V_c \right) \qquad \text{...(17.7.5)}$$

$$V_{a1} = \frac{1}{3}\left(V_a + \alpha V_b + \alpha^2 V_c \right) \qquad \text{...(17.7.6)}$$

$$V_{a2} = \frac{1}{3}\left(V_a + \alpha^2 V_b + \alpha V_c \right) \qquad \text{...(17.7.7)}$$

Usually the symmetrical components of phase *a* are found. The components of other phases are found from these results. Equations (17.6.1) to (17.6.6) can be used to find V_{b0}, V_{b1}, V_{b2}, V_{c0}, V_{c1}, V_{c2}.

Equations (17.7.5) to (17.7.7) represent the symmetrical components of line-to-neutral voltage phasors. Symmetrical components of line-to-line voltages may also be found by using the same technique.

Equations (17.7.5) to (17.7.7) may also be deduced without the use of matrices. Addition of Eqs. (17.67) to (17.6.9) gives

$$V_a + V_b + V_c = 3V_{a0} + (1 + \alpha^2 + \alpha) V_{a1} + (1 + \alpha^2 + \alpha) V_{c2}$$

Since $\qquad 1 + \alpha + \alpha^2 = 0$

$$V_a + V_b + V_c = 3V_{a0}$$

$$V_{a0} = \frac{1}{3}\left(V_a + V_b + V_c \right) \qquad \text{...(17.7.5)}$$

Multiplication of Eq. (17.6.8) by α and Eq. (17.6.9) by α^2 and adding the products to Eq. (17.6.7) gives

$$V_a + \alpha V_b + \alpha^2 V_c = (1 + \alpha + \alpha^2) V_{a0} + (1 + \alpha^3 + \alpha^3) V_{a1} + (1 + \alpha^2 + \alpha^4) V_{a2}$$

Since $\qquad 1 + \alpha + \alpha^2 = 0, \qquad \alpha^3 = 1, \qquad \alpha^4 = \alpha$

$$1 + \alpha^2 + \alpha^4 = 0$$

$$V_a + \alpha V_b + \alpha^2 V_c = 3V_{a1}$$

$$V_{a1} = \frac{1}{3}\left(V_a + \alpha V_b + \alpha^2 V_c \right) \qquad \text{...(17.7.6)}$$

Multiplying Eq. (17.6.8) by α^2 and Eq. (17.6.9) by α and adding the products to Eq. (17.6.7) gives

$$V_a + \alpha^2 V_b + \alpha V_c = (1 + \alpha + \alpha^2) V_{a0} + (1 + \alpha^2 + \alpha^4) V_{a1} + (1 + \alpha^3 + \alpha^3) V_{a2} = 3V_{a2}$$

$$\therefore \quad V_{a2} = \frac{1}{3}(V_a + \alpha^2 V_b + \alpha V_c) \qquad\qquad\qquad …(17.7.7)$$

17.8 GRAPHICAL METHOD OF DETERMINING SEQUENCE COMPONENTS

For a graphical method consider the three unbalanced voltage phasors shown in Fig. 17.3 (a). Equations (17.7.5) to (17.7.7) are used to determine the sequence components.

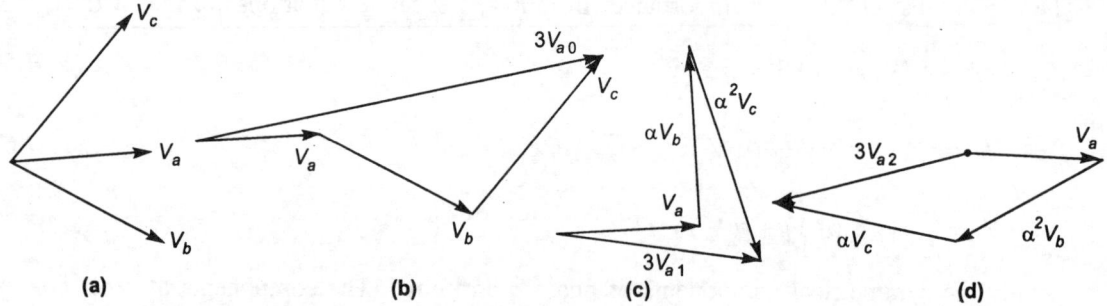

Fig. 17.3. Graphical method of determining sequence components.

Zero Sequence Component V_{a0}

Equation (17.7.5) is used to determine the zero sequence component V_{a0}. It is obtained by direct phasor addition of phasors V_a, V_b and V_c and then dividing the phasor which forms the closing side of the resulting polygon by 3 [Fig. 17.3 (b)].

Positive Sequence Component V_{a1}

Equation (17.7.6) is used to determine the positive sequence component V_{a1}. It is found by adding V_a to αV_b, i.e. V_b rotated through 120°. To the sum of V_a and αV_b is added $\alpha^2 V_c$, i.e., V_c rotated through 240°. The component V_{a1} is thus obtained by dividing the phasor which forms the closing side of the resulting polygon by 3 [Fig. 17.3 (c)].

Negative Sequence Component V_{a2}

Equation (17.7.7) is used to determine the negative sequence component V_{a2}. It is determined by adding V_a to $\alpha^2 V_b$, i.e., V_b rotated through 240°. To their sum is added αV_c, i.e., V_c rotated through 120°. The component V_{a2} is obtained by taking one-third of the sum [Fig. 17.3 (d)].

17.9 SYMMETRICAL COMPONENTS OF CURRENT PHASORS

The method of symmetrical components is applicable to any unbalanced set of n-phase quantities. We have found the symmetrical components of unbalanced voltages. Identical relations can be written for unbalanced current phasors. Thus, for a system of unbalanced, three-phase currents I_a, I_b, and I_c we can determine nine symmetrical components I_{a0}, I_{b0}, I_{c0}, I_{a1}, I_{b1}, I_{c1}, I_{a2}, I_{b2}, and I_{c2}.

We have

$$\mathbf{I}_{abc} = \mathbf{A}\mathbf{I}_{012}$$

$$\mathbf{I}_{012} = \mathbf{A}^{-1}\,\mathbf{I}_{abc}$$

where $\mathbf{I}_{abc} = \begin{bmatrix} I_a \\ I_b \\ I_c \end{bmatrix} \triangleq$ phase current vector

and $\mathbf{I}_{012} = \begin{bmatrix} I_{a0} \\ I_{a1} \\ I_{a2} \end{bmatrix} \triangleq$ symmetrical sequence current vector

In the form of separate equations

$$I_a = I_{a0} + I_{a1} + I_{a2}$$

$$I_b = I_{a0} + \alpha^2 I_{a1} + \alpha I_{a2}$$

$$I_c = I_{a0} + \alpha I_{a1} + \alpha^2 I_{a2}$$

$$I_{a0} = \frac{1}{3}(I_a + I_b + I_c)$$

$$I_{a1} = \frac{1}{3}(I_a + \alpha I_b + \alpha^2 I_c)$$

$$I_{a2} = \frac{1}{3}(I_a + \alpha^2 I_b + \alpha I_c)$$

17.10 ZERO SEQUENCE COMPONENTS OF VOLTAGE

Consider the case of a balanced system. In such a system

$$V_b = \alpha^2 V_a, \ V_c = \alpha V_a$$

$$V_{a0} = \frac{1}{3}(V_a + V_b + V_c) = \frac{1}{3}(V_a + \alpha^2 V_a + \alpha V_a) = \frac{V_a}{3}(1 + \alpha^2 + \alpha) = 0$$

$$V_{a1} = \frac{1}{3}(V_a + \alpha V_b + \alpha^2 V_c) = \frac{1}{3}(V_a + \alpha^3 V_a + \alpha^3 V_a) = V_a$$

$$V_{a2} = \frac{1}{3}(V_a + \alpha^2 V_b + \alpha V_c) = \frac{1}{3}(V_a + \alpha^4 V_a + \alpha^2 V_a) = \frac{V_a}{3}(1 + \alpha + \alpha^2) = 0.$$

Thus, the zero and negative sequence components of voltage in a balanced three-phase system are zero, and only positive sequence component is present. It is equal to the balanced phasor.

Also, in a three-phase system where the three phasors do not constitute a balanced system, yet if their resultant is zero, i.e., if the phasor sum $(V_a + V_b + V_c)$ is zero the zero sequence component will disappear since

$$V_{a0} = \frac{1}{3}(V_a + V_b + V_c) = 0$$

It follows that in a delta-connected system there are no zero-sequence components of voltage.

17.11 GENERAL NATURE OF ZERO-SEQUENCE CURRENTS

The zero-sequence currents are given by

$$I_{a0} = I_{b0} = I_{c0} = \frac{1}{3}(I_a + I_b + I_c)$$

(a) Delta-Connected Winding

Fig. 17.4 shows a delta-connected winding. The zero-sequence currents of phases a, b, and c are equal in magnitude and in phase with one another. Therefore, the zero-sequence currents can circulate in the phase windings of the delta connection. Hence,

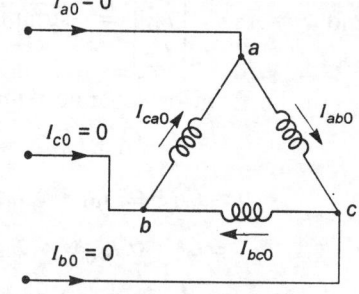

$$I_{ab0} = I_{bc0} = I_{ca0} = I_0 \text{ (say)}$$

Such currents are produced due to the existence of the zero-sequence voltages.

By KCL at node a of Fig. 17.4,

$$I_{a0} + I_{ca0} = I_{ab0}$$

Fig. 17.4. Delta-connected winding and zero-sequence currents.

$$I_{a0} = I_{ab0} - I_{ca0} = I_0 - I_0 = 0$$

Similarly, by applying KCL at nodes b and c, we have

$$I_{b0} = 0, \ I_{c0} = 0$$

Hence we conclude that *no zero-sequence currents can flow in the lines connected to a delta-connected system because of the absence of the return path for these currents.*

Since a delta-connected circuit provides no path for zero-sequence currents in the line, the zero-sequence impedance as viewed from P (Fig. 17.5) towards delta is infinite. This infinite impedance is shown by an open circuit (a break) at P in the single-phase equivalent zero sequence network for a delta-connected circuit with zero-sequence impedance Z_0 in Fig. 17.5.

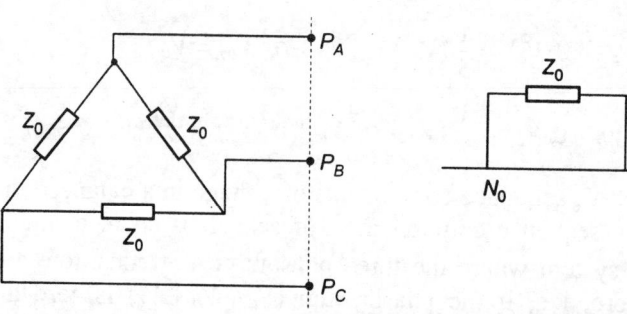

Fig. 17.5. Zero-sequence equivalent circuit of a delta-connected circuit.

However, there is a closed path for zero-sequence currents within the delta circuit. This is indicated by connecting the zero sequence impedance Z_0 to the zero-sequence neutral bus N_0 in Fig. 17.5.

(b) Star-connected winding with neutral isolated from ground

Consider a star-connected winding without a neutral return as shown in Fig. 17.6. In this case

$$I_a + I_b + I_c = I_n = 0$$

$$I_{a0} = I_{b0} = I_{c0} = \frac{1}{3}(I_a + I_b + I_c) = 0$$

This result shows that *for a three-phase three-wire system without neutral return, the zero-sequence line currents are zero.*

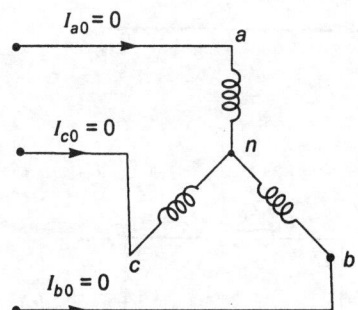

Fig. 17.6. Star-connected winding without neutral wire.

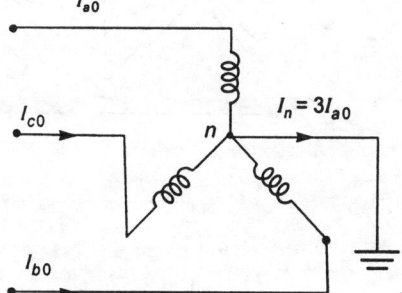

Fig. 17.7. Star-connected winding with neutral grounded.

(c) Star-connected winding with neutral return

Fig. 17.7 shows a star connected winding with neutral grounded. Here,

$$I_n = I_a + I_b + I_c$$

$$I_{a0} = I_{b0} = I_{c0} = \frac{1}{3}(I_a + I_b + I_c) = \frac{1}{3}I_n$$

Therefore, $I_n = 3I_{a0}$

Hence we conclude that *for a three-phase grounded system or three phase with neutral return, zero-sequence currents will flow both in the phase windings as well as in the lines.*

Fig. 17.8 shows the zero-sequence equivalent circuits for star-connected systems.

Example 17.1 In a three-phase four-wire system the currents in the lines a, b, and c under abnormal conditions of loading were as follows :

$$\mathbf{I}_a = 100\,\underline{/30°}\ \text{A}, \quad \mathbf{I}_b = 50\,\underline{/300°}\ \text{A}, \quad \mathbf{I}_c = 30\,\underline{/180°}\ \text{A}.$$

Calculate the zero, positive and negative phase sequence currents in line a and the return current in the neutral conductor.

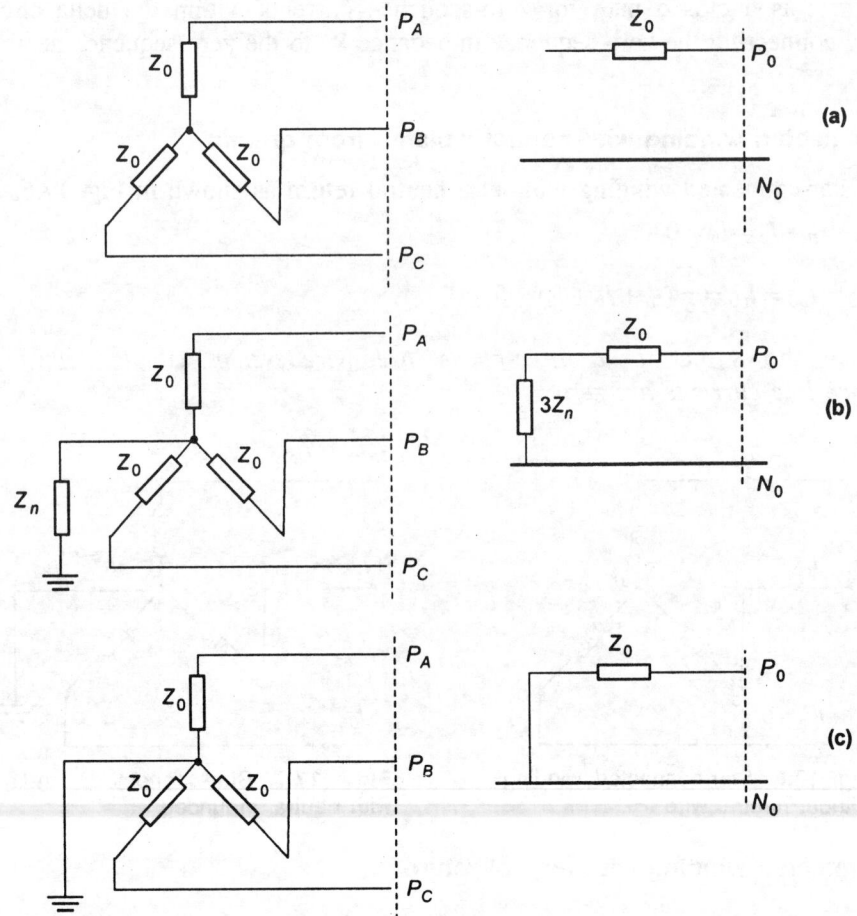

Fig. 17.8. Zero-sequence equivalent networks for symmetrically star-connected systems with (a) neutral isolated from ground (b) neutral grounded through an impedance Z_n, (c) solidly grounded neutral.

Solution

The zero-sequence component of current in line a is given by

$$\mathbf{I}_{a0} = \frac{1}{3} (\mathbf{I}_a + \mathbf{I}_b + \mathbf{I}_c) = \frac{1}{3} (100 \underline{/30°} + 50 \underline{/300°} + 30 \underline{/180°})$$

$$= \frac{1}{3} [(86.60 + j\,50) + (25 - j\,43.30) + (-30 + j\,0)] = \frac{1}{3} (81.6 + j\,6.7) = 27.29 \underline{/4.69°} \text{ A}$$

The positive-sequence component of current in line a is given by

$$\mathbf{I}_{a1} = \frac{1}{3} (\mathbf{I}_a + \alpha\, \mathbf{I}_b + \alpha^2\, \mathbf{I}_c) = \frac{1}{3} [100 \underline{/30°} + (1 \underline{/120°}) (50 \underline{/300°}) + (1 \underline{/240°}) (30 \underline{/180°})]$$

$$= \frac{1}{3} (100 \underline{/30°}) + 50 \underline{/420°} + 30 \underline{/420°})$$

$$= \frac{1}{3}[(86.6 + j\,50) + (25 + j\,43.30) + (15 + j\,25.98)]$$

$$= \frac{1}{3}(126.6 + j\,119.28) = 57.98\,\underline{/43.29°}\ A$$

The negative-sequence component of current in line a is given by

$$I_{a2} = \frac{1}{3}(I_a + \alpha^2\,I_b + \alpha\,I_c) = \frac{1}{3}[100\,\underline{/30°} + (1\,\underline{/240°})\,(50\,\underline{/300°}) + (1\,\underline{/120°})\,(30\,\underline{/180°})]$$

$$= \frac{1}{3}(100\,\underline{/30°} + 50\,\underline{/540°} + 30\,\underline{/300°}) = \frac{1}{3}[(86.6 + j\,50) + (-50 + j\,0) + (15 - j\,25.98)]$$

$$= \frac{1}{3}(51.6 + j\,24.02) = 18.97\,\underline{/24.96°}\ A$$

The return current in the neutral conductor is given by

$$I_n = I_a + I_b + I_c = 3I_{a0} = 81.87\,\underline{/4.69°}\ A$$

Example 17.2 Determine the symmetrical components of currents in a three-phase system, the original phasors of which are

$$I_a = 12 + j\,6, \quad I_b = 12 - j\,12, \quad I_c = -15 + j\,10.$$

Solution

$$I_a = 12 + j\,6, \quad I_b = 12 - j\,12 = 16.97\,\underline{/-45°}, \quad I_c = -15 + j\,10 = 18.028\,\underline{/146.31°}$$

(a) *Calculation of zero-sequence currents*

$$I_{a0} = I_{b0} = I_{c0} = \frac{1}{3}(I_a + I_b + I_c)$$

$$= \frac{1}{3}(12 + j\,6 + 12 - j\,12 - 15 + j\,10) = 3 + j\,1.333 = 3.283\,\underline{/23.96°}$$

(b) *Calculation of positive-sequence currents*

$$I_{a1} = \frac{1}{3}(I_a + \alpha\,I_b + \alpha^2\,I_c)$$

$$= \frac{1}{3}[(12 + j\,6) + (1\,\underline{/120°})\,(16.97\,\underline{/-45°}) + (1\,\underline{/240°})\,(18.028\,\underline{/146.31°})]$$

$$= \frac{1}{3}[(12 + j\,6) + 16.97\,\underline{/75°} + 18.028\,\underline{/386.31°}]$$

$$= \frac{1}{3}(12 + j\,6 + 4.392 + j\,16.392 + 16.16 + j\,7.99)$$

$$= \frac{1}{3}(32.55 + j\,30.382) = 14.842\,\underline{/43.02°}$$

$$I_{b1} = \alpha^2\,I_{a1} = (1\,\underline{/240°})\,(14.842\,\underline{/43.02°}) = 14.842\,\underline{/283.02°} = 3.343 - j\,14.46$$

$$I_{c1} = \alpha\,I_{a1} = (1\,\underline{/120°})\,(14.842\,\underline{/43.02°}) = 14.842\,\underline{/163.02°} = -14.19 + j\,4.33$$

(c) *Calculation of negative-sequence currents*

$$I_{a2} = \frac{1}{3}(I_a + \alpha^2 I_b + \alpha I_c)$$

$$= \frac{1}{3}[(12 + j\,6) + (1\,\underline{/240^\circ})\,(16.97\,\underline{/-45^\circ}) + (1\,\underline{/120^\circ})\,(18.028\,\underline{/146.31^\circ})]$$

$$= [(12 + j\,6)\,16.97\,\underline{/195^\circ} + 18.028\,1\,\underline{/266.31^\circ}]$$

$$= \frac{1}{3}(12 + j\,6 - 16.392 - j\,4.392 - 1.16 - j\,17.99)$$

$$= -1.85 - j\,5.46 = 5.766\,\underline{/-108.72^\circ}$$

$$I_{b2} = \alpha\,I_{a2} = (1\,\underline{/120^\circ})\,(5.766\,\underline{/-108.72^\circ}) = 5.766\,\underline{/11.28^\circ} = 5.65 + j\,1.127$$

$$I_{c2} = \alpha^2\,I_{a2} = (1\,\underline{/240^\circ})\,(5.766\,\underline{/-108.72^\circ}) = 5.766\,\underline{/131.28^\circ} = -3.80 + j\,4.33$$

Example 17.3 A balanced star-connected load takes 150 A from a balanced 3-phase 4-wire supply. If the fuses in two of the lines are removed, find the symmetrical components of the line currents before and after the fuses are removed.

Solution

(a) *Before fuse removal*

$$I_a = 150\,\underline{/0^\circ}\ \text{A},\ I_b = 150\,\underline{/-120^\circ}\ \text{A},\ I_c = 150\,\underline{/120^\circ}\ \text{A}$$

$$I_{a0} = \frac{1}{3}(I_a + I_b + I_c) = \frac{1}{3}(150\,\underline{/0^\circ} + 150\,\underline{/-120^\circ} + 150\,\underline{/120^\circ}) = 0$$

$$I_{a1} = \frac{1}{3}(I_a + \alpha\,I_b + \alpha^2\,I_c)$$

$$= \frac{1}{3}(150\,\underline{/0^\circ} + 1\,\underline{/120^\circ} \cdot 150\,\underline{/-120^\circ} + 1\,\underline{/-120^\circ} \cdot 150\,\underline{/120^\circ})$$

$$= \frac{150}{3}(1\,\underline{/0^\circ} + 1\,\underline{/0^\circ} + 1\,\underline{/0^\circ}) = 50\,\underline{/0^\circ}\ \text{A}$$

$$I_{a2} = \frac{1}{3}(I_a + \alpha^2\,I_b + \alpha\,I_c)$$

$$= \frac{1}{3}(150\,\underline{/0^\circ} + 1\,\underline{/240^\circ} \cdot 150\,\underline{/-120^\circ} + 1\,\underline{/120^\circ} \cdot 150\,\underline{/120^\circ}$$

$$= \frac{150}{3}(1\,\underline{/0^\circ} + 1\,\underline{/120^\circ} + 1\,\underline{/240^\circ}) = 0$$

(b) *After fuse removal in two lines*

Let the fuses be removed in lines *b* and *c*

$$\therefore \qquad I_b = 0, \qquad I_c = 0$$

$$I_{a0} = (1/3)(I_a + I_b + I_c) = (1/3)(150\,\underline{/0^\circ} + 0 + 0) = 50\,\underline{/0^\circ}\ \text{A}$$

$$I_{a1} = (1/3)(I_a + \alpha\,I_b + \alpha^2\,I_c) = (1/3)(150\,\underline{/0^\circ} + 0 + 0) = 50\,\underline{/0^\circ}\ \text{A}$$

$$I_{a2} = (1/3)(I_a + \alpha^2\,I_b + \alpha\,I_c) = (1/3)(150\,\underline{/0^\circ} + 0 + 0) = 50\,\underline{/0^\circ}\ \text{A}$$

Example 17.4 Prove that the three line voltage phasors $\mathbf{V}_{ab}$, $\mathbf{V}_{bc}$, and $\mathbf{V}_{ca}$ will have no zero-sequence component.

Solution

The three line voltages are given by

$$\mathbf{V}_{ab} = \mathbf{V}_a - \mathbf{V}_b$$
$$\mathbf{V}_{bc} = \mathbf{V}_b - \mathbf{V}_c$$
$$\mathbf{V}_{ca} = \mathbf{V}_c - \mathbf{V}_a$$

Addition of these equations gives, $\mathbf{V}_{ab} + \mathbf{V}_{bc} + \mathbf{V}_{ca} = 0$

$$\mathbf{V}_{ab0} = \mathbf{V}_{bc0} = \mathbf{V}_{ca0} = \frac{1}{3}(\mathbf{V}_{ab} + \mathbf{V}_{bc} + \mathbf{V}_{ca}) = 0$$

Hence the phasor set of line voltages will have no zero-sequence component.

Example 17.5 Prove that neutral current can flow only if zero-sequence currents exist.

Solution

$$\mathbf{I}_{a1} + \mathbf{I}_{b1} + \mathbf{I}_{c1} = \mathbf{I}_{a1} + \alpha^2 \mathbf{I}_{a1} + \alpha \mathbf{I}_{a1} = (1 + \alpha^2 + \alpha) I_{a1} = 0 \times I_{a1} = 0$$

$$\mathbf{I}_{a2} + \mathbf{I}_{b2} + \mathbf{I}_{c2} = \mathbf{I}_{a2} + \alpha \mathbf{I}_{a2} + \alpha^2 \mathbf{I}_{a2} = (1 + \alpha + \alpha^2) I_{a2} = 0 \times I_{a2} = 0$$

It is seen that neither the positive nor the negative sequence currents will contribute any neutral current.

Now, $\mathbf{I}_n = \mathbf{I}_a + \mathbf{I}_b + \mathbf{I}_c = 3\mathbf{I}_{a0} = \mathbf{I}_{a0} + \mathbf{I}_{b0} + \mathbf{I}_{c0}$

This result shows that the neutral current is equal to the sum of the zero-sequence currents in each phase.

17.12 POWER IN TERMS OF SYMMETRICAL COMPONENTS

In a single-phase system voltamperes S is given by

$$\mathbf{S}_{1\,\varphi} = P + j\,Q = \mathbf{V}\mathbf{I}^* \qquad\qquad\qquad ...(17.12.1)$$

where $\mathbf{V}$ and $\mathbf{I}$ are phase voltage and phase current and $\mathbf{I}^*$ is the conjugate of $\mathbf{I}$. Similarly, in a three-phase system

$$\mathbf{S}_{abc} = \mathbf{S}_{3\,\varphi} = \mathbf{V}_a \mathbf{I}_a^* + \mathbf{V}_b \mathbf{I}_b^* + \mathbf{V}_c \mathbf{I}_c^* \qquad\qquad ...(17.12.2)$$

where $\mathbf{V}_a$, $\mathbf{V}_b$, and $\mathbf{V}_c$ are the phase voltages and $\mathbf{I}_a^*$, $\mathbf{I}_b^*$ and $\mathbf{I}_c^*$ are the conjugates of phase currents $\mathbf{I}_a$, $\mathbf{I}_b$ and $\mathbf{I}_c$.

In matrix form, Eq. (7.12.2) may be written as

$$\mathbf{S}_{abc} = [\mathbf{V}_a \ \mathbf{V}_b \ \mathbf{V}_c] \begin{bmatrix} \mathbf{I}_a \\ \mathbf{I}_b \\ \mathbf{I}_c \end{bmatrix}^* = \begin{bmatrix} \mathbf{V}_a \\ \mathbf{V}_b \\ \mathbf{V}_c \end{bmatrix}^T \begin{bmatrix} \mathbf{I}_a \\ \mathbf{I}_b \\ \mathbf{I}_c \end{bmatrix}^* = \mathbf{V}_{abc}^T \mathbf{I}_{abc}^*$$

where the subscript T stands for transpose.

Now, $\mathbf{V}_{abc} = \mathbf{A}\mathbf{V}_{012}$

$\mathbf{I}_{abc} = \mathbf{A}\mathbf{I}_{012}$

$\therefore$ $\mathbf{S}_{abc} = [\mathbf{A}\mathbf{V}_{012}]^T [\mathbf{A}\mathbf{I}_{012}]^* = \mathbf{V}_{012}^T \mathbf{A}^T \mathbf{A}^* \mathbf{I}_{012}^*$ ...(17.12.3)

because the transpose of the product of two matrices is equal to the product of the transpose of the matrices in the reverse order. The transpose of the transformation matrix $\mathbf{A}$ is obtained by putting the columns of $\mathbf{A}$ as the rows of $\mathbf{A}^T$.

$$\mathbf{A} = \begin{bmatrix} 1 & 1 & 1 \\ 1 & \alpha^2 & \alpha \\ 1 & \alpha & \alpha^2 \end{bmatrix}$$

$$\mathbf{A}^T = \begin{bmatrix} 1 & 1 & 1 \\ 1 & \alpha^2 & \alpha \\ 1 & \alpha & \alpha^2 \end{bmatrix} = \mathbf{A}$$

$$\mathbf{A}^* = \begin{bmatrix} 1 & 1 & 1 \\ 1 & \alpha^{2*} & \alpha^* \\ 1 & \alpha^* & \alpha^{2*} \end{bmatrix}$$

But $\alpha^* = \alpha^2, \qquad \alpha^{2*} = \alpha$

$$\mathbf{A}^* = \begin{bmatrix} 1 & 1 & 1 \\ 1 & \alpha & \alpha^2 \\ 1 & \alpha^2 & \alpha \end{bmatrix}$$

$$\mathbf{A}^T \mathbf{A}^* = \begin{bmatrix} 1 & 1 & 1 \\ 1 & \alpha^2 & \alpha \\ 1 & \alpha & \alpha^2 \end{bmatrix} \begin{bmatrix} 1 & 1 & 1 \\ 1 & \alpha & \alpha^2 \\ 1 & \alpha^2 & \alpha \end{bmatrix} = \begin{bmatrix} 3 & 0 & 0 \\ 0 & 3 & 0 \\ 0 & 0 & 3 \end{bmatrix} = 3 \begin{bmatrix} 1 & 0 & 0 \\ 0 & 1 & 0 \\ 0 & 0 & 1 \end{bmatrix} = 3\mathbf{U}$$

where $\mathbf{U} = \begin{bmatrix} 1 & 0 & 0 \\ 0 & 1 & 0 \\ 0 & 0 & 1 \end{bmatrix} =$ the unit matrix

Substituting the value of $\mathbf{A}^T \mathbf{A}^*$ in Eq. (17.12.3) we get

$$\mathbf{S}_{abc} = 3\mathbf{V}_{012}^T \mathbf{U}\mathbf{I}_{012}^* = 3\mathbf{V}_{012}^T \mathbf{I}_{012}^* = 3\mathbf{S}_{012} = 3 \begin{bmatrix} \mathbf{V}_{a0} & \mathbf{V}_{a1} & \mathbf{V}_{a2} \end{bmatrix} \begin{bmatrix} \mathbf{I}_{a0} \\ \mathbf{I}_{a1} \\ \mathbf{I}_{a2} \end{bmatrix}^*$$

$$= 3\mathbf{V}_{a0} \mathbf{I}_{a0}^* + 3\mathbf{V}_{a1} \mathbf{I}_{a1}^* + 3\mathbf{V}_{a2} \mathbf{I}_{a2}^*$$ ...(17.12.4)

Equation (17.12.4) shows that the total complex power in the unbalanced system is equal to the sum of the complex powers of the three symmetrical components. Hence we can say that the symmetrical component transformation is power invariant.

Example 17.6 Prove that

$$| \mathbf{I}_a |^2 + | \mathbf{I}_b |^2 + | \mathbf{I}_c |^2 = 3 \, (| \mathbf{I}_{a0} |^2 + | \mathbf{I}_{a1} |^2 + | \mathbf{I}_{a2} |^2).$$

Solution

We know that

$$| \mathbf{I}_a |^2 + | \mathbf{I}_b |^2 + | \mathbf{I}_c |^2 = \mathbf{I}_a \cdot \mathbf{I}_a^* + \mathbf{I}_b \cdot \mathbf{I}_b^* + \mathbf{I}_c \cdot \mathbf{I}_c^* = [\mathbf{I}_a \, \mathbf{I}_b \, \mathbf{I}_c] \begin{bmatrix} \mathbf{I}_a^* \\ \mathbf{I}_b^* \\ \mathbf{I}_c^* \end{bmatrix} = \mathbf{I}_{abc}^T \, \mathbf{I}_{abc}^*$$

$$\mathbf{I}_{abc}^T = [\mathbf{AI}_{012}]^T = \mathbf{I}_{012}^T \, \mathbf{A}^T = [I_{a0} \, I_{a1} \, I_{a2}] \begin{bmatrix} 1 & 1 & 1 \\ 1 & \alpha^2 & \alpha \\ 1 & \alpha & \alpha^2 \end{bmatrix}$$

$$\mathbf{I}_{abc}^* = [\mathbf{AI}_{012}]^* = \mathbf{A}^* \, \mathbf{I}_{012}^* = \begin{bmatrix} 1 & 1 & 1 \\ 1 & \alpha & \alpha^2 \\ 1 & \alpha^2 & \alpha \end{bmatrix} \begin{bmatrix} I_{a0}^* \\ I_{a1}^* \\ I_{a2}^* \end{bmatrix}$$

Therefore,

$$| \mathbf{I}_a |^2 + | \mathbf{I}_b |^2 + | \mathbf{I}_c |^2 = \mathbf{I}_{abc}^T \, \mathbf{I}_{abc}^* = [I_{a0} \, I_{a1} \, I_{b2}] \begin{bmatrix} 1 & 1 & 1 \\ 1 & \alpha^2 & \alpha \\ 1 & \alpha & \alpha^2 \end{bmatrix} \begin{bmatrix} 1 & 1 & 1 \\ 1 & \alpha & \alpha^2 \\ 1 & \alpha^2 & \alpha \end{bmatrix} \begin{bmatrix} I_{a0}^* \\ I_{a1}^* \\ I_{a2}^* \end{bmatrix}$$

$$= [I_{a0} \, I_{a1} \, I_{a2}] \begin{bmatrix} 3 & 0 & 0 \\ 0 & 3 & 0 \\ 0 & 0 & 3 \end{bmatrix} \begin{bmatrix} I_{a0}^* \\ I_{a1}^* \\ I_{a2}^* \end{bmatrix}$$

$$= 3 \, (I_{a0} \, I_{a0}^* + I_{a1} \, I_{a1}^* + I_{a2} \, I_{a2}^*) = 3 \, (| I_{a0} |^2 + | I_{a1} |^2 + | I_{a2} |^2)$$

Example 17.7 The resolution of a set of three-phase unbalanced voltages into symmetrical components gave the following results :

$$\mathbf{V}_{a0} = 30 \, \underline{/- 30°} \text{ V}, \quad \mathbf{V}_{a1} = 450 \, \underline{/0°} \text{ V}, \quad \mathbf{V}_{a2} = 225 \, \underline{/40°} \text{ V}.$$

The component currents are

$$\mathbf{I}_{a0} = 10 \, \underline{/190°} \text{ A}, \quad \mathbf{I}_{a1} = 6 \, \underline{/20°} \text{ A}, \quad \mathbf{I}_{a2} = 5 \, \underline{/50°} \text{ A}.$$

Determine the complex power represented by these voltages and currents by (a) symmetrical components (b) unbalanced phase components.

Solution

(a) *Calculation of complex power with symmetrical components*

$$\begin{aligned}
\mathbf{S}_{3\,\varphi} &= 3\mathbf{V}_{a0} \, \mathbf{I}_{a0}^* + 3\mathbf{V}_{a1} \, \mathbf{I}_{a1}^* + 3\mathbf{V}_{a2} \, \mathbf{I}_{a2}^* \\
&= 3 \, [(30 \, \underline{/- 30°}) \, (10 \, \underline{/- 190°}) + (450 \, \underline{/0°}) \, (6 \, \underline{/- 20°}) + (225 \, \underline{/40°}) \, (5 \, \underline{/- 50°})] \\
&= 3 \, (330 \, \underline{/- 220°} + 2700 \, \underline{/- 20°} + 1125 \, \underline{/- 10°}) \\
&= 3 \, (- 229.81 + j \, 192.83 + 2537.17 - j \, 923.45 + 1107.9 - j \, 195.35) \\
&= 3 \, (3415.26 - j \, 925.97) = (10245.78 - j \, 2777.9) \text{ VA}
\end{aligned}$$

(b) *Calculation of complex power with unbalanced phase components*

The phase voltages V_a, V_b and V_c are calculated as follows :

$$V_a = V_{a0} + V_{a1} + V_{a2}$$
$$= 30\underline{/-30°} + 450\underline{/0°} + 225\underline{/40°}$$
$$= 25.98 - j\,15 + 450 + j\,0 + 172.36 + j\,144.62$$
$$= 648.34 + j\,129.62 = 661.17\underline{/11.3°}\ \text{V}$$

$$V_b = V_{b0} + V_{b1} + V_{b2} = V_{a0} + \alpha^2\,V_{a1} + \alpha\,V_{a2}$$
$$= 30\underline{/-30°} + (1\underline{/240°})\,(450\underline{/0°}) + (1\underline{/120°})\,(225\underline{/40°})$$
$$= 30\underline{/-30°} + 450\underline{/240°} + 225\underline{/160°}$$
$$= 25.98 - j\,15 - 225 - j\,389.7 - 211.43 + j\,76.95$$
$$= -410.45 - j\,327.75 = 525.25\underline{/-141.39°}\ \text{V}$$

$$V_c = V_{c0} + V_{c1} + V_{c2} = V_{a0} + \alpha\,V_{a1} + \alpha^2\,V_{a2}$$
$$= 30\underline{/-30°} + (1\underline{/120°})\,(450\underline{/0°}) + (1\underline{/240°})\,(225\underline{/40°})$$
$$= 30\underline{/-30°} + 450\underline{/120°} + 225\underline{/280°}$$
$$= 25.98 - j\,15 - 225 + j\,389.71 + 39.07 - j\,221.58$$
$$= -159.95 + j\,153.13 = 221.43\underline{/136.24°}\ \text{V}$$

The phase currents, I_a, I_b and I_c are calculated as follows :

$$I_a = I_{a0} + I_{a1} + I_{a2} = 10\underline{/190°} + 6\underline{/20°} + 5\underline{/50°}$$
$$= -9.848 - j\,1.736 + 5.638 + j\,2.052 + 3.21 + j\,3.83$$
$$= -1.0 + j\,4.146 = 4.265\underline{/103.56°}\ \text{A}$$

$$I_b = I_{b0} + I_{b1} + I_{b2} = I_{a0} + \alpha^2\,I_{a1} + \alpha\,I_{a2}$$
$$= 10\underline{/190°} + (1\underline{/240°})\,(6\underline{/20°}) + (1\underline{/120°})\,(5\underline{/50°})$$
$$= 10\underline{/190°} + 6\underline{/260°} + 5\underline{/170°}$$
$$= -9.848 - j\,1.736 - 1.042 - j\,5.9088 - 4.924 + j\,0.868$$
$$= -15.814 - j\,6.777 = 17.20\underline{/-156.8°}\ \text{A}$$

$$I_c = I_{c0} + I_{c1} + I_{c2} = I_{a0} + \alpha\,I_{a1} + \alpha^2\,I_{a2}$$
$$= 10\underline{/190°} + (1\underline{/120°})\,(6\underline{/20°}) + (1\underline{/240°})\,(5\underline{/50°})$$
$$= 10\underline{/190°} + 6\underline{/140°} + 5\underline{/290°}$$
$$= -9.848 - j\,1.736 - 4.596 + j\,3.856 + 1.71 - j\,4.698$$
$$= -12.73 - j\,2.578 = 12.988\underline{/-168.55°}\ \text{A}$$

The complex three-phase power is given by

$$S_{3\,\varphi} = V_a\,I_a^* + V_b\,I_b^* + V_c\,I_c^*$$
$$= (661.17\underline{/11.3°})\,(4.265\underline{/-103.56°}) + (525.25\underline{/-141.39°})\,(17.2\underline{/156.8°})$$
$$+ (221.43\underline{/136.24°})\,(12.988\underline{/168.55°})$$
$$= 2819\underline{/-92.26°} + 9034\underline{/15.41°} + 2875.9\underline{/304.79°}$$
$$= -111.2 - j\,2817.7 + 8709.5 + j\,2400.6 + 1640.9 - j\,2361.8 = 10239.2 - j\,2778.3$$

17.13 POTENTIAL OF NEUTRAL

The neutral may or may not be at the same potential as ground in the cases of ungrounded neutrals or neutrals grounded through impedances. Consider the network shown in Fig. 17.9 in which the neutral is grounded through an impedance Z_n. If the zero-sequence currents exist in the circuit, then a current will flow through Z_n. Consequently, there will be a potential difference V_{gn} between ground and neutral. Let V_a, V_b and V_c be the voltages to ground at P. Let us also assume that

V_{gn} = voltage between ground and neutral

V'_a = voltage between n and P for phase a

V'_b = voltage between n and P for phase b

V'_c = voltage between n and P for phase c

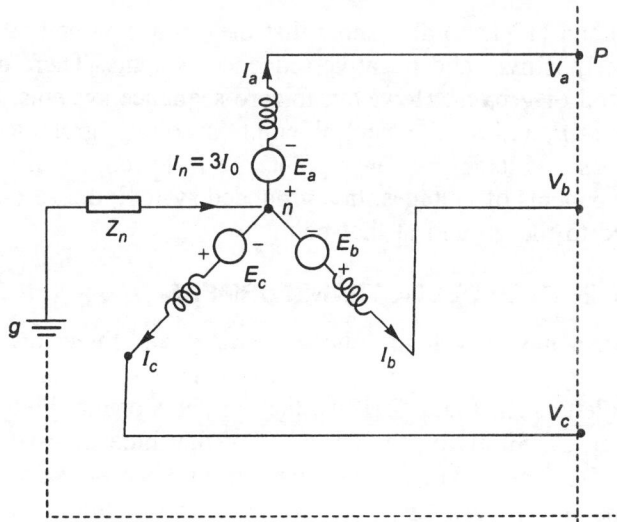

Fig. 17.9. Potential of neutral.

By KVL we can write

$$V_a = V_{gn} + V'_a \; ; \; V_b = V_{gn} + V'_b \; ; \; V_c = V_{gn} + V'_c$$

We shall determine the phase sequence components for V_{gn} of all the three phase voltages.

$$V_{a0} = \frac{1}{3}(V_{gn} + V_{gn} + V_{gn}) = V_{gn} \qquad \qquad ...(17.13.1)$$

$$V_{a1} = \frac{1}{3}(V_{gn} + \alpha V_{gn} + \alpha^2 V_{gn}) = \frac{1}{3}(1 + \alpha + \alpha^2)V_{gn} = 0 \qquad ...(17.13.2)$$

$$V_{a2} = \frac{1}{3}(V_{gn} + \alpha^2 V_{gn} + \alpha V_{gn}) = \frac{1}{3}(1 + \alpha^2 + \alpha)V_{gn} = 0 \qquad ...(17.13.3)$$

Equations (17.13.2) and (17.13.3) show that the positive and negative sequence voltages between the neutral and ground are zero, while Eq. (17.13.1) shows that the voltage between ground and neutral represents the zero-sequence voltage. If

Z'_0 = zero-sequence impedance from n to P

Z_0 = total zero-sequence impedance from g to P

the total zero-sequence voltage drop from ground to P is given by

$$Z_n I_n + Z'_0 I_0 = Z_0 I_0 \qquad \qquad ...(17.13.4)$$

Since $I_n = 3I_0$, Eq. (17.13.4) becomes

$$(3Z_n + Z'_0) I_0 = Z_0 I_0$$

or $\qquad Z_0 = Z'_0 + 3Z_n \qquad \qquad ...(17.13.5)$

If the neutral is solidly grounded, $Z_n = 0$

then $\qquad Z_0 = Z'_0$

Equations (17.13.4) and (17.13.5) also show that the voltage to neutral or voltage to ground are the same in case of positive- and negative-sequence systems. Therefore, the zero-potential bus may be either neutral or *ground*. However, in zero-sequence systems, the zero-potential bus is *ground* and *not neutral*, unless the neutral point is solidly grounded at the point under consideration. The phase voltages at any point are referred to neutral in positive and negative-sequence components of voltages. In a grounded system, the zero-sequence components of voltages are referred to the ground at that point.

17.14 PHASE SHIFT IN STAR-DELTA TRANSFORMERS

A three-phase transformer has three high-voltage windings and three low-voltage windings. By convention, the low-voltage side is denoted by lowercase letters a, b, and c and the high-voltage side by the upper case letters A, B and C. Thus, the terminals of the three low-voltage windings are a_1, a_2, b_1, b_2 and c_1, c_2. Similarly, the terminals of the three high-voltage windings are A_1, A_2, B_1, B_2, and C_1, C_2. The polarities are marked in such a way that the voltages are instantaneously in time phase between terminals with corresponding markings. Thus, at the instant when the potential of A_1 is at a positive maximum value with respect to A_2, the potential of a_1 will also be at a positive maximum value with respect to a_2.

In a Y-Y or Δ-Δ transformer, it is possible to label the phases in such a way that there is no phase shift between corresponding primary and secondary quantities. However, in a Y-Δ or Δ-Y transformer, it is impossible to label the phases in such a way that there is no phase shift between corresponding quantities.

Fig. 17.10 (a) shows a Δ-Y transformer in which the low-voltage side is connected in Δ and the high-voltage side in Y. For the positive sequence a, b, c, the phasor diagram is shown in Fig. 17.10 (b), where V_{ab}, V_{bc}, and V_{ca} are the phase voltages on the low-voltage (delta) side. The corresponding voltages on the high-voltage (star) side are V_{AN}, V_{BN}, and V_{CN}. It is to be noted that $\mathbf{V}_{ab}$ and $\mathbf{V}_{AN}$ are in phase because they are produced by the same flux in transformer a-A. Also, $\mathbf{V}_{AN}$ is shown of greater length than V_{ab} because V_{AN} represents the high voltage. The line voltage on high-voltage side is V_{AB}. By KVL

$$\mathbf{V}_{AB} = \mathbf{V}_{AN} + \mathbf{V}_{NB}$$

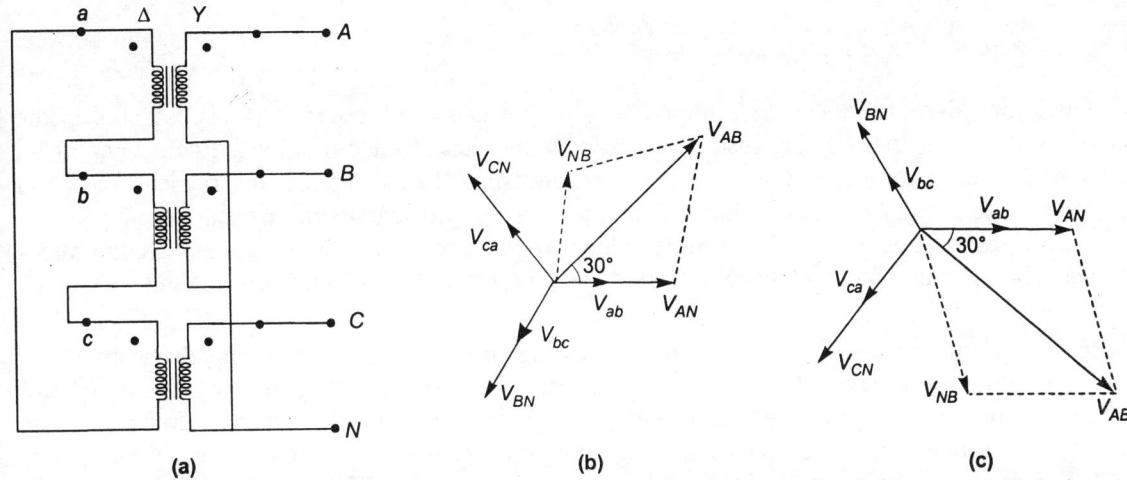

Fig. 17.10. Illustrating the 30° displacement between line voltages in a star-delta connection of transformer (a) phase coil connections with the *hv* side in star; (b) the phasor diagram for the positive-sequence set; (c) the phasor diagram for the negative-sequence set.

It is also seen that $|\mathbf{V}_{AB}| = \sqrt{3}\,|\mathbf{V}_{AN}|$ and $\mathbf{V}_{AB}$ leads $\mathbf{V}_{AN}$ by 30°. Since $\mathbf{V}_{AN}$ and $\mathbf{V}_{ab}$ are in time phase, it follows that the line voltage between A and B on the star side *leads* the line voltage on the delta side by 30°. Similarly, the line voltage $\mathbf{V}_{BC}$ leads $\mathbf{V}_{bc}$ by 30° and $\mathbf{V}_{CA}$ leads $\mathbf{V}_{ca}$ by 30°. Also, the positive-sequence line currents on the high-voltage side lead their corresponding positive-sequence line currents on the low-voltage side by 30°.

Fig. 17.10 (c) shows the phasor diagram for the negative-sequence voltages. In this case, the line voltage on the star side *lags* the line voltage on the delta side by 30°. As a result of 30° phase shift between the line voltages of the Y-Δ transformer, the following standard convention is generally adopted :

The phases of star-delta or delta-star transformers are labelled in such a way that the positive-sequence quantity on the high voltage side *leads* the corresponding positive-sequence quantity on the low-voltage side by 30°. Similarly, the negative-sequence quantity on the high-voltage side *lags* the corresponding negative-sequence quantity on the low-voltage side by 30°.

17.15 SEQUENCE IMPEDANCES

The main application of the theory of symmetrical components is to the analysis of three-phase networks which are subjected to unsymmetrical faults. In such networks the impedance of each phase *upto the fault* are equal. For a given phase sequence the ratio of phase sequence voltage to phase sequence current is defined as the impedance to phase sequence current or simply the phase sequence impedance.

Thus,   positive sequence impedance $Z_1 = \dfrac{V_1}{I_1}$

negative sequence impedance $Z_2 = \dfrac{V_2}{I_2}$

and zero sequence impedance $Z_0 = \dfrac{V_0}{I_0}$

The knowledge of the sequence impedances of power system components is essential in the study of the behaviour of the system under asymmetrical fault conditions. Each component, whether static (transmission lines, transformers, and static loads), or rotating (synchronous and induction machines), has three values of impedances, one for each symmetrical components of current. In some cases two or all the three values are the same. For example, in static equipments Z_1 and Z_2 are the same and have different Z_0. For rotating machines Z_1, Z_2 and Z_0 are all different.

In general, the positive sequence current flowing in positive sequence impedance produces positive-sequence voltage drop, negative sequence current flowing in negative sequence impedance produces negative sequence voltage drop, and zero sequence current flowing in zero sequence impedance produces zero sequence voltage drop. In other words, generally there is no mutual impedance between various symmetrical components. Thus, each sequence can be considered separately. This simplifies the calculation of asymmetrical fault currents considerably.

17.16 SEQUENCE NETWORKS FOR FAULT CALCULATIONS

The presence of a fault at any point in a system means generally that the system is put into an unbalanced state of operation. We know from symmetrical component theory that this unbalanced condition can be replaced by a symmetrically balanced positive-sequence set, a symmetrically balanced negative sequence set, and a single-phase zero-sequence set. Thus, when the fault occurs, we can assume that these three sequences sets are injected into the system replacing the unbalanced set of voltages and or currents. Hence, in order to calculate the postfault voltages and currents, we have to determine the response of the system to each component set. For this purpose, we have to draw the three sequence networks of the system.

Since currents of any one sequence produce voltage drops of that sequence only, each current may be considered to flow in an independent network containing the impedance to the circuit of that sequence only. Also, the currents of each sequence are balanced, the per phase basis can be adopted for calculation work. The single-phase equivalent network containing the impedance to current of any one sequence only is called the *sequence network* for that particular sequence. Thus, a *sequence network* may be defined as an equivalent network for the balanced power system under an imagined operating condition such that only one sequence component of voltages and currents is present in the system.

Each sequence network, is replaced by a Thevenin's equivalent circuit between two points, i.e., each sequence network can be reduced to a single voltage and single impedance. One point is the fault point F and the other is the zero potential of reference bus N. Sequence networks are represented by boxes in which fault point F, the zero potential or reference bus N, and the Thevenin voltage and impedance are shown. The sequence networks are shown in Fig. 17.11.

The Thevenin voltage in the positive sequence network is the open-circuit voltage V_f at the fault point. The voltage V_f is the prefault voltage of phase a at the fault point F. This voltage is sometimes represented by E_{a1}. Thevenin voltages in negative and zero sequence networks are zero because the negative and zero sequence voltages at the fault point are zero in a balanced system.

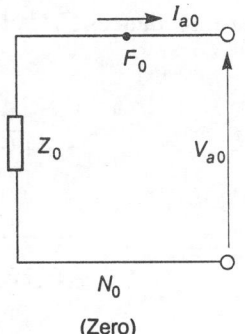
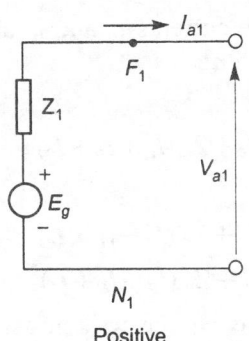
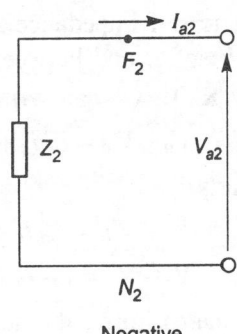

Fig. 17.11. Sequence networks.

Since I_a is the current flowing from the system into the fault, its components I_{a0}, I_{a1}, I_{a2} flow away from the fault point F. It is to be noted that the unbalanced connection is to be attached at F and the current flows from the balanced system to this connection.

The symmetrical components of voltages at the fault may be written as

$$V_{a0} = -Z_0 I_{a0}$$
$$V_{a1} = V_f - Z_1 I_{a1}$$
$$V_{a2} = -Z_2 I_{a2}$$

where Z_0, Z_1 and Z_2 are the total equivalent impedances of the zero, positive, and negative sequence networks upto the fault point.

Similarly, for a fault on the unloaded generator with excitation voltage E_g the following will be the sequence voltage drops :

$$V_{a0} = -Z_0 I_{a0}$$
$$V_{a1} = E_g - Z_1 I_{a1}$$
$$V_{a2} = -Z_2 I_{a2}$$

17.17 SEQUENCE IMPEDANCES OF A TRANSMISSION LINE

Consider a three-phase transmission line (Fig. 17.12) with self impedance Z_s per phase and equal mutual impedances Z_m between phases. It is assumed that Z_m is the same for each pair of conductors. Let I_a, I_b, and I_c be the line currents. If the system voltages are unbalanced, we have a neutral current I_n flowing through the neutral (ground) having impedance Z_n. Thus,

$$I_n = I_a + I_b + I_c$$

The phase voltages at the two ends of the line are V_a, V_b, V_c and V'_a, V'_b, V'_c respectively. The voltage drop in line a is caused by the current

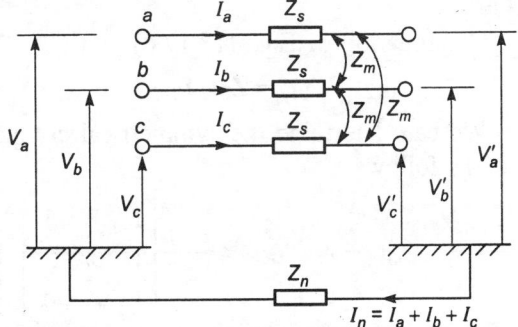

Fig. 17.12. Unsymmetrically loaded transmission line.

I_a due to self impedance Z_s and by the currents I_b and I_c due to mutual impedances Z_m between lines b and a and between lines c and a.

By KVL, we can write

$$V_a - V'_a = Z_s I_a + Z_m I_b + Z_m I_c + Z_n (I_a + I_b + I_c) \qquad \dots(17.17.1)$$

Similarly,

$$V_b - V'_b = Z_m I_a + Z_s I_b + Z_m I_c + Z_n (I_a + I_b + I_c) \qquad \dots(17.17.2)$$

$$V_c - V'_c = Z_m I_a + Z_m I_b + Z_s I_c + Z_n (I_a + I_b + I_c) \qquad \dots(17.17.3)$$

In matrix form, the above equations can be written as

$$\begin{bmatrix} V_a \\ V_b \\ V_c \end{bmatrix} - \begin{bmatrix} V'_a \\ V'_b \\ V'_c \end{bmatrix} = \begin{bmatrix} Z_s + Z_n & Z_m + Z_n & Z_m + Z_n \\ Z_m + Z_n & Z_s + Z_n & Z_m + Z_n \\ Z_m + Z_n & Z_m + Z_n & Z_s + Z_n \end{bmatrix} \begin{bmatrix} I_a \\ I_b \\ I_c \end{bmatrix} \qquad \dots(17.17.4)$$

In a more compact form we can write

$$\mathbf{V}_{abc} - \mathbf{V}'_{abc} = \mathbf{Z}_{abc} \mathbf{I}_{abc} \qquad \dots(17.17.5)$$

These equations show that the currents of one phase produce voltage drops in other phases also. These impedances are said to be *coupled* so far as phase voltages and phase currents are concerned.

Applying symmetrical transformation to Eq. (17.17.5) we get

$$\mathbf{A} \mathbf{V}_{012} - \mathbf{A} \mathbf{V}'_{012} = \mathbf{Z}_{abc} \mathbf{A} \mathbf{I}_{012}$$

Premultiplying both the sides by $\mathbf{A}^{-1}$ we get

$$\mathbf{V}_{012} - \mathbf{V}'_{012} = \mathbf{A}^{-1} \mathbf{Z}_{abc} \mathbf{A} \mathbf{I}_{012} \qquad \dots(17.17.6)$$

Let us define

$$\mathbf{Z}_{012} \triangleq \mathbf{A}^{-1} \mathbf{Z}_{abc} \mathbf{A} \qquad \dots(17.17.7)$$

$\mathbf{Z}_{012}$ is called the *symmetrical components impedance matrix*.

Equation (17.17.7) is used to convert the transmission line matrix $\mathbf{Z}_{abc}$ into a form that is consistent with the use of zero-, positive-, and negative-sequence voltages ($\mathbf{V}_{abc}$) and currents ($\mathbf{I}_{abc}$).

Combination of Eqs. (17.17.6) and (17.17.7) gives

$$\mathbf{V}_{012} - \mathbf{V}'_{012} = \mathbf{Z}_{012} \mathbf{I}_{012} \qquad \dots(17.17.8)$$

We can determine the symmetrical components impedance matrix $\mathbf{Z}_{012}$ for the transmission line as follows :

$$\mathbf{Z}_{012} = \mathbf{A}^{-1} \mathbf{Z}_{abc} \mathbf{A} = \frac{1}{3} \begin{bmatrix} 1 & 1 & 1 \\ 1 & \alpha & \alpha^2 \\ 1 & \alpha^2 & \alpha \end{bmatrix} \begin{bmatrix} Z_s + Z_n & Z_m + Z_n & Z_m + Z_n \\ Z_m + Z_n & Z_s + Z_n & Z_m + Z_n \\ Z_m + Z_n & Z_m + Z_n & Z_s + Z_n \end{bmatrix} \begin{bmatrix} 1 & 1 & 1 \\ 1 & \alpha^2 & \alpha \\ 1 & \alpha & \alpha^2 \end{bmatrix}$$

For the sake of convenience, let us assume, that

$$Z_s + Z_n = x, \qquad Z_m + Z_n = y$$

$$\mathbf{Z}_{abc}\,\mathbf{A} = \begin{bmatrix} x & y & y \\ y & x & y \\ y & y & x \end{bmatrix} \begin{bmatrix} 1 & 1 & 1 \\ 1 & \alpha^2 & \alpha \\ 1 & \alpha & \alpha^2 \end{bmatrix} = \begin{bmatrix} x+y+y & x+y\,(\alpha+\alpha^2) & x+y\,(\alpha+\alpha^2) \\ y+x+y & x+\alpha^2 x+\alpha\,y & y+\alpha\,x+\alpha^2\,y \\ y+y+x & y+\alpha^2 y+\alpha\,x & y+\alpha\,y+\alpha^2\,x \end{bmatrix}$$

$$= \begin{bmatrix} x+2y & x-y & x-y \\ x+2y & \alpha^2\,(x-y) & \alpha\,(x-y) \\ x+2y & \alpha\,(x-y) & \alpha^2\,(x-y) \end{bmatrix} = \begin{bmatrix} Z_s+2Z_m+3Z_n & Z_s-Z_m & Z_s-Z_m \\ Z_s+2Z_m+3Z_n & \alpha^2\,(Z_s-Z_m) & \alpha\,(Z_s-Z_m) \\ Z_s+2Z_m+3Z_n & \alpha\,(Z_s-Z_m) & \alpha^2\,(Z_s-Z_m) \end{bmatrix}$$

Let us put $Z_s+2Z_m+3Z_n = p$　and　$Z_s-Z_m = q$

$$\mathbf{Z}_{abc}\,\mathbf{A} = \begin{bmatrix} p & q & q \\ p & \alpha^2\,q & \alpha\,q \\ p & \alpha\,q & \alpha^2\,q \end{bmatrix}$$

$$\mathbf{A}^{-1}\,\mathbf{Z}_{abc}\,\mathbf{A} = \frac{1}{3} \begin{bmatrix} 1 & 1 & 1 \\ 1 & \alpha & \alpha^2 \\ 1 & \alpha^2 & \alpha \end{bmatrix} \begin{bmatrix} p & q & q \\ p & \alpha^2\,q & \alpha\,q \\ p & \alpha\,q & \alpha^2\,q \end{bmatrix}$$

$$= \frac{1}{3} \begin{bmatrix} p+p+p & q\,(1+\alpha^2+\alpha) & q\,(1+\alpha+\alpha^2) \\ p\,(1+\alpha+\alpha^2) & q\,(1+\alpha^3+\alpha^3) & q\,(1+\alpha^2+\alpha^4) \\ p\,(1+\alpha+\alpha^2) & q\,(1+\alpha^4+\alpha^2) & q\,(1+\alpha^3+\alpha^3) \end{bmatrix}$$

Since $1+\alpha+\alpha^2 = 0$,　$\alpha^3 = 1$　and　$\alpha^4 = \alpha$, we have

$$\mathbf{Z}_{012} = \mathbf{A}^{-1}\,\mathbf{Z}_{abc}\,\mathbf{A} = \frac{1}{3} \begin{bmatrix} 3p & 0 & 0 \\ 0 & 3q & 0 \\ 0 & 0 & 3q \end{bmatrix} = \begin{bmatrix} p & 0 & 0 \\ 0 & q & 0 \\ 0 & 0 & q \end{bmatrix}$$

$$= \begin{bmatrix} Z_s+2Z_m+3Z_n & 0 & 0 \\ 0 & Z_s-Z_m & 0 \\ 0 & 0 & Z_s-Z_m \end{bmatrix} \qquad \ldots(17.17.9)$$

Let us define the following sequence impedances for the line :

$Z_0 \overset{\Delta}{=} Z_s+2Z_m+3Z_n$ = zero-sequence impedance

$Z_1 \overset{\Delta}{=} Z_s-Z_m$ = positive-sequence impedance

$Z_2 \overset{\Delta}{=} Z_s-Z_m$ = negative-sequence impedance.

Therefore,

$$\mathbf{Z}_{012} = \begin{bmatrix} Z_0 & 0 & 0 \\ 0 & Z_1 & 0 \\ 0 & 0 & Z_2 \end{bmatrix} \qquad \ldots(17.17.10)$$

and　$$\begin{bmatrix} V_{a0} \\ V_{a1} \\ V_{a2} \end{bmatrix} - \begin{bmatrix} V'_{a0} \\ V'_{a1} \\ V'_{a2} \end{bmatrix} = \begin{bmatrix} Z_0 & 0 & 0 \\ 0 & Z_1 & 0 \\ 0 & 0 & Z_2 \end{bmatrix} \begin{bmatrix} I_{a0} \\ I_{a1} \\ I_{a2} \end{bmatrix} \qquad \ldots(17.17.11)$$

In the expanded form we can write

$$\left. \begin{array}{l} V_{a0} - V'_{a0} = Z_0 \, I_{a0} \\ V_{a1} - V'_{a1} = Z_1 \, I_{a1} \\ V_{a2} - V'_{a2} = Z_2 \, I_{a2} \end{array} \right\} \qquad \qquad ...(17.17.12)$$

These relations show that in *symmetrical* circuits with or without mutual coupling, currents of a given sequence produce voltage drops of the same sequence only. The sequence impedances are said to be *uncoupled*. This fact indicates that the voltage drops due to zero-, positive-, and negative-sequences may be treated separately.

The impedance 'seen' by the positive and negative-sequence waves is the same as we move along the transmission line. That is $Z_1 = Z_2$. This is logical also due to the fact that the reversal of the phase sequence of voltages applied to a static component (the transmission line or a transformer) does not experience a different impedance because these waves encounter the same geometry of the lines regardless of the sequence.

It is also observed that the zero-sequence impedance Z_0 of the line has a different value and is usually 2 to 3.5 times greater than Z_1 and Z_2.

Equations (17.17.12) can be represented in network form as shown in Fig. 17.13.

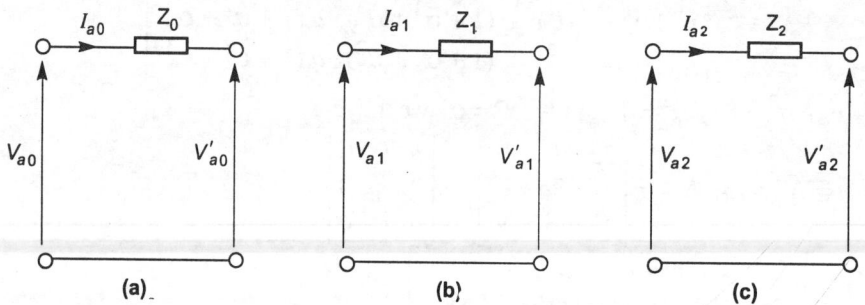

Fig. 17.13. Sequence networks of a transmission line. (a) Zero-sequence network; (b) Positive-sequence network; (c) Negative-sequence network.

17.18 SEQUENCE IMPEDANCES OF SYNCHRONOUS MACHINES

It is usual to neglect resistance and therefore, all sequence impedances for a synchronous machine are reactive. The positive-, negative-, and zero-sequence impedances have in general different values.

Positive-Sequence Impedance Z_1

Depending on the time interval of interest, one of the three reactances may be used :

1. For the subtransient interval, we use the subtransient reactance

   $Z_1 = j \, X''_d$.

2. For the transient interval, we use the transient reactance

   $Z_1 = j \, X'_d$.

3. If the steady-state value is of interest, we have

   $Z_1 = j \, X_d$.

Negative-Sequence Impedance Z_2

Negative sequence currents produce a rotating magnetic field rotating at the same speed but in a direction opposite to that produced by positive sequence currents. Hence the negative-sequence impedance Z_2 is different from the positive-sequence impedance Z_1. The negative-sequence impedance Z_2 for a synchronous machine is often defined as

$$Z_2 = j \frac{X''_d + X''_q}{2}$$

Zero-Sequence Impedance Z_0

Zero-sequence currents are all in phase and therefore, do not produce any rotating field. The zero-sequence impedance Z_0 depends upon the type of grounding and the zero-sequence impedance per phase of the generator.

17.19 SEQUENCE NETWORKS OF SYNCHRONOUS MACHINES

A three-phase synchronous machine (generator or motor) with its neutral grounded through an impedance Z_n is shown in Fig. 17.14. The generator is not supplying any load, but because of a fault at the generator terminals, currents I_a, I_b, and I_c, flow through the lines a, b and c. Depending upon the type of fault, one or two line currents may be zero. For example, $I_b = 0$ and $I_c = 0$ when only phase a terminal is grounded. Let the generator induced voltages be E_a, E_b, and E_c in the three phases. For any unbalanced condition, the phase currents can be resolved into symmetrical components.

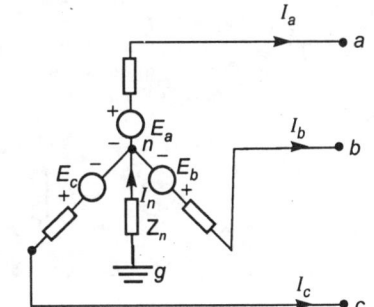

Fig. 17.14. Three-phase synchronous generator with grounded neutral.

Positive-Sequence Network

The windings of a synchronous machine are symmetrical. Thus the generated voltages are of positive sequence only. Fig. 17.15 (a) shows the three-phase positive-sequence network model of a synchronous generator. It consists of a voltage equal to no-load terminal voltage in series with the positive sequence impedance Z_1 of the machine. Fig. 17.15 (a) shows the paths for positive sequence currents. Since the phasor sum of I_{a1}, I_{b1}, and I_{c1} is zero, no positive sequence current flows through Z_n. For this reason, Z_n does not appear in the model. Since it is a balanced network it can be represented by a single-phase network model as shown in Fig. 17.15 (b). The reference bus for the positive sequence network is the neutral of the generator. Further, since no current flows from ground to neutral, the neutral is at ground potential. From Fig. 17.15 (b), the positive-sequence voltage of terminal a with respect to the reference bus is given by

$$V_{a1} = E_a - Z_1 I_{a1} \qquad \qquad ...(17.19.1)$$

Negative-Sequence Network

A synchronous machine does not generate any negative sequence voltage. Negative-sequence network models of a synchronous machine on a three-phase and single-phase basis are shown

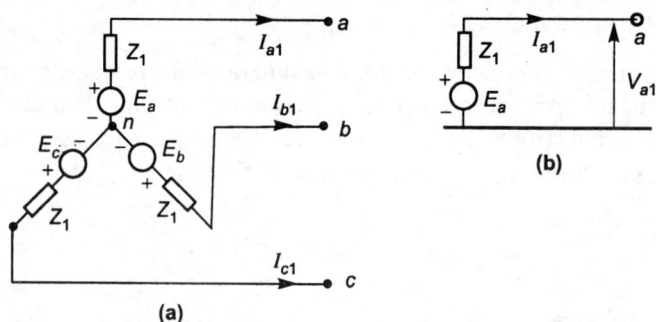

Fig. 17.15. Positive-sequence network of a synchronous machine. (a) Three-phase model (b) Single-phase model.

in Fig. 17.16 (a) and (b) respectively. The reference bus for the negative sequence network is at neutral potential which is the same as the ground potential From Fig. 17.16 (b), the negative-sequence voltage of terminal a with respect to reference bus is given by

$$V_{a2} = -Z_2 I_{a2} \qquad \qquad ...(17.19.2)$$

where Z_2 is the negative-sequence impedance of the generator.

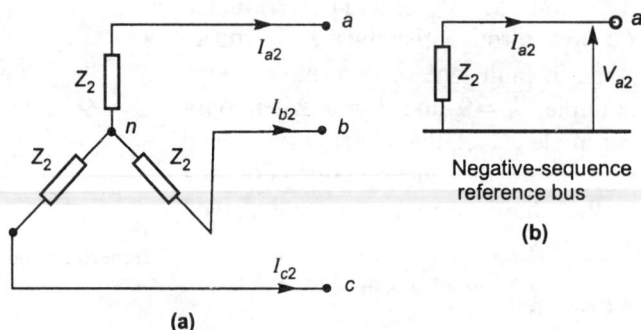

Fig. 17.16. Negative-sequence network of a synchronous machine. (a) Three-phase model; (b) Single-phase model.

Zero-Sequence Network

Let Z_{0g} be the zero-sequence impedance per phase of the generator. The zero-sequence network model on a three-phase basis is shown in Fig. 17.17 (a). Current flowing through Z_n is given by

$$I_n = I_{a0} + I_{b0} + I_{c0} = 3I_{a0}$$

If V_{a0} is the zero-sequence voltage of terminal a with respect to the reference bus (which is at ground potential), then by KVL

$$V_{gn} + V_{na} + V_{ag} = 0$$

$$-Z_n I_n - Z_{0g} I_{a0} - V_{a0} = 0$$

$$V_{a0} = -3Z_n I_{a0} - Z_{0g} I_{a0}$$

or $\qquad V_{a0} = -(3Z_n + Z_{0g}) I_{a0} \qquad \qquad ...(17.19.3)$

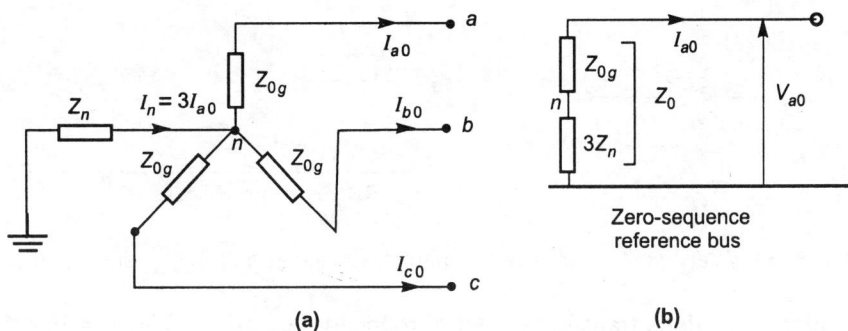

Fig. 17.17. Zero-sequence network of a synchronous machine. (a) Three-phase model; (b) Single-phase model.

Since the single-phase model of zero-sequence network carries only per phase zero-sequence current I_{a0}, this network must have a total zero-sequence impedance given by

$$Z_0 = 3Z_n + Z_{0g} \qquad\qquad ...(17.19.4)$$

From Fig. 17.17 (b), the zero-sequence voltage of terminal a with respect to the reference bus is given by

$$V_{a0} = -Z_0 I_{a0} \qquad\qquad ...(17.19.5)$$

17.20　ZERO-SEQUENCE NETWORKS OF TRANSFORMERS

The zero-sequence network of a system depends on the nature of the connections of three-phase windings for each of the system components. It is to be noted that zero-sequence current is a single-phase component and, therefore, its existence is dependent upon a closed path that must be completed through reference ground. If there is no ground return, zero-sequence currents cannot flow, and in the corresponding zero-sequence networks these networks may be replaced by open circuits.

When magnetising current in a transformer is neglected, the primary ampere-turns balance the secondary ampere-turns. Hence current can flow in the primary only if there is a current in secondary. The zero-sequence currents can flow through the winding connected in star only if the star point is grounded. It the star point is isolated from ground (that is, ungrounded) zero-sequence currents cannot flow in the winding. Since the delta-connected winding has no intentional physical connection to ground, no return path is available for zero-sequence currents. Hence zero-sequence currents cannot flow in the *lines* connected to a delta-connected winding. However, zero-sequence currents can circulate in the legs of the delta-connected winding. The currents flowing round a delta winding may be due to the presence of zero-sequence voltages.

The above rules are to be kept in mind in determining the zero-sequence networks for any three-phase transformer.

Consider the star-star transformer with any one neutral grounded as shown in Fig. 17.18. In this figure, there is a ground return for the primary zero-sequence currents, but this is not enough, because there is no path for the resulting secondary currents. Hence an open circuit exists in the zero-sequence network between a, and a', that is, between the two paths of the system connected by the transformer as shown in Fig. 17.18.

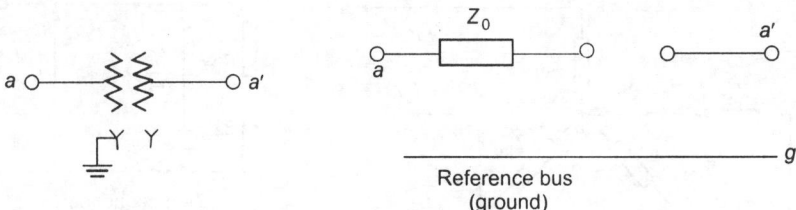

Fig. 17.18. Star/star transformer with one neutral grounded and it zero-sequence network.

Now consider a star-delta transformer with grounded neutral, as shown in Fig. 17.19. In this figure, the primary currents can flow because there is a ground return and paths for the resulting secondary currents. The balancing zero sequence currents can flow in delta legs, but no zero sequence currents flow in the line on the delta side. Thus the zero-sequence network has a path from the line on the star side through the zero-sequence impedance of the transformer to the reference bus. On the other hand, looking into the secondary circuit we see an open circuit (no ground return). It is to be noted that if the neutral point of the star connection is grounded through an impedance Z_n an impedance $3Z_n$ appears in series with Z_0 in the zero-sequence network.

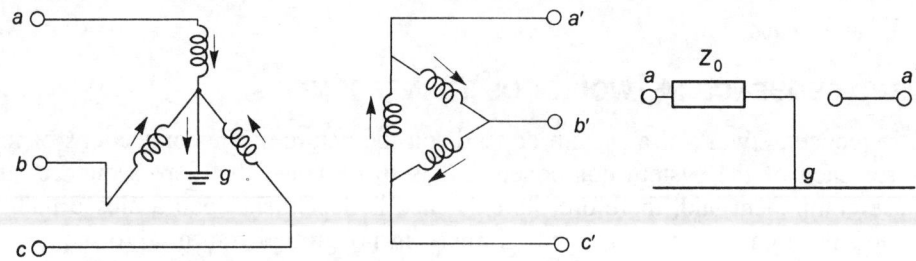

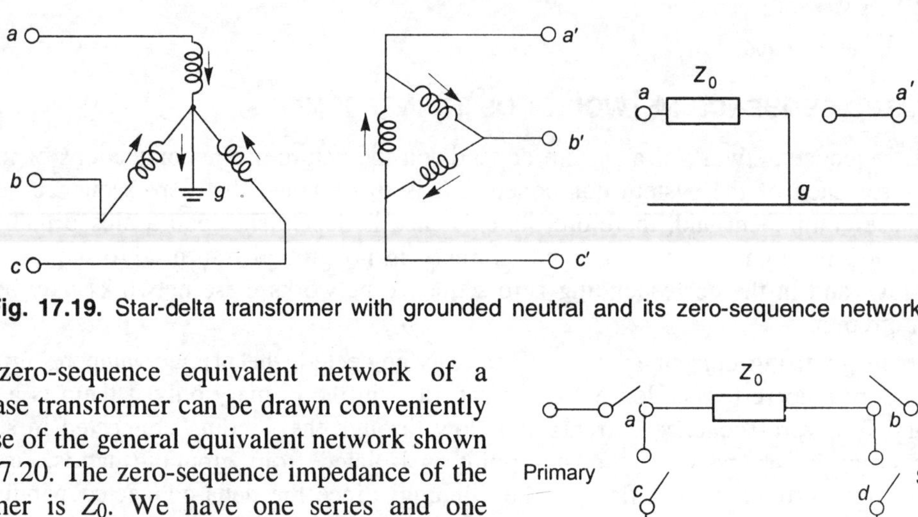

Fig. 17.19. Star-delta transformer with grounded neutral and its zero-sequence network.

The zero-sequence equivalent network of a three-phase transformer can be drawn conveniently by the use of the general equivalent network shown in Fig. 17.20. The zero-sequence impedance of the transformer is Z_0. We have one series and one shunt switch for the each side. The general rule to determine the correct zero-sequence network of the transformer is as follows :

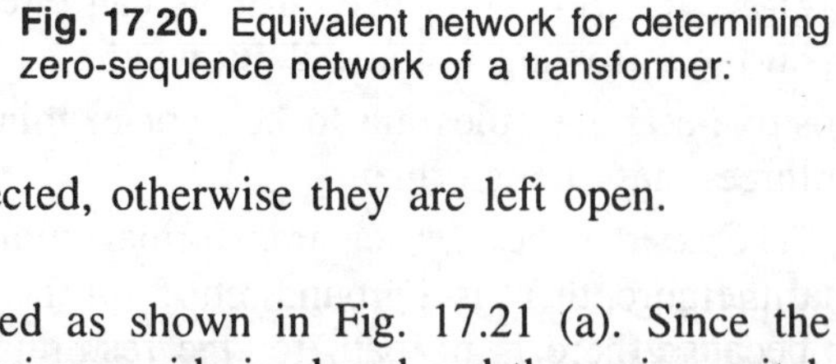

The series switch of a particular side is closed if it is star connected with neutral grounded, and the shunt switch is closed if that side is delta connected, otherwise they are left open.

Fig. 17.20. Equivalent network for determining zero-sequence network of a transformer.

In the final network the switches are not shown.

Consider a Δ/Y transformer with star ungrounded as shown in Fig. 17.21 (a). Since the primary is delta connected, the shunt switch of the primary side is closed and the series switch is left open. The secondary is star connected with neutral ungrounded. Therefore, the series

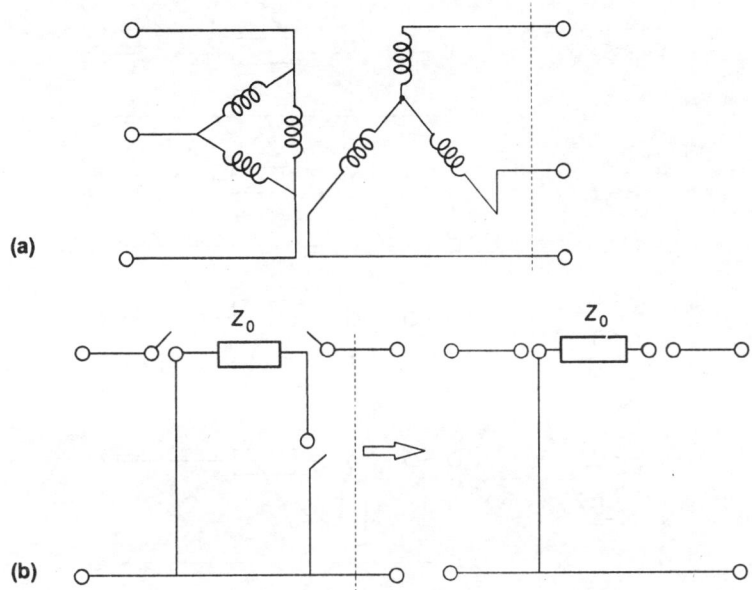

Fig. 17.21. (a) Delta/star transformer with ungrounded star. (b) Zero-sequence equivalent network.

switch is left open and the shunt switch is also left open as shown in Fig. 17.21 (b). Fig. 17.22 shows some typical transformer connections and their zero-sequence networks.

The following problem illustrates a number of different practical possibilities that exist in deriving paths for zero-sequence currents.

Example 17.8 Draw the zero-sequence network for the system shown in Fig. 17.23 (a).

Solution

The zero-sequence, network is shown in Fig. 17.23 (b) in which all reactances and resistances are in per unit.

17.21 ASSEMBLY OF SEQUENCE NETWORKS OF A POWER SYSTEM

A power system network consists of synchronous machines, transformers and transmission lines. We have already discussed the sequence networks of all these components in the proceeding sections. Complete sequence networks of a power system can be drawn by combining these component networks. The positive-sequence network is the same as the single-line impedance diagram drawn for calculation of symmetrical fault currents. The reference bus for positive-sequence network is the system neutral. It should be noted that positive sequences voltages are present in synchronous generators and synchronous motors only.

The negative-sequence network is similar to the positive-sequence network with the exception that the negative-sequence network does not contain any voltage source. The negative-sequence impedances for static components (transmission line and transformers) are the same as positive-sequence impedances. The negative sequence impedance of a synchronous machine may be different from its positive sequence impedance. Any impedance connected between a neutral

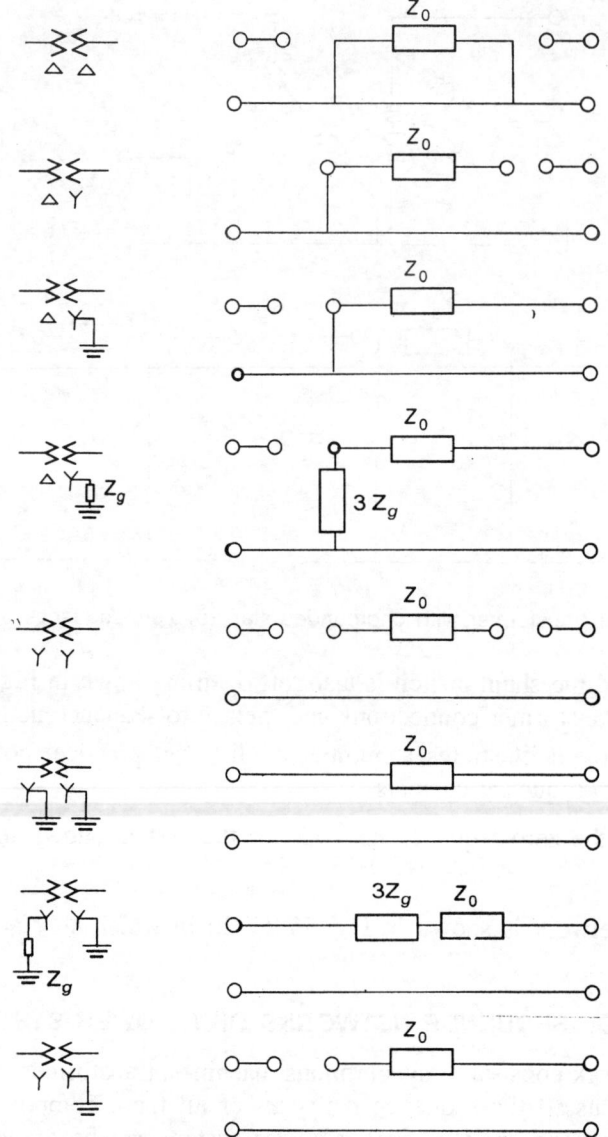

Fig. 17.22. Zero-sequence equivalent networks for transformers.

and ground is not included in the positive and negative-sequence networks because the positive and negative-sequence currents cannot flow in such an impedance.

Complete zero-sequence network of a power system can be obtained by combining the zero-sequence networks of the various components of the system. The zero-sequence network also does not contain any voltage source. The zero-sequence impedances should be connected with sufficient care. Zero-sequence networks of all possible transformer connections have been discussed in Section 17.20. It should be noted that impedance of the ground device in generator or transformer becomes three times it actual value in a zero-sequence network.

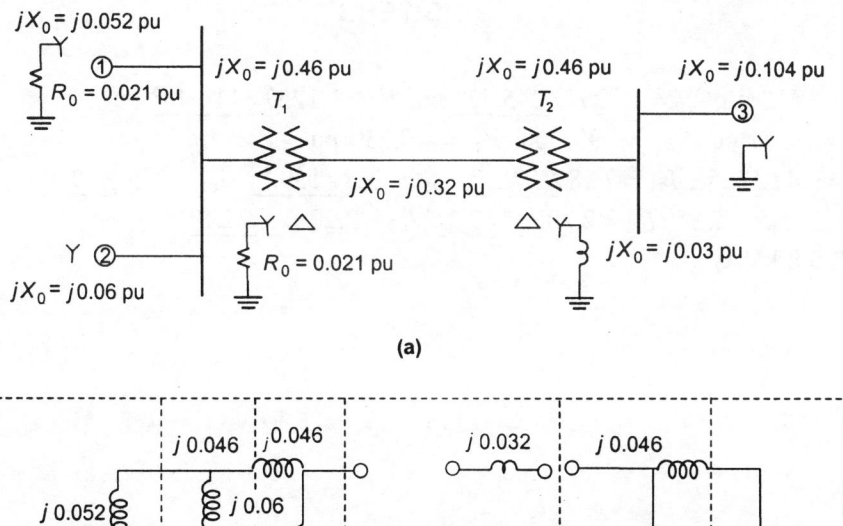

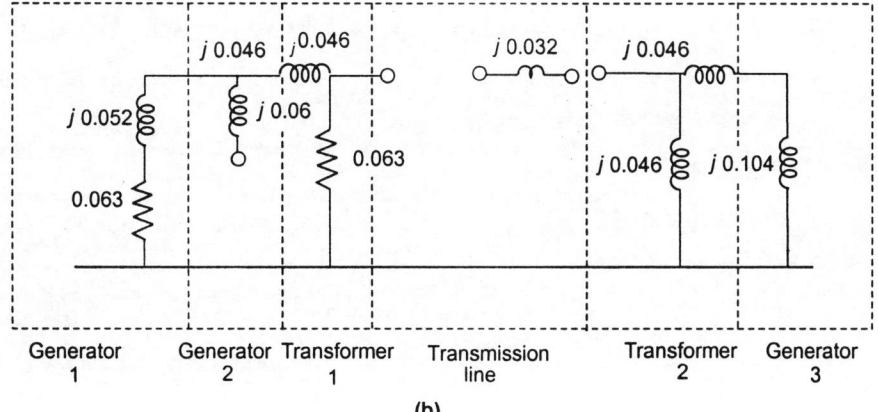

Fig. 17.23. (a) Network of Example 17.8. (b) Zero-sequence network of Example 17.8.

EXERCISES

1. The three sequence components of bus voltage for a given bus are :
   $$V_0 = -0.105 \text{ pu}, \quad V_1 = 0.953 \text{ pu}, \quad V_2 = -0.230 \text{ pu}.$$

   Determine (a) the three-phase voltages and (b) the three sequence components of phase voltages obtained in (a).

2. A set of unbalanced line currents in a three phase, four-wire system is as follows :
   $$\mathbf{I}_a = -j6 \text{ A}, \quad \mathbf{I}_b = (-8 + j5) \text{ A} \quad \text{and} \quad \mathbf{I}_c = 7 \text{ A}.$$

   Determine the zero-, positive-, and negative-sequence components of the current.

3. The unbalanced set of line-to-neutral voltages at a fault point in a power system are given by
   $$\mathbf{V}_a = 7.5 \underline{/-30^\circ} \text{ kV} ; \quad \mathbf{V}_b = 7.5 \underline{/-150^\circ} \text{ kV} ; \quad \mathbf{V}_c = 0$$
   $$\mathbf{I}_a = 2500 \underline{/-90^\circ} \text{ A} ; \quad \mathbf{I}_b = 2500 \underline{/90^\circ} \text{ A} ; \quad \mathbf{I}_c = 0$$

   Determine the complex power calculated with (a) symmetrical components (b) unbalanced phase components.

ANSWERS

1. $V_a = 0.618$ pu, $V_b = 1.1257 \underline{/245.52°}$ pu, $V_c = 1.1257 \underline{/114.48°}$ pu
   $V_{a0} = 0.105$ pu, $V_{a1} = 0.953$ pu, $V_{a2} = -0.230$ pu
2. $I_{a0} = 0.47 \underline{/225°}$, $I_{a1} = 7.28 \underline{/259.9°}$, $I_{b1} = 7.28 \underline{/139.9°}$, $I_{c1} = 7.28 \underline{/19.9°}$,
   $I_{a2} = 2.205 \underline{/42.9°}$, $I_{b2} = 2.205 \underline{/162.9°}$, $I_{c2} = 2.205 \underline{/-77.1°}$
3. 32475.8 kVAr

18

Unsymmetrical Faults

18.1 INTRODUCTION

The term unsymmetrical fault is used to mean an unbalanced condition. It is a connection or situation which causes an unbalance among the three phases. If an unbalanced connection is attached at a point F on a balanced system then F is called the *fault point* of the system. Thus, if there is a single line-to-ground fault at bus M then bus M becomes the fault point. Similarly, if an unbalanced three-phase load is connected at bus N then bus N becomes the fault point.

An unsymmetrical *shunt fault* is an unbalance between phases or between phase and ground. A *series fault* is an unbalance in the line impedances. It does not involve any connection between lines or between line and ground at fault point.

Shunt faults in a three-phase system can be classified under the following headings :

1. Single line-to-ground (LG) fault.

2. Line-to-line (LL) fault.

3. Double line-to-ground (LLG) fault.

4. Three-phase short circuit (LLL) fault.

5. Three-phase-to-ground (LLLG) fault.

In general, a single line-to-ground (LG) fault on a transmission system occurs when one conductor falls to ground or comes in contact with the neutral conductor. A line-to-line (LL) fault occurs when two conductors are short circuited. A double line-to-ground (LLG) fault occurs when two conductors fall and connected through ground or when two conductors come in contact with the neutral of a three-phase grounded system.

Faults of the types LG, LL, and LLG are unsymmetrical faults, while LLL and LLLG are symmetrical faults. In case of a symmetrical fault, the system remains symmetrical i.e., balanced even after the fault. For unsymmetrical faults the voltages and currents become unbalanced after the fault.

These faults may occur at the terminals of the generators or any part of the power system, and may take place either through zero impedance or through an impedance. This fault impedance may arise on account of the resistance of the arc between the conductors or due to the tower footing resistance. The resistance of arc is usually negligible on high voltage circuits. The resistance of tower footing ranges from 5 to 20 ohms but is usually neglected giving results for the most severe conditions.

18.2 ASSUMPTIONS

The following assumptions are made in the analysis of unsymmetrical faults :
1. In most of the cases load currents in the system are negligible in comparison to fault currents.
2. The network impedances upto the fault are balanced, so that the phase sequence components are independent of one another.
3. By convention fault current is taken positive when directed out of the fault point.
4. Resistance of various elements are neglected.

To simplify the interconnection of sequence networks the fault is assumed to be symmetrical with respect to the reference phase, otherwise the sequence networks will require to be interconnected by phase shifting devices to represent the operators α and α^2. Usually phase a is selected the reference phase in a system of phase sequence abc.

18.3 SEQUENCE VOLTAGES OF A GENERATOR

Consider a symmetrically designed three-phase generator. Let E_a, E_b, and E_c be the generated voltages and E_{a0}, E_{a1} and E_{a2} the zero, positive and negative sequence generated voltages of phase a respectively.

$$E_b = \alpha^2 E_a, \qquad E_c = \alpha E_a$$

$$E_{a0} = \frac{1}{3}(E_a + E_b + E_c) = \frac{1}{3}(E_a + \alpha^2 E_a + \alpha E_a) = \frac{E_a}{3}(1 + \alpha^2 + \alpha) = 0 \qquad \text{...(18.3.1)}$$

$$E_{a1} = \frac{1}{3}(E_a + \alpha E_b + \alpha^2 E_c) = \frac{1}{3}(E_a + \alpha^3 E_a + \alpha^3 E_a) = \frac{E_a}{3}(1 + 1 + 1) = E_a \qquad \text{...(18.3.2)}$$

$$E_{a2} = \frac{1}{3}(E_a + \alpha^2 E_b + \alpha E_c) = \frac{1}{3}(E_a + \alpha^4 E_a + \alpha^2 E_a) = \frac{E_a}{3}(1 + \alpha + \alpha^2) = 0 \qquad \text{...(18.3.3)}$$

These relations show that a symmetrically designed generator generates voltage of positive sequence only. The zero and negative sequence generated voltages are zero.

18.4 SEQUENCE VOLTAGES AT A FAULT POINT

Let V_a, V_b, and V_c be the prefault phase voltages of phases a, b, and c, at the fault point. V_f is the prefault voltage of phase a at the fault point. Suppose that V_{a0}, V_{a1}, and V_{a2} denote the zero, positive, and negative sequence voltages of phase a.

For a balanced system

$$V_b = \alpha^2 V_a, \qquad V_c = \alpha V_a$$

$$V_{a0} = \frac{1}{3}(V_a + V_b + V_c) = \frac{1}{3}(V_a + \alpha^2 V_a + \alpha V_a) = \frac{V_a}{3}(1 + \alpha^2 + \alpha) = 0 \qquad \ldots (18.4.1)$$

$$V_{a1} = \frac{1}{3}(V_a + \alpha V_b + \alpha^2 V_c) = \frac{1}{3}(V_a + \alpha^3 V_a + \alpha^3 V_a) = \frac{V_a}{3}(1 + 1 + 1) = V_a \qquad \ldots (18.4.2)$$

$$V_{a2} = \frac{1}{3}(V_a + \alpha^2 V_a + \alpha V_c) = \frac{1}{3}(V_a + \alpha^4 V_a + \alpha^2 V_a) = \frac{V_a}{3}(1 + \alpha + \alpha^2) = 0 \qquad \ldots (18.4.3)$$

These relations show that for a balanced system the zero and negative sequence voltages at the fault point are zero and the line-to-neutral voltage of phase a at the fault will be positive sequence voltage only. This voltage is denoted by V_f. This is prefault voltage.

$$\left.\begin{array}{l} V_{a1} = V_f \\ V_{a0} = 0 \\ V_{a2} = 0 \end{array}\right\} \qquad \ldots (18.4.4)$$

$\therefore$

18.5 GENERAL PROCEDURE

The general procedure adopted in the analysis of various types of fault is outlined below :

1. Circuit Diagram

A circuit diagram of the fault showing all phase connections to the fault is drawn. The directions of currents and the polarities of voltages are marked properly.

In Fig. 18.1, a, b, c are the phases of the original balanced system. I_a, I_b and I_c are the fault currents flowing from the original balanced system to the fault point. V_a, V_b and V_c denote the phase voltages at the fault point. V_f is the pre-fault voltage of phase a at the fault point. Since the system is balanced V_f is the positive sequence voltage.

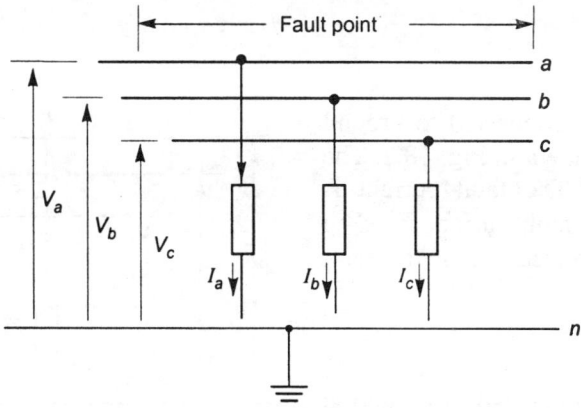

Fig. 18.1. Circuit diagram of the fault.

2. Boundary Conditions

For a given type of fault the relations between known phase voltages and currents at the fault point are written. These conditions at this fault point are called *boundary conditions*.

3. Transformation

The equations obtained in step 2 are solved to find the symmetrical components of voltages and currents.

This process is called the *transformation* from the *a-b-c* to 0-1-2 system. The transformation is carried out with the help of relations :

$$\mathbf{V}_{012} = \mathbf{A}^{-1}\,\mathbf{V}_{abc}$$

4. Sequence Currents and Sequence Voltages

The sequence currents and sequence voltages obtained in step 3 are examined to determine the interconnection of sequence networks. Required impedances may be added in the sequence networks.

5. Interconnection of Sequence Networks

The sequence networks are interconnected in such a way that the equations describing the fault conditions are satisfied and that the interconnection represents the constraints impressed on the system by the fault.

6. Informations from Sequence Networks

Voltage of phase *a* and components of current are found from the sequence network. The voltages and currents for phases *b* and *c* are then found by the known angular relationships of the balanced sets.

18.6 SINGLE LINE-TO-GROUND (LG) FAULT

The single line-to-ground fault occurs when one conductor falls to ground or contacts the neutral conductor.

Circuit Diagram

Suppose that phase *a* is connected to ground at the fault point *F* as shown in Fig. 18.2. The fault impedance is Z_f. The fault current is $I_{af} = I_a$. By convention, fault current is taken to be positive when directed *out* of the fault point.

Boundary Conditions

Since only phase *a* is connected to ground at the fault, phases *b* and *c* are open circuited and carry no current. That is, $I_b = 0$, $I_c = 0$, and the fault current is I_a. The voltage above ground at fault point *F* is $V_a = Z_f I_a$.

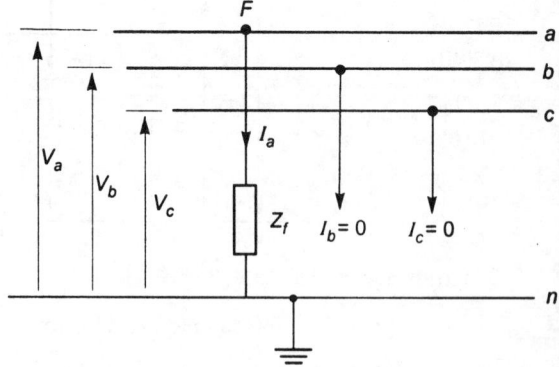

Fig. 18.2. Single line-to-ground fault.

Transformation

The symmetrical components of fault current in phase a at the fault point can be written as

$$I_{a0} = \frac{1}{3}(I_a + I_b + I_c) = \frac{1}{3}(I_a + 0 + 0) = \frac{1}{3}I_a$$

$$I_{a1} = \frac{1}{3}(I_a + \alpha I_b + \alpha^2 I_c) = \frac{1}{3}(I_a + 0 + 0) = \frac{1}{3}I_a$$

$$I_{a2} = \frac{1}{3}(I_a + \alpha^2 I_b + \alpha I_c) = \frac{1}{3}(I_a + 0 + 0) = \frac{1}{3}I_a$$

$$\therefore \qquad I_{a0} = I_{a1} = I_{a2} = \frac{1}{3}I_a \qquad\qquad\qquad ...(18.6.1)$$

This relation can also be found by matrix method as follows :

$$\mathbf{I}_{012} = \mathbf{A}^{-1}\,\mathbf{I}_{abc}$$

$$\begin{bmatrix} I_{a0} \\ I_{a1} \\ I_{a2} \end{bmatrix} = \frac{1}{3}\begin{bmatrix} 1 & 1 & 1 \\ 1 & \alpha & \alpha^2 \\ 1 & \alpha^2 & \alpha \end{bmatrix}\begin{bmatrix} I_a \\ I_b \\ I_c \end{bmatrix} = \frac{1}{3}\begin{bmatrix} 1 & 1 & 1 \\ 1 & \alpha & \alpha^2 \\ 1 & \alpha^2 & \alpha \end{bmatrix}\begin{bmatrix} I_a \\ 0 \\ 0 \end{bmatrix} = \frac{I_a}{3}\begin{bmatrix} 1 \\ 1 \\ 1 \end{bmatrix}$$

$$I_{a0} = I_{a1} = I_{a2} = \frac{1}{3}I_a \qquad\qquad\qquad ...(18.6.1)$$

Hence we conclude that in the case of a single line-to-ground fault, the sequence currents are equal.

The sequence voltages at the fault point are found as follows :

$$V_{a0} = E_{a0} - Z_{a0}\,I_{a0}$$

$$V_{a1} = E_{a1} - Z_{a1}\,I_{a1}$$

$$V_{a2} = E_{a2} - Z_{a2}\,I_{a2}$$

where E_{a0}, E_{a1} and E_{a2} are the sequence generated voltages of phase a, and Z_{a0}, Z_{a1}, Z_{a2} are the sequence impedances to the flow of currents I_{a0}, I_{a1}, and I_{a2} respectively. For a balanced system

$$E_{a0} = 0, \quad E_{a2} = 0, \quad E_{a1} = V_f$$

We know that

$$V_a = V_{a0} + V_{a1} + V_{a2}$$

$$Z_f I_a = -Z_{a0}\,I_{a0} + V_f - Z_{a1}\,I_{a1} - Z_{a2}\,I_{a2} \qquad\qquad ...(18.6.2)$$

Combination of Eqs. (18.6.1) and (18.6.2) gives

$$Z_f I_a = V_f - \frac{I_a}{3}(Z_{a0} + Z_{a1} + Z_{a2})$$

$$I_a = \frac{V_f}{Z_f + \frac{1}{3}(Z_{a0} + Z_{a1} + Z_{a2})} \qquad\qquad ...(18.6.3)$$

Since all the impedances and the voltage V_f at the fault point are known, the fault current can be determined from Eq. (18.6.3). By combining Eqs. (18.6.1) and (18.6.3) we get the sequence current as

$$I_{a0} = I_{a1} = I_{a2} = \frac{V_f}{Z_{a0} + Z_{a1} + Z_{a2} + 3Z_f} \qquad \ldots(18.6.4)$$

The phase voltages V_a, V_b and V_c at the fault point F can be found from the following relations :

$$V_a = V_{a0} + V_{a1} + V_{a2} = Z_f I_a$$

$$V_b = V_{a0} + \alpha^2 V_{a1} + \alpha V_{a2}$$

$$V_c = V_{a0} + \alpha V_{a1} + \alpha^2 V_{a2}$$

Interconnection of Sequence Networks

Since the sequence currents are equal, the sequence networks should be connected in series to satisfy this condition. If an external impedance of $3Z_f$ is connected as shown in Fig. 18.3 (a). Eq. (18.6.4) is completely satisfied. Fig. 18.3 (a) is an equivalent circuit showing the interconnection of sequence networks to simulate LG fault for phase a. With the help of such an equivalent circuit all the required equations can be written by inspection.

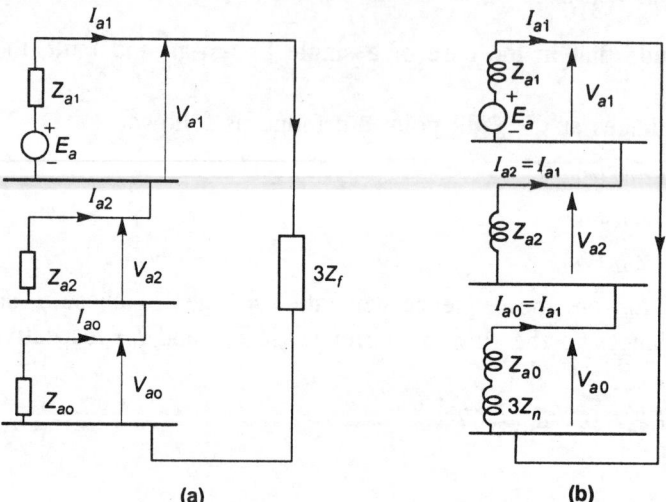

(a)   (b)

Fig. 18.3. (a) Interconnection of sequence networks to simulate LG fault for phase a. (b) Equivalent circuit.

The phase currents can be written as

$$I_a = \frac{3V_f}{Z_{a0} + Z_{a1} + Z_{a2} + 3Z_f} \qquad \ldots(18.6.5)$$

$$I_b = 0, \qquad I_c = 0 \qquad \ldots(18.6.6)$$

Special Case

If $Z_f = 0$, then $I_{a0} = I_{a1} = I_{a3} = \dfrac{V_f}{Z_{a0} + Z_{a1} + Z_{a2}}$

The equivalent circuit showing interconnection of sequence networks is given in Fig. 18.3 (b).

18.7 LINE-TO-LINE FAULT

A line-to-line fault occurs when two conductors are short circuited. Fig. 18.4 shows a three-phase system with a line-to-line (LL) fault between phases b and c. The fault impedance is assumed to be Z_f. The LL fault is placed between lines b and c in order that the fault be symmetrical with respect to the reference phase a which is unfaulted.

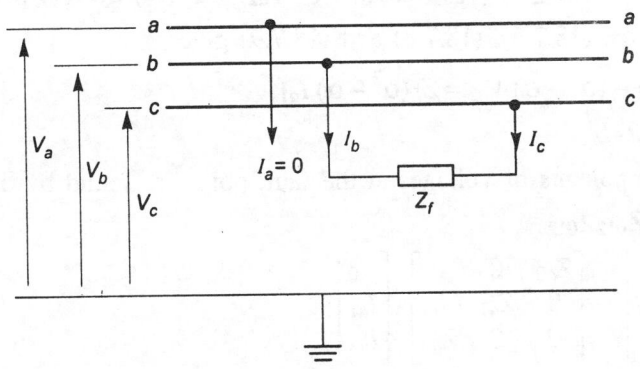

Fig. 18.4. Line-to-line fault.

Boundary Conditions

The boundary conditions are

$$I_a = 0 \qquad\qquad \text{...(18.7.1)}$$

$$I_b = -I_c \qquad\qquad \text{...(18.7.2)}$$

$$V_b - V_c = Z_f I_b \qquad\qquad \text{...(18.7.3)}$$

Fault current, $I_f = I_b$ \qquad\qquad ...(18.7.4)

Transformation

The symmetrical components of fault current in phase a at the fault point can be written as

$$I_{a0} = \frac{1}{3}(I_a + I_b + I_c) = \frac{1}{3}(0 + I_b - I_b) = 0 \qquad\qquad \text{...(18.7.5)}$$

$$I_{a1} = \frac{1}{3}(I_a + \alpha I_b + \alpha^2 I_c) = \frac{1}{3}(0 + \alpha I_b - \alpha^2 I_b) = \frac{1}{3}(\alpha - \alpha^2) I_b \qquad\qquad \text{...(18.7.6)}$$

$$I_{a2} = \frac{1}{3}(I_a + \alpha^2 I_b + \alpha I_c) = \frac{1}{3}(0 + \alpha^2 I_b - \alpha I_b) = -\frac{1}{3}(\alpha - \alpha^2) I_b \qquad\qquad \text{..:(18.7.7)}$$

$$I_{a1} = -I_{a2} = \frac{1}{3}(\alpha - \alpha^2) I_b \qquad\qquad \text{...(18.7.8)}$$

The sequence currents can also be found by matrix method as follows :

$$\mathbf{I}_{012} = \mathbf{A}^{-1} \mathbf{I}_{abc}$$

$$\begin{bmatrix} I_{a0} \\ I_{a1} \\ I_{a2} \end{bmatrix} = \frac{1}{3} \begin{bmatrix} 1 & 1 & 1 \\ 1 & \alpha & \alpha^2 \\ 1 & \alpha^2 & \alpha \end{bmatrix} \begin{bmatrix} I_a \\ I_b \\ I_c \end{bmatrix} = \frac{1}{3} \begin{bmatrix} 1 & 1 & 1 \\ 1 & \alpha & \alpha^2 \\ 1 & \alpha^2 & \alpha \end{bmatrix} \begin{bmatrix} 0 \\ I_b \\ -I_b \end{bmatrix}$$

$\therefore \qquad I_{a0} = 0, \quad \text{and} \quad I_{a1} = -I_{a2}$

Expressing V_b, V_c and I_b in terms of their sequence components, Eq. (18.7.3) can be written as

$$(V_{a0} + \alpha^2 V_{a1} + \alpha V_{a2}) - (V_{a0} + \alpha V_{a1} + \alpha^2 V_{a2}) = Z_f (I_{a0} + \alpha^2 I_{a1} + \alpha I_{a2}) \qquad \ldots(18.7.9)$$

Combination of Eqs. (18.7.5), (18.7.8) and (18.7.9) gives

$$(\alpha^2 - \alpha) V_{a1} - (\alpha^2 - \alpha) V_{a2} = Z_f (\alpha^2 - \alpha) I_{a1}$$

or $\qquad V_{a1} - V_{a2} = Z_f I_{a1} \qquad \qquad \ldots(18.7.10)$

The sequence components of voltages at the fault point are found by the relations given by

$$\mathbf{V}_{012} = \mathbf{E} - \mathbf{Z}_{012} \mathbf{I}_{012}$$

$$\begin{bmatrix} V_{a0} \\ V_{a1} \\ V_{a2} \end{bmatrix} = \begin{bmatrix} 0 \\ V_f \\ 0 \end{bmatrix} - \begin{bmatrix} Z_{a0} & 0 & 0 \\ 0 & Z_{a1} & 0 \\ 0 & 0 & Z_{a2} \end{bmatrix} \begin{bmatrix} I_{a0} \\ I_{a1} \\ I_{a2} \end{bmatrix}$$

$V_{a0} = -Z_{a0} I_{a0} \qquad \qquad \ldots(18.7.11)$

$V_{a1} = V_f - Z_{a1} I_{a1} \qquad \qquad \ldots(18.7.12)$

$V_{a2} = -Z_{a2} I_{a2} \qquad \qquad \ldots(18.7.13)$

From Eqs. (18.7.12) and (18.7.13)

$V_{a1} - V_{a2} = V_f - Z_{a1} I_{a1} + Z_{a2} I_{a2} \qquad \qquad \ldots(18.7.14)$

Combination of Eqs. (18.7.10), (18.7.14) and (18.7.8) gives

$Z_f I_{a1} = V_f - Z_{a1} I_{a1} - Z_{a2} I_{a2}$

$(Z_{a1} + Z_{a2} + Z_f) I_{a1} = V_f$

$$I_{a1} = \frac{V_f}{Z_{a1} + Z_{a2} + Z_f} \qquad \qquad \ldots(18.7.15)$$

The fault current is given by

$I_f = I_b = -I_c$

$$I_b = I_{a0} + \alpha^2 I_{a1} + \alpha I_{a2} = 0 + \alpha^2 I_{a1} - \alpha I_{a1} = (\alpha^2 - \alpha) I_{a1} = \frac{(\alpha^2 - \alpha) V_f}{Z_{a1} + Z_{a2} + Z_f} \qquad \ldots(18.7.16)$$

$$= \frac{-j\sqrt{3}\, V_f}{Z_{a1} + Z_{a2} + Z_f} \qquad \qquad \ldots(18.7.17)$$

Equation (18.7.5) shows that for a line-to-line fault the zero-sequence component of current I_{a0} is equal to zero. This is an expected result since there is no earth return path. Equation

(18.7.8) shows that the positive-sequence component of current is equal in magnitude but opposite in phase to the negative-sequence component of current.

Interconnection of Sequence Networks

Since $I_{a0} = 0$, the zero-sequence network is open (unconnected). Equations (18.7.8) and (18.7.10) suggest parallel connection of positive- and negative-sequence networks through a series impedance Z_f as shown in Fig. 18.5.

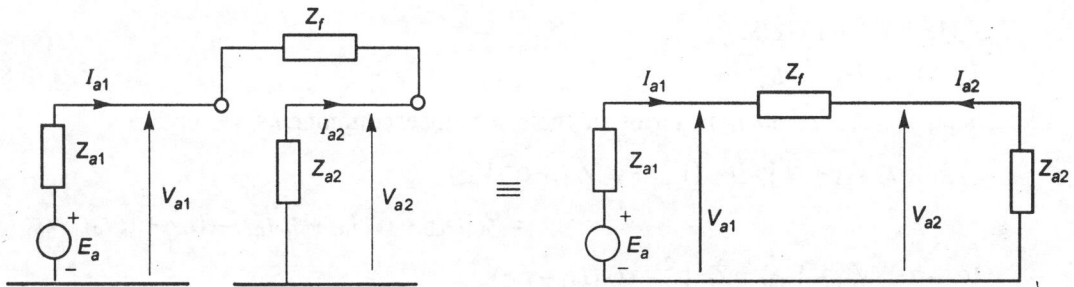

Fig. 18.5. Interconnection of sequence networks for a line-to-line (LL) fault.

Special Case

If $Z_f = 0$ (solid fault)

$$I_{a1} = \frac{V_f}{Z_{a1} + Z_{a2}} \; ; \qquad I_f = I_b = \frac{(\alpha^2 - \alpha) V_f}{Z_{a1} + Z_{a2}} = \frac{-j \sqrt{3} \, V_f}{Z_{a1} + Z_{a2}}$$

18.8 DOUBLE LINE-TO-GROUND (LLG OR DLG) FAULT

The double line-to-ground fault is shown in Fig. 18.6. For the sake of symmetry with respect to phase a, phase b and phase c are faulted. Phase a is unfaulted. Phase b has fault impedance of Z_f, phase c has fault impedance of Z_f and the common line-to-ground fault impedance is Z_g.

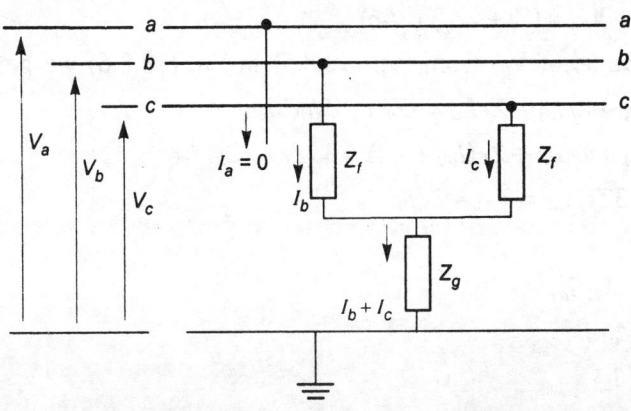

Fig. 18.6. Double line-to-ground (LLG) fault.

Boundary Conditions

The boundary conditions at the fault point F are as follows :

$$I_a = 0 \qquad \qquad \ldots(18.8.1)$$

$$V_b = Z_f I_b + Z_g (I_b + I_c) = (Z_f + Z_g) I_b + Z_g I_c \qquad \ldots(18.8.2)$$

$$V_c = Z_f I_c + Z_g (I_b + I_c) = (Z_f + Z_g) I_c + Z_g I_b \qquad \ldots(18.8.3)$$

Transformation

From Eqs. (18.8.2) and (18.8.3),

$$V_b - V_c = Z_f (I_b - I_c) \qquad \ldots(18.8.4)$$

Expressing V_b, V_c, I_b and I_c in terms of their sequence components we obtain

$$(V_{a0} + \alpha^2 V_{a1} + \alpha V_{a2}) - (V_{a0} + \alpha V_{a1} + \alpha^2 V_{a2})$$

$$= Z_f [(I_{a0} + \alpha^2 I_{a1} + \alpha I_{a2}) - (I_{a0} + \alpha I_{a1} + \alpha^2 I_{a2})]$$

$$(\alpha^2 - \alpha)(V_{a1} - V_{a2}) = Z_f (\alpha^2 - \alpha)(I_{a1} - I_{a2})$$

$$V_{a1} - V_{a2} = Z_f (I_{a1} - I_{a2})$$

$$V_{a1} - Z_f I_{a1} = V_{a2} - Z_f I_{a2} \qquad \ldots(18.8.5)$$

From Eqs. (18.8.2) and (18.8.3)

$$V_b + V_c = (Z_f + 2Z_g)(I_b + I_c)$$

Expressing V_b, V_c, I_b and I_c in terms of their sequence components we get

$$(V_{a0} + \alpha^2 V_{a1} + \alpha V_{a2}) + (V_{a0} + \alpha V_{a1} + \alpha^2 V_{a2})$$

$$= (Z_f + 2Z_g)[(I_{a0} + \alpha^2 I_{a1} + \alpha I_{a2}) + (I_{a0} + \alpha I_{a1} + \alpha^2 I_{a2})]$$

$$2V_{a0} + (\alpha + \alpha^2) V_{a1} + (\alpha + \alpha^2) V_{a2} = (Z_f + 2Z_g)[2I_{a0} + (\alpha + \alpha^2) I_{a1} + (\alpha + \alpha^2) I_{a2}]$$

$$2V_{a0} - V_{a1} - V_{a2} = (Z_f + 2Z_g)(2I_{a0} - I_{a1} - I_{a2})$$

Since $\quad I_a = 0, \qquad I_{a0} + I_{a1} + I_{a2} = I_a = 0$

$$2V_{a0} - V_{a1} - V_{a2} = (Z_f + 2Z_g)(3I_{a0}) \qquad \ldots(18.8.6)$$

Substituting the value of V_{a2} from Eq. (18.8.5) in Eq. (18.8.6) we get

$$2V_{a0} - 2V_{a1} + Z_f I_{a1} - Z_f I_{a2} = (Z_f + 2Z_g)(3I_{a0})$$

$$2V_{a0} - 2V_{a1} + Z_f I_{a1} + Z_f (I_{a0} + I_{a1}) = 3(Z_f + 2Z_g) I_{a0}$$

$$V_{a0} - (Z_f + 3Z_g) I_{a0} = V_{a1} - Z_f I_{a1} \qquad \ldots(18.8.7)$$

We know that

$$V_{a1} = E_{a1} - Z_{a1} I_{a1}$$

or $\qquad V_{a1} = V_f - Z_{a1} I_{a1}$

$$V_{a2} = - Z_{a2} I_{a2}$$

$$V_{a0} = - Z_{a0} I_{a0}$$

Substitution of the values of V_{a0}, and V_{a1} in Eq. (18.8.7) gives

$$V_f - Z_{a1} I_{a1} - Z_f I_{a1} = - Z_{a0} I_{a0} - (Z_f + 3Z_g) I_{a0}$$

$$V_f - (Z_{a1} + Z_f) I_{a1} = - (Z_{a0} + Z_f + 3Z_g) I_{a0} \qquad ...(18.8.8)$$

Substituting the value of V_{a1} in (18.8.5) we get

$$V_{a1} - Z_f I_{a1} = - Z_{a2} I_{a2} - Z_f I_{a2}$$

$$V_{a1} - Z_f I_{a1} = - (Z_{a2} + Z_f) I_{a2} \qquad ...(18.8.9)$$

Combining Eqs. (18.8.8) and (18.8.9) we get

$$V_f - (Z_{a1} + Z_f) I_{a1} = - (Z_{a2} + Z_f) I_{a2}$$

$$= - (Z_{a0} + Z_f + 3Z_g) I_{a0} \qquad ...(18.8.10)$$

Interconnection of Sequence Networks

The relationship $I_{a0} + I_{a1} + I_{a2} = 0$ requires that the reference points N_0, N_1 and N_2 be connected at one point. According to Eq. (18.8.5) the voltage across the positive- and negative-sequence networks are equal if an external impedance Z_f is connected in series with each network. Equation (18.8.7) suggests that an external impedance of $(Z_f + 3Z_g)$ be connected to the zero-sequence network. The interconnection sequence networks satisfying all these conditions is shown in Fig. 18.7. It is seen from Eq. (18.8.10) that the sequence networks are connected in parallel.

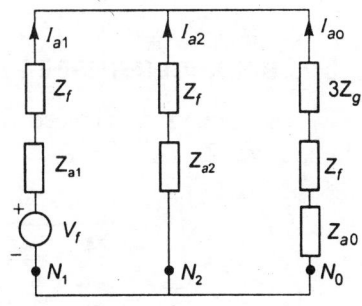

Fig. 18.7. Interconnection of sequence networks to simulate LLG fault.

From the equivalent network we have the positive-sequence current

$$I_{a1} = \cfrac{V_f}{Z_{a1} + Z_f + \left[\dfrac{(Z_{a2} + Z_f)(Z_{a0} + Z_f + 3Z_g)}{Z_{a2} + Z_{a0} + 2Z_f + 3Z_g)}\right]} \qquad ...(18.8.11)$$

The negative-sequence current is

$$I_{a2} = - I_{a1} \left(\frac{Z_{a0} + Z_f + 3Z_g}{Z_{a2} + Z_{a0} + 2Z_f + 3Z_g}\right) \qquad ...(18.8.12)$$

The zero-sequence current is found from

$$I_{a0} = - (I_{a1} + I_{a2}) \qquad ...(18.8.13)$$

Special Case

If $Z_f = 0$

$$I_{a1} = \cfrac{V_f}{Z_{a1} + \left(\dfrac{Z_{a2}(Z_{a0} + 3Z_f)}{Z_{a2} + Z_{a0} + 3Z_g}\right)} \qquad ...(18.8.14)$$

Also, if $Z_g = 0$,

$$I_{a1} = \frac{V_f}{Z_{a1} + \dfrac{Z_{a0} Z_{a2}}{Z_{a0} + Z_{a2}}} \qquad \qquad ...(18.8.15)$$

In our previous discussion, we have assumed that the single line-to-ground fault occurred on phase a, and the line-to-line faults occurred between phases b and c. These choices were not arbitrary; it is simpler to deal with these 'canonical cases'. Suppose that phase b is now connected to ground at the fault point. That is, there is an LG fault on phase b. To tackle this problem we relabel the phases so that phase b is now labelled a, phase c is now labelled b, and phase a is now labelled c. Similar type of relabelling is done for other types of faults to reduce them to the canonical forms. Care should be taken to ensure that (originally) positive-sequence sources remain positive-sequence after relabelling. This is done by labelling the conductors in the same cyclic order. Thus, the conductors labelled abc are labelled bca and cab.

18.9 BALANCED THREE-PHASE FAULT

Fig. 18.8 shows a balanced three-phase fault on phases a, b and c, all through the same fault impedance Z_f.

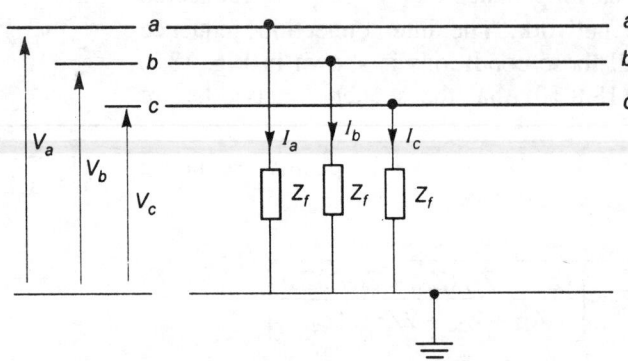

Fig. 18.8. Balanced three-phase fault.

Boundary Conditions

The phase voltages at the fault point are given by

$$V_a = Z_f I_a \qquad \qquad ...(18.9.1)$$
$$V_b = Z_f I_b \qquad \qquad ...(18.9.2)$$
$$V_c = Z_f I_c \qquad \qquad ...(18.9.3)$$

Transformation

The positive-sequence voltage is given by

$$V_{a1} = \frac{1}{3}(V_a + \alpha V_b + \alpha^2 V_c) \qquad \qquad ...(18.9.4)$$

Substituting the values of V_a, V_b and V_c we obtain

$$V_{a1} = \frac{1}{3}(I_a + \alpha I_b + \alpha^2 I_c) Z_f$$

or $\qquad V_{a1} = Z_f I_{a1}$ ...(18.9.5)

Similarly,

$\qquad V_{a2} = Z_f I_{a2}$ ...(18.9.6)

$\qquad V_{a0} = Z_f I_{a0}$ ...(18.9.7)

For a balanced system,

$\qquad V_{a1} = E_{a1} - Z_{a1} I_{a1}$ ...(18.9.8)

$\qquad V_{a2} = 0 - Z_{a2} I_{a2}$ ...(18.9.9)

$\qquad V_{a0} = 0 - Z_{a0} I_{a0}$ ...(18.9.10)

Combination of Eqs. (18.9.5) and (18.9.8) gives

$\qquad V_{a1} = E_{a1} - Z_{a1} I_{a1} = Z_f I_{a1}$ ...(18.9.11)

$\therefore \qquad I_{a1} = \dfrac{V_{a1}}{Z_{a1} + Z_f}$ ...(18.9.12)

Combination of Eqs. (18.9.6) and (18.9.9) gives

$\qquad V_{a2} = - Z_{a2} I_{a2} = Z_f I_{a2}$

$\therefore \qquad I_{a2} = 0$ ...(18.9.13)

Combination of Eqs. (18.9.7) and (18.9.10) gives

$\qquad V_{a0} = - Z_{a0} I_{a0} = Z_f I_{a0}$

$\therefore \qquad I_{a0} = 0$ ...(18.9.14)

Equations (18.9.13) and (18.9.14) show that the negative and zero sequence components of current are zero in the case of a balanced three-phase fault. Only the positive-sequence current is present.

Example 18.1 An 11 kV, 25 MVA synchronous generator has positive, negative and zero sequence reactances of 0.12, 0.12 and 0.08 per unit, respectively. The generator neutral is grounded through a reactance of 0.03 per unit. A single line-to-ground fault occurs at the generator terminals. Determine the fault current and the line-to-line voltages. Assume that the generator was unloaded before the fault.

Solution

Since the generator was unloaded before the fault, the internal generator voltage before the fault is equal to the terminal voltage. Let the line-to-neutral voltage (phase voltage) at the fault point before the fault be $\mathbf{E} = 1\underline{/0°}$ pu. The sequence network for a single line to ground (L-G) fault is shown in Fig. 18.3 (b).

Calculation of sequence current components

For a line to ground fault,

$$\mathbf{I}_{a1} = \frac{E_a}{Z_{a0} + Z_{a1} + Z_{a2} + 3Z_f} = \frac{1\underline{/0°}}{j\,0.08 + j\,0.12 + j\,0.12 + j\,3 \times 0.03} = \frac{1\underline{/0°}}{j\,0.41} = -j\,2.439 \text{ pu}$$

For a single line-to-ground fault,

$$\mathbf{I}_{a0} = \mathbf{I}_{a1} = \mathbf{I}_{a2}$$

$$\therefore \qquad \mathbf{I}_{a0} = \mathbf{I}_{a1} = \mathbf{I}_{a2} = -j\,2.439 \text{ pu}$$

The phase current in a is given by

$$\mathbf{I}_a = \mathbf{I}_{a0} + \mathbf{I}_{a1} + \mathbf{I}_{a2} = 3\mathbf{I}_{a1} = -j\,7.317 \text{ pu}$$

Calculation of base current

Let the base quantities be 25 MVA, and 11 kV.

$$\text{Base current} = \frac{\text{base VA}}{\sqrt{3} \times \text{base voltage}} = \frac{25 \times 10^6}{\sqrt{3} \times 11 \times 10^3} = 1312 \text{ A}$$

Calculation of phase currents

Therefore, the fault current in amperes is given by

$$I_a = 7.312 \times 1312 = 9593 \text{ A}$$

Since only phase a is connected to ground in a single L-G fault, phases b and c are open-circuited, therefore,

$$I_b = 0, \qquad I_c = 0$$

Calculation of sequence voltage components

$$\mathbf{V}_{a1} = \mathbf{E}_a - \mathbf{Z}_{a1}\,\mathbf{I}_{a1} = 1 + j\,0 - (j\,0.12)(-j\,2.439) = 1 - 0.29268 = 0.70732 \text{ pu}$$

$$\mathbf{V}_{a2} = -\mathbf{Z}_{a2}\,\mathbf{I}_{a2} = -(j\,0.12)(-j\,2.439) = -0.29268 \text{ pu}$$

$$\mathbf{V}_{a0} = -\mathbf{Z}_{a0}\,\mathbf{I}_{a0} = -(j\,0.08 + j\,0.09)(-j\,2.439) = -0.41463 \text{ pu}$$

Calculation of phase voltages

The next step is to convert the sequence voltage components to phase voltages.

$$\mathbf{V}_a = \mathbf{V}_{a0} + \mathbf{V}_{a1} + \mathbf{V}_{a2} = -0.41463 + 0.70732 - 0.29268 = 0$$

$$\mathbf{V}_b = \mathbf{V}_{a0} + \alpha^2\,\mathbf{V}_{a1} + \alpha\,\mathbf{V}_{a2} = -0.41463 + (1\,\underline{/240°})\,(0.70732) + (1\,\underline{/120°})\,(-0.29268)$$

$$= -0.41463 - 0.35366 - j\,0.612557 + 0.14634 - j\,0.2534683$$

$$= (-0.62195 - j\,0.866) \text{ pu}$$

$$\mathbf{V}_c = \mathbf{V}_{a0} + \alpha\,\mathbf{V}_{a1} + \alpha^2\,\mathbf{V}_{a2} = -0.41463 + (1\,\underline{/120°})\,(0.70732) + (1\,\underline{/240°})\,(-0.29268)$$

$$= -0.41463 - 0.35366 + j\,0.612557 + 0.14634 + j\,0.2534683$$

$$= (-0.62195 + j\,0.866) \text{ pu}$$

Calculation of line voltages

$$\mathbf{V}_{ab} = \mathbf{V}_a - \mathbf{V}_b = 0 - (-0.62196 - j\,0.866) = 0.62195 + j\,0.866 = 1.0661978\,\underline{/54.32°} \text{ pu}$$

$$\mathbf{V}_{bc} = \mathbf{V}_b - \mathbf{V}_c = -0.62195 - j\,0.866 - (-0.62195 + j\,0.866) = -j\,1.732 = 1.732\,\underline{/270°} \text{ pu}$$

$$\mathbf{V}_{ca} = \mathbf{V}_c - \mathbf{V}_a = -0.62195 + j\,0.866 = 1.0662\,\underline{/125.68°} \text{ pu}$$

We have assumed that voltage per phase $E = 1\,\underline{/0°}$ pu. Therefore,

$$1.0 \text{ pu voltage} = \frac{11}{\sqrt{3}} = 6.35085 \text{ kV}$$

$\therefore$ $V_{ab} = (1.0661978 \,\underline{/54.32°})\,(6.35085) = 6.7712 \,\underline{/54.32°}$ kV

       $V_{bc} = (1.732 \,\underline{/270°})\,(6.35085) = 10.9997 \,\underline{/270°} \simeq 11 \,\underline{/270°}$ kV

       $V_{ca} = (1.0662 \,\underline{/125.68°})\,(6.35085) = 6.7713 \,\underline{/125.68°}$ kV

Example 18.2 Draw the sequence networks for the system shown in Fig. 18.9. Determine the fault current when (a) LLG and (b) LL fault occurs at point F. The per unit reactances all referred to the same base are as follows :

	X_0	X_1	X_2
Generator G_1	0.05	0.3	0.2
Generator G_2	0.03	0.25	0.15
Line 1	0.70	0.3	0.3
Line 2	0.70	0.3	0.3
Transformer T_1	0.12	0.12	0.12
Transformer T_2	0.10	0.1	0.1

Both the generators are generating 1 pu voltage.

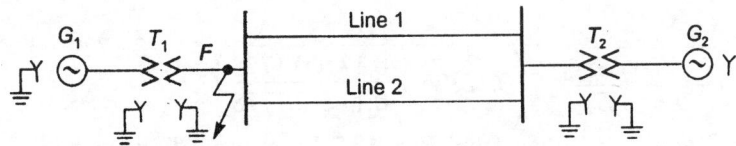

Fig. 18.9. Network of Example 18.2.

Solution

The sequence networks are shown in Fig. 18.10.

From the positive-sequence network of Fig. 18.10 (a), the equivalent impedance upto the point of the fault is given by

$$Z_1 = j\left[(0.3 + 0.12)\;\middle|\middle|\;\left(0.25 + 0.1 + \frac{0.3}{2}\right)\right] = j\,[(0.42)\;||\;(0.5)] = j\,\frac{0.42 \times 0.5}{0.42 + 0.5} = j\,0.22826 \text{ pu}$$

From the negative-sequence network of Fig. 18.10 (b), the equivalent impedance upto F is given by

$$Z_2 = j\left[(0.2 + 0.12)\;||\;\left(0.15 + 0.1 + \frac{0.3}{2}\right)\right] = j\,(0.32)\;||\;(0.4) = j\,\frac{0.32 \times 0.4}{0.32 + 0.4} = j\,0.1778 \text{ pu}$$

From the zero-sequence network of Fig. 18.10 (c), the equivalent impedance upto F is given by

$$Z_0 = j\,(0.05 + 0.12) = j\,0.17 \text{ pu}$$

(a) LLG Fault at F

If phase a is assumed to be the reference phasor and phases b and c are shorted at the fault, then from Eq. (18.8.15)

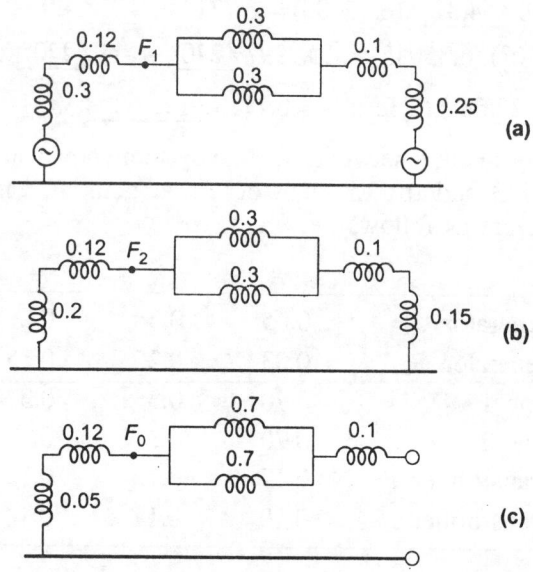

Fig. 18.10. Sequence networks of Example 18.2.

$$\mathbf{I}_{a1} = \frac{\mathbf{V}_f}{\mathbf{Z}_{a1} + \dfrac{\mathbf{Z}_{a0}\,\mathbf{Z}_{a2}}{\mathbf{Z}_{a0} + \mathbf{Z}_{a2}}} = \frac{1\,/0°}{j\left(0.22826 + \dfrac{0.17 \times 0.1778}{0.17 + 0.1778}\right)} = -j\,3.1729 \text{ pu} = 3.1729\,/-90° \text{ pu}$$

If we put $Z_f = 0$ and $Z_g = 0$ then from Fig. 18.7 by current division rule

$$\mathbf{I}_{a0} = -\mathbf{I}_{a1}\,\frac{\mathbf{Z}_{a2}}{\mathbf{Z}_{a0} + \mathbf{Z}_{a2}} = j\left[3.1729 \times \frac{0.1778}{0.17 + 0.1778}\right] = j\,1.622 \text{ pu} = 1.622\,/90° \text{ pu}$$

$$\mathbf{I}_{a2} = -\mathbf{I}_{a1}\,\frac{\mathbf{Z}_{a0}}{\mathbf{Z}_{a0} + \mathbf{Z}_{a2}} = j\left[3.1729 \times \frac{0.17}{0.17 + 0.1778}\right] = j\,1.551 \text{ pu} = 1.551\,/90° \text{ pu}$$

Check

$$\mathbf{I}_a = \mathbf{I}_{a0} + \mathbf{I}_{a1} + \mathbf{I}_{a2} = j\,1.622 - j\,3.1729 + j\,1.551 \simeq 0$$

$$\mathbf{I}_b = \mathbf{I}_{a0} + \alpha^2\,\mathbf{I}_{a1} + \alpha\,\mathbf{I}_{a2} = j\,1.622 + (1\,/240°)\,(3.1729\,/-90°) + (1\,/120°)\,(1.551\,/90°)$$

$$= j\,1.622 + 3.1729\,/150° + 1.551\,/210°$$

$$= j\,1.622 - 2.7478 + j\,1.5864 - 1.3432 - j\,0.7755$$

$$= -4.091 + j\,2.4729$$

$$\mathbf{I}_c = \mathbf{I}_{a0} + \alpha\,\mathbf{I}_{a1} + \alpha^2\,\mathbf{I}_{a2} = j\,1.622 + (1\,/120°)\,(3.1729\,/-90°) + (1\,/240°)\,(1.551\,/90°)$$

$$= j\,1.622 + 2.7478 + j\,1.5864 + 1.3432 - j\,0.7755 = 4.091 + j\,2.4729$$

$$|\mathbf{I}_b| = |\mathbf{I}_c| = \sqrt{(4.091)^2 + (2.4729)^2} = 4.78 \text{ pu}$$

(b) LL Fault at F

If the line-to-line fault is between phases b and c, then from Eq. (18.7.15)

$$\mathbf{I}_{a1} = \frac{\mathbf{V}_f}{\mathbf{Z}_{a1} + \mathbf{Z}_{a2}} = \frac{1\,/0°}{j\,(0.22826 + 0.1778)} = -j\,2.4627 \text{ pu} = 2.4627\,/-90° \text{ pu}$$

The phase a negative-sequence current is given by

$$\mathbf{I}_{a2} = -\mathbf{I}_{a1} = +j\,2.4627 \text{ pu} = 2.4627\,\underline{/90°}\ \text{pu}$$

The phase a fault current

$$\mathbf{I}_a = \mathbf{I}_{a0} + \mathbf{I}_{a1} + \mathbf{I}_{a2} = 0 - j\,2.4627 + j\,2.4627 = 0$$

The phase b fault current

$$\mathbf{I}_b = \mathbf{I}_{a0} + \alpha^2\,\mathbf{I}_{a1} + \alpha\,\mathbf{I}_{a2} = 0 + (1\,\underline{/240°})\,(2.4627\,\underline{/-90°}) + (1\,\underline{/120°})\,(2.4627\,\underline{/90°})$$

$$= 2.4627\,\underline{/150°} + 2.4627\,\underline{/210°} = -2.133 + j\,1.231 - 2.133 - j\,1.231 = -4.266\ \text{pu}$$

The phase c fault current

$$\mathbf{I}_c = \mathbf{I}_{a0} + \alpha\,\mathbf{I}_{a1} + \alpha^2\,\mathbf{I}_{a2} = 0 + (1\,\underline{/120°})\,(2.4627\,\underline{/-90°}) + (1\,\underline{/240°})\,(2.4627\,\underline{/90°})$$

$$= 2.4627\,\underline{/30°} + 2.4627\,\underline{/330°}) = 2.133 + j\,1.231 + 2.133 - j\,1.231 = 4.266\ \text{pu}$$

Thus, it is calculated that

$$\mathbf{I}_b = -\mathbf{I}_c = -4.266\ \text{pu}$$

Example 18.3 A three phase 10 MVA, 11 kV synchronous generator with a solidly grounded neutral point supplies a feeder. The relevant impedances of the generator and feeder are as follows :

	Generator	Feeder
To positive sequence currents, Z_1	$j\,1.2\ \Omega$	$j\,1.0\ \Omega$
To negative sequence currents, Z_2	$j\,0.9\ \Omega$	$j\,1.0\ \Omega$
To zero sequence currents, Z_0	$j\,0.4\ \Omega$	$j\,3.0\ \Omega$

A fault from one phase to ground occurs at the far end of the feeder. Determine the voltage to neutral of the faulty phase at the terminals of the gender.

Solution

Rated phase e.m.f. $= \dfrac{11000}{\sqrt{3}} = 6350.8\ \text{V}$

Let the fault occur on phase a.

Total impedance per phase upto the far end of the feeder to different sequence currents are :

$$Z_{a0} = j\,0.4 + j\,3.0 = j\,3.4\ \Omega$$

$$Z_{a1} = j\,1.2 + j\,1.0 = j\,2.2\ \Omega$$

$$Z_{a2} = j\,0.9 + j\,1.0 = j\,1.9\ \Omega$$

$$I_{a0} = I_{a1} = I_{a2} = \frac{V_f}{Z_{a0} + Z_{a1} + Z_{a2}} = \frac{6350.8}{j\,3.4 + j\,2.2 + j\,1.9}$$

$$= -j\,846.77\ \text{A}$$

Fault current

$$I_f = I_a = 3I_{a0} = -j\,(3 \times 846.77) = -j\,2540.3\ \text{A}$$

The voltage to neutral of the faulty phase of the generator is

$$V_a = E_a - (Z_{ga0} I_{a0} + Z_{ga1} I_{a1} + Z_{ga2} I_{a2}) = E_a - I_{a0}(Z_{ga0} + Z_{ga1} + Z_{ga2})$$

$$= 6350.8 - (-j\,846.77)(j\,0.4 + j\,1.2 + j\,0.9) = 6350.8 - 2116.9 = 4233.9 \text{ V}$$

18.10 COMPARISON OF SINGLE LINE-TO-GROUND AND THREE-PHASE FAULT CURRENTS

Suppose that the sequence impedances Z_0, Z_1 and Z_2 are purely reactive and $Z_f = 0$. Therefore,

$$Z_0 = j\,X_0,\ Z_1 = j\,X_1,\ Z_2 = j\,X_2$$

For a three-phase fault,

$$(I_a)_{3L} = \frac{V_f}{j\,X_1} \qquad\qquad \ldots(18.10.1)$$

For a line-to-ground fault

$$(I_a)_{LG} = \frac{3V_f}{j\,(X_0 + X_1 + X_2)} \qquad\qquad \ldots(18.10.2)$$

We shall consider three important cases.

(a) Fault at Generator Terminals with Generator Neutral Solidly Grounded

For a generator $X_0 \ll X_1$. Let us assume that $X_1 = X_2$. Substituting these values in Eqs. (18.10.1) and (18.10.2) we get

$$|I_a|_{3L} = \frac{|V_f|}{X_1} = \frac{3|V_f|}{3X_1} \qquad\qquad \ldots(18.10.3)$$

$$|I_a|_{LG} = \frac{3|V_f|}{2X_1 + X_0} \qquad\qquad \ldots(18.10.4)$$

Comparison of Eqs. (18.10.3) and (18.10.4) shows that

$$|I_a|_{LG} > |I_a|_{3L} \qquad\qquad \ldots(18.10.5)$$

Hence we conclude that *a line-to-ground fault at the terminals of a generator with solidly grounded neutral is more severe than a 3L fault.*

(b) Fault at Generator Terminals with Generator Neutral Grounded through a Reactance X_n

For a 3L fault, the method of neutral grounding does not affect the fault current. Therefore,

$$|I_a|_{3L} = \frac{|V_f|}{X_1} \qquad\qquad \ldots(18.10.6)$$

For a line-to-ground fault,

$$|I_a|_{LG} = \frac{3|V_f|}{X_0 + X_1 + X_2 + 3X_n} \qquad\qquad \ldots(18.10.7)$$

Equation (18.10.7) shows that the line-to-ground fault current depends on the value of X_n. If X_n is large $|I_a|_{LG} < |I_a|_{3L}$. If X_n is very small, $|I_a|_{LG}$ may be greater than $|I_a|_{3L}$. For the two fault currents to be equal

$$\frac{|V_f|}{X_1} = \frac{3|V_f|}{X_0 + X_1 + X_2 + 3X_n}$$

$$X_0 + 2X_1 + 3X_n = 3X_1$$

$$X_n = \frac{1}{3}(X_1 - X_0) \qquad\qquad ...(18.10.8)$$

If $X_n < \frac{1}{3}(X_1 - X_0)$, a single line-to-ground fault is more severe than a 3L fault. If $X_n > \frac{1}{3}(X_1 - X_0)$, a 3L fault is more severe than an LG fault.

(c) Fault on Transmission Line

For a transmission line $X_0 \gg X_1$ and $X_1 = X_2$, so that for a line fault at a point which is at a large distance from the generator, 3L fault will be more severe than an LG fault.

Example 18.4 A 50 MVA, 11 KV, three-phase synchronous generator was subjected to different types of faults. The fault currents are as follows :

LG fault — 4200 A ;   LL fault — 2600 A ;   LLL fault — 2000 A.

The generator neutral is solidly grounded. Find the per unit values of the three sequence reactances of the generator.

Solution

For a three-phase (3L) fault

$$I_a = \frac{|V_f|}{X_1} ; \qquad 2000 = \frac{11000/\sqrt{3}}{X_1}$$

$$X_1 = \frac{11000}{\sqrt{3} \times 2000} = 3.175 \ \Omega$$

For a line-to-line fault

$$I_a = \frac{\sqrt{3}\,|V_f|}{X_1 + X_2} ; \qquad 2600 = \frac{\sqrt{3} \cdot 11000/\sqrt{3}}{X_1 + X_2}$$

$$X_1 + X_2 = \frac{11000}{2600} = 4.231 \ \Omega$$

$$X_2 = 4.231 - 3.175 = 1.056 \ \Omega$$

For an LG fault

$$I_a = \frac{3|V_f|}{X_0 + X_1 + X_2} ; \qquad 4200 = \frac{3 \times 11000/\sqrt{3}}{X_0 + X_1 + X_2}$$

$$X_0 + X_1 + X_2 = \frac{\sqrt{3} \times 11000}{4200} = 4.536 \ \Omega$$

$$\therefore \qquad X_0 + 4.231 = 4.536$$

$$X_0 = 4.536 - 4.231 = 0.305 \ \Omega$$

Base impedance,

$$Z_b = \frac{(kV_l)_b^2}{(MVA)_b} = \frac{(11)^2}{50} = 2.42 \ \Omega$$

$$X_{0\,pu} = \frac{X_{0\,\Omega}}{Z_b} = \frac{0.305}{2.42} = 0.126 \ pu$$

$$X_{1\,pu} = \frac{X_{1\,\Omega}}{Z_b} = \frac{3.175}{2.42} = 1.312 \ pu$$

$$X_{2\,pu} = \frac{X_{2\,\Omega}}{Z_b} = \frac{1.056}{2.42} = 0.436 \ pu$$

18.11 OPEN CONDUCTOR FAULTS

A break in one or two conductors in a 3-phase circuit leads to an open conductor fault. An open conductor fault is in series with the line. The method of symmetrical components is used to analyze the open conductor faults.

18.11.1 One Conductor Open

Fig. 18.11(a) shows a 3-phase line with one conductor a open. The ends of the system on sides of the fault are shown as FF' and the conductor ends are shown as aa', bb', cc'.

The boundary conditions are:

$$I_a = 0 \qquad\qquad \text{...(18.11.1)}$$

$$V_{bb'} = V_{cc'} = 0 \qquad\qquad \text{...(18.11.2)}$$

Since $I_a = 0$,

$$I_{a0} + I_{a1} + I_{a2} = I_a = 0 \qquad\qquad \text{...(18.11.3)}$$

$$V_{aa'0} = V_{aa'1} = V_{aa'2} = \frac{1}{3}V_{aa'} \qquad\qquad \text{...(18.11.4)}$$

Equations (18.11.3) and (18.11.4) suggest a parallel connection of sequence networks as shown in Fig. 18.11(b).

18.11.2 Two Conductors Open

Fig. 18.12(a) shows the fault at FF' with conductors b and c open. The boundary conditions are:

$$V_{aa'} = 0 \qquad\qquad \text{...(18.11.5)}$$

$$I_b = I_c = 0 \qquad\qquad \text{...(18.11.6)}$$

In terms of symmetrical components:

$$V_{aa'0} + V_{aa'1} + V_{aa'2} = 0 \qquad\qquad \text{...(18.11.7)}$$

$$I_{a0} + I_{a1} = I_{a2} = \frac{1}{3}I_a \qquad\qquad \text{...(18.11.8)}$$

Equations (18.11.7) and (18.11.8) show that the sequence networks are connected in series. This is shown in Fig. 18.12(b).

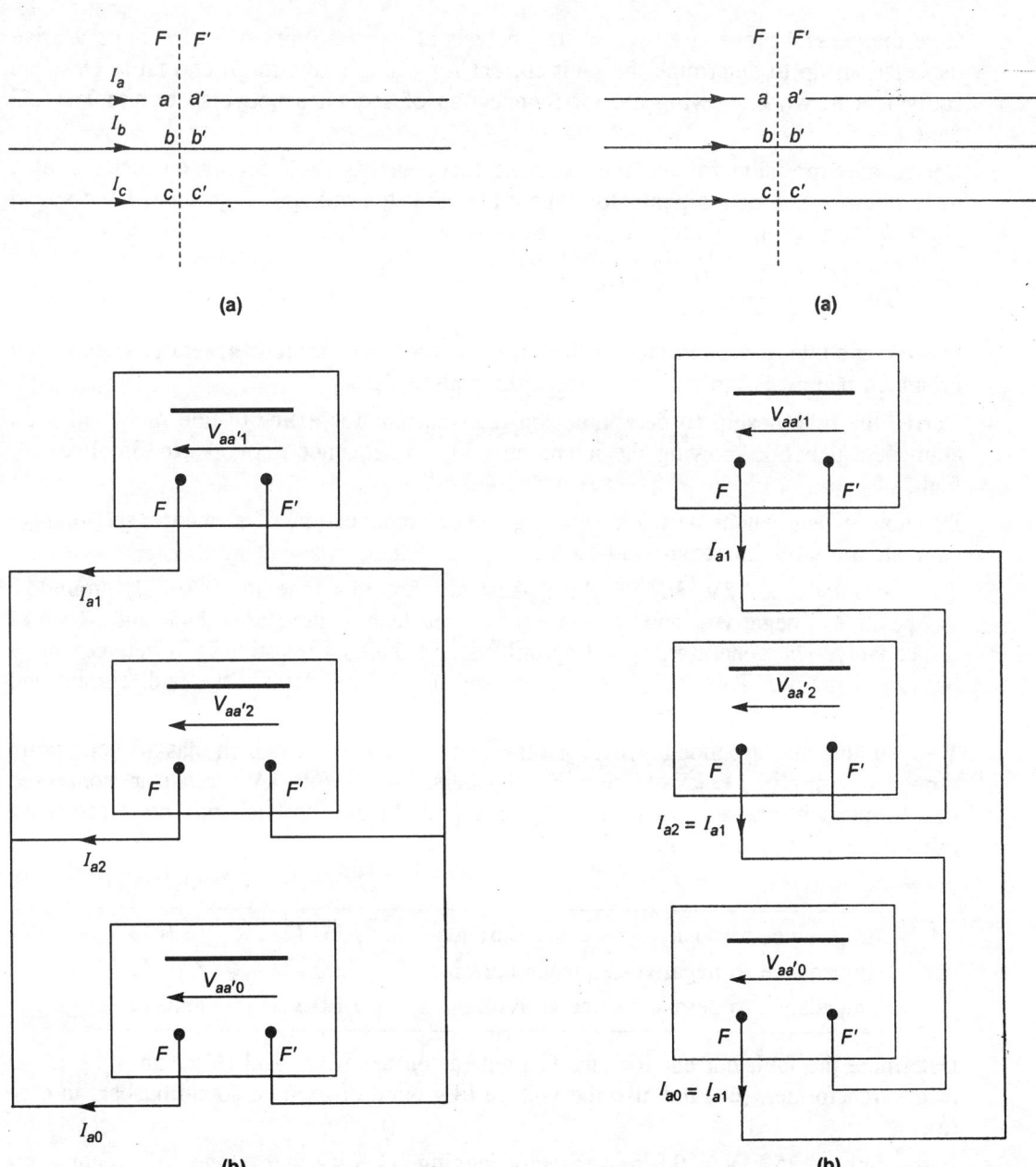

Fig. 18.11. One conductor open. (a) Circuit diagram, (b) sequence network.

Fig. 18.12. Two conductors open. (a) Circuit diagram, (b) sequence network.

EXERCISES

1. Give the general procedure used in the analysis of various types of shunt faults. Derive the relationship to determine the fault current for a single line-to-ground fault. Draw an equivalent network showing the interconnection of sequence networks to simulate LG fault.

2. Derive an expression for the fault current when an LG fault occurs on phase a of a three-phase synchronous generator. Show also that the voltage to ground of the sound phase b at the point of fault is given by

$$V_b = \frac{Z_0\,(\alpha^2 - 1) + Z_2\,(\alpha^2 - \alpha)}{Z_0 + Z_1 + Z_2}$$

3. Prove that a line-to-ground fault at the terminals for a synchronous generator with solidly grounded neutral is more severe than a three-phase fault.

4. Derive the relationship to determine the fault current for a line-to-line fault. Draw an equivalent network showing the interconnection of sequence networks to simulate L-L fault.

5. Develop an equivalent network showing the interconnection of sequence networks to simulate a double line-to-ground fault.

6. The star point of a 3 kV, 3 MVA three-phase synchronous generator is solidly grounded. Its positive-, negative- and zero-sequence reactances are 2.4, 0.45 and 0.30 Ω respectively. The generator, operating unloaded, sustains a resistive fault between the a phase and ground. This fault has a resistance of 1.2 Ω. Calculate the fault current and the voltage to ground of the a phase.

7. Two similar star-grounded synchronous generators, one of which has its star point grounded, supply 11 kV three-phase busbars. An 11/66 kV delta/star connected transformer with the star point grounded is supplied from the busbars. The impedances, referred to 11 kV, are as given below :

	Generator	Transformer
Impedance to positive-sequence currents	$j\,3.0\ \Omega$	$j\,3.0\ \Omega$
Impedance to negative-sequence currents	$j\,2.0\ \Omega$	$j\,3.0\ \Omega$
Impedance to zero-sequence currents	$j\,1.0\ \Omega$	$j\,3.0\ \Omega$

Determine the fault current for an LG fault (a) on a busbar, and (b) at an hv terminal of the transformer. (c) Find also the voltage to ground of the two sound busbars in case (a).

8. A three-phase 75 MVA, 0.8 power factor lagging, 11.8 kV star connected synchronous generator which has its star point solidly grounded supplies a feeder. The relevant per unit impedances based on the rated phase voltage and phase current of the generator are as follows :

	Generator	Feeder
Positive-sequence impedance (pu)	$j\,1.70$	$j\,0.10$
Negative-sequence impedance (pu)	$j\,0.18$	$j\,0.10$
Zero-sequence impedance (pu)	$j\,0.12$	$j\,0.30$

Determine the fault current and the line-to-neutral voltages at the generator terminals for an LG fault occuring at the distant end of the feeder.

9. A star-connected three-phase synchronous generator has a solidly grounded neutral and feeds a line upon which there is a fault. The total impedances upto the fault are as follows :

To positive-sequence currents, $Z_1 = j\,2.2\ \Omega$

To negative-sequence currents, $Z_2 = j\,1.9\ \Omega$

To zero-sequence currents, $Z_0 = j\,1.4\ \Omega$.

If the generator induced voltage is 11 kV, calculate the fault currents for (a) an LG fault, (b) a double line-to-ground fault, (c) a line-to-line fault. Also calculate the voltages of the lines at the fault.

ANSWERS

6. $I_a = 1086\,\underline{/-41.18°}$ A, $V_a = 1303.5\,\underline{/-41.18°}$ V

7. (a) $I_a = 5443\,\underline{/-90°}$ A, (b) $I_a = 1656\,\underline{/-90°}$ A,
   (c) $V_b = 5443.5\,\underline{/-120°}$ V, $V_c = 5443.5\,\underline{/120°}$ V

8. $I_a = 8175\,\underline{/-90°}$ A, $V_a = 1362$ V, $V_b = 2593\,\underline{/-116.86°}$ V, $V_c = 2593\,\underline{/116.86°}$ V

9. (a) $I_a = 3464\,\underline{/-90°}$ A, $V_a = 0$, $V_b = 5737.5\,\underline{/-115°}$ V, $V_c = 5737.5\,\underline{/115°}$ V
   (b) $I_a = 0$, $I_b = 3181\,\underline{/145°}$ A, $I_c = 3181\,\underline{/35°}$ A, $V_a = 5109$ V, $V_b = 0$, $V_c = 0$
   (c) $I_f = I_b = 2682.9$ A, $I_c = -2682.9$ A, $V_a = 5886.4$ V, $V_b = V_c = -2943.2$ V

Power System Stability

19.1 INTRODUCTION

A large power system consists of a number of synchronous machines operating in synchronism. It is necessary that they should maintain perfect synchronism under all steady-state conditions. When the system is subjected to some form of disturbance, there is a tendency for the system to develop force, to bring it to a normal or stable condition.

The ability of a system to reach a normal or stable condition after being disturbed is called *stability*. The stability problem is concerned with the behaviour of synchronous machines after a disturbance. Under stable conditions the system stays in synchronism.

Synchronous stability may be divided into two main categories depending upon the magnitude of the disturbance.

Steady-state stability refers to the ability of the power system to regain synchronism after *small* and slow disturbance, such as gradual power changes.

The **transient stability** is the ability of the system to regain synchronism after a *large* disturbance. The large disturbance can occur due to sudden changes in application or removal of large loads, line switching operations, faults on the system, sudden outage of a line, or loss of excitation. Transient stability studies are needed to ensure that the system can withstand the transient conditions following a major disturbance. Frequently, such studies are conducted when new generating and transmitting facilities are planned.

Steady-state stability is subdivided into static stability and dynamic stability.

Static stability refers to *inherent* stability that prevails without the aid of automatic control devices such as governors and voltage regulators.

Dynamic stability, on the other hand, denotes *artificial* stability given to an inherently unstable system by automatic control devices. Dynamic stability is concerned with small disturbances lasting for times of the order of 10 to 30 seconds with the inclusion of automatic control devices.

Stability studies are helpful for the following purposes:

* Determination of critical clearing time of circuit breakers.
* Investigation of schemes of protective relaying.
* Determination of voltage levels.
* Transfer capability between systems.

Analysis of power system stability is complex and nonlinear. Consequently, final designs are generally based on computer simulations. For approximate purposes simplified calculations are used. Simplified calculations provide a starting point for, and check of, computer simulations. They are also useful in studying the factors that influence the power system stability. Invariably stability studies of power systems are carried out on a digital computer. In the following, special cases to illustrate certain principles and basic concepts are presented.

19.2 STABILITY LIMITS AND POWER TRANSMISSION CAPABILITY

The *stability limit* is the maximum power that can be transferred in a network between sources and loads without loss of synchronism. The *steady-state limit* is the maximum power that can be transferred without the system becoming unstable when the load is increased gradually under steady-state conditions. *Transient limit* is the maximum power that can be transferred without the system becoming unstable when a sudden or large disturbance occurs.

The system experiences a shock by sudden and large power changes and violent fluctuations of voltage occur. Consequently, individual machines or group of machines may go out of step. The rapidity of the application of a large disturbances is responsible for the loss of stability, otherwise it may be possible to maintain stability if the same large load is applied gradually. Thus, the transient stability limit is lower than the steady-state limit.

Power Transmission Capability

The power transmission capability of a line is limited by the thermal loading limit and the stability limit. The real power loss increases the conductor temperature. This will increase the sag of the conductors between the transmission towers. The thermal limit is specified by the current-carrying capacity of the conductor and is available in the manufacturer's data.

Let $I_{thermal}$ = current–carrying capacity

$\quad$ $S_{thermal}$ = thermal loading limit of the line

$\quad\quad$ V_p = rated phase voltage

$\quad\quad$ $S_{thermal} = 3V_pI_{thermal}$

19.3 INFINITE BUS

In a power system normally more than two generators operate in parallel. The machines may be located at different places. A group of machines located at one place may be treated as a single large machine. Also, the machines not connected to the same bus but separated by lines of low reactance, may be grouped into one large machine. The operation of one machine connected in parallel with such a large system comprising many other machines is of great interest. The capacity of the system is so large that its voltage and frequency may be taken constant. The connection or disconnection of a single small machine on such a system would not affect the magnitude and phase of the voltage and frequency. Such a system of constant voltage and constant frequency regardless of the load is called *infinite busbar system* or simply *infinite bus*. Physically it is not possible to have a perfect infinite bus. An infinite bus is an ideal voltage source.

19.4 SYNCHRONOUS GENERATOR CONNECTED TO AN INFINITE BUS

Consider a simple system consisting of a synchronous generator connected to an infinite bus through a network represented by the **ABCD** parameters as shown in Fig. 19.1.

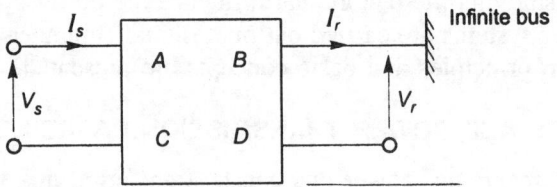

Fig. 19.1. One generator connected to an infinite bus through a two-port network.

The sending-end and receiving-end voltages are assumed as

$$\mathbf{V}_s = V_s \underline{/\delta}, \qquad \mathbf{V}_r = V_r \underline{/0^\circ}$$

We have $\mathbf{V}_s = \mathbf{A}\mathbf{V}_r + \mathbf{B}\mathbf{I}_r$ 　　　　　　　　　　　...(19.4.1)

$$\mathbf{I}_r = \frac{\mathbf{V}_s}{\mathbf{B}} - \frac{\mathbf{A}}{\mathbf{B}}\mathbf{V}_r = \frac{V_s\underline{/\delta}}{B\underline{/\beta}} - \frac{A\underline{/\alpha}}{B\underline{/\beta}} V_r\underline{/0^\circ} = \frac{V_s}{B}\underline{/(\delta-\beta)} - \frac{AV_r}{B}\underline{/(\alpha-\beta)} \qquad ...(19.4.2)$$

Complex power received at the infinite bus

$$\mathbf{S}_r = \mathbf{V}_r \mathbf{I}_r^* = P_r + j Q_r \qquad ...(19.4.3)$$

where $\mathbf{I}_r^*$ is the complex conjugate of $\mathbf{I}_r$.

$$\mathbf{I}_r^* = \frac{V_s}{B}\underline{/(\beta-\delta)} - \frac{AV_r}{B}\underline{/(\beta-\alpha)} \qquad ...(19.4.4)$$

$$P_r + j Q_r = \mathbf{V}_r \mathbf{I}_r^* = \frac{V_s V_r}{B}\underline{/(\beta-\delta)} - \frac{AV_r^2}{B}\underline{/(\beta-\alpha)}$$

$$= \frac{V_s V_r}{B}[\cos(\beta-\delta) + j\sin(\beta-\delta)] - \frac{AV_r^2}{B}[\cos(\beta-\alpha) + j\sin(\beta-\alpha)]$$

Equating real and imaginary parts, we get

$$P_r = \frac{V_s V_r}{B}\cos(\beta-\delta) - \frac{AV_r^2}{B}\cos(\beta-\alpha) \qquad ...(19.4.5)$$

$$Q_r = \frac{V_s V_r}{B}\sin(\beta-\delta) - \frac{AV_r^2}{B}\sin(\beta-\alpha) \qquad ...(19.4.6)$$

The power received is a maximum when $\delta = \beta$. Therefore,

$$P_{r\,\text{max}} = \frac{V_s V_r}{B} - \frac{AV_r^2}{B}\cos(\beta-\alpha) \qquad ...(19.4.7)$$

19.5 POWER-ANGLE CURVE

Fig. 19.2 shows a synchronous machine connected to an infinite bus through a transmission line of reactance X_l. Let us assume that line resistance and capacitance are neglected.

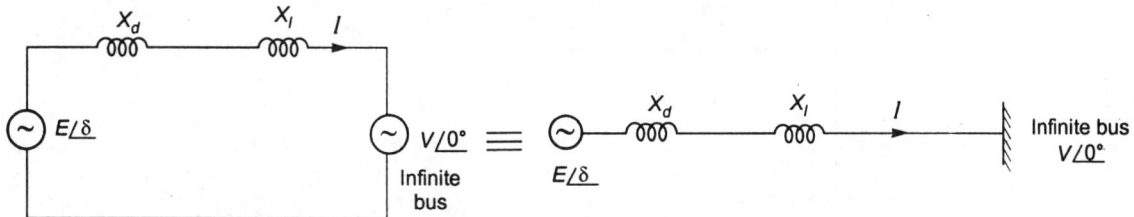

Fig. 19.2. Synchronous machine connected to infinite bus through a transmission line of series reactance X_l.

Let, $\mathbf{V} = V\,\underline{/0^\circ}$ = voltage of infinite bus

$\mathbf{E} = E\,\underline{/\delta}$ = voltage behind direct axis synchronous reactance of the machine

X_d = synchronous/transient reactance of the machine.

The complex power delivered by the generator to the system is

$$\mathbf{S} = \mathbf{VI}^* = V\left[\frac{E\,\underline{/\delta} - V\,\underline{/0^\circ}}{j\,(X_d + X_l)}\right]^* \qquad\qquad \text{...(19.5.1)}$$

Let $X_d + X_l = X$ \qquad\qquad\qquad\qquad\qquad\qquad\qquad\qquad ...(19.5.2)

$$\mathbf{S} = V\left[\frac{E\,\underline{/\delta}}{X\,\underline{/90^\circ}} + j\,\frac{V}{X}\right] = \frac{EV}{X}\,\underline{/(90-\delta)} - j\,\frac{V^2}{X} = \frac{EV}{X}\sin\delta + j\,\frac{EV}{X}\cos\delta - j\,\frac{V^2}{X}$$

$$P_e + j\,Q_e = \frac{EV}{X}\sin\delta + j\left(\frac{EV}{X}\cos\delta - \frac{V^2}{X}\right) \qquad\qquad \text{...(19.5.3)}$$

Active power transferred to the system

$$P_e = \frac{EV}{X}\sin\delta \qquad\qquad\qquad\qquad\qquad\qquad \text{...(19.5.4)}$$

The reactive power transferred to the system

$$Q_e = \frac{EV}{X}\cos\delta - \frac{V^2}{X} \qquad\qquad\qquad\qquad \text{...(19.5.5)}$$

The maximum steady-state power transfer occurs when $\delta = 90^\circ$. From Eq. (19.5.4)

$$P = \frac{EV}{X}\sin 90^\circ = \frac{EV}{X} \qquad\qquad\qquad\qquad \text{...(19.5.6)}$$

$\therefore$ \qquad $P_e = P_{e\,\text{max}}\sin\delta$ \qquad\qquad\qquad\qquad\qquad\qquad ...(19.5.7)

The graphical representation of power P_e and the load angle δ is called the *power-angle diagram* or *power-angle curve*. Such a diagram is widely used in power-system stability studies. A power angle diagram is shown in Fig. 19.3.

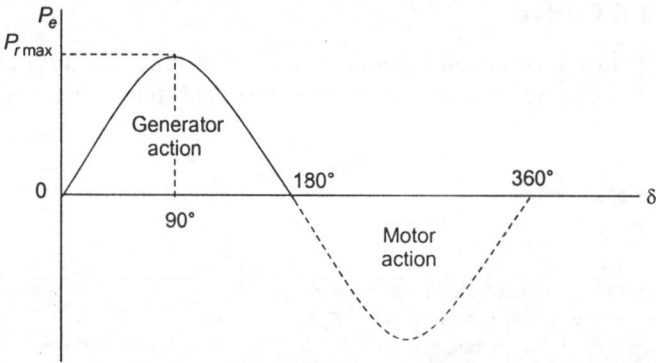

Fig. 19.3. Power-angle diagram.

Maximum power is transferred when $\delta = 90°$. As δ is increased beyond $90°$, P_e decreases and becomes zero at $\delta = 180°$. Beyond $\delta = 180°$, P_e becomes negative which implies that the power flow direction is reversed and the power is supplied from the infinite bus to the generator, the value of $P_{e\,max}$ is often called the *pull-out power*. It is also called the *steady-state limit*.

The total reactance X between two voltage sources V and E is called the *transfer reactance*. It is seen that the maximum power limit is inversely proportional to the transfer reactance. Equation (19.5.4) is valid for both steady-state and transient conditions. For steady-state conditions we use synchronous reactance and we take E as the e.m.f. behind synchronous reactance. For transient conditions, the transient reactance X'_d is used and E is taken as the e.m.f. behind transient reactance.

19.6 REPRESENTATION OF A T-CIRCUIT WITH A SERIES CIRCUIT

The T-circuit shown in Fig. 19.4 (a) can be replaced by a series circuit shown in Fig. 19.4 (b) having the modified receiving-end voltage given by

$$V'_2 = \frac{Z_3}{Z_2 + Z_3} V_2$$

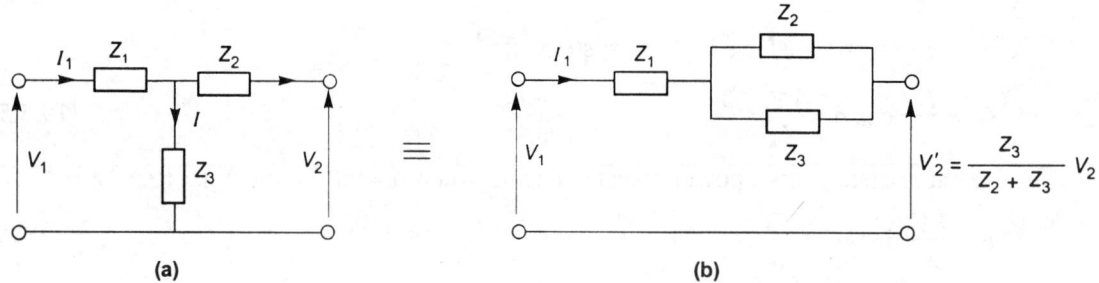

Fig. 19.4. Representation of a T-circuit with series circuit.

Proof
From circuit in Fig. 19.4 (a), by KVL

$$V_1 = Z_1 I_1 + Z_3 I$$

$$\therefore \qquad I = \frac{V_1 - Z_1 I_1}{Z_3}$$

Again, by KVL

$$V_1 = Z_1 I_1 + Z_2 (I_1 - I) + V_2 = Z_1 I_1 + Z_2 \left(I_1 - \frac{V_1 - Z_1 I_1}{Z_3} \right) + V_2$$

$$V_1 \left(1 + \frac{Z_2}{Z_3} \right) = \left[Z_1 \left(1 + \frac{Z_2}{Z_3} \right) + Z_2 \right] I_1 + V_2$$

$$V_1 \left(\frac{Z_2 + Z_3}{Z_3} \right) = \left[Z_1 \frac{(Z_2 + Z_3)}{Z_3} + Z_2 \right] I_1 + V_2$$

$$V_1 = \left(Z_1 + \frac{Z_2 Z_3}{Z_2 + Z_3} \right) I_1 + \frac{Z_3}{Z_2 + Z_3} V_2 \qquad \qquad ...(19.6.1)$$

From circuit in Fig. 19.4 (b), by KVL

$$V_1 = \left(Z_1 + \frac{Z_2 Z_3}{Z_2 + Z_3} \right) I_1 + V'_2 \qquad \qquad ...(19.6.2)$$

Comparison of Eqs. (19.6.1) and (19.6.2) shows that

$$V'_2 = \frac{Z_3}{Z_2 + Z_3} V_2 \qquad \qquad ...(19.6.3)$$

19.7 POWER-ANGLE RELATIONS FOR GENERAL NETWORK CONFIGURATIONS

In general, active power flow from bus i to bus j of an a.c. network when resistances are neglected is given by

$$P_{ij} = \frac{V_i V_j}{X_{ij}} \sin \delta_{ij} \qquad \qquad ...(19.7.1)$$

where $P_{ij} =$ active power flow from bus i to bus j

$V_i =$ voltage at bus i

$V_j =$ voltage at bus j

$\delta_{ij} =$ angle between bus i and bus j with bus i taken as reference

$X_{ij} =$ equivalent transfer reactance between buses i and j .

The reactive power flow is given by

$$Q_{ij} = \frac{V_i V_j}{X_{ij}} \cos \delta_{ij} = \frac{V_i^2}{X_{ij}} \qquad \qquad ...(19.7.2)$$

where $Q_{ij} =$ reactive power flow from bus i to bus j .

Let us determine the power-angle relation for the network configuration shown in Fig. 19.5. Here, $E_G =$ generator voltage

$V =$ infinite bus voltage

$X_G =$ generator reactance

X_s = system reactance

X_f = fault reactance

P_e = machine electrical power output.

We can replace the T-circuit of Fig. 19.5 (a) by a series circuit shown in Fig. 19.5 (b).

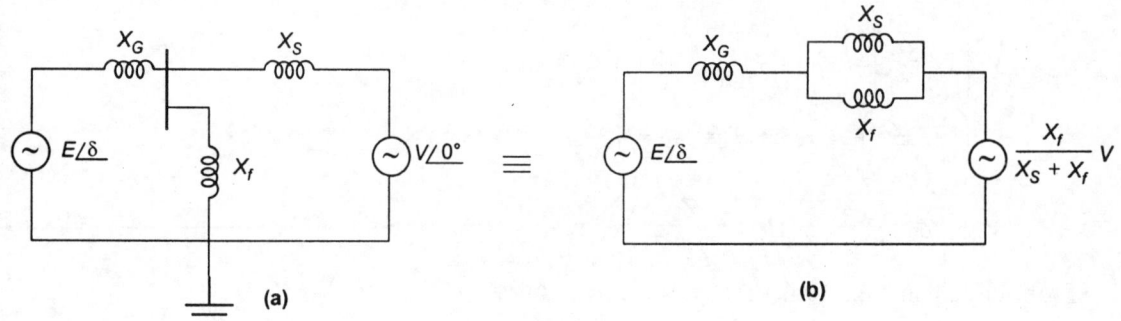

Fig. 19.5.

From Eq. (19.7.1)

$$P_e = \frac{E\left(\dfrac{X_f}{X_s + X_f}\,V\right)}{X_G + \dfrac{X_s X_f}{X_s + X_f}} \sin \delta = \frac{EV}{X_G\left(\dfrac{X_s + X_f}{X_f}\right) + X_s} \sin \delta = \frac{EV}{X_G + X_s + \dfrac{X_G X_s}{X_f}} \sin \delta \quad \ldots(19.7.3)$$

Equation (19.7.3) gives the power-angle relation for the network configuration shown in Fig. 19.5 (a).

Special Cases

(a) For the network of Fig. 19.6 (b), $X_f = \infty$. Therefore, the power-angle relation is given by

$$P_e = \frac{EV}{X_G + X_s} \sin \delta \quad \ldots(19.7.4)$$

(b) For the network of Fig. 19.6 (c), $X_f = 0$

$\therefore \qquad P_e = 0 \qquad \qquad \ldots(19.7.5)$

19.8 TRANSFER IMPEDANCE

For the 2-port network shown in Fig. 19.7,

$\qquad E_1 = AE_2 + BI_2$

$\qquad I_1 = CE_2 + DI_2$

when, $\quad E_2 = 0, \quad B = \dfrac{E_1}{I_2}$

B is called the *short-circuit transfer impedance*. Consider a power system reduced to the unsymmetrical T-circuit equivalent to a general 2-port network shown in Fig. 19.8 (a).

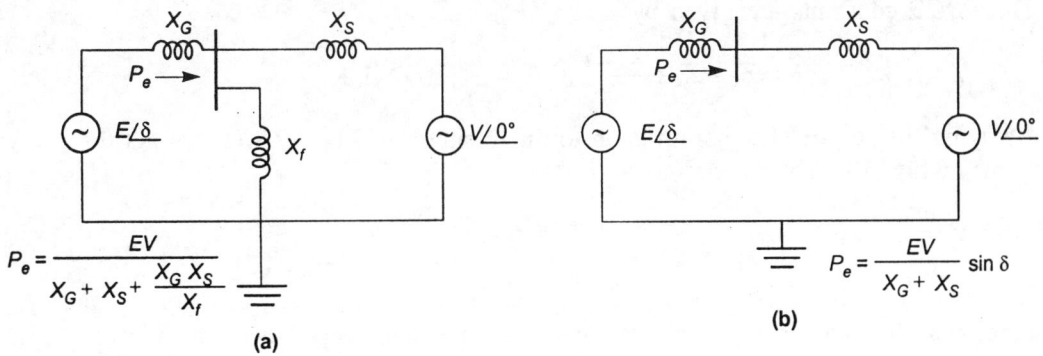

$$P_e = \frac{EV}{X_G + X_S + \frac{X_G X_S}{X_f}}$$

(a)

$$P_e = \frac{EV}{X_G + X_S} \sin \delta$$

(b)

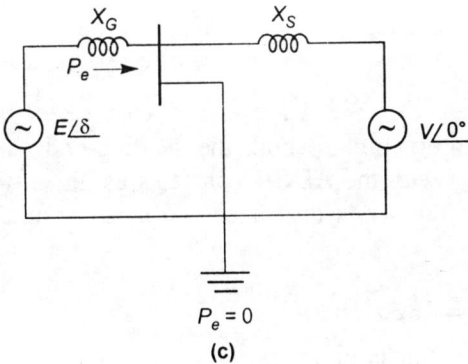

$P_e = 0$

(c)

Fig. 19.6.

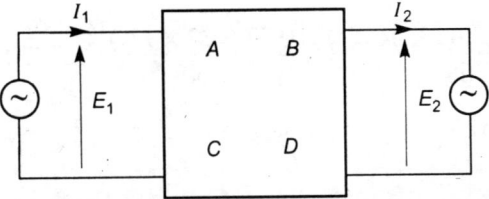

Fig. 19.7. Two-port network.

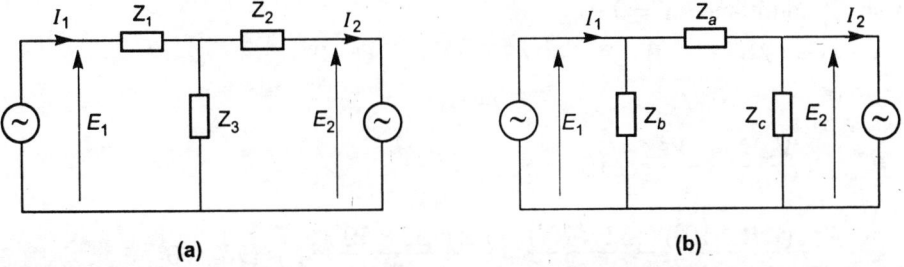

(a)   (b)

Fig. 19.8. (a) Equivalent T and (b) equivalent π networks of a power system.

The *ABCD* constants are given by

$$A = 1 + \frac{Z_1}{Z_3}, \quad B = Z_1 + Z_2 + \frac{Z_1 Z_2}{Z_3}, \quad C = \frac{1}{Z_3}, \quad D = 1 + \frac{Z_2}{Z_3} \qquad \ldots(19.8.1)$$

Applying the star-delta transformation to the T-circuit of Fig. 19.8 (a) we get the equivalent π circuit of Fig. 19.8 (b) where

$$Z_a = \frac{1}{Z_3}(Z_1 Z_2 + Z_2 Z_3 + Z_3 Z_1) = Z_1 + Z_2 + \frac{Z_1 Z_2}{Z_3} = B \qquad \ldots(19.8.2)$$

Equation (19.8.2) shows that Z_a is the transfer impedance between ports 1 and 2. It is observed that the transfer impedance is the single equivalent impedance which directly connects the two voltage sources.

Example 19.1 The **ABCD** constants of a nominal π network representing a three phase transmission line are

$$\mathbf{A} = \mathbf{D} = 0.97 \underline{/0.6°}, \quad \mathbf{B} = 60 \underline{/70°} \; \Omega, \quad \mathbf{C} = 0.001 \underline{/91°} \; \text{S}.$$

Find the steady-state stability limit if both the sending-end and receiving-end voltages are held constant at 132 kV : (a) with the **ABCD** constants as given, (b) with the shunt admittance neglected, and (c) with both the series resistance and shunt admittance neglected.

Solution

(a) $\qquad V_s = V_r = 132 \text{ kV} = 132 \times 10^3 \text{ V}$

The steady-state stability limit is given by

$$P_{r \max} = \frac{V_s V_r}{B} - \frac{A V_r^2}{B} \cos (\beta - \alpha)$$

$$= \frac{(132 \times 10^3)(132 \times 10^3)}{60} - \frac{0.97 (132 \times 10^3)^2}{60} \cos (70° - 0.6°)$$

$$= \frac{(132)^2}{60} \times 10^6 (1 - 0.97 \cos 69.4°) = 191.29 \times 10^6 \text{ W} = 191.29 \text{ MW}$$

(b) For a nominal π network

$$\mathbf{A} = \mathbf{D} = 1 + \frac{ZY}{2}, \qquad \mathbf{B} = \mathbf{Z}, \qquad \mathbf{C} = \mathbf{Y}\left(1 + \frac{ZY}{4}\right)$$

When shunt admittance $\mathbf{Y} = 0$

$$\mathbf{A} = \mathbf{D} = 1 \underline{/0°}, \qquad \mathbf{B} = \mathbf{Z} = 60 \underline{/70°} \; \Omega, \qquad \mathbf{C} = 0$$

$$\alpha = 0°, \quad \beta = 70°$$

$$P_{r \max} = \frac{V_s V_r}{B} - \frac{A V_r^2}{B} \cos (\beta - \alpha)$$

$$= \frac{(132 \times 10^3)(132 \times 10^3)}{60} - \frac{1 \times (132 \times 10^3)^2}{60} \cos (70° - 0°)$$

$$= 191.077 \times 10^6 \text{ W} = 191.077 \text{ MW}$$

(c) When both the series resistance and shunt admittances are neglected

$$A = 1 \underline{/0^\circ}, \quad D = A = 1 \underline{/0^\circ}, \quad C = 0, \quad B = Z = R + jX = 60 \underline{/70^\circ} \; \Omega$$

$\therefore \qquad R + jX = 60 \cos 70^\circ + j \, 60 \sin 70^\circ$

If the resistance is neglected

$$B = jX = j \, 60 \sin 70^\circ = j \, 56.38 = 56.38 \underline{/90^\circ}$$

$\therefore \qquad \beta = 90^\circ$

$$P_{r\,max} = \frac{V_s \, V_r}{B} - \frac{A V_r^2}{B} \cos (\beta - \alpha)$$

$$= \frac{(132 \times 10^3)\,(132 \times 10^3)}{56.38} - \frac{1 \times (132 \times 10^3)^2}{56.38} \cos (90^\circ - 0^\circ)$$

$$= 309.046 \times 10^6 \; \text{W} = 309.046 \; \text{MW}$$

Example 19.2 Develop an expression for the maximum steady-state power which can be transmitted over the line if the voltage at each end is maintained constant. Neglect the effect of capacitance of the line.

Show that if the reactance X of the line could be varied, the resistance remaining constant, the maximum steady state power that could be transmitted over the line would be greatest when $X = \sqrt{3} \, R$.

Solution

When the effect of capacitance is neglected

$$A = D = 1 \underline{/0^\circ}, \quad C = 0, \quad B = Z = R + jX$$

$$B \underline{/\beta} = R + jX, \qquad \cos \beta = \frac{R}{Z}$$

$$P_{r\,max} = \frac{V_s \, V_r}{B} - \frac{A V_r^2}{B} \cos (\beta - \alpha) = \frac{V_s \, V_r}{Z} - \frac{1 \times V_r^2}{Z} \cos (\beta - 0)$$

$$= \frac{V_s \, V_r}{Z} - \frac{V_r^2}{Z} \cdot \frac{R}{Z} = \frac{V_s \, V_r}{Z} - \frac{V_r^2}{Z^2} R$$

If X is varied and R remains constant, $P_{r\,max}$ would be greatest when

$$\frac{d}{dX} P_{r\,max} = 0$$

$$\frac{d}{dX} \left[\frac{V_s \, V_r}{Z} - \frac{V_r^2}{Z^2} R \right] = 0$$

$$\frac{d}{dX} \left[\frac{V_s \, V_r}{(R^2 + X^2)^{1/2}} - \frac{V_r^2 \, R}{(R^2 + X^2)} \right] = 0$$

$$V_s \, V_r \frac{d}{dX} (R^2 + X^2)^{-1/2} - R V_r^2 \frac{d}{dX} (R^2 + X^2)^{-1} = 0$$

$$V_s V_r \left(-\frac{1}{2}\right)(R^2 + X^2)^{-3/2}(2X) - RV_r^2(-1)(R^2 + X^2)(2X) = 0$$

$$(R^2 + X^2)^{1/2} = \frac{2V_r R}{V_s}$$

$$R^2 + X^2 = 4R^2\left(\frac{V_r}{V_s}\right)^2$$

$$X^2 = R^2\left[4\left(\frac{V_r}{V_s}\right)^2 - 1\right]$$

If, $V_s = V_r$

$$X^2 = R^2(4-1) = 3R^2$$

∴ $X = \sqrt{3}\, R$ is the condition for $P_{r\,max}$ to be greatest.

Example 19.3 For the system shown in Fig. 19.9 (a), an inductor of reactance 0.6 pu per phase is connected at the mid-point of transmission line. Determine the steady-state power limit under the following conditions :

(a) with inductor switch S open;

(b) with inductor switch S closed;

(c) with inductor replaced by a capacitor of the same per unit reactance;

(d) with a capacitor of the same per unit reactance connected in series with the line at the same point instead of shunt capacitor;

(e) with the inductor replaced by a resistor of resistance 1.5.

Solution

(a) *When switch S is open*

Transfer reactance, $X = X_{dg} + X_T + \frac{1}{2}X_l + \frac{1}{2}X_l = 0.8 + 0.1 + 0.3 + 0.3 = 1.5$ pu

Therefore, the steady-state power limit

$$P_{r\,max} = \frac{EV}{X} = \frac{1.2 \times 1.0}{1.5} = 0.8 \text{ pu}$$

(b) *When switch S is closed*

With switch S closed, the system is represented by the equivalent network shown in Fig. 19.9 (b). The transformer impedance is found by using the relation

$$\mathbf{B} = \mathbf{Z}_1 + \mathbf{Z}_2 + \frac{\mathbf{Z}_1 \mathbf{Z}_2}{\mathbf{Z}_3}$$

Here, $\mathbf{Z}_1 = j\left(X_{dg} + X_T + \frac{1}{2}X_l\right) = j(0.8 + 0.1 + 0.3) = j\,1.2$ pu

$$\mathbf{Z}_2 = j\frac{1}{2}X_l = j\,0.3 \text{ pu}$$

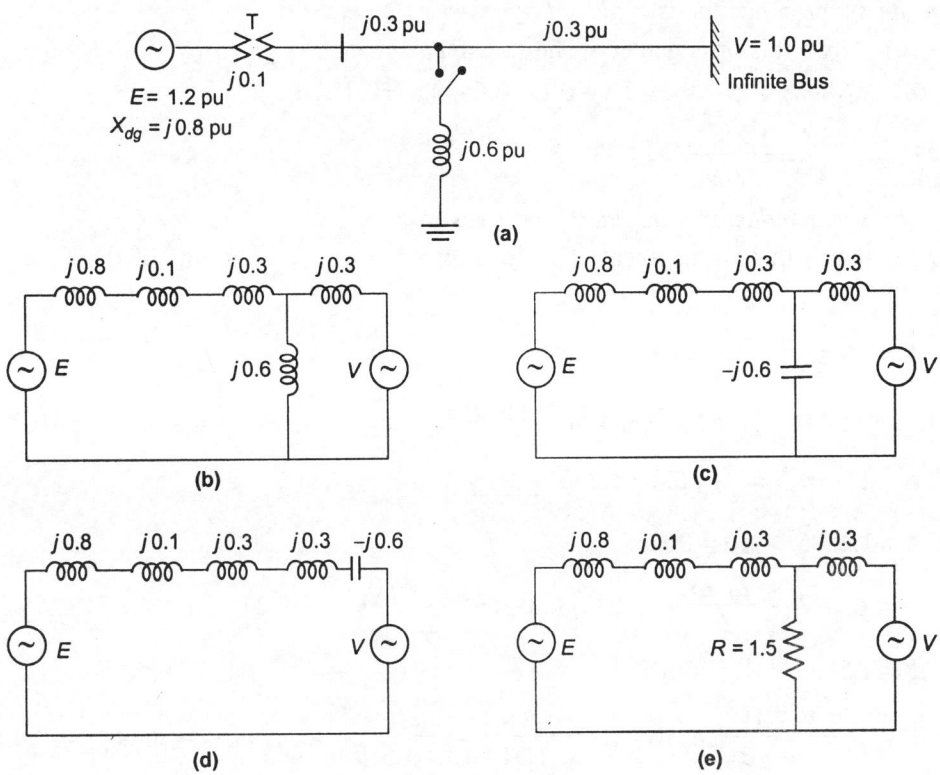

Fig. 19.9. Illustrating Example 19.3.

$$\mathbf{Z}_3 = j\, X_r = j\, 0.6 \text{ pu}$$

where X_r = inductive reactance of the reactor.

$$\therefore \qquad \mathbf{B} = j\, X = j\, 1.2 + j\, 0.3 + \frac{(j\, 1.2) \times (j\, 0.3)}{j\, 0.6} = j\, 2.1 \text{ pu}$$

$$|\mathbf{B}| = X = 2.1 \text{ pu}, \quad \beta = 90°$$

$$P_{\max} = \frac{EV}{X} = \frac{1.2 \times 1.0}{2.1} = 0.5714 \text{ pu}$$

(c) *When the shunt reactor is replaced by a shunt capacitor*

Under this condition the equivalent network is shown in Fig. 19.9 (c).

$$\mathbf{B} = \mathbf{Z}_1 + \mathbf{Z}_2 + \frac{\mathbf{Z}_1\, \mathbf{Z}_2}{\mathbf{Z}_3} = j\, 1.2 + j\, 0.3 + \frac{(j\, 1.2) \times (j\, 0.3)}{(-j\, 0.6)} = j\, 0.9 = 0.9 \underline{/90°}$$

$$|\mathbf{B}| = X = 0.9, \quad \beta = 90°$$

$$P_{\max} = \frac{EV}{X} = \frac{1.2 \times 1.0}{0.9} = 1.333 \text{ pu}$$

(d) *When the shunt capacitor is replaced by a series capacitor*

Fig. 19.9 (d) shows this arrangement. The transfer reactance is given by

$$X = X_{dg} + X_T + X_l - X_c = 0.8 + 0.1 + 0.6 - 0.6 = 0.9$$

$$P_{max} = \frac{EV}{X} = \frac{1.2 \times 1.0}{0.9} = 1.333 \text{ pu}$$

(e) *When the shunt inductor is replaced by a resistor*

Fig. 19.9 (e) shows this arrangement. The generalized **A** and **B** constants of the T-network are given by

$$\mathbf{A} = 1 + \frac{\mathbf{Z}_1}{\mathbf{Z}_3}$$

$$\mathbf{Z}_1 = j\,0.8 + j\,0.1 + j\,0.3 = j\,1.2, \qquad \mathbf{Z}_3 = 1.5$$

∴ $$\mathbf{A} = 1 + \frac{j\,1.2}{1.5} = 1 + j\,0.8 = 1.2806\,\underline{/36.66^\circ}$$

$$A = 1.2806, \qquad \alpha = 36.66^\circ$$

$$\mathbf{B} = \mathbf{Z}_1 + \mathbf{Z}_2 + \frac{\mathbf{Z}_1\,\mathbf{Z}_2}{\mathbf{Z}_3} = j\,1.2 + j\,0.3 + \frac{j\,(1.2) \times j\,0.3}{1.5} = -0.24 + j\,1.5 = 1.519\,\underline{/99.1^\circ}$$

$$B = 1.519, \qquad \beta = 99.1^\circ$$

$$P_{max} = \frac{EV}{B} - \frac{AV^2}{B}\cos(\beta - \alpha) = \frac{1.2 \times 1}{1.519} - \frac{1.2806 \times 1}{1.519}\cos(99.1^\circ - 36.66^\circ) = 0.4 \text{ pu}$$

19.9 POWER-ANGLE CHARACTERISTIC OF A SALIENT-POLE SYNCHRONOUS MACHINE

Let X_d = direct–axis synchronous reactance

X_q = quadrature–axis synchronous reactance

E_f = generated voltage

V = terminal voltage

I_a = output current

φ = phase angle between V and I_a

I_d = direct–axis component of I_a

I_q = quadrature–axis component of I_a

The resistance R_a of the armature has negligible effect on the relationship between the power output of a synchronous machine and its torque angle δ. It may, therefore, be neglected. The phasor diagram at lagging pf for a salient-pole synchronous generator, neglecting R_a, is shown in Fig. 19.10.

The torque angle characteristic of a salient-pole machine may be derived from the phasor diagram of Fig. 19.10.

The complex power output per phase

$$S_{1\,\varphi} = VI_a^* \qquad \qquad \qquad ...(19.9.1)$$

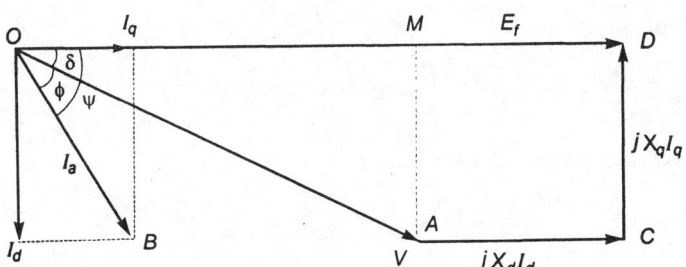

Fig. 19.10. Phasor diagram at lagging pf of a salient-pole generator, neglecting R_a.

Taking E_f as the reference phasor.

$\therefore$ $\mathbf{V} = V\underline{/-\delta} = V\cos\delta - jV\sin\delta$

   $\mathbf{I_a} = I_q - jI_d$

   $\mathbf{I_a^*} = I_q + jI_d$

   $S_{1\,\varphi} = \mathbf{VI_a^*} = (V\cos\delta - jV\sin\delta)(I_q + jI_d)$ ...(19.9.2)

From the phasor diagram of Fig. 19.10

   $X_qI_q = CD = AM = V\sin\delta$

$\therefore$ $I_q = \dfrac{V\sin\delta}{X_q}$ ...(19.9.3)

   $X_dI_d = AC = MD = OD - OM = E_f - V\cos\delta$

$\therefore$ $I_d = \dfrac{E_f - V\cos\delta}{X_d}$ ...(19.9.4)

Substituting the values of I_q and I_d in Eq. (19.9.2), we get

$$S_{1\,\varphi} = (V\cos\delta - jV\sin\delta)\left(\frac{V\sin\delta}{X_q} + j\frac{E_f - V\cos\delta}{X_d}\right)$$

$$= \left(\frac{V^2}{X_q}\sin\delta\cos\delta + \frac{VE_f}{X_d}\sin\delta - \frac{V^2}{X_d}\sin\delta\cos\delta\right) + j\left(\frac{VE_f}{X_d}\cos\delta - \frac{V^2}{X_d}\cos^2\delta - \frac{V^2}{X_q}\sin^2\delta\right)$$

$$= \left[\frac{VE_f}{X_d}\sin\delta + \frac{V^2}{2}\left(\frac{1}{X_q} - \frac{1}{X_d}\right)\sin 2\delta\right] + j\left[\frac{VE_f}{X_d}\cos\delta - \frac{V^2}{2X_d}(1+\cos 2\delta) - \frac{V^2}{2X_q}(1-\cos 2\delta)\right]$$

$$= \left[\frac{VE_f}{X_d}\sin\delta + \frac{V^2}{2}\left(\frac{1}{X_q} - \frac{1}{X_d}\right)\sin 2\delta\right] + j\left[\frac{VE_f}{X_d}\cos\delta - \frac{V^2}{2X_dX_q}\left\{(X_d+X_q) - (X_d-X_q)\cos 2\delta\right\}\right]$$

    ...(19.9.5)

Also, $S_{1\,\varphi} = P_{1\,\varphi} + jQ_{1\,\varphi}$ ...(19.9.6)

Therefore, the real power per phase in watts is

$$P_{1\,\varphi} = \frac{VE_f}{X_d}\sin\delta + \frac{V^2}{2}\left(\frac{1}{X_q} - \frac{1}{X_d}\right)\sin 2\delta$$ ...(19.9.7)

Total real power in watts

$$P_{3\,\phi} = 3P_{1\,\phi} = \frac{3VE_f}{X_d}\sin\delta + \frac{3V^2}{2}\left(\frac{1}{X_q} - \frac{1}{X_d}\right)\sin 2\delta \qquad \qquad ...(19.9.8)$$

The reactive power per phase in vars is

$$Q_{1\,\phi} = \frac{VE_f}{X_d}\cos\delta - \frac{V^2}{2X_dX_q}[(X_d + X_q) - (X_d - X_q)\cos 2\delta] \qquad ...(19.9.9)$$

Total reactive power in vars

$$Q_{3\,\phi} = 3Q_{1\,\phi} = \frac{3VE_f}{X_d}\cos\delta - \frac{3V^2}{2X_dX_q}[(X_d + X_q) - (X_d - X_q)\cos 2\delta] \qquad ...(19.9.10)$$

For an infinite bus system V is constant and E_f can be made constant by keeping excitation constant. Hence it is seen from Eq. (19.9.7) that real power P is a function only of angle δ.

The first term on the right hand side of Eq. (19.9.7) is called *excitation power* and the second term the *reluctance power*. Thus,

$$\text{excitation power} = \frac{VE_f}{X_d}\sin\delta \qquad \qquad ...(19.9.11)$$

$$\text{reluctance power} = \frac{V^2}{2}\left(\frac{1}{X_q} - \frac{1}{X_d}\right)\sin 2\delta \qquad \qquad ...(19.9.12)$$

The reluctance power depends on the **saliency** defined by the quantity $\left[\dfrac{1}{X_q} - \dfrac{1}{X_d}\right]$. The saliency disappears when $X_d = X_q$ (that is, for a cylindrical rotor). It is to be noted that if there is no field excitation, $E_f = 0$, and excitation power is zero. However, the reluctance power exists even when there is no field current. That is, the machine still has some P generation capability. It is impractical to operate a synchronous generator without field excitation on a power system, because it would supply only about 25% or less of its real power rating. Also, it would absorb an excessive amount of reactive power.

The reluctance power is of the order of 10 to 20 percent of the excitation power. The reluctance power is usually neglected in the steady-state stability studies.

For a non-salient pole (cylindrical rotor) machine $X_q = X_d$ and

$$P_{1\,\phi} = \frac{VE_f}{X_d}\sin\delta \qquad \qquad ...(19.9.13)$$

Equation (19.9.11) shows that the excitation power is proportional to $\sin\delta$ and Eq. (19.9.12) shows that the reluctance power is proportional to $\sin 2\delta$. The power-angle (P–δ) curve for a salient-pole machine is shown in Fig. 19.11. It is to be noted that the peak power or steady-state limit occurs at a value of δ less than 90°. Positive P, indicating generator action, is obtained for $\delta > 0$, that is, for E_f leading **V**. Negative P, corresponding to motor action, is obtained for negative δ.

For a cylindrical rotor machine, $X_d = X_q$, reactive power per phase in vars is

$$Q_{1\,\phi} = \frac{VE_f}{X_d}\cos\delta - \frac{V^2}{X_d} \qquad \qquad ...(19.9.14a)$$

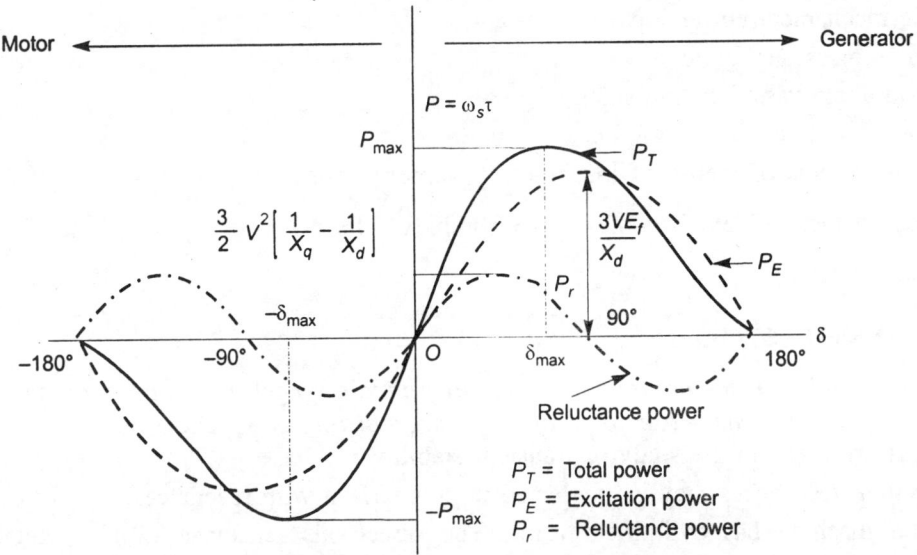

Fig. 19.11. Power-angle curve of a salient-pole machine.

or $\qquad Q1 \; \varphi = \dfrac{V}{X_d}(E_f \cos \delta - V)$ $\qquad\qquad\qquad\qquad$ ...(19.9.14b)

Equation (19.9.14b) shows that when $E_f \cos \delta = V$, that is under normal excitation, $Q = 0$, and the generator operates at unity power factor.

When $E_f \cos \delta > V$, that is, the generator is overexcited, Q is positive. Therefore, the generator supplies reactive power to the busbars. When $E_f \cos \delta < V$, that is, the alternator is underexcited, Q is negative. Hence the generator absorbs or consumes reactive power.

Equation (19.9.9) may be applied to both synchronous generators and synchronous motors. Motor action results when δ is negative, that is $\mathbf{V}$ leading E_f. The reactive power Q is independent of sign of δ which shows that, in general, an overexcited generator or motor supplies reactive power to the busbars, and an underexcited generator or motor consumes or absorbs reactive power from the busbars.

19.10 STEADY-STATE STABILITY CRITERION

The rate $\dfrac{dP}{d\delta}$, i.e., the differential power increase obtained per differential load angle increase is called the *synchronizing power coefficient* or *electrical stiffness* of a synchronous machine. It is taken as the measure of the stability of a system. The direct-state synchronous stability criterion for a simple system is $\dfrac{dP}{d\delta} > 0$, i.e., the synchronizing coefficient is positive. The steady-state stability limit is reached when $\dfrac{dP}{d\delta} = 0$ and if $\dfrac{dP}{d\delta} < 0$, the system is unstable.

The criterion of a stability holds only under conditions satisfying the following assumptions :

(a) generators are represented by constant impedances in series with the no-load voltages;

    (b) the mechanical power input is constant;

    (c) damping is negligible;

    (d) load angle variations are small;

    (e) speed variations are negligible.

When the effects of inertia of machines, governor action and automatic voltage regulators are considered the problem becomes more complex. The criterion $\dfrac{dP}{d\delta} = 0$ alone gives a low result which is safe.

19.11 TRANSIENT STABILITY

The transient stability is the ability of a system to maintain synchronous operation and to reach a stable state or the one close to it after a large disturbance. The following simplifying *assumptions* are made in the study of transient stability :

    (a) System resistances may be neglected in comparison with reactances.

    (b) The machine has cylindrical rotor. The direct-axis reactance (X_d) is equal to the quadrature-axis reactance (X_q).

    (c) The system may be reduced to an equivalent two-machine system.

    (d) Each machine may be assumed to supply an infinite bus.

    (e) Direct axis transient reactance (X_d) is used for machine representation.

    (f) The shaft input power may be assumed constant for few seconds after occurrence of a disturbance. This assumption may be valid on the grounds that the mechanical system involving governors, steam valves etc., is relatively sluggish in operation as compared to rapidly changing electrical quantities. With fast acting valves the assumption of constant input will not be true.

The problem of stability revolves around the determination of whether or not the torque angle δ will stabilize after a sudden disturbance. In case δ continues to increase after a disturbance the machine will lose synchronism.

In a synchronous generator the input is the mechanical or shaft torque and the output is the electromagnetic torque. Both these torques are assumed positive in the following discussion. For a synchronous motor, the input is the electromagnetic torque and the output is the shaft torque. Based upon the sign conventions adopted for synchronous generators, the values of the shaft torque and electromagnetic torque are taken as negative for motor action.

Let T_e = electromagnetic torque

    T_s = shaft torque

If the losses are neglected the difference between the shaft torque and the electromagnetic torque is equal to the accelerating or decelerating torque.

For a generator, when $T_s > T_e$, then T_a is positive and the rotor accelerates. In case of synchronous motors T_a is positive only when $T_e > T_s$ since T_s and T_e are both negative.

19.12 SWING EQUATION

The behaviour of a synchronous machine during transients is described by the swing equation. Let θ be the angular position of the rotor at any instant t. However, θ is continuously changing with time. It is convenient to measure θ with respect to reference axis that is rotating at synchronous speed. If δ is the angular displacement of the rotor in electrical degrees from the synchronously rotating reference axis and ω_s the synchronous speed in electrical radians, then θ can be expressed as the sum of (1) time varying angle $\omega_s t$ on the rotating reference axis, plus (2) the torque angle δ of the rotor with respect to the rotating reference axis. In other words,

$$\theta = \omega_s t + \delta \text{ electrical radians} \qquad \qquad \qquad ...(19.12.1)$$

Differentiating Eq. (19.12.1) with respect to t we get

$$\frac{d\theta}{dt} = \omega_s + \frac{d\delta}{dt} \qquad \qquad \qquad ...(19.12.2)$$

Differentiation of Eq. (19.2.2) gives

$$\frac{d^2\theta}{dt^2} = \frac{d^2\delta}{dt^2} \qquad \qquad \qquad ...(19.12.3)$$

Angular acceleration of rotor

$$\alpha = \frac{d^2\theta}{dt^2} = \frac{d^2\delta}{dt^2} \text{ elec. rad/s}^2 \qquad \qquad \qquad ...(19.12.4)$$

If damping is neglected the accelerating torque, T_a in a synchronous generator is equal to the difference of input mechanical or shaft torque T_s and the output electromagnetic (electrodynamic) torque T_e. That is,

$$T_a = T_s - T_e \qquad \qquad \qquad ...(19.12.5)$$

Let, $\omega =$ synchronous speed of the rotor

$J =$ moment of inertia of the rotor

$M =$ angular momentum of the rotor

$P_s =$ mechanical power input

$P_e =$ electrical power output

$P_a =$ accelerating power.

Now $M = J\omega$ $\qquad \qquad \qquad ...(19.12.6)$

Multiplying both the sides of Eq. (19.12.5) by ω we get

$$\omega T_a = \omega T_s - \omega T_e$$

$$P_a = P_s - P_e$$

But $J\dfrac{d^2\theta}{dt^2} = T_a$; $J\dfrac{d^2\delta}{dt^2} = T_a$

$$\omega J \frac{d^2\delta}{dt^2} = \omega T_a$$

$$M \frac{d^2\delta}{dt^2} = P_a = P_s - P_e \qquad \qquad \qquad ...(19.12.7)$$

Equation (19.12.7) gives the relation between the accelerating power and angular acceleration. It is called the *swing equation*. It is a non-linear differential equation of the second order. With this differential equation we can discuss stability in a quantitative way, because it describes *swings* in the power angle δ during transients.

19.13 SWING CURVES

A graph of δ (usually in electrical radians) *versus* time in seconds is called the *swing curve*. Swing curves (Fig. 19.12) provide information regarding stability. They show any tendency of δ to oscillate and/or increase beyond the point of return. If δ increases continuously with time the system is unstable. While if δ starts decreasing after reaching a maximum value it is inferred that the system will remain stable.

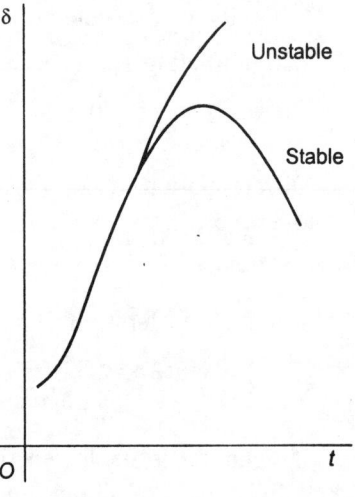

Fig. 19.12. Swing curves.

Swing curves are useful in determining the adequacy of relay protection on power systems with regard to the clearing of faults before one or more machines become unstable and fall out of synchronism. The critical clearing time is found to specify the correct speed of the circuit breaker.

The solution of swing equation involves elliptic integrals. Step-by-step (or point-by-point) method may be used for numerical solution of swing equation. At present digital computer is used for solving swing equation.

19.14 *M* AND *H* CONSTANTS

The transient conditions of synchronous machines depend, in part, on the mechanical constants of the rotor and load or prime mover.

Let ω = synchronous speed of the rotor in rad/s

   m = mass of the rotor in kg

   r = radius of gyration in m

   J = moment of inertia of the rotor in kg–m^2

   M = angular momentum of the rotor in Js/rad

   W = kinetic energy of the rotor in J

   f = system frequency in Hz

   T = torque in N–m

   P = power in watts

   α = angular acceleration of the rotor.

19.14.1 *M* Constant

Now, $J = mr^2$

   $W = \dfrac{1}{2} J \omega^2$

$$M = J\,\omega = \frac{2W}{\omega}$$

$\omega = 2\pi f$ rad/s $= 360 f$ elec. deg/s

$T = J\,\alpha$

$P = \omega\,T = \omega\,J\,\alpha = M\,\alpha$

$$M = \frac{P}{\alpha}$$

Thus, M constant may be defined as the power in MW required to produce unit angular acceleration.

19.14.2 *H* Constant or Per Unit Inertia Constant

Another constant H, called the per unit inertia constant, is more frequently used by the manufacturers. The *per unit inertia constant H* is defined as the kinetic energy stored in the rotating parts of the machine at *synchronous speed* per unit megavoltamperes (MVA) of the machine. Thus

$$H \overset{\Delta}{=} \ = \frac{\text{kinetic energy in MJ at rated speed}}{\text{machine rating in MVA}}$$

It is expressed in MJ/MVA.

If W is the stored energy in megajoules (MJ) and S is the rating of the machine in MVA, then

$$H \overset{\Delta}{=} \ \frac{W}{S} = \frac{\omega\,M}{2S} = \frac{2\pi fM}{2S} = \frac{\pi fM}{S} \qquad \qquad \ldots(19.14.1)$$

$$M = \frac{HS}{\pi f} \ \text{MJs/elec. radian} \qquad \qquad \ldots(19.14.2)$$

or $\qquad M = \frac{HS}{180 f} \ \text{MJs/elec. degree} \qquad \qquad \ldots(19.14.3)$

The value of angular momentum M varies over a wide range of MVA for a given type of machine and prime mover, but the value of H is fairly constant. Hence H is more convenient to use. Typical values are:

Cylindrical-rotor alternators	4-10
Salient-pole alternators	2-4
Synchronous compensators	1-2
Salient-pole synchronous motors	0.5-2

The swing equation can be written as

$$\frac{HS}{180 f}\,\frac{d^2\delta}{dt^2} = P_s - P_e \qquad \qquad \ldots(19.14.4)$$

If we combine Eqs. (19.14.3) and (19.14.4) and divide by S, we obtain the per unit swing equation as

$$\frac{H}{180 f}\,\frac{d^2\delta}{dt^2} = P_{s_{pu}} - P_{e_{pu}} = P_{a_{pu}} \qquad \qquad \ldots(19.14.5)$$

19.14.3 *H* Constant on a Common Base

An inertia constant H_{mach} based on a machine's own MVA rating may be converted to a value H_{syst} relative to the system base S_{syst} with the formula

$$H_{syst} = H_{mach}\frac{S_{mach}}{S_{syst}} \qquad ...(19.14.6)$$

A convenient system base value is 100 MVA.

19.15 EQUIVALENT SYSTEM

Suppose that a number of generators are connected in parallel to the same busbars.

Let, $S_1, S_2, ..., S_n$ = MVA rating of individual machines

$\qquad\qquad S_e$ = MVA rating of the equivalent machine

$\qquad\qquad S_b$ = base MVA

$H_1, H_2, ..., H_n$ = inertia constants of individual machines

$\qquad\qquad H_e$ = inertia constant of a single equivalent machine or overall inertia constant

$\qquad\qquad S$ = total rating of the machines.

Energy stored by the equivalent machines = sum of the energies stored by individual machines

$$W = W_1 + W_2 + ... + W_n \qquad ...(19.15.1)$$

$$S_e H_e = S_1 H_1 + S_2 H_2 + ... + S_n H_n \qquad ...(19.15.2)$$

$$S_e = S_1 + S_2 + ... + S_n \qquad ...(19.15.3)$$

If the base MVA is equal to the combined MVA rating of the individual machines, i.e., $S_b = S_e$ Eq. (19.15.2) becomes

$$H_e = H_1\left(\frac{S_1}{S_b}\right) + H_2\left(\frac{S_2}{S_b}\right) + ... + H_n\left(\frac{S_n}{S_b}\right)$$

Thus, the equivalent inertia constant is the sum of the individual constants when these are referred to the total rating of the machines.

If the machines are identical

$$S_1 = S_2 = ... = S_n = S \text{ (say)} \qquad ...(19.15.4)$$

$$H_1 = H_2 = ... = H_n = H \text{ (say)} \qquad ...(19.15.5)$$

$$S_b = S_e = nS \qquad ...(19.15.6)$$

$$H_e = \frac{nHS}{nS} = H \qquad ...(19.15.7)$$

Thus, the equivalent H constant of several identical machines operating in parallel is the same as that of any one of the machines.

19.16 EQUIVALENT *M* CONSTANT OF TWO MACHINES

Two synchronous machines connected by a reactance can be replaced by one equivalent machine connected by the reactance to an infinite busbar. Let suffixes 1 and 2 be used for the two machines.

For one machine connected to infinite busbars,

$$M \frac{d^2\delta}{dt^2} = P_s - P_e \qquad\qquad\qquad ...(19.16.1)$$

$$\therefore \qquad M_1 \frac{d^2\delta_1}{dt^2} = P_{s1} - P_{e1} \qquad\qquad\qquad ...(19.16.2)$$

and $\qquad M_2 \frac{d^2\delta_2}{dt^2} = P_{s2} - P_{e2} \qquad\qquad\qquad ...(19.16.3)$

Let δ be the relative angle between the rotors of the two machines

$$\delta = \delta_1 - \delta_2 \qquad\qquad\qquad ...(19.16.4)$$

$$\frac{d^2\delta}{dt^2} = \frac{d^2\delta_1}{dt^2} - \frac{d^2\delta_2}{dt^2} = \frac{1}{M_1} (P_{s1} - P_{e1}) - \frac{1}{M_2} (P_{s2} - P_{e2})$$

Multiplying both the sides of the above equation by $\dfrac{M_1 M_2}{M_1 + M_2}$

$$\frac{M_1 M_2}{M_1 + M_2} \cdot \frac{d^2\delta}{dt^2} = \frac{M_2}{M_1 + M_2} (P_{s1} - P_{e1}) - \frac{M_1}{M_1 + M_2} (P_{s2} - P_{e2})$$

$$= \frac{M_2 P_{s1} - M_1 P_{s2}}{M_1 + M_2} - \frac{M_2 P_{e1} - M_1 P_{e2}}{M_1 + M_2} \qquad\qquad ...(19.16.5)$$

Equation (19.16.5) can be represented in the form

$$M' \frac{d^2\delta}{dt^2} = P'_s - P'_e \qquad\qquad\qquad ...(19.16.6)$$

where $\qquad M' = \dfrac{M_1 M_2}{M_1 + M_2} \qquad\qquad\qquad ...(19.16.7)$

$$P'_s = \frac{M_2 P_{s1} - M_1 P_{s2}}{M_1 + M_2} \qquad\qquad\qquad ...(19.16.8)$$

$$P'_e = \frac{M_2 P_{e1} - M_1 P_{e2}}{M_1 + M_2} \qquad\qquad\qquad ...(19.16.9)$$

It is seen that Eq. (19.16.6) is similar to Eq. (19.16.1). Thus, two interconnected machines can be represented as a single source. The quantities M'_s, P'_s and P'_e represent the equivalent values of the inertia constant, the input at the shaft and the electromagnetic output respectively. The load angle δ of the equivalent machine is given by Eq. (19.16.4).

Multi-Machine Systems

The swing equation of a machine is given by

$$M \frac{d^2\delta}{dt^2} = P_s - P_e \qquad\qquad\qquad ...(19.16.10)$$

In a system with n machines, the rotor of each machine will respond in accordance with Eq. (19.16.10) so that we have

$$M_1 \frac{d^2\delta_1}{dt^2} = P_{s_1} - P_{e_1}$$

$$M_2 \frac{d^2\delta_2}{dt^2} = P_{s_2} - P_{e_2}$$

... ...

$$M_n \frac{d^2\delta_n}{dt^2} = P_{s_n} - P_{e_n}$$

It is seen that there is a separate swing equation for each machine. Consequently, it is the relative displacement between power angles of the machines that is essential in determining the system stability. In a multi-machine system, a single machine is usually chosen as a reference and the rotor swings (changes in power angles) of the remaining machines are determined relative to the reference machine. Stability is maintained if the machine rotors return to a stable operating state relative to each other.

Example 19.4 A two-pole 50 Hz 60 MVA turbogenerator has a moment of inertia of 9×10^3 kg-m^2. Calculate

(a) the kinetic energy in MJ at rated speed,

(b) the inertia constants M and H,

(c) the inertia constant on 50 MVA base.

Solution

$$N = \frac{120f}{P} = \frac{120 \times 50}{2} = 3000 \text{ rpm.}$$

(a)   Kinetic energy stored $= \frac{1}{2} J \omega^2 = \frac{1}{2} J \left(\frac{2\pi N}{60}\right)^2 = \frac{1}{2} \times 9 \times 10^3 \left(\frac{2\pi \times 3000}{60}\right)^2$

$$= 444 \times 10^6 \text{ joules} = 444 \text{ MJ}$$

(b) $H = \dfrac{\text{kinetic energy stored in MJ}}{\text{MVA rating}} = \dfrac{444}{60} = 7.4 \text{ MJ/MVA}$

$M = \dfrac{\text{kinetic energy stored in MJ}}{180f} = \dfrac{444}{180 \times 50} = 0.0493 \text{ MJs/elec. degree}$

(c)   Per unit inertia constant $= \dfrac{M}{\text{base MVA}} = \dfrac{0.0493}{50} = 9.86 \times 10^{-4} \text{ pu.}$

Example 19.5. A 500 MVA synchronous machine has $H_1 = 4.6$ MJ/MVA, and a 1500 MVA machine has $H_2 = 3.0$ MJ/MVA. The two machines operate in parallel in a power station. What is the equivalent H constant for the two, relative to a 100 MVA base?

Solution

The total kinetic energy of the two machines $= S_1 H_1 + S_2 H_2$

$$= 4.6 \times 500 + 3 \times 1500 = 6800 \text{ MJ}$$

The equivalent H relative to a 100 MVA base $= \dfrac{6800}{100} = 68 \text{ MJ/MVA}$

19.17 EQUAL-AREA CRITERION OF STABILITY

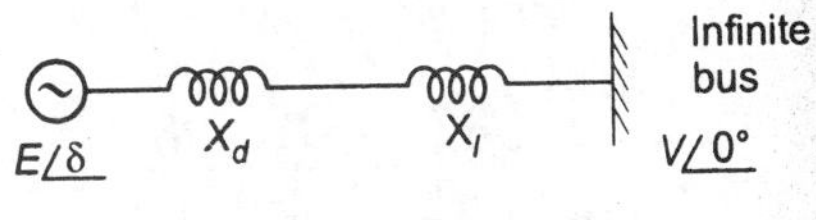

Equal area criterion may be used to assess the transient stability of a two-machine system or one machine connected to an infinite bus without actually solving the swing equation. Consider a loss-free synchronous generator supplying an infinite bus through a purely reactive transmission line of reactance X_l as shown in Fig. 19.13. The electrical power transferred is given by Eq. (19.5.7) which is reproduced here for convenience.

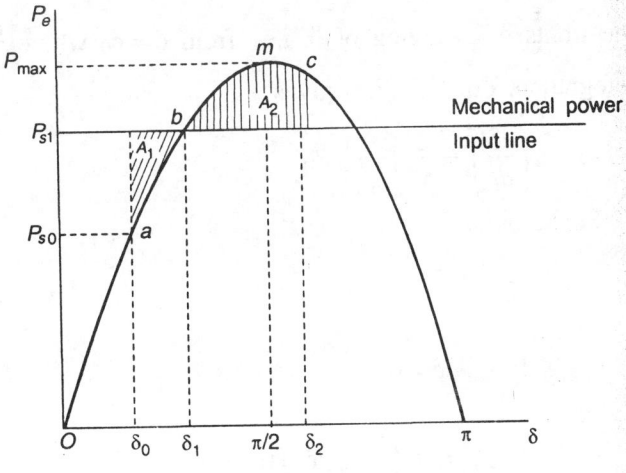

Fig. 19.13.

$$P_e = P_{e\,\max} \sin \delta \qquad \qquad \ldots(19.17.1)$$

The power-angle curve is shown in Fig. 19.14. Suppose that initially the mechanical input (shaft power) is P_{s0} at load angle δ_0. It is represented by point a on the power-angle curve. Let the mechanical input power suddenly increase to P_{s1}. With the sudden increase of shaft power there is momentarily more shaft input than electrical output. The increase in power $(P_{s1} - P_{s0})$ accelerates the rotor so that it is advanced with respect to the initial position with the result that load angle is increased. Let this new load angle be δ_1 corresponding to P_{s1}. Since the rotor is in acceleration and running slightly

Fig. 19.14.

above synchronous speed the load angle goes on increasing overshooting point b. When the load angle is more than δ_1 the rotor retards since the power transferred to the busbar is greater than input power P_{s1}. The rotor decelerates until it reaches some maximum point c, where it is again running at synchronous speed. The rotor swing starts in the reverse direction. The load angle goes on decreasing until it is equal to δ_0 where again the rotor is running at synchronous speed. The cycle is repeated. The rotor will oscillate for some time about b, before finally coming to rest at b.

Here, we have made an important assumption that the first swing or oscillation of the rotor does not make the system unstable. In practice the system is more likely to be stable during subsequent swings, particularly if it is stable for the subsequent steady-state condition. This assumption may be justified by the fact that the losses of the system progressively damp the amplitude of the swing.

Consider the swing equation

$$\frac{d^2\delta}{dt^2} = \frac{P_a}{M} \qquad \qquad \text{...(19.17.2)}$$

Multiplying both the sides by $2\frac{d\delta}{dt}$

$$2\frac{d\delta}{dt}\left(\frac{d^2\delta}{dt^2}\right) = \frac{2P_a}{M} \cdot \frac{d\delta}{dt}$$

$$\frac{d}{dt}\left(\frac{d\delta}{dt}\right)^2 = 2\frac{P_a}{M} \cdot \frac{d\delta}{dt} \qquad \qquad \text{...(19.17.3)}$$

The time rate of change of load angle $d\delta/dt$ is the speed of the machine with respect to the synchronously revolving reference frame. For the stability, this speed must become zero at some time after disturbance. That is, $d\delta/dt = 0$.

Since the condition $\left(\frac{d\delta}{dt}\right) = 0$ implies synchronous running, Eq. (19.17.3) is integrated between the limits of swinging of δ, i.e., from $\delta = \delta_0$ where $\frac{d\delta}{dt} = 0$ to $\delta = \delta_2$ where again $\frac{d\delta}{dt} = 0$. On integration, Eq. (19.17.3) gives

$$\left(\frac{d\delta}{dt}\right)^2 = \frac{2}{M}\int_{\delta_0}^{\delta_2} P_a \, d\delta$$

For stability

$$\frac{d\delta}{dt} = 0$$

$$\therefore \qquad \int_{\delta_0}^{\delta_2} P_a \, d\delta = 0 \qquad \qquad \text{...(19.17.4)}$$

$$\int_{\delta_0}^{\delta_1} P_a \, d\delta + \int_{\delta_1}^{\delta_2} P_a \, d\delta = 0$$

$$\int_{\delta_0}^{\delta_1} P_a \, d\delta = -\int_{\delta_1}^{\delta_2} P_a \, d\delta \qquad \qquad \text{...(19.17.5)}$$

$$A_1 = -A_2 \qquad \qquad \text{...(19.17.6)}$$

where $A_1 = \int_{\delta_0}^{\delta_1} P_a \, d\delta$ = positive or accelerating area

= the amount of work done on the rotor to move it from point a to point b in increasing the kinetic energy of the rotor

$A_2 = \int_{\delta_1}^{\delta_2} P_a \, d\delta$ = negative or decelerating area

= the amount of work done on the rotor to move it from point b to point c when the rotor returns its energy to the circuit.

Since the positive or accelerating area A_1 is equal to the negative or decelerating area A_2, it is called the *equal-area criterion of stability*. Thus, for stability area A_1 is equal to area A_2.

The equal area criterion of stability provides the following informations:

1. It is an easy means of finding the maximum angle of swing.
2. An estimate of whether synchronism will be maintained.
3. The maximum amount of disturbance that can be allowed without losing synchronism.

The equal area criterion is applicable only to a two-machine system or one machine connected to infinite bus. It is not applicable to multi-machine systems.

19.17.1 Application to Sudden Increase in Mechanical Power Input

The equal-area criterion is used to determine the maximum additional power P_s which can be applied for stability to be maintained. With a sudden change in power input, the stability is maintained only if area A_2 is at least equal to A_1 can be located above P_s. If area A_2 is less than area A_1, the accelerating momentum can never be overcome. For the system to remain stable it is possible to find angle δ_2 such that $A_2 = A_1$. As P_{s1} is increased a limiting condition is finally reached when A_1 equals the area above the P_{s1} line as shown in Fig. 19.15. The limit of stability occurs when δ_{max} is at the intersection of line P_s and the power-angle curve for $90° < \delta < 180°$, as shown in Fig. 19.15.

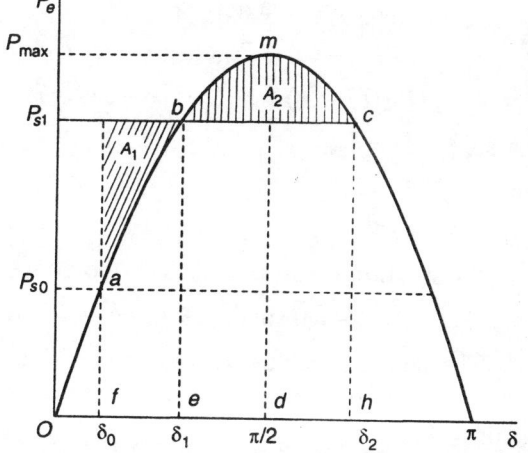

Fig. 19.15. Limiting case of transient stability with mechanical input suddenly increased.

Under this condition δ_2 acquires the maximum value δ_{max} such that

$$\delta_2 = \delta_{max} = 180° - \delta_1$$

Any further increase in P_{s1} means that the area available for A_2 is less than the area A_1, so that excess kinetic energy causes δ to increase beyond point c and the retarding power changes over to accelerating power with the system consequently become unstable.

It may also be noted from Fig. 19.15 that the system will remain stable even though the rotor may oscillate beyond $\delta = 90°$ so long as equal-area criterion is met. The condition $\delta = 90°$ is meant for use in steady-state stability only, and does not apply to the transient stability.

Applying the equal-area criterion to Fig. 19.15, we have

area A_1 = area A_2

area agb = area bmc

area A_1 = area agb

= area of the rectangle $gbef$ – area $abef$ under the sine curve

$$= P_{s1}(\delta_1 - \delta_0) - \int_{\delta_0}^{\delta_1} P_m \sin \delta \, d\delta$$

$$= P_{s1}(\delta_1 - \delta_0) + P_m(\cos \delta_1 - \cos \delta_0)$$

Area A_2 = area bmc

= area $bmche$ under the sine curve – area of rectangle $bche$

$$= \int_{\delta_1}^{\delta_2} P_m \sin \delta \, d\delta - P_{s1}(\delta_2 - \delta_1)$$

$$= P_m(\cos \delta_1 - \cos \delta_2) - P_{s1}(\delta_2 - \delta_1)$$

By equal area criterion

area A_1 = area A_2

$$P_{s1}(\delta_1 - \delta_0) + P_m(\cos \delta_1 - \cos \delta_0) = P_m(\cos \delta_1 - \cos \delta_2) - P_{s1}(\delta_2 - \delta_1)$$

or $\quad P_{s1}(\delta_2 - \delta_0) = P_m(\cos \delta_0 - \cos \delta_2)$

Also, $\quad \delta_2 = \delta_{max}$

and $\quad P_{s1} = P_m \sin \delta_{max}$

at point c of the sine curve.

Substitution of these values in the above equation gives

$$(\delta_{max} - \delta_0) \sin \delta_{max} + \cos \delta_{max} = \cos \delta_0$$

The above nonlinear algebraic equation can be solved by trial and error method for δ_{max}. Once δ_{max} is obtained, the maximum permissible power or the transient stability limit is found from

$$P_{s1} = P_m \sin \delta_1$$

where $\quad \delta_1 = \pi - \delta_{max}$

Example 19.6 A loss-free generator supplies 50 MW to an infinite bus, the steady-state limit of the system being 100 MW. Determine whether the generator will remain in synchronism if the prime mover input is abruptly increased by 30 MW.

Solution

The power-angle curve is shown in Fig. 19.16. The equation of the power-angle curve is

$$P_e = P_{max} \sin \delta \qquad \text{...(E19.6.1)}$$

The initial operating point is at a where $P_e = 50$ MW. The prime mover input is abruptly increased by 30 MW. The desired operating point is at b so that $ag = 30$ MW. The maximum swing

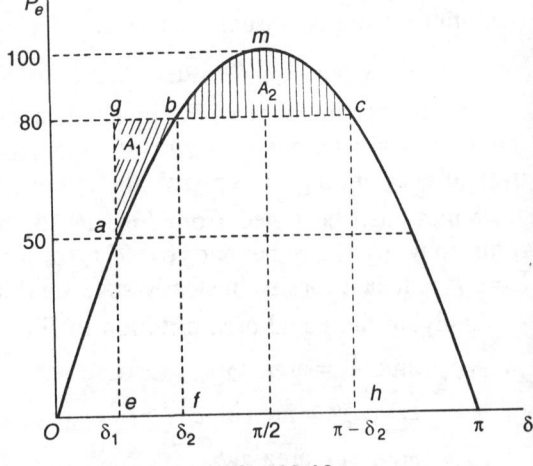

Fig. 19.16.

of the rotor can be upto point c. The system will be stable if the area agb is less than area bmc. The load angles corresponding to points a and b can be found from Eq. (E19.6.1).

For point a, $P_e = 50$, $\delta = \delta_1$

Substituting these values in Eq. (E19.6.1) we get

$$50 = 100 \sin \delta_1$$

$$\sin \delta_1 = \frac{50}{100}, \; \delta_1 = 30° = 0.523 \text{ rad}$$

For point b, $P_e = 80$, $\delta = \delta_2$

$$80 = 100 \sin \delta_2$$

$$\sin \delta_2 = \frac{80}{100}, \; \delta_2 = 53.2° = 0.927 \text{ rad}$$

A_1 = area agb = area of rectangle $gbfe$ – area $abfe$ under the power angle sine curve

$$= (eg \times ef) - \int_{\delta_1}^{\delta_2} P_e \, d\delta = 80 \, (\delta_2 - \delta_1) - \int_{\delta_1}^{\delta_2} 100 \sin \delta \, d\delta$$

$$= 80 \, (0.927 - 0.523) + 100 \, (\cos \delta_2 - \cos \delta_1) = 32.32 + 100 \, (0.600 - 0.866)$$

$$= 5.72 \text{ MW rad}$$

A_2 = area bmc = area $bmchf$ under the sine curve – area of the rectangle $bchf$

$$= \int_{\delta_2}^{\pi - \delta_2} P_e \, d\delta - 80 \, (\pi - \delta_2 - \delta_2) = \int_{\delta_2}^{\pi - \delta_2} 100 \sin \delta \, d\delta - 80 \, (\pi - 2\delta_2)$$

$$= -100 \, [\cos (\pi - \delta_2) - \cos \delta_2] - 80 \, (\pi - 2 \times 0.927)$$

$$= 200 \cos \delta_2 - 103 = 200 \times 0.6 - 103 = 117 \text{ MW rad}.$$

Since area agb (A_1) is less than area bmc (A_2) the system is stable.

Example 19.7 A generator with constant excitation supplies 30 MW through a step-up transformer and a high voltage line to an infinite busbar. If the steady-state stability limit of the system is 60 MW, estimate the maximum permissible sudden increase of generator output (resulting from a sudden increase in prime mover input) if the stability is to be maintained. The resistances of the generator, transformer and line may be neglected.

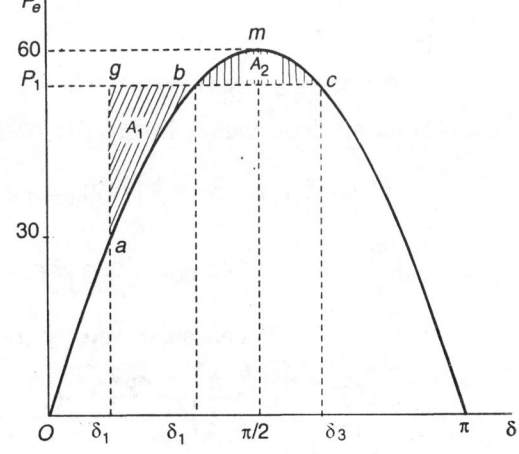

Fig. 19.17.

Solution

The power-angle curve is shown in Fig. 19.17. The initial operating point is at a where $P_e = 30$ MW. With the sudden increase of load to P_1 the rotor advances corresponding to the point b. The maximum swing of the rotor can be upto point c. For the system to remain stable the accelerating area A_1 = area agb should be less than or equal to the retardation area A_2 = area bmc.

The equation of the power-angle curve is

$$P_e = P_{max} \sin \delta$$

or $\qquad P_e = 60 \sin \delta \qquad\qquad\qquad$ ...(E19.7.1)

$$A_1 = \text{area } agb = \int_{\delta_1}^{\delta_2} (P_1 - 60 \sin \delta) \, d\delta$$

$$A_2 = \text{area } bmc = \int_{\delta_2}^{\delta_3} (60 \sin \delta - P_1) \, d\delta$$

For stability, $A_1 = A_2 \quad$ or $\quad A_1 - A_2 = 0$

$$\int_{\delta_1}^{\delta_2} (P_1 - 60 \sin \delta) \, d\delta - \int_{\delta_2}^{\delta_3} (60 \sin \delta - P_1) \, d\delta = 0$$

$$\int_{\delta_1}^{\delta_2} (P_1 - 60 \sin \delta) \, d\delta + \int_{\delta_2}^{\delta_3} (P_1 - 60 \sin \delta) \, d\delta = 0$$

$$\int_{\delta_1}^{\delta_3} (P_1 - 60 \sin \delta) \, d\delta = 0$$

$$\Big[P_1 \delta + 60 \cos \delta \Big]_{\delta_1}^{\delta_3} = 0$$

$$P_1 (\delta_3 - \delta_1) + 60 (\cos \delta_3 - \cos \delta_1) = 0 \qquad\qquad \text{...(E19.7.2)}$$

At point a, $P_e = 30$, $\delta = \delta_1$

Substituting these values in Eq. (E19.7.1), we get

$$30 = 60 \sin \delta_1$$

$\therefore \qquad \sin \delta_1 = 0.5, \quad \delta_1 = \dfrac{\pi}{6}$ radians, $\quad \cos \delta_1 = 0.866$

At point b, $P_1 = 60 \sin \delta_2$

Also, $\qquad \delta_3 = \pi - \delta_2, \quad \cos \delta_3 = -\cos \delta_2$

Substituting these values in Eq. (E19.7.2), we get

$$(60 \sin \delta_2) \left(\pi - \delta_2 - \frac{\pi}{6} \right) + 60 \, (- \cos \delta_2 - 0.866) = 0$$

$$\left(\frac{5\pi}{6} - \delta_2 \right) \sin \delta_2 = \cos \delta_2 + 0.866 \qquad\qquad \text{...(E19.7.3)}$$

Equation (E19.7.3) can be solved by trial and error method. Put $\delta_2 = \pi/3$ radians $= 60°$

$$\text{L.H.S.} = \left(\frac{5\pi}{6} - \frac{\pi}{3} \right) \sin \frac{\pi}{3} = 1.36035$$

$$\text{R.H.S.} = \cos \frac{\pi}{3} + 0.866 = 1.366$$

Let us put $\delta_2 = 60.4° = 1.0541789$ radians

$$\text{L.H.S.} = \left(\frac{5\pi}{6} - \frac{60.4 \times \pi}{180}\right) \sin 60.4° = 1.35973$$

$$\text{R.H.S.} = (\cos 60.4°) + 0.866 = 1.35994$$

$\therefore$ L.H.S. = R.H.S. (approx.)

If $\delta_2 = 60.4°$

$$P_1 = 60 \sin \delta_2 = 60 \sin 60.4° = 52.17 \text{ MW}$$

The maximum permissible sudden increase of load $= P_1 - 30 = 52.17 - 30 = 22.17$ MW

Equation (E19.7.3) can also be solved by graphical method. The left-hand side and right-hand side of this equation are plotted for arbitrary values of δ_2 which lie between δ_1 and 90°. The intersection of the two curves gives the value of δ_2.

19.17.2 One of the Parallel Lines Suddenly Switched Off

Consider a system (Fig. 19.18) consisting of a synchronous generator feeding an infinite bus through a double-circuit line. The two circuits are operating in parallel. If one of the circuits is switched off suddenly, the system may become unstable inspite of the fact that the load could be supplied over by the other circuit under steady-state conditions.

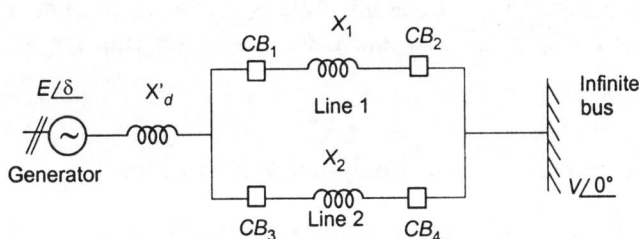

Fig. 19.18. A synchronous generator feeding an infinite bus through a double-circuit line.

We shall use the equal-area criterion to study the transient stability of the system when one of the lines is switched off. When both the lines are operating in parallel, the power transfer is given by

$$P_{e1} = \frac{EV}{X_A} \sin \delta$$

where X_A is the transfer reactance when both the lines are operating in parallel.

$$X_A = X'_d + X_1 \parallel X_2 = X'_d + \frac{X_1 X_2}{X_1 + X_2}$$

$\therefore$ $$P_{e1} = \frac{EV}{X'_d + \dfrac{X_1 X_2}{X_1 + X_2}} \sin \delta$$

or $P_{e1} = P_{\max 1} \sin \delta$ ...(19.17.7)

where $$P_{\max 1} = \frac{EV}{X'_d + \dfrac{X_1 X_2}{X_1 + X_2}}$$

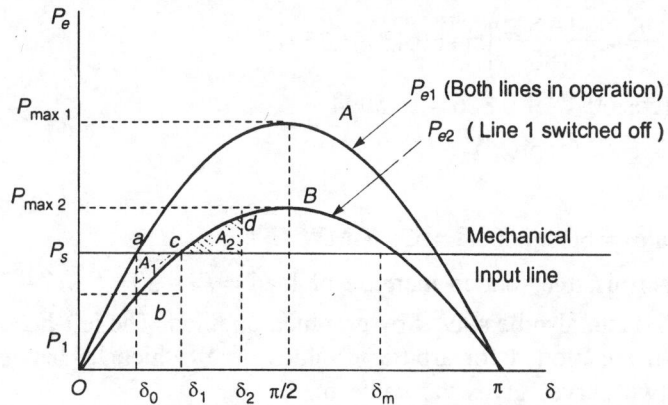

Fig. 19.19. Power-angle curves. Curve A for two lines in parallel and curve B for line 2.

The power-angle curve given by Eq. (19.17.7) is shown in Fig. 19.19 as curve A.

Let the mechanical power input to the generator be P_s corresponding to a point a on the power-angle curve A of the two lines in parallel. At point a the load angle is δ_0 and the input power P_s at the shaft is equal to the output power P_e of the generator.

When power P_s is being transferred at an angle δ_0, suppose that line 1 is suddenly switched off by opening the circuit breakers CB_1 and CB_2. The power transfer is given by

$$P_{e2} = \frac{EV}{X_B} \sin \delta$$

where X_B is the transfer reactance when only line 2 is in operation.

$$X_B = X'_d + X_2$$

$\therefore \qquad P_{e2} = \frac{EV}{X'_d + X_2} \sin \delta$

$$P_{e2} = P_{\max 2} \sin \delta \qquad\qquad\qquad ...(19.17.8)$$

where $\qquad P_{\max 2} = \dfrac{EV}{X'_d + X_2}$

The power-angle curve given by Eq. (19.17.8) is shown as curve B in Fig. 19.19. Immediately on switching off line 1, the load angle δ_0 cannot change instantaneously due to rotor inertia. Since the load angle is still δ_0, the output power has reduced to P_1. The operating point shifts to point b on the new operating curve B. This sudden change in generator output is not immediately detected by the governor of the prime mover. Thus the shaft power P_s is not changed. Now, the power input to the generator from the shaft is P_s and power output of the generator into the line 2 is P_1. The power output P_1 is less than the power input P_s. The difference $(P_s - P_1)$ accelerates the rotor. With a slight increase in speed the load angle δ_0 increases. The operating point moves along the curve B from b to c. At point c, $\delta = \delta_1$ and the accelerating power is zero, but the rotor is running slightly above synchronous speed. It will therefore, continue to advance upto δ_2 due to rotor inertia. However, in the region between δ_1 and δ_2 the electrical output P_e is more than the the shaft input P_s, with the result the rotor slows down. At point d the rotor relative speed (w.r.t. the synchronous speed) becomes zero. When $\delta = \delta_2$, the area A_1 is equal to the area A_2, and the rotor starts swinging back towards δ_1. The rotor will oscillate about c and the oscillations go on diminishing due to damping. The operating point finally comes at c where the input power is equal to the output power again, that is $P_{e2} = P_{\max 2} \sin \delta_1$ as shown in Fig. 19.19.

If the initial input power P_s is increased (so that the input line P_s is shifted upwards) a limit is reached beyond which the retarding area A_2, cannot be equal to the accelerating area A_1. If $A_2 < A_1$, the rotor will overshoot past δ_m and the machine will lose synchronism. Thus, the maximum value which δ_2 can attain without loss of stability is δ_m and is given by

$$\delta_2 = \delta_m = (\pi - \delta_1) \text{ elec. radians}$$

This is shown in Fig. 19.20. The value of P_s corresponding to this condition is called the *transient stability limit*.

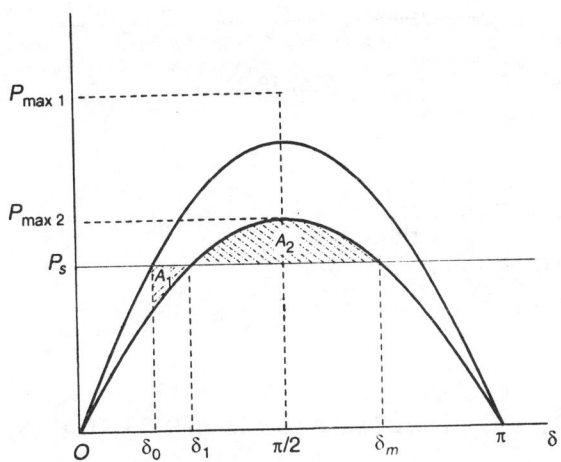

Fig. 19.20. Transient stability limit for the system in Fig. 19.18.

Example 19.8 A double-circuit three-phase feeder connects a single generator to a large network. The power corresponding to the limit of steady-state stability for each circuit is 100 MW. The line is transmitting 80 MW when one of the circuits is suddenly switched out. Determine with reference to appropriate diagram whether the generator is likely to remain in synchronism.

Solution

The equation of the power-angle curve is

$$P_e = P_{\max} \sin \delta$$

For curve A in Fig. 19.21,

$$80 = 200 \sin \delta_1, \quad \delta_1 = \sin^{-1} \frac{80}{200}$$

$$\delta_1 = 23.578° = 0.4115 \text{ rad}$$

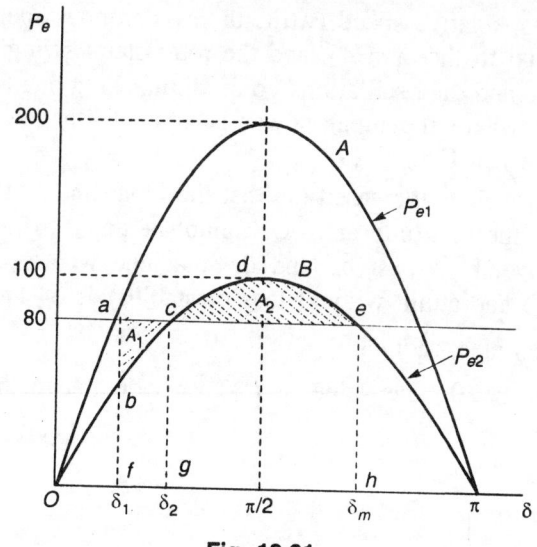

Fig. 19.21.

For curve B,

$$80 = 100 \sin \delta_2, \quad \delta_2 = \sin^{-1} \frac{80}{100}$$

$$\delta_2 = 53.13° = 0.9273 \text{ rad}$$

Area A_1 = area abc

= area of rectangle $acgf$ – area $bcgf$ under the power–angle sine curve B

$$= 80 \,(\delta_2 - \delta_1) - \int_{\delta_1}^{\delta_2} 100 \sin \delta \, d\delta = 80 \,(0.9273 - 0.4115) + [100 \cos \delta]_{\delta_1}^{\delta_2}$$

$$= 41.26 + 100 \,(\cos \delta_2 - \cos \delta_1) = 41.26 + 100 \,(0.6 - 0.9165) = 9.6 \text{ MW radians}$$

Area A_2 = area cde = area $cdehg$ under the sine curve B – area of rectangle $cehg$

$$= \int_{\delta_2}^{\delta_m} 100 \sin \delta \, d\delta - 80 \,(\delta_m - \delta_2) = [-100 \cos \delta]_{\delta_2}^{\delta_m} - 80 \,(\delta_m - \delta_2)$$

$$\delta_m = \pi - \delta_2$$

∴ $$A_2 = -100 \,[\cos (\pi - \delta_2) - \cos (\delta_2)] - 80 \,(\pi - \delta_2 - \delta_2)$$

$$= 100 \times 2 \cos 53.13° - 80 \,(\pi - 2 \times 0.9273) = 120 - 102.96 = 17.04 \text{ MW radians}$$

Since area A_1 is less than area A_2, the system is stable.

19.18 SYSTEM FAULT AND SUBSEQUENT CIRCUIT ISOLATION

Consider a synchronous generator supplying power to an infinite bus through a double-circuit line as shown in Fig. 19.22. Suppose that some type of fault (say line-to-ground fault) occurs in the middle of line 2. Let us further assume that the fault is not sustained but it is cleared after some time by opening of the circuit breakers at both the ends of the faulted line. The fault

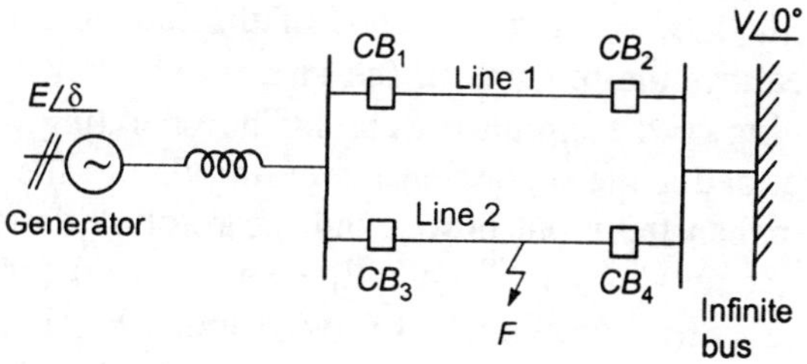

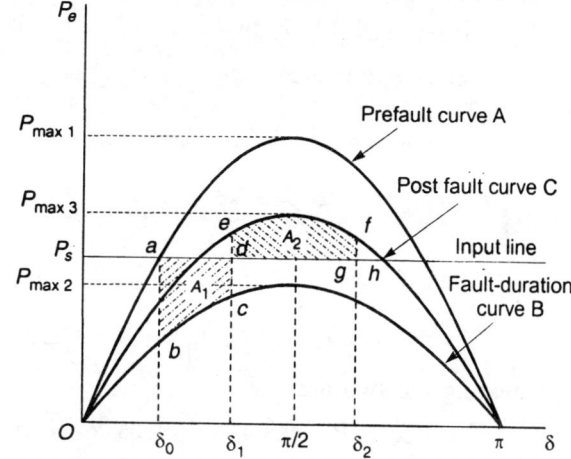

Fig. 19.22. Fault in the middle of line 2.

produces a transient change which may render the system unstable. However, if the circuit breakers clear the fault in time it is possible to maintain stability. The maximum value of time allowed for protective gear to operate without loss of stability is called the *critical clearing time*. The torque angle corresponding to this time is called *critical clearing angle*.

The conditions of operation are shown by three power-angle curves A, B and C. Curve A is called the *prefault curve*. It represents the prefault condition when both the lines are in healthy condition. Curve B is called the *fault-duration curve*. It represents the condition of the system during fault when one healthy line and one faulted line are in the circuit. Curve C is called the *post-fault curve*. This curve represents the condition of the system after the faulted line has been switched out and only one healthy line is in the circuit. The three power-angle curves are drawn from the following equations :

For prefault condition,

$$P_{e1} = \frac{EV}{X_A} \sin \delta = P_{\max 1} \sin \delta$$

For the condition during fault,

$$P_{e2} = \frac{EV}{X_B} \sin \delta = P_{\max 2} \sin \delta$$

For the post-fault condition,

$$P_{e3} = \frac{EV}{X_C} \sin \delta = P_{\max 3} \sin \delta$$

where X_A = transfer reactance prior to fault

X_B = transfer reactance during fault

X_C = transfer reactance for the post–

fault condition.

Fig. 19.23. Power-angle curves when fault occurs on one line.

The three power-angle curves are shown in Fig. 19.23. The input line is given by $P_e = P_s$.

Before the occurrence of the fault the system was operating at point *a* on the prefault curve A. The initial load angle δ_0 is obtained by the intersection of the input line $P_e = P_s$ and the prefault curve A. When the fault occurs the transfer reactance is changed and the power output is reduced. The operating point shifts from point *a* to point *b* corresponding to load angle δ_0 on the fault-duration curve B. At this point, the input P_s is greater than the electrical output, with the result that the rotor accelerates and the operating point moves to point *c*. At the point *c* the circuit breakers CB_3 and CB_4 at the two ends of the faulted line 2 open and the fault is cleared. Thus, only one healthy line remains in the circuit. The transfer reactance changes to a new value due to the opening of the circuit breakers. The operating point moves to point *e* on the post-fault curve C. The load angle δ goes on increasing due to the inertia of the rotor. Now, the output power is greater than the input power and the rotor starts retarding till the point *f* is reached. At the point *f*, the angle δ is δ_2 and the speed of the rotor with respect to the synchronous speed becomes zero. The extent of overshoot, that is, the value of δ_2, can be determined by equating the areas *defg* and *abcd*. If the area included between the curve C and the line $P_e = P_s$, bounded by $\delta = \delta_1$ is less than area *abcd*, the machine will lose synchronism after the operation of the circuit breakers. The system will be stable if the retardation area A_2 (= area *defg*) is equal to the acceleration area A_1 (= area *abcd*). It is to be noted that the acceleration area A_1 depends upon the clearing angle δ_1. When δ_1 increases, area A_1 also increases, and to find $A_2 = A_1$, δ_2 is increased till it has a value δ_m. This is the maximum allowable value of δ for stability. Such a condition is shown in Fig. 19.24a where $\delta_1 = \delta_c$. The angle δ_c is called the *critical clearing angle*. For the system to be stable, the clearing angle should be less than the critical clearing angle. If the actual clearing angle is greater than the critical clearing angle, the system becomes unstable. Thus, more rapidly the fault is cleared, the smaller will the accelerating area be, and the greater the chance of stable operation being restored.

For simplicity, let us take $P_{\max 1} = P_{m1}$, $P_{\max 2} = P_{m2}$ and $P_{\max 3} = P_{m3}$.

For transient stability limit,

$$\text{area } abcd = \text{area } defh$$

$$\text{area } abcd = \text{area } adkg - \text{area } bckg = P_s (\delta_c - \delta_0) - \int_{\delta_0}^{\delta_c} P_{m2} \sin \delta \, d\delta$$

$$= P_s (\delta_c - \delta_0) - [- P_{m2} \cos \delta]_{\delta_0}^{\delta_c} = P_s (\delta_c - \delta_0) + P_{m2} \cos \delta_c - P_{m2} \cos \delta_0$$

$$\text{area } defh = \text{area } kdefhm - \text{area } dhmk = \int_{\delta_c}^{\delta_m} P_{m3} \sin \delta \, d\delta - P_s (\delta_m - \delta_c)$$

$$= [- P_{m3} \cos \delta]_{\delta_c}^{\delta_m} - P_s (\delta_m - \delta_c) = P_{m3} \cos \delta_c - P_{m3} \cos \delta_m - P_s (\delta_m - \delta_c)$$

Equating the two areas we get,

$$P_s (\delta_c - \delta_0) + P_{m2} \cos \delta_c - P_{m2} \cos \delta_0 = P_{m3} \cos \delta_c - P_{m3} \cos \delta_m - P_s (\delta_m - \delta_c) \quad \ldots(19.18.1)$$

Also, $P_s = P_{m1} \sin \delta_0$ at point *a* on curve A

$P_s = P_{m3} \sin \delta_q$ at point *q* on curve C

$$\delta_m = \pi - \delta_q = \pi - \sin^{-1} \left(\frac{P_s}{P_{m3}} \right) \text{ radians}$$

From Eq. (19.18.1)

$$\cos \delta_c = \frac{P_s (\delta_m - \delta_0) - P_{m2} \cos \delta_0 + P_{m3} \cos \delta_m}{P_{m3} - P_{m2}} \quad \ldots(19.18.2)$$

Equation (19.18.2) can be used to determine critical clearing angle. The angles in this equation are in radians. If the angles are in degrees, then Eq. (19.18.2) becomes

$$\cos \delta_c = \frac{\dfrac{\pi}{180} P_s (\delta_m - \delta_0) - P_{m2} \cos \delta_0 + P_{m3} \cos \delta_m}{P_{m3} - P_{m2}} \qquad \text{...(19.18.3)}$$

19.18.1 System Fault, Circuit Isolation and Reclosing

Most of the faults on the system are of transient nature. Automatic quick reclosing circuit breakers are used with transmission lines. When a fault occurs, the faulted line is disconnected. After an interval, the circuit breakers of the faulted line are reclosed automatically. The input is P_s and the initial angle is δ_0. When a fault occurs, the operation shifts to the point b on the fault duration curve B as shown in Fig. 19.24(b). When the load angle is δ_c, the faulted line is isolated and the operation shifts to the post-fault curve C. When the load angle is δ_r the circuit breakers reclose and the operation shifts to the prefault curve A.

For stable operation the accelerating area A_1 (= area *abcd*) should be equal to the decelerating area A_2 (= area *defghk*). The maximum angle to which the rotor angle swings is δ_2. It is less than δ_m (that is, the maximum permissible rotor swing if stability is to be maintained).

Example 19.9 A generator is delivering 0.6 of maximum power to an infinite bus through a transmission line. A fault occurs such that the reactance between the generator and the infinite bus is increased to three times its prefault value. When the fault is cleared, the maximum power that can be delivered is 0.8 of the original maximum value. Determine the critical clearing angle.

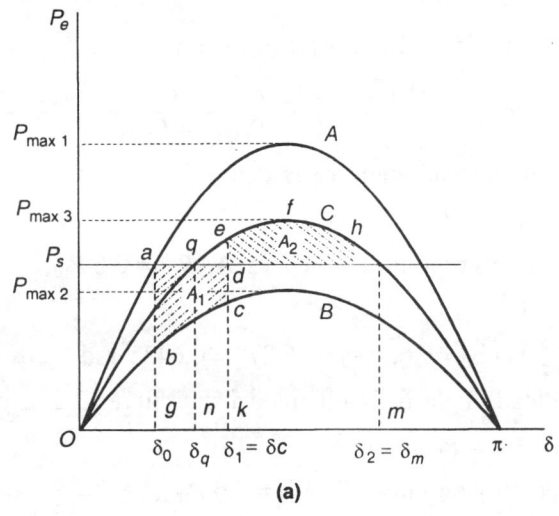

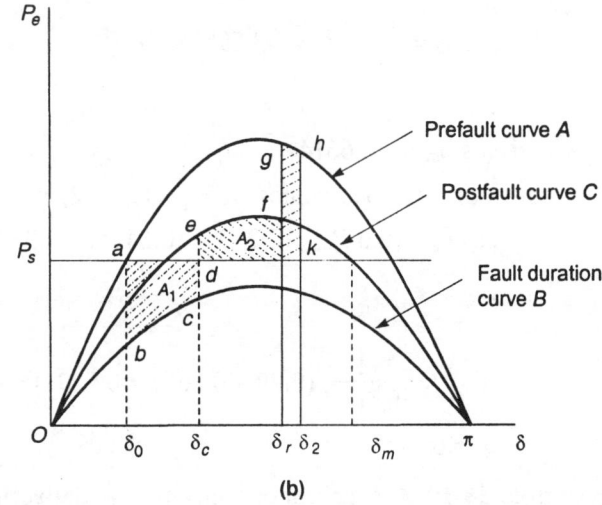

Fig. 19.24. (a) Determination of critical clearing angle. (b) Equal-area criterion for system fault, circuit isolation and reclosing.

Solution

The power-angle curves are shown in Fig. 19.24 (a).

Here, $P_{\max 1} = P_{\max}$

$$P_{\max 2} = \frac{1}{3} P_{\max} = 0.3333 \, P_{\max}$$

$$P_{\max 3} = 0.8 \, P_{\max}$$

$$P_s = 0.6 \, P_{\max}$$

Equation (19.18.2) is used to determine δ_c.

$$\cos \delta_c = \frac{P_s \, (\delta_m - \delta_0) - P_{\max 2} \cos \delta_0 + P_{\max 3} \cos \delta_m}{P_{\max 3} - P_{\max 2}}$$

The initial load angle δ_0 is determined from the prefault curve A. For curve A,

$$P_{e1} = P_{\max 1} \sin \delta_0$$

At operating point a on curve A, $P_{e1} = 0.6 \, P_{\max}$, $\delta = \delta_0$

∴ $0.6 \, P_{\max} = P_{\max} \sin \delta_0$

$\sin \delta_0 = 0.6$, $\delta_0 = 36.87° = 0.6435$ rad, $\cos \delta_0 = 0.8$

The load angle δ_q is obtained from curve C. For curve C,

$$P_{e3} = P_{\max 3} \sin \delta$$

At point q on curve C, $P_{e3} = 0.6 \, P_{\max}$, $\delta = \delta_q$

∴ $0.6 \, P_{\max} = 0.8 \, P_{\max} \sin \delta_q$

$$\sin \delta_q = \frac{0.6}{0.8}, \quad \delta_q = 48.59°$$

$\delta_m = 180° - 48.59° = 131.41° = 2.2935$ rad

$\cos \delta_m = -0.6614$

Substituting various values in Eq. (19.18.2) we get

$$\cos \delta_c = [0.6 \, P_{\max} (2.2935 - 0.6435) - 0.3333 \, P_{\max} \times 0.8 + 0.8 \, P_{\max} (-0.6614)]$$

$$\times \frac{1}{0.8 \, P_{\max} - 0.3333 \, P_{\max}}$$

$$= \frac{1}{0.4667} (0.99 - 0.2667 - 0.5291) = 0.4161$$

∴ $\delta_c = 65.41°$

Example 19.10 A synchronous machine is delivering 1.0 pu power to an infinite bus through a double-circuit transmission line shown in Fig. 19.25 (a). The direct-axis transient reactance of the generator is 0.3 pu. The per unit reactance for each line is 0.5. All reactances are given to a base of the machine rating. One of the transmission lines experiences a solid three phase fault to ground, during which occurrence the system reactances are as shown in Fig. 19.25 (c). Determine the critical clearing angle before which the circuit breakers of the faulted line should operate if the stability is to be maintained.

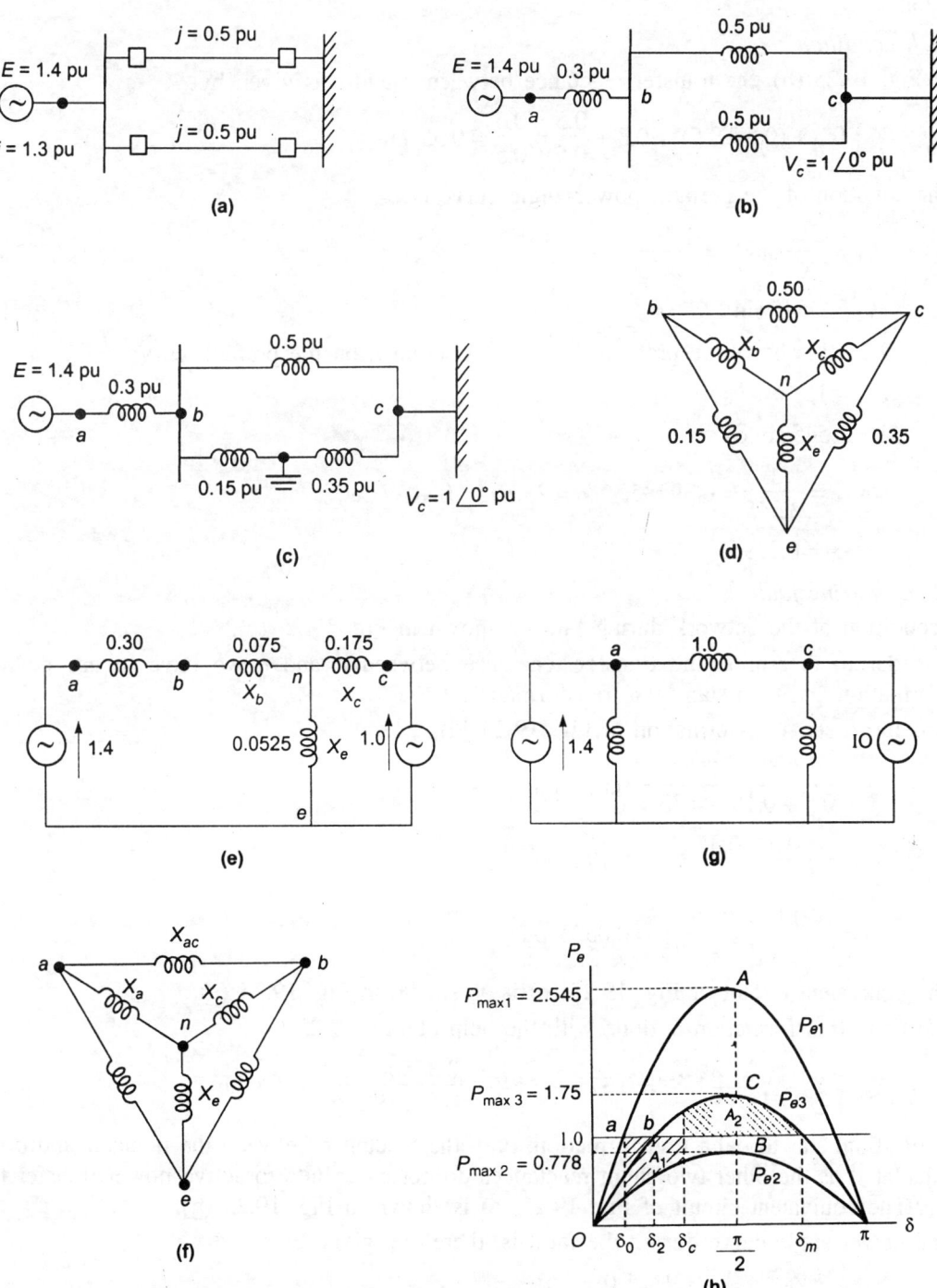

Fig. 19.25. Illustrating Example 19.9.

Solution

Prefault condition

From Fig. 19.25 (b), the transfer reactance between a and c is given by

$$X = 0.3 + (0.5 \parallel 0.5) = 0.3 + \frac{0.5 \times 0.5}{0.5 + 0.5} = 0.55 \text{ pu}$$

The equation of the prefault power-angle curve is

$$P_{e1} = \frac{EV_c}{X} \sin \delta = \frac{1.4 \times 1}{0.55} \sin \delta \qquad \qquad \text{...(E19.10.1)}$$

or $\qquad P_{e1} = 2.545 \sin \delta$

Fig. 19.25 (h) shows the prefault curve A. At point a on the prefault curve,

$$P_{e1} = 1, \quad \delta = \delta_0$$

$\therefore \qquad 1 = 2.545 \sin \delta_0$

$$\sin \delta_0 = \frac{1}{2.545} = 0.39285, \quad \delta_0 = 23.13° = 0.4037 \text{ rad}$$

$$\cos \delta_0 = 0.9196$$

Condition during fault

The condition of the network during fault is shown in Fig. 19.25 (c).

In order to determine the transfer reactance between a and c we have to use delta-star transformation and then star-delta transformation.

In delta to star transformation in Fig. 19.25 (d)

$$X_b = \frac{0.5 \times 0.15}{0.5 + 0.15 + 0.35} = 0.075 \text{ pu}$$

$$X_c = \frac{0.5 \times 0.35}{0.5 + 0.15 + 0.35} = 0.175 \text{ pu}$$

$$X_e = \frac{0.15 \times 0.35}{0.5 + 0.15 + 0.35} = 0.0525 \text{ pu}$$

The equivalent circuit of Fig. 19.25 (c) is as shown in Fig. 19.25 (e).

Star-delta transformation is done with the help of Fig. 19.25 (f).

$$X_{ac} = X_a + X_c + \frac{X_a X_c}{X_e} = 0.375 + 0.175 + \frac{0.375 \times 0.175}{0.0525} = 1.8 \text{ pu}$$

In the final star to delta transformation, only the reactance between the nodes a and c need be calculated as the other two shunt reactances do not contribute to active power transfer from a to c. The equivalent circuit of Fig. 19.25 (e) is shown in Fig. 19.25 (g).

The power-angle curve during the fault is, therefore, given by

$$P_{e2} = \frac{EV_c}{X_{ac}} \sin \delta = \frac{1.4 \times 1.0}{1.8} \sin \delta$$

or $\qquad P_{e2} = 0.778 \sin \delta$ $\qquad \qquad \text{...(E19.10.2)}$

Equation (19.7.3) can also be used to determine the relationship given by Eq. (E19.10.2).

Post-fault condition

The post-fault power-angle curve can be derived from Fig. 19.25 (b) with one line out of circuit. The transfer reactance in this case is

$$X = 0.3 + 0.5 = 0.8 \text{ pu}$$

$$\therefore \qquad P_{e3} = \frac{EV_c}{X} \sin \delta = \frac{1.4 \times 1.0}{0.8} \sin \delta$$

$$P_{e3} = 1.75 \sin \delta \qquad \qquad \qquad \text{...(E19.10.3)}$$

The power angle diagrams for the three conditions are shown in Fig. 19.25 (h). At point *b* of curve *C*, $P_{e3} = 1.0$, $\delta = \delta_2$. Substituting these values in Eq. (E19.10.3), we get

$$1 = 1.75 \sin \delta_2, \quad \delta_2 = 34.85°$$

$$\delta_m = 180° - 34.85° = 145.5° = 2.5333 \text{ rad}$$

$$\cos \delta_m = -0.82065$$

For stability, $A_1 = A_2$

$$A_1 = P_s (\delta_c - \delta_0) - \int_{\delta_0}^{\delta_c} P_{\max 2} \sin \delta \, d\delta = P_s (\delta_c - \delta_0) + P_{\max 2} (\cos \delta_c - \cos \delta_0)$$

$$A_2 = \int_{\delta_c}^{\delta_m} P_{\max 3} \sin \delta \, d\delta - P_s (\delta_m - \delta_c) = P_{\max 3} (\cos \delta_c - \cos \delta_m) - P_s (\delta_m - \delta_c)$$

Equating the two areas and simplifying we get

$$\cos \delta_c = \frac{P_s (\delta_m - \delta_0) - P_{\max 2} \cos \delta_0 + P_{\max 3} \cos \delta_m}{P_{\max 3} - P_{\max 2}}$$

$$= \frac{[1 (2.5333 - 0.4037) - 0.778 \times 0.9196 + 1.75 (-0.82065)]}{(1.75 - 0.778)}$$

$$= \frac{(2.1296 - 0.71545 - 1.4361)}{0.972} = -0.02258$$

$$\delta_c = 91.3°$$

19.19 NUMERICAL SOLUTION OF SWING EQUATION (POINT-BY-POINT SOLUTION)

The equal-area criterion may be used to determine the critical fault clearing angle. The point-by-point (or step-by-step) method is used to determine the critical fault clearing time associated with this angle. A swing curve (δ vs. t) is plotted for each machine to know its tendency to remain stable. We shall illustrate a general method for one machine connected to an infinite bus. However, it can be applied to every machine of a multimachine system.

The following *assumptions* are made in the solution of swing equation by point-by-point method.

1. The accelerating power P_a and angular acceleration α are constant from the middle of the preceding interval to the middle of the present interval considered. Both of these values are calculated at the beginning of this interval.

2. The angular velocity ω, computed at the middle of an interval, remains constant from the beginning to the end of the interval considered.

These assumptions are not strictly correct, since δ is changing continuously and both P_a and ω depend upon δ. However, if time increment Δt is made smaller, the computed curve approximates the actual curve.

The accelerating at $t = (n - 1)$ is α_{n-1}. Over the region of constant acceleration α_{n-1} there is an increment of angular velocity from $\omega_{n-3/2}$ to $\omega_{n-1/2}$,

$$\therefore \qquad \alpha_{n-1} = \frac{\omega_{n-1/2} - \omega_{n-3/2}}{\Delta t} \qquad\qquad ...(19.19.1)$$

$$\text{Also,} \qquad \alpha_{n-1} = \frac{P_{a(n-1)}}{M} \qquad\qquad ...(19.19.2)$$

$$\therefore \qquad \omega_{n-1/2} - \omega_{n-3/2} = \frac{P_{a(n-1)}}{M} (\Delta t) \qquad\qquad ...(19.19.3)$$

Again, the angular velocity $\omega_{n-1/2}$ remains constant from $t = n - 1$ to $t = n$. The displacement angle increases from δ_{n-1} to δ_n during the time interval Δt.

From the velocity/time curve [Fig. 19.26 (b)]

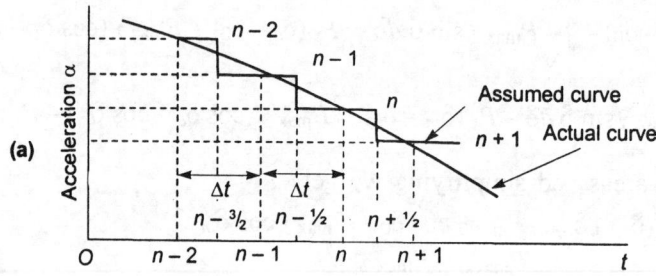

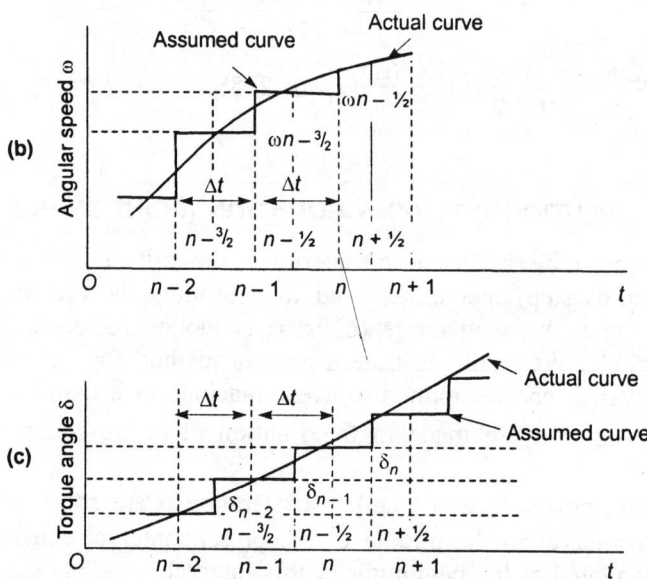

Fig. 19.26. (a) Acceleration time curve. (b) Angular speed-time curve. (c) Torque angle-time curve.

$$\omega_{n-1/2} = \frac{\delta_n - \delta_{n-1}}{\Delta t}$$

$\therefore$ $\delta_n - \delta_{n-1} = \omega_{n-1/2} (\Delta t)$ ...(19.19.4)

Similarly,

$$\omega_{n-3/2} = \frac{1}{\Delta t} (\delta_{n-1} - \delta_{n-2})$$

$\therefore$ $\delta_{n-1} - \delta_{n-2} = \omega_{n-3/2} (\Delta t)$ ...(19.19.5)

The change in δ over any interval = value of ω for that interval $\times$ time of interval

But the change in δ over nth interval, that is from $t = n$ to $t = n-1$ is given by

$(\Delta \delta)_n = \delta_n - \delta_{n-1} = \omega_{n-1/2} (\Delta t)$ ...(19.19.6)

The change in δ over $(n-1)$th interval, that is from $t = (n-1)$ to $t = (n-2)$ is given by

$(\Delta \delta)_{n-1} = \delta_{n-1} - \delta_{n-2} = \omega_{n-3/2} (\Delta t)$ ...(19.19.7)

$\therefore$ $(\Delta \delta)_n - (\Delta \delta)_{n-1} = \omega_{n-1/2} (\Delta t) - \omega_{n-3/2} (\Delta t) = \dfrac{P_{a(n-1)}}{M} (\Delta t)^2$

$\therefore$ $(\Delta \delta)_n = (\Delta \delta)_{n-1} + \dfrac{P_{a(n-1)}}{M} (\Delta t)^2$ ...(19.19.8)

Equation (19.19.8) shows that the change in torque angle during a given interval is equal to the change in torque angle during the preceding interval plus the accelerating power at the beginning of the interval times $(\Delta t)^2/M$. This equation forms the basis of the numerical solution by point-by-point method. The accelerating power is calculated at the beginning of each new interval. The subsequent steps are written in a similar manner and the points are obtained for plotting a curve between δ and t. More accurate results are obtained by decreasing Δt and the assumed curve approaches the actual swing curve. However, the amount of labour required in calculations goes on increasing by making Δt smaller. In practice, Δt is usually taken of the order of 0.05 s as the usual periodic time of swing curve is of the order of 0.5 to 2 s. If δ continues to increase the system is unstable and if δ decreases after attaining a maximum value, the system is stable.

The occurrence or removal of a fault, or a switching operation causes a discontinuity in the accelerating power P_a. If a discontinuity occurs at the beginning of an interval, then the average of the values of P_a just before and just after the disturbance is taken as the accelerating power.

Thus, for calculating the increment in δ occurring during the first interval after a fault is applied at $t = 0$, Eq. (19.19.8) becomes

$$(\Delta \delta)_1 = \frac{P_{a0+}}{2} \frac{(\Delta t)^2}{M}$$

where P_{a0+} is the accelerating power immediately after the occurrence of the fault. Immediately before the occurrence of the fault, the system is in steady state so that $P_{a0-} = 0$ and δ_0 is known. If the fault is cleared at the beginning of the mth interval, the value $\frac{1}{2} [P_{a(m-1)-} + P_{a(m-1)+}]$ should be used instead of $P_{a(m-1)-}$ where $P_{a(m-1)+}$ is the accelerating power immediately before clearing the fault and $P_{a(m-1)+}$ is the accelerating power immediately after clearing the fault. If a discontinuity occurs at the middle of an interval there is no need of special procedure. The increment in δ during such an interval is calculated in the usual manner from the value of P_a at the beginning of the interval.

Example 19.11 A 50 MVA, 50 Hz generator delivers 50 MW over a double circuit line to an infinite bus. The generator has an inertia constant $H = 2.7$ MJ/MVA. The generator has transient reactance $X'_d = 0.3$ pu. Each transmission line has a reactance of 0.5 pu on 50 MVA base. $|E| = 1.4$ pu and $V = 1.0 \underline{/0^\circ}$ pu. A solid three-phase fault to ground occurs on line 2. The system reactances during the fault are shown in Fig. 19.25 (c). Plot the swing curve for a sustained fault upto a time of 0.4 s.

Solution

Let us take base MVA = 50, $S = 1.0$ pu

$$M = \frac{HS}{180f} = \frac{2.7 \times 1}{180 \times 50}$$

$$= 3 \times 10^{-4} \text{ MJs/elec. deg.}$$

From Example 19.10 we have

$$\delta_0 = 23.13^\circ$$

$$P_{e1} = 2.545 \sin \delta$$

$$P_{e2} = 0.778 \sin \delta$$

$$P_{e3} = 1.75 \sin \delta$$

We shall use the following relations to plot the swing curve.

$$P_{a(n-1)} = P_s - P_{max} \sin \delta_{(n-1)}$$

$$(\Delta \delta)_n = (\Delta \delta)_{n-1} + \frac{P_{a(n-1)}}{M} (\Delta t)^2$$

$$\delta_n = \delta_{n-1} + (\Delta \delta)_n$$

Since there is a discontinuity at $t = 0$, therefore, the average value of P_a is to be used.

At $t = 0_-$,

$$P_{max} = 2.545, \quad P_a = 0$$

At $t = 0_+$,

$$P_{max} = 0.778$$

$$P_a = P_s - P_{e2} = P_s - P_{max\,2} \sin \delta_0$$

$$= 1.0 - 0.778 \sin 23.13^\circ = 0.6944$$

$$P_{a(avg)} = \frac{1}{2} (0 + 0.6944) = 0.3472 \text{ pu}$$

Let us take an interval $\Delta t = 0.05$ s

$$\frac{P_{a(n-1)}}{M} (\Delta t)^2 = \frac{(0.05)^2}{3 \times 10^{-4}} P_{a(n-1)} = 8.333 \, P_{a(n-1)}$$

For first interval

$$(\Delta \delta)_1 = (\Delta \delta)_0 + \frac{(\Delta t)^2}{M} P_{a0} = 0 + 8.333 \times 0.3472 = 2.89°$$

$$\delta_1 = \delta_0 + (\Delta \delta)_1 = 23.13° + 2.89° = 26.02°$$

For second interval

$$P_{a(1)} = P_s - 0.778 \sin \delta_1$$

$$= 1 - 0.778 \sin 26.02° = 0.6587 \text{ pu}$$

$$(\Delta \delta)_2 = (\Delta \delta)_1 + \frac{(\Delta t)^2}{M} P_{a(1)}$$

$$= 2.89 + 8.333 \times 0.6587 = 8.38°$$

$$\delta_2 = \delta_1 + (\Delta \delta)_2 = 26.02 + 8.38 = 34.4°$$

Calculations for other intervals may be carried out in a similar manner. The details of calculations are shown in Table 19.1.

Table 19.1   Point-by-Point Solution of Swing Equation for Sustained Fault

t	P_{max}	$\sin \delta$	P_e	P_a	$\frac{(\Delta t)^2}{M} P_a$	$\Delta \delta$	δ
s	pu		pu	pu	elec. deg.	elec. deg.	elec. deg.
			$= 0.778 \sin \delta$	$(1 - P_e)$	$8.333 P_a$		
0_	2.545	0.3928	1	0	0	0	23.13
0+	0.778	0.3928	0.3056	0.6944	–	–	
0avg					0.3472 → 2.89	→ 2.89	
0.05	0.778	0.4387 →	0.3413 →	0.6587 →	5.49 →	8.38	26.02
0.10	0.778	0.5649 →	0.4395 →	0.5605 →	4.67 →	13.05 →	34.4
0.15	0.778	0.73669 →	0.5731 →	0.4269 →	3.56 →	16.61	47.45
0.20	0.778	0.899	0.6996	0.3004	2.50	19.11	64.06
0.25	0.778	0.9929	0.7724	0.2276	1.89	21.0	83.17
0.30	0.778	0.9965	0.7543	0.2456	2.047	23.05	104.17
0.35	0.778	0.7963	0.6195	0.3805	3.17	26.22	127.22
0.40	0.778	0.4471	0.3478	0.6522	5.43	31.65	153.09

In Table 19.1, the staggering of the $(\Delta \delta)$ figures with respect to the others is to be noted carefully. The arrows indicate the order in which the figures are obtained. The swing curve for a sustained fault upto a time of 0.4 s is shown in Fig. 19.27.

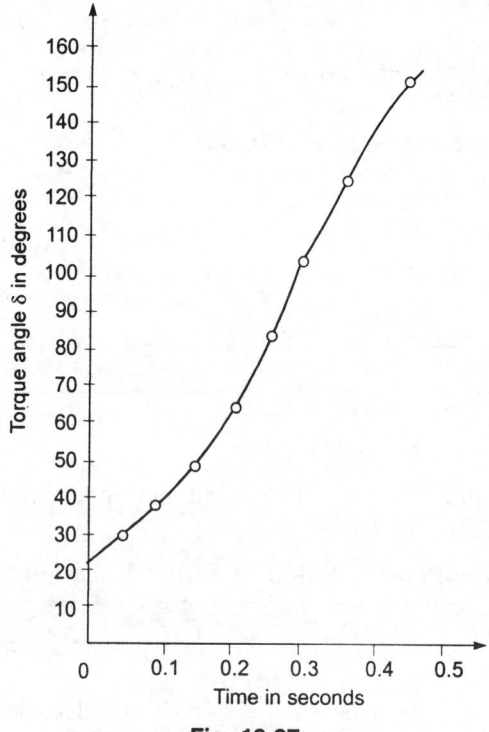

Fig. 19.27.

19.20 METHODS OF IMPROVING STABILITY

Fig. 19.24 shows that when the maximum power limit of various power-angle curves is raised, the accelerating area decreases and decelerating area increases for a given clearing angle. Consequently δ_0 is decreased and δ_m is increased. This means that by increasing $P_{\max}$ the rotor can swing through a larger angle from its original position before it reaches a critical clearing angle. Thus, raising the value of $P_{\max}$ increases the critical clearing time and improves stability.

The steady-state power limit is given by

$$P_{\max} = \frac{EV}{X}$$

It can be seen from this expression that $P_{\max}$ can be increased by increasing either or both V and E and reducing the transfer reactance. The following methods are available for reducing the transfer reactance :

(a) Use of Double-Circuit Lines

The impedance of a double-circuit line is less than that of a single-circuit line. A double-circuit line doubles the transmission capability. An additional advantage is that the continuity of supply is maintained over one line with reduced capacity when the other line is out of service for maintenance or repair. But the provision of additional line can hardly be justified by stability consideration alone.

(b) Use of Bundled Conductors

Bundling of conductors reduces to a considerable extent the line reactance and so increases the power limit of the line.

(c) Series Compensation of the Lines

The inductive reactance of a line can be reduced by connecting static capacitors in series with the line.

It is to be noted that any measure to increase the steady-state limit P_{max} will improve the transient stability limit. The use of generators of high inertia and low reactance improves the transient stability, but generators with these characteristics are costly. In practice, only those methods are used which are economical.

High-speed Excitation Systems

High-speed excitation helps to maintain synchronism during a fault by quickly increasing the excitation voltage. High-speed governors help by quickly adjusting the generator inputs.

Fast Switching

Rapid isolation of faults is the principal way of improving transient stability. The fault should be cleared as fast as possible. It should be noted that the time required for fault removal is the sum of relay response time plus the circuit breaker operating time. Therefore, high speed relaying and circuit breaking are commonly used to improve stability during fault conditions. It has now become possible to isolate the fault in less than two cycles (that is, 0.04 s for 50 Hz system). System stability can be further improved by making circuit-breaker reclosure automatic, as many faults do not re-establish themselves after restoration of supply. The time interval between line removal and reclosure should be reduced keeping in mind that the line must remain de-energized for a certain minimum time in order that the line insulation should recover fully.

Some of the *recent methods* to maintain stability are given below :

Braking Resistors

In this method an artificial electric load in the form of shunt resistors is temporarily connected at or near the generator bus. Such resistors partially compensate the reduction of load on a generator following a fault. The acceleration of the generator rotor is therefore, reduced. For this reason, these resistors are called braking resistors. This method is also known as *dynamic braking*. A control scheme connects the braking resistors to the generators through circuit breakers. The scheme also determines the amount of resistance to be connected and the duration of its connection. The braking resistors are connected immediately following the fault and remain in the circuit for few cycles. They are disconnected at the moment of reclosure when the system voltage has recovered.

Turbine Fast Valving or Bypass Valving

In the event of a fault, the generator output is reduced resulting in a high accelerating power. If the mechanical input power to the turbine could be momentarily reduced, the acceleration could be reduced. Fast valving is a means of reducing the mechanical input power to the turbine during the fault. Certain steam valves are rapidly closed (in 0.1 to 0.2 s) and immediately reopened. This procedure increases the critical clearing time.

Single-Pole Switching

Majority of the line faults are single line-to-ground (LG) faults. In *single-pole switching* (also called *independent pole operation*), the three phases of the circuit breaker are closed or opened independently of each other. In the event of an LG fault, the circuit breaker pole corresponding to the faulty line is opened and the remaining two healthy phases continue to transfer power. Since most of the faults are transitory, this phase can be reclosed after it has been open for a predetermined time. The system should not be operated for long periods with one phase open. Therefore, provision should be made to trip the whole line if one phase remains open for a predetermined time.

HVDC Links

High voltage direct current (HVDC) links are helpful in maintaining stability due to the following advantages :
  (a) A d.c. tie line provides a loose coupling between two a.c. systems to be interconnected.
  (b) A d.c. link may interconnect two a.c. systems at different frequencies.
  (c) There is no transfer of fault energy from one a.c. system to another if they are interconnected by a d.c. tie line.

Load Shedding

If there is insufficient generation to maintain system frequency, some of the generators are disconnected during or immediately after a fault. Thus, the stability of the remaining generators is improved. The unit to be disconnected is provided with a large steam bypass system. When the system recovers from the shock of the fault, the disconnected unit is resynchronized and reloaded. Extra cost of a large steam bypass system is the limitation of this method. Disconnection of some consumers, that is, *load shedding* (removal of load), is also helpful in improving transient stability.

EXERCISES

  1. Define the terms (a) steady-state stability, (b) transient stability, (c) steady-state limit, (d) transient limit.
  2. Distinguish between steady-state and transient stability of a power system and discuss the factors on which these depend.
  3. Explain the reason why the H constant, expressed in MW-s/MVA is more frequently used than the inertia constant M, expressed in MW-s^2/electrical degree, in power system stability studies.

4. Define the H constant relating to the stored energy of a synchronous machine used in power system stability studies. Discuss its applications and field of usefulness. (I.E.E.)

5. Show that two synchronous generating sources of inertia constants M_1 and M_2 respectively, and interconnected by means of a transmission system, may be regraded for purposes of stability studies as a single generator, of inertia constant $(M_1 M_2)/(M_1 + M_2)$, connected through the same transmission system to an infinite busbar.

6. Explain briefly the equal-area criterion and how it may be used to study the stability of a two-machine system. List the factors determining the stability limit and indicate how it may be improved.

7. Two large synchronous systems are interconnected by a transmission line over which there is a power transfer. State the 'two-machine' simplifying assumptions and then derive the equal-area stability criterion. Show how this can be applied to the following cases which might cause instability : (a) sudden changes in load, (b) sudden circuit changes.   (I.E.E.)

8. Explain the equal-area criterion for the stability of an alternator supplying infinite busbars via an inductive interconnector.

9. What is meant by swing curve and how is it determined? What information is supplied by it?

10. Explain the equal-area criterion as applied to the power-angle diagram for assessing the transient stability of a transmission line acting as an interconnector between two constant voltage networks. Determine, graphically or otherwise, the maximum additional load that can be suddenly applied to such an interconnector already carrying 50 MW if the power-angle diagram is given by the equation $P = 100 \sin \alpha$, where P is the power transmitted in megawatts, and α is the displacement between the voltage phasors at the ends of the line.   (L.U.)

11. A generator supplies a large power system via a dual circuit line. With one circuit in service the maximum power transfer is 150 MW. Find approximate magnitude of the minimum angle between terminal voltage phasors occurring after switching in one circuit while 50 MW is being delivered by the other.

12. A double-circuit interconnected line has a power-angle diagram represented by $P = 100 \sin \alpha$ where P = total power (in MW), and α is the angle between the sending-end and receiving-end voltage phasors. A fault occurs on one line when a total of 50 MW is being transmitted and reduces the diagram to $P = 30 \sin \alpha$. If the fault is cleared by disconnection of one line by circuit breaker operation, until what angle may this operation be delayed if the system is to remain stable with one line in circuit? The power-angle diagram in this latter condition is $P = 70 \sin \alpha$.

13. . Show with detailed diagram how the equal area criterion may be employed to assess the stability of a two-machine system. A three-phase power system consists of a synchronous machine connected through a lossless transmission line to an infinite busbar. A fault occurs on the transmission line. The maximum power transfer of this system when unfaulted is $5\,P$, and immediately prior to the instant of the fault of power transfer is $2.5\,P$. The power/angle curves during fault and post fault conditions have peak values

of $2P$ and $4P$ respectively. Determine the permissible increase in the angular displacement between the voltages at the two ends of the system beyond which he circuit breakers could not clear the fault in time for the system to remain in synchronism.

(I.E.E.)

14. Describe the difference between transient and steady-state stability of a synchronous system. Describe a tabular method of computing of transient stability by a step-by-step process, and indicate how a model network may be used to assist the computation.

(I.E.E.)

15. Explain the methods of improving power system stability.

16. The **ABCD** constants of a nominal π-network representing a three-phase transmission line are

$$\mathbf{A} = \mathbf{D} = 0.950 \underline{/1.27°}, \quad \mathbf{B} = 92.4 \underline{/76.87°} \ \Omega, \quad \mathbf{C} = 0.0006 \underline{/90°} \ \text{S}.$$

Find the steady-state stability limit if both the sending-end and receiving-end voltages are held at 138 kV : (a) with the **ABCD** constants as given, (b) with the shunt admittance neglected, and (c) with both the series resistance and the shunt admittance neglected.

17. For the system shown in Fig. P 19.1, determine the maximum steady-state power transfer when (a) reactor is disconnected, (b) reactor is connected. All the reactances shown in the diagram are based on the generator rating.

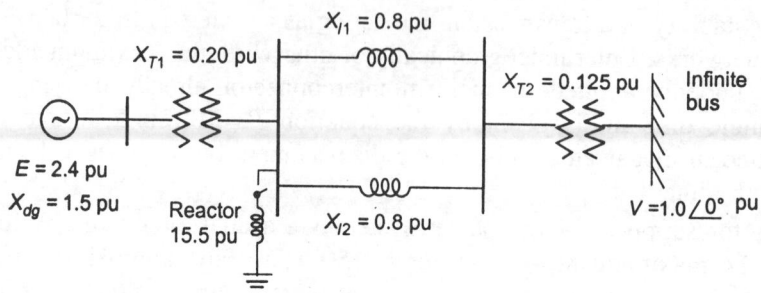

Fig. P. 19.1.

18. Explain the terms critical clearing angle and critical clearing time in connection with the transient stability of a power system.

A synchronous generator is delivering 1.0 pu active power to an infinite bus of 1.0 pu voltage through a transmission line of 0.2 pu reactance and negligible resistance. The generator reactance is 0.3 pu and the voltage behind the reactance is 1.3 pu. A three-phase short-circuit fault occurs close to the generator terminals. Determine the torque angle before which the fault must be cleared by a circuit breaker if the stability is to be maintained.

19. A synchronous generator capable of developing 500 MW operates at a load angle of 8°. Determine by how much the input shaft power can be increased suddenly if the stability is to be maintained.

20. A synchronous generator is delivering 0.5 of maximum power to an infinite bus through a transmission line. A fault occurs such that the new maximum power is 0.3 of the

original. When the fault is cleared, the maximum power that can be delivered is 0.8 of the original maximum value. Determine the critical clearing angle. If the fault is cleared at $\delta = 75°$, find the maximum value of δ for which the machine swings around its new equilibrium position.

21. A synchronous generator is delivering 0.25 of maximum power to an infinite bus through a transmission line. A fault occurs such that the reactance between the generator and the bus is increased to two times its prefault value. Determine the maximum value of the δ swing for a sustained fault.

22. A 100 MVA, 2-pole, 60-Hz generator has a moment of inertia of 50×10^3 kgm^2. Determine (a) the energy stored in the rotor at the rated speed, (b) the angular momentum M, (c) the inertia constant H.

23. The kinetic energy stored in the rotor of a 50 MVA, 6-pole, 60-Hz synchronous machine is 200 MJ. The input to the machine is 25 MW at a developed power of 22.5 MW. Calculate the accelerating power and the acceleration.

24. The **ABCD** constants for the nominal π-circuit representation of a transmission line are:

$$\mathbf{A} = \mathbf{D} = 0.9 \underline{/0.3°}$$
$$\mathbf{B} = 82.5 \underline{/76°} \ \Omega$$
$$\mathbf{C} = 0.0005 \underline{/90°} \ S$$

What is the maximum power that can be transmitted over the line without making the system unstable if

$$|\mathbf{V_s}| = |\mathbf{V_r}| = 110 \text{ kV } ?$$

25. A synchronous motor develops 30 percent of its rated power for a certain load. The load on the motor is suddenly increased by 150 percent of the original value. Neglecting all losses, calculate the maximum power angle on the swing curve.

26. A motor delivers 0.25 pu of its rated power while operating from an infinite bus. If the load on the motor is suddenly doubled, determine the maximum power angle δ_m. Neglect losses.

27. A synchronous generator, capable of developing 1500 MW, operates at a power angle of 8°. By how much can the input shaft power be increased suddenly without loss of stability? Assume that the generator field current and terminal voltages do not change such that P_{max} will remain constant.

28. A synchronous generator is operating at an infinite bus and supplying 0.45 pu of its maximum power capacity. A fault occurs, and the reactance between the generator and the line becomes four times its value before the fault. The maximum power that can be delivered after the fault is cleared is 70 percent of the original maximum value. Determine the critical clearing angle.

29. A 100 MVA synchronous generator supplies 62.5 MVA at 0.8 lagging power factor. The reactance between the load and the generator is normally 1.0 pu, but it increases to 3.0 pu because of a sudden three-phase short circuit. The fault is subsequently cleared and the generator then supplies 43.75 MVA at 0.8 lagging power factor. Determine the critical clearing angle.

ANSWERS

10. 37 MW   11. 9°

12. 66°   13. $\delta_2 = 90°$

16. (a) 157.4 MW, (b) 159.3 MW, (c) 211.6 MW

17. (a) 1.00786 pu, (b) 1.0514 pu

18. 91°   19. 316.2 MW

20. $\delta_c = 79.96°$, $\delta_m = 118.9°$ 21. 46.3°

22. (a) 3553 J, (b) $M = 0.329$ MJ rad/s, (c) $H = 35.53$ MJ/MVA

23. 2.5 MW, 2.356 rad/s^2 24. 114.09 MW

25. 40°   26. 45°

27. 940.3 MW   28. 72.2°

29. 68.58°

20

Travelling Waves

20.1 INTRODUCTION

When the current, voltage, or power change in a circuit, there is a transient state between two steady-state conditions. The transient behaviour of transmission lines differs from its steady-state behaviour. The transient period, although of a very short duration of the order of a few microseconds, is sufficient to cause much disturbance in the line. The transients are set up in the transmission lines mainly due to switching, faults and lightning. A *surge* is a movement of charge along the conductor. Such surges are characterized by a sudden very steep rise in voltage (the *surge front*) followed by a gradual decay in voltage (the *surge tail*). The surges produced on the line due to lightning are of particular interest due to their great magnitudes and different wave-shapes. The surges reach the terminal apparatus such as cable boxes, transformers or switchgear, and may damage them if they are not properly protected. As the waves travel along the line their wave-shapes and magnitudes are also modified. The study of travelling waves, therefore, plays an important role in knowing the voltages and currents at all points in a power system. It helps in the design of insulators, protective equipment, the insulation of the terminal equipment, and overall insulation coordination.

20.2 WAVE EQUATION

It has been already pointed out in Chapter 7 that the line parameters are uniformly distributed along the line. It may be assumed that the line is made up of short sections of length dx (Fig. 20.1). Suppose that

  R = the resistance of line per unit length

  L = the inductance of line per unit length

  C = the capacitance of line per unit length

and G = the shunt conductance of line per unit length.

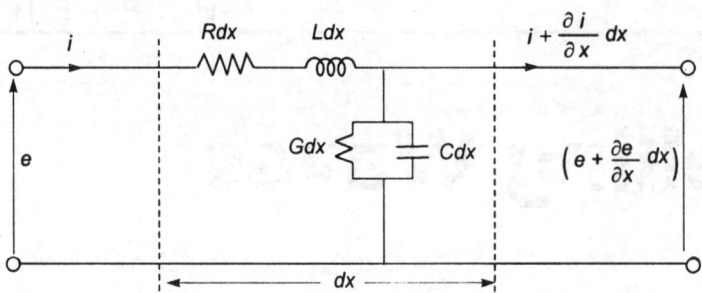

Fig. 20.1. Short section of a two-wire transmission line.

For convenience, the resistance and inductance of both line conductors are shown lumped in one conductor only. The resistance, inductance, capacitance, and shunt conductance for the short section of length dx are Rdx, Ldx, Cdx and Gdx respectively.

It should be appreciated that as the number of the series or lumps is increased and their sizes decreased, the line with distributed parameters approaches nearer to the actual smooth line. It will be assumed that the line parameters are constant throughout the line. They are independent of voltage, current and frequency. Actually, due to skin effect both inductance and resistance vary with frequency. At the front of a steep wave the resistance is greater and the inductance less. During corona formation the shunt conductance varies considerably. The constancy of line parameters is assumed on the grounds so as to make the analysis of travelling waves mathematical.

Consider a short section of the line of length dx. The instantaneous values of voltage and current are functions of both distance x and time t. At a distance x from the sending end they can be represented as $e\,(x, t)$ and $i\,(x, t)$ respectively. The voltage at a neighbouring point distance $(x + dx)$ from the sending end is $e\,(x + dx)$. By Taylor's theorem,

$$e\,(x + dx) = e\,(x) + \frac{\partial e}{\partial x}\,dx \qquad \qquad \ldots(20.2.1)$$

The current i flowing through the resistance Rdx causes a voltage drop $i\,(Rdx)$ in it. It changes at the rate of $\partial i/\partial t$ in the inductance Ldx to produce a voltage drop equal to $\left(L\dfrac{\partial i}{\partial t}\,dx\right)$ in it. The difference in the voltage between the ends of the section is obviously due to the voltage drops in the resistance Rdx and inductance Ldx. Expressed mathematically,

$$e - \left(e + \frac{\partial e}{\partial x}\,dx\right) = (Rdx)\,i + (Ldx)\,\frac{\partial i}{\partial t}$$

or $\qquad -\dfrac{\partial e}{\partial x}\,dx = (Rdx)\,i + (Ldx)\,\dfrac{\partial i}{\partial t}$

Cancelling dx from both the sides gives

$$-\frac{\partial e}{\partial x} = Ri + L\,\frac{\partial i}{\partial t} \qquad \qquad \ldots(20.2.2)$$

Also, the difference of the currents between the two ends of the section is equal to the sum of the currents in the conductance Gdx due to the voltage e impressed on it and the current through the capacitance Cdx due to the voltage changing at the rate of $\dfrac{\partial e}{\partial t}$. Expressing mathematically,

$$i - \left(i + \frac{\partial i}{\partial x}\,dx\right) = (Gdx)\,e + (Cdx)\,\frac{\partial e}{\partial t}$$

or $\qquad -\dfrac{\partial t}{\partial x}\,dx = (Gdx)\,e + (Cdx)\,\dfrac{\partial e}{\partial t}$

or $\qquad -\dfrac{\partial i}{\partial x} = Ge + C\,\dfrac{\partial e}{\partial t}$ $\hspace{4cm}$ ...(20.2.3)

The solution of the Eqs. (20.2.2) and (20.2.3) is complicated. Since the losses in the line are much smaller than the energy travelling along the line, they can be neglected. For a lossless line, $R = 0$ and $G = 0$. With this assumption the calculations become simpler, and at the same time the results obtained are safer and satisfactory. Equations (20.2.2) and (20.2.3) then take the forms

$$\frac{\partial e}{\partial x} = -L\,\frac{\partial i}{\partial t} \hspace{5cm} \text{...(20.2.4)}$$

and $\qquad \dfrac{\partial i}{\partial x} = -C\,\dfrac{\partial e}{\partial t}$ $\hspace{4.5cm}$ ...(20.2.5)

Differentiating Eq. (20.2.4) partially with respect to distance x and Eq. (20.2.5) with respect to time t gives

$$\frac{\partial^2 e}{\partial x^2} = -L\,\frac{\partial^2 i}{\partial x \partial t} \hspace{4.5cm} \text{...(20.2.6)}$$

and $\qquad \dfrac{\partial^2 i}{\partial t \partial x} = -C\,\dfrac{\partial^2 e}{\partial t^2}$ $\hspace{4cm}$ ...(20.2.7)

Also, $\qquad \dfrac{\partial^2 i}{\partial t \partial x} = \dfrac{\partial^2 i}{\partial x \partial t}$

Substituting the value of $\dfrac{\partial^2 i}{\partial x \partial t}$ in Eq. (20.2.6) from Eq. (20.2.7) we get

$$\frac{\partial^2 e}{\partial x^2} = LC\,\frac{\partial^2 e}{\partial t^2} \hspace{4.5cm} \text{...(20.2.8)}$$

Similarly, differentiating Eq. (20.2.4) with respect to t and Eq. (20.2.5) with respect to x

$$\frac{\partial^2 e}{\partial x \partial t} = -L\,\frac{\partial^2 i}{\partial t^2} \hspace{4.5cm} \text{...(20.2.9)}$$

and $\qquad \dfrac{\partial^2 i}{\partial x^2} = -C\,\dfrac{\partial^2 e}{\partial x \partial t}$ $\hspace{3.8cm}$ ...(20.2.10)

Substituting the value of $\dfrac{\partial^2 e}{\partial x \partial t}$ from Eq. (20.2.9) in Eq. (20.2.10),

$$\frac{\partial^2 i}{\partial x^2} = LC \frac{\partial^2 i}{\partial t^2} \qquad \qquad \text{...(20.2.11)}$$

Equations (20.2.8) and (20.2.11) are identical in form and give similar solutions. They are called the *wave equations*. They represent the distribution of voltage and current along the line in terms of time and distance.

Equation (20.2.8), when solved, gives the wave equations in the form

$$e = f_1\left(x - \frac{1}{\sqrt{LC}} t\right) + f_2\left(x + \frac{1}{\sqrt{LC}} t\right) \qquad \qquad \text{...(20.2.12)}$$

where f_1 is an arbitrary single-valued function of $\left(x - \dfrac{1}{\sqrt{LC}} t\right)$ and f_2 that of $\left(x + \dfrac{1}{\sqrt{LC}} t\right)$. Now,

any function of $\left(x - \dfrac{1}{\sqrt{LC}} t\right)$ represents a distribution of voltage travelling along the line in the

direction of increasing x. As long as x is increased with velocity $\dfrac{1}{\sqrt{LC}}$, the quantity

$\left(x - \dfrac{1}{\sqrt{LC}} t\right)$ remains constant. Its function will, therefore, remain constant, which means that the

voltage travels along the line at a velocity $\dfrac{1}{\sqrt{LC}}$ without changing its magnitude or form. In

other words, there is no attenuation or distortion in the travelling wave.

On similar lines it may be shown that any function of $\left(x + \dfrac{1}{\sqrt{LC}} t\right)$ is a wave travelling in

the direction of decreasing x with a velocity $\dfrac{1}{\sqrt{LC}}$ without a change in amplitude or form.

Now, we shall prove that functions f_1 and f_2 satisfy Eq. (20.2.8) independently. Since Eq. (20.2.8) is linear, the sum of the functions f_1 and f_2 will also satisfy Eq. (20.2.8). To verify this consider the forward wave represented by

$$e = f_1\left(x - \frac{1}{\sqrt{LC}} t\right) = f_1 (x - ut)$$

where $u = \dfrac{1}{\sqrt{LC}}$

Differentiation with respect to x and t gives

$$\frac{\partial e}{\partial x} = \frac{\partial f_1}{\partial (x - ut)} \cdot \frac{\partial (x - ut)}{\partial x} = \frac{\partial f_1}{\partial (x - ut)} \qquad \qquad \text{...(20.2.13)}$$

and $\quad \dfrac{\partial e}{\partial t} = \dfrac{\partial f_1}{\partial (x - ut)} \cdot \dfrac{\partial (x - ut)}{\partial t} = - u \dfrac{\partial f_1}{\partial (x - ut)} \qquad \qquad \text{...(20.2.14)}$

Differentiating again with respect to x and t gives

$$\frac{\partial^2 e}{\partial x^2} = \frac{\partial^2 f_1}{\partial (x - ut)^2} \cdot \frac{\partial (x - ut)}{\partial x} = \frac{\partial^2 f_1}{\partial (x - ut)^2} \qquad \qquad \text{...(20.2.15)}$$

and $\quad \dfrac{\partial^2 e}{\partial t^2} = -u \dfrac{\partial^2 f_1}{\partial (x-ut)^2} \cdot \dfrac{\partial (x-ut)}{\partial t} = u^2 \dfrac{\partial^2 f_1}{\partial (x-ut)^2}$ $\qquad$ ...(20.2.16)

Comparing Eq. (20.2.15) and (20.2.16),

$$\dfrac{\partial^2 e}{\partial x^2} = \dfrac{1}{u^2} \dfrac{\partial^2 e}{\partial t^2}$$

or $\qquad \dfrac{\partial^2 e}{\partial x^2} = LC \dfrac{\partial^2 e}{\partial t^2}$

which is Eq. (20.2.8).

This shows that any function of the form $f_1(x-ut)$ satisfies Eq. (20.2.8). The assumed solution is, therefore, correct.

The solution of Eq. (20.2.11) gives a similar result for current, i.e.,

$$i = f_3 \left(x - \dfrac{1}{\sqrt{LC}} t \right) + f_4 \left(x + \dfrac{1}{\sqrt{LC}} t \right) \qquad \text{...(20.2.17)}$$

Equation (20.2.17) also represents two waves travelling in opposite directions, the former being the *incident* and the latter the *reflected* wave.

The forward and reflected waves travel in opposite directions. They superimpose to produce a new type of waves which appear stationary along the line. These waves are called *stationary waves* or *standing waves*. The wave pattern along the line is called the standing wave pattern.

20.3 CHARACTERISTIC IMPEDANCE

The voltage and current waves are dependent upon each other. We shall establish a relation between them. Consider the forward voltage represented by

$$e = f_1 \left(x - \dfrac{1}{\sqrt{LC}} t \right)$$

Integrating both the sides of the Eq. (20.2.5),

$$\int \dfrac{\partial i}{\partial x} dx = -\int C \dfrac{\partial e}{\partial t} dx$$

From Eqs. (20.2.13) and (20.2.14)

$$\dfrac{\partial e}{\partial t} = -u \dfrac{\partial e}{\partial x} \qquad \text{...(20.3.1)}$$

Using this result in the above expression

$$\int \dfrac{\partial i}{\partial x} dx = -C(-u) \int \dfrac{\partial e}{\partial x} dx$$

$$i = C u e$$

Thus, the ratio of the magnitudes of voltages and associated current surges is given by

$$\dfrac{e}{i} = \dfrac{1}{Cu} = \dfrac{1}{C} \sqrt{LC} = \sqrt{\dfrac{L}{C}} \qquad \text{...(20.3.2)}$$

The quantity $\sqrt{\dfrac{L}{C}}$ is called the *characteristic impedance* for a loss-free line. It is denoted by Z_0. This is also known as *surge impedance*, *wave impedance* or *natural impedance* of line. The term surge impedance is more common in power system work. For a loss-free line Z_0 has the dimension of a pure resistance. It is, therefore, expressed in ohms.

This may be seem rather surprising that Z_0 has the units expressed in ohms. This is, however, true.

$$Z_0 = \left(\frac{L}{C}\right)^{1/2} = \left(\frac{\text{henries/length}}{\text{farads/length}}\right)^{1/2} = \left(\frac{2\pi f \text{ henries}}{2\pi f \text{ farads}}\right)^{1/2}$$

$$= [(\text{inductive reactance in ohms}) \times (\text{capacitive reactance in ohms})]^{1/2}$$

The quantity under radical sign has got the dimensions of ohms. Z_0 is independent of the length of the line. The reciprocal of surge impedance is called the *surge admittance*. It is denoted by Y_0.

$$Y_0 = \frac{1}{Z_0} \qquad\qquad\qquad ...(20.3.3)$$

The surge impedance for overhead lines is usually between 200 and 500 ohms. For cables, it is of the order of 30-50 ohms, and the transformers have several thousand ohms as their surges impedance.

The expressions for velocity of surge propagation and characteristic impedance of the line can easily be deduced as follows :

For a short section, dx, of the line, total inductance $= Ldx$.

Change in flux linkage in time dt is $iLdx$.

Rate of change of flux linkage $= iL\dfrac{dx}{dt}$ = induced voltage

But this induced voltage is equal to the voltage e.

$$e = iL\frac{dx}{dt} = iLu \qquad\qquad\qquad ...(20.3.4)$$

Total capacitance of the section $= Cdx$

The charge, dq, delivered to the section $= eCdx$

$$\therefore \qquad i = \frac{dq}{dt} = eC\frac{dx}{di} = eCu \qquad\qquad\qquad ...(20.3.5)$$

Multiplication of Eqs. (20.3.4) and (20.3.5) gives

$$ei = eiLCu^2$$

$$u = \pm\sqrt{\frac{1}{LC}} \qquad\qquad\qquad ...(20.3.6)$$

The double sign indicates that the surge is split into two components which travel along the line in opposite directions. From Eqs. (20.3.4), (20.3.5) and (20.3.6)

$$\frac{e}{i} = Lu = \sqrt{\frac{L}{C}} = Z_0 \qquad\qquad\qquad ...(20.3.7)$$

20.4 ENERGY AND POWER OF A SURGE

Every surge comprises a definite amount of energy which is partly in the form of electrostatic energy and partly in the form of electromagnetic energy.

Electrostatic energy per unit length of line $= \dfrac{1}{2} Ce^2$

Electromagnetic energy per unit length of line $= \dfrac{1}{2} Li^2$

Total energy per unit length of line $= \dfrac{1}{2} Ce^2 + \dfrac{1}{2} Li^2$

But, $\dfrac{e}{i} = \sqrt{\dfrac{L}{C}}$

$$Ce^2 = Li^2 \qquad\qquad\qquad\qquad\qquad\qquad ...(20.4.1)$$

Thus, the total energy of the surge is equally divided between its electric and magnetic fields.

Total surge energy $= \displaystyle\int \left(\dfrac{1}{2} Ce^2 + \dfrac{1}{2} Li^2 \right) dx = C \int e^2 \, dx = L \int i^2 \, dx$

The power of the surge flowing along the line

$$p = ei = i^2 Z_0 \qquad\qquad\qquad\qquad\qquad ...(20.4.2)$$

For each element of the line, power loss of the surge is $-dp$.

Now, $-dp = -2iZ_0 \, di$

Also, power loss in elementary length, dx, of the line $= (i^2 R dx + Ge^2 dx)$
Equating these values

$$-2iZ_0 \, di = (i^2 R + Ge^2) \, dx = i^2 (R + GZ_0^2) \, dx$$

$$\dfrac{di}{i} = -\dfrac{1}{2} \left(\dfrac{R}{Z_0} + GZ_0 \right) dx$$

Let, $\dfrac{1}{2} \left(\dfrac{R}{Z_0} + GZ_0 \right) = a \qquad\qquad\qquad\qquad ...(20.4.3)$

$$\dfrac{di}{i} = -a \, dx \qquad\qquad\qquad\qquad\qquad\qquad ...(20.4.4)$$

Integration of Eq. (20.4.4) gives

$$\ln i = -ax + D$$

where D is a constant.

If, $i = i_0$ for $x = 0$

$$D = \ln i_0, \qquad \ln i = -ax + \ln i_0$$

$\therefore$ $i = i_0 \, \varepsilon^{-ax} \qquad\qquad\qquad\qquad\qquad\qquad ...(20.4.5)$

Similarly, $e = e_0 \, \varepsilon^{-ax}$

Hence, power per unit length at a point distance x from the wave-front :

$$p = ei = e_0 i_0 \, \varepsilon^{-2ax} \qquad \qquad ...(20.4.6)$$

Equations (20.4.4) and (20.4.5) show that R and G produce an exponential damping or attenuation of both the current and voltage amplitudes of the surge. Since R, G and Z_0 are all constants of the line, attenuation is independent of the surge amplitude and is determined only by the distance travelled by the surge. The surges retain their shape without distortion. The velocity of propagation is independent of R and G.

For high voltage lines G can be neglected in comparison with R.

$$\therefore \qquad \frac{e}{e_0} = \frac{i}{i_0} = \varepsilon^{\frac{-Rx}{2Z_0}} \qquad \qquad ...(20.4.7)$$

Equation (20.4.7) shows that damping is directly proportional to R and inversely proportional to Z_0. Thus, damping is much higher in cables than in overhead lines,

If, t = time required by the surge to travel the distance x

$$x = ut$$

$$\frac{Rx}{2Z_0} = \frac{Rut}{2Z_0} = \frac{R}{2L} t$$

Thus, $\qquad \dfrac{e}{e_0} = \dfrac{i}{i_0} = \varepsilon^{(-R/2L)\,t} = \varepsilon^{-t/\tau}$ $\qquad \qquad ...(20.4.8)$

where τ is the time constant. The phenomena caused by the surge propagation occur only during a very short time interval, of the order of 1 ms, following a switching process.

20.5 EVALUATION OF SURGE IMPEDANCE

Surge impedance is evaluated here for the following cases :

(a) Single-phase overhead line with air as dielectric

Loop inductance, $L = 4 \times 10^{-7} \ln \dfrac{D}{r}$ H/m

Capacitance, $C = \dfrac{1}{36 \times 10^9 \ln \dfrac{D}{r}}$ F/m

Surge impedance, $Z_0 = \sqrt{\dfrac{L}{C}} = \left[\left(4 \times 10^{-7} \ln \dfrac{D}{r} \right) \times \left(36 \times 10^9 \ln \dfrac{D}{r} \right) \right]^{1/2} = 120 \ln \dfrac{D}{r}$ ohms

$$...(20.5.1)$$

(b) Single-core cable

If R = inner sheath radius

$\quad r$ = radius of conductor

$\quad \varepsilon_r$ = relative permittivity

$$L = 2 \times 10^{-7} \ln \frac{R}{r} \text{ H/m}$$

$$C = \frac{\varepsilon_r}{18 \times 10^9 \ln \frac{R}{r}} \text{ F/m}$$

$$Z_0 = \sqrt{\frac{L}{C}} = \left[\left(2 \times 10^{-7} \ln \frac{R}{r} \right) \times \left(\frac{18 \times 10^9}{\varepsilon_r} \ln \frac{R}{r} \right) \right]^{1/2} = \frac{60}{\sqrt{\varepsilon_r}} \ln \frac{R}{r} \text{ ohms} \qquad \ldots(20.5.2)$$

(c) Single-conductor line with ground return

If h = height of conductor above ground

$\quad$ r = radius of conductor

then,$\quad$ $L = 2 \times 10^{-7} \ln \frac{2h}{r} \text{ H/m}$

$$C = \frac{1}{18 \times 10^9 \ln \frac{2h}{r}} \text{ F/m}$$

$$Z_0 = \sqrt{\frac{L}{C}} = \left[\left(2 \times 10^{-7} \ln \frac{2h}{r} \right) \times \left(18 \times 10^9 \ln \frac{2h}{r} \right) \right]^{1/2} = 60 \ln \frac{2h}{r} \text{ ohms} \qquad \ldots(20.5.3)$$

20.6 VELOCITY OF TRAVELLING WAVES

A surge travelling along a line has a definite velocity determined by the inductance and capacitance of the line. The velocity of travelling waves is found here for a single-phase line. Similar, calculations can be made for a three-phase line. The result remains the same.

For a single-phase overhead line with air as dielectric

$$L = 2 \times 10^{-7} \ln \frac{D}{r} \text{ H/m}$$

$$C = \frac{1}{18 \times 10^9 \ln \frac{D}{r}} \text{ F/m}$$

Therefore, the velocity of travelling waves for overhead lines

$$u = \sqrt{\frac{1}{LC}} = \left(\frac{18 \times 10^9 \ln \frac{D}{r}}{2 \times 10^{-7} \ln \frac{D}{r}} \right)^{1/2} = 3 \times 10^8 \text{ m/s} \qquad \ldots(20.6.1)$$

Thus, it is found that velocity of travelling waves on overhead lines with air as dielectric is the same as the speed of light in free space.

For a single-phase, two-wire cable of relative permittivity ε_r,

$$L = 2 \times 10^{-7} \ln \frac{D}{r} \text{ H/m}$$

$$C = \frac{\varepsilon_r}{18 \times 10^9 \ln \frac{D}{r}} \text{ F/m}$$

The velocity of propagation in a single-phase, two-wire cable is, therefore, given by

$$u = \sqrt{\frac{1}{LC}} = \left(\frac{18 \times 10^9 \ln \frac{D}{r}}{\varepsilon_r \times 2 \times 10^{-7} \ln \frac{D}{r}} \right) = \frac{3 \times 10^8}{\sqrt{\varepsilon_r}} \text{ m/s} \qquad \ldots(20.6.2)$$

Since for a cable the relative permittivity, ε_r is greater than unity, the value of velocity of propagation in it is usually between one-half and two-thirds of the velocity of light.

20.7 INCIDENT AND REFLECTED WAVES

For a wave travelling in the backward direction given by

$$e = f_2 (x + ut)$$

The ratio of the voltage to current is

$$\frac{e}{i} = \sqrt{\frac{L}{C}} = -Z_0$$

and the value of the current corresponding to voltage moving in backward direction is

$$i = -\frac{e}{Z_0} \qquad \ldots(20.7.1)$$

We, therefore, arrive at the conclusion that if there is any change in current, voltage, or a constant parameter at any point in a line, voltage surges are set up. These voltage surges travel along the line at a very high speed. They are accompanied by current waves. The current and voltage waves travelling in forward direction have got the same sign. If they travel in backward direction, they are opposite in sign. To distinguish between the two waves travelling in opposite directions, use is made of suffix f to represent a wave travelling from the source towards the load and suffix r to represent the wave in opposite direction. Thus, e_f and i_f are used to represent the voltage and current waves travelling in the forward direction, and e_r and i_r, the waves in the opposite direction. The waves travelling from the source towards the load are known as *incident waves* and those from load to the source are called *reflected waves*.

These waves may superpose each other and the resultant wave at any point is the sum of incident and reflected waves there. Thus, if e_R and i_R denote the resultant voltage and current respectively at any point on the line.

$$e_R = e_f + e_r \qquad \ldots(20.7.2)$$

and $\quad i_R = i_f + i_r \qquad \ldots(20.7.3)$

Fig. 20.2 illustrates the two waves travelling in opposite directions.

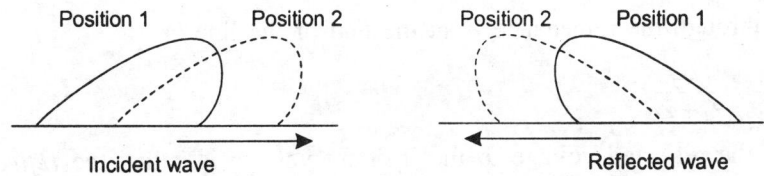

Fig. 20.2. Two waves travelling in opposite directions.

20.8 REFLECTION OF WAVES

If the uniformity of the line is disturbed at a point, the surge impedance changes there. There will be a partial refection and partial transmission of the wave when it reaches the point of discontinuity i.e., the incident wave gives rise to two waves, one transmitted forward and the other reflected back. The discontinuity in the line may occur as a consequence of the change in the characteristics of the circuit, e.g., junction of an overhead line and a cable, the end of the line, forked line, the junction of the line with terminal apparatus, etc.

Consider a general case of a line which is terminated in an impedance Z_t, which is a pure non-inductive resistance. By Ohm's law,

$$\frac{\text{the resultant voltage at the receiving end}}{\text{the resultant current at the receiving end}} = \text{impedance at the receiving end}$$

i.e., $$\frac{e_R}{i_R} = Z_t \qquad\qquad ...(20.8.1)$$

Substituting the values of e_R and i_R from Eq. (20.7.2) and (20.7.3) in (20.8.1)

$$\frac{e_f + e_r}{i_f + i_r} = Z_t \qquad\qquad ...(20.8.2)$$

For incident wave,

$$e_f = i_f Z_0 \qquad\qquad ...(20.8.3)$$

and for reflected wave,

$$e_r = - i_r Z_0 \qquad\qquad ...(20.8.4)$$

The values of the currents obtained from these equations when substituted in the expression for Z_t give

$$\frac{e_f + e_r}{e_f / Z_0 - e_r / Z_0} = Z_t \qquad\qquad ...(20.8.5)$$

Normally, the value of the incident voltage being known and, therefore, the reflected voltage e_r, is given by

$$e_r = e_f \frac{Z_t - Z_0}{Z_t + Z_0} \qquad\qquad ...(20.8.6)$$

The voltage at the end of the line is

$$e_R = e_f + e_r = 2 e_f \frac{Z_t}{Z_t + Z_0}$$

The current through the impedance Z_t at the end of the line is

$$i_R = \frac{e_R}{Z_t} = \frac{2e_f}{Z_t + Z_0}$$

The ratio of the reflected voltage to the incident voltage is called the *reflection coefficient* or the *reflection factor* for voltage. It is denoted by ρ_v.

$$\rho_v \triangleq \frac{e_r}{e_f} = \frac{Z_t - Z_0}{Z_t + Z_0} \qquad \qquad \text{...(20.8.7)}$$

This implies that a voltage wave is reflected at the termination and its amplitude becomes ρ_v times the incident wave. The ratio of the reflected current to the incident current is

$$\rho_i \triangleq \frac{i_r}{i_f} = \frac{-e_r/Z_0}{e_f/Z_0} = -\frac{e_r}{e_f} = -\rho_v \qquad \qquad \text{...(20.8.8)}$$

$$\therefore \qquad \rho_v = -\rho_i = \rho \text{ (say)} \qquad \qquad \text{...(20.8.9)}$$

The physical meaning of the negative reflection coefficient needs an explanation. The current is assumed to be positive in the direction of the incident voltage wave, i.e., from the sending end to the receiving end. It is negative in the direction of the reflected voltage wave. The reflected voltage is positive. Therefore, a positive voltage at the receiving end will cause a current to flow from the receiving end to the sending end. But it has been assumed that the current is taken positive in the direction from the sending end to the receiving end. The current in the direction from the receiving end to the sending end should, therefore, be taken negative.

20.9 TRANSMISSION OR REFRACTION OF WAVES

It has been noticed earlier that there is a partial reflection and partial transmission of a wave at a point of discontinuity. The ratio of the voltage at the termination to the incident voltage is called the *transmission coefficient* or *refraction coefficient*. It is denoted by τ.

$$\frac{e_R}{e_f} = \tau \qquad \qquad \text{...(20.9.1)}$$

$$\frac{e_R}{e_f} - 1 = \tau - 1 \; ; \qquad \frac{e_R - e_f}{e_f} = \tau - 1$$

i.e., $\qquad \rho = \tau - 1 \qquad \qquad \text{...(20.9.2)}$

The values of reflected and resultant current and voltage for their incident values of i_f and e_f can be conveniently written as :

Reflected voltage, $e_r = \rho \, e_f$ $\qquad \qquad \text{...(20.9.3)}$

Reflected current, $i_r = -\rho \, i_f$ $\qquad \qquad \text{...(20.9.4)}$

Resultant voltage at termination, $e_R = e_f \dfrac{2Z_t}{Z_t + Z_0} = \tau \, e_f$ $\qquad \qquad \text{...(20.9.5)}$

Resultant current at termination, $i_R = i_f \dfrac{2Z_0}{Z_t + Z_0}$ $\qquad \qquad \text{...(20.9.6)}$

In the above discussion the value of the terminating impedance is assumed to be a general one. In practice, it can have any value. But here only those cases will be considered which are of particular interest. They are as follows :

(a) When $Z_t = \infty$, i.e., the line is open-circuited.

(b) When $Z_t = 0$, i.e., the line is short-circuited.

(c) When $Z_t = Z_0$, i.e., the line is terminated in its characteristic impedance.

20.10 OPEN-CIRCUITED LINE

When the line is open at the receiving end, Z_t becomes infinite and

$$\rho = \frac{Z_t - Z_0}{Z_t + Z_0} = \frac{1 - Z_0/Z_t}{1 + Z_0/Z_t} = +1 \qquad \qquad \ldots(20.10.1)$$

Also, $\tau = 2$

The amplitude of the reflected wave is equal to that of the incident wave, the two waves being of the same sign. The resultant voltage at the receiving end

$$e_R = e_f + e_r = e_f + e_f = 2e_f \qquad \qquad \ldots(20.10.2)$$

Also, for an open circuit,

$$i_R = 0$$

i.e., $i_f + i_r = 0$

$$i_r = -i_f \qquad \qquad \ldots(20.10.3)$$

The direction of the reflected current is opposite to that of the incident current. The voltage at the receiving end of an open circuited line thus becomes equal to two times the incident voltage, and the current there is zero.

The voltage and current distribution along the line at various instants may be conveniently followed with the help of diagrams shown in Fig. 20.3. Here, it is assumed that d.c. supply with negligible internal resistance is available so that on closing the switch the voltage builds up suddenly from zero to e and a rectangular wavefront of incident voltage travels along the line to the termination in a time $t = l/u$ (Fig. 20.3).

(a) The voltage wave has completed half of its total distance to reach the receiving end. Its peak value is e.

(b) The voltage wave has reached the receiving end. There it has been doubled and now it is travelling in the direction shown.

(c) The wave has practically reached its sending end and the whole line is at a voltage $2e$.

(d) After reaching the sending end the wave is reflected there and travels back along the line. But the voltage at the sending end is held constant at e.

(e) The wave reaches the open end where its value is e.

(f) The voltage at the receiving end becomes zero and the wave travels from right to left.

(g) When $t = 4l/u$ the voltage becomes zero. The cycle is completed. The line then takes its original state for a moment. After this the repetition of the whole cycle occurs.

From this it is observed that for a cycle to repeat itself the wave will have to travel a distance equal to four times the length of the line.

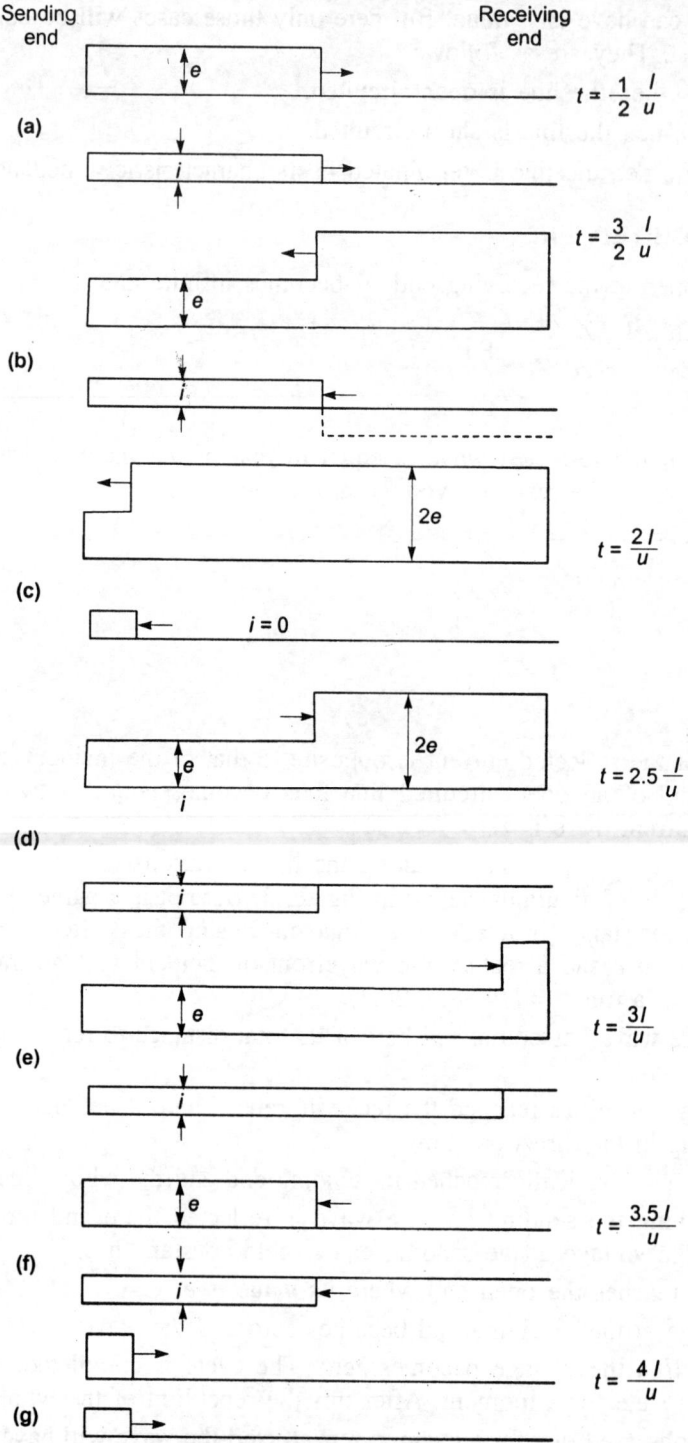

Fig. 20.3. Voltage and current distributions in an open-circuited line.

Time for a cycle, $t = \dfrac{4l}{u}$

Frequency of the wave $= \dfrac{1}{t} = \dfrac{u}{4l} = \dfrac{1}{4l\sqrt{LC}}$

The distribution of the associated current is also shown in Fig. 20.3.

20.11 SHORT-CIRCUITED LINE

For a line which is short circuited at the receiving end, $Z_t = 0$. Therefore, the voltage there will be zero, i.e.,

$$e_f + e_r = e_R = 0$$

Thus, $e_f = -e_r$ and $\rho = -1$, $\tau = 0$

Also, $i_f = i_r$ and $i_R = i_f + i_r = 2i_f$

These equations show that at the receiving point the reflected voltage will be equal to incident voltage but of opposite sign. The reflected current is also equal to incident current but of the same sign. The resultant voltage is zero at the termination, while the resultant current is double there.

The distributions of voltage and current are shown in Fig. 20.4. The line voltage becomes zero periodically, but the current is increased by $\dfrac{e}{Z_0}$ at each subsequent reflection. The current thus goes on increasing indefinitely in steps. This is an ideal case. In actual practice, these current steps become smaller and smaller in magnitude until a steady current $\dfrac{e}{R_l}$ is established in the circuit, where R_l is the resistance of the whole line.

20.12 LINE TERMINATED IN Z_0

If the line be terminated in its characteristic impedance Z_0, then $Z_t = Z_0$.

$$\rho = \dfrac{Z_t - Z_0}{Z_t + Z_0} = 0$$

$$\tau = 1, \quad e_r = 0, \quad i_r = 0$$

which shows that there will be no reflected wave, i.e., no reflection occurs. The incident waves of voltage and current will be entirely absorbed by the load. Such a line with the receiving-end impedance equal to its characteristic impedance is said to be *correctly terminated* or *matched*. It is also called a *flat line* or an *infinite line*.

20.13 SERIES REACTIVE TERMINATION

Suppose that a wave travels along the line of surge impedance Z_0 having a termination of a resistor R in series with an inductor L (Fig. 20.5).

$$e_f + e_r = e = iR + L\dfrac{di}{dt} \qquad\qquad\qquad ...(20.13.1)$$

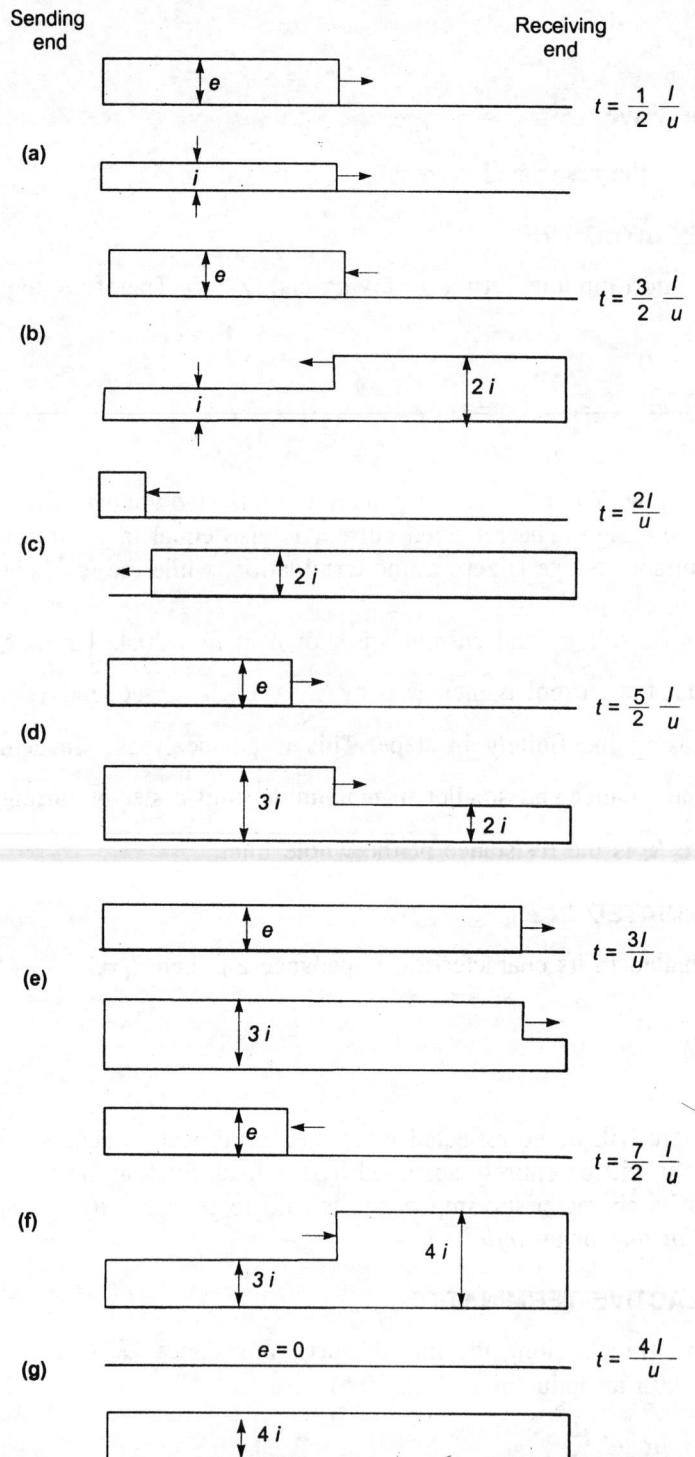

Fig. 20.4. Voltage and current distributions in a short-circuited line.

$$i_f + i_r = i \qquad \qquad \ldots(20.13.2)$$

$$i_f = \frac{e_f}{Z_0}, \qquad i_r = -\frac{e_r}{Z_0}$$

$$i_f - i_r = \frac{1}{Z_0}(e_f + e_r) = \frac{1}{Z_0}\left(iR + L\frac{di}{dt}\right) \qquad \qquad \ldots(20.13.3)$$

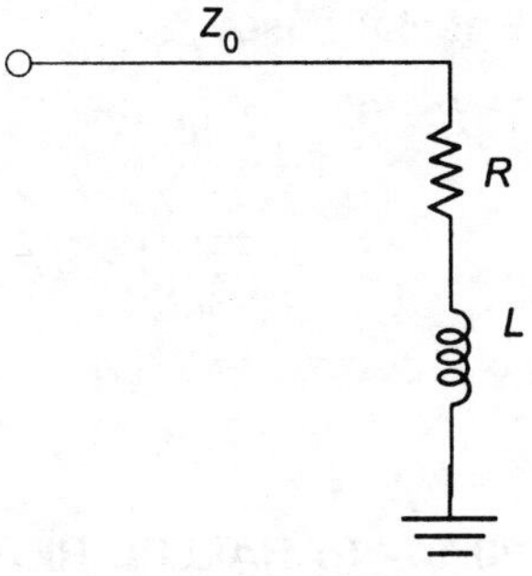

Addition of Eqs. (20.13.2) and (20.13.3) gives

$$2i_f = \left(1 + \frac{R}{Z_0}\right)i + \frac{L}{Z_0}\frac{di}{dt}$$

$$\frac{di}{dt} + \frac{R + Z_0}{L}i = \frac{2Z_0}{L}i_f \qquad \qquad \ldots(20.13.4)$$

Fig. 20.5. *R-L* termination of a line.

Equation (20.13.4) is a linear differential equation of first order of the form

$$\frac{di}{dt} + ai = b \qquad \qquad \ldots(20.13.5)$$

where $a = \dfrac{R + Z_0}{L}, \quad b = \dfrac{2Z_0}{L}i_f$

$$\frac{b}{a} = \frac{2Z_0}{R + Z_0}i_f = \frac{2e_f}{R + Z_0}$$

$$\therefore \qquad i = \frac{b}{a} + A\,\varepsilon^{-at}$$

At $t = 0, \quad i = 0, \quad A = -\dfrac{b}{a}$

$$i = \frac{b}{a}(1 - \varepsilon^{-at}) \qquad \qquad \ldots(20.13.6)$$

$$i = \frac{2e_f}{R + Z_0}(1 - \varepsilon^{-\frac{t}{\tau_L}}) \qquad \qquad \ldots(20.13.7)$$

where $\tau_L = \dfrac{1}{a} = \dfrac{L}{R + Z_0} \qquad \qquad \ldots(20.13.8)$

$$e = iR + L\frac{di}{dt} = \frac{b}{a}R(1 - \varepsilon^{-at}) + \frac{b}{a}La\varepsilon^{-at} = \frac{b}{a}[R - (R - aL)\varepsilon^{-at}]$$

$$= \frac{2e_f}{R + Z_0}(R + Z_0\,\varepsilon^{-\frac{t}{\tau_L}}) \qquad \qquad \ldots(20.13.9)$$

$$e_r = e - e_f = \frac{e_f}{R + Z_0}(R - Z_0 + 2Z_0\,\varepsilon^{-t/\tau_L}) \qquad \qquad \ldots(20.13.10)$$

The reflected voltage changes exponentially, with time constant τ_L, to a final value $\dfrac{R - Z_0}{R + Z_0}e_f$, which may be described as a reflection from a pure resistance termination. Thus, in a time τ_L the effect of inductance disappears.

Special case

If $R = 0$, i.e., the line is terminated in a pure inductor,

$$i = \frac{2e_f}{Z_0}\left(1 - \varepsilon^{\frac{-Z_0}{L}t}\right) \qquad\qquad ...(20.13.11)$$

$$e = 2e_f\varepsilon^{\frac{-Z_0}{L}t}\,) \qquad\qquad ...(20.13.12)$$

20.14 PARALLEL REACTIVE TERMINATION

Suppose that a wave travels along the line of surge impedance Z_0 having at termination of a resistor R in parallel with a capacitor C (Fig. 20.6).

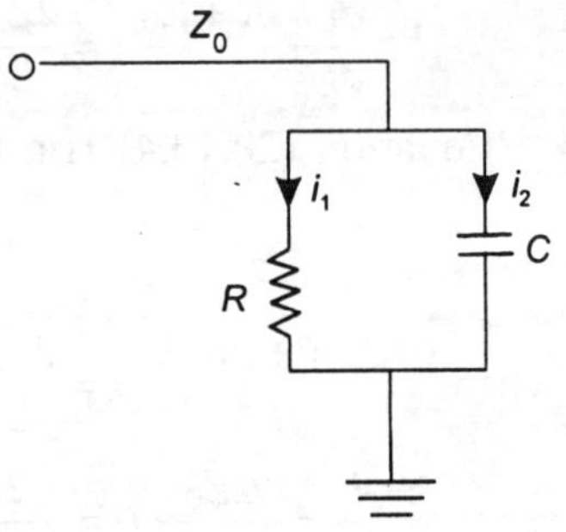

$$e_f + e_r = e \qquad\qquad ...(20.14.1)$$

$$i_f + i_r = i = i_1 + i_2 \qquad\qquad ...(20.14.2)$$

$$i_1 = \frac{e}{R}, \qquad i_2 = C\frac{de}{dt}, \qquad i_f = \frac{e_f}{Z_0}, \qquad i_r = -\frac{e_r}{Z_0}$$

Substitution of the values of i_f, i_r, i_1 and i_2 in Eq. (20.14.2) gives

Fig. 20.6. *R-C* termination of a line.

$$\frac{e_f}{Z_0} - \frac{e_r}{Z_0} = \frac{e}{R} + C\frac{de}{dt}$$

$$e_f - e_r = \frac{Z_0}{R}e + CZ_0\frac{de}{dt} \qquad\qquad ...(20.14.3)$$

Addition of Eqs. (20.14.1) and (20.14.3) gives

$$2e_f = \left(1 + \frac{Z_0}{R}\right)e + CZ_0\frac{de}{dt}$$

$$\frac{de}{dt} + \frac{R + Z_0}{CRZ_0}e = \frac{2e_f}{CZ_0} \qquad\qquad ...(20.14.4)$$

This equation is of the form

$$\frac{de}{dt} + ae = b \qquad\qquad ...(20.14.5)$$

where $a = \dfrac{R + Z_0}{CRZ_0}$, $\quad b = \dfrac{2e_f}{CZ_0}$.

The solution of Eq. (20.14.5) is

$$e = \frac{b}{a} + A\,\varepsilon^{-at}$$

Assuming that the capacitor cannot charge instantaneously, so that

$$e = 0 \text{ at } t = 0$$

$\therefore$ $\qquad A = -\dfrac{b}{a}$ and $e = \dfrac{b}{a}(1 - \varepsilon^{-at})$

$$e = \dfrac{2R}{R + Z_0} e_f (1 - e^{-t/\tau_c}) \qquad\qquad\qquad ...(20.14.6)$$

where $\tau_c = \dfrac{1}{a} = \dfrac{CRZ_0}{R + Z_0}$ = time constant.

Equation (20.14.6) shows that the voltage across the load increases exponentially, with time constant τ_c, to a final value $\dfrac{2R}{R + Z_0} e_f$. This value is obtained in the absence of the capacitor. Thus, the effect of capacitor, is to prevent the instantaneous change of voltage at the load. Initially, the capacitor behaves as short circuited.

20.15   LINE TERMINATED IN A CAPACITOR

Here, $\quad e_f + e_r = e$ $\qquad\qquad\qquad\qquad\qquad\qquad\qquad ...(20.15.1)$

$\qquad\quad i_f + i_r = i$ $\qquad\qquad\qquad\qquad\qquad\qquad\qquad ...(20.15.2)$

$$\dfrac{e_f}{Z_0} + \dfrac{e_r}{Z_0} = \dfrac{e}{Z_0} \qquad\qquad\qquad\qquad ...(20.15.3)$$

$$\dfrac{e_f}{Z_0} - \dfrac{e_r}{Z_)} = C\dfrac{de}{dt} \qquad\qquad\qquad\qquad ...(20.15.4)$$

$$\dfrac{2e_r}{Z_0} = \dfrac{e}{Z_0} + C\dfrac{de}{dt}$$

$$\dfrac{de}{dt} + \dfrac{1}{CZ_0}e = \dfrac{2}{CZ_0}e_f \qquad\qquad\qquad ...(20.15.5)$$

The equation is of the form $\quad \dfrac{de}{dt} + ae = b$

The solution is $\quad e = \dfrac{b}{a} + A\,\varepsilon^{-at}$

Assuming that the capacitor cannot charge instantaneously, so that $e = 0$ at $t = 0$,

$\therefore$ $\qquad A = -\dfrac{b}{a}$

and $\qquad e = \dfrac{b}{a}(1 - \varepsilon^{-at})$

or $\qquad e = 2e_f\left(1 - \varepsilon^{-\frac{t}{CZ_0}}\right) \qquad\qquad\qquad ...(20.15.6)$

$\qquad\qquad e_r = e - e_f$

$$e_r = e_f\left(1 - 2\varepsilon^{-\frac{t}{CZ_0}}\right) \qquad\qquad\qquad ...(20.15.7)$$

$$i = C\dfrac{de}{dt} = \dfrac{2e_f}{Z_0}\varepsilon^{-\frac{t}{CZ_0}} \qquad\qquad\qquad ...(20.15.8)$$

At $t = 0$, $e = 0$, $\qquad e_r = -e_f$, $\quad i = \dfrac{2e_f}{Z_0} = 2i_f$

Thus, if the line is terminated in a capacitor a charging current flows through it. At the moment of the arrival of the surge the capacitor behaves as short circuited so that the terminal voltage is zero. The reflected voltage is negative and the current at the load is momentarily doubled. The terminal voltage increases with the charging of the capacitor. Finally, when the capacitor is fully charged it behaves as an open circuit. The terminal current becomes zero and the terminal voltage is doubled.

20.16 JUNCTION OF TWO DISSIMILAR LINES

If a surge travelling along a line A of surge impedance Z_1 meets a junction of a line B of surge impedance Z_2, there will be a partial reflection of wave at the junction and partially the surge will be transmitted into the line B.

Let e_f, i_f = incident waves of the voltage and current

$\quad e_r$, i_r = reflected waves of the voltage and current

$\quad e_t$, i_t = voltage and current waves transmitted into the line.

Then $\qquad \left. \begin{aligned} i_f &= \frac{e_f}{Z_1} \\[2mm] i_t &= \frac{e_t}{Z_2} \\[2mm] i_r &= -\frac{e_r}{Z_1} \end{aligned} \right\}$ $\qquad\qquad$ ...(20.16.1)

Since the total current in the two lines at the junction is the same,

$$i_t = i_f + i_r \qquad\qquad \text{...(20.16.2)}$$

Also, the total voltage at the junction will be equal to the algebraic sum of the incident and reflected voltages, i.e.,

$$e_t = e_f + e_r \qquad\qquad \text{...(20.16.3)}$$

Substituting the values of currents obtained from Eq. (20.16.1) in Eq. (20.16.2) we get

$$\frac{e_t}{Z_2} = \frac{e_f}{Z_1} - \frac{e_r}{Z_1} \qquad\qquad \text{...(20.16.4)}$$

Substituting for e_r from Eq. (20.16.3) in Eq. (20.16.4) we obtain

$$\frac{e_t}{Z_2} = \frac{e_f}{Z_1} - \frac{e_t - e_f}{Z_1}$$

$$\frac{e_t}{Z_2} + \frac{e_t}{Z_1} = \frac{2e_f}{Z_1}$$

$$\left(\frac{Z_1 + Z_2}{Z_1 Z_2} \right) e_t = \frac{2e_f}{Z_1}$$

Therefore, the voltage transmitted in B is

$$e_t = \frac{2Z_2}{Z_1 + Z_2}\, e_f = \tau_v\, e_f \qquad \qquad ...(20.16.5)$$

where, τ_v = transmission factor for voltage.

$$\therefore \qquad \tau_v \overset{\Delta}{=} \frac{e_t}{e_f} = \frac{2Z_2}{Z_1 + Z_2} \qquad \qquad ...(20.16.6)$$

The current transmitted in line B is

$$i_t = \frac{e_t}{Z_2} = \frac{2e_f}{Z_1 + Z_2} = \frac{2Z_1}{Z_1 + Z_2}\, i_f = \tau_i\, i_f$$

where τ_i = transmission factor for current.

$$\therefore \qquad \tau_i \overset{\Delta}{=} \frac{i_t}{i_f} = \frac{2Z_1}{Z_1 + Z_2} \qquad \qquad ...(20.16.7)$$

Voltage reflected in line A is given by

$$e_r = e_t - e_f = \frac{2Z_2}{Z_1 + Z_2}\, e_f - e_f$$

$$e_r = \frac{Z_2 - Z_1}{Z_1 + Z_2}\, e_f = \rho_v\, e_f \qquad \qquad ...(20.16.8)$$

where ρ_v = reflection factor for voltage.

$$\therefore \qquad \rho_v \overset{\Delta}{=} \frac{e_r}{e_f} = \frac{Z_2 - Z_1}{Z_1 + Z_2} \qquad \qquad ...(20.16.9)$$

Current reflected in line A is given by

$$i_r = i_t - i_f = \frac{Z_1 - Z_2}{Z_1 + Z_2}\, i_f = \rho_i\, i_f$$

where ρ_i = refection factor for current.

$$\therefore \qquad \rho_i \overset{\Delta}{=} \frac{i_r}{i_f} = \frac{Z_1 - Z_2}{Z_1 + Z_2} \qquad \qquad ...(20.16.10)$$

It is seen that

$$\rho_i = -\rho_v \qquad \qquad ...(20.16.11)$$

Example 20.1 An overhead line is connected in series with a cable. The overhead line has an inductance of 2 mH/km and capacitance of 0.01 μF/km. The cable has an inductance of 0.25 mH/km and capacitance of 0.102 μF/km. If a surge having a maximum value of 100 kV travels along the overhead line towards its junction with the cable, calculate :

(a) the surge impedances of the line and the cable,

(b) the velocities of wave propagation in the line and the cable,

(c) the reflected and transmitted waves of voltage and current at the junction.

If the 100 kV surge originates in the cable, calculate the reflected and transmitted waves of voltage and current at the junction.

Solution

$$\text{Surge impedance of overhead line, } Z_1 = \sqrt{\frac{L_1}{C_1}} = \left(\frac{2 \times 10^{-3}}{0.01 \times 10^{-6}}\right)^{1/2} = 447 \ \Omega$$

$$\text{Surge impedance of the cable, } Z_2 = \sqrt{\frac{L_2}{C_2}} = \left(\frac{0.25 \times 10^{-3}}{0.102 \times 10^{-6}}\right)^{1/2} = 49.5 \ \Omega$$

$$\text{Velocity of the wave in the overhead line, } u_1 = \frac{1}{\sqrt{L_1 \ C_1}} = \frac{1}{(2 \times 10^{-3} \times 0.01 \times 10^{-6})^{1/2}}$$
$$= 2.24 \times 10^5 \text{ km/s}$$

$$\text{Velocity of the wave in cable, } u_2 = \frac{1}{\sqrt{L_2 \ C_2}} = \frac{1}{(0.25 \times 10^{-3} \times 0.102 \times 10^{-6})^{1/2}}$$
$$= 1.98 \times 10^5 \text{ km/s}$$

Waves originating in overhead line

When the waves originate in the overhead line and travel towards the cable we shall take $Z_1 = 447 \ \Omega$, $Z_2 = 49.5 \ \Omega$.

Incident values

Incident voltage in the overhead line, $e_f = 100$ kV

$$\text{Incident current in the overhead line, } i_f = \frac{e_f}{Z_1} = \frac{100 \times 1000}{447} = 223.7 \text{ A}$$

Transmitted values

$$\text{Line-to-cable transmission factor for voltage, } \tau_{v1} = \frac{2Z_2}{Z_1 + Z_2} = \frac{2 \times 49.5}{447 + 49.5} = 0.1994$$

Transmitted voltage in the cable, $e_t = \tau_{v1} \ e_f = 0.1994 \times 100 = 19.94$ kV

$$\text{Line-to-cable transmission factor for current, } \tau_{i1} = \frac{i_t}{i_f} = \frac{2Z_1}{Z_1 + Z_2} = \frac{2 \times 447}{447 + 49.5} = 1.8006$$

Transmitted current in the cable, $i_t = \tau_{i1} \ i_f = 1.8006 \times 223.7 = 402.8$ A

$$\text{Alternatively, } i_t = \frac{e_t}{Z_2} = \frac{19.94 \times 1000}{49.5} = 402.8 \text{ A}$$

Reflected values

$$\text{Reflection factor for voltage, } \rho_{v1} = \frac{Z_2 - Z_1}{Z_1 + Z_2} = \frac{49.5 - 447}{447 + 49.5} = -0.8006$$

Voltage reflected the overhead line, $e_r = \rho_{v1} \ e_f = -0.8006 \times 100 = -80.06$ kV

$$\text{Reflection factor for current, } \rho_{i1} = \frac{Z_1 - Z_2}{Z_1 + Z_2} = -\rho_{v1} = 0.8006$$

Current reflected back in the overhead line, $i_r = \rho_{i1} \ i_f = 0.8006 \times 223.7 = 179.1$ A

$$\text{Alternatively, } i_r = -\frac{e_r}{Z_1} = \frac{80.06 \times 1000}{447} = 179.1 \text{ A}$$

Check

$$e_f + e_r = e_t$$

$$e_t = 100 + (- 80.06) = 19.94 \text{ kV}$$

$$i_r = i_t - i_f = 402.8 - 223.7 = 179.1 \text{ A}$$

The voltage and current distributions of the wave reaching the junction of the line and cable are shown in Fig. 20.7.

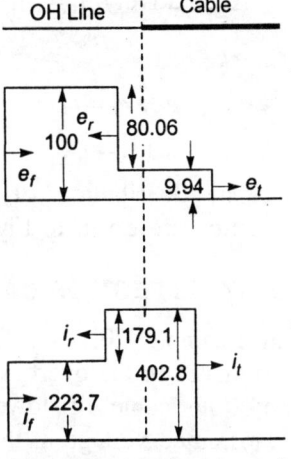

Fig. 20.7. Voltage and current distributions.

Waves originating in cable

When the waves originate in the cable we shall take $Z_1 = 49.5 \ \Omega$, $Z_2 = 447 \ \Omega$

Incident values

Incident voltage in the cable, $e_f = 100 \text{ kV}$

Incident current in the cable, $i_f = \dfrac{e_f}{Z_1} = \dfrac{100 \times 1000}{49.5} = 2020 \text{ A}$

Transmitted values

Cable-to-line transmission factor for voltage, $\tau_{v2} = \dfrac{2Z_2}{Z_1 + Z_2}$

$$= \dfrac{2 \times 447}{49.5 + 447} = 1.8006$$

The voltage transmitted into the overhead line after its arrival at the junction

$$e_t = \tau_{v2} \, e_f = 1.8006 \times 100 = 180.06 \text{ kV}$$

Cable-to-line transmission factor for current, $\tau_{i2} = \dfrac{2Z_1}{Z_1 + Z_2} = \dfrac{2 \times 49.5}{49.5 + 447}$

$$= 0.1994$$

Current transmitted into the overhead line, $i_t = \tau_{i2} \, i_f = 0.1994 \times 2020$

$$= 402.8 \text{ A}$$

Alternatively, $i_t = \dfrac{e_t}{Z_2} = \dfrac{180.06 \times 1000}{447} = 402.8 \text{ A}$

Reflected values

Reflection factor for voltage, $\rho_{v2} = \dfrac{Z_2 - Z_1}{Z_2 + Z_1} = \dfrac{447 - 49.5}{447 + 49.5}$

$$= 0.8006$$

Reflected voltage into the cable, $e_r = \rho_{v2} \, e_f = 0.8006 \times 100$

$$= 80.06 \text{ kV}$$

Reflection factor for current, $\rho_{i2} = \dfrac{Z_1 - Z_2}{Z_1 + Z_2} = - \rho_{v2}$

$$= - 0.8006$$

Reflected current in the cable, $i_r = \rho_{i2} \, i_f = - 0.8006 \times 2020$

$$= - 1617.2 \text{ A}$$

Alternatively, $i_r = -\dfrac{e_r}{Z_1} = -\dfrac{80.06 \times 1000}{49.5} = -1617.3$ A

Check

$$e_f + e_r = e_t; \qquad 100 + 80.06 = 180.06$$

$$i_f + i_r = i_t; \qquad 2020 - 1617.2 = 402.8$$

The distribution of voltage and current waves reaching the junction are given in Fig. 20.8.

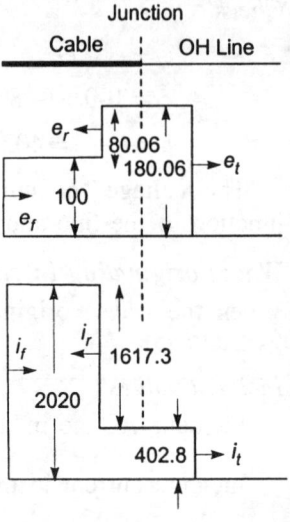

Fig. 20.8. Voltage and current distribution.

20.17 EFFECT OF CABLE ON SURGE

In Example 20.1 we have seen that in the first case the surge originates in the overhead line of higher surge impedance and enters into the cable of lower surge impedance. It is found that the transmitted voltage in the cable is much reduced and the transmitted current is increased. In the second case the surge travels from the cable to the overhead line; the transmitted voltage is increased and the transmitted current reduced. The connection of a short length of a cable at the termination of an overhead line with a substation serves to protect the terminal equipment there from dangerous values of voltage rises. The length of the cable in the circuit is also important. Unless the cable length is more than the length of the incident wave, the peak value of the wave at the junction will be higher than that of the incident wave. This is due to the repeated reflections at the junction. Thus, instead of absorbing the surge, the cable of short length shall prove more dangerous in the matter of building up of voltage which is greater than the original incident voltage wave. The cable reduces the front steepness of a surge, and reduces its peak value depending on the ratio of the propagation time in the cable to the duration of the surge.

Example 20.2 An overhead line of surge impedance 400 ohms is connected in series with a cable of surge impedance 40 ohms. The cable is 5 km in length and is open at its far end. A surge of peak value 100 kV travels along the overhead line towards the junction of the line and the cable. Find the voltage distribution at the instants 1 and 50 microseconds after the surge reaches the junction. The velocities of propagation in the overhead line and the cable are 3×10^5 km/s and 1.4×10^5 km/s respectively.

Solution

In the problem

$$Z_1 = 400 \ \Omega, \ Z_2 = 40 \ \Omega$$

Line-to-cable reflection coefficient, $\rho_1 = \dfrac{Z_2 - Z_1}{Z_2 + Z_1} = \dfrac{40 - 400}{40 + 400} = -0.818$

Reflected voltage in overhead line, $e_r = \rho_1 \ e_f = -0.818 \times 100 = -81.8$ kV

Transmission factor for voltage, $\tau_v = \dfrac{2Z_2}{Z_1 + Z_2} = \dfrac{2 \times 40}{400 + 40} = 0.182$

Therefore the voltage transmitted into the cable, $e_t = \tau_v \ e_f = 0.182 \times 100 = 18.2$ kV

Alternatively, $e_t = e_f + e_r = 100 + (-81.8) = 18.2$ kV

The voltage at the junction just after the surge strikes the junction is, therefore, 18.2 kV. The distance travelled by the transmitted wave into the cable in one microsecond

$$= 1.4 \times 10^8 \times 10^{-6} = 140 \text{ m}$$

The distance travelled by the reflected wave into the overhead line in one microsecond

$$= 3 \times 10^8 \times 10^{-6} = 300 \text{ m}$$

The voltage distribution after 1 μs is given in Fig. 20.9.

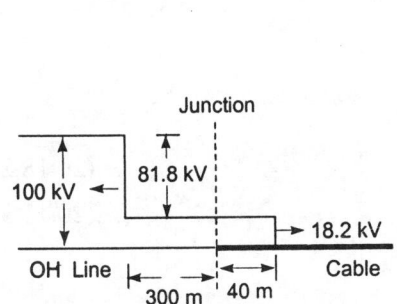

Fig. 20.9. Voltage distribution after 1 μs.

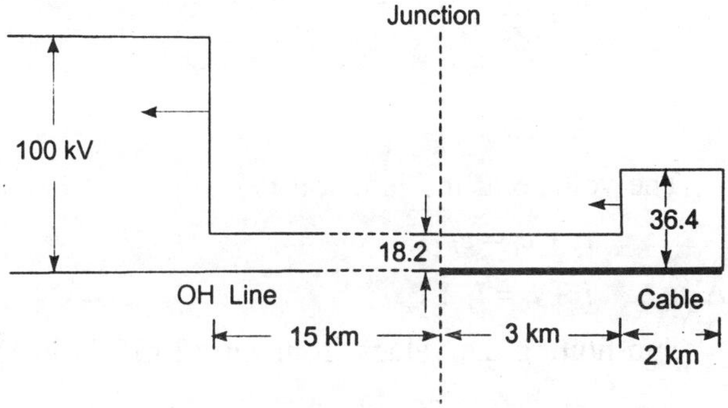

Fig. 20.10. Voltage distribution after 50 μs.

Time required by the transmitted wave to reach the open end of the cable

$$= \frac{5 \times 1000}{1.4 \times 10^5 \times 10^3} \text{ s} = \frac{5 \times 10^6}{1.4 \times 10^5} = 35.7 \text{ μs}$$

When this voltage reaches the open end of the cable it is reflected and is doubled. In the remaining time of $(50 - 35.7)$, i.e., 14.3 μs the distance covered by the wave reflected at the open end is $(14.3 \times 10^{-6} \times 1.4 \times 10^5)$, i.e., 2 km from the free end of the cable towards the junction.

The distance covered by the reflected wave in the overhead line is 50 μs

$$= 3 \times 10^5 \times 50 \times 10^{-6} = 15 \text{ km}$$

The voltage distribution after 50 μs the original surge reaches the junction of the line and the cable is shown in Fig. 20.10.

20.18 JUNCTION OF SEVERAL LINES

Now consider the case of a surge travelling along a line of surge impedance Z_1 reaching a junction of several lines of surge impedance Z_2, Z_3, Z_4, etc. The wave will be partially reflected on the line of surge impedance Z_1 and partially transmitted in other branches. For the sake of simplicity, let a line OA be bifurcated into two branches AB and AC (Fig. 20.11) of surge impedances

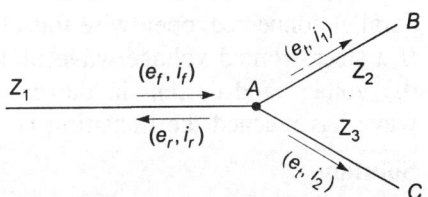

Fig. 20.11. Forked line.

Z_2 and Z_3 respectively. Since AB and AC are in parallel, the voltages transmitted in them will be the same.

Let i_1 and i_2 be the currents transmitted in branches AB and AC. The following relations may be written on the basis of the arguments in the previous sections :

$$\left.\begin{array}{l} i_f = \dfrac{e_f}{Z_1} \\[3mm] i_r = \dfrac{-e_r}{Z_1} \\[3mm] i_1 = \dfrac{e_t}{Z_2} \\[3mm] i_2 = \dfrac{e_t}{Z_3} \end{array}\right\} \qquad \qquad \text{...(20.18.1)}$$

The voltage at the junction is

$$e_f + e_r = e_t \qquad \qquad \text{...(20.18.2)}$$

Also, $i_f + i_r = i_1 + i_2$ 　　　　　　　　　　...(20.11.3)

Substituting the values from Eq. (20.18.1) in Eq. (20.18.3).

$$\frac{e_f}{Z_1} - \frac{e_r}{Z_1} = \frac{e_t}{Z_2} + \frac{e_t}{Z_3} \qquad \qquad \text{...(20.18.4)}$$

Putting the values of e_r from Eq. (20.18.2) in Eq. (20.18.4)

$$\frac{e_f}{Z_1} - \frac{e_t - e_f}{Z_1} = \frac{e_t}{Z_2} + \frac{e_t}{Z_3}$$

which on simplification gives

$$e_t = \frac{\dfrac{2e_f}{Z_1}}{\dfrac{1}{Z_1} + \dfrac{1}{Z_2} + \dfrac{1}{Z_3}} = 2e_f \frac{\dfrac{1}{Z_1}}{\dfrac{1}{Z_1} + \dfrac{1}{Z_2} + \dfrac{1}{Z_3}} \qquad \qquad \text{...(20.18.5)}$$

The reflected voltage, $e_r = e_t - e_f = e_f \dfrac{\dfrac{1}{Z_1} - \dfrac{1}{Z_2} - \dfrac{1}{Z_3}}{\dfrac{1}{Z_1} + \dfrac{1}{Z_2} + \dfrac{1}{Z_3}}$ 　　　　...(20.18.6)

Example 20.3 A cable with a surge impedances of 100 ohms is terminated in two parallel-connected, open-wire lines having surge impedances of 600 and 1000 ohms respectively. If a steep-fronted voltage wave of 1000 V travels along the cable, find from the first principles the voltage and current in the cable and the open-wire lines immediately after the travelling wave has reached the transition point. The line may be assumed to be of infinite length.

Solution

In the problem

$$Z_1 = 100 \ \Omega, \ Z_2 = 600 \ \Omega, \ Z_3 = 1000 \ \Omega \text{ and } e_f = 1000 \text{ V.}$$

$$i_f + i_r = i_1 + i_2$$

$$\frac{e_f}{Z_1} - \frac{e_r}{Z_1} = \frac{e_t}{Z_2} + \frac{e_t}{Z_3}$$

$$e_f - e_r = e_t \left(\frac{1}{Z_2} + \frac{1}{Z_3} \right) Z_1 = e_t \left(\frac{1}{600} + \frac{1}{1000} \right) \times 100 = \frac{4}{15} e_t$$

Also, $\quad e_f + e_r = e_t$

By addition

$$2e_f = \frac{4}{15} e_t + e_t = \frac{19}{15} e_t$$

$$e_t = \frac{15 \times 2}{19} e_f = \frac{30}{19} \times 1000 = 1579 \text{ V}$$

Reflected voltage, $e_r = e_t - e_f = 1579 - 1000 = 579 \text{ V}$

Current transmitted in the 600 ohm line, $i_1 = \dfrac{e_t}{Z_2} = \dfrac{1579}{600} = 2.63 \text{ A}$

Current transmitted in the 1000 ohm line, $i_2 = \dfrac{e_t}{Z_3} = \dfrac{1579}{1000} = 1.58 \text{ A}$

Current in the cable $= i_1 + i_2 = 2.63 + 1.58 = 4.21 \text{ A}$

Reflected current, $i_r = -\dfrac{e_r}{Z_1} = -\dfrac{579}{1000} = -5.79 \text{ A}$

Incident current, $i_f = \dfrac{e_f}{Z_1} = \dfrac{1000}{100} = 10 \text{ A}$

Current in the cable, $i_f + i_r = 10 - 5.79 = 4.21 \text{ A}$

This result is the same as obtained above. It serves as a check.

20.19 REPEATED REFLECTIONS

Surges on a power system are subject to repeated reflected at terminations, junctions, towers, and similar discontinuities. We shall consider the case of a short cable connected to an overhead line.

Let Z_1 = surge impedance of the overhead line

$\quad Z_2$ = surge impedance of the cable

$\quad \rho_1$ = line–to–cable reflection factor

$\quad \tau_1$ = line–to–cable transmission factor

$\quad \rho_2$ = cable–to–line reflection factor

$\quad \tau_2$ = cable–to–line transmission factor

$\quad \rho_c$ = reflection factor at the far end of the cable

$\quad \tau_c$ = transmission factor at the far end of the cable.

Then, $\quad \rho_1 = \dfrac{Z_2 - Z_1}{Z_2 + Z_1}$ ...(20.19.1)

$$\tau_1 = \frac{2Z_2}{Z_1 + Z_2} \qquad \qquad \ldots(20.19.2)$$

$$\rho_2 = \frac{Z_1 - Z_2}{Z_1 + Z_2} \qquad \qquad \ldots(20.19.3)$$

$$\tau_2 = \frac{2Z_1}{Z_1 + Z_2} \qquad \qquad \ldots(20.19.4)$$

If the far end of the cable is open $\rho_c = 1$ and $\tau_c = 0$

Suppose that a rectangular wave of amplitude e_1 originates in the overhead line. On reaching the junction of the line and the cable, the wave is divided into two parts. One of the waves is reflected back into the line and the other is transmitted into the cable.

The transmitted wave in the cable travels along the cable and it reaches its far end, where it is reflected back. After covering a distance of twice the length of the cable the wave reaches back the junction. At the junction once again this wave is split up into two parts, one transmitted along the line and the other reflected back into the cable. This process continues. The amplitudes of the successive waves in the cable go on decreasing.

Bewley's lattice diagram provides a simple and convenient method to study the effects of these multiple or repeated reflections. It gives the picture of the position and direction of every incident, reflected, and transmitted wave on the system at every instant. Lattice diagram is a distance-time graph. The distance along the system in the direction of original surge is taken horizontally. The time scale is taken vertically downwards. The slopes of the line give the propagation velocities in the system. The sloping lines thus form a lattice in the distance-time plane. The amplitude of each wave is marked on its line in terms of the amplitude of original wave and the coefficients of reflection and transmission.

Fig. 20.12 shows the lattice or zig-zag diagram. In the diagram AB represents the original wave of amplitude e_1. The slope of AB gives the velocity of propagation along the overhead line. When the surge reaches the junction B it splits up into waves BC and BD. The wave $\rho_1 e_1$ represented by BC is reflected back in the overhead line and the wave $\tau_1 e_1$ represented

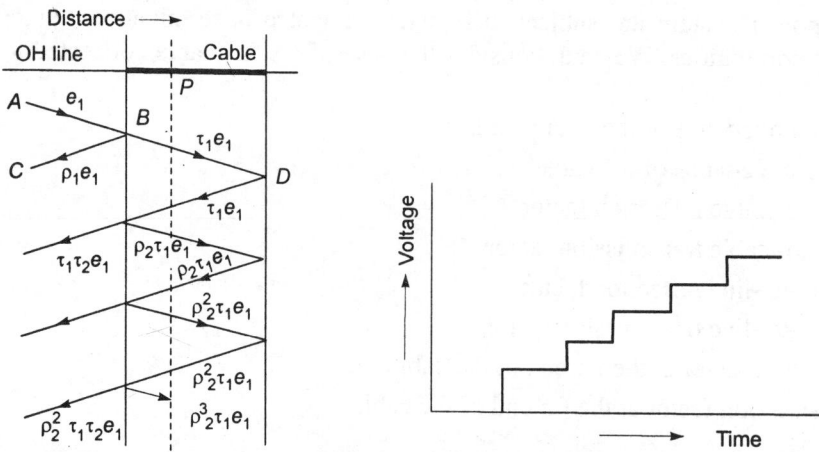

Fig. 20.12. Lattice diagram.

by *BD* in the diagram is transmitted into the cable. The slope of *BD* gives velocity in the cable. The process is repeated. The voltage or current at any distant can be found from the diagram.

The incremental waves are projected on the time scale. The voltage at any point *P* is shown in the diagram to the right of the lattice diagram. The lattice diagram for current can also be drawn in a similar manner.

Example 20.4 Explain the terms travelling wave and standing wave as applied to transmission lines.

Two long transmission lines each having a surge impedance of 400 Ω are connected by a cable having surge impedance of 50 Ω. If a short pulse of magnitude 10 kV travels along the first line towards the junction, determine from first principles the magnitudes of the first and second pulses entering the second line. State any assumption made.   (I.E.E.)

Solution

The problem may be worked out from first principles as given in Section 20.16. It is therefore, left to the reader to solve it by that method. Here it shall be solved with the help of Bewley's lattice diagram shown in Fig. 20.13.

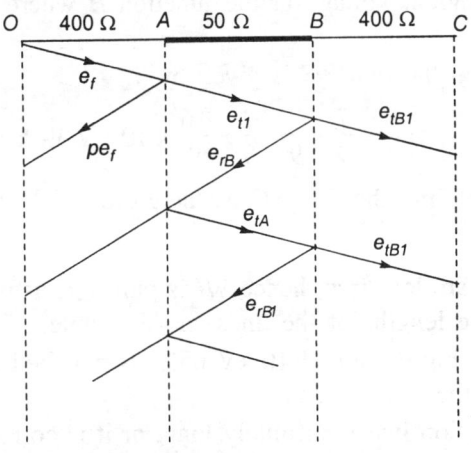

Fig. 20.13.

Let the two lines *OA* and *BC* be joined by the cable *AB*. The surge impedances of *OA*, *AB* and *BC* are 400, 50 and 400 ohms respectively.

Transmission coefficient for voltage at junction *A* , $\tau_1 = \dfrac{2Z_2}{Z_1 + Z_2} = \dfrac{2 \times 50}{400 + 50} = \dfrac{2}{9}$

Reflected coefficient for voltage at junction *A* , $\rho_1 = \dfrac{Z_2 - Z_1}{Z_2 + Z_1} = \dfrac{50 - 400}{50 + 400} = -\dfrac{7}{9}$

Transmission coefficient for voltage at junction *B* , $\tau_2 = \dfrac{2Z_3}{Z_3 + Z_2} = \dfrac{2 \times 400}{400 + 50} = \dfrac{16}{9}$

Reflection coefficient for voltage at junction *B* , $\rho_2 = \dfrac{Z_3 - Z_2}{Z_3 + Z_2} = \dfrac{400 - 50}{400 + 50} = \dfrac{7}{9}$

At junction A, the incident wave e_f is partially reflected in the transmission line OA, and partially it is transmitted in the cable AB.

The voltage transmitted in $AB = e_{t1} = \tau_1 e_f$

e_{t1} now becomes the incident voltage in the cable towards the junction B. At B the voltage e_{t1} is partially transmitted in the line BC and partially reflected in the cable.

The transmitted part which is the first pulse entering the line BC is e_{tB}.

$$e_{tB} = \tau_2 e_{t1} = \tau_2 \tau_1 e_f = \frac{16}{9} \times \frac{2}{9} \times 10 = 3.95 \text{ kV}$$

The reflected component of e_{t1} is e_{rB}. It is given by

$$e_{rB} = \rho_2 e_{t1}$$

e_{rB} now becomes the incident voltage in the cable towards the junction A. At A the voltage e_{rB} is partially transmitted in the line OA and partially it is reflected in the cable. The reflected voltage e_{tA} is given by

$$e_{tA} = \rho_1 e_{rB} = \rho_1 \rho_2 e_{t1} = \rho_1 \rho_2 \tau_1 e_f$$

e_{tA} now becomes the incident voltage at the junction B where it is partially reflected and partially transmitted.

The voltage transmitted in the line BC is given by

$$e_{tB1} = \tau_2 e_{tA} = \rho_1 \rho_2 \tau_1 \tau_2 e_f = \frac{7}{9} \times \frac{7}{9} \times \frac{2}{9} \times \frac{16}{9} \times 10 = 2.39 \text{ kV}$$

The second pulse e_{tB1} entering the line BC is, therefore, 2.39 kV.

Assumptions

1. The lines and cable are loss-free, hence V/I is purely resistive and equal to Z_0 at all the points throughout the lengths of the lines and the cable.
2. The generator supplying the initial 10 kV pulse is matched to the line, so that there are no reflections from the sending end.
3. The second transmission line is infinitely long, or it is correctly terminated. Thus, there are no reflections from the transmission line towards the junction B.

Example 20.5 Show from basic considerations that the ratio of voltage to current in a surge on a uniform, loss-free, transmission line of inductance L henrys and capacitance C farads per metre length is given by

$$\frac{v}{i} = Z_0 = \sqrt{\frac{L}{C}}$$

Find also an expression for the velocity of propagation.

A long-tailed unit function 500 kV surge voltage on an overhead line of surge impedance 400 Ω arrives at a point where the line continues into a cable, 8 km long and having a total inductance of 265 μH and total capacitance of 0.165 μF. At the end of the cable connection is made to a transformer of surge impedance 1000 Ω. Find the surge voltage distribution 12 μs after the surge arrives at the line cable junction. (I.E.E.)

Solution

The surge impedance of the cable is

$$Z_2 = \sqrt{\frac{L}{C}} = \left(\frac{265 \times 10^{-6}}{0.165 \times 10^{-6}}\right)^{1/2} = 40 \ \Omega$$

Velocity of the surge in the cable

$$u_2 = \frac{1}{\sqrt{LC}} = \frac{1}{(265 \times 10^{-6} \times 0.165 \times 10^{-6} \times 10^{-3} \times 10^{-3})^{1/2}} \ \text{m/s}$$

$$= 0.151 \times 10^6 \ \text{km/s}$$

The distance travelled by the wave in 12 µs in the cable $= 12 \times 0.151 = 1.812$ km.

Thus, the wave reaches the end C of the cable and is reflected. The reflected wave travels a distance CD equal to 0.812 km towards the line-cable junction.

The incident voltage in the cable

$$e_{t1} = \frac{2Z_2}{Z_1 + Z_2} \ e_f = \frac{2 \times 40}{400 + 40} \times 500 = 91 \ \text{kV}$$

The reflected voltage at the transformer end of the cable

$$e_{rB} = \rho_2 \ e_{t1} = \frac{Z_3 - Z_2}{Z_3 + Z_2} \times e_{t1} = \frac{1000 - 40}{1000 + 40} \times 91 = 84 \ \text{kV}$$

Hence the voltage in CD = incident voltage + reflected voltage $= 91 + 84 = 175$ kV

From the point D upto B, the voltage is equal to 91 kV. Fig. 20.14 shows the surge voltage distribution.

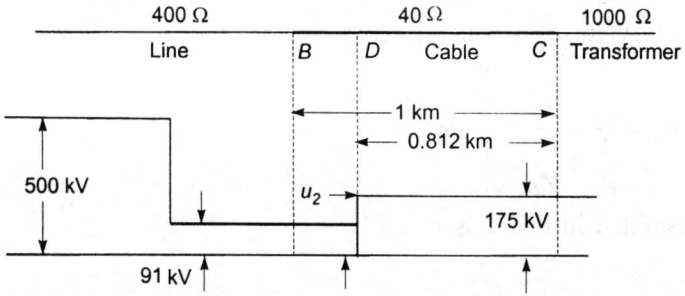

Fig. 20.14. Surge voltage distribution.

Example 20.6 Two single lines A and B with earth return are connected in series, and at the junction a resistance of 2000 Ω is connected between the lines and the earth. The surge impedance of line A is 400 Ω and of line B is 600 Ω. A rectangular wave having an amplitude of 100 kV travels towards the junction from line A.

Develop expressions for, and determine the magnitude of, the voltage and current waves reflected from and transmitted beyond the junctions. What value of resistance at the junction makes the magnitude of the transmitted wave 100 kV.   (L.U.)

Solution

Since the voltage in both the lines is the same,

$$e_1 = e_2$$

$$e_{f1} + e_{r1} = e_{f2} \qquad \qquad \qquad \dots(E20.6.1)$$

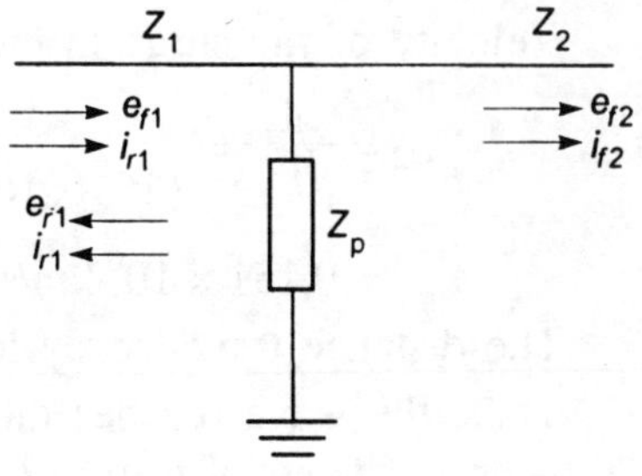

The currents in the two lines differ by the amount of current flowing in Z_p.

$$i_1 = i_2 + \frac{e_2}{Z_p}$$

$$i_{f1} + i_{r1} = i_{f2} + \frac{e_{f2}}{Z_p}$$

$$\frac{e_{f1}}{Z_1} - \frac{e_{r1}}{Z_1} = \frac{e_{f2}}{Z_p} + \frac{e_{f2}}{Z_2}$$

$$e_{f1} - e_{r1} = Z_1 \left(\frac{1}{Z_2} + \frac{1}{Z_p} \right) e_{f2} \qquad \qquad \dots(E20.6.2)$$

Fig. 20.15.

Addition of Eqs. (E20.6.1) and (E20.6.2) gives

$$2e_{f1} = \left[1 + \left(\frac{1}{Z_2} + \frac{1}{Z_p} \right) Z_1 \right] e_{f2}$$

Therefore, the forward travelling surge voltage in line 2 is given by

$$e_{f2} = \frac{\dfrac{2}{Z_1}}{\dfrac{1}{Z_1} + \dfrac{1}{Z_2} + \dfrac{1}{Z_p}} \, e_{f1} \qquad \qquad \dots(E20.6.3)$$

The reflected surge voltage in line 1 is given by

$$e_{r1} = e_{f2} - e_{f1} = \frac{\dfrac{1}{Z_1} - \dfrac{1}{Z_2} - \dfrac{1}{Z_p}}{\dfrac{1}{Z_1} + \dfrac{1}{Z_2} + \dfrac{1}{Z_p}} \, e_{f1} \qquad \qquad \dots(E20.6.4)$$

The current transmitted in line 2 is given by

$$i_{f2} = \frac{e_{f2}}{Z_2}$$

The reflected current in the line 1 is given by

$$i_{r1} = \frac{e_{r1}}{Z_1}$$

The voltage transmitted in line 2 is

$$e_{f2} = \frac{\dfrac{2}{Z_1}}{\dfrac{1}{Z_1} + \dfrac{1}{Z_2} + \dfrac{1}{Z_p}} \, e_{f1} = \frac{\dfrac{2}{400}}{\dfrac{1}{400} + \dfrac{1}{600} + \dfrac{1}{2000}} \times 100 = \frac{0.5 \times 100}{\dfrac{1}{4} + \dfrac{1}{6} + \dfrac{1}{20}} = \frac{50 \times 60}{15 + 10 + 3} = 107 \text{ kV}$$

The voltage reflected in line 1 is

$$e_{r1} = e_{f2} - e_{f1} = 107 - 100 = 7 \text{ kV}$$

The current reflected in line 1 is

$$i_{r1} = \frac{e_{r1}}{Z_1} = \frac{7 \times 1000}{400} = 17.5 \text{ A}$$

Calculation of Z_p

Here $e_{f2} = 100 \text{ kV}$, $e_{f1} = 100 \text{ kV}$, $Z_1 = 400 \ \Omega$, $Z_2 = 600 \ \Omega$

Substituting the values in Eq. (20.6.3)

$$100 = \frac{\dfrac{2}{400}}{\dfrac{1}{400} + \dfrac{1}{600} + \dfrac{1}{Z_p}} \times 100 \, ; \qquad \frac{2}{400} = \frac{1}{400} + \frac{1}{600} + \frac{1}{Z_p}$$

$$\frac{1}{Z_p} = \frac{1}{400} - \frac{1}{600} = \frac{3-2}{1200}$$

$$Z_p = 1200 \ \Omega.$$

Example 20.7 An overhead transmission line with a surge impedance of 500 Ω has a load comprising a 10 kΩ resistor in parallel with a 0.005 μF capacitor connected across the far end. A surge voltage of 10 kV magnitude and unit function form travels along the line. Determine an expression for the time variation of the voltage across the load, and calculate this voltage 5 μ sec after the arrival of the wave front of the surge. State any assumption made.   (I.E.E.)

Solution

From Eq. (20.14.6)

$$e = \frac{2R \, e_f}{Z_0 + R} (1 - \varepsilon^{-t/\tau})$$

where $\tau = \dfrac{CR Z_0}{Z_0 + R}$

$$\frac{1}{\tau} = \frac{Z_0 + R}{CR Z_0} = \frac{500 + 10 \times 10^3}{0.005 \times 10^{-6} \times 10 \times 10^3 \times 500} = 0.42 \times 10^6$$

$$\frac{2R \, e_f}{Z_0 + R} = \frac{2 \times 10 \times 10^3}{500 + 10 \times 10^3} \times 10 \text{ kV} = 19.04 \text{ kV}$$

$$e = 19.04 \, (1 - \varepsilon^{-0.42 \times 10^6 \, t}) \text{ kV}$$

After 5 μ sec

$$t/\tau = 0.42 \times 10^6 \times 5 \times 10^{-6} = 2.1$$

$$e = 19.04 \, (1 - \varepsilon^{-2.1}) = 19.04 \, (1 - 0.1225) = 16.71 \text{ kV}$$

Assumptions

1. The line is lossless.
2. The generator impedance is matched to the line so that there are no reflections from the sending end.
3. The capacitor C is initially uncharged.

Example 20.8 An overhead line has a surge impedance of 400 Ω. A surge voltage $v = 250\,(e^{-0.05\,t} - e^{-t})$ kV, where t is in μs, travels along the line. The termination of the line is connected to two parallel overhead line transformer feeders. The surge impedance of each feeder is 300 Ω. The transformers are protected by surge diverters, each of surge impedance 50 Ω. Determine the maximum voltage which would initially appear across the feeder-end windings of each transformer due to the surge. Assume the transformer to have infinite surge impedance.

Solution

Fig. 20.16 shows the circuit. Since AB and AC are in parallel, the voltage transmitted in them will be the same. The transmitted voltage in AB or AC is given by Eq. (20.18.5) as

$$e_t = 2e_f \frac{\dfrac{1}{Z_1}}{\dfrac{1}{Z_1} + \dfrac{1}{Z_2} + \dfrac{1}{Z_3}} = 2v\,\frac{\dfrac{1}{400}}{\dfrac{1}{400} + \dfrac{1}{300} + \dfrac{1}{300}} = \frac{6}{11}\,v$$

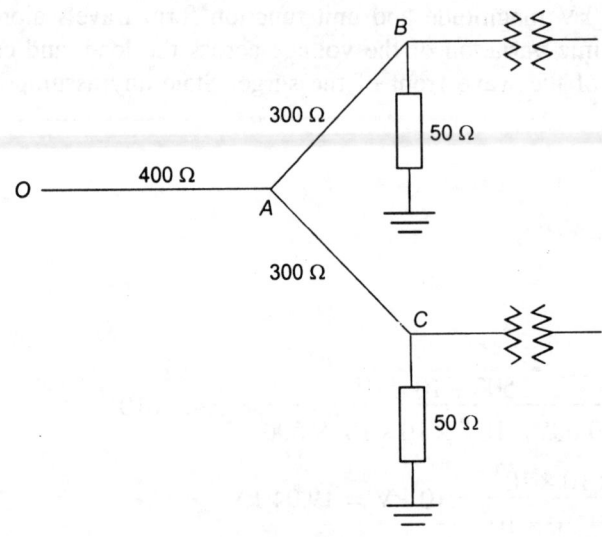

Fig. 20.16.

The conditions at junctions B and C are identical. The voltage transmitted at B or C

$$e_{t1} = e_t \times \frac{2Z_d}{Z_2 + Z_d}$$

where Z_d = surge impedance of a surge diverter = 50 Ω

$$e_{t1} = \frac{6}{11} \, v \times \frac{2 \times 50}{300 + 50} = \frac{12}{77} \, v$$

Now, $v = 250 \, (e^{-0.05\,t} - e^{-t})$

For maximum voltage to appear $\dfrac{dv}{dt}$ is zero, i.e.,

$$\frac{d}{dt} \, [250 \, (e^{-0.05\,t} - e^{-t}) = 0$$

$$- 0.05 \, e^{-0.05\,t} + e^{-t} = 0 \; ; \qquad e^{-t} = 0.05 \, e^{-0.05\,t} \; ; \qquad e^{0.95\,t} = 20$$

$\therefore$ $0.95 \, t = \ln 20, \; t = 3.1534 \; \mu s$

Putting this value in expression for v gives the maximum transmitted voltage.

Therefore maximum voltage appearing across the feeder-end windings of each transformer

$$= \frac{12}{77} \, v = \frac{12}{77} \times 250 \, (e^{-0.05 \times 3.1534} - e^{-3.1534}) = \frac{12}{77} \times 250 \, (0.8541 - 0.0427) = 31.61 \; kV.$$

EXERCISES

1. Explain what is meant by the surge impedance of a line and show upon what factors it depends.

   An overhead line of surge impedance 500 Ω terminated in a transformer of surge impedance 3500 Ω. Find the amplitudes of current and voltage surge transmitted to the transformer due to an incident voltage of 30 kV.

2. Obtain an expression for the surge impedance of a transmission line and for the velocity of propagation of electric waves in terms of the line inductance and capacitance.

   A cable having an inductance of 0.3 mH per km and a capacitance of 0.4 μF per km is connected in series with a transmission line having an inductance of 1.5 mH per km and a capacitance of 0.012 μF per km. A surge of peak value 50 kV originates in the line and progresses towards the cable. Find the voltage transmitted into cable.

3. Derive an expression for the surge impedance of a transmission line.

   A transmission line has a capacitance of 0.012 μF per km and an inductance of 1.8 mH per km. This overhead line is continued by an underground cable with a capacitance of 0.45 μF per km and an inductance of 0.3 mH per km. Calculate the maximum voltage occurring at the junction of line and cable when a 20 kV surge travels along the cable towards the overhead line.

4. A cable with a characteristic impedance of 80 ohms is joined in series with an open-wire line having a characteristic impedance of 700 ohms. If, as a consequence of connecting a direct voltage to the cable, a steep-fronted voltage wave of 1.2 kV travels along it, determine the voltage and current in the cable and open-wire line immediately after the travelling wave has reached the junction. Assume the line to be loss-free.   (I.E.E.)

5. A surge of 20 kV magnitude travels along a cable towards its junction with an overhead line. The inductance and capacitance of the cable are 0.8 mH and 1.0 μF, and of the overhead line, 6 mH and 0.05 μF respectively. Calculate the voltage rise at the junction due to the surge. (C & G)

6. Explain what is meant by the surge impedance of a transmission line and derive its value in terms of the line constants. Derive expressions for the values of transmitted and reflected waves of current and voltage relative to those of incident waves at a point where the surge impedance changes from Z_1 to Z_2.

   A rectangular wave of 200 kV amplitude travels along a line having a surge impedance of 500 ohms to a transition point where it is connected to a line of 50 ohms surge impedance. Determine the values of the transmitted and reflected voltage and current waves. (L.U.)

7. An underground cable having an inductance of 0.3 mH per km and a capacitance of 0.4 μF per km is connected in series with an overhead line having an inductance of 2.0 mH per km and a capacitance of 0.014 μF per km. Calculate the values of reflected and transmitted waves of voltage and current at the junction due to a voltage surge of 100 kV travelling to the junction (a) along the cable, and (b) along the overhead line.

   (L.U.)

8. An overhead transmission line having a surge impedance of 500 Ω is connected at one end to two underground cable lines, one having a surge impedance of 40 Ω and the other 60 Ω. A rectangular wave having a value of 100 kV travels along the overhead line to the junction. Deduce expressions for, and determine the magnitude of the voltage and current waves reflected from and transmitted beyond the junction. (L.U.)

9. An overhead transmission line has a surge impedance of 700 Ω and a voltage wave of 10000 V travelling along it. The wave is assumed to be of infinite length and the wave-front is vertical. At a certain point the overhead line terminates and the circuit is connected by two cables in parallel. The surge impedance of one cable is 100 Ω, and that of the other is 200 Ω. Calculate the voltage and current in the overhead line and in the two cables immediately after the travelling wave has reached the fork. (L.U.)

10. Two stations are connected together by an underground cable having a capacitance of 1.15 μF/km and an inductance of 0.35 mH/km joined to an overhead line having a capacitance of 0.01 μF/km and an inductance of 2 mH/km. If a surge having a steady value of 100 kV travels along the cable towards the junction with an overhead line, determine the values of the reflected and transmitted waves of voltage and current at the junction. (L.U.)

11. Two lengths of overhead line of surge impedance 400 Ω have a 2 km length of underground cable, *AB*, between them. The surge impedance of the cable is 60 Ω and the velocity of surge propagation along it is 2×10^8 m/s. Develop expressions for, and calculate the values of, the transmission and reflection coefficients of surges arriving at each of the junctions *A* and *B*.

    Plot against time a curve of voltage at junction *A* for 30 μs subsequent to the arrival of a 500 kV unit function surge from the overhead line connected to *A*. Neglect reflection from the remote ends of the overhead lines. (I.E.E.)

ANSWERS

1. 15 A, 52.5 kV
2. 7.2 kV
3. 37.5 kV
4. 2.154 kV, 3.08 A
5. 11.78 kV
6. 36.4 kV, 163.6 kV, 728 A, 327.2 A
7. (a) 86 kV, 186 kV, – 3.16 kA, 492 A; (b) – 86.5 kV, 13.5 kV, 230 A, 494 A
8. 90.8 kV, 182 A, 9.16 kV, 229 A, 153 A
9. Voltage at junction = 1.74 kV; Current in 100 Ω line = 17.4 A
   Current in 200 Ω line = 8.7 A; Current in overhead line = 26 A
10. 81 kV, 181 kV, 1.57 kA, 404 A
11. Transmission factor (line-to-cable) = 0.26; Reflection factor (cable-to-line) = 0.74
    A surge of 130 kV enters cable, 96 kV reflected from B after 10 μs, arrives at A
    after 20 μs, causes voltage to rise by 200 kV. No further change during first 30 μs.

CHAPTER 21

Overvoltage Protection

21.1 INTRODUCTION

Power systems may be subjected to overvoltages which may be of transient or persistent nature. Overvoltages stress the insulation of the equipment beyond safe limits. Their study, therefore, serves as a guide in the choice of insulation level and the requirements for protective devices so as to minimise damage of equipment. The main causes due to which overvoltages are produced in the power systems may be conveniently grouped into two categories, namely, internal and external.

21.2 INTERNAL OVERVOLTAGES

The internal overvoltages have got their origin within the system itself. They are further classified as transient, dynamic and stationary. The transient overvoltages persist for short durations. They may be caused by sudden changes in the system parameters, usually as a result of switching-in or tripping-off of highly inductive or capacitive components of the system such as transformers shunt reactors and long transmission lines. The magnitude of the switching surges depends on the system voltage, the design of the circuit breakers, and the circuit constants. For a well designed system their magnitude normally does not exceed three to four times the working voltage. The nature of these transient voltages is often oscillatory and their wave-forms vary considerably. They are propagated along the line as travelling waves and reach the terminal equipment. The insulation of the line and terminal equipment is thus subjected to high voltages. Switching surges have relatively greater importance at high voltages.

Overvoltage is also produced due to arcing faults on systems having insulated neutrals. An *insulated neutral system* is a system which has no intentional connection to earth except through indicating, measuring or protective devices of very high impedance. On such a system if one of the phase breaks down to earth due to failure of insulation, the voltage between sound conductors and earth may build up to high values. This is due to the fact that the faulty conductor comes in contact with earth, the voltage between sound conductors and earth becomes the line voltage. An arc is formed between earth and faulty conductors at the point where the insulation failed. With the formation of the arc the voltage across it becomes zero and thereby the arc is extinguished, the potential of the faulty conductor is restored and a formation of second arc takes place. This phenomenon is called *arcing ground*. The successive extinction and re-ignition of the charging current flowing in the arc may increase the potential of the other two healthy conductors appreciably due to setting up of high frequency oscillations. The overvoltage in healthy conductors may damage the insulation at some other point and thus resulting in a short circuit on the system. For this reason, it is generally avoided to operate high voltage systems with insulated neutrals.

Overvoltages produced at normal system frequency and persisting only for a few seconds are called *dynamic overvoltages*. Such voltages are encountered in generators of hydro-power plants which overspeed on disconnection or removal of a large part of the load from them.

Stationary overvoltages are also produced at system frequency and persist for long periods, perhaps for hours. These overvoltages are produced on healthy conductors under sustained earth fault conditions on one conductor in the system which is earthed through an arc suppression coil.

21.3 EXTERNAL OVERVOLTAGES

Lightning is a very serious cause of overvoltages on the lines. It is advisable at this stage to know the nature of these voltages. Lightning denotes the electrical discharge phenomena in the atmosphere which are essentially associated with thunderstorms. The discharge may take place between the thundercloud and earth, between different thunderclouds, or even between different charges in the same thunder cloud. Only the discharges from the clouds to earth are of interest to electrical engineers because such discharges affect the transmission lines and equipment.

21.4 MECHANISM OF THE LIGHTNING DISCHARGE

Many theories have been put forward to account for the accumulation of charge in clouds. The subject is still controversial. But it has been established, irrespective of the process of charge accumulation, that during thunderstorms there is a concentration of negative charge in the lower layers of the cloud, while the upper zone becomes positively charged (Fig. 21.1)

When the voltage gradient in the cloud is much greater than that at the earth's surface beneath, the discharge originates in the cloud. When the potential gradient, which is not uniform in a cloud, becomes of the order of 10-30 kV per cm in any part of the cloud, an initial discharge, called the *pilot streamer* or *pilot leader*, moves slowly from it towards the earth. The first

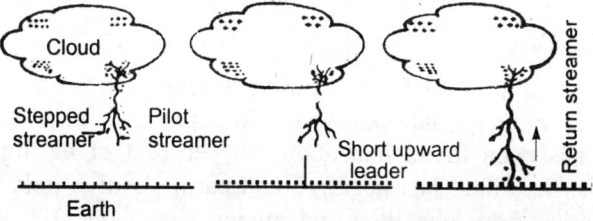

Fig. 21.1. Mechanism of a lightning stroke.

discharge moves to earth in steps of about 50 metres each and is, therefore, termed the *stepped leader* (Fig. 21.1). The pilot leader carries a charge with it, and the potential gradient at its tip is very high. It ionises the air and provides a path or channel for the pilot leader. The channel also becomes charged. The pilot leader carries with it the secondary streamers, which branch out from it. As the pilot streamer approaches the earth, the electric field intensity increases, and due to this a charge of opposite polarity in the form of a short streamer rises from the earth to meet the tip of downward leader. When a contact is made between the pilot leader and the short upward streamer, a return streamer travels from the earth to cloud along the ionised channel formed by the pilot leader (Fig. 21.1). The return streamer moves very fast and produces the well-known intensely luminous lightning flash.

The potential of the portion of the cloud where the discharge originated is lowered by the passage of the charge through the ionised channel to the earth. But other portions of the cloud remain charged. Therefore, a high potential develops between original charge centre and another charge centre in the cloud. The charge of the second charge centre is first transferred to the first one, and then it passes to the earth through the ionised channel made by the first discharge. The second discharge is unbranched and without steps. Its velocity is therefore, more than that of the pilot leader. This is known as *dart leader* and it is also followed by a return stroke.

Other charges are similarly discharged to the earth in the form of leader and return strokes along the same ionised channel. Lightning strokes with many discharges are known as *multiple* or *repetitive strokes*. But only a single flash is seen with an unaided eye. As many as forty component strokes have been recorded in such multiple strokes.

The lightning strokes are referred to as *hot* or *cold* depending upon the magnitude and duration of the stroke current. The stroke of hot lightning has a low current but long duration. It causes fire when it strikes an object. The stroke of cold lightning has a high current but is of short duration. It causes an explosion when it strikes an object.

21.5 FREQUENCY OF LIGHTNING FLASHOVERS

The frequency of lightning flashovers to overhead lines at a place is important to know the probability of outages due to lightning. In a particular region there is a wide variation in the frequency of thunderstorms, but the average frequency is somewhat constant over a period of years. The number of days per year on which thunder is heard is defined as the *isokeraunic level*. This number gives an idea of the severity of lightning at a particular place. The isokeraunic level does not give any information about duration and number of strokes reaching ground. The isokeraunic level in Great Britain is about 15, and in India it is generally between 20 to 35 with the exception of extreme South (Cape Comorin) and regions of Assam and Bay of Bengal. In tropical zones the values are higher.

21.6 STROKE CURRENTS

The instruments commonly employed for measuring lightning stroke current are klydonograph, magnetic link, surge crest ammeter, laboratory oscillographs and fulchronograph. With the help of these instruments it has been found that about 90 per cent of the lightning strokes have a negative polarity, that is, the cloud has negative and earth positive polarity. The magnitude of the lightning current varies from one thousand amperes to about 200 kA or even more; the average value is about 20 kA.

21.7   DIRECT AND INDUCED STROKES

Lightning overvoltages are produced in the lines either by a direct stroke or by an induced stroke. When lightning strikes a line conductor, the current divides and flows in each direction along the line. The line is subjected to a voltage given by the product of the lighting current and half the surge impedance of the line. This voltage causes a flashover of the line insulation to earth. It has been observed that the risk of lightning flashovers goes on diminishing with the increase in operation voltages.

Overvoltages can also be induced on overhead lines by strokes to earth or to an earthed object, near the line. Such strokes are known as *induced strokes*. Induced strokes are also called *indirect strokes*. The voltages due to induced strokes are very much less as compared to those produced by direct strokes.

21.8   WAVE SHAPES OF STROKE CURRENTS

The wave shape of stroke current is also important. It varies widely. For testing electrical equipment arbitrary wave shapes having the nature of uni-directional impulses have been standardized. Such a wave shape can be produced by impulse generator. The wave shape consists of a portion showing the steep rise of voltage upto a peak or crest value called the *wavefront*, and the other portion showing the decay of voltage called the *wavetail*. Such a wave is shown in Fig. 21.2. This may be represented as the difference of two exponentials, thus

$$e = E\,(\varepsilon^{-\alpha t} - \varepsilon^{-\beta t})$$

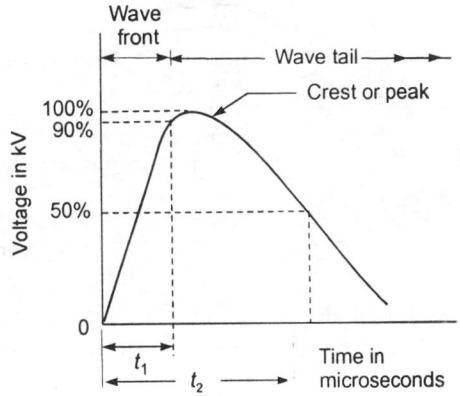

Fig. 21.2. Impulse waveform.

where α and β are constants which determine the shape. The wave is defined by times t_1 and t_2 in microseconds. The time to reach the impulse current or voltage to its maximum amplitude is denoted by t_1, while t_2 denotes the time when the current or voltage has fallen to one-half of its peak value. Both the timings are measured from the start of the wave. This is designated as t_1/t_2 or $t_1 \times t_2$. In Europe 1/50 microsecond, while in the USA 1.5/40 microsecond waveshapes have been standardized for testing purpose. The International Electrotechnical Commission recommends 1×50 microsecond wave as standard. In India 1.2/50 microsecond is the standard.

21.9   OVERVOLTAGE PROTECTION

During their travel the voltage surges suffer a reduction in magnitude and steepness. This may result in a reduced stress on insulation. Even after attenuation these overvoltages may cause damage to insulators and substation equipment. It is, therefore, necessary to provide means to protect the insulators and other apparatus from the harmful effects of the surge voltages. A number of devices are available to reduce the amplitude and front steepness of surges. Of the various devices adopted for overvoltage protection the following will be described here :

    1. Overhead earth wire.

2. Rod gap.

3. Surge diverter.

21.10 OVERHEAD EARTH WIRE

An *overhead earth wire* or *ground wire* is one of the most common devices used to protected the line against lightning. It is a wire carried by the line supports and runs over the phase conductors (Fig. 21.3a). The purpose of the earth wire is to intercept direct lightning strokes, which would otherwise strike the phase conductors. It has no effect on switching surges. When lightning strikes an earth wire at mid span, waves are produced which travel in opposite

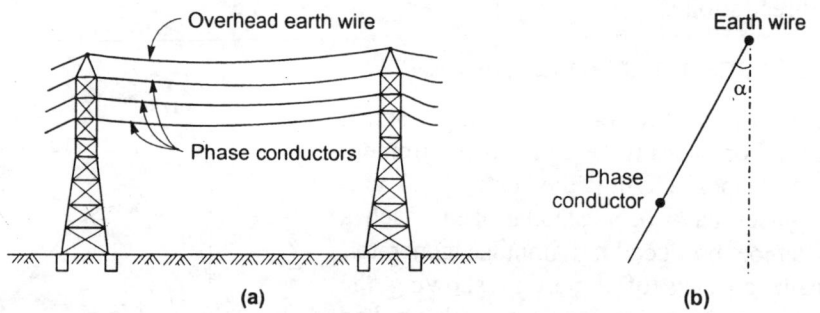

Fig. 21.3. (a) Overhead earth wire. (b) Shielding angle.

directions along the line. The waves reach the adjoining towers which pass them to earth safely. It should be noted that the earth wire is effective only when the resistance between the tower footing and earth is sufficiently low. In case the resistance is not low, the earth wire or tower struck by lightning will be raised to a vary high potential, which will cause a flashover from the tower to one or more phase conductors. Such a flashover is known as *back flashover*. It occurs only when the product of tower current and tower impedance exceeds the insulation level of the line. Back flashovers on the line can be minimised by reducing tower footing resistances using driven rods and *counterpoises* where soil resistivity is high. A *counterpoise* is a conductor buried in the ground. Counterpoise wires are usually made of galvanized steel. The counterpoise for an overhead transmission line consists of a special grounding terminal that reduces the surge impedance of the ground connection and increases the coupling between the ground wire and the conductors. Two types of counterpoises namely, parallel type and ring type, are used for transmission lines. The *parallel* (or *continuous*) *counterpoise* is made up of one or more counterpoise wires buried under the transmission line throughout its length. The counterpoise wires are connected to the overhead earth wire at all towers or poles. The *radial type counterpoise* is made up of number of wires extending radially from the tower legs. The number and length of the wires are determined by the tower location and soil conditions. Fig. 21.4 shows the two types of counterpoises. The depth of counterpoises may be 0.5 m to 1.0 m below the surface of earth.

Tower footing resistance of 30 Ω for 132 kV lines and less than 20 Ω for 275 and 400 kV lines to avoid risk of back flashovers have generally proved satisfactory in most cases.

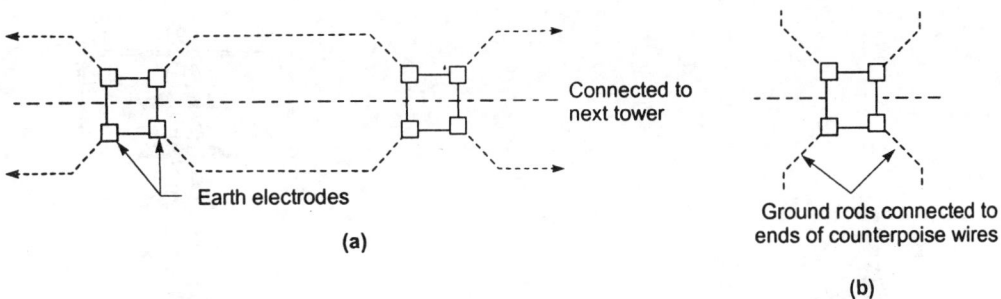

Fig. 21.4. Types of counterpoises (a) Parallel; (b) Radial.

The efficacy of an earth wire to shield a conductor depends on its shielding angle. The *shielding* or *protective* angle α is the angle between the vertical through the earth wire and phase conductor to be protected (Fig. 21.3b). Usually the angle included between the vertical through the earth wire and the line joining the earth wire to the outer most phase conductor is taken as the shielding angle. For effective shielding the protective angle should be kept as small as possible Adequate shielding is achieved provided that this angle does not exceed 40°. The angles between 20° - 30° are quite safe.

Two earth wires are used in modern high voltage systems with wider spacing between the conductors. The protection afforded by two earth wires is definitely better than a single earth wire. Moreover, the surge impedance is low for two earth wires and the coupling effect with the conductor is increased.

The earth wire arrangement is useful in front of high voltage stations for one or two kilometres on the line for extra protection. Both electrical and mechanical considerations are made to select an earth wire. It should have a current capacity to carry heavy short circuit currents or the currents when the phase conductors are earthed. It should also be sufficiently strong mechanically. Steelcored aluminium conductors or alumoweld conductors may be used as earth wires.

21.11 ROD GAP

The rod gap is one of the most common forms of protective devices. It is an air gap between ends of two rods. The rod gaps are generally used on transformer bushings and at the ends of strings of suspension insulators. They are shown in Figs. 21.5. and 21.6. The gap setting should be such that it should break under all conditions before the equipment to be protected is affected. The chief merits of this device are simplicity, reliability and cheapness. Although rod gaps provide a certain degree of protection against overvoltages they have got the following limitations :

    (a) They are unable to prevent the flow of power follow current which flows in the gap after the breakdown. The flow of this current sets up an arc across the gap which is extinguished by the operation of the protective gear resulting in the supply interruption.

    (b) The rod gap has a bent sparkover voltage-time characteristic. If it is set to protect the equipment at the long times it would not do so for short times. If it is set for protecting the equipment at the short times, its minimum spark-over voltage will be too low and cause frequent flashovers. The supply will, therefore, be interrupted even due to harmless surges. A compromise is, therefore, made in gap setting.

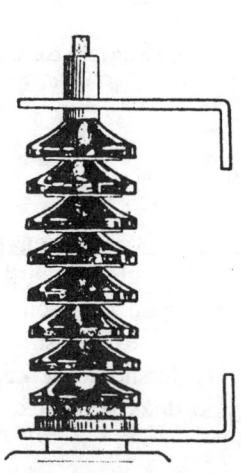

Fig. 21.5. Road gap on a transformer bushing.

Fig. 21.6. Road gap on an insulator string.

(c) Atmospheric conditions such as humidity, temperature and barometric pressure affect the performance of gaps.

(d) The arc setup may produce sufficient heat to damage the rod gap by melting it and thus alter the gap setting.

(e) The polarity of the surge also affects the performance of the gap.

The rod gaps may be used in practice where the continuity of supply is not of much importance. In such cases they are used with automatic reclosing circuit breakers, which will operate as soon as the arc current has been interrupted. Despite the drawbacks rod gaps are used with surge diverters so as to protect the equipment to some degree in case of failure of surge diverters.

21.12 SURGE DIVERTERS

Surge diverter or lightning arrester is a device used for diverting abnormal high voltages to ground without affecting the continuity of supply. The name surge diverter appears to be more correct than lightning arrester. Three types of surge diverters are described here.

1. Expulsion-type surge diverter
2. Valve-type surge diverter
3. Metal-oxide surge diverter

21.13 EXPULSION-TYPE SURGE DIVERTER

The expulsion-type diverter (Fig. 21.7) consists essentially of a tube made of fibre or other gas producing material. The metal electrodes are provided at the ends of the tube forming a spark gap inside the tube. The electrode at the upper end is separated from the line conductor to form an external series gap with the conductor. Under the normal working conditions this gap avoids the leakage currents through the tube. The bottom electrode, which is hollow, is connected to earth.

When a high voltage appears on the conductor both external and internal gaps spark-over. A gas is produced by the vaporization of the fibre tube walls. The great pressure developed in the tube flows out the arc through the vent formed by the bottom electrode. The expulsion of the arc takes place so violently that the arc is interrupted when the current passes through zero point. If the rate of recovery of dielectric strength of the gap is more than the rate of rise restriking voltage, the restriking of the arc is prevented, and thus there is no interruption in supply.

Although, it has been possible to design the expulsion gap to extinguish the arc at the smallest earth fault currents in high voltage systems, there is an upper limit for the current to be interrupted.

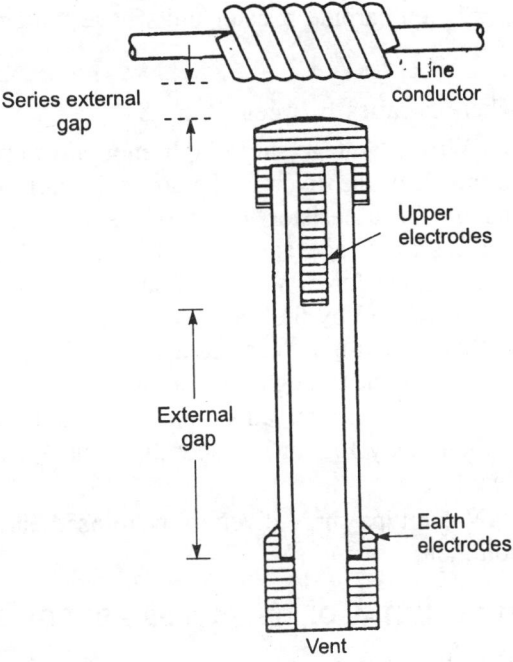

Fig. 21.7. Expulsion-type surge diverter.

If that limit is exceeded it may generate excessive gas which may produce sufficient pressure to cause mechanical failure of the tube, or the insulating material may be eroded to a considerable amount.

The expulsion tube is an improvement over the rod gap. The impulse sparkover characteristic is better than that of the rod gap. The expulsion tubes are mostly used on transmission lines. They may be used for protecting distribution type of transformers where economy and continuity of supply are the main considerations. The use is mainly restricted to system with a low fault level.

21.14 VALVE-TYPE SURGE DIVERTER

Valve-type surge diverter is also known as conventional gap surge diverter or silicon carbide (SiC) surge diverter with series gaps. This type of diverter is commonly known by trade names such as thyrite, miurite, magnavalve, etc.

The valve-type surge diverter (Fig. 21.8) consists of a multiple spark gap assembly in series with resistor elements of nonlinear voltampere (V-I) characteristic. Each spark gap has two electrodes. In order to have uniform voltage distribution, between the gaps, nonlinear resistors are connected in parallel across each gap. Resistor elements are made of silicon carbide (SiC) with inorganic binders. The material may be in the form of pellets, discs, or blocks. The whole arrangement is enclosed in a sealed porcelain or pyrex glass housing filled with nitrogen gas or SF_6 gas.

Silicon carbide resistor units have voltampere (*V-I*) characteristic given by the relationship

$$I = kV^n$$

where *n* varies between 5 and 8.

When a voltage surge high enough to endanger the equipment insulation appears across the terminals of the diverter, the air gaps spark over, and the current is discharged to ground through the nonlinear resistor which offers very small resistance. After the passage of the surge, the impressed voltage across the diverter falls and the diverter resistance increases until normal voltage is restored. When the surge disappears a small current at power frequency flows in the path produced by the flashover. This current is known as *power-follow current*. The magnitude of this current is reduced to a value which can be interrupted by the spark gaps as they recover their dielectric strength. The power-follow current is extinguished at the first current zero. The supply therefore, remains uninterrupted. The normal conditions of the diverter is restored and it is then ready for further operation with incoming surges. This is called resealing of the surge diverter.

Valve type surge diverters were used earlier in AC systems. At present their use is not very common.

21.15 METAL-OXIDE GAPLESS SURGE DIVERTERS

With the development of metal-oxide resistors, the need for series spark gaps in diverters has disappeared. At present gapless surge diverters are very widely used. Modern zinc oxide (ZnO) diverters provide protection against all types of AC and DC overvoltages. They are used for overvoltage protection at all voltage levels in a power system.

Metal-oxide surge diverters are also known as *gapless surge diverters*, or *zinc oxide* (ZnO) *diverters*. A ZnO diverter is shown in Fig. 21.9. The base material used in manufacturing metal oxide resistors is zinc oxide (ZnO). It is a semiconducting material of *N*-type. It is pulverized and finely grained. About 10 or more doping materials are added in the form of fine powders of insulating oxides such as Bi_2O_3, Sb_2O_3, CoO, MnO_2, Cr_2O_3. The actual composition differs from one manufacturer to the other. After treating the powder with a number of processes, the mixture is spray dried to obtain dry granulates. The granulates are compressed into disc shaped blocks. The blocks are sintered to obtain a dense polycrystalline ceramic. Each disc metal oxide resistor is coated with a conducting compound in order to protect the discs from undesirable environmental effects. The conducting coating also provides proper contact and uniform current distribution. The discs are then enclosed in a porcelain housing filled with nitrogen gas or SF_6 gas. Silicon rubber is used to keep the discs in position. It also helps in heat transfer from discs to the porcelain housing. The discs are held under pressure using suitable springs.

The ZnO diverter material has extremely nonlinear voltage-current characteristic given by

$$I = kV^n$$

where *n* may vary between 35 and 50. The *V-I* characteristic of a typical ZnO disc is shown in Fig. 21.10.

The leakage current in a ZnO element at normal voltage is of the order of 100 µA. The same element can carry thousands of amperes at twice the normal operating voltage. This property makes it possible to eliminate series spark gaps in the diverter. It is also seen from

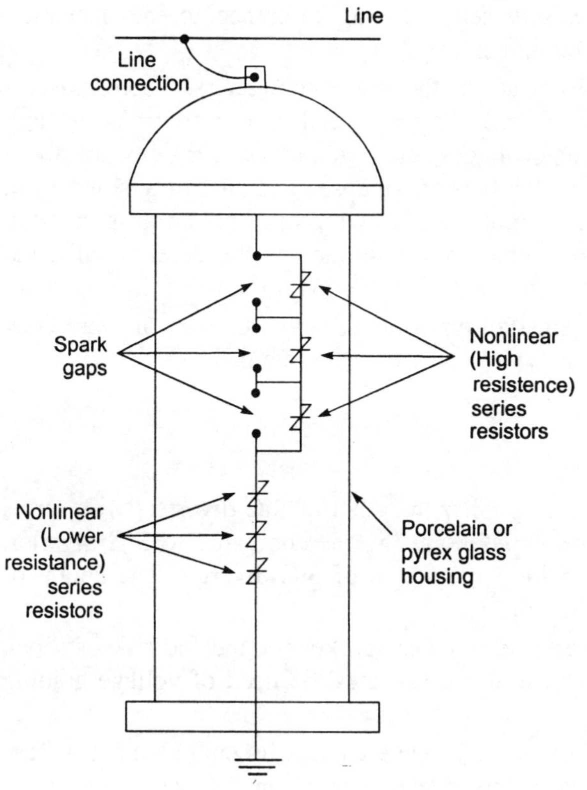

Line
Line connection
Spark gaps
Nonlinear (High resistence) series resistors
Nonlinear (Lower resistance) series resistors
Porcelain or pyrex glass housing

Fig. 21.8. Valve-type surge diverter.

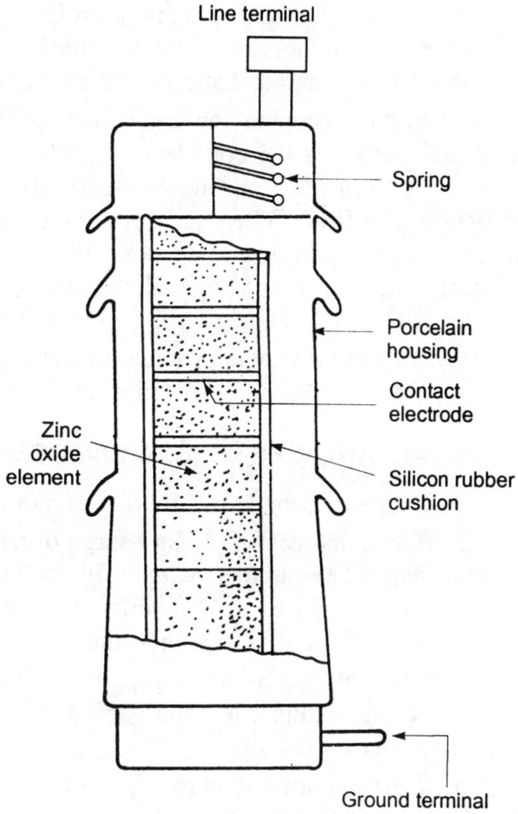

Line terminal
Spring
Porcelain housing
Contact electrode
Zinc oxide element
Silicon rubber cushion
Ground terminal

Fig. 21.9. ZnO surge diverter.

SiC
ZnO
Voltage (kV)
1
2
3
4
5
6
10^{-3} 10^{-2} 10^{-1} 10^{-0} 10^{1} 10^{2} 10^{3} 10^{4} 10^{5} 10^{6}
Current (A)

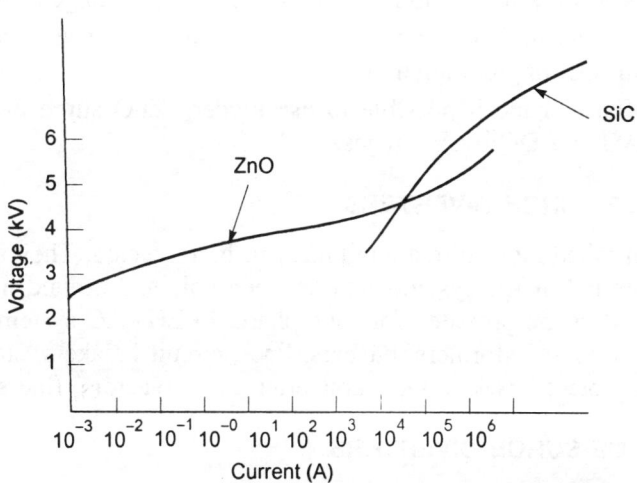

Fig. 21.10. Typical *V-I* characteristic of a ZnO element.

the *V-I* curve that above certain voltage, the characteristic is almost flat. Therefore, the characteristic of the ZnO diverter can be precisely selected with reference to the insulation withstand level characteristic of the protected equipment.

The voltage drop in the ZnO diverter takes place at the grain boundaries. There exists a potential barrier at the boundary of each grain of ZnO. This potential barrier controls the flow of current from one grain to the next. At normal voltages, the potential barrier does not allow the current to flow. When an overvoltage occurs, the barrier collapses and sharp transition from insulating to conducting state takes place. The current starts flowing and the surge is diverted to ground. After the passage of the surge, the impressed voltage across the diverter falls, the current is reduced to a negligible value by the resistor units and there is no power-follow current.

Modern ZnO diverters provide protection against all types of AC and DC overvoltages. They are used for overvoltage protection at all voltage levels in a power system.

21.16 ADVANTAGES OF ZnO DIVERTERS

The ZnO diverters have the following merits :
1. The construction of ZnO surge diverters is simpler than that of SiC diverters with series gaps. The number of parts in ZnO diverter is about 35% as compared to SiC diverter. The weight of ZnO diverter is about 45% of the weight of SiC diverter. The height of ZnO diverter is less than that of SiC diverter.
2. The absence of spark gaps in ZnO diverter eliminates sparkovers and the risk of shock to the system when the gap breaks down. It also eliminates the need of voltage grading system.
3. During normal voltage operating conditions the leakage current in ZnO element is very low of the order the 100 µA while it is of the order of 100 A in SiC valve resistor.
4. There is no power-follow current in ZnO diverter.
5. ZnO diverters have higher energy absorbing capability.
6. ZnO diverters possess high stability during and after prolonged discharges.
7. It is possible to control dynamic overvoltages in addition to switching surges. This results in economic insulation coordination.

These advantages have made it possible to use modern ZnO surge diverters for protection against all types of AC and DC overvoltages.

21.17 LOCATION OF SURGE DIVERTERS

Surge diverters are installed close to the equipment to be protected. They are usually connected between phase and ground in AC systems and between pole and ground in DC systems. In AC systems, separate diverters are provided for each phase. In EHVAC systems, surge diverters are used to protect generators, transformers, busbars, lines, circuit breakers, etc. In HVDC systems, the diverters used to protect buses, valves, converter units, reactors, filters, etc.

21.18 SELECTION OF SURGE DIVERTERS

The important factors to be considered in the selection of surge diverters are as follows :
1. Continuous operating voltage (COV)

2. Temporary overvoltage (TOV)
3. Rated voltage
4. Switching surge discharge voltage
5. Energy dissipation capability.

21.19 LIGHTNING PERFORMANCE OF LINES

The lightning performance of EHV lines is dependent on many factors. The basic considerations made in the design of transmission line based on direct lightning strokes are as follows :

1. The lines must be provided with earth wires. The shielding angle should lie between 20° and 30°.
2. Surge diverters should be located properly.
3. Adequate clearance should be provided between phase conductors and tower (ground).
4. There should be sufficient clearance between phase conductors and earth wire specially at midspan. This would eliminate flashovers between conductors and earth wire during conductor swings, dancing, galloping, etc.
5. The tower footing resistance should be as low as economically justified.

21.20 BASIC IMPULSE INSULATION LEVEL OR BASIC INSULATION LEVEL (BIL)

Basic impulse insulation level (BIL) can be defined as reference level expressed in impulse crest (peak) voltage of a standard wave not longer than 1.2/50 microsecond wave, according to Indian standards.

Apparatus insulation levels as demonstrated by suitable tests should be equal to or greater than the BIL.

The basic insulation level measures the ability to withstand test voltages without disruptive discharge. The BIL for a system is selected such that the system should be protected with a suitable lightning protective device, for example, a surge diverter. The margin between the BIL and the surge diverter should be selected such that it provides good protection economically. The BIL of the equipment insulation chosen must be higher than the maximum expected surge voltage across the surge diverter (selected for protection).

Critical Flashover (CFO) Voltage

Critical flashover (CFO) voltage is the peak voltage for a 50% probability of flashover or disruptive discharge.

Impulse Ratio

Impulse ratio (for flashover or puncture of insulation) is the impulse peak voltage divided by the peak value of the power frequency voltage to cause flashover or puncture.

Each equipment is normally tested for both the normal ac frequency and impulse strength.

21.21 VOLT-TIME CHARACTERISTIC (CURVE)

The volt-time curve is graph of the crest flashover voltages plotted against time to flashover for a series of impulse applications of a given waveshape.

For plotting volt-time curve of a given insulation, a series of application of impulse voltages of the same waveshape and polarity but different peak values is carried out. Suitable test waveshape and polarity is applied through an impulse voltage generator. This impulse voltage peak is increased in constant steps of voltage. The largest value of peak voltage at which the given insulation does not flashover is called *full wave impulse withstand voltage*. If the peak voltage is further increased such that the insulation under test fails to flashover on 50% of applications and flashes over the tail of the wave on the other 50 percent of the applications, the crest value of this voltage is known as *critical flashover voltage*. If the insulation under test flashes over exactly at crest value of the test wave, then it is known as *crest flashover voltage*. The test voltage at which the test insulation flashes over the front of the wave is called *front flashover voltage*. The construction of the volt-time curve and the terminology associated with impulse testing is shown in Fig. 21.11.

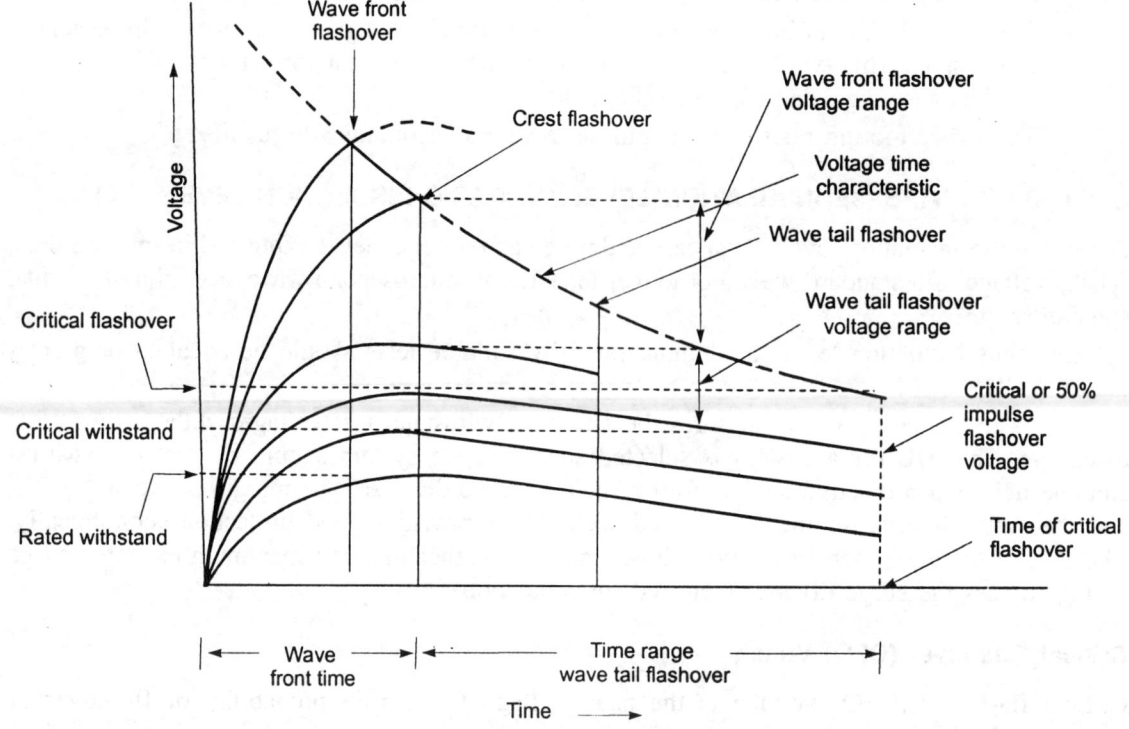

Fig. 21.11. Construction of volt-time curve.

The values of voltages and the time required for flashovers if plotted on *V-t* axes give volt-time curves for the test insulation.

Rated insulated voltage is always chosen smaller than the critical withstand voltage depending upon the factor of safety required and economy. Insulation coordination helps in adjusting the insulation levels of different items in a power system.

21.22 INSULATION COORDINATION

The equipment used in a power system comprises items having different breakdown or withstand voltages and different volt-time characteristics. Insulation coordination aims at correlating the insulations of these various items with the characteristics of protective devices that the equipment is protected from excessive voltages.

Insulation coordination involves the following basic steps:

1. The determination of the line insulation level.
2. Selection of standard insulation level (BIL) and insulation level of the equipment.
3. Making sure that every equipment has a breakdown strength equal to or higher than the basic impulse level.
4. The selection of proper protective devices that will give the apparatus as good protection as can be justified economically.

Thus, the insulation coordination is the matching of the volt-time flashover and breakdown characteristics of the equipment and protective devices. Proper insulation should ensure that the volt-time curve of the weakest equipment in the system will lie above the volt-time curve of the protective device such as surge diverter, over the whole range of the volt-time curve.

Fig. 21.12 shows volt-time curves of various items of the system. The volt-time curve of the substation bus is highest located such that it can even withstand direct lightning strokes.

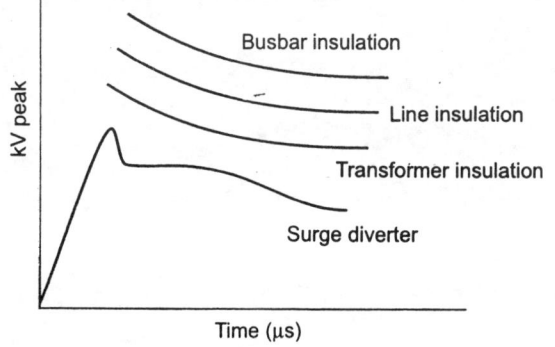

Fig. 21.12. Volt-time curves of various items of system.

Magnitudes and waveshapes of overvoltages occurring on a power system vary over a wide range. This variation is totally a random phenomenon. Designing of system insulation that could withstand the voltages upto the highest level and providing 100 percent protection is very uneconomical. Insulation failure may occur even with 100 percent protection. For this reason, practical insulation designs use probabilistic and statistical methods considering both economy and safety margins specially for high voltage insulation requirements.

EXERCISES

1. What are the causes of overvoltages in a power system?
2. Explain the mechanism of the lightning discharge. What are direct and indirect strokes' Draw and explain the waveshape of stroke currents.

3. What are various devices used for overvoltage protection?

4. Explain the use of an overhead earth wire for protection of the line against lightning. What is shielding angle? Explain its significance.

5. Briefly explain the working of rod gap.

6. What are the various types of surge diverters?

7. Describe the construction and working of expulsion-type surge diverter.

8. Describe the construction and working of valve-type surge diverter.

9. Describe the construction and working of metal-oxide surge diverter.

10. What are the various types of surge diverters? Explain the working of the gapless surge diverter.

11. Why is insulation coordination needed in a power system? What is meant by basic insulation level?

22

Corona

22.1 INTRODUCTION

The use of higher voltages has become necessary in order to meet the rapidly increasing demand for power. With the coming into prominence of EHV transmission lines of more than 230 kV, the corona characteristics of conductors have assumed great importance. Under this, the size of the conductor is mainly determined by corona loss and radio interference considerations. Investigations on the basis of series of experiments carried out in many countries reveal that it is now possible to predict with a fair degree of accuracy the extent of corona performance of a line under different operating conditions.

22.2 THE PHENOMENON OF CORONA

The name *corona* has been derived from the glow surrounding the conductors when the operating voltage is sufficiently high. For an overhead transmission system the atmospheric air, which is the dielectric medium, behaves practically like a perfect insulator when the potential difference between the conductors is small. If the voltage impressed between the conductors is of alternating nature, sustained charging current will flow due to the capacitance of the line. With the increase of the voltage, there is a corresponding increase in the electric field intensity.

As long as the air is subjected to a uniform electric field intensity of peak value less than 3×10^6 V/m (3000 kV/m or 30 kV/cm), the flow of current between the two conductors of the line is negligibly small for practical purposes. But when the electric field intensity (voltage gradient) reaches this critical value of 3×10^6 V/m, the air in the immediate vicinity of conductors no more remains a dielectric but it ionizes and becomes conducting. This electric breakdown is accompanied by the following phenomena :

    1. A faint glow appears around the conductors which is visible in the dark.

    2. There is an accoustical noise.

3. There is a tendency in the conductors to vibrate.
4. Ozone and oxides of nitrogen are produced.
5. There is a loss of power.
6. There is radio interference.

If the voltage gradient is increased further, the size and the brightness of the luminous envelope goes on increasing until finally a spark or arc is established between the conductors because of complete breakdown of the insulating properties of the air between them. The spark may occur before any corona formation if the conductors are very close together. The effect of corona is more pronounced at the protruding points of the conductors due to local higher field intensity there.

22.3 THEORY OF CORONA FORMATION

The electrons and ions are always present to a small extent in the atmospheric air due to the random action of the ionizing sources such as cosmic rays, ultra-violet radiations from the sun, radioactivity of the soil, etc. When an electric field of uniform intensity is established between the conductors, these ions and free electrons experience forces upon them proportional to the product of the field intensity and the charge on them. Due to this force, the ions and free electrons get accelerated and move in opposite directions. The charged particles during their motion collide with one another and also with the very slow moving uncharged molecules. The velocity acquired by a charged particle at collision depends on its mass, charge and the *mean free path*. By mean free path is meant the average distance travelled by a particle without making any collision. The distance travelled by a particle without making any collision is purely a matter of chance. The mean free path depends upon the number of molecules contained in a unit volume of the air and thus upon the size of molecules, the temperature and pressure of air.

Experimentally it has been seen that at normal conditions of atmosphere pressure of 760 mm of mercury and normal temperature of 25°C when the electrical field intensity reaches a critical value of about 3000 kV/metre, the velocity attained by a charged ion or electron is sufficiently high to tear off one or more electrons from the uncharged molecule with which it makes collision, thereby leaving one or more free electrons and a new positive ion behind. This process is called *ionization*. These newly formed free electrons and ions in turn are accelerated by a field and collide with other air molecules to ionize them further. Thus, the number of charged particles goes on increasing rapidly. If a uniform field is assumed between the conductors such conditions are produced everywhere in the gap. As a result of this, the saturation is reached. The air becomes conducting and a complete breakdown occurs. At this stage, a discharge or arc is established between the conductors.

22.4 THE CALCULATION OF POTENTIAL GRADIENT

From the preceding considerations, it is obvious that corona performance depends upon the electric field intensity or potential gradient. This section will, therefore, be devoted to the calculation of potential gradient for any arrangement of conductors of a transmission line.

(a) Single-phase line

Fig. 22.1 shows a single-phase line. It has two conductors A and B, each of radius r and the

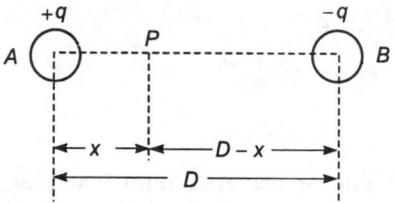

Fig. 22.1. Single-phase line.

distance between the conductors is D. The voltage gradient at a point P situated at a distance x from conductor A is given by

$$G = G_a - G_b$$

where G_a = voltage gradient at P due to a charge $+ q$ on conductor A

G_b = voltage gradient at P due to a charge $- q$ on conductor B.

Now, $G_a = \dfrac{q}{2\pi\varepsilon x}$ V/m   ...(22.4.1)

$$G_b = \dfrac{-q}{2\pi\varepsilon\,(D - x)}\ \text{V/m}$$ ...(22.4.2)

$$G = \dfrac{q}{2\pi\varepsilon}\left[\dfrac{1}{x} + \dfrac{1}{D - x}\right]\text{V/m}$$ ...(22.4.3)

The voltage gradient will be a maximum at a point where $x = r$, that is, at the surface of the conductor. Substituting $x = r$ in Eq. (22.4.3), we get,

$$G_{\max} = \dfrac{q}{2\pi\varepsilon}\left[\dfrac{1}{r} + \dfrac{1}{D - r}\right]\text{V/m}$$ ...(22.4.4)

Since the inter-axial distance D between the conductors is assumed to be fairly large in comparison with their radii r, the second term in Eq. (22.4.4) may be neglected. The maximum value of voltage gradient is, therefore, given by

$$G_{\max} = \dfrac{q}{2\pi\varepsilon r}\ \text{V/m}$$ ...(22.4.5)

The r.m.s. value of voltage gradient is given by

$$G_0 = \dfrac{G_{\max}}{\sqrt{2}}$$

Also, the potential difference E_{ab} between the conductors of a single phase line can be written from Eq. (7.19.5) as

$$E_{ab} = \dfrac{q}{\pi\varepsilon}\ln\dfrac{D}{r}\ \text{V}$$ ...(22.4.6)

If E_n be the voltage from conductor to neutral

$$E_n = \dfrac{1}{2}\,E_{ab} = \dfrac{q}{2\pi\varepsilon}\ln\dfrac{D}{r}\ \text{V}$$ ...(22.4.7)

Combining Eqs. (22.4.5) and (22.4.7) we get

$$G_{max} = \frac{E_n}{r \ln \dfrac{D}{r}} \text{ V/m} \qquad \qquad ...(22.4.8)$$

Equation (22.4.8) gives the value of the maximum potential gradient in terms of voltage to neutral, the inter-axial distance and the radii of the conductors for single-phase line. The variation of potential gradient is shown in Fig. 22.2.

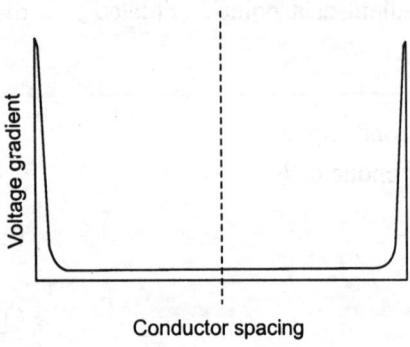

Fig. 22.2. Variation of potential gradient between two conductors.

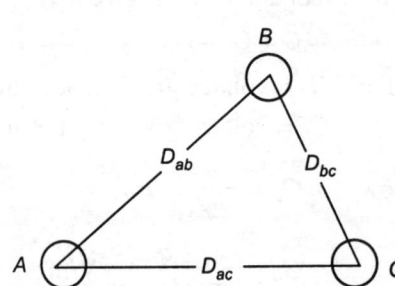

Fig. 22.3. Unsymmetrical three-phase line.

(b) Three-phase line

(i) *Unsymmetrical spacing*

Fig. 22.3 illustrates the three conductors, A, B and C of a three-phase line with unequal triangular spacing. The radius of each conductor is r. The spacing between the conductors is also known in Fig. 22.3.

Let G_a, G_b and G_c denote the r.m.s. values of the voltage gradients in V/m at the surface of conductors A, B and C respectively. Then,

$$G_a = \frac{q_a}{2\pi\varepsilon_0 r} \qquad \qquad ...(22.4.9)$$

$$G_b = \frac{q_b}{2\pi\varepsilon_0 r} \qquad \qquad ...(22.4.10)$$

and $\quad G_c = \dfrac{q_c}{2\pi\varepsilon_0 r} \qquad \qquad ...(22.4.11)$

where q_a, q_b and q_c are the respective r.m.s. charges per metre of conductor.

$$\therefore \quad \frac{q_a}{2\pi\varepsilon_0} = r\, G_a \qquad \qquad ...(22.4.12)$$

$$\frac{q_b}{2\pi\varepsilon_0} = r\, G_b \qquad \qquad ...(22.4.13)$$

and $\quad \dfrac{q_c}{2\pi\varepsilon_0} = r\, G_c \qquad \qquad ...(22.4.14)$

The voltage between the conductors A and B is given by Eq. (7.18.3) as

$$\mathbf{E}_{ab} = \frac{1}{2\pi\varepsilon_0}\left[\mathbf{q}_a \ln\frac{D_{ab}}{D_{aa}} + \mathbf{q}_b \ln\frac{D_{bb}}{D_{ba}} + \mathbf{q}_c \ln\frac{D_{cb}}{D_{ca}}\right] = \frac{1}{2\pi\varepsilon_0}\left[\mathbf{q}_a \ln\frac{D_{ab}}{r} - \mathbf{q}_b \ln\frac{D_{ab}}{r} + \mathbf{q}_c \ln\frac{D_{bc}}{D_{ca}}\right]$$

$$= \frac{1}{2\pi\varepsilon_0}\left[(\mathbf{q}_a - \mathbf{q}_b) \ln\frac{D_{ab}}{r} + \mathbf{q}_c \ln\frac{D_{bc}}{D_{ca}}\right] = \left(\frac{\mathbf{q}_a}{2\pi\varepsilon_0} - \frac{\mathbf{q}_b}{2\pi\varepsilon_0}\right)\ln\frac{D_{ab}}{r} + \frac{\mathbf{q}_c}{2\pi\varepsilon_0}\ln\frac{D_{bc}}{D_{ca}}$$

$$= (r\,\mathbf{G}_a - r\,\mathbf{G}_b) \ln\frac{D_{ab}}{r} + r\,\mathbf{G}_c \ln\frac{D_{bc}}{D_{ca}} = r\left[(\mathbf{G}_a - \mathbf{G}_b)\ln\frac{D_{ab}}{r} + \mathbf{G}_c \ln\frac{D_{bc}}{D_{ca}}\right]\mathbf{V}$$

$$\ldots(22.4.15)$$

By Eq. (7.18.3) the voltage between the conductors A and C can be written as

$$\mathbf{E}_{ac} = \frac{1}{2\pi\varepsilon_0}\left[\mathbf{q}_a \ln\frac{D_{ac}}{r} + \mathbf{q}_b \ln\frac{D_{bc}}{D_{ba}} + \mathbf{q}_c \ln\frac{r}{D_{ca}}\right] = \frac{1}{2\pi\varepsilon_0}\left[\mathbf{q}_a \ln\frac{D_{ac}}{r} + \mathbf{q}_b \ln\frac{D_{bc}}{D_{ab}} - \mathbf{q}_c \ln\frac{D_{ac}}{r}\right]$$

$$= \frac{1}{2\pi\varepsilon_0}\left[(\mathbf{q}_a - \mathbf{q}_c) \ln\frac{D_{ac}}{r} + \mathbf{q}_b \ln\frac{D_{bc}}{D_{ab}}\right] = r\left[(\mathbf{G}_a - \mathbf{G}_c)\ln\frac{D_{ac}}{r} + \mathbf{G}_b \ln\frac{D_{bc}}{D_{ab}}\right]\mathbf{V}$$

$$\ldots(22.4.16)$$

Let E_{an} = phase voltage of phase a. This phase voltage is the line-to-neutral voltage of phase a. If $\mathbf{E}_{an}$ be taken as the reference phasor,

$$\mathbf{E}_{an} = E_{an}\,\underline{/0°}$$

$$\mathbf{E}_{ab} = \sqrt{3}\ E_{an}\,\underline{/30°} \qquad\qquad\qquad \ldots(22.4.17)$$

$$\mathbf{E}_{ac} = -\mathbf{E}_{ca} = \sqrt{3}\ E_{an}\,\underline{/-30°} \qquad\qquad \ldots(22.4.18)$$

The phasor diagram is shown in Fig. 22.4.

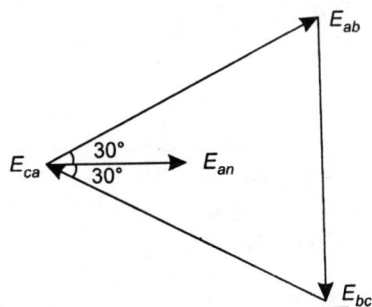

Fig. 22.4. Phasor diagram of voltages.

Also, for a balanced three-phase circuit,

$$\mathbf{q}_a + \mathbf{q}_b + \mathbf{q}_c = 0 \qquad\qquad\qquad\qquad \ldots(22.4.19)$$

$$2\pi\varepsilon r\mathbf{G}_a + 2\pi\varepsilon r\mathbf{G}_b + 2\pi\varepsilon r\mathbf{G}_c = 0$$

or $$\mathbf{G}_a + \mathbf{G}_b + \mathbf{G}_c = 0 \qquad\qquad\qquad\qquad \ldots(22.4.20)$$

By combining the above equations we get the following relations :

$$E_{ab} = \sqrt{3}\ E_{an}\ \underline{/30°} = r\left[(G_a - G_b)\ \ln\frac{D_{ab}}{r} + G_c\ \ln\frac{D_{bc}}{D_{ca}}\right] \qquad \text{...(22.4.21)}$$

$$E_{ac} = \sqrt{3}\ E_{an}\ \underline{/-30°} = r\left[(G_a - G_c)\ \ln\frac{D_{ac}}{r} + G_b.\ln\frac{D_{bc}}{D_{ca}}\right] \qquad \text{...(22.4.22)}$$

Equations (22.4.20), (22.4.21) and (22.4.22) are sufficient to calculate the three voltage gradients if the interaxial distances, radii voltage between the conductors are known.

(ii) *Equilateral spacing*

For equilateral spacing (Fig. 7.5)

$$D_{ab} = D_{bc} = D_{ca} = D\ \text{(say)}$$

From Eqs. (22.4.21) and (22.4.22) we get

$$\sqrt{3}\ E_{an}\ \underline{/30°} = r\ (G_a - G_b)\ \ln\frac{D}{r}$$

$$\sqrt{3}\ E_{an}\ \underline{/-30°} = r\ (G_a - G_c)\ \ln\frac{D}{r}$$

Addition of the above equations gives

$$\sqrt{3}\ E_{an}\ (\cos 30° + j\sin 30° + \cos 30° - j\sin 30°) = r\ [2G_a - (G_b + G_c)]\ \ln\frac{D}{r}$$

$$3\ E_{an} = r\ [2G_a - (-G_a)]\ \ln\frac{D}{r} = 3G_a\ r\ \ln\frac{D}{r}$$

$$G_a = \frac{E_{an}}{r\ \ln\dfrac{D}{r}} \qquad \text{...(22.4.23)}$$

Similarly,

$$G_b = \frac{E_{bn}}{r\ \ln\dfrac{D}{r}} \qquad \text{...(22.4.24)}$$

$$G_c = \frac{E_{cn}}{r\ \ln\dfrac{D}{r}} \qquad \text{...(22.4.25)}$$

Thus, we find that surface voltage gradient is the same for every phase and can be written in a generalised from as

$$G = \frac{E_n}{r\ \ln\dfrac{D}{r}}\ \text{V/m} \qquad \text{...(22.4.26)}$$

where E_n = line–to–neutral voltage in volts = phase voltage in volts

r = radius of the conductor in m

D = spacing between the conductors in m.

(iii) *Flat spacing*

The potential gradients for every phase can also be found for a flat configuration of the line (Fig. 7.7). This is again a particular case, where $D_{ab} = D_{bc} = D$ and $D_{ac} = 2D$.

It should be noted that in case of unsymmetrically spaced three-phase lines the surface voltage gradient differs for every conductor. The transposition of conductor equalizes the loss due to corona per phase for the complete length of the line. However, it does not affect the corona loss of a specific phase in a specific section. The surface voltage gradient at any point is fixed. It is not altered by any transposition of conductors. For a flat horizontal configuration of conductors the value of surface voltage gradient for the middle conductor is more than the values for the outer counter when compared with a triangular configuration.

It is interesting to compare the voltage gradients for single-phase and three-phase equilaterally spaced lines having the same line voltage, conductor diameter, and spacing between conductors.

It is seen that the following relations holds good for both the cases :

$$G = \frac{E_n}{r \ln \frac{D}{r}}$$

where, E_n is the voltage to neutral in V.

For single-phase, $E_n = \frac{1}{2} \times$ line voltage $= \frac{E_l}{2}$

and for three-phase, $E_n = \frac{1}{\sqrt{3}} \times$ line voltage $= \frac{E_l}{\sqrt{3}}$

where, E_l = line-to-line voltage

$$\therefore \quad \frac{\text{voltage gradient for three–phase line}}{\text{voltage gradient for single–phase line}} = \frac{E_l/\sqrt{3}}{E_l/2} = \frac{2}{\sqrt{3}} = 1.155$$

which shows that the surface voltage gradient for a three-phase line is 15.5 per cent greater than that for a single-phase line, provided it is assumed that both the lines operate at the same line voltage and that the radii of the conductors and the spacing between them are equal for both the lines. In other words, on a three-phase line corona starts at a voltage of 15.5 per cent lower than that for a single-phase line having same r, D, and E_l.

22.5 FACTORS AFFECTING CORONA

Due to the inherent complexity and predominance of weather conditions, the theoretical analysis of corona has always lagged behind, and its studies are mainly based upon experimental evidence. Large number of full-scale experimental lines had been constructed. The main purpose of these lines was to determine the factors affecting corona loss and radio interference so as to form a basis for economical and satisfactory performance of the existing EHV/UHV lines and the economical designs of the new ones.

Corona loss depends upon a large number of factors, the important being broadly classified in the following way :

1. Conductor surface gradient,

2. Condition of conductor surface,
3. Atmospheric conditions, and
4. Air density factor.

1. Conductor Surface Gradient

Corona loss on a conductor is a function of potential gradient at the conductor surface. The surface gradient is given by Eq. (22.4.8), which is applicable to single conductors only when the effects of stranding, surface condition, and air density are not considered. The field strength will be different for bundled conductor. In such cases, the surface gradients become non-uniform by the distortion produced by induction.

From the equations of surface gradients for single conductors and multiple conductor lines, it is seen that radius and spacing of the conductors affect the surface gradients and therefore, the voltage at which corona starts. In addition to this, the surface gradient also depends on the configuration of the line and the height of the conductors above the ground. For example, in case of a line with flat spacing the surface gradient at the middle phase conductor is higher than the outers. The corona will, therefore, start first on the middle conductor.

Corona loss also depends upon the frequency and waveform of the supply voltage.

2. Condition of Conductor Surface

The voltage at which corona starts is also affected by the condition of conductor surface. If the conductor is smooth, the electric field will be more uniform as compared to that when its surface is rough. The roughness of conductor may be caused by stranding, die burrs, scratches, or deposition of dust or other airborne particles on its surface. Due to the concentration of field intensity at the rough spots local corona discharges occur there. Therefore, corona starting voltage is lower in case of conductors with rough surface than for those with smooth surfaces.

It has been seen that after an operation period of six months or so, the conductor surface becomes smooth due to weathering effects. This has been explained by the fact that during corona discharge ozone (O_3) and oxides of nitrogen are produced. The are rather unstable to give atomic oxygen easily, which is a very active oxidizing agent. With the action of atomic oxygen at the pronounced points, they become blunted making the surface smooth. When the conductors get *aged* in this way, the corona loss is brought from the initial high value to a comparatively smaller stable value.

3. Atmospheric Conditions

Atmospheric conditions due to weather affect the corona loss to a great extent. The corona loss is fairly large during *foul* weather. Foul weather reduces the uniformity of electric field and lowers the corona formation voltage. The increase of corona loss due to rainfall is due to the fact that the rain drops settle on the conductor surface to form protruding points, which become the sources of local discharges because of field intensity concentration at those points.

When the rate of rainfall increases, it is observed that the corona loss for a given conductor at any voltage increases. But the rate of increase of loss is lesser than that of rainfall. The loss becomes practically constant when the rate of rainfall is very high.

4. Air Density Factor

Air density is also one of the main factors to affect the corona loss. The *air density factor* δ is defined as the ratio of density of air at any given barometric pressure, and temperature to that at the conditions 25°C (77°F) and 760 mm (29.9 inches) of mercury barometric pressure. The value of air density is taken to be equal to unity at standard barometric pressure of 760 mm of mercury and 25°C.

The value of δ at a pressure p mm of mercury and t °C can be found out from the well-known gas equation, which states that the product of pressure and volume divided by the absolute temperature of a gas is constant

$$\delta = \frac{(273+25)}{760} \times \frac{p}{273+t} = \frac{0.392\,p}{273+t} \qquad \qquad ...(22.5.1)$$

Humidity also affects corona voltage. An increase of humidity decreases corona starting voltage. With the increase of the altitude the air density is reduced and corona starting voltage will be lower. However, lower average temperatures of air at high altitudes compensate this effect by increasing the corona starting voltage.

22.6 DISRUPTIVE CRITICAL VOLTAGE

The minimum voltage at which the breakdown of the insulating properties of air occurs and corona starts is called the *disruptive critical voltage*. The potential gradient corresponding to this value of the voltage is known as *disruptive critical voltage gradient*.

(a) *Disruptive critical voltage for a single-phase line*

Let E = voltage between the conductors = line–to–line voltage in volts

r = radius of each conductor in m

D = spacing between the conductors in m.

From Eq. (7.19.5).

$$E = \frac{q}{\pi\varepsilon} \ln \frac{D}{r} \text{ V} \qquad \qquad ...(22.6.1)$$

The electric field intensity or voltage gradient at the conductor surface of radius r is given by Eq. (7.17.2) as

$$G_r = \frac{q}{2\pi\varepsilon r} \qquad \qquad ...(22.6.2)$$

where q is the charge on the conductor per metre length and ε is the permittivity of the medium. Combination of Eqs. (22.6.1) and (22.6.2) gives

$$E = 2r\,G_r \ln \frac{D}{r} \qquad \qquad ...(22.6.3)$$

According to F.W. Peek, for air under conditions near sea level at 25°C and at atmospheric pressure of 760 mm of mercury without impurities, the maximum value of voltage gradient at which ionization of air starts is about 3×10^6 V/m (30 kV/cm). That is, the maximum value of the voltage gradient is the disruptive critical voltage gradient. The voltage corresponding to this voltage gradient is the disruptive critical voltage.

Let $E_{o\,max}$ = maximum value of disruptive critical voltage in volts

$\quad$ $G_{o\,max}$ = maximum value of disruptive critical voltage gradient in V/m.

Substituting these values in Eq. (22.6.3) gives

$$E_{o\,max} = 2rG_{o\,max}\ln\frac{D}{r} \qquad\qquad ...(22.6.4)$$

But $\qquad G_{o\,max} = 3\times10^6$ V/m

$\therefore \qquad E_{o\,max} = 2r\times3\times10^6\ln\frac{D}{r} \qquad\qquad ...(22.6.5)$

The r.m.s. value of the disruptive critical voltage for a single-phase line is given by

$$E_o = \frac{1}{\sqrt{2}}E_{o\,max} = \frac{1}{\sqrt{2}}\times6\times10^6\,r\ln\frac{D}{r}\ \text{V} \qquad\qquad ...(22.6.6)$$

where r and D are in metres.

(b) *Disruptive critical voltage for a three-phase line*

Let E_n = line-to-neutral voltage in volts = phase voltage in volts

$\quad$ D_{eq} = equivalent spacing between the conductors in m = $(D_{ab}\,D_{bc}\,D_{ca})^{1/3}$

$\quad$ r = radius of each conductor in m.

Then $\qquad E_n = \dfrac{q}{2\pi\varepsilon}\ln\dfrac{D_{eq}}{r} \qquad\qquad ...(22.6.7)$

Combining Eqs. (22.6.2) and (22.6.7) we get

$$E_n = rG_r\ln\frac{D_{eq}}{r} \qquad\qquad ...(22.6.8)$$

When G_r becomes $G_{o\,max}$, E_n becomes $E_{o\,max}$ per phase. Therefore, Eq. (22.6.8) becomes

$$E_{o\,max} = rG_{o\,max}\ln\frac{D_{eq}}{r} \qquad\qquad ...(22.6.9)$$

The r.m.s. value of the disruptive critical voltage for a three-phase line is given by

$$E_o = \frac{1}{\sqrt{2}}\times E_{o\,max} = \frac{3\times10^6}{\sqrt{2}}\,r\ln\frac{D_{eq}}{r}\ \text{V/phase} \qquad\qquad ...(22.6.10)$$

where r and D_{eq} are in metres.

Peek concluded from his limited experimental data that the disruptive critical voltage is directly proportional to the air density factor δ over a considerable range. Therefore, a correction factor should be introduced for calculation of disruptive critical voltage in Eq. (22.6.10) at conditions other than standard ones. The formula for disruptive critical voltage then takes the form

$$E_o = \frac{3\times10^6}{\sqrt{2}}\,r\,\delta\ln\frac{D_{eq}}{r}\ \text{V/phase} \qquad\qquad ...(22.6.11)$$

In all the above expressions for disruptive critical voltage the conductors have been assumed to be smooth and clean. The disruptive critical voltage is somewhat smaller when the conductor surface is rough and dirty. To take this effect into account, the disruptive critical voltage expressions given above should be multiplied by a factor m_0 known as the *irregularity factor, surface factor* or *roughness factor*. The approximate values of m_0 given by F.W. Peek are as follows :

$m_0 = 1.00$ for smooth, polished conductors

$m_0 = 0.93$ to 0.98 for rough and weathered conductors

$m_0 = 0.80$ to 0.87 for stranded conductors.

The complete formulae for disruptive critical voltages can, therefore, be written as follows : For single-phase lines

$$E_0 = 2G_0\, m_0\, r\, \delta \ln \frac{D}{r} \text{ r.m.s. volts to neutral} \qquad \qquad ...(22.6.12)$$

If $\qquad G_0 = \dfrac{3 \times 10^6}{\sqrt{2}}$ V/m,

then $\qquad E_0 = \dfrac{6 \times 10^6}{\sqrt{2}}\, m_0\, r\, \delta \ln \dfrac{D}{r}$ volts r.m.s. to neutral $\qquad \qquad ...(22.6.13)$

where r and D are in metres.

For a three-phase line with equivalent spacing D_{eq}, the r.m.s. value of the critical disruptive voltage is given by

$$E_0 = G_0\, m_0\, r\, \delta \ln \frac{D_{eq}}{r} \text{ V/phase} \qquad \qquad ...(22.6.14)$$

If $\qquad G_0 = \dfrac{3 \times 10^6}{\sqrt{2}}$ V/m,

then $\qquad E_0 = \dfrac{3 \times 10^6}{\sqrt{2}}\, m_0\, r\, \delta \ln \dfrac{D_{eq}}{r}$ r.m.s. volts/phase. $\qquad \qquad ...(22.6.15)$

Peterson showed that the critical disruptive voltage varied as $\delta^{2/3}$.

22.7 VISUAL CRITICAL VOLTAGE

Visual glow of corona occurs at a voltage higher than the disruptive critical voltage. The minimum voltage at which the visual corona begins is termed as *visual critical voltage*.

According to F.W. Peek, the actual visual corona does not start at the disruptive critical value of voltage. The maximum value of voltage gradient responsible for starting of corona is 3×10^6 V/m. But this value of the voltage gradient at the surface of conductor will not ionize the air. The maximum value of 3×10^6 V/m will cause ionization when this value is reached at a distance of $(r + 0.0301 \sqrt{r})$ from the conductor axis, where r is in metres. The reason being that some energy is required by the charged ions to start corona. If the maximum voltage gradient at the surface of conductor be 3×10^6 V/m, the value of the maximum voltage gradient at any other point away from the centre would be less than this and, thus there will be corona discharge at that point.

Under the standard conditions of temperature and pressure, the values of the visual critical voltages for single-phase and three-phase lines can be written as follows when the effects of irregularity of the surface of the conductor and air density factor are considered.

For a single-phase line

$$E_v = 2G_0 \, m_v \, r \, \delta \left(1 + \frac{0.0301}{\sqrt{r \, \delta}}\right) \ln \frac{D}{r} \qquad \qquad ...(22.7.1)$$

If $\quad G_0 = \dfrac{3 \times 10^6}{\sqrt{2}}$ V/m,

then $\quad E_v = \dfrac{6 \times 10^6}{\sqrt{2}} \, r \, \delta \, m_v \left(1 + \dfrac{0.0301}{\sqrt{r \, \delta}}\right) \ln \dfrac{D}{r}$ r.m.s. volts $\qquad \qquad ...(22.7.2)$

For a three-phase line

$$E_v = G_0 \, m_v \, r \, \delta \left(1 + \frac{0.0301}{\sqrt{r \, \delta}}\right) \ln \frac{D_{eq}}{r} \text{ V/phase} \qquad \qquad ...(22.7.3)$$

If $\quad G_0 = \dfrac{3 \times 10^6}{\sqrt{2}}$ V/m,

then $\quad E_v = \dfrac{3 \times 10^6}{\sqrt{2}} \, m_v \, r \, \delta \left(1 + \dfrac{0.0301}{\sqrt{r \, \delta}}\right) \ln \dfrac{D_{eq}}{r}$ r.m.s. volts/phase $\qquad \qquad ...(22.7.4)$

It should be noted that r, D, and D_{eq} are to be taken in metres. In Eqs. (22.7.1) and (22.7.2), m_v has been taken as another roughness factor or irregularity factor.

$m_v = 1.00$ for smooth conductors

$m_v = 0.70$ to 0.75 for local corona when the effect is first visible at some places along the line

$m_v = 0.80$ to 0.85 for decided or general corona along the whole length of the conductor.

Example 22.1 A three-phase, 132 kV, 50 Hz, 150 km long transmission line consists of three stranded aluminium conductors spaced triangularly at 3.8 m between centres. Each conductor has a diameter of 19.53 mm. The surrounding air is at a temperature of 30°C and at the barometric pressure of 750 mm of mercury. If the breakdown strength of air is 21.1 kV (rms) per cm and the surface factor is 0.85, determine the disruptive critical voltage. Also, determine the visual critical voltages for local and general corona if the surface factors are 0.72 and 0.82 for visual corona (local) and visual corona (general) respectively.

Solution

$$d = 19.53 \text{ mm}, \quad r = 0.5 \, d = 0.5 \times 19.53 \text{ mm} = 9.765 \times 10^{-3} \text{ m}$$

Air density factor, $\delta = \dfrac{0.392 \, p}{273 + t} = \dfrac{0.392 \times 750}{273 + 30} = 0.9703$

$$D = 3.8 \text{ m}, \quad m_0 = 0.85$$

$$G_0 = 21.1 \text{ kV (r.m.s.) per cm} = 21.1 \times 1000 \times 100 \text{ V/m} = 2.11 \times 10^6 \text{ V/m}$$

The disruptive critical r.m.s. voltage per phase is given by

$$E_0 = G_0 \, m_0 \, r \, \delta \ln \frac{D}{r}$$

$$= 2.11 \times 10^6 \times 0.85 \times 9.765 \times 10^{-3} \times 0.9703 \ln \frac{3.8}{9.765 \times 10^{-3}}$$

$$= 101.347 \times 10^3 \ \text{V} = 101.347 \ \text{kV}$$

For local corona, $m_v = 0.72$

The visual critical r.m.s. voltage per phase for local corona is given by

$$E_v = G_0 \, m_v \, r \, \delta \left(1 + \frac{0.0301}{\sqrt{r \, \delta}} \right) \ln \frac{D}{r}$$

$$= 2.11 \times 10^6 \times 0.72 \times 9.765 \times 10^{-3} \times 0.9703 \left[1 + \frac{0.0301}{(9.765 \times 10^{-3} \times 0.9703)^{1/2}} \right]$$

$$\times \ln \frac{3.8}{9.765 \times 10^{-3}}$$

$$= 112.393 \times 10^3 \ \text{V} = 112.393 \ \text{kV}$$

For decided corona, $m_v = 0.82$

The visual critical r.m.s. voltage per phase for decided (general) corona is given by

$$E_v = 2.11 \times 10^6 \times 0.82 \times 9.765 \times 10^{-3} \times 0.9703 \left[1 + \frac{0.0301}{(9.765 \times 10^{-3} \times 0.9703)^{1/2}} \right]$$

$$\times \ln \frac{3.8}{9.765 \times 10^{-3}}$$

$$= 128 \times 10^3 \ \text{V} = 128 \ \text{kV}$$

It is to be noted that there is no corona under normal working conditions since the actual operating voltage to neutral is $\frac{132}{\sqrt{3}}$, that is, 76.21 kV.

22.8 CORONA POWER LOSS

The power dissipated in the system due to corona discharges is called *corona loss*. It is very difficult to estimate corona loss accurately because of its extremely variable nature. A number of empirical formulae have been developed by various investigators. Such formulae are applicable under fair-weather conditions. To estimate the losses under foul weather conditions rather arbitrary annual loss factors are applied to take account of the losses during precipitations. It has been found that in the fair weather the corona loss is of small significance and it is negligible as compared to the losses found for widely different weather patterns.

According to F.W. Peek, Jr., the corona loss for single-phase and equilaterally spaced three-phase lines under fair-weather conditions is given by the formula

$$P_c = \frac{244}{\delta} (f + 25) (E_n - E_0)^2 \sqrt{\frac{r}{D}} \times 10^{-5} \ \text{kW/km/phase} \qquad \qquad ...(22.8.1)$$

In the Peek's formula

P_c = corona power loss

f = frequency of supply in Hz

δ = air density factor

E_n = r.m.s. phase voltage (line–to–neutral voltage) in kV

E_0 = disruptive critical voltage per phase in kV (r.m.s.)

r = radius of conductor in metres

D = spacing (or equivalent spacing) between conductors in metres.

It is also to be noted that for a single-phase line

$$E_n = \frac{1}{2} \times \text{line voltage}$$

and for a three phase line

$$E_n = \frac{1}{\sqrt{3}} \times \text{line voltage}$$

The total loss of the line is the sum of the losses due to the three conductors. Peek's formula is applicable for decided visual corona. The formula given in Eq. (22.8.1) is, therefore, only approximate one. Experimental results indicate that corona takes place even when the voltage between the conductor is well below the disruptive critical voltage.

Peek's formula gives inaccurate results when the losses are low and E_n/E_0 is less than 1.8. It is superseded by Peterson's formula.

Peterson's empirical formula for fair-weather loss is

$$P_c = 2.1 f F \frac{E_n^2}{\left(\log_{10} \dfrac{D}{r}\right)^2} \times 10^{-5} \text{ kW/km/conductor} \qquad \ldots(22.8.2)$$

where P_c = corona power loss

f = frequency of supply in Hz

E_n = voltage per phase (line–to–neutral voltage in kV (r.m.s.)

r = radius of conductor in metres

D = spacing (or equivalent spacing) between conductors in metres.

The factor F is called the *corona-loss function*. It varies with the ratio (E_n/E_0). It should be noted that E_0 is not given by Eq. (22.6.14) but it is given by

$$E_0 = G_0 m_0 r \, \delta^{2/3} \ln \frac{D_{eq}}{r} \text{ V/phase (r.m.s.)} \qquad \ldots(22.8.3)$$

In order to determine the ratio, E_n and E_0 should be expressed in the same units (either both in volts or both in kilovolts). Table 22.1 gives the values, of F.

Table 22.1   Values of Corona Loss Functions

E_n/E_0	F	E_n/E_0	F	E_n/E_0	F	E_n/E_0	F
1.00	0.037	1.26	0.120	1.52	1.80	1.78	4.72
1.02	0.039	1.28	0.136	1.54	1.33	1.80	4.95
1.04	0.042	1.30	0.154	1.56	1.59	1.82	5.17
1.06	0.045	1.32	0.176	1.58	1.88	1.84	5.39
1.08	0.048	1.34	0.200	1.60	2.20	1.86	5.60
1.10	0.052	1.36	0.228	1.62	2.52	1.88	5.81
1.12	0.057	1.38	0.260	1.64	2.83	1.90	6.01
1.14	0.063	1.40	0.30	1.66	3.13	1.92	6.21
1.16	0.069	1.42	0.38	1.68	3.42	1.94	6.41
1.18	0.075	1.44	0.48	1.70	3.70	1.96	6.61
1.20	0.082	1.46	0.60	1.72	3.97	1.98	6.81
1.22	0.092	1.48	0.74	1.74	4.23	2.00	7.00
1.24	0.105	1.50	0.90	1.76	4.48	–	–

The empirical formulae do not give accurate results. It was found better to replace the use of formulae by the curves drawn over a wide range and under different conditions.

Example 22.2 Estimate the corona loss for a three-phase, 110 kV, 50 Hz, 150 km long transmission line consisting of three conductors each of 10 mm diameter and spaced 2.5 m apart in an equilateral triangle formation. The temperature of air is 30°C and the atmospheric pressure is 750 mm of mercury. Take the irregularity factor as 0.85. Ionization of air may be assumed to take place at a maximum voltage gradient of 30 kV/cm.

Solution

$$d = 10 \text{ mm}, \quad r = 0.5 \, d = 5 \text{ mm} = 5 \times 10^{-3} \text{ m}$$

Air density factor, $\delta = \dfrac{0.392 \, p}{273 + t} = \dfrac{0.392 \times 750}{273 + 30} = 0.9703$

$$D = 2.5 \text{ m}, \quad m_0 = 0.85$$

$$E_0 = G_0 \, m_0 \, r \, \delta \ln \frac{D}{r} \text{ V/phase} = \frac{3 \times 10^6}{\sqrt{2}} \times 0.85 \times 5 \times 10^{-3} \times 0.9703 \times \ln \frac{2.5}{5 \times 10^{-3}}$$

$$= 54.372 \times 10^3 \text{ V} = 54.372 \text{ kV/phase}$$

$$E_n = 110/\sqrt{3} = 63.508 \text{ kV/phase}$$

According to Peek, corona power loss under fair weather conditions is given by

$$P_c = \frac{244}{\delta} \, (f + 25) \, (E_n - E_0)^2 \, \sqrt{\frac{r}{D}} \times 10^{-5} \text{ kW/km/phase}$$

$$= \frac{244}{0.9703} (50 + 25) (63.508 - 54.372)^2 \sqrt{\frac{5 \times 10^{-3}}{2.5}} \times 10^{-5} \times 150 \text{ kW/phase}$$

$$= 105.6 \text{ kW/phase}$$

Total corona loss $= 3 P_c = 3 \times 105.6 = 316.8$ kW.

22.9 RADIO AND TELEVISION INTERFERENCE (RI)

The corona discharges emit radiations which may introduce noise signals in the communication channels, radio and television receivers in the vicinity. This is called *radio interference* (RI).

Radio noise from overhead power lines is caused by corona on conductors and fittings, surface discharges on insulators and poor contacts in fittings and insulator strings.

The importance of radio interference (RI) problem in the present age can hardly be over-emphasized. Great attention is paid to reduce the level of RI from EHV lines to tolerable limits.

According to Trichel a discharge from a negative point contains fast and regular pulses which increase in frequency with the increase in voltage. A corona discharge, which builds up and extinguishes itself in a short period of time of the order of fraction of a micro-second, is called a *corona pulse*. These pulses from a negative source are known as *Trichel pulses*. Trichel also found the presence of pulses in positive point corona. These pulses are of two forms. They are called *burst* and *streamer pulses*. The main cause of radio interference (RI) from high voltage transmission lines is due to the presence of above type of pulses in the corona discharge. The streamer type pulses produce RI much higher than the negative half cycle RI. When the field intensity is below its critical value, the interference is mainly due to negative half-cycle pulses, but above the critical value it appears in the positive half-cycle also.

The radio and TV interference increases with the voltage more rapidly than the corona loss and, like the corona loss, it is very much influenced by the weather. Extremely foul weather conditions can be expected to increase RI levels as much as 20 times. Ten to fifteen times increase is more general in foul weather. A thorough investigation of the problems for standardization of permissible levels of radio and television interference has been taken up in various projects in many countries.

22.10 MINIMIZING CORONA

The undesirable effects of corona are :

1. Audible noise and radio interference.
2. Power loss in the line.
3. Corrosion due to ozone formation, etc.
4. Harmonic current flow resulting from corona formation.

It is, therefore, necessary to minimize corona. The following factors may be considered to control corona :

(a) Voltage of the line;
(b) Spacing between the conductors;
(c) Conductor diameter.

The voltage of transmission is fixed by economic considerations. To increase the disruptive voltage, the spacing of conductors is to be increased, but this method has also got its limitations. With the increase of spacing of conductors the cost of the towers increases and also there is an increase in the voltage drop in the line due to the increase in the inductive reactance X_L. The method of increasing the diameter of conductors has been found to be very effective in reducing corona. The diameter of the conductors can be increased by either using hollow conductors or steel cored aluminium conductors (ACSR) or expanded ACSR conductors. The description of these conductors is given in Chapter 3.

22.10.1 Bundled Conductors

The present practice is to use bundled conductors for reducing corona. The use of bundled conductors has become so common for transmission of larger blocks of power over longer distances at EHV and UHV that the bundled conductors may be considered as an integral part of EHV and UHV transmission lines.

A *bundled conductor* consists of two or more parallel subconductors at a spacing of several diameters. These groups of subconductors form the phase conductors. The number of subconductors used for each phase of ac lines or each pole of dc lines may be 2, 3, 4, 6, 8. At present commercial lines of 1150-1200 kV use upto 8 subconductors, but a maximum of 18 subconductors have been tried on experimental lines. Bundled conductors are also called *grouped* or *multiple* conductors.

The bundle acts as far as the electric field is concerned, like a conductor of diameter much larger than that of the component conductors. This reduces the voltage gradient. In other words, a higher voltage can be used for permissible levels of RI. The GMR of a bundle is high and therefore the inductive reactance of the line is low. The bundled conductors have higher capacitance and therefore lower surge impedance as compared to single conductors of equivalent diameters. The lower value of inductive reactance helps in reducing the cost of series capacitors which are used to increase the transient stability limit of very long lines. Due to higher capacitance of bundled conductors total generated reactive power capacity of the lines is also reduced. The power transmission capability of a bundled conductor line depends upon the spacing of subconductors in a bundle, the size of subconductors, number of subconductors, and the phase separation. Bundled conductors have noise level 5-6 dB less than that of a single conductor operating at the same voltage gradient. Compared with a single conductor of the same cross-sectional area, bundled conductors, having a larger surface area exposed to the air, are better cooled. Thus, higher currents may be allowed to flow without exceeding thermal limits.

Bundled conductors have certain *limitations* also. Bundled conductor lines are subject to ice and wind loads. The bundled conductors have the disadvantage from the point of view of their complex mechanical structure, increased clearance requirements at structures, and higher initial costs. The advantage of higher transmission capability from the point of view of stability outweighs all disadvantages. The economic use of bundled conductors is limited to very high voltages above 220 kV.

Corona is helpful in one respect, namely, it reduces the effect of surges and acts as a relief valve for them. This is so because the surges are partially dissipated as corona.

EXERCISES

1. Give a short account of corona on hv transmission lines and derive a formula for the disruptive critical voltage between two smooth circular conductors, assuming the breakdown strength of air to be 30 kV/cm.

2. Assuming uniformly distributed charges, develop an expression for the voltage gradient at the surface of two very long parallel cylindrical conductors of radius r, spaced d between centres and with a steady potential difference of V volts between them.

   Discuss the phenomenon of 'corona' and 'corona loss' and indicate the circumstances under which they are likely to occur.

   Compare the single- and twin-conductor arrangements for an overhead line with reference to their surface voltage gradients and their corona loss.

3. Explain the formation of corona on hv transmission line and develop an expression in terms of conductor radius, conductor spacing and breakdown voltage of air, for the voltage at which the corona effect might be expected to appear on a symmetrically-spaced smooth conductor three-phase transmission line. Indicate other factors that might influence the value of this voltage. Discuss the importance of corona in the design and operation of the line.

4. Explain the occurrence of corona on hv lines. What is meant by the terms disruptive critical voltage and the visual critical voltage? How does corona affect the design of hv line?

5. Explain the factors which determine the formation of corona on overhead lines. What are the advantages and limitations of corona and how can it be minimised?

6. Calculate the disruptive critical voltage for a three-phase overhead line which has three smooth round conductors of 12.70 mm diameter arranged in a 3 m delta. The barometric pressure is 750 mm of mercury and the air temperature is $-1°C$. The breakdown strength of air is 21 kV (r.m.s.) per cm.

7. A three-phase, 220 kV, 50 Hz, 100 km long overhead line consists of three stranded aluminium conductors spaced 5 m apart in equilateral formation. The surrounding air is at a temperature of 27°C and a pressure of 740 mm of mercury. The breakdown strength of air is 3000 V/mm (max.). Determine (a) the disruptive critical voltage, (b) the visual critical voltage, and (c) the corona power loss of the line. Assume the conductor diameter of 21 mm, the irregularity factor of 0.9 for the disruptive critical voltage and 0.8 for the visual critical voltage.

8. A three-phase 132 kV line with 19.53 mm diameter conductors is built so that corona takes place if the line voltage exceeds 210 kV (rms). If the value of the potential gradient at which ionisation occurs is 3000 V/mm (max.), find the spacing between the conductors.

9. Each conductor of a three-phase overhead line has a diameter of 21.00 mm. The conductors are arranged in equilateral formation. Find the minimum spacing between the conductors if the maximum value of breakdown strength of air is 30 kV/cm, the disruptive critical voltage 230 kV (rms), air density factor 0.98 and irregularity factor 0.95.

10. Explain the importance of bundled conductors in EHV/UHV lines.

11. A three-phase, 400 kV, 50 Hz, 250 km long transmission line consists of three stranded aluminium conductors spaced triangularly at 5.5 m between centres. Each conductor has a diameter of 30 mm. The surrounding air is at a temperature of 35°C and at the barometric pressure of 740 mm of mercury. If the breakdown strength of air is 21.1 kV (r.m.s.) per cm and the irregularity factor is 0.84, determine the disruptive critical voltage. Also, determine the visual critical voltages for local and general corona if the surface factors are 0.72 and 0.82 for visual corona (local) and visual corona (general) respectively.

ANSWERS

 6. 89.18 kV

 7. (a) 119.53 kV (b) 137.99 kV (c) 145.97 kW

 8. 3.401 m   **9.** 6.34 m

11. 147.843 kV/phase; 165.43 kV/phase; 185.283 kV/phase

High Voltage Direct Current (HVDC) Transmission

23.1 INTRODUCTION

The growing trend towards the rapid increase of load and present development systems have made it necessary to transmit more power over longer distances to cope with the demands. The reliability and flexibility of power transmission by AC is indisputable. That is why AC power transmission at high voltages is in use in most of the countries of the world. Today's designs, although technically safe, are not necessarily the most economical. The inherent simplicity and economic considerations have urged the power engineers to rethink in terms of DC at high voltage for power transmission. This will enable them to make the best use of hydro-power sources that are usually located far from load centres.

It has been proved that AC is better from generation and utilisation points of view while DC is preferable for transmission over very long distances. In a combined AC and DC system, generated AC voltage is converted into DC voltage at the sending end. Then the DC voltage is inverted to AC voltage at the receiving end for distribution purpose. Thus, in a combined system conversion and inversion equipment is also needed at the two ends of the line. The operation of conversion and inversion equipment is made reversible by suitable converter control.

Transmission of power by DC is by no means new. From the very beginning electrical energy was generated, distributed and utilised as DC. The need of high voltages and ease of transmission of voltage from one level to another led to the general adoption of AC. In the beginning of the century the French engineer Thury used DC to transmit 20 MW at a voltage of 125 kV from Moutiers to Lyons over a distance of 220 km (36 km of cable). A large number of DC generators of small output voltage were connected in series to give constant current. The system worked quite satisfactorily until 1937. Since then continuous attempts are being made to develop HVDC system. The successful operation of 100 km of DC cable link to carry 20 MW at 100 kV between Swedish mainland and the Baltic island of Gottland in 1954 has revived an interest in many countries to use HVDC as a means of transporting large amount of power over long distances.

A number of HVDC schemes have come in operational use throughout the world since that time. Transmission voltages and currents have increased from 100 kV to ±600 kV and from 200 to 3000 A. The power rating of a typical system has increased to 6300 MW.

Table 23.1 shows some major HVDC systems in existence or under construction.

23.2 CLASSIFICATION OF HVDC SYSTEMS

The HVDC links are classified as follows:

(a) Monopolar Link

A monopolar link (Fig. 23.1) has a single conductor usually of negative polarity and uses earth or sea for the return path of current. Sometimes metallic return is also used. Earthing of poles is done by earth electrodes located about 15 to 55 km away from the respective terminal stations.

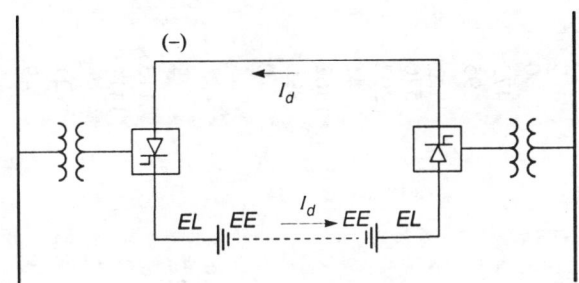

Fig. 23.1. A monopolar link.

(b) Bipolar Link

A bipolar link (Fig. 23.2) has two conductors, one positive and the other negative with respect to earth. The mid-points of converters at each terminal station are earthed via electrode lines and earth electrodes. The voltage between the conductors is equal to two times the voltage between either of the conductors and earth. Since one conductor is at positive polarity with respect to earth and the other is at negative polarity with respect to earth, a bipolar HVDC system is described as say ±500 kV. A bipolar system is advantageous in the sense that when one pole goes out of operation, the system may be changed to monopolar mode with ground return. Thus, the other pole continues to supply half the rated power through ground return. Bipolar links are most commonly used in all high power HVDC systems.

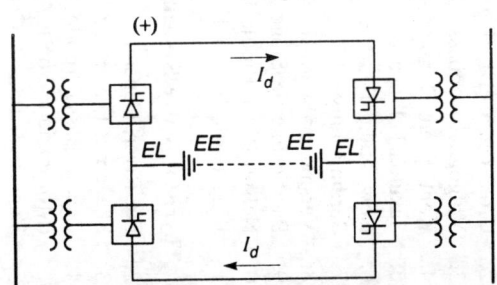

Fig. 23.2. A bipolar link.

Table 23.1 Major HVDC Systems in Operation

No.	HVDC system	Capacity MW	Rated DC voltage (kV) × No. of circuits	Transmission distance (km) Overhead line	Cable	Commissioning date	Remarks
(1)	(2)	(3)	(4)	(5)	(6)	(7)	(8)
A. Mercury-Arc Valve Systems							
1.	Gottland-Swedish mainland	30	150	0	96	1954/70	
2.	Cross Channel I, England-France	160	±100	0	65	1961	
3.	Volgograd-Donbass, USSR	720	±400	472	0	1962-65	
4.	Konti-Skan, Denmark-Sweden	250	250	95	85	1965	
5.	Sakuma I, Japan	300	125 × 2	–	–	1965	
6.	New Zealand	600	±250	570	39	1965	
7.	Sardinia-Italy	200	200	292	121	1967	
8.	Vancouver Stage III, Canada	312	±260	41	33	1968/69	
9.	Pacific Intertie Stage I, USA	1600	±400	1362	0	1970	
10.	Nelson River Bipole I, Canada	1620	±450	890	0	1973/77	
11.	Kingsnorth, England	640	±266	0	82	1975	
B. Thyristor Valve Systems							
12.	Eel River, Canada	320	80 × 2	–	–	1972	Asynchronous Tie
13.	Skagerrak I, Norway-Denmark	500	±250	113	127	1976/77	
14.	Cabora Bassa-Apollo (Mozambique-South Africa)	1920	±533	1414	0	1979	
15.	Vancouver Pole 2, Canada	370	-280	41	32	1977/79	
16.	Square Butte, USA	500	±250	749	0	1977	
17.	Shin-Shinano, Japan	300	125 × 2	–	–	1977	50/60 Hz Tie
18.	Nelson River Bipole 2, Canada	900	±250	930	0	1978	
19.	CU, US	1000	±400	656	–	1979	
20.	Hokkaido-Honshu, Japan	300	250	124	44	1979/80	
21.	Vyborg, (USSR-Finland)	170	±85 × 3	–	–	1982	Asynchronous Tie
22.	Inga Shaba, Zaire	560	±500	1700	–	1982	
23.	Gottland-Swedish mainland	130	150	7	91	1983	

(1)	(2)	(3)	(4)	(5)	(6)	(7)	(8)
24.	Eddy Co. (USA)	200	82	—	—	1983	Asynchronous Tie
25.	Itaipu, Brazil	1575	±300	783/806	0	1984	Asynchronous Tie
26.	Chateauguary (Canada)	1000	140	—	—	1984	
27.	Itaipu, Brazil	6300	±600 × 2	783/806	0	1985-87	Asynchronous Tie
28.	Oklaunion (USA)	200	82	—	—	1984	
29.	Pacific Intertie (USA)	400	±500	—	—	1985	
30.	Corsica Tap (France)	50	200	—	—	1986	
31.	Madawaska (Canada)	350	144	—	—	1985	Asynchronous Tie
32.	Miles City (USA)	200	82	—	—	1985	Asynchronous Tie
33.	Walker Co. (USA)	500-1500	±400	256	0	1985	Asynchronous Tie
34.	Cross Channel II, England-France	2000	±270 × 2	0	72	1985/86	
35.	Konti-Skan, Denmark-Sweden	270	250	95	85	1988/89	
36.	Ekibastus-Centre (USSR)	6000	±250	2400	0	1985-88	
37.	Store Baelt (Denmark)	350	280	35	30	1989-90	
38.	Skagerrak II (Denmark-Norway)	320	300	113	127	1988-89	
39.	Intermountain (USA)	1600	±500	794	0	1987	
40.	Liberty-Mead (USA)	1600/2200	±364/±500	400	0	1989-90	
41.	Nelson River Bipole 3, Canada	2000	±500	930	0	1992/97	
42.	Chicoasen (Mexico)	900/1800	±500	720	0	1985/90	
43.	Quebec-New England	690/2070	±450	175/375	0	1986/92	
44.	Des Cantons-Camerford	690	±450	175	0	1986	
45.	Blackwater (USA)	200	56	—	—	1985	Asynchronous Tie
46.	Highgate (USA)	200	56	—	—	1985	Asynchronous Tie
47.	SACOI-2 (Italy)	300	200				
48.	Gezhouba-Nan Qio (China)	1200	±500	1080	—	1987-91	
49.	Rihand-Delhi (India)	1500	±500	810	—	1987	
50.	Vindhyachal (India)	250 × 2	70	—	—	1989	Asynchronous Tie
51.	Gottland 3-Swedish mainland	130	150	—	98	1987	
52.	South Finland-East Sweden	420	350	35	185		

(c) Homopolar Link

A homopolar link is shown in Fig. 23.3. It has two conductors of the same polarity (usually negative), and always operates with ground or metallic return. This system is not used presently.

The choice between AC and DC transmission is based upon the following factors :

(a) Technical performance

(b) Economic consideration

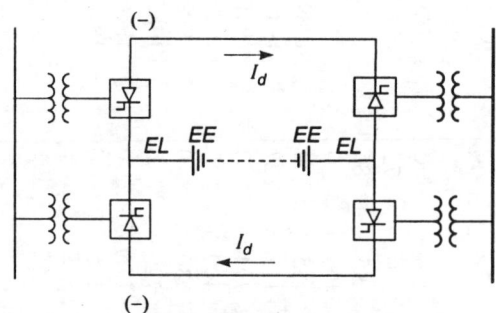

Fig. 23.3. A homopolar link.

The DC transmission provides many technical advantages which are not offered by AC transmission. The main problems associated with AC transmission are interconnection, stability and charging current. These are described below :

23.3 AC INTERCONNECTION

When two AC systems are to be interconnected by an AC link it is necessary that there should be sufficiently close frequency control on each of the two systems. For a 50 Hz system, the frequency should lie between 48.5 Hz and 51.5 Hz. Such an interconnection is known as *synchronous interconnection* or *synchronous tie*. Thus, an AC link provides a *rigid connection* between two AC systems to be interconnected.

23.3.1 Limitations of AC Interconnection

Even with the coordinated control of two interconnected AC systems, the operation of AC link suffers from the following problems :

1. An interconnection of two AC networks with an AC line is a synchronous tie. Frequency disturbances in one system are transferred to the other system.

2. Power swings in one system affect the other system. Large power swings may result in frequent tripping. Major faults in any of the networks may lead to the complete failure of the whole interconnected system.

3. There is an increase in the fault level if an existing AC system is interconnected with another AC system by an AC tie line. This is due to the fact that the additional parallel lines reduce the equivalent reactance of the interconnected system. However, if two AC systems are interconnected by a DC tie line, the fault level of each AC system remains unchanged.

23.3.2 Advantages of DC Interconnection

A DC interconnection or DC tie provides a *loose coupling* between two AC systems to be interconnected. The DC tie between two AC systems is *non-synchronous* (*asynchronous*). The main advantages of DC interconnection are as follows :

1. Since the DC tie between two AC systems is asynchronous, the two systems to be interconnected may either operate at the same frequency or may have different frequencies. A DC link thus provides a very important advantage of interconnecting two AC systems at different frequencies to enable them to operate independently and to maintain their frequency standards. In Japan 50 and 60 Hz systems have been interlinked. Direct current tie interconnects two systems at 25 and 60 Hz in the USA.

2. HVDC link provides fast and reliable control of magnitude and direction of power flow by controlling firing angles of the converters. The rapid control of power flow increases the limit of transient stability.

3. Power swings in the interconnected AC networks can be damped rapidly by modulating the power flow through the DC tie. Thus, the stability of the system is increased.

23.4 STABILITY LIMITS

The power transfer in AC lines depends upon the difference of the phase angles between the voltages at the two ends. For a given power level, this angle increases with the length of the line. The maximum power transfer is limited by the considerations of steady state and transient stability. The need of maintaining stability imposes a serious technical limit to the distance over which power can be transmitted on a simple AC line. It becomes more difficult to maintain stability as the length and, therefore, the reactance of the line increases. The reactive power is to be injected at regular intervals to limit the reactive voltage drop which is the main cause of instability. The stability problems do not arise with DC transmission and, therefore, there is no longer a length limitation with DC. Thus, the power transfer capability of DC lines is unaffected by the distance of transmission.

23.5 HVDC CABLE TRANSMISSION

The charging current becomes a problem at very high voltages and long distances. There is a continuous loss of power on no load in an AC system, but in a DC system the charging current has to be supplied only at the time of switching.

The EHAC cables take continuous charging currents. These charging currents become more important for longer lengths of cables and result in dielectric heating. Consequently, thermal limit is reached even on no load in AC system. Hence the power transfer capability of long EHVAC power cables is low. The length of EHVAC power cables is, therefore, limited by charging current and temperature rise. The charging MVAr is particularly important in transferring power by underground or submarine cables. For a three-phase cable line, it is of the order of 1.25 MVAr per circuit/km at 132 kV, 3.0 MVAr per circuit/km at 220 kV, and 10 MVAr per circuit/km at 400 kV.

The reactive compensation is to be done at intermediate points at regular intervals. It may be possible with underground systems, but long-distance transmission by submarine (underwater) cables is either impracticable or uneconomical with AC. In submarine cables, it is not possible to connect shunt reactors at suitable intervals at the bottom of the sea. Thus, there is a distance limit over which AC transmission by cables is feasible. The limits of distance may be taken as 60 km for 132 kV, 40 km for 220 kV, and 25 km for 400 kV. Such limitations may be avoided with the use of DC. No intermediate switching stations are required. Although DC converter stations require reactive power supply depending upon the active power loading of the line, the line itself does not require reactive power.

The current carrying capacity of HVDC cables is considerably high due to reduced dielectric losses. Higher dielectric stresses are permissible in HVDC cables due to lesser temperature rise. Hence HVDC cables are more compact and cheaper than EHVAC cables. Their installation cost is smaller and it is easier to repair them.

23.6 ECONOMIC COMPARISON

The cost of a transmission line consists of two parts, namely, the fixed cost and the operational cost. The fixed cost includes the cost of Right of Way (ROW), transmission towers, conductors, insulators and terminal equipment. The operational costs include mainly the cost of losses. The economic comparison shall be based on two principles, namely, a new DC line is to be provided instead of a three-phase line, or converting an existing three-phase double circuit line to a three-circuit DC line (three parallel DC circuits). The two methods are discussed separately.

In the first instance, let us compare a 2-wire mid-point earthed DC system (bipolar link) with a three-phase AC earthed central system. The two systems are shown in Figs. 23.4 and 23.5 respectively. In comparing the two systems, the assumptions made are :

1. same power transmitted;
2. same power losses; and
3. same conductor size.

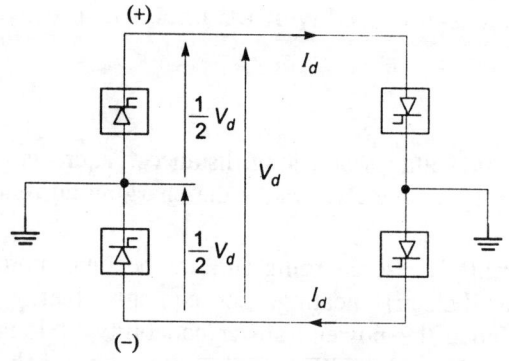

Fig. 23.4. Two-wire mid-point earthed DC system.

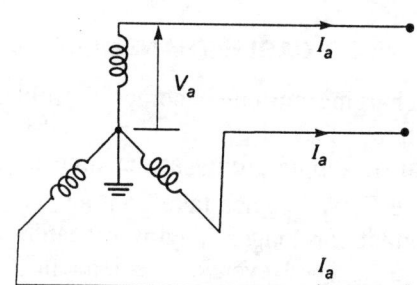

Fig. 23.5. Three-phase AC system.

Let I_d = direct current

V_d = voltage between conductors in DC system.

Since the DC line under discussion carries two conductors, one positive and the other negative with respect to earth, the line to earth voltage will be $\frac{1}{2} V_d$.

Also, let I_a = rms alternating current

       V_a = rms alternating voltage to earth of each conductor

       R = resistance of each conductor

  $\cos \varphi$ = power factor.

Power transmitted by DC = $V_d I_d$

Power transmitted by AC = $3 V_a I_a \cos \varphi = 3 V_a I_a$ (assuming the power factor, $\cos \varphi = 1$)

Power loss in DC system = $2 I_d^2 R$

Power loss in AC system = $3 I_a^2 R$

For AC power loss to be equal to DC power loss

$$3 I_a^2 R = 2 I_d^2 R$$

$\therefore$ $\dfrac{I_a}{I_d} = \sqrt{\dfrac{2}{3}}$ ...(23.6.1)

For the same power transmitted by both the systems,

$$3 V_a I_a = V_d I_d$$

$\therefore$ $\dfrac{V_d}{V_a} = 3 \left(\dfrac{I_a}{I_d} \right) = 3 \sqrt{\dfrac{2}{3}} = \sqrt{6}$ ...(23.6.2)

The DC insulation level $= \dfrac{1}{2} V_d$

The AC insulation level = peak value of alternating voltage to earth = $\sqrt{2}\, V_a$

$\therefore$ $\dfrac{\text{DC insulation level}}{\text{AC insulation level}} = \dfrac{V_d}{2\sqrt{2}\, V_a} = \dfrac{\sqrt{6}}{2\sqrt{2}} = 0.867$ ...(23.6.3)

Thus, for the same power to be transmitted, the insulation level of a DC line is 86.7 per cent of that of an AC line. Moreover, the number of conductors will be two in a DC line as compared to three for an AC line.

23.7 CONVERSION OF THREE-PHASE AC LINE TO DC LINE

Another method of comparing two systems is based on converting an existing three-phase double circuit AC line to three-circuit DC line. Each circuit of DC system shall comprise two conductors, one positive and the other negative to earth. The voltage of one will be $+V_d/2$ while that of the other $-V_d/2$ to earth.

(a) Consider the first case when the current and insulation levels are assumed to be the same in both the systems. For current to be the same, $I_a = I_d$.

For equality of the insulation levels, $\sqrt{2}\, V_a = \dfrac{1}{2} V_d$

Power transmitted by three-phase, double-circuit AC line = $2 \times (3 V_a I_a \cos \varphi) = 6 V_a I_a$

                                           (assuming $\cos \varphi = 1$)

Power transmitted by three-circuit DC line = $3 V_d I_d$

$$\therefore \quad \frac{\text{power transmitted by DC}}{\text{power transmitted by AC}} = \frac{3V_d I_d}{6V_a I_a} = \frac{3\,(2\sqrt{2}\,V_a)\,I_a}{6V_a I_a} = \sqrt{2} = 1.414 \qquad \ldots(23.7.1)$$

That is, power transmitted by DC $= 1.414 \times$ power transmitted by AC

Also, percentage loss by DC $= \dfrac{\text{DC loss}}{\text{DC power}} \times 100$

and percentage loss by AC $= \dfrac{\text{AC loss}}{\text{AC power}} \times 100$

$$\therefore \quad \frac{\text{percentage loss by DC}}{\text{percentage loss by AC}} = \frac{\text{DC loss}}{\text{DC power}} \times \frac{\text{AC power}}{\text{AC loss}} = \frac{1}{\sqrt{2}} = 0.707 \qquad \ldots(23.7.2)$$

The above result shows that the conversion of an existing three-phase, double-circuit line to three DC lines increases the power transmission capability by 41.4 per cent and the line losses are reduced by 29.3 per cent.

(b) We shall now compare the two systems if percentage losses and insulation level remain the same in both of them. For the same insulation level

$$\sqrt{2}\,V_a = \frac{1}{2}\,V_d$$

and if the same percentage losses are assumed

$$\frac{\text{AC loss}}{\text{AC power}} \times 100 = \frac{\text{DC loss}}{\text{DC power}} \times 100$$

$$\frac{6I_a^2 R}{2 \times 3V_a I_a} = \frac{6I_d^2 R}{3V_d I_d} = \frac{6I_d^2 R}{3\,(2\sqrt{2}\,V_a)\,I_d}$$

$$I_d = \sqrt{2}\,I_a \qquad \ldots(23.7.3)$$

$$\therefore \quad \frac{\text{power transmitted by DC}}{\text{power transmitted by AC}} = \frac{3V_d I_d}{2 \times 3V_a I_a} = \frac{3\,(2\sqrt{2}\,V_a)\,\sqrt{2}\,I_a}{6V_a I_a} = 2 \qquad \ldots(23.7.4)$$

Thus, the converted DC line can transmit twice the power of a three-phase, double-circuit system without modification of insulation and with the same percentage loss.

Since there is no skin effect with DC, power losses are reduced marginally. The dielectric losses in HVDC cables are small and, therefore, the current carrying capacity of HVDC cables is considerably large.

The other factors affecting the line costs are the costs of compensation and terminal equipment. Line compensation is not required in DC lines but the cost of terminal equipment is increased due to the presence of converters and filters.

23.8 ADVANTAGES OF HVDC TRANSMISSION

1. Lesser number of conductors and insulators and therefore reduced conductor and insulator costs.
2. Lesser phase-to-phase clearance and lesser phase-to-ground clearance.
3. Lighter and cheaper towers.

4. Reduction in right-of-way (ROW): The DC line corridor, being extremely compact, results in reduced right-of-way requirement. For example, the total requirement of ROW for ±500 kV DC line is about half of that for 400 kV AC line for the same quantum of power to be transmitted.

These factors indicate that DC line construction is simpler and cheaper than AC line construction.

5. Lesser corona loss and reduced radio and television interference. Since the corona effects are less, the choice of economic size of conductors with HVDC is possible. It is to be noted that EHVDC lines have bundled conductors to minimize corona effects.

6. The power losses are also reduced with DC as there are two conductors for a bipolar HVDC line.

7. Earth return can be used in HVDC transmission. In the event of any fault on any pole in an HVDC bipolar line, the other pole with earth return acts as an independent circuit and continues to supply power. Thus, there is a flexibility of operation.

8. It forms an asynchronous connection between two AC stations connected through an HVDC link, that is, the transmission of power is independent of the sending-end and receiving-end system frequencies. For this reason, one of the major uses of HVDC is to interconnect two regions which are usually operating at different frequencies.

9. There is no stability problem with DC line and hence there is no longer a length limitation with DC.

10. Due to the absence of the frequency factor on DC link, there is no skin effect with DC. Thus, complete cross-section of the conductor can be used effectively and more power can be transmitted on the same size of the conductor.

11. Higher dielectric stresses are permissible in HVDC cables due to lesser temperature rise. Hence HVDC cables are more compact and cheaper than EHVAC cables.

12. Since a DC transmission line does not generate or absorb any reactive power, it helps to increase the capability of the link to transmit large quantities of power over long distances. A DC line requires no reactive compensation.

13. Additional EHVAC parallel lines result in a higher fault level at receiving end due to reduced equivalent impedance. When an existing AC system is interconnected with another AC system by an AC tie, the fault level of both the systems increases. However, when the two AC systems are interconnected by a DC tie, the fault level of each system remains unchanged.

14. The power flow through a DC link can be controlled rapidly and accurately under steady-state as well as transient conditions.

23.9 ECONOMIC DISTANCES FOR HVDC TRANSMISSION

The total capital cost of a transmission system is equal to the sum of the capital cost of substations plus capital cost of lines. DC lines and HVDC cables are cheaper than AC lines and HVAC cables. However, the cost of DC line terminal equipment is very high. Fig. 23.6 shows the variation of costs of transmission with line length for AC and DC transmission. The abscissa of the point of intersection of the two curves is called the *breakeven distance*. When the transmission distance is greater than the breakeven distance, HVDC is more economical than

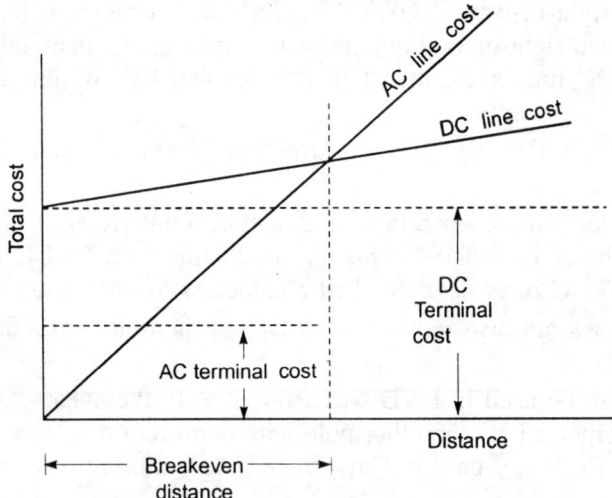

Fig. 23.6. Comparison of the costs of AC and DC transmission.

equivalent HVAC. The breakeven distance is different for different projects. The breakeven distance usually varies from 500 to 900 km in overhead lines. For submarine cables the breakeven distance varies from 25 to 50 km, and for underground cables the range is between 50 to 100 km. Thus, except for interconnection, HVDC link is not preferred below the breakeven distance of the line.

23.10 COMPONENTS OF AN HVDC TRANSMISSION SYSTEM

An HVDC transmission has the following main components :
   1. AC substation and HVDC substation at each terminal
   2. Interconnecting HVDC line(s)
   3. Electrode lines and earth (ground) electrodes.

AC substation is of the conventional type having busbars, AC switchgear, CTs, VTs, surge diverters, etc. The HVDC substation is called the converter station.

23.11 CONVERTER STATION

The terminal substation which converts AC to DC is called *rectifier terminal* while the terminal substation which converts DC to AC is called *inverter terminal*. In general, every terminal is designed to operate in both rectifier and inverter modes. Therefore, each terminal is called a *converter terminal* or *converter station*. A two-terminal HVDC system has only two terminals and one HVDC line. A MTDC transmission system has three or more terminal stations.

The major components of a typical HVDC converter station are shown in Fig. 23.7.

23.12 CONVERTER UNIT

The conversion from AC to DC and vice versa is done in HVDC converter stations by using three-phase bridge converters. Fig. 23.8 shows a six-pulse bridge converter. This bridge circuit is also called Graetz circuit.

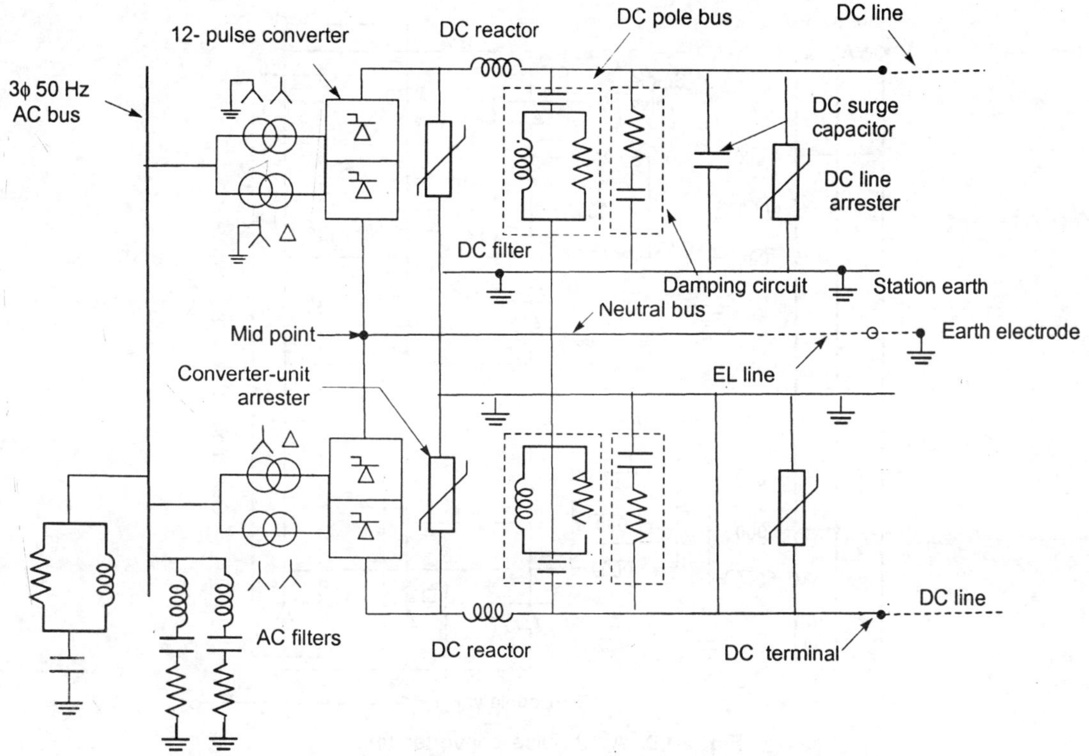

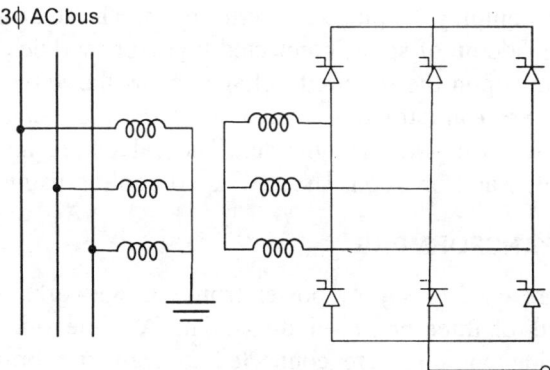

Fig. 23.7. Schematic diagram of a typical HVDC converter station.

Fig. 23.8. A 6-pulse converter unit (Graetz circuit).

In HVDC transmission, a 12-pulse bridge converter (Fig. 23.9) is used. Such a converter is obtained by connecting two 6-pulse bridges in series. The two series connected 6-pulse bridges are identical except that AC supply voltages to the two bridges are shifted in phase by 30°. This phase shift is usually obtained by supplying one 6-pulse bridge by a star-star connected three-phase transformer and the other bridge by a star-delta connected three-phase transformer. The use of 12-pulse converter is preferred over the 6-pulse converter because of the reduced filtering requirements.

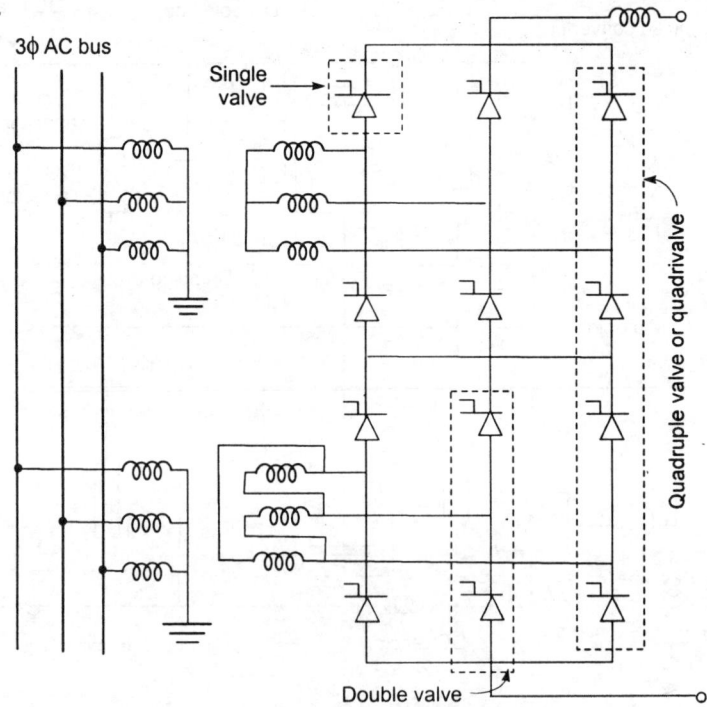

Fig. 23.9. A 12-pulse converter unit.

23.13 CONVERTER VALVES

Modern HVDC converters employ 12-pulse converter units. The total number of valves in each unit is 12. Each valve is made up of series connected thyristor modules. The number of thyristor modules per valve depends upon the required voltage across the valve. Usually four valves are assembled to form one vertical structure called a *quadruple valve* or *quadrivalve*. Three quadrivalves are needed for one pole of converter. The valves are installed in valve halls and they are cooled by air, oil, water or freon. Usually liquid cooling using deionised water is used.

23.14 CONVERTER TRANSFORMERS

Converter transformers are used to supply power from AC networks to DC converters or vice versa. They have two sets of three-phase windings. The AC line side winding is connected to AC busbar. The valve-side windings are connected to converter bridge. These windings are connected in star for one transformer and delta for the other. The AC side windings of the two three-phase transformers are connected in star with their neutrals grounded. For one 12-pulse converter bridge, the following converter transformer configurations may be used :

   (a) 3-phase 2-winding transformers = 2

   (b) 1-phase 3-winding transformers = 3

   (c) 1-phase 2-winding transformers = 6.

Converter transformers are specially designed power transformers differing in many respects from the usual power transformers in AC substations. The valve side transformer windings is designed to withstand alternating voltage stress (from induced voltage) and direct voltage stress from valve bridge. There are increased eddy current losses due to harmonic currents. The magnetization in the core of a converter transformer is due to (a) alternating voltage from AC network containing fundamental and several harmonies, (b) direct voltage from valve side terminals having some harmonics also.

23.15 FILTERS

Both AC and DC harmonics are generated in HVDC converters. AC harmonics are injected into the AC system and DC harmonics are injected into the DC line. These harmonics have the following harmful effects :

1. Telephone interference in adjacent telephone lines.
2. Extra power losses in machines and capacitors connected in the system.
3. Some harmonics may produce resonances in AC circuits resulting in overvoltages.
4. Instability of converter controls.

Harmonics are minimised by using filters. The following types of filters are used :

(a) AC filters
(b) DC filters
(c) High frequency filters.

23.15.1 AC Filters

AC filters are RLC circuits connected between phase and earth. They offer low impedance to harmonic frequencies. Thus, AC harmonic currents are passed to earth. Both tuned and damped filter arrangements are used. The AC harmonic filters also provide reactive power required for satisfactory operation of converters.

23.15.2 DC Filters

DC filters are similar to AC filters. A DC filter is connected between pole bus and neutral bus. It diverts DC harmonics to earth and prevents them from entering DC lines. Such a filter does not supply reactive power as DC line does not require reactive power.

23.15.3 High Frequency Filters

HVDC converters may produce electrical noise in the carrier frequency band from 20 kHz to 490 kHz. They also generate radio interference (RI) noise in the mega Hertz range of frequencies. High frequency (PLC-RI) filters are used to minimize noise and interference with power line carrier communication. Such filters are connected between the converter transformer and the station AC bus.

23.16 REACTIVE POWER SOURCE

Reactive power is required for satisfactory operation of converters. AC harmonic filters provide reactive power partly. Additional supply may be obtained from shunt (switched) capacitors, synchronous phase modifiers and static var systems. The choice depends upon the speed of control desired.

23.17 SMOOTHING REACTOR

Smoothing reactor is an oil-filled, oil-cooled reactor having a large inductance (0.27 H to 1.5

H). It is connected in series with the converter before the DC filter. It can be located either on the line side or neutral side.

Smoothing reactors serve the following purposes :

1. They smooth the ripples in the direct current.
2. They decrease the harmonic voltages and currents in the DC line.
3. They limit the fault current in the DC line.
4. They reduce commutation failure in inverters in case of low voltages caused by faults in AC systems.
5. Consequent commutation failures in inverters are prevented by smoothing reactors by reducing the rate of rise of DC in the bridge when the direct voltage of another series-connected bridge collapses.
6. Smoothing reactors reduce steepness of voltage and current surges coming from DC line. Thus, the stresses on converter valves and valve surge diverters are reduced.

23.18 HVDC SYSTEM POLE

HVDC system pole (abbreviated pole) is a part of an HVDC system consisting of all the equipment in the HVDC substation and interconnecting transmission lines (if any) which during normal operating conditions exhibit a common direct polarity with respect to earth (associated converter transformers are included). Thus the word *pole* refers to the path of DC which has the same polarity with respect to earth. The total pole includes substation pole and transmission line pole.

23.19 GROUND ELECTRODES

The grounding of neutral points in a bipolar DC line is not done near the converter stations in order to prevent (a) galvanic corrosion of underground pipes, buried cables, structures, (b) the interference effects of ground currents in converter transformers. The ground electrodes are usually located at a distance of 15-55 km away from the respective terminal stations. The line connecting the neutral to the ground electrode is called the *electrode line*. This line usually consists of two sections in parallel. Such an arrangement helps in detecting faults in electrode lines by differential protection. Electrode line is either a cable or an overhead line. The electrodes are placed in soil of low resistivity or sea. A land electrode is always kept moist. The electrodes are designed for either continuous operation or intermittent operation (during monopolar operation of a bipolar line with ground return).

23.20 BACK-TO-BACK HVDC STATION (HVDC COUPLING STATION)

An HVDC system which transfers energy between AC buses at the same location is called *back-to-back system* or an *HVDC coupling system*. In a back-to-back HVDC station, the converters and rectifiers are installed in the same station. Thus, there is no DC transmission line in a back-to-back system. Such a system gives an asynchronous interconnection between two adjacent independently controlled AC networks without transferring frequency disturbances. Back-to-back DC links reduce he overall conversion costs and improve the reliability of DC system. The back-to-back coupling stations are generally designed for bipolar operation.

Vindhyachal back-to-back HVDC link forms an asynchronous connection between UP (Northern) and MP (Western) power grids in India. The link was put into commercial operation in June 1989. It is in the form of two blocks of 250 MW capacity each with capability to transmit power in either direction. There are several back-to-back systems in operational use and several more under construction in different countries.

23.21 TWO TERMINAL HVDC SYSTEM

An HVDC system with two terminals (converter stations) and one HVDC transmission line is called a two terminal DC system (2 TDC system). It is also known as point-to-point system. In a 2 TDC system there are no parallel HVDC lines and no intermediate tappings. HVDC circuit breakers are also not required for 2 TDC systems. Normal and abnormal currents are interrupted by effective converter control. Majority of the present HVDC systems are 2 TDC systems.

23.22 MULTITERMINAL DC (MTDC) SYSTEM

A multiterminal DC (MTDC) system has more than two converter stations (terminals) and interconnecting DC transmission lines. Some of the converter stations operate as rectifiers while others operate as inverters. The total power taken from rectifier stations is equal to the total power supplied by the inverter station. There are two types of MTDC systems :

(a) Series MTDC system

(b) Parallel MTDC system.

In a series MTDC system, converters are connected in series while in a parallel MTDC system, converters are connected in parallel. At present parallel MTDC systems are commonly used. The parallel connected MTDC systems may be operated without the use of HVDC circuit breakers.

The first application of a MTDC system is the Sardinia-Corsica-Italy system. The Italy to Sardinia monopolar 2 TDC system existed since 1976. In 1986, a parallel tap has been commissioned in Corsica. The first 5 TDC system is under construction (1988-1996) to transfer bulk power from Quebec (Canada) to New England (USA). Three more bipolar terminals will be added to the existing 2 TDC link between Des Cantons (Canada) to Camerford (USA). This scheme does not use HVDC circuit breakers. It is expected that the application of MTDC systems will increase in future.

23.22.1 Advantages of MTDC Systems

The following are the advantages of MTDC systems over equivalent separate 2 TDC systems :

1. A MTDC is more flexible and economical.
2. The frequency oscillations in the interconnected AC networks can be damped quickly.
3. A MTDC system has inherent overload capability. Hence the transient stability of the system is increased without an increase in the installed capacity. Thus, the overall stability of the interconnected AC networks is improved.
4. Heavily loaded AC networks can be reinforced by using MTDC systems.

23.22.2 Application of MTDC Systems

1. Bulk power transfer from several remote generating sources to several load centres.
2. Cable transmission between three or more AC systems.

3. System interconnections between three or more AC systems by radial MTD systems.

4. Reinforcing heavily loaded urban AC networks by MTD systems.

23.23 DC CIRCUIT BREAKERS

In an AC system, the current can be interrupted at the zero point in the circuit.

The main problem in the circuit interruption in a DC circuit is that DC is a steady unidirectional current and there is no zero point in a DC system. Hence the interruption of a DC circuit requires that the steady or fault current should be forced down to zero by some method. At the same time, an arrangement should be made to dissipate the large amount of energy released by such an operation. Thus, in the early days of HVDC transmission only point-to-point (2 TDC) transmission was possible due to lack of efficient DC circuit breakers.

Converter control is used for protection of DC lines and converters. In the event of a fault on DC side, the thyristor valves are blocked quickly and the line current is brought to a low value. The AC circuit breakers on the AC side of the converter transformers at both terminals are made to trip. The DC pole is then isolated. Thus in a 2 TDC system there is no need of DC circuit breakers. Similarly, in the present MTDC systems, DC circuit breakers are not needed due to lack of economical HVDC circuit breakers and the protection against faults in a DC line is most commonly done by converter control. However, the lack of suitable HVDC circuit breakers presented the following problems :

(a) More complex and costly protection and control systems.

(b) Limited growth of MTDC systems.

Recently a 500 kV DC circuit breaker with current interrupting capability upto 4000 A has been developed. The availability of cheap HVDC circuit breakers would provide the following advantages :

1. There is a scope of converting the present 2 TDC systems to MTDC system for interconnecting three or more AC systems.

2. The designs of MTDC systems would be simplified.

3. Operation and control of MTDC systems would become more flexible.

At present HVDC circuit breakers are used in 2 TDC and MTDC systems for transferring from ground to metallic return.

23.24 LIMITATIONS OF HVDC TRANSMISSION

In spite of the fact that DC is economical and offers some operation advantages not obtained by AC, it has got its limitations also. The scope of application of HVDC transmission is limited by the following factors :

1. **Circuit Breakers.** Circuit breaking is difficult in DC circuits. Therefore, the cost of DC circuit breakers is high.

2. **Transformers.** DC system does not have step-up or step-down transformers to change the voltage levels.

3. **Cost of Converter Stations.** The cost of terminal stations is very high.

4. **Generation of Harmonics.** Both AC and DC harmonics are generated. AC and DC filters are required to minimise harmonics. Thus, the cost of converter stations is increased.

5. **Cooling of HVDC Substations.** HVDC substations have additional losses in converters and transformers. Effective cooling system is required to dissipate heat.

6. **HVDC System Control.** The system control in a HVDC link is quite complex.

23.25 DEVELOPMENTS IN DC TECHNOLOGY

The following are the main developments made in recent years in DC technology :

1. Higher rating of thyristor valves.
2. Twelve-pulse operation of converters.
3. Application of digital electronics and fiber optics in the control of converters.
4. Development of HVDC circuit breakers.
5. Use of metal oxide gapless surge diverters.
6. Better cooling techniques.
7. Suspension of quadrivalve assembly from ceiling to withstand seismic forces.

These developments have improved the system reliability and reduced the cost of HVDC substations. Recently, a 500 kV DC circuit breaker with current interrupting capacity upto 4000 A has been developed.

23.26 APPLICATIONS OF HVDC TRANSMISSION

Considering all the advantages of DC, it is seen that DC transmission is in no way competitive with AC transmission. Both systems have their own advantages and limitations. It is indisputable that generation by AC, transmission by HVDC, and distribution by AC are most economical. Thus, AC and DC are complementary. The special difficulties encountered in one may be solved completely by the other. But there are certain fields of application where DC seems more attractive. These include the following :

1. Long distance bulk power transmission by overhead lines.
2. Underground or underwater cables.
3. Interconnection of AC systems operating at different frequencies.
4. Back-to-back HVDC coupling stations.
5. MTDC asynchronous interconnections between three or more AC networks.
6. Control and stabilisation of power flows in AC interconnections of large interconnected systems.

23.27 HVDC SYSTEMS IN INDIA

A 400 kV, 500 MW, Singrauli (UP)–Vindhyachal back-to-back HVDC link is in operation since 1989. The Rihand–Vidyutnagar bipole HVDC system is the first commercial long-distance system in India. It was commissioned in December 1990. It is transmitting 1500 MW at ±500 kV over a distance of 915 km. It is being maintained and operated by Power Grid Corporation of India. The corporation has commissioned on 14 Feb. 2003 a 2,000 MW Talcher–Kolar ±500 kV, HVDC bipole transmission system enabling excess power to flow from East to South. It is expected that 7,000 circuit km of ±500 kV HVDC would be in operation by the year 2012.

EXERCISES

1. An existing three-phase, double-circuit AC line is to be converted to three-circuit DC line. Assuming the same insulation level and unity power factor in the AC systems, show that (a) the ratio of power transmitted by DC to that by AC is equal to $\sqrt{2}$, and (b) the ratio of percentage loss by DC to that by AC is equal to $1/\sqrt{2}$.

2. A new bipolar DC transmission system is compared with a three-phase AC system transmitting the same power and having the same losses and size of conductor. If the direct voltage for breakdown of an insulator string is equal to the peak value of the alternating voltage to cause breakdown, show that

$$\frac{\text{DC insulation level}}{\text{AC insulation level}} = 0.867 \cos \varphi$$

where $\cos \varphi$ is the power factor in the AC system.

3. A new bipolar DC system is compared with a three-phase AC system. If

$$\frac{\text{power transmitted by DC}}{\text{power transmitted by AC}} = k_1,$$

$$\frac{\text{power loss in DC system}}{\text{power loss in AC system}} = k_2,$$

$$\frac{\text{AC line resistance}}{\text{DC line resistance}} = k_3$$

and $\cos \varphi$ is the power factor in AC system, show that

$$\frac{\text{DC insulation level}}{\text{AC insulation level}} = \frac{0.867 \, k_1 \cos \varphi}{\sqrt{k_2 k_3}}.$$

4. Discuss the advantages and limitations of HVDC systems.

5. What are the problems associated with AC interconnection? How HVDC interconnections are technically superior to HVAC interconnections?

6. Explain with diagrams the various types of HVDC transmission systems.

7. Write brief notes on the following :
   (a) Converter valves
   (b) Converter transformers
   (c) Smoothing reactors
   (d) Harmonic filters
   (e) MTDC systems
   (f) Ground electrodes
   (g) Back-to-back HVDC station
   (h) HVDC cable transmission.

8. What are the applications of HVDC transmission?

C H A P T E R **24**

System Neutral Grounding

24.1 INTRODUCTION

A *system neutral ground* is a connection to ground from the neutral point or points of a system or rotating machine or transformer. The neutral grounding is an important aspect of power system design because the performance of the system in terms of short circuits, stability, protection, etc., is greatly affected by the condition of the neutral. A three-phase system can be operated in two possible ways :

1. With ungrounded neutral
2. With grounded neutral.

24.2 UNGROUNDED NEUTRAL SYSTEM

In an ungrounded neutral system, the neutral is not connected to ground. In other words, the neutral is isolated from the ground. Therefore, this system is also known as *isolated neutral system* or *free neutral system*. It is shown in Fig. 24.1.

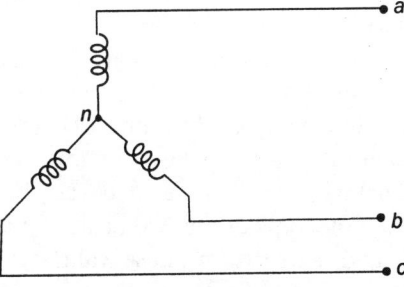

Fig. 24.1. Isolated neutral.

The line conductors have distributed capacitances between one another and to ground. The former are delta connected while the latter are star connected as shown in Fig. 24.2. The delta-connected capacitances have little effect on grounding characteristics of the system and therefore can be neglected.

In a perfectly transposed line, will have the same capacitance to ground. Thus, $C_a = C_b = C_c = C$. Under normal operating balanced conditions, the charging current of each phase is the same. They form a balanced set of currents spaced 120° apart. Therefore, their resultant is zero and no current flows to ground. The potential of the neutral is the same as the

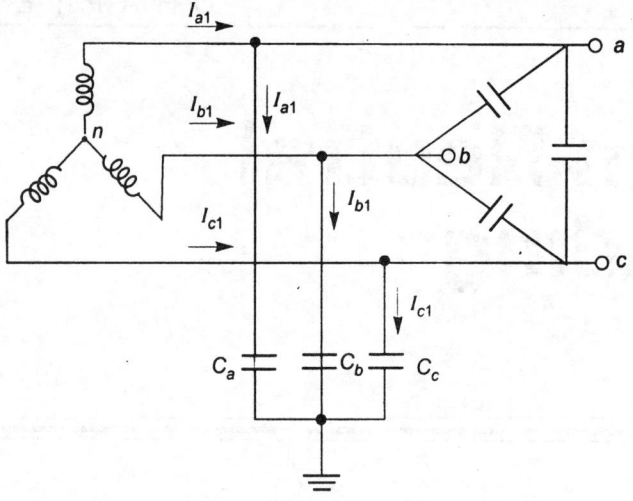

Fig. 24.2.

ground potential. Let the phase voltages (line-to-neutral voltages) be V_{an}, V_{bn} and V_{cn}. It is to be noted that the charging currents I_{a1}, I_{b1} and I_{c1} lead their respective phase voltages by 90°. Fig. 24.3 shows the phasor diagram under normal conditions.

The magnitude of each charging current is given by

$$|\mathbf{I}_{a1}| = |\mathbf{I}_{b1}| = |\mathbf{I}_{c1}| = \frac{V_p}{X_C}$$

where V_p = phase voltage

X_C = capacitive reactance of the line to ground.

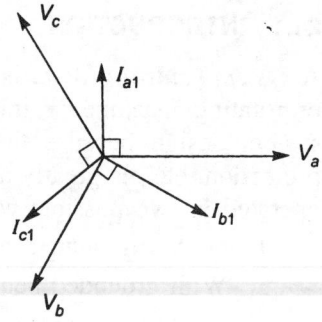

Fig. 24.3. Phasor diagram under normal conditions.

Now consider a line-to-ground fault in line a at a point F as shown in Fig. 24.4. As a result of this LG fault, the potential of phase a becomes equal to ground potential, thus short circuiting the capacitance of this line. Hence no charging current flows in this phase. That is, $I_{a2} = 0$. It is to be noted that the potentials of the two healthy phases b and c rise from phase values to the line values. The capacitance currents now become unbalanced and the fault current $\mathbf{I}_f$ is the phasor sum of the currents $\mathbf{I}_{b2}$ and $\mathbf{I}_{c2}$ which flow through the capacitances C_b and C_c under the potential differences of V_{ba} and V_{ca} respectively. These currents lead their respective voltages by 90°. The phasor diagram is shown in Fig. 24.5.

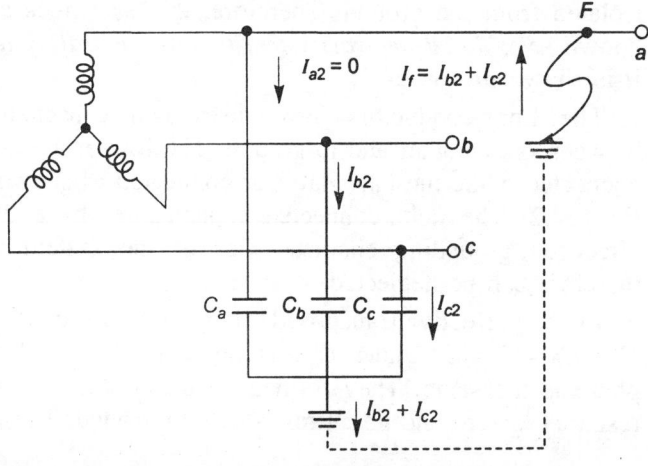

Fig. 24.4. Charging current during LG fault on line a.

$$I_{b2} = \frac{V_{ba}}{X_{cb}} = \frac{\sqrt{3}\ V_p}{X_C}$$

$$I_{c2} = \frac{V_{ca}}{X_{cc}} = \frac{\sqrt{3}\ V_p}{X_C}$$

$\therefore \qquad I_{b2} = I_{c2}$

$\qquad \mathbf{I}_f = \mathbf{I}_{b2} + \mathbf{I}_{c2}$

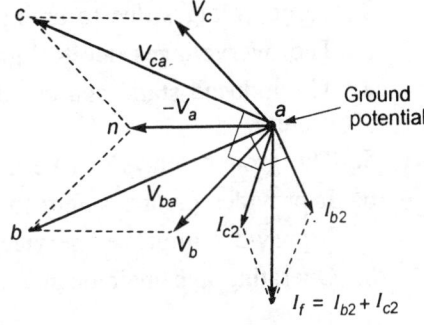

From the phasor diagram in Fig. 24.5,

$$I_f = \sqrt{3}\ I_{b2} = \sqrt{3}\ I_{c2} = \frac{\sqrt{3} \times 3\ V_p}{X_C} = \frac{3\ V_p}{X_C}$$

Fig. 24.5. Phasor diagram with LG fault on line a.

$\therefore \qquad I_f = 3\ I_{b1} = 3\ I_{c1}$

Thus, we arrive at the following *conclusions* in the event of a single line-to-ground fault on a system with isolated neutral :

1. The potential of the faulted phase becomes equal to the ground potential while the voltages of the remaining two healthy phases rise from their normal phase voltages (line-to-neutral voltages) to full line-to-line voltages.

2. The charging current (capacitance current) in the two healthy phases increases to $\sqrt{3}$ times their normal values.

3. The fault current becomes three times the normal per phase charging current under balanced conditions.

24.3 ARCING GROUNDS

The insulation of all equipment connected to the lines is subjected to the high voltage. If it exists for a very short duration, the insulation may be able to withstand it, otherwise it may be damaged. For operating the protective devices, it is necessary that the current should be sufficient in magnitude. However, in case of a single line-to-ground fault on an isolated neutral system, the resultant capacitive current is usually not large enough to operate the protective device. Further, a current of over 4 or 5 A flowing through the fault may give rise to an arc in the ionised part of the fault. With the formation of the arc, the voltage across it becomes zero and, therefore, the arc is extinguished. The potential of the faulty conductor is restored and the formation of second arc takes place. This phenomenon of intermittent arcing is called the *arcing ground*. The successive extinction and reignition of the charging current flowing in the arc may increase the potential of the other two healthy conductors appreciably due to setting up of high-frequency oscillations. These high-frequency oscillations are superimposed on the system and produce surge voltages as high as six times the normal value. The overvoltages in healthy conductors may damage the insulation at some other point of the system.

24.4 ADVANTAGES OF NEUTRAL GROUNDING

Because of the problems associated with ungrounded neutral systems, the neutrals are grounded in most of the modern high-voltage systems. Some of the advantages of neutral grounding are as follows :

1. Voltages of phases are limited to the line-to-ground voltages.

2. Surge voltages due to arcing grounds are eliminated.
3. The overvoltages due to lightning are discharged to ground.
4. The induced static charges do not produce any disturbance since they are conducted to ground.
5. The ground relays can be used to protect against the ground faults.
6. It provides greater safety to personnel and equipment.
7. It provides improved service reliability.
8. Operating and maintenance expenditure are reduced.

24.5 METHODS OF NEUTRAL GROUNDING

The methods commonly used for grounding the system neutral are :
1. Solid grounding (or effective grounding)
2. Resistance grounding
3. Reactance grounding
4. Peterson-coil grounding (or resonant grounding).

The selection of the type of grounding depends on the size of the unit, system voltage and protection scheme to be used.

24.6 SOLID GROUNDING

The term effectively grounded is now used instead of the old term 'solidly grounded'. A power system is said to be effectively grounded when the neutral of a generator, power transformer, or grounding transformer are directly connected to ground through a conductor of negligible resistance and negligible reactance as shown in Fig. 24.6. A system or a portion of a system is defined to be effectively grounded when $R_0 \leq X_1$ and $X_0 \leq 3X_1$, and such relationships exist at any point in the system for any condition of operation and for any amount of generator capacity.

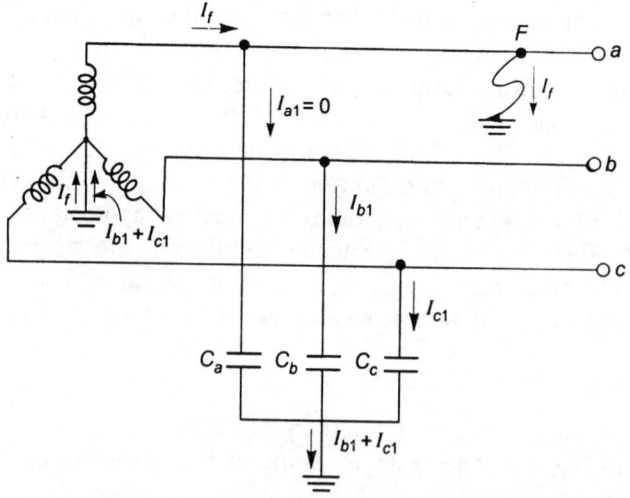

Fig. 24.6. Solidly grounded system.

Here R_0 = zero–sequence resistance

$\quad X_0$ = zero–sequence reactance

$\quad X_1$ = positive–sequence reactance.

Consider a line-to-ground fault in line a at a point F as shown in Fig. 24.6. As a result of this fault the line-to-ground voltage of phase a becomes zero. However, the remaining two phases b and c will still have the same voltages as before as shown in Fig. 24.7.

It should be noted that in this system, in addition to charging currents, the power source also feeds the fault current I_f. As the generator or transformer have their own reactances in series with the neutral circuit, solid grounding does not make a zero-impedance circuit. If the impedance of the generator is too low, solid grounding of the generator without any external impedance may cause the single line-to-ground fault current from the generator to exceed the maximum three-phase fault current which the generator can deliver, and this may exceed the short-circuit current for which its windings are braced. If the reactance of the generator or transformer is very large, then also the purpose of grounding is defeated. For solidly grounded systems, it is necessary that the ground fault current should not exceed 80% of the 3-phase fault current to prevent the production of surge voltages.

Solid grounding is usually used where the circuit impedance is sufficiently high so as to keep the fault current within safe limits.

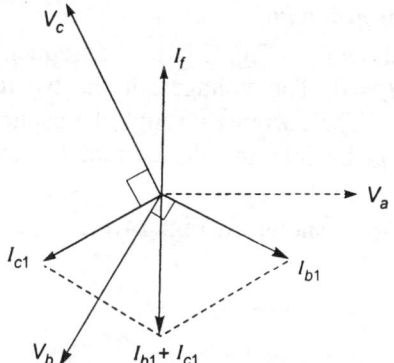

Fig. 24.7. Phasor diagram of a solidly grounded system.

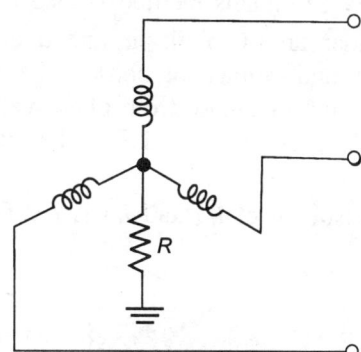

Fig. 24.8. Resistance grounding.

24.7 RESISTANCE GROUNDING

In a resistance-grounded system, the system neutral is connected to ground through one or more resistors as shown in Fig. 24.8. Resistance grounding limits the ground-fault currents. A system that is properly grounded by resistance is not subject to dangerous transient overvoltages. Resistance grounding reduces the arcing ground hazards and permits ground-fault protection.

The value of resistance to be used in the neutral to be grounded should neither be very low nor very high. A very low resistance, makes the system similar to the solidity grounded system. If the grounding resistance is high, the system conditions become similar to an isolated (ungrounded) neutral system. The value of resistance is chosen such that the ground-fault current is limited but still sufficient ground current flows to permit the operation of ground fault protection. In general, the ground-fault current may be limited to 5% to 20% of that which would occur with a three-phase fault.

24.8 REACTANCE GROUNDING

In this system, a reactance is inserted between the neutral and ground to limit the fault current as shown in Fig. 24.9. In order to minimize transient overvoltages, the ground-fault current in a reactance grounded system should not be less than 25% of the 3-phase fault current. This is considerably more than the minimum current desirable in resistance-grounded systems.

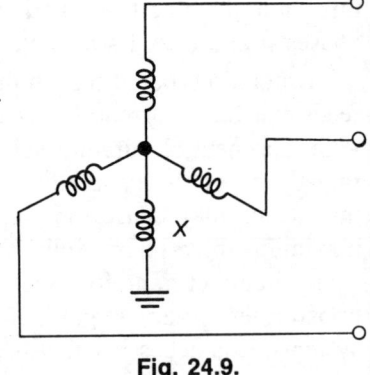

Fig. 24.9.

24.9 PETERSON-COIL GROUNDING

Peterson coil is an iron-cored reactor connected between neutral and ground. It is provided with tappings so that it can be *tuned* with the capacitance of the system. The reactance is selected so that the current through the reactor is equal to the small line-charging current which would flow into the line-to-ground fault if the system were operated with the neutral ungrounded. The reactance of the reactor balances the system ground capacitance so that the resultant ground-fault current is practically zero. This method is also known as *resonant grounding*.

Consider an LG fault in line *a* at a point F as shown in Fig. 24.10. Consequently the line-to-ground voltage of phase *a* becomes zero and $I_a = 0$. The voltages of the two healthy phases *b* and *c* increase from phase values to line values. The currents through the capacitances C_b and C_c become I_{b2} and I_{c2}. The current I_{b2} leads V_{ba} by 90° and the current I_{c2} leads V_{ca} by 90°.

The resultant of I_{b2} and I_{c2} is I_C. The phasor diagram is shown in Fig. 24.11.

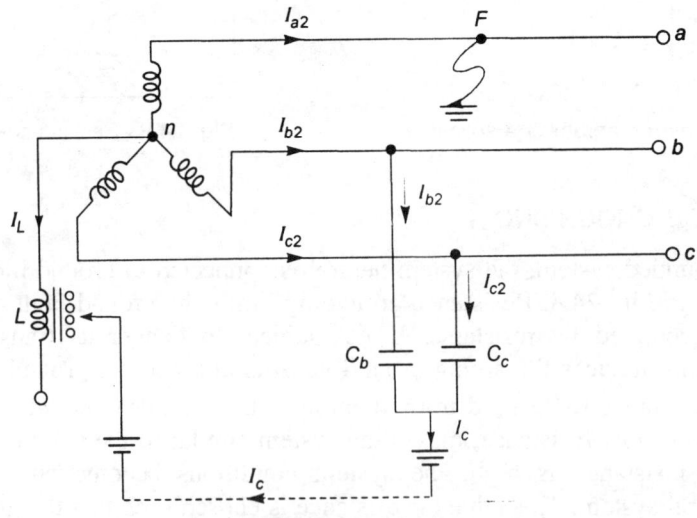

Fig. 24.10. LG fault on line *a*.

$$I_{b2} = \frac{V_{ba}}{X_{cb}} = \frac{\sqrt{3}\, V_p}{X_C}$$

$$I_{c2} = \frac{V_{ca}}{X_{cc}} = \frac{\sqrt{3}\, V_p}{X_C}$$

∴ $I_{b2} = I_{c2}$

$\mathbf{I}_C = \mathbf{I}_{b2} + \mathbf{I}_{c2}$

From the phasor diagram in Fig. 24.11,

$$I_C = \sqrt{3}\, I_{b2} = \sqrt{3}\, I_{c2} = \frac{\sqrt{3} \times \sqrt{3}\, V_p}{X_C} = \frac{3V_p}{X_C}$$

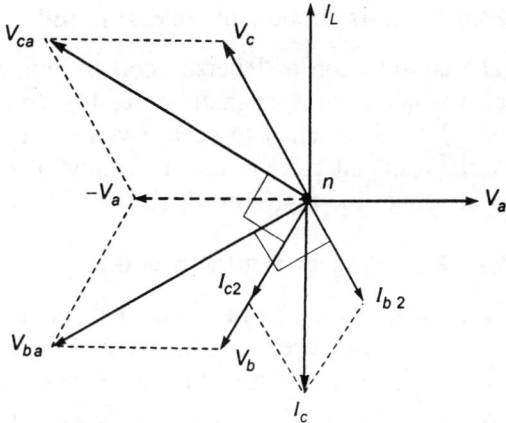

The current I_L through the inductances L of the Peterson coil is given by

Fig. 24.11. Phasor diagram with LG fault on line a.

$$|I_L| = \left| \frac{V_{na}}{\omega L} \right| = \frac{V_p}{\omega L}$$

But V_{na} is equal to $-V_a$. The current I_L lags behind V_{na} by 90°. The current I_C is in phase opposition to I_L. Hence if I_L is equal to I_C, there will be no current through the ground and there will be no tendency of the arcing grounds to occur. With the use of Peterson coil, arc current is reduced to such a small value that it is usually self extinguishing. Therefore, Peterson coil is also known as *ground fault neutralizer* or *arc suppression coil*.

The inductance value for the Peterson coil can be calculated as follows :

For no current through the ground fault,

$$|I_L| = |I_C|$$

$$\frac{V_p}{\omega L} = \frac{3V_p}{X_C} = 3V_p \omega C$$

∴ $$L = \frac{1}{3\omega^2 C}$$ ...(24.9.1)

Equation (24.9.1) can also be written as

$$\omega = \frac{1}{\sqrt{(3C)L}}$$

But this relationship is also the condition for resonance at the supply frequency ω rad/s, when a capacitance of $3C$ farad is connected in series with an inductance of L henry. Hence the required condition for the quenching of the arc at the fault point F is that the reactance ωL shall be in series resonance with the capacitance $3C$ farad, where C farad is the capacitance of each line to ground.

It is seen from Eq. (24.9.1) that the Peterson coil depends for its operation on such an adjustment of its reactance that

$$\omega L = \frac{1}{\omega \times (3C)}$$ ...(24.9.2)

In other words, this condition states that the inductance must be "tuned" to the capacitance if the Peterson coil is to fulfil its function.

24.9.1 Construction of Peterson coil

The construction of Peterson coil is similar to that of an oil immersed transformer. It consists of a winding on a magnetic core. The coil is provided with tappings so that its inductance can be adjusted or tuned to such a value that the current passing through it is equal and opposite to the resultant current passing through the ground capacitances of the healthy lines in the event of a single line-to-ground fault.

24.9.2 Rating of Peterson coil

Peterson coil may be rated either for a short-time duty of about 5 minutes, or it is designed to carry its rated current continuously. A short-time rated coil is provided with an automatic circuit breaker, which by the action of a relay shorts the coil after a definite time and connects the neutral directly to ground. A short-time rated coil clears transient faults without supply interruption. Sustained faults are cleared by isolating the faulty section. A continuously rated coil allows a fault to remain in the system till it is located and removed.

24.9.3 Advantages of Peterson-coil grounding

1. The Peterson-coil grounding is an effective method of clearing both transient faults due to lightning and sustained single line-to-ground faults.
2. There is no tendency of arcing grounds to occur and the arcs are usually self extinguishing.
3. Voltage drops to single line-to-ground faults are minimized.

24.9.4 Limitations of Peterson-coil grounding

1. There is a need for retuning after any network modification.
2. The lines should be transposed.
3. There is an increase in corona and radio interference (RI) in the event of a double line-to-ground fault.

24.10 GROUNDING TRANSFORMER

If the system neutral is not available for some reason, for example, when a system is delta connected, the grounding can be done as follows :

(a) By using a zigzag grounding transformer or (b) by using a delta star grounding transformer.

(a) Zigzag Transformer

The zigzag transformer has no secondary winding. It is a three-limbed transformer. Each limb has got two identical windings. One set of windings is connected in star to provide the neutral point. The other ends of this set of windings are connected to the second set of windings as shown in Fig. 24.12. The directions of current in the two windings on each limb are opposite to each other.

Under normal operating conditions the total flux in each limb is negligibly small. Therefore, the transformer draws very small magnetizing current. Under fault conditions the impedance of

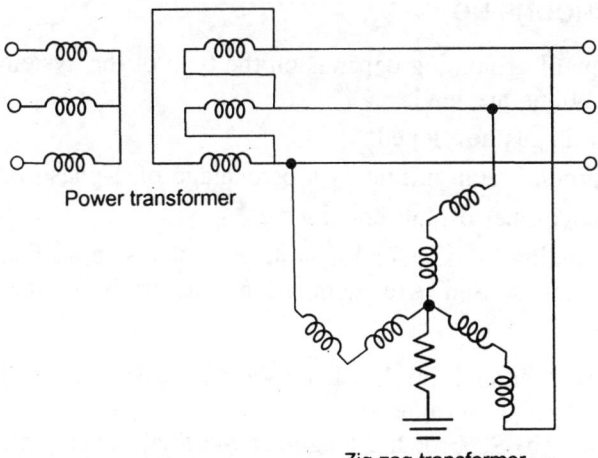

Fig. 24.12. Zig-zag transformer for neutral grounding.

the grounding transformer is very low. In order to limit the fault current, a resistor is connected in series with the neutral grounding connection as shown in Fig. 24.12.

(b) Delta-Star Grounding Transformer

In case of delta-star grounding transformer, (Fig. 24.13) the delta side is closed to provide a path for zero-sequence current. The star winding must be of the same voltage rating as the circuit that is to be grounded, whereas the delta voltage rating can be chosen to be any standard voltage level.

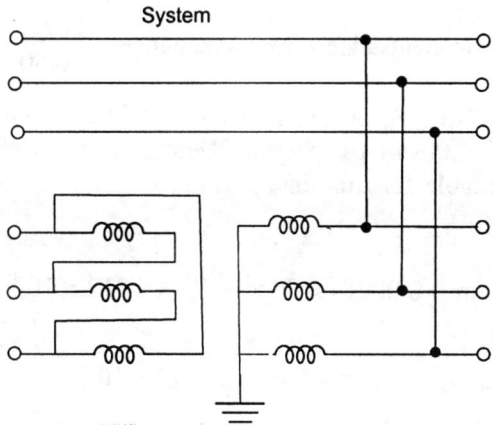

'**Fig. 24.13.** Delta-star transformer grounding.

The zigzag grounding transformer is designed for a short-time kVA rating equal to the rated phase voltage multiplied by the rated neutral current. The grounding transformer is designed to carry the rated current for very short time of 10 seconds or a maximum of one minute. The grounding transformer is therefore much smaller than an ordinary 3-phase transformer of the same rated kVA capacity.

24.11 CHOICE OF GROUNDING

The selection of the type of grounding depends on the type of the system and its voltage levels. The following considerations are made :

1. Transient overvoltages developed.
2. Magnitude of ground-fault current as a percentage of 3-phase fault current.
3. Dip in line voltage due to fault conditions.

Generally, solid grounding is used for low-voltage systems (upto 600 V). For voltages upto 11 kV resistance grounding is used. Arc suppression coils are best suited for high-voltage over head lines.

Example 24.1 A 50 Hz overhead line has line-to-earth capacitance of 1.2 μF. It is decided to use an earth fault neutralizer. Determine the reactance to neutralize the capacitance of (a) 100% of the length of the line, (b) 90% of the length of the line, and (c) 80% of the length of the line.

Solution

(a) The inductive reactance to neutralize 100% of the capacitance is given by

$$\omega L = \frac{1}{3\omega C} = \frac{1}{3 \times 2\pi \times 50 \times 1.2 \times 10^{-6}} = 884.19 \ \Omega$$

(b) The capacitance of 90% of the length of line $= 0.90 \times 1.2 = 1.08$ μF.

The inductive reactance to neutralize 90% capacitance

$$= \frac{1}{3\omega \times 0.9C} = \frac{1}{3 \times 2\pi \times 50 \times 0.9 \times 1.2 \times 10^{-6}} = 982.43 \ \Omega$$

(c) The inductive reactance to neutralize 80% capacitance $= \dfrac{884.19}{0.80} = 1105.24 \ \Omega$

Example 24.2 A 132 kV, 3-phase, 50 Hz transmission line 80 km long has a capacitance to earth of each line equal to 0.00914 μF per km. Determine the inductance and kVA rating of the arc suppression coil suitable for this line.

Solution

$$C = 0.00914 \ \mu\text{F/km} = 0.00914 \times 80 \times 10^{-6} = 0.7312 \times 10^{-6} \ \text{F}$$

$$L = \frac{1}{3\omega^2 C} = \frac{1}{3 \, (2\pi f)^2 C} = \frac{1}{3 \, (2\pi \times 50)^2 \times 0.7312 \times 10^{-6}} = 4.619 \ \text{H}$$

$$I_L = \frac{V_{ph}}{X_L} ; \qquad V_{ph} = \frac{V_l}{\sqrt{3}} = \frac{132 \times 10^3}{\sqrt{3}} \ \text{V}$$

$$X_L = 2\pi f L = 2\pi \times 50 \times 4.619$$

$$\therefore \qquad I_L = \frac{132 \times 10^3}{\sqrt{3} \times 2\pi \times 50 \times 4.619} = 52.5188 \ \text{A}$$

Rating of the a.c. suppression coil

$$S_{PC} = V_{ph}I_L = \frac{132 \times 10^3}{\sqrt{3}} \times 52.5188 = 4002.59 \times 10^3 \text{ VA} = 4002.59 \text{ kVA}$$

Alternative Method

Rating of the arc suppression coil

$$S_{PC} = V_{ph}I_L ; \qquad I_L = \frac{V_{ph}}{X_L}, \ X_L = \omega L = \frac{1}{3\omega C}$$

$\therefore$
$$S_{PC} = V_{ph}^2 \times 3\omega C = \left(\frac{V_l}{\sqrt{3}}\right)^2 \cdot 3\omega C = V_l^2 \omega C = V_l^2 \ (2\pi f)C$$

$$= (132 \times 10^3)^2 \times 2\pi \times 50 \times 0.7312 \times 10^{-6} \text{ VA}$$

$$= 4002.52 \times 10^3 \text{ VA} = 4002.52 \text{ kVA}$$

EXERCISES

1. Explain the phenomenon of *arcing grounds*. How does neutral grounding eliminate arcing grounds?

2. Explain the advantages of grounding power system neutrals.

3. What are the various methods of neutral grounding? Compare their performance with respect to (a) protective relaying, (b) fault levels, (c) stability, and (d) voltage levels of power systems.

4. Explain the following statement :

   In the event of a SLG fault on a system with isolated neutral the voltages of the remaining two healthy phases rise from their normal phase voltages to full line-to-line voltages.

5. Explain with the help of circuit and phasor diagrams the function of a Peterson coil in a 3-phase system.

6. Derive an expression for the inductance of the Peterson coil in terms of the capacitance of the protected line.

7. A 33 kV, 3-phase, 50 Hz, 60 km overhead line has a capacitance to earth of each line equal to 0.01 µF/km. Determine the inductance and kVA rating of the arc suppression coil.

8. A 50 Hz transmission line has a capacitance of 0.1 µF per phase. Determine the inductance of Peterson coil to neutralize the effect of capacitance of (a) complete length of line, (b) 95% of the length of line, (c) 85% of the length of the line.

9. A 132 kV, 50 Hz, 3-phase, 75 km long transmission line has a capacitance of 0.015 µF/km per phase. Determine the inductive reactance and MVA rating of the arc suppression coil suitable for the line to eliminate arcing ground phenomenon.

10. A 132 kV, 3-phase, 50 Hz overhead line of 100 km length has a capacitance to earth of 0.012 μF/km. Determine the inductance and MVA rating of the arc suppression coil suitable for this line.

ANSWERS

7. 5.629 H ; 205.27 kVA   **8.** (a) 33.77 H (b), 35.55 H (c) 39.73 H

9. 943.14 Ω ; 6.158 MVA   **10.** 2.814 H ; 6.5687 MVA

25

Tariffs

25.1 INTRODUCTION

By *tariff* is meant the schedule of rates framed by the supplier for the supply of electrical energy to various types of consumers. Thus, different methods of charging a consumer or rates of payments by the consumer, for consuming electrical energy are known as tariffs. The following factors are taken into account to decide the tariff :

- Type of load (domestic, commercial, or industrial)
- Maximum demand
- Time at which load is required
- Power factor of the load
- Amount of energy used.

The manner in which consumers pay for electrical energy varies according to their requirements. Industrial consumers use more energy for longer times than domestic consumers. Tariff should be such as to cover total cost of producing and supplying electrical energy plus a reasonable profit.

The cost of electrical energy supplied by a generating station depends on the installed capacity of the plant and kWh generated. Increase in maximum demand increases the installed capacity of the generating station.

The time at which maximum demand occurs is also important in plant economics. If the maximum demand of the consumer coincides with the maximum demand on the system, additional plant capacity is required. However, if the maximum demand of the consumer occurs during off-peak hours, the load factor is improved and no extra plant capacity is needed. Thus, the overall cost per kWh generated is reduced.

Power factor is also important from the point of view of plant economics. At a low power factor, the load current is large. Consequently, the current to be supplied from the generating station is also large. This large current is responsible for greater I^2R losses in the system and larger voltage drops. Thus, the regulation becomes poor; In order to supply the consumer's voltage within permissible limits, power factor correction equipment is to be installed. Therefore, the cost of generation increases.

The cost of electrical energy is reduced by using a large amount of energy for longer periods.

25.2 TYPES OF TARIFFS

Several types of tariffs have been proposed. All these tariffs have been derived from following relationship :

$$C = Ax + By + D$$

where C = total charge for a certain period (say one month)

x = maximum demand during the period (kW or kVA)

y = total energy consumed during the period (kWh)

A = cost per kW or per kVA of maximum demand

B = cost per kWh of energy consumed

D = fixed charge during each billing period.

Thus, the total bill of the consumer has three parts, namely, fixed charge D, semi-fixed charge Ax and running charge By. This is known as a *three-part tariff*. It is generally applied to big consumers.

Some of the important types of tariff are as follows :

1. Flat demand rate tariff
2. Straight-line meter rate tariff
3. Block meter rate tariff
4. Two-part tariff
5. Power-factor tariff
6. Seasonal rate tariff
7. Peak load tariff
8. Three-part tariff.

25.2.1 Flat demand rate tariff

The flat demand rate tariff is expressed in the form, $C = Ax$. Here the bill depends only on the maximum demand. It is independent of the energy consumed. This system is used in street lighting, sign lighting, signal system and irrigation tube wells. In all such systems the amount of connected load and hours of their use are known and the rates of charges are made accordingly. Thus, metering is not required in this system of tariff.

25.2.2 Straight-line meter rate tariff

This type of tariff is given by the relationship $C = By$. In this system the bill depends only on the amount of energy consumed. In this system, different types of consumers are charged at different rates. The rate for each type of consumption is decided by taking into consideration the load factor and diversity factor of the load. For example, the flat rate for light and fan loads is higher than that for power loads. For this purpose, separate energy meters are to be installed for light and power loads.

25.2.3 Block meter rate tariff

In this system, the energy consumption is divided into blocks and the price per unit is fixed in each block. The price per unit in the first block is the highest and it decreases for the succeeding blocks. Generally the price and energy consumption are divided into three blocks. The first few units of energy at a certain rate, the next few at a slightly lower rate and the remaining units at a still lower rate. Theoretically, this form of tariff does not encourage energy conservation. The energy crisis in India has compelled several states to adopt the reverse form of this tariff. In order to discourage to use more energy, the first few kWh are charged at a certain rate, the next few at a slightly higher rate, the next few at a still higher rate, and so on. For example, the first 100 units or less may be charged at the rate of Re. 1.43 per unit, the next 100 units at the rate of Rs 1.63 per unit, and the remaining additional units may be charged at the rate of Rs 1.83 pr unit. This system of tariff is used for majority of domestic and commercial consumers.

25.2.4 Two-part tariff

In two-part tariff the total charge to be made from the consumer is split into two components. The first component is the fixed charge which is dependent upon the maximum demand. the second component is the running charge which depends upon the energy consumed.

This tariff can be expressed as

$$C = Ax + By$$

or $\quad C = A \text{ (kW)} + B \text{ (kWh)}$

where A = charge per kW of maximum demand

$\quad B$ = charge per kWh of energy consumed.

The factors A and B may be constant or may vary according to some sliding scale.

The maximum demand can either be taken as a certain fraction of the connected load or measured by a maximum demand indicator. The maximum demand indicator records the maximum load in kilowatts during any 15 minutes or 30 minutes in a month or some other period. The two-part tariff is used for industrial consumers.

25.2.5 Power factor tariff

The tariff in which power factor of the consumer's load is taken into consideration is called power factor tariff. The following are the important types of power factor tariff in use :

1. kVA maximum demand tariff

$$\text{Total charges} = A \text{ (kVA)} + B \text{ (kWh)}$$

This is also a 2-part tariff. In this system the consumer is compelled to improve the p.f. of his load since low p.f. will increase the kVA.

2. kWh and kVArh tariff

Total charges $= A_1 \,(\text{kWh}) + B_1 \,(\text{kVArh})$

Since kVArh decreases with the increase in p.f., this type of tariff will induce the consumer to improve the p.f.

3. Sliding Scale or average power factor tariff

In this type of tariff an average power factor, say 0.8 lagging may be taken as the reference. If the power factor of the consumer is below this reference value, a suitable additional charge is realized from the consumer for each step decrease of 0.01 below 0.8. Similarly, if the power factor is above this reference value, a suitable discount is allowed to the consumer for each step rise of 0.01 above 0.8. The sliding-scale power tariffs are however, very rare.

25.2.6 Seasonal rate tariff

The seasonal rate tariff specifies higher prices per kWh used during the season of the year in which the system peak occurs. This is known as the on-peak season. The prices per kWh are lower during the season of the year in which the usage is lowest. This is known as off-peak season.

25.2.7 Peak-load tariff

The peak load tariff is similar to seasonal rate tariff. It specifies higher prices per kWh used during the peak period of the day and lower prices during the off-peak period of the day.

The seasonal rate tariff and peak load tariff are both designed to reduce the systems peak load and therefore, reduce the system idle standby capacity.

25.2.8 Three-part tariff

As discussed earlier, the three-part tariff is of the form

$C = Ax + By + D$

It is generally applied to big consumers.

Example 25.1 A consumer has a maximum demand of 80 kW at 0.45 load factor. If the tariff is Rs 750 per kW of maximum demand plus Rs 1.10 per kWh, determine the overall cost per kWh.

Solution

Annual energy consumption $= 80 \times 0.45 \times 8760 = 315360$ kWh

Cost of annual energy consumption, $C_E = $ Rs $1.10 \times 315360 = $ Rs 346896

Fixed charges per year, $C_F = $ Rs $750 \times 80 = $ Rs 60000

Total annual charges $= C_E + C_F = $ Rs $(346896 + 60000) = $ Rs 406896

Overall cost per kWh $= $ Rs $\dfrac{406896}{315360} = $ Rs 1.29.

Example 25.2 An industrial consumer has a connected load of 50 kW. The maximum demand is 40 kW. On an average each machine works for 8 hours a day for 300 working days in a year. The tariff is (Rs 5000 + Rs 800 per kW of maximum demand + Rs 1.15 per kWh of energy consumed). Calculate the annual bill of the consumer.

Solution

Energy consumed per year $= 50 \times 8 \times 300 = 120000$ kWh

Annual bill of the consumer $= \text{Rs } (5000 + 800 \times 40 + 120000 \times 1.15) = \text{Rs } 175000.$

Example 25.3 The output of a generating station is 960×10^6 kWh per annum. The average load factor is 0.55. If the annual fixed charges are Rs 1000 per kW of the installed capacity and the running charges are Rs 0.40 per kWh, determine the cost of energy per kWh at the busbars.

Solution

$$\text{Maximum demand} = \frac{\text{kWh generated per year}}{\text{load factor} \times 8760} = \frac{960 \times 10^6}{0.55 \times 8760} = 199.2528 \text{ MW}$$

Annual demand charges $= \text{Rs } 1000 \times 199.2528 \times 10^3 = \text{Rs } 199252.8 \times 10^3$

Annual energy charges $= \text{Rs } 960 \times 10^6 \times 0.40 = \text{Rs } 384 \times 10^6$

Total annual charges $= \text{Rs } (199252.8 \times 10^3 + 432 \times 10^6) = \text{Rs } 631.258 \times 10^6$

$$\text{Cost of energy per kWh at the busbars} = \frac{\text{total annual cost}}{\text{total units generated per year}}$$

$$= \frac{631.258 \times 10^6}{960 \times 10^6} = \text{Rs } 0.6575$$

Example 25.4 The daily load of an industrial concern is as follows : 100 kW for 9 hours; 125 kW for 6 hours; 50 kW for 7 hours; 5 kW for 2 hours. The tariff is Rs 800 per kW of maximum demand per year plus Rs 1.30 per kWh. Determine the energy consumption per year (for 365 days) and the yearly bill.

Solution

Energy consumption per day $= 100 \times 9 + 125 \times 6 + 50 \times 7 + 5 \times 2 = 2010$ kWh

Annual energy consumption $= 2010 \times 365 = 733650$ kWh

Annual cost of energy consumption $= \text{Rs } 733650 \times 1.30 = \text{Rs } 953745$

Maximum demand $= 125$ kW

Annual maximum demand charges $= \text{Rs } 125 \times 800 = \text{Rs } 100000$

Total annual charges $= \text{Rs } (953745 + 100000) = \text{Rs } 1053745.$

Example 25.5 A consumer has an annual consumption of 200000 kWh at a load factor of 40%. The tariff is Rs 750 per kW of maximum demand per year plus Rs 1.15 per kWh. Determine the saving in energy cost if the load factor is improved to 60%.

Solution

$$\text{Average load} = \frac{\text{energy consumed per year}}{8760} = \frac{200000}{8760} = 22.831 \text{ kW}$$

$$\text{Maximum demand} = \frac{\text{average load}}{\text{load factor}} = \frac{22.831}{0.4} = 57.0776 \text{ kW}$$

If the load factor is improved to 60%, then maximum demand $= \dfrac{22.831}{0.6} = 38.05166$ kW

Reduction in maximum demand $= 57.0776 - 38.05166 = 19.0259$ kW

Saving in demand charges $=$ Rs $19.0259 \times 750 =$ Rs 14269.45.

Example 25.6 An industrial consumer has a maximum demand of 150 kW at a load factor of 0.65. The tariff is Rs 900 per kVA of maximum demand per year plus Rs 1.30 per kWh of energy consumed. If the average power factor is 0.82 lagging, calculate the total energy consumed per year and the total yearly electricity bill.

Solution

$$\text{Maximum kVA demand} = \frac{\text{maximum kW demand}}{\text{power factor}} = \frac{150}{0.82} = 182.9268 \text{ kVA}$$

Annual energy consumption $=$ (maximum kW demand) $\times$ load factor $\times 8760$
$$= 150 \times 0.65 \times 8760 = 854100 \text{ kWh}$$

Annual cost of energy consumption $=$ Rs $854100 \times 1.30 =$ Rs 1110330

Annual demand charges $=$ Rs $900 \times 182.9268 =$ Rs 164634

Annual electricity bill $=$ Rs $(1274964 + 1110330) =$ Rs 1274964.

Example 25.7 An industrial consumer has a maximum demand of 100 kW. Two alternative tariffs are available :

(a) A fixed charge of Rs 800 per kW of maximum demand per year plus a running charge of Rs 1.30 per kWh of energy consumed.

(b) A flat rate of Rs 1.83 per kWh.

Which tariff is economical if the factory runs for 3600 hours per year with a load factor of 0.8?

Solution

Average demand $=$ (maximum demand $\times$ load factor) $= 100 \times 0.8 = 80$ kW

Annual energy consumption $=$ average demand $\times$ working hours per year
$$= 80 \times 3600 = 288000 \text{ kWh}$$

Annual bill with first tariff $=$ Rs $(800 \times 100 + 1.3 \times 288000) =$ Rs 454400

Annual bill with second tariff $=$ Rs $1.83 \times 288000 =$ Rs 527040.

Since the annual bill with two-part tariff is lower than that with flat-rate tariff, the two-part tariff is economical.

Example 25.8 Two systems of tariff are available for a factory, working 8 hours a day for 300 working days in a year.

(a) High-voltage supply at Rs 75 per month per kVA of maximum demand plus Rs 1.15 per kWh consumed.

(b) Low-voltage supply at Rs 80 per month per kVA of maximum demand plus Rs 1.43 per kWh consumed.

The factory has an average load of 400 kW at 0.8 power factor and a maximum demand of 500 kW at the same power factor. The cost of high voltage equipment is Rs 900 per kVA of maximum demand and losses can be taken as 4%. The interest and depreciation charges on high voltage equipment are 15 per cent. Calculate the annual bill for both the systems.

Solution

(a) *High-voltage supply*

Maximum kVA demand of the factory $= \dfrac{\text{maximum kW demand}}{\text{power factor}} = \dfrac{500}{0.8} = 625$

kVA capacity of the high-voltage equipment $= \dfrac{625}{0.96} = 651.04$

Cost of hv equipment $= \text{Rs } 900 \times 651.04$

Annual interest and depreciation $= \text{Rs } \dfrac{15}{100} \times 900 \times 651.04 = \text{Rs } 87890.62$

Annual maximum demand charges $= \text{Rs } 75 \times 12 \times 625 = \text{Rs } 562500$

Energy consumption per year $=$ (load in kW) (working hours per year)

$$= \dfrac{400}{0.96} \times (300 \times 8) = 1000000 \text{ kWh}$$

Cost of energy consumption per year $= \text{Rs } 1000000 \times 1.15 = \text{Rs } 1150000$

Total charges per year with hv supply $= \text{Rs } (562500 + 1150000) = \text{Rs } 1712500.$

(b) *Low-voltage supply*

Annual maximum demand charges $= \text{Rs } 80 \times 12 \times 625 = \text{Rs } 600000$

Annual energy consumption $= 400 \times 300 \times 8 = 960000 \text{ kWh}$

Cost of energy consumption per year $= \text{Rs } 960000 \times 1.43 = \text{Rs } 1372800$

Total charges per year with lv supply $= \text{Rs } (600000 + 1372800) = \text{Rs } 1972800.$

Example 25.9 The following rates are offered to a consumer :

(a) Rs 400 plus Rs 1.43 per kWh;

(b) A flat rate of Rs 1.83 per kWh.

Find the number of units consumed for which the first tariff is economical.

Solution

Let x be the number of units (kWh).

Charge due to first tariff, $C_1 = \text{Rs } (400 + 1.43x)$

Charge due to second tariff, $C_2 = \text{Rs } 1.83x$

If the first tariff is economical

$$C_1 < C_2$$

That is, $400 + 1.43x < 1.83x$

or $1.83x - 143x > 400$

or $0.40x > 400$

or $\qquad x > \dfrac{400}{0.4}$

or $\qquad x > 1000.$

Therefore, the first tariff is economical if the energy consumption is greater than 1000 kWh.

Example 25.10 The following tariffs are offered to a consumer :
  (a) Rs 500 per year plus Rs 0.90 per kWh;
  (b) Rs 1.43 for the first 100 units per month and Rs 1.63 for next 100 units and Rs 1.83 for all the additional units.

Find the energy consumed per year for which the charges due to both tariffs become equal.

Solution

Let x be the number of units consumed, where $x > 200.$
  Annual charges due to first tariff,
$$C_1 = \text{Rs } (500 + 0.90x)$$
  Annual charges due to second tariff,
$$C_2 = \text{Rs } (1.43 \times 100) + \text{Rs } (1.63 \times 100) + \text{Rs } (1.83) \, (x - 200)$$
$$= \text{Rs } (143 + 163 + 1.83x - 366) = \text{Rs } (1.83x - 60)$$
If $\qquad C_1 = C_2$
$$500 + 0.90x = 1.83x - 60$$
$$(1.83 - 0.90) \, x = 560$$
$$x = \dfrac{560}{0.93} = 602 \text{ kWh}$$

Example 25.11 Determine the generation cost per kWh from the following data :
  Installed capacity = 500 MW
  Capital cost = Rs 35000 per kW
  Interest and depreciation = 12%
  Fuel consumption = 0.85 kg/kWh
  Fuel cost = Rs 800 per 1000 kg
  Other operating costs = 25% of fuel cost
  Peak load = 475 MW
  Load factor = 0.82.

Solution

  Average load = peak load × load factor = $475 \times 0.82 = 389.5$ MW
  Energy generated per year = (average load in kW) × number of hours per year
$$= 389.5 \times 10^3) \times 8760 \text{ kWh}$$
  Total investment = Rs $(500 \times 10^3) \times 35000 = \text{Rs } 1.75 \times 10^{10}$

Annual interest and depreciation, $C_{ID} = \text{Rs } \dfrac{12}{100} \times 1.75 \times 10^{10} = \text{Rs } 2.1 \times 10^9$

Fuel consumption per year $= 0.85 \times 389.5 \times 10^3 \times 8760 \text{ kg}$

Annual fuel cost, $C_{AF} = \text{Rs } \dfrac{800}{1000} \times 0.85 \times 389.5 \times 10^3 \times 8760 = \text{Rs } (2.32017 \times 10^9)$

Other operating costs per year, $C_{AO} = 25\%$ of fuel cost $= \dfrac{25}{100} \times C_{AF} = 0.25 \, C_{AF}$

Annual plant cost, $C_{AP} =$ annual fixed cost $+$ annual operating costs
$$= C_{ID} + C_{AF} + C_{AO} = C_{ID} + 1.25 \, C_{AF}$$
$$= \text{Rs } (2.1 \times 10^9 + 1.25 \times 2.32017 \times 10^9) = \text{Rs } 5.00021 \times 10^9$$

Generation cost per kWh $= \dfrac{C_{AP}}{\text{energy generated per year}} = \text{Rs } \dfrac{5.00021 \times 10^9}{389.5 \times 10^3 \times 8760} = \text{Rs } 1.47.$

EXERCISES

1. What is meant by tariff? Name the factors taken into account to decide the tariff.
2. Describe some of the types of tariff commonly used in practice.
3. Write short notes on the following :
   (a) Two-part tariff
   (b) Three-part tariff
   (c) Power factor tariff.

26

Power Factor Improvement

26.1 INTRODUCTION

The power factor is defined as the ratio of the active power (P) and voltamperes (S).

$$\text{power factor} = \frac{P}{S} = \frac{P}{VI}$$

For sinusoidal waveforms the power factor is the cosine of the phase angle ϕ between voltage and current.

$$\text{power factor} = \cos \varphi$$

$\therefore \qquad \cos \varphi = \dfrac{P}{VI}$

$P = VI \cos \varphi$

$I = \dfrac{P}{V \cos \varphi}$ \hfill ...(26.1.1)

Equation (26.1.1) shows that the current is affected by the power factor. The supply voltage V is kept fairly constant. Hence, for a given power P required by the load, the current I taken by the load varies inversely as the load power factor $\cos \varphi$. Thus, a given load takes more current at a low power factor than it does at a high power factor.

26.2 DISADVANTAGES OF A LOW POWER FACTOR

The undesirable effects of operating a load at a low power factor are due to larger current required at low power factor. The important disadvantages of low power factor are :

1. Higher currents require larger cables, switchgears, transformers, alternators, etc. Thus, the capital cost of the equipment is increased. This is, uneconomical from the supplier's point of view.

2. Higher currents give rise to higher copper ,losses in the system and therefore the efficiency of the system is reduced. Also, the cost of energy loss (that is, running cost) in the system is increased.

3. Higher currents produce larger voltage drop in cables and other apparatus. This results in poor voltage regulation. The *voltage regulation* is defined as the change in voltage between the no-load and full-load conditions. To keep the consumer's voltage within the prescribed limits, the cost of the apparatus for compensating the voltage drop increases.

Since both the capital and running costs are increased, the operation of a system at a low power factor (whether it is lagging or leading) is uneconomical from the supplier's point of view. Because of this, the supplier's point of view. Because of this, the supplier usually takes extra charges from the consumers who operate their loads at low power factors. Such consumers are offered special tariffs which persuade them to operate at higher power factors. The tariff of a large consumer consists of two parts.

Total cost = cost for maximum demand + cost for energy used

$$C = C_1 \, (\text{kVA}) + C_2 \, (\text{kWh}) \qquad \qquad \qquad ...(26.2.1)$$

where C_1 = cost per maximum demand in kVA

C_2 = cost per kWh used

C = total cost paid by the consumer.

kW = kVA cos φ

$$\text{kVA} = \text{kW}/\cos \varphi \qquad \qquad \qquad ...(26.2.2)$$

This relation shows that if cos φ is large, kVA is small. Hence, to reduce kVA taken from the supply, the consumer is compelled to operate at a high power factor which is as near to unity as possible. In practice, the power factor is rarely corrected to unity because the cost of equipment required to improve the power factor is usually greater than the saving on tariff and plant. The most economical power factor usually lies between 0.9 to 0.95 lagging.

26.3 CAUSES OF LOW POWER FACTOR

The usual cause of low power factor is due to inductive loads. The current in an inductive load lags behind the voltage. The power factor is therefore lagging. The important inductive loads responsible for low power factor are as follows :

1. Most of the ac motors are of induction type. Three-phase induction motors operate at a power factor of about 0.8 lagging at full load. At light loads these motors work at a very small power factor of the order of 0.2 to 0.3 lagging. Single-phase motors operate at power factor of around 0.6.

2. A transformer draws magnetizing current from the supply. At normal load, this current does not affect the power factor much but at light loads the primary current power factor is low.

3. Arc lamps, electric discharge lamps, industrial heating furnaces, welding equipment operate at low lagging power factors.

26.4 POWER FACTOR IMPROVEMENT

If the power factor is low (or poor) it is necessary to improve (or correct) it. The basic principle of power factor improvement is to inject a leading current into the circuit so as to neutralize the effect of lagging current. The power factor may be improved by the following methods :

1. By using static capacitors
2. By using synchronous motors.

26.5 POWER FACTOR CORRECTION BY STATIC CAPACITORS

Consider an inductive load consisting of a resistor R and an inductor L connected to an ac supply. The circuit and phasor diagrams are shown in Fig. 26.1.

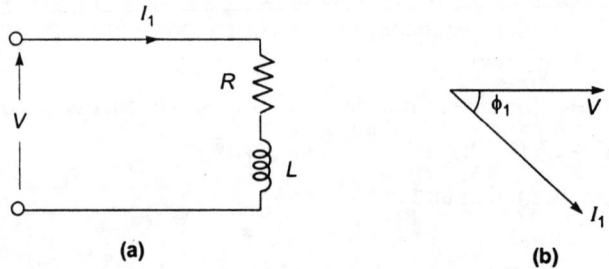

(a)   **(b)**

Fig. 26.1. (a) Circuit diagram (b) Phasor diagram.

Let V = supply voltage

I_1 = load current

φ_1 = phase angle by which the current I_1 lags behind the voltage V

cos φ_1 = original power factor.

Let a capacitor C be placed in parallel with the load. It will take a leading current I_C from the supply. The circuit and phasor diagrams are shown in Fig. 26.2.

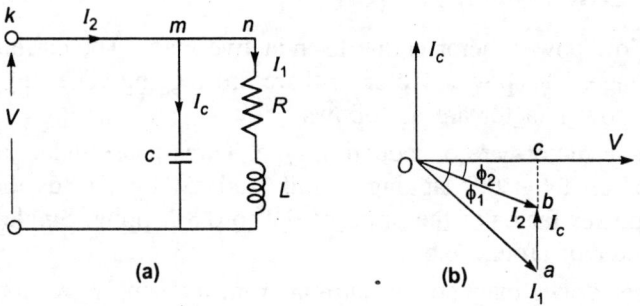

(a)   **(b)**

Fig. 26.2. (a) Circuit diagram (b) Phasor diagram.

The total current $\mathbf{I}_2$ drawn from the supply will be equal to the phasor sum of $\mathbf{I}_1$ and $\mathbf{I}_C$, that is,

$$\mathbf{I}_2 = \mathbf{I}_1 + \mathbf{I}_C \qquad \qquad \qquad \qquad \text{...(26.5.1)}$$

Conclusions

1. The phase angle of I_2 is φ_2. It is seen from the phasor diagram that φ_2 is less than φ_1 and hence $\cos \varphi_2$ is greater than $\cos \varphi_1$. In other words, the power factor is improved from $\cos \varphi_1$ to $\cos \varphi_2$.

2. From the phasor diagram [Fig. 26.2 (b)]

$$Oc = I_1 \cos \varphi_1 = I_2 \cos \varphi_2$$

Therefore, the new supply current is given by

$$I_2 = I_1 \frac{\cos \varphi_1}{\cos \varphi_2}$$

Since $\cos \varphi_2 > \cos \varphi_1$, $\qquad I_2 < I_1$

Hence the new current drawn from the supply is less than the load current I_1. This is also seen from the phasor diagram.

3. Again, $I_2 \cos \varphi_2 = I_1 \cos \varphi_1$

$$\therefore \qquad VI_2 \cos \varphi_2 = VI_1 \cos \varphi_1$$

or $\qquad P_2 = P_1$

This relation shows that the power taken from the supply has not altered.

Hence by connecting a capacitor in parallel (shunt capacitor) with an inductive load, the power factor is improved and the current taken from the supply is reduced without altering either the current or power taken by the load.

26.6 CAPACITOR RATING CALCULATIONS

From the phasor diagram [Fig. 26.2 (b)]

$$ab = ac - bc$$

$$I_C = I_1 \sin \varphi_1 - I_2 \sin \varphi_2 \qquad \qquad \text{...(26.6.1)}$$

Multiplying both the sides of Eq. (26.6.1) by V we obtain

$$VI_C = VI_1 \sin \varphi_1 - VI_2 \sin \varphi_2 \qquad \qquad \text{...(26.6.2)}$$

or $\qquad Q_C = Q_1 - Q_2 \qquad \qquad \text{...(26.6.3)}$

where $\qquad Q_1 = VI_1 \sin \varphi_1 \qquad \qquad \text{...(26.6.4)}$

$\qquad \qquad$ = reactive voltamperes taken by the load

$\qquad Q_2 = VI_2 \sin \varphi_2 \qquad \qquad \text{...(26.6.5)}$

$\qquad \qquad$ = reactive voltamperes taken from the supply

$\qquad Q_C = VI_C \qquad \qquad \text{...(26.6.6)}$

$\qquad \qquad$ = leading reactive voltamperes drawn by the capacitor from the supply.

The power diagram is shown in Fig. 26.3. Let P be the power received by the load. In Fig. 26.3,

$$OD = P$$

$$OA = S_1 = VI_1$$

$$AD = Q_1$$

$$OB = S_2 = VI_2$$

$$BD = Q_2$$

$$Q_1 = P \tan \varphi_1$$

$$Q_2 = P \tan \varphi_2$$

$$Q_C = Q_1 - Q_2$$

$$Q_C = P (\tan \varphi_1 - \tan \varphi_2) \qquad \qquad \qquad \qquad \text{...(26.6.7)}$$

Equation (26.6.7) is used to determine the VAr rating of the capacitor. If P is expressed in kW, then Q_C is obtained in kVAr.

Calculation of C

The value of C can also be calculated.

Since $Q_C = VI_C$

and $I_C = \dfrac{V}{X_C} = V\omega C$

$\therefore$ $Q_C = V^2 \omega C$

and $C = \dfrac{Q_C}{\omega V^2} \qquad \qquad \qquad \qquad \qquad \qquad \text{...(26.6.8)}$

Equation (26.6.8) shows that the capacitance required for power factor correction is inversely proportional to V^2.

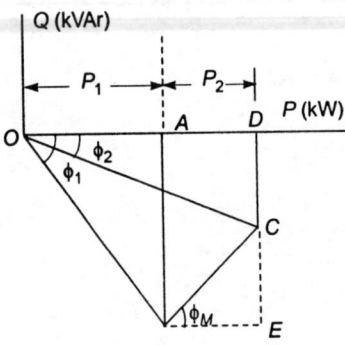

Fig. 26.3. Power diagram.

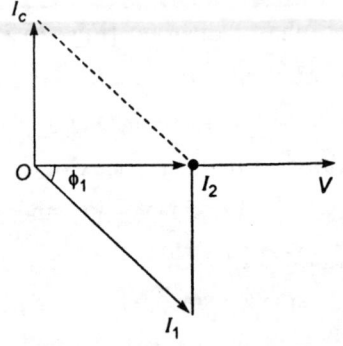

Fig. 26.4. Phasor diagram for improving the power factor to unity.

Special case

If the overall power factor is to be improved to unity

$$\varphi_2 = 0, \quad \sin \varphi_2 = 0, \quad \tan \varphi_2 = 0$$

The phasor diagram is shown in Fig. 26.4.

From Eq. (26.6.1), since $\sin \varphi_2 = 0$

$$I_C = I_1 \sin \varphi_1 \qquad \qquad \qquad \qquad \qquad \text{...(26.6.9)}$$

But $\qquad I_C = \dfrac{V}{X_C} = V\omega C$

$\therefore \qquad C = \dfrac{I_C}{\omega V} = \dfrac{I_1 \sin \varphi_1}{\omega V}$ ...(26.6.10)

From Eq. (26.6.7), since $\tan \varphi_2 = 0$

$\qquad Q_C = P \tan \varphi_1$ ...(26.6.11)

Example 26.1 A fluorescent lamp takes a current of 0.75 A when connected across a 240 V, 50 Hz ac supply. The power consumed by the lamp is 80 W. Calculate the value of the capacitance to be connected in parallel with the lamp to improve the power factor to (a) unity (b) 0.95 lagging.

Solution

$\qquad I_1 = 0.75$ A, $\quad V = 240$ V, $\quad P = 80$ W

$\qquad VI_1 \cos \varphi_1 = P$

$\qquad I_1 \cos \varphi_1 = \dfrac{P}{V} = \dfrac{80}{240} = \dfrac{1}{3} ; \qquad \cos \varphi_1 = \dfrac{1}{3 I_1} = \dfrac{1}{3 \times 0.75} = 0.444$

$\therefore \qquad \varphi_1 = 63.61°, \ \tan \varphi_1 = 2.0155$

(a) $\qquad \cos \varphi_2 = 1, \quad \varphi_2 = 0, \quad \tan \varphi_2 = 0$

$\qquad I_C = I_1 \cos \varphi_1 (\tan \varphi_1 - \tan \varphi_2) = \dfrac{1}{3} (2.0155 - 0) = 0.6718$ A

$\qquad C = \dfrac{I_C}{V\omega} = \dfrac{0.6718}{240 \times 2\pi \times 50} = 8.91 \times 10^{-6}$ F $= 8.91 \ \mu$F

(b) $\qquad \cos \varphi_2 = 0.95, \quad \varphi_2 = 18.19°, \quad \tan \varphi_2 = 0.3287$

$\qquad I_C = I_1 \cos \varphi_1 (\tan \varphi_1 - \tan \varphi_2) = \dfrac{1}{3} (2.0155 - 0.3287) = 0.5623$ A

$\qquad C = \dfrac{I_C}{V\omega} = \dfrac{0.5623}{240 \times 2\pi \times 50} = 7.457 \times 10^{-6}$ F $= 7.457 \ \mu$F

Example 26.2 A single-phase 50 Hz motor takes 20 A at 0.75 power factor lagging from a 230 V sinusoidal supply. Calculate the kVAr and capacitance of a capacitor to be connected in parallel to raise the power factor to 0.9 lagging. What is the new supply current?

Solution

$\qquad I_1 = 20$ A, $\quad f = 50$ Hz

$\qquad \cos \varphi_1 = 0.75, \quad \varphi_1 = 41.4°, \quad \tan \varphi_1 = 0.8819$

$\qquad \cos \varphi_2 = 0.90, \quad \varphi_2 = 25.84°, \quad \tan \varphi_2 = 0.4843$

$\qquad I_C = I_1 \cos \varphi_1 (\tan \varphi_1 - \tan \varphi_2) = 20 \times 0.75 (0.8819 - 0.4843) = 5.9637$ A

$\qquad C = \dfrac{I_C}{\omega V} = \dfrac{5.9637}{2\pi \times 50 \times 230} = 82.53 \times 10^{-6}$ F $= 82.53 \ \mu$F

$$Q_C = VI_C = 230 \times 5.9637 = 1371.65 \text{ VAr} = 1.3716 \text{ kVAr}$$

Let I_2 be the new supply current. Since the active component of supply current remains unchanged,

$$I_2 \cos \varphi_2 = I_1 \cos \varphi_1$$

$$I_2 = I_1 \frac{\cos \varphi_1}{\cos \varphi_2} = 20 \times \frac{0.75}{0.9} = 16.67 \text{ A}$$

26.7 POWER FACTOR CORRECTION IN THREE-PHASE SYSTEMS

The problems of power factor improvement in 3-phase systems are solved in a manner similar to the single-phase problems. In a 3-phase system it is convenient to use the per-phase basis.

We can compare the capacitance required per phase when capacitors are connected first in delta and then in star. The delta and star-connected banks are shown in Fig. 26.5.

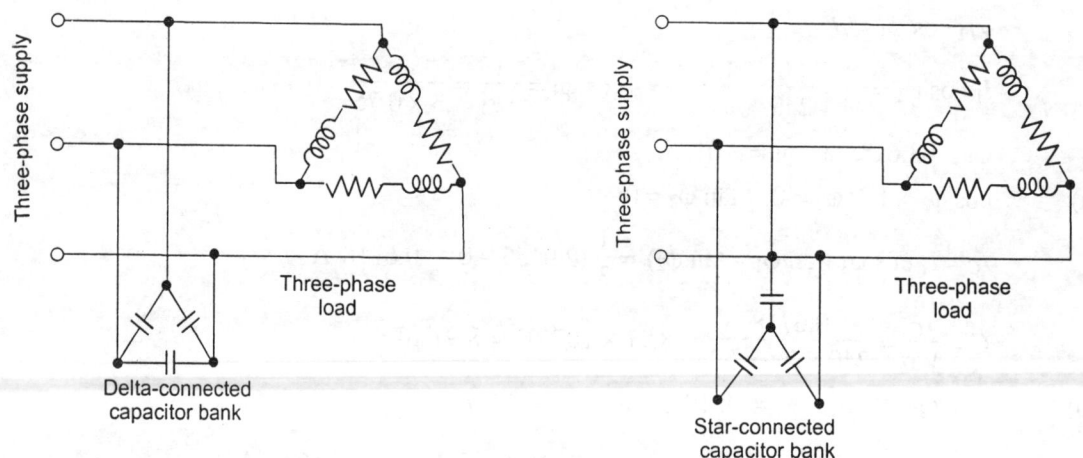

Fig. 26.5. Delta and star-connected capacitor banks.

Let V_L = line voltage

V_p = phase voltage

C_Δ = capacitance per phase when the capacitors are connected in delta

C_Y = capacitance per phase when the capacitors are connected in star

Q_C = VAr rating of each capacitor.

Delta connection

$$V_p = V_L$$

From Eq. (26.6.7)

$$C_\Delta = \frac{Q_C}{\omega V_p^2} = \frac{Q_C}{\omega V_L^2} \qquad \qquad ...(26.7.1)$$

Star connection

$$V_p = \frac{1}{\sqrt{3}} V_L$$

From Eq. (26.6.7)

$$C_Y = \frac{Q_C}{\omega V_p^2} = \frac{Q_C}{\omega (V_L/\sqrt{3})^2} = \frac{3Q_C}{\omega V_L^2} \qquad \qquad ...(26.7.2)$$

Combining Eqs. (26.7.1) and (26.7.2)

$$C_Y = 3C_\Delta \qquad \qquad ...(26.7.3)$$

Hence the capacitance required per phase in star connection is equal to three times the capacitance required per phase when the capacitors are connected in delta. Also, the capacitors for the star-connected bank have a working voltage equal to $1/\sqrt{3}$ times that for the delta-connected bank. For these reasons, the capacitors are connected in delta in a 3-phase system for power factor improvement. Delta connection is also better if the capacitors are designed for higher working voltage.

Example 26.3 A 3-phase load of 750 kW at 400 V, 50 Hz, operates at a power factor of 0.7 lagging. Calculate the kVAr rating per phase and the capacitance per phase of a mesh connected capacitor bank to improve the power factor to 0.95 lagging.

Solution

Power per phase, $P_p = \dfrac{750}{3} = 250$ kW

Original power factor, $\cos \varphi_1 = 0.7$

∴ $\qquad \varphi_1 = 45.57°$ and $\tan \varphi_1 = 1.02$

Improved power factor, $\cos \varphi_2 = 0.95$

∴ $\qquad \varphi_2 = 18.19°$, $\quad \tan \varphi_2 = 0.32868$

$\qquad Q_{Cp} = P_p (\tan \varphi_1 - \tan \varphi_2) = 250 (1.02 - 0.32868) = 172.83$ kVAr/phase

$\qquad Q_C = VI_C$; $\qquad I_C = \dfrac{V}{X_C} = V\omega C$

∴ $\qquad Q_C = V^2 \omega C$

$\qquad C = \dfrac{Q_C}{V^2 \omega}$

$\qquad C_\Delta = \dfrac{Q_{Cp}}{V_p^2 (2\pi f)} = \dfrac{172.83 \times 10^3}{(400)^2 \times 2\pi \times 50} = 3438 \times 10^{-6}$ F $= 3438$ μF

Example 26.4 A 400 V, 40 kW, 3-phase, 50 Hz induction motor runs at a power factor of 0.72 lagging with an efficiency of 85%. Find the capacitance per phase of a mesh-connected capacitor bank necessary to raise the power factor of the supply to 0.98 lagging.

Solution

$$\text{Input to motor} = \frac{\text{output}}{\text{efficiency}} = \frac{40}{0.85} = 47.0588 \text{ kW}$$

$$\text{Input power per phase, } P_p = \frac{47.0588}{3} = 15.686 \text{ kW}$$

Original power factor, $\cos \varphi_1 = 0.72$

$\therefore \qquad \varphi_1 = 43.945°$, $\tan \varphi_1 = 0.96385$

Improved power factor, $\cos \varphi_2 = 0.98$

$\therefore \qquad \varphi_2 = 11.478°$, $\tan \varphi_2 = 0.20306$

$$Q_{Cp} = P_p (\tan \varphi_1 - \tan \varphi_2) = 15.686 (0.96385 - 0.20306) = 11.9338 \text{ kVAr}$$

$$Q_{Cp} = V_p I_C = V_p \times \frac{V_p}{X_C} = V_p^2 \cdot 2\pi f C$$

$$11.9338 \times 10^3 = (400)^2 \times 2\pi \times 50 \times C_\Delta$$

$$C_\Delta = \frac{11.9338 \times 10^3}{(400)^2 \times 2\pi \times 50} = 2.374 \times 10^{-4} \text{ F} = 237.4 \text{ μF}$$

Example 26.5 A 230 V, 7.5 kW, 50 Hz single-phase motor working at full load with an efficiency of 87% has a power factor of 0.75 lagging.

(a) Calculate the current supplied to the motor at full load.

(b) If a capacitor is connected across the motor terminals to raise the overall power factor to unity, find

    (i) the current through the capacitor,

    (ii) the value of the capacitor in microfarads.

Solution

(a) Input to motor, $P = \dfrac{\text{output}}{\text{efficiency}} = \dfrac{7.5}{0.87} = 8.6207 \text{ kW}$

$$P = VI \cos \varphi_1$$

Current supplied to the motor, $I = \dfrac{P}{V \cos \varphi_1} = \dfrac{8.6207 \times 10^3}{230 \times 0.75} = 49.975 \text{ A}$

(b) $\qquad Q_C = P (\tan \varphi_1 - \tan \varphi_2)$

$\qquad \cos \varphi_1 = 0.75$, $\quad \tan \varphi_1 = 0.8819$

$\qquad \cos \varphi_2 = 1$, $\quad \tan \varphi_2 = 0$

$\therefore \qquad Q_C = 8.6207 (0.8819 - 0) = 7.6029 \text{ kVAr}$

Also, $\quad Q_C = VI_C$

$\therefore \qquad I_C = \dfrac{Q_C}{V} = \dfrac{7.6029 \times 10^3}{230} = 33.056 \text{ A}$

$$C = \frac{Q_C}{V^2 (2\pi f)} = \frac{7.6029 \times 10^3}{(230)^2 \times 2\pi \times 50} = 457.48 \times 10^{-6} \text{ F} = 457.48 \text{ μF}$$

Example 26.6 A 3-phase induction motor takes 50 kW at 415 V, 50 Hz, and a power factor of 0.72 lagging. Determine the kVAr rating of a capacitor bank to improve the power factor to 0.9 lagging. What capacitance per phase is required if the capacitor bank is connected (a) in delta (b) in star?

Solution

Power per phase, $P_p = \dfrac{50}{3}$ kW

Original power factor, $\cos \varphi_1 = 0.72$ $\varphi_1 = 43.95°$, $\tan \varphi_1 = 0.9638$

Improved power factor, $\cos \varphi_2 = 0.9$ $\varphi_2 = 25.84°$, $\tan \varphi_2 = 0.4843$

$$Q_{Cp} = P_p (\tan \varphi_1 - \tan \varphi_2) = \frac{50}{3} (0.9638 - 0.4843) = 7.991 \text{ kVAr/phase}$$

(a) *Delta connection of capacitor bank*

Phase voltage of capacitor bank = line voltage

$V_p = V_L = 415$

$Q_C = V_p I_C$

$I_C = \dfrac{Q_C}{V_p} = \dfrac{7.991 \times 1000}{415} = 19.256$ A/phase

Capacitance per phase, $C_\Delta = \dfrac{I_C}{\omega V_p} = \dfrac{19.256}{2\pi \times 50 \times 415} = 147.69 \times 10^{-6}$ F $= 147.69$ μF

(b) *Star connection of capacitor bank*

Phase voltage of capacitor bank, $V_p = \dfrac{V_L}{\sqrt{3}} = \dfrac{415}{\sqrt{3}} = 239.6$ V

Phase current, $I_C = \dfrac{Q_C}{V_p} = \dfrac{7.991 \times 1000}{239.6} = 33.35$ A

Capacitance per phase, $C_Y = \dfrac{I_C}{\omega V_p} = \dfrac{I_C}{2\pi f V_p} = \dfrac{33.35}{2\pi \times 50 \times 239.6} = 443 \times 10^{-6}$ F $= 443$ μF

$\dfrac{C_Y}{C_\Delta} = \dfrac{443}{147.69} = 3$

It is seen that for star connection the capacitance is three times that required for delta connection.

26.8 ADVANTAGES AND LIMITATIONS OF STATIC CAPACITORS

Some of the advantages of using static capacitors for improvement of power factor are :

1. They are robust.
2. They are easy to install, occupy little space and do not require any special foundation.
3. They are practically loss free.
4. Their maintenance is easy.
5. They can be manufactured in very small sizes.
6. They are efficient and give trouble free service.

The limitation with the static capacitors is that they tend to over compensate on light loads unless arrangements for automatic switching off the capacitor bank are provided.

26.9 LOCATION OF CAPACITORS

From Fig. 26.2 (a) it is seen that the current I_2 flows from k to m, that is, upto the point of connection of the capacitor. Hence, it is better to improve the power factor of each load by connecting a separate capacitor to it. Such an arrangement reduces the current in all parts of the installation and therefore, the sizes of the circuit cables may be reduced. An alternative method of improving the power factor of a group of loads of a large installation is to connect bank of capacitors at the supply terminals. The banks may be connected or disconnected either manually or automatically with the variation of the load. The method of individual power factor correction is more expensive than bulk power factor correction. For large installations, it is better to apply individual correction to large motors and bulk correction to group of small motors.

26.10 POWER FACTOR CORRECTION BY SYNCHRONOUS MOTORS

Power factor can be corrected with the help of a synchronous phase modifier. It is a synchronous motor running without a mechanical load. It is connected in parallel with the load. It can generate or absorb reactive voltamperes (VAr) by varying the excitation of its field winding. It can be made to take leading current with over excitation of the field winding. It is used to correct power factor in bulk. The output of a synchronous phase modifier can be varied smoothly. It has the disadvantage of being relatively costly. Its installation, maintenance and operation are not easy.

Example 26.7 A synchronous motor improves the power factor of a load of 500 kW from 0.707 lagging to 0.95 lagging. Simultaneously the motor carries a load of 100 kW. Find (i) the leading kVAr supplied by the motor, (ii) kVA rating of the motor, and (iii) power factor at which the motor operates.

Solution

Load, $P_1 = 500$ kW

Motor load, $P_2 = 100$ kW

Power factor of the given load, $\cos \varphi_1 = 0.707$ (lag)

Power factor of combined load, $\cos \varphi_2 = 0.95$ (lag)

Total load, $P = P_1 + P_2 = 500 + 100 = 600$ kW

In Fig. 26.6, Δ OAB is the power triangle for the given load. Here,

$$OA = P_1 = 500 \text{ kW}, \quad \angle AOB = \varphi_1$$

Δ OCD is the power triangle for the combined load. Here,

$$OD = OA + AD = 500 + 100 = 600 \text{ kW}, \quad \angle COD = \varphi_2.$$

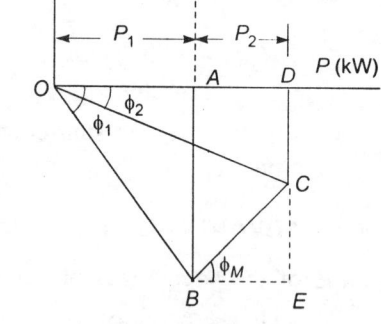

Fig. 26.6.

The power triangle for the synchronous motor is Δ BEC. In this triangle,

$$BE = P_2 = 100 \text{ kW}$$

$$EC = \text{leading kVAr supplied by the motor}$$

$BC = $ kVA rating of the synchronous motor

$\angle CBE = \varphi_M = $ power factor angle at which the motor operates.

(i) Leading kVAr supplied by the motor $= EC = DE - DC = AB - DC$

$$= P_1 \tan \varphi_1 - (P_1 + P_2) \tan \varphi_2$$

$$= 500 \tan (\cos^{-1} 0.707) - 600 \tan (\cos^{-1} 0.95)$$

$$= 500 \times 1 - 600 \times 0.32868 = 302.79 \text{ kVAr}$$

(ii) kVA rating of the motor $= BC = \sqrt{(BE)^2 + (EC)^2} = \sqrt{100^2 + (302.79)^2}$

$$= 318.875 \text{ kVA}$$

(iii) Power factor of the motor, $\cos \varphi_M = \dfrac{\text{motor kW}}{\text{motor kVA}} = \dfrac{BE}{BC} = \dfrac{100}{318.875}$

$$\doteq 0.3136 \text{ (leading)}$$

Alternative method

Let $S_1 = $ kVA of the load

$S_M = $ kVA of the synchronous motor

$S_T = $ total kVA supplied.

$$S_1 = P_1 - j\, Q_1 = P_1 - j\, P_1 \tan \varphi_1$$

$$S_M = P_M + j\, Q_M = P_2 + j\, P_2 \tan \varphi_M$$

$$S_T = P_T - j\, Q_T = (P_1 + P_2) - j\, (P_1 + P_2) \tan \varphi_2$$

It is to be noted that the j term in the expression for S_M is positive because the synchronous motor supplies leading kVAr.

Now $S_T = S_1 + S_M$

$$(P_1 + P_2) - j\, (P_1 + P_2) \tan \varphi_2 = P_1 - j\, P_1 \tan \varphi_1 + P_2 + j\, P_2 \tan \varphi_M$$

$$(P_1 + P_2) \tan \varphi_2 = P_1 \tan \varphi_1 - P_2 \tan \varphi_M$$

$$\tan \varphi_M = \frac{1}{P_2} [P_1 \tan \varphi_1 - (P_1 + P_2) \tan \varphi_2]$$

The p.f. at which the motor operates is $\cos \varphi_M$ (leading).

Leading kVAr supplied by the motor, $Q_M = P_M \tan \varphi_M$

In this problem

$$\tan \varphi_M = \frac{1}{100} [500 \tan (\cos^{-1} 0.707) - (500 + 100) \tan (\cos^{-1} 0.95)]$$

$$= \frac{1}{100} (500 \times 1 - 600 \times 0.32868) = 3.0279$$

Power factor of the motor, $\cos \varphi_M = 0.3136$ leading.

Leading kVAr supplied by the motor, $Q_M = P_M \tan \varphi_M = 100 \times 3.0279 = 302.79 \text{ kVAr}$

$$S_M = P_M + j\, Q_M = P_M + j\, P_M \tan \varphi_M$$

$$= 100 + j\, 302.79 = \sqrt{(100)^2 + (302.79)^2} = 318.875 \text{ kVA}.$$

Example 26.8 A 400 V, 3 φ installation takes a current of 36 A at 0.8 p.f. (lagging). A synchronous motor improves the overall p.f. to 0.92 (lagging). Simultaneously the synchronous motor drives a 15 hp (metric) load at an efficiency of 0.85. Determine (a) the power factor of the synchronous motor, (b) the leading kVAr supplied by the motor, and (c) the kVA rating of the motor.

Solution

Active power requirement of the installation

$$P_1 = \sqrt{3} \; V_L I_L \cos \varphi_1 = \sqrt{3} \times 400 \times 36 \times 0.8 = 19953 \text{ W} = 19.953 \text{ kW}$$

Input to the synchronous motor

$$P_2 = \frac{\text{output}}{\text{efficiency}} = \frac{15 \times 735.5}{0.85} = 12979 \text{ W} = 12.979 \text{ kW}$$

(a) As in Example 26.7

$$\tan \varphi_M = \frac{1}{P_2} [P_1 \tan \varphi_1 - (P_1 + P_2) \tan \varphi_2]$$

$$= \frac{1}{12.979} [19.953 \tan (\cos^{-1} 0.8) - (19.953 + 12.979) \tan (\cos^{-1} 0.92)]$$

$$= \frac{1}{12.979} [19.953 \times 0.75 - 32.932 \times 0.426] = \frac{1}{12.979} (14.9647 - 14.0289) = 0.0721$$

∴ $\cos \varphi_M = 0.9974$ (leading)

(b) Leading kVAr supplied by the motor

$$Q_M = P_M \tan \varphi_M = 12.979 \times 0.0721 = 0.9358 \text{ kVAr}$$

(c) kVA rating of the motor

$$S_M = P_M + j \, Q_M = 12.979 + j \, 0.9358$$

$$S_M = \sqrt{(12.979)^2 + (0.9358)^2} = 13.0127 \text{ kVA}$$

Example 26.9 The total load taken from an ac supply consists of :
(a) a heating loading of 15 kW,
(b) a motor load of 40 kVA at 0.6 power factor lagging,
(c) a load of 20 kW at 0.8 power factor lagging.

Calculate the total load from the supply in kW and kVA and its power factor. What would be the kVAr rating of a capacitor to bring the power factor to unity and how would the capacitor be connected?

Solution

It is more convenient to solve the problems for parallel loads by constructing a table. The kVA, kW and kVAr are tabulated for each load. The following results are used :

$$kW = kVA \cos \varphi ; \qquad kVAr = kVA \sin \varphi$$

$$kVA = \sqrt{(kW)^2 + (kVAr)^2} ; \qquad \cos \varphi = \frac{kW}{kVA}$$

By convention, the lagging kVAr is taken positive and the leading kVAr is taken negative. The total kW is found by adding *algebraically* the kW of individual loads. The total kVAr is found by the *algebraic* addition of kVAr of the individual loads. The total kVA is found from the relation

$$\text{total kVA} = \sqrt{(\text{total kW})^2 + (\text{total kVAr})^2}$$

The table for this problem is given.

Load	kVA	cos φ	sin φ	kW	kVAr
(a)	15	**1.0**	0	**15**	0
(b)	40	**0.6** (lag)	0.8	24	+32
(c)	25	**0.8** (lag)	0.6	**20**	+15
Totals				59	47

In the above table the boldfaced values are known and the other values are calculated values.

Total kVA $= \sqrt{(\text{total kW})^2 + (\text{total kVAr})^2} = \sqrt{(59)^2 + (47)^2} = 75.432$ kVA

Overall power factor, $\cos\varphi = \dfrac{\text{total kW}}{\text{total kVA}} = \dfrac{59}{75.432} = 0.782$ lagging

In order to improve the power factor to unity the lagging kVAr of +47 is to be cancelled by the leading kVAr by connecting a capacitor in parallel with the load.

Rating of the capacitor = 47 kVAr leading.

Alternative method

$$S_a = P_a + j\,Q_a = S\cos\varphi_a + j\,S\sin\varphi_a$$

∴ $\quad S_a = 15 + j\,0$

$$S_b = P_b - j\,Q_b = S\cos\varphi_b - j\,S\sin\varphi_b = 40\times0.6 - j\,40\times0.8 = 24 - j\,32$$

$$S_c = P_c - j\,Q_c = P_c - j\,P_c\tan\varphi_c = 20 - j\,20\tan(\cos^{-1}0.8) = 20 - j\,15$$

$$S_T = S_a + S_b + S_c = 15 + j\,0 + 24 - j\,32 + 20 - j\,15 = 59 - j\,47$$

∴ $\quad$ total active power = 59 kW

Total kVA $= \sqrt{(59)^2 + (47)^2} = 75.432$

Overall power factor, $\cos\varphi = \dfrac{P_T}{S_T} = \dfrac{59}{75.432} = 0.782$ (lagging).

As before, rating of the capacitor = 47 kVAr (leading).

26.11 ECONOMICS OF POWER FACTOR IMPROVEMENT

When the power factor is improved, there is a reduction in maximum kVA demand. Hence there will be annual saving over the maximum kVA demand charges. However, when power factor is improved, it involves capital investment on the power factor correction equipment. Therefore, an expenditure is to be made every year in the form of interest and depreciation etc., on the initial cost of power factor correction equipment. The total annual saving will be equal to the annual saving on the maximum kVA demand charges minus annual expenditure on p.f. correction equipment.

The *most economical power factor* is that value of power factor at which net annual saving is maximum.

26.11.1 Most Economical Power Factor When kW Demand is Constant

Consider an installation having an active power requirement of P kW at a power factor of $\cos \varphi_1$. Suppose that the p.f. is improved to $\cos \varphi_2$ by installing power factor correction equipment. Thus, there is a reduction in kVA from S_1 to S_2.

Fig. 26.7 shows the power triangles. The power triangle at the original power factor $\cos \varphi_1$ is OAB and the power triangle for the improved power factor is OAC.

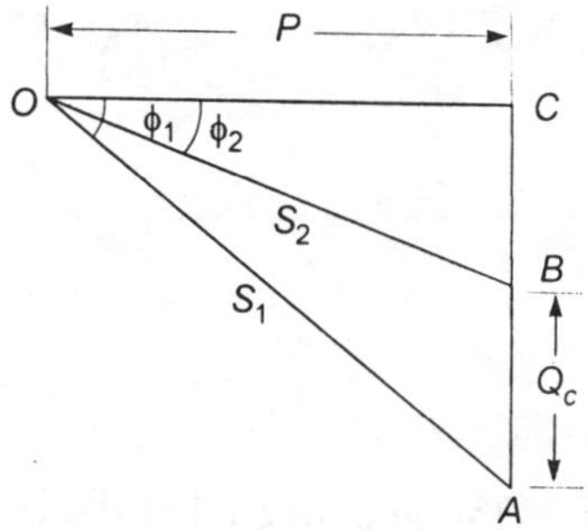

Original kVA demand, $S_1 = OA$

Final kVA demand, $S_2 = OB$

Reduction in kVA demand when the power factor is improved from $\cos \varphi_1$ to $\cos \varphi_2$ is $(S_1 - S_2)$.

Fig. 26.7.

$$S_1 - S_2 = OA - OB = P \sec \varphi_1 - P \sec \varphi_2 = P (\sec \varphi_1 - \sec \varphi_2)$$

If x = annual cost per kVA of maximum demand then annual saving in the kVA demand charges

$$C_D = x (S_1 - S_2) = xP (\sec \varphi_1 - \sec \varphi_2)$$

Reactive kilovoltamperes (kVAr) of the power factor correction equipment

$$Q_C = AB = AC - BC = P \tan \varphi_1 - P \tan \varphi_2$$

or $\qquad Q_C = P (\tan \varphi_1 - \tan \varphi_2)$

If y = annual cost per kVAr of the power factor correction equipment then annual cost of the power factor correction equipment

$$C_{PF} = yQ_C = yP (\tan \varphi_1 - \tan \varphi_2)$$

Total annual saving, $C_S = C_D - C_{PF}$

or $\qquad C_S = xP (\sec \varphi_1 - \sec \varphi_2) - yP (\tan \varphi_1 - \tan \varphi_2)$

It should be noted that x, y, P, and φ_1 are constants. The only variable is φ_2.

For net saving to be maximum,

$$\frac{dC_S}{d\varphi_2} = 0 \quad \text{and} \quad \frac{d^2C_S}{d\varphi_2^2} < 0$$

Now, $\qquad \dfrac{dC_S}{d\varphi_2} = xP (0 - \sec \varphi_2 \tan \varphi_2) - yP (0 - \sec^2 \varphi_2) = 0$

$$- x \times \frac{1}{\cos \varphi_2} \times \frac{\sin \varphi_2}{\cos \varphi_2} + y \frac{1}{\cos^2 \varphi_2} = 0$$

$$x \sin \varphi_2 = y$$

$$\sin \varphi_2 = \frac{y}{x} \qquad \qquad ...(26.11.1)$$

$$= \frac{\text{annual cost per kVAr of p.f. correction equipment}}{\text{annual cost per kVA of maximum demand}}$$

Also, $\quad \dfrac{d^2 C_S}{d\varphi_2^2} = \dfrac{d}{d\varphi_2}(-xP \sec \varphi_2 \tan \varphi_2 + yP \sec^2 \varphi_2)$

$$= -xP(\sec \varphi_2 \times \sec^2 \varphi_2 + \tan \varphi_2 \times \sec \varphi_2 \tan \varphi_2) + yP(2 \sec \varphi_2 \times \sec \varphi_2 \tan \varphi_2)$$

$\therefore \qquad \dfrac{d^2 C_S}{d\varphi_2^2} < 0$

Thus, the expression given by Eq. (26.11.1) gives the condition for maximum annual saving. In other words, maximum saving is obtained when power factor is improved to $\cos \varphi_2$ where φ_2 is given by Eq. (26.11.1).

The most economical power factor is given by

$$\cos \varphi_2 = \sqrt{1 - \sin^2 \varphi_2} = \sqrt{1 - \left(\frac{y}{x}\right)^2} \qquad \qquad ...(26.11.2)$$

Equation (26.11.2) shows that the most economical p.f. depends upon x and y and is independent of original p.f. $\cos \varphi_1$. It is of the order of 0.95.

26.11.2 Most Economical Power Factor When kVA Demand is Constant

Consider an installation having an active power requirement of P_1 kW at a p.f. of $\cos \varphi_1$. The power triangle at original p.f. $\cos \varphi_1$ is OAB as shown in Fig. 26.8.

Here, $\quad OB = P_1, \quad OA = S_1, \quad \angle AOB = \varphi_1$.

Suppose that power factor is improved to $\cos \varphi_2$ by supplying leading kVAr Q_C by installing p.f. correction equipment. The kVA output S is to be kept constant. Therefore, the output increases from P_1 to P_2. The power triangle corresponding to this condition is OCD as shown in Fig. 26.8.

Here, $\quad OD = P_2, \quad OC = S_2, \quad \angle COD = \varphi_2$.

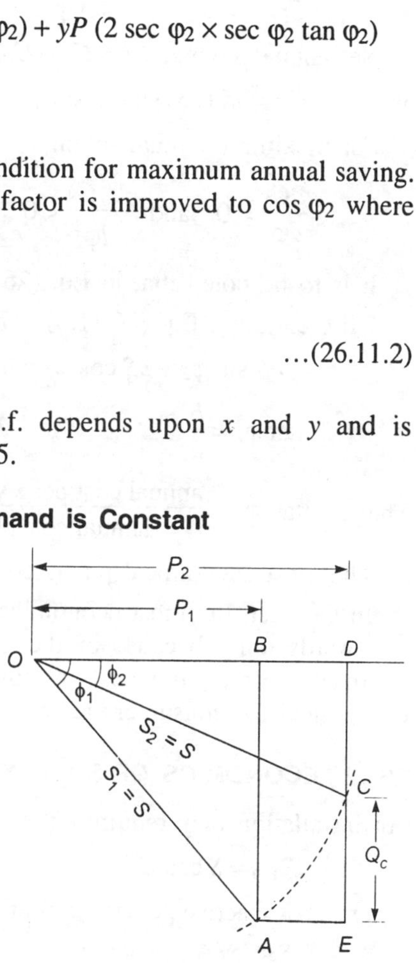

Since the original kVA is equal to the final kVA

$$OA = OC$$

or $\quad S_1 = S_2 = S$ (say)

Fig. 26.8.

Increase in active power output of the installation due to improved power factor

$$= OD - OB = P_2 - P_1$$

If a = annual cost per kW of installation then the annual saving due to increased power output is given by

$$C_p = a\,(P_2 - P_1) = a\,(OD - OB) = a\,(S_2 \cos \varphi_2 - S_1 \cos \varphi_1) = aS\,(\cos \varphi_2 - \cos \varphi_1)$$

Reactive kilovoltampere (kVAr) of the p.f. correction equipment is given by

$$Q_C = EC = ED - CD = AB - CD = S_1 \sin \varphi_1 - S_2 \sin \varphi_2 = S\,(\sin \varphi_1 - \sin \varphi_2)$$

If b = annual cost per kVAr of the p.f. correction equipment then the annual cost of the power factor correction equipment is given by

$$C_{PF} = bQ_C = bS\,(\sin \varphi_1 - \sin \varphi_2)$$

Net annual saving, $C_S = C_P - C_{PF}$

or $\qquad C_S = aS\,(\cos \varphi_2 - \cos \varphi_1) - bS\,(\sin \varphi_1 - \sin \varphi_2)$ $\hfill ...(26.11.3)$

For maximum annual saving

$$\frac{dC_S}{d\varphi_2} = 0 \quad \text{and} \quad \frac{d^2C_S}{d\varphi_2^2} < 0$$

It is to be noted that in Eq. (26.11.3), a, b, S, φ_1 are constants and φ_2 is variable. Differentiating Eq. (26.11.3) w.r.t. φ_2 and equating it to zero we get

$$- aS \sin \varphi_2 + bS \cos \varphi_2 = 0$$

$$\tan \varphi_2 = \frac{b}{a} \hfill ...(26.11.4)$$

That is, $\quad \tan \varphi_2 = \dfrac{\text{annual cost per kVAr of the p.f. equipment}}{\text{annual cost per kW of installation}}$

The most economical p.f. is $\cos \varphi_2$, where φ_2 is given by Eq. (26.11.4).

In the past, the most economical value of power factor was around 0.95. The present value is towards unity because of the high costs of plant and fuel. However, as the power factor approaches unity, the cost of power correction equipment becomes more. Hence, it is not economical for consumers to raise power factor to unity.

26.12 ECONOMICS OF SUPPLY OF INCREASED POWER DEMAND

An installation may require extra active power for its future expansion. We have,

$$P_{1\,\varphi} = S \cos \varphi$$

where $P_{1\,\varphi}$ = active power requirement in kW per phase

$\qquad S$ = kVA per phase

$\qquad \cos \varphi$ = power factor.

The increase in active power demand can be obtained by either of the following two methods :

(a) By increasing the kVA capacity of the generating plant from S_1 to S_2 at the original power factor say $\cos \varphi_1$. This will require extra cost for the installation of extra generating plant.

(b) By improving the power factor of the system from $\cos \varphi_1$ to $\cos \varphi_2$ without increasing the kVA capacity S_1 of the system. This will also involve extra cost for the power factor correction equipment.

We shall make economical comparison of the two methods.

Consider an installation having an original active power demand of P_1 kW at a power factor $\cos \varphi_1$. If S_1 is the original kVA at power factor $\cos \varphi_1$ then

$$P_1 = S_1 \cos \varphi_1$$

Suppose that the active power requirement of the system is to be increased from P_1 to P_2 for its future expansion.

(a) *Cost of extra kVA of the generating plant*

Fig. 26.9 shows the power diagram. In power triangle *OLM*

$$OM = P_1, \ OL = S_1, \ \angle LOM = \varphi_1$$

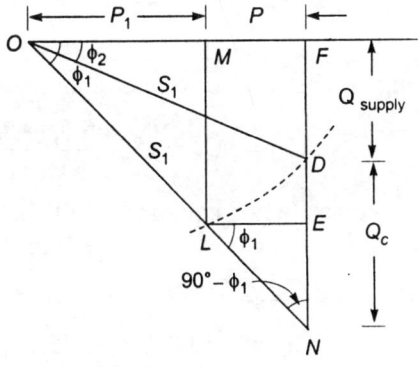

Fig. 26.9.

If the power factor $\cos \varphi_1$ is kept constant, the generating capacity is to be increased from *OL* to *ON* to meet the active power demand *OF*.

In power triangle *OFN*

OF = new active power demand = P_2

ON = new kVA = S_2

$\angle NOF = \varphi_1$

MF = increase in active power demand = p

LN = additional kVA required to meet additional active power demand.

Draw *LE* perpendicular to *NF*.

From Fig. 26.9,

$$MF = LE$$

$$OF - OM = LN \cos \angle NLE$$

$$OD \cos \varphi_2 - OL \cos \varphi_1 = LN \cos \varphi_1$$

But $\quad OD = OL = S_1$

$\therefore \quad S_1 \cos \varphi_2 - S_1 \cos \varphi_1 = LN \cos \varphi_1$

or $\quad LN = \dfrac{S_1 \, (\cos \varphi_2 - \cos \varphi_1)}{\cos \varphi_1}$

If x = annual cost per kVA of the extra capacity installed, then annual cost of extra kVA of the generating plant = $x \, (LN)$

$$C_{GP} = \frac{xS_1}{\cos \varphi_1} \, (\cos \varphi_2 - \cos \varphi_1) \qquad \qquad \dots(26.12.1)$$

(b) *Cost of power factor correction equipment*

Let S_1 be kept constant by installation of power factor correction equipment.

Reactive kilovoltamperes (kVAr) of the power factor correction equipment to improve the power factor from $\cos \varphi_1$ to $\cos \varphi_2$.

$$Q_C = DN$$

By using sine formula in triangle ODN, we have

$$\frac{DN}{\sin \angle DON} = \frac{OD}{\sin \angle OND}$$

$$\frac{DN}{\sin (\varphi_1 - \varphi_2)} = \frac{S_1}{\sin (90 - \varphi_1)}$$

$$\therefore \qquad DN = S_1 \frac{\sin (\varphi_1 - \varphi_2)}{\cos \varphi_1}$$

Let y = annual cost per kVAr rating of the power factor correction equipment. Therefore, annual cost of the p.f. correction equipment

$$C_{PF} = yQ_C = yS_1 \frac{\sin (\varphi_1 - \varphi_2)}{\cos \varphi_1} \qquad \qquad ...(26.12.2)$$

Power factor equipment will be cheaper if annual cost on p.f. correction equipment is less than the annual cost of increasing the generating plant capacity. That is,

$$C_{PF} < C_{GP}$$

or $\qquad yS_1 \dfrac{\sin (\varphi_1 - \varphi_2)}{\cos \varphi_1} < \dfrac{xS_1}{\cos \varphi_1} (\cos \varphi_2 - \cos \varphi_1)$

or $\qquad y \sin (\varphi_1 - \varphi_2) < x (\cos \varphi_2 - \cos \varphi_1)$

In the limiting case, the maximum cost of p.f. correction equipment which would justify its installation, should be equal to the cost of increasing the generating plant capacity. That is,

$$C_{PF} = C_{GP}$$

or $\qquad yS_1 \sin (\varphi_1 - \varphi_2) = xS_1 (\cos \varphi_2 - \cos \varphi_1)$

so that, $\quad y = x \dfrac{\cos \varphi_2 - \cos \varphi_1}{\sin (\varphi_1 - \varphi_2)}$ $\qquad \qquad ...(26.12.3)$

Example 26.10 A consumer has a constant load of 2000 kW at a power factor of 0.8 lagging. The tariff is Rs 900 per kVA of maximum demand plus 90 paise per kWh. If the power factor is improved to 0.96 lagging by installing power factor correction equipment, calculate the annual saving. The cost of the power factor correction equipment is Rs 1950 per kVAr and the annual interest and depreciation is 15%.

Solution

Saving in maximum demand charges per year

$$C_{MD} = xP \left(\frac{1}{\cos \varphi_1} - \frac{1}{\cos \varphi_2} \right) = 900 \times 2000 \left(\frac{1}{0.8} - \frac{1}{0.96} \right)$$

$$= \text{Rs } 375000$$

Annual cost of p.f. correction equipment

$$C_{PF} = yP \ (\tan \varphi_1 - \tan \varphi_2)$$

$$= \left(\frac{15}{100} \times 1950\right) \times 2000 \ [\tan (\cos^{-1} 0.8) - \tan (\cos^{-1} 0.96)] = \text{Rs} \ 268125$$

Total annual saving $= C_{MD} - C_{PF} = 375000 - 268125 = \text{Rs} \ 106875.$

Example 26.11 A consumer takes a load of 900 kW at 0.65 p.f. lagging. The tariff is Rs 1000 per kVA of maximum demand annually + 80 paise per kWh. The cost of installation of the power factor correction equipment is Rs 2000 per kVAr. The annual interest and depreciation is 15 per cent. Determine (a) the most economical p.f. (b) the kVAr rating of the power factor correction equipment to improve the p.f. to this value.

Solution

$$P = 900 \ \text{kW}, \quad \cos \varphi_1 = 0.65$$

Maximum demand charges per kVA per year, $x = \text{Rs} \ 1000.$

Annual charges per kVAr of the power factor correction equipment,

$$y = 15\% \ \text{of Rs} \ 2000 = \text{Rs} \ 300.$$

(a) *Most economical power factor*

$$\cos \varphi_2 = \sqrt{1 - \left(\frac{y}{x}\right)^2} = \sqrt{1 - \left(\frac{300}{1000}\right)^2} = 0.9539 \ \text{(lagging)}$$

(b) *Rating of the p.f. correction equipment*

$$Q_C = P \ (\tan \varphi_1 - \tan \varphi_2) = 900 \ [\tan (\cos^{-1} 0.65) - \tan^{-1} (\cos^{-1} 0.9539)]$$
$$= 900 \ (1.1691 - 0.3146) = 769 \ \text{kVAr}$$

Example 26.12 A consumer has a constant load of 800 kW at a lagging power factor of 0.72 for 6520 hours a year. The tariff is Rs 500 per kVA of maximum demand annually and 50 paise per kWh. The annual cost of power factor correction equipment is Rs 160 per kVAr. Calculate the annual saving if the power factor is improved to most economical limit.

Solution

$$P = 800 \ \text{kW}, \quad \cos \varphi_1 = 0.72$$

Annual maximum demand charges per kVA, $x = \text{Rs} \ 500.$

Cost of power factor correction equipment per kVAr, $y = \text{Rs} \ 160.$

Therefore, most economical power factor

$$\cos \varphi_2 = \sqrt{1 - \left(\frac{y}{x}\right)^2} = \sqrt{1 - \left(\frac{160}{500}\right)^2} = 0.9474 \ \text{(lagging)}$$

$$\varphi_2 = 18.663°$$

∴ $\tan \varphi_2 = 0.33776$

Since $\cos \varphi_1 = 0.72, \quad \varphi_1 = 43.945°, \quad \tan \varphi_1 = 0.96385$

Leading kVAr supplied by the power factor correction equipment

$$Q_C = P (\tan \varphi_1 - \tan \varphi_2) = 800 (0.96385 - 0.33776)$$

$$= 500.87 \text{ kVAr}$$

Annual saving in demand charges

$$C_{MD} = xP (\sec \varphi_1 - \sec \varphi_2) = xP \left(\frac{1}{\cos \varphi_1} - \frac{1}{\cos \varphi_2} \right)$$

$$= 500 \times 800 \left(\frac{1}{0.72} - \frac{1}{0.9474} \right) = \text{Rs } 133347.41$$

Annual cost of p.f. correction equipment, $C_{PF} = yQ_C = 160 \times 500.87 = \text{Rs } 80139.20$.

Net annual saving $= C_{MD} - C_{PF} = 133347.41 - 80139.20 = \text{Rs } 53208.21$.

Example 26.13 A system is working at its maximum kVA capacity with a lagging power factor of 0.707. An anticipated increase of load could be met by (a) raising the power factor of the system to 0.866 by installing power factor correction equipment, or (b) installing extra generating plant to meet the increased power demand. The total cost of the generating plant is Rs 8000 per kVA. Estimate the limiting cost per kVAr of the p.f. correction equipment which would justify its installation.

Solution

$$\cos \varphi_1 = 0.707, \quad \varphi_1 = 45°$$

$$\cos \varphi_2 = 0.866, \quad \varphi_2 = 30°$$

$$x = \text{Rs } 8000$$

$$y = x \frac{\cos \varphi_2 - \cos \varphi_1}{\sin (\varphi_1 - \varphi_2)} = 8000 \frac{0.866 - 0.707}{\sin 15°}$$

$$\therefore \quad y = \text{Rs } 4914.63.$$

EXERCISES

1. Show, with the aid of a phasor diagram, how the power factor of a load can be improved by connecting a capacitor in parallel with it.

   A coil dissipates 3 kW when connected to a 415 V, 50 Hz supply and takes a current of 10 A. Calculate the capacitance required to bring the power factor to unity.

2. Define power factor and explain why, in general, it should be kept as high as possible in power supply systems.

   A single-phase motor takes 50A at a power factor of 0.6 lagging from a 250 V, 50 Hz supply. What value of capacitance must a shunting capacitor have to raise the overall power factor to 0.9 lagging? How does the installation of the capacitor affect the line and motor currents?

3. A single-phase load of 5 kW operates at a power factor of 0.6 lagging. It is proposed to improve the power factor to 0.95 lagging by connecting a capacitor across the load. Calculate the kVAr rating of the capacitor.

4. A 415 V, 3-phase, 50 Hz factory installation load is 450 kVA at a power factor of 0.6 lagging. What is the rating in kVAr of the capacitor bank required to improve the power factor to 0.95 lagging?

5. A 415 V, 3-phase, 50 Hz induction motor having an output of 74.6 runs on full load at a power factor of 0.7 lagging with an efficiency of 85%. Find the capacitance per phase of a mesh connected capacitor bank to raise the power factor : (a) to unity (b) to 0.9 lagging.

6. The power supply to a 415 V, 50 Hz, 3-phase induction motor is 50 kW at 0.72 power factor lagging. A bank of capacitors is connected in delta across the line to improve the overall power factor. Calculate the capacitance per phase in order to raise the power factor to 0.9 lagging.

7. A substation supplies the following loads :

   (a) Induction motors : total kW 84.77, power factor 0.8 lagging; (b) Synchronous motor : total kW 52.66; (c) Heating load 25 kW.

   At what power factor must the synchronous motor work so that the substation power factor is (i) unity (ii) 0.95 lagging?

8. A 3-phase induction motor is connected in parallel with a load of 250 kW at 0.8 power factor (lagging) and its raises the total power factor to 0.9 lagging. If the mechanical load on the motor including losses is 60 kW, calculate (a) the power factor at which motor operates, (b) the leading kVAr supplied by the motor, and (c) the kVA rating of the motor.

9. The load taken by a factory is 400 kVA at 0.8 power factor lagging. A synchronous motor is installed to supply 110 kW output at 90% efficiency and to raise the overall power factor to 0.95 lagging. Find the power factor at which the motor must operate.

10. A consumer has constant load of 100 kW at power factor 0.8 lagging. The tariff is Rs 1600 per kVA plus 90 paise per kWh. Find the economical power factor to which he could improve by installing power factor correction equipment costing Rs 3000 per kVA. The rate of interest per year is 12%.

11. For increasing the kW capacity of a power plant working at 0.7 power factor lagging, the necessary increase in power can be obtained by raising the power factor to 0.9 lagging or by installing additional generating plant. What is the maximum cost per kVAr if the power factor correction equipment to make its use more economical than the additional generating plant at Rs 8000 per kVA.

12. A factory has a maximum demand of 1200 kVA at power factor 0.75 lagging. There is a further demand of the load by the factory which could be met either by raising the power factor of the system to 0.9 lagging by installing power factor correction equipment or by installing extra generating plant. Estimate the maximum cost per kVAr of the power factor correction equipment which could be justified for use if the capital cost of generating plant is Rs 5000 per kVA. The interest and depreciation charges are 15 per cent in both cases.

ANSWERS

1. 53.02 μF
2. 324 μF
3. 5.0232 kVAr
4. 90.418 kVAr/phase
5. (a) 551.5 μF (b) 289.69 μF
6. 147.7 μF
7. (i) 0.6379 (leading) (ii) 0.9818 (leading)
8. (a) 0.8489 (leading) (b) 37.36 kVAr (c) 70.68 kVA
9. 0.7895 (leading)
10. 0.9743 (lagging)
11. Rs 4739.94
12. Rs 2796.30

Voltage Stability

27.1 INTRODUCTION

Voltage stability is concerned with the ability of a power system to maintain acceptable voltages at all buses in the system under normal conditions and after being subjected to a disturbance. A power system at a given operating state is voltage stable if, following a disturbance, voltages near loads are identical or close to the pre-disturbance values. A power system is said to have entered a state of voltage instability when a disturbance results in a progressive and uncontrollable decline in voltage. Following voltage instability, a power system may undergo voltage collapse, if the post-disturbance equilibrium voltages near loads are below acceptable limits. *Voltage collapse* is also defined as a process by which voltage instability leads to very low voltage profile in a significant part of the system. Voltage collapse may be total (blackout) or partial. Voltage stability is sometimes also called *load stability*. The terms voltage instability and voltage collapse are often used interchangeably.

In recent years, voltage instability has resulted in several major system failures in Florida, France, Belgium, Sweden, Japan, India and Washington DC.

The present transmission networks are getting more and more stressed due to economic and environmental constraints. Voltage stability problems normally occur in heavily stressed systems. With growing size of power system networks, greater emphasis is being given to the study of voltage stability. The voltage stability can be studied either on static (slow time-frame) or dynamic (long time-frame) considerations. Depending upon the disturbance and system dynamics, voltage stability may be regarded as a slow or fast phenomenon.

27.2 CLASSIFICATION OF VOLTAGE STABILITY

Voltage stability may be classified into two categories:
- Large-disturbance voltage stability;
- Small-disturbance voltage stability.

27.2.1 Large-Disturbance Voltage Stability

It is concerned with a system's ability to control voltages following large disturbances such as system faults, loss of load, or loss of generation. Determination of this form of stability requires the examination of the dynamic performance of the system over a period of time sufficient to capture the interactions of such devices as underload tap changing transformers, and generator field current limiters. Large-disturbance voltage stability can be studied by using nonlinear time-domain simulations which include proper modelling. It may be further subdivided into transient and long-term time-frames.

27.2.2 Small-Disturbance Voltage Stability

A power system at a given operating state is *small-disturbance voltage stable* if, following any small disturbance, voltage near loads do not change or remain close to the predisturbance values. The concept of small-disturbance stability is related to steady-state stability and can be analysed using small-signal (linearised) model of the system.

27.3 ROTOR-ANGLE STABILITY AND VOLTAGE STABILITY

The problem of power system stability is discussed in Chapter 19. To make the distinction clear, the term rotor angle (or synchronous) stability may be used in place of power system stability. Voltage stability and rotor angle stability are more or less interlinked. Transient voltage stability is often interlinked with transient rotor angle stability, and slower forms of voltage stability are interlinked with small-disturbance rotor angle stability. Rotor angle stability, as well as voltage stability, are affected by reactive power control. Voltage stability problems arise mainly in the event of faults. Rotor angle stability problems also arise during and after faults. The voltage stability and rotor angle stability are, to some extent, interlinked. However, in some aspects they are quite different. Voltage stability is concerned with load areas and load characteristics. For rotor angle stability, we are often concerned with integrating remote power plants to a large system over long transmission lines. Voltage stability is basically *load stability*, and rotor angle stability is basically *generator stability*.

In a large interconnected system voltage instability of a load area may occur, but angle stability is still maintained without loss of synchronism of any generators.

Transient voltage stability is usually closely associated with transient rotor angle stability. Longer term voltage stability is less interlinked with rotor angle stability.

It can be said that, if there is voltage collapse at a point in a transmission system remote from loads, it is an angle stability problem. If the voltage collapses in a load area, it is probably mainly a voltage instability problem.

27.4 VOLTAGE STABILITY LIMIT

The *voltage stability limit* can be defined as the limiting stage in a power system beyond which no amount of reactive power injection will elevate the system voltage to its nominal state. The system voltage can only be adjusted by reactive power injections till the system voltage stability is maintained.

The power transfer over a lossless line is given by:

$$P = \frac{V_s V_r}{X} \sin \delta \qquad \qquad ...(27.4.1)$$

where P = power transferred per phase

V_s = sending–end phase voltage

V_r = receiving–end phase voltage

X = transfer reactance per phase

δ = phase angle between V_s and V_r

Since the line is lossless

$$P_r = P_s = P$$

$$\therefore \qquad \frac{dP_r}{dV_s} = \frac{V_r}{X} \sin \delta + \frac{V_s V_r}{X} \cos \delta \frac{d\delta}{dV_s} \qquad \qquad ...(27.4.2)$$

Assuming the power generation to be constant,

$$\frac{dP_r}{dV_s} = 0 \qquad \qquad ...(27.4.3)$$

$$\therefore \qquad \frac{V_r}{X} \sin \delta + \frac{V_s V_r}{X} \cos \delta \frac{d\delta}{dV_s} = 0$$

or $\qquad \dfrac{d\delta}{dV_s} = -\dfrac{\tan \delta}{V_s} \qquad \qquad ...(27.4.4)$

For maximum power transfer:

$\delta = 90°$, so that as $\delta \to \infty$

$$\frac{d\delta}{dV_s} \to -\infty \qquad \qquad ...(27.4.5)$$

Equation (27.4.5) gives the location of critical point on the δ versus V_s curve. It is assumed that receiving-end voltage is constant.

Similar result can be obtained assuming the sending-end voltage constant and analysing the system taking V_r as a variable parameter. In this case the resulting equation is

$$\frac{d\delta}{dV_r} = -\frac{\tan \delta}{V_r} \qquad \qquad ...(27.4.6)$$

The reactive power expression at the receiving-end bus may be written as

$$Q_r = \frac{V_s V_r}{X} \cos \delta - \frac{V_r^2}{X} \qquad \qquad ...(27.4.7)$$

Therefore,

$$\frac{dQ_r}{dV_r} = \frac{1}{X}\left[V_s \cos \delta - V_s V_r \sin \delta \frac{d\delta}{dV_r} - 2V_r \right] \qquad \qquad ...(27.4.8)$$

Substituting the value of dS/dV_r from Eq. (27.4.6) into Eq. (27.4.7), we get

$$\frac{dQ_r}{dV_r} = \frac{1}{X}\left[V_s \cos \delta + V_s V_r \sin \delta \left(\frac{\tan \delta}{V_r}\right) - 2V_r \right]$$

$$= \frac{1}{X}\left[V_s \cos \delta + V_s \frac{\sin^2 \delta}{\cos \delta} - 2V_r \right]$$

$$= \frac{1}{X}\left[\frac{V_s}{\cos \delta} (\cos^2 \delta + \sin^2 \delta) - 2V_r \right]$$

or $\qquad \dfrac{dQ_r}{dV_r} = \dfrac{1}{X}\left(\dfrac{V_s}{\cos \delta} - 2V_r \right)$ $\hfill$ ...(27.4.9)

At the steady-state power angle stability, $\delta = 90°$ so that as $\delta \to 90°$

$$\frac{dQ_r}{dV_r} \to \infty \hspace{4cm} ...(27.4.10)$$

Equation (27.4.10) represents steady-state voltage stability limit. It shows that, at steady-state stability limit, the reactive power sensitivity (dQ_r/dV_r) becomes infinite. This means that (dV_r/dQ_r) becomes zero. Hence the rotor angle stability limit under steady-state conditions is coincident with steady-state voltage stability limit.

The voltage stability at the receiving-end bus is also affected by the load. In the above analysis the load effect is neglected. In a rigorous analysis, the load effect must be taken into account to analyse the voltage stability of a power line.

27.5 VOLTAGE STABILITY ANALYSIS

The analysis of voltage stability problem requires examination of wide range of system conditions and a large number of contingencies. Following a disturbance, power simulations provide a method of study of voltage instability problem. Two sets of graphs are used to study voltage instability. They are P-V curves and V-Q curves.

Consider a transmission having voltages V_s and V_r at the two ends and feeding a load $(P_r + j Q_r)$ at the receiving end. The line is assumed to be lossless and has a total series inductive reactance X.

27.5.1 P-V Curves

P-V curves are plotted between receiving-end power and receiving-end voltage for different values of load power factor with sending-end voltage V_s constant. In practice, normalized values are used in preparation of these curves. Thus, the receiving-end power P_r is represented in terms of p where $p = P_r \dfrac{X}{V_s^2}$. The receiving-end voltage is represented by v where $v = \dfrac{V_r}{V_s}$. The values of v can be found for different values of p and power factor from line equations. For a large system, the values are found from load flow study. The P-V curve shows a plot of bus voltage v as a function of p for a specified power factor. The P-V curves are plotted for various power factors, lagging as well as leading, as shown in Fig. 27.1.

Two values of load voltage exist for each value of load. For each value of power factor, the higher voltage indicates stable voltage case, while the lower voltage lies in the unstable voltage

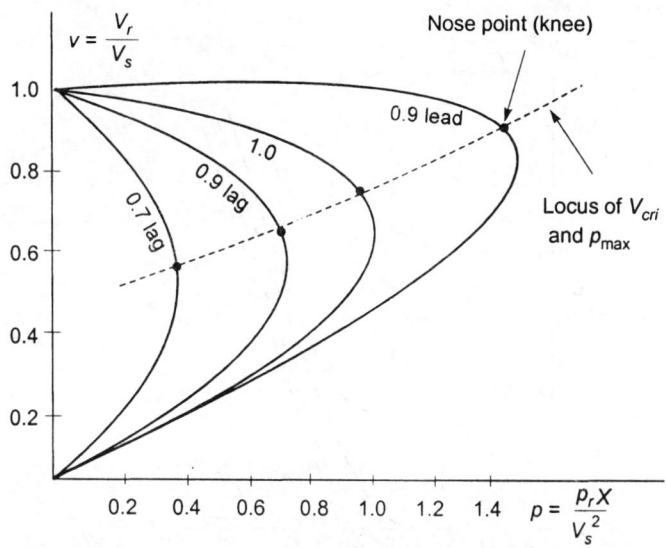

Fig. 27.1. *P-V* curves for different load power factors.

operation zone. The changeover occurs at $v_{critical}$ and p_{max}. The dotted line in Fig. 27.1 shows the locus of $v_{critical} - p_{max}$ points for various power factors. Only the operating points above the critical points represent satisfactory operating conditions. At the 'knee' of the *P-V* curve, the voltage drops rapidly with the increase in load. It may be noted that critical voltage is high for high values of p. For leading power factors the maximum power and critical voltage are higher. *P-V* curves are useful for conceptual analysis of voltage stability especially for radial systems.

27.5.2 V-Q Curves

V-Q curves are plotted between receiving-end voltage V_r and reactive power Q_r for different values of P_r. In practice, normalized values are used in plotting these curves. Thus, $q = \dfrac{Q_r X}{V_s^2}$ and $v = \dfrac{V_r}{V_s}$. The *v-q* curves can be obtained from the *p-v* curves. These curves are shown in Fig. 27.2. These curves have several advantages for study of certain aspects of voltage stability.

27.6 METHODS OF IMPROVING VOLTAGE STABILITY

Power system voltage instability can be improved by using the following methods:
1. Reactive power compensation
2. Load shedding during contingencies
3. Use of LTC (load tap changing transformers)
4. Erection of additional transmission lines

27.6.1 Reactive Power Compensation

Voltage instability is basically the effect of reactive power imbalance between generation and

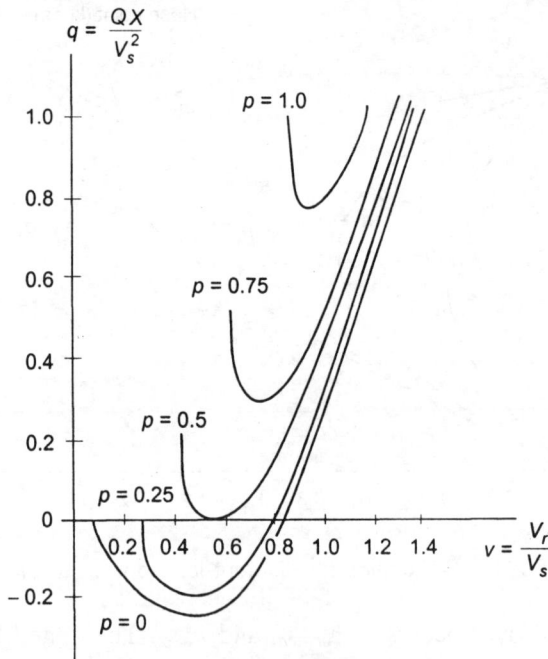

Fig. 27.2. Normalized v-q curves.

demand. Load bus is most susceptible to voltage instability. Therefore, localised reactive support may be used to improve voltage stability. The following methods are used to give the reactive power support:

- Shunt reactors
- Shunt capacitors
- Series capacitors
- Static VAR systems (SVS)
- Synchronous phase modifiers

Reactive power compensation is often the most effective way to improve both power transfer capability and voltage stability. These methods have already been discussed in Chapter 13.

27.6.2 Load Shedding During Contingencies

The risk of collapse due to voltage instability can be reduced by load shedding. Undervoltage load shedding during contingencies is the cheapest way to prevent voltage instability. Load shedding may be manual or automatic. Different steps involved in load shedding are as follows:

(a) Determination of the amount of load to be shed.

(b) Selection of load to be shed.

(c) Determination of time steps of load shedding.

(d) Determination of voltage levels at which shedding begins.

System planners conduct numerous studies using V-Q curves and other analytical methods to determine the amount of load that needs to be shed to retain voltage stability during contingencies. Voltage collapse is most probable under heavy load conditions. Therefore, the amount of load to be shed depends on system load peak and generation sources.

Load shedding schemes should be designed so as to differentiate between faults, transient voltage dips and low voltage conditions leading to voltage collapse.

27.6.3 Use of LTC (Load Tap Changing Transformers)

Load tap changing (LTC) transformers are frequently used for voltage and reactive power control at the load-end bus in power system. It is to be noted that LTC transformers only adjust the reactive power flow between the buses. They do not produce any reactive power to assist the system when the system faces contingencies. Tap changers can be controlled either locally or centrally. Microprocessor-based LTC controls offer greater flexibility. LTC operation offers limited assistance. Therefore, LTC transformers, except for small range of voltage control, provide low remedial action to solve the voltage stability problem.

EXERCISES

1. Explain the term 'voltage stability'. Why does voltage instability occur in power systems.
2. Define the term 'voltage collapse'. How is voltage stability classified?
3. Differentiate between rotor angle stability and voltage stability.
4. What is voltage stability limit? Deduce the condition for steady-state voltage stability limit. State the assumptions made.
5. Draw p-v and v-q curves for a line. What is the use of these curves?
6. State and explain the methods of improving power system voltage instability.

28

Flexible AC Transmission Systems (FACTS)

28.1 INTRODUCTION

In general, flexible AC transmission systems (FACTS) is a new technology. It has the principal role of enhancing controllability and power transfer capability in ac systems.

With the increasing demand of power, the power system networks are becoming more complex from the point of view of operation and control. The existing networks are mostly mechanically controlled. Microelectronics, computers and high-speed communications are widely used for protection and control of transmission systems. However, when operating signals are sent to power circuits, where the final control action takes place, the switching devices are mechanical. These devices are slow in operation. Another problem associated with mechanical switching devices is that control cannot be initiated frequently. This is due to the fact that mechanical devices wear out quickly as compared to static devices. Consequently, both steady-state and dynamic operations of the system are practically uncontrolled.

Since 1990, a number of control devices were developed under the term FACTS technology. FACTS technology opens up new opportunities for controlling power and enhancing the usable capacity of existing and new lines. It also enables a line to carry power closer to its thermal rating. FACTS devices can be effectively used for power flow control, load sharing along parallel corridors, voltage regulation, enhancement of transient stability and mitigation of system oscillations.

Flexible AC transmission system (FACTS) is defined as alternating current transmission systems incorporating power electronics-based and other static controllers to enhance controllability and increase power transfer capability.

FACTS controller is a power electronics-based system or other static equipment that provides control of one or more AC transmission system parameters.

28.2 BASIC TYPES OF FACTS CONTROLLERS

In general, FACTS controllers can be divided into four categories:

- Series controllers
- Shunt controllers
- Combined series-series controllers
- Combined series-shunt controllers

Fig. 28.1(a) shows the general symbol for a FACTS controller: a thyristor arrow inside a box.

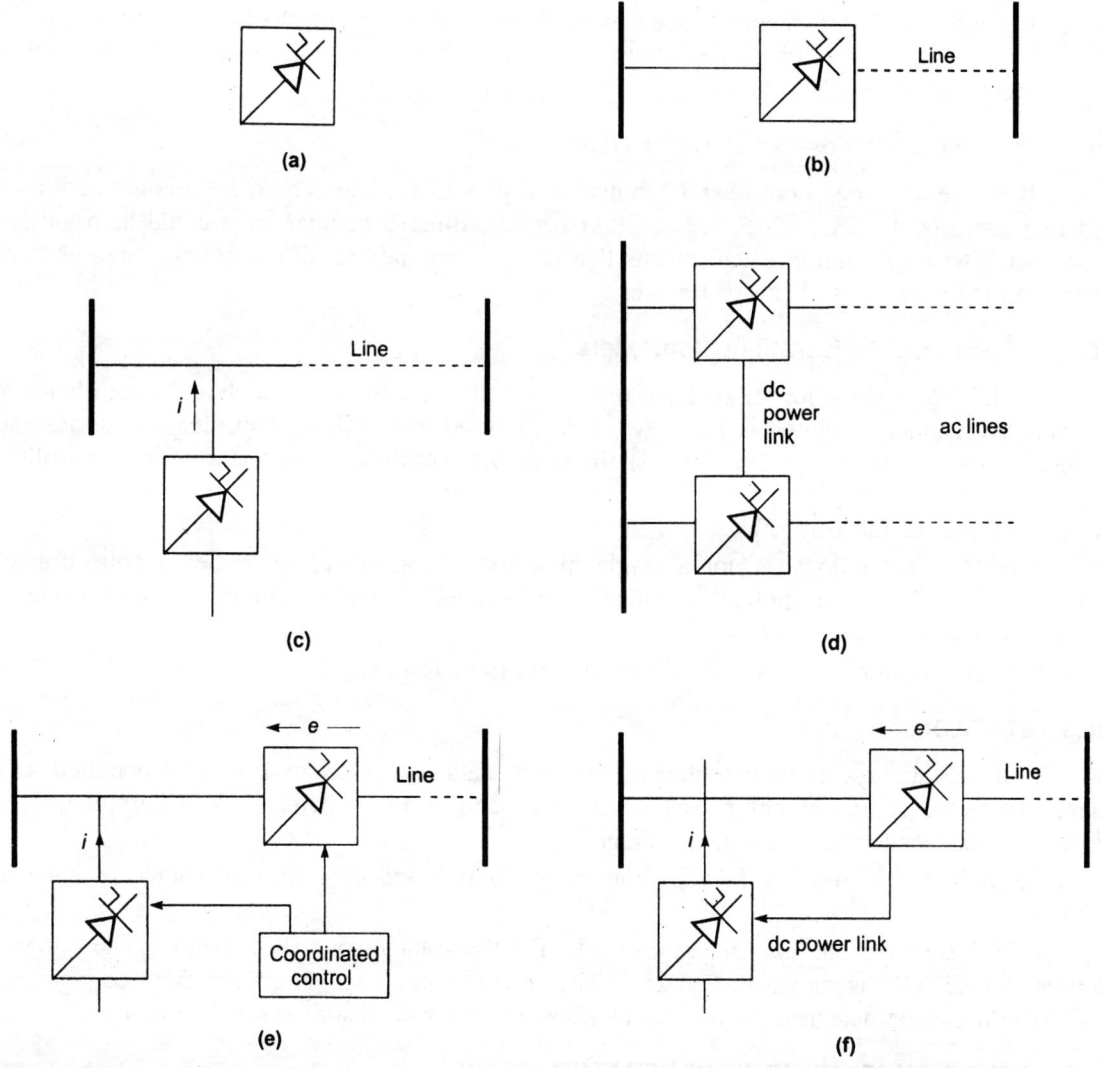

Fig. 28.1. Basic types of FACTS controllers. (a) General symbol for FACTS controller, (b) series controller, (c) shunt controller, (d) unified series-series controller, (e) coordinated series and shunt controller, (f) unified series-shunt controller.

28.2.1 Series Controllers

The series controller is shown in Fig. 28.1(b). It could be a variable impedance, such as capacitor, reactor etc., or a power electronics-based variable source. In principle, all series controllers inject voltage in series with the line. As long as the voltage is in phase quadrature with the line current, the series controller only supplies or consumes variable reactive power. Any other phase relationship will involve handling of real power also.

28.2.2 Shunt Controllers

A shunt controller is shown in Fig. 28.1(c). It may be a variable impedance, variable source, or a combination of these. In principle, all shunt controllers inject current into the system at the point of connection. If the injected current is in phase quadrature with the line voltage, the shunt controller only supplies or consumes variable reactive power. Any other phase relationship will involve real power also.

28.2.3 Combined Series-Series Controllers

A combined series-series controller is shown in Fig. 28.1(d). This could be a combination of separate series controllers, which are controlled in coordinated manner, or it could be a unified controller. The term 'unified' here means that the dc terminals of all controller converters are connected together for real power transfer.

28.2.4 Combined Series-Shunt Controllers

This could be a combination of separate series and shunt controllers, which are controlled in a coordinated manner as shown in Fig. 28.1(e) or a unified power flow controller with series and shunt elements as shown in Fig. 28.1(f). In principle, combined shunt and series controllers inject current into the system with shunt part of the controller and voltage in series in the line with series part of the controller.

The FACTS technology is not a single high-power controller, but rather a collection of controllers, which can be applied individually or in combination with others to control one or more of the system parameters.

A brief description of some FACTS controllers is as follows:

28.3 STATCOM

STATCOM (Static synchronous compensator) is a static synchronous generator operated as a shunt connected static var compensator whose capacitive or inductive output current can be controlled independent of ac system voltage.

Statcom is one of the key FACTS controllers. It is based on a voltage source or current-sourced converter.

Fig. 28.2 shows a one-line diagram of STATCOM voltage-sourced converter. The current-sourced STATCOM is shown in Fig. 28.3. The voltage-source converters are more economical. STATCOM can be designed to also act as an active filter to absorb system harmonics.

28.4 STATIC SYNCHRONOUS GENERATOR (SSG)

A combination of Statcom and any energy source to supply or absorb power is called static

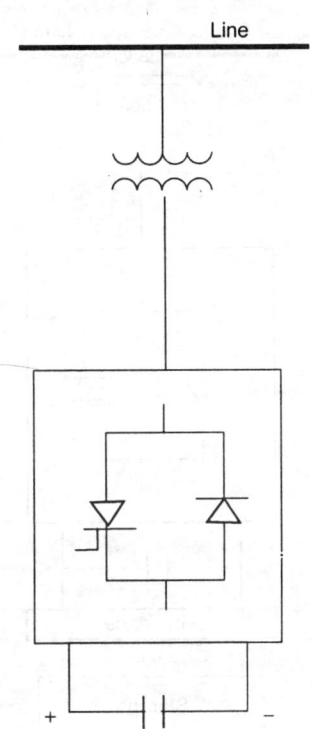

Fig. 28.2. Statcom based on voltage-sourced converter.

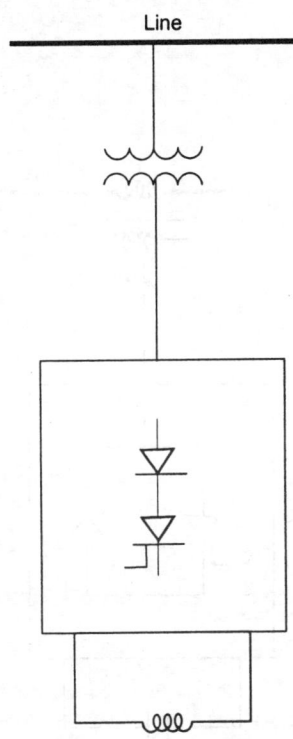

Fig. 28.3. Statcom based on current-sourced converter.

synchronous generator (SSG). Energy source may be a battery flywheel, superconducting magnet, large dc storage capacitor, another rectifier/inverter, etc.

28.5 STATIC SYNCHRONOUS SERIES COMPENSATOR (SSSC)

It is a static synchronous generator operated without an external energy source as a series compensator whose output voltage is in quadrature with, and controllable independently of, the line current for the purpose of increasing or decreasing the overall reactive voltage drop across the line and thereby controlling the transmitted electric power.

SSSC is one of the most important FACTS controllers. It is like a Statcom, except that the output ac voltage is in series with the line. A voltage-sourced converter SSSC is shown in Fig. 28.4. Battery storage or superconducting magnetic storage can also be connected in series controller (Fig. 28.5) to inject a voltage phasor of variable angle in series with the line.

28.6 INTERLINE POWER FLOW CONTROLLER (IPFC)

The IPFC is a recently introduced controller. It is a combination of two or more static synchronous series compensators which are coupled via a common dc link to facilitate bi-directional flow of real power between the ac terminals of the SSSCs, and are controlled to provide independent reactive compensation for adjustment of real power flow in each line and maintain the desired distribution of reactive power flow among the lines.

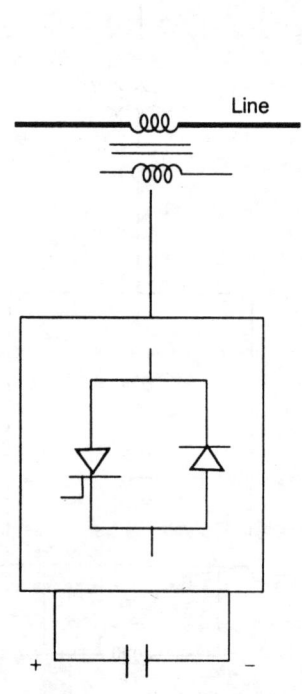

Fig. 28.4. Static synchronous series compensator (SSSC).

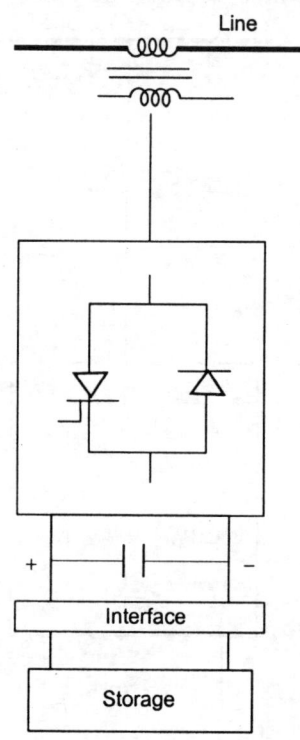

Fig. 28.5. SSSC with storage.

28.7 UNIFIED POWER FLOW CONTROLLER (UPFC)

It is a combination of STATCOM and SSSC which are coupled via a common dc link to allow bidirectional flow of real power between the series output terminals of the SSSC and the shunt output terminals of the STATCOM. These are controlled to provide concurrent real and reactive series line compensation without an external energy source. The UPFC, by means of angularly unconstrained series voltage injection, is able to control, concurrently or selectively, the transmission line voltage, impedance, and angle or, alternatively, the real and reactive power flow in the line. The UPFC may also provide independently controllable shunt reactive compensation. A unified power flow controller (UPFC) is shown in Fig. 28.6.

This is a complete controller for controlling active and reactive power control through the line, as well as line voltage control.

28.8 THYRISTOR-CONTROLLED PHASE SHIFTING TRANSFORMER (TCPST)

It is a phase shifting transformer, adjusted by thyristor switches to provide a rapidly variable phase angle. This controller is also known as thyristor-controlled phase angle regulator (TCPAR).

28.9 INTERPHASE POWER CONTROLLER (IPC)

It is a series-connected controller of active and reactive power consisting, in each phase, of inductive and capacitive branches subjected to separately phase-shifted voltages. The active and

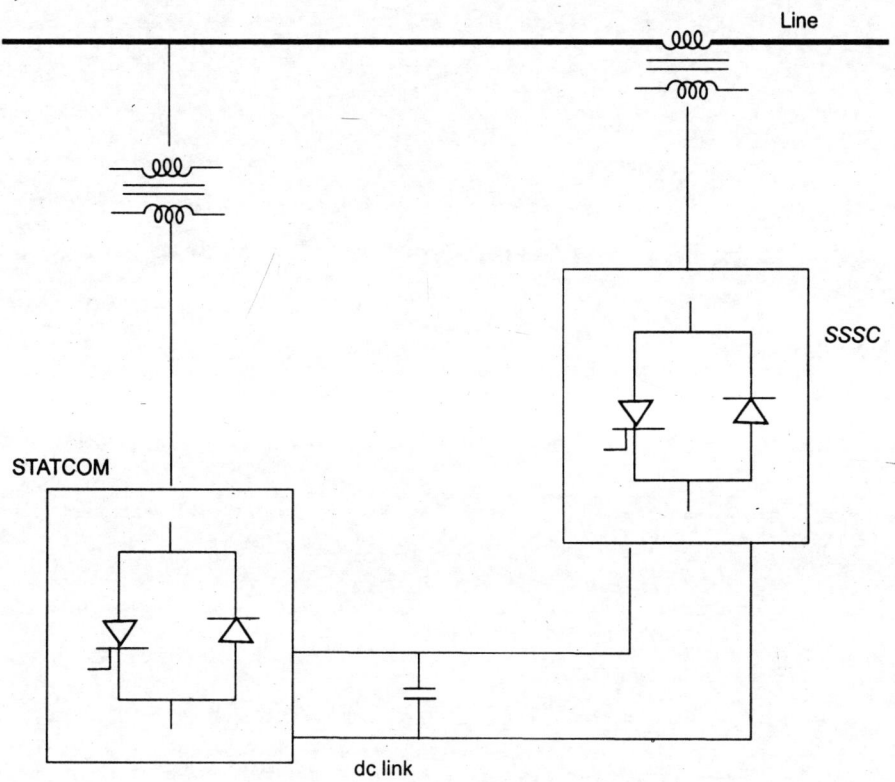

Fig. 28.6. Unified power flow converter (UPFC).

reactive power can be set independently by adjusting the phase shifts and/or the branch impedance, using mechanical or electronic switches.

28.10 THYRISTOR CONTROLLED VOLTAGE LIMITER (TCVL)

It is a thyristor-switched metal-oxide varistor (MOV) used to limit the voltage across its terminal during transient conditions.

APPENDIX

ALUMINIUM CONDUCTOR STEEL REINFORCED (A.C.S.R.)
Based on IS 398 : (1961)

		Electrical Characteristics				Mechanical Characteristics										
Code Word	Nominal Copper Area (sq. mm)	Calculated Equivalent Area of Aluminium (sq. mm)	Resistance at 20°C (Ohms/km)	Approx. Current Carrying Capacity At 40°C ambient temperature (A)	At 45°C ambient temperature (A)	No. of Wires Al	No. of Wires St	Diameter of Wires Al (mm)	Diameter of Wires St (mm)	Diameter of Conductor (mm)	Weight of Conductor Total (kg/km)	Weight Al (kg/km)	Weight St (kg/km)	Approx. Ultimate Tensile Strength of Conductor (kg)	Modulus of Elasticity approx. (kg/cm² ×10⁶)	Coefficient of Linear Expansion approx. (per °C ×10⁻⁶)
Squirrel	13	20.71	1.374	115	107	6	1	2.11	2.11	6.33	85	57.6	27.4	771	0.8090	18.99
Gopher	16	25.91	1.098	133	123	6	1	2.36	2.36	7.08	106	71.9	34.1	952	0.8090	18.99
Weasel	20	31.21	0.9116	150	139	6	1	2.59	2.59	7.77	128	86.8	41.2	1136	0.8090	18.99
Ferret	25	41.87	0.6795	181	168	6	1	3.00	3.00	9.00	171	116.1	54.9	1503	0.8090	18.99
Rabbit	30	52.21	0.5449	208	193	6	1	3.35	3.35	10.05	214	144.8	69.2	1860	0.8090	18.99
Mink	40	62.32	0.4565	234	217	6	1	3.66	3.66	10.98	255	172.6	82.4	2207	0.8090	18.99
Beaver	45	74.07	0.3841	267	242	6	1	3.99	3.99	11.97	303	205.4	97.6	2613	0.8090	18.99
Raccoon	48	77.83	0.3656	270	250	6	1	4.09	4.09	12.27	318	215	103	2746	0.8090	18.99
Otter	50	82.85	0.3434	281	260	6	1	4.22	4.22	12.66	339	229.2	109.8	2923	0.8090	18.99
Cat	55	94.21	0.3020	305	283	6	1	4.50	4.50	13.50	385	260.9	124.1	3324	0.8090	18.99
Dog	65	103.6	0.2745	324	300	6	7	4.72	1.57	14.15	394	288.0	106.0	3299	0.7350	19.53
Leopard	80	129.7	0.2193	375	348	6	7	5.28	1.76	15.84	493	359.6	133.4	4137	0.7350	19.53
Coyote	80	128.5	0.2214	375	348	26	7	2.54	1.90	15.86	521	365.0	156.0	4638	0.7730	18.99
Tiger	80	128.1	0.2221	382	354	30	7	2.36	2.36	16.52	604	362.5	241.5	5758	0.7870	17.73
Wolf	95	154.3	0.1844	430	398	30	7	2.59	2.59	18.13	727	436.0	291.0	6880	0.7870	17.73
Lynx	110	179.0	0.1589	475	440	30	7	2.79	2.79	19.53	844	506.0	338.0	7950	0.7870	17.73
Panther	130	207.0	0.1375	520	482	30	7	3.00	3.05	21.00	976	586.0	390.0	9127	0.7870	17.73
Lion	140	232.5	0.1223	555	515	30	7	3.18	3.18	22.26	1097	659	438.0	10210	0.7870	17.73
Bear	160	258.1	0.1102	595	552	30	7	2.35	3.35	23.43	1228	738.0	491.0	11310	0.7870	17.73
Goat	185	316.5	0.08989	680	630	30	7	3.71	3.71	25.97	1492	896.0	596.0	13780	0.7870	17.73
Sheep	225	366.1	0.07771	745	690	30	7	3.99	3.99	27.93	1726	1036.0	690.0	15910	0.7870	17.73
Deer	260	419.3	0.06786	806	747	30	7	4.27	4.27	29.89	1977	1188.0	789.0	18230	0.7870	17.73
Elk	300	465.7	0.06110	860	796	30	7	4.50	4.50	31.50	2196	1320.0	876.0	20240	0.7870	17.73
Moose	325	515.7	0.05517	900	835	54	7	3.53	3.53	31.77	2002	1463.0	539.0	16250	0.6860	19.35
* Sparrow	20	33.16	0.8578	—	—	6	1	2.67	2.67	8.01	135	92.0	43.0	1208	0.8090	18.99
* Fox	22	36.21	0.7857	165	153	6	1	2.79	2.79	8.37	148.9	100.8	48.1	1313	0.8090	18.99
* Guinea	49	78.56	0.3620	—	—	12	7	2.92	2.92	14.60	590.0	224.0	366.0	6664	1.167	15.88
* Lark	125	196.1	0.1451	—	—	30	7	2.92	2.92	20.44	922.0	556.0	366.0	8659	0.7870	17.73
* Zebra	260	418.6	0.0680	795	736	54	7	3.18	3.18	28.62	1623	1185.0	438.0	13316	0.6860	19.35

Index